VECTOR IDENTITIES

TRIPLE PRODUCTS

(1) $\mathbf{A} \cdot (\mathbf{B} \times \mathbf{C}) = \mathbf{B} \cdot (\mathbf{C} \times \mathbf{A}) = \mathbf{C} \cdot (\mathbf{A} \times \mathbf{B})$

(2) $\mathbf{A} \times (\mathbf{B} \times \mathbf{C}) = \mathbf{B}(\mathbf{A} \cdot \mathbf{C}) - \mathbf{C}(\mathbf{A} \cdot \mathbf{B})$

PRODUCT RULES

(3) $\nabla(fg) = f(\nabla g) + g(\nabla f)$

(4) $\nabla(\mathbf{A} \cdot \mathbf{B}) = \mathbf{A} \times (\nabla \times \mathbf{B}) + \mathbf{B} \times (\nabla \times \mathbf{A}) + (\mathbf{A} \cdot \nabla)\mathbf{B} + (\mathbf{B} \cdot \nabla)\mathbf{A}$

(5) $\nabla \cdot (f\mathbf{A}) = f(\nabla \cdot \mathbf{A}) + \mathbf{A} \cdot (\nabla f)$

(6) $\nabla \cdot (\mathbf{A} \times \mathbf{B}) = \mathbf{B} \cdot (\nabla \times \mathbf{A}) - \mathbf{A} \cdot (\nabla \times \mathbf{B})$

(7) $\nabla \times (f\mathbf{A}) = f(\nabla \times \mathbf{A}) - \mathbf{A} \times (\nabla f)$

(8) $\nabla \times (\mathbf{A} \times \mathbf{B}) = (\mathbf{B} \cdot \nabla)\mathbf{A} - (\mathbf{A} \cdot \nabla)\mathbf{B} + \mathbf{A}(\nabla \cdot \mathbf{B}) - \mathbf{B}(\nabla \cdot \mathbf{A})$

SECOND DERIVATIVES

(9) $\nabla \cdot (\nabla \times \mathbf{A}) = 0$

(10) $\nabla \times (\nabla f) = 0$

(11) $\nabla \times (\nabla \times \mathbf{A}) = \nabla(\nabla \cdot \mathbf{A}) - \nabla^2\mathbf{A}$

FUNDAMENTAL THEOREMS

Gradient Theorem: $\int_a^b (\nabla f) \cdot d\mathbf{l} = f(\mathbf{b}) - f(\mathbf{a})$

Divergence Theorem: $\int_{\text{volume}} (\nabla \cdot \mathbf{A})\, d\tau = \int_{\text{surface}} \mathbf{A} \cdot d\mathbf{a}$

Curl Theorem: $\int_{\text{surface}} (\nabla \times \mathbf{A}) \cdot d\mathbf{a} = \oint_{\text{line}} \mathbf{A} \cdot d\mathbf{l}$

INTRODUCTION TO ELECTRODYNAMICS

Second Edition

DAVID J. GRIFFITHS
Department of Physics
Reed College

PRENTICE HALL, Englewood Cliffs, New Jersey 07632

Griffiths, David J. (David Jeffrey), (date)
Introduction to electrodynamics / David J. Griffiths.—2nd ed.
p. cm.
Includes index.
ISBN 0-13-481367-7
1. Electrodynamics. I. Title.
QC680.G74 1989
537.6—dc19 88-36566
CIP

Editorial/production supervision: Zita de Schauensee and Cristina Ferrari
Cover design: 20/20 Services, Inc.
Manufacturing Buyer: Paula Massenaro

A Division of Simon & Schuster
Englewood Cliffs, New Jersey 07632

Printed in the United States of America

10 9 8 7 6 5 4 3 2 1

ISBN 0-13-481367-7

Prentice-Hall International (UK) Limited, *London*
Prentice-Hall of Australia Pty. Limited, *Sydney*
Prentice-Hall Canada Inc., *Toronto*
Prentice-Hall Hispanoamericana, S.A., *Mexico*
Prentice-Hall of India Private Limited, *New Delhi*
Prentice-Hall of Japan, Inc., *Tokyo*
Simon & Schuster Asia Pte. Ltd., *Singapore*
Editora Prentice-Hall do Brasil, Ltda., *Rio de Janeiro*

CONTENTS

PREFACE

For this edition the original text has been modified in a number of small ways, most of them suggested by instructors who used the book in its earlier form. Several technical errors (and typos) have been corrected; some examples have been changed; a few problems have been eliminated, and many new ones added. (The problems have been distinguished more clearly from the text, which should make them easier to find. I resisted, however, a recommendation that they all be moved to the end of each chapter. Many of them have a specific purpose and should be worked immediately after reading the section to which they pertain; I think it is best if their pedagogical role is indicated by their placement on the page.) In several cases the solution to a problem is used later in the text; such problems are indicated with a bullet (•) in the left margin. Difficult or lengthy problems are flagged with an exclamation point (!). Many people have asked that the answers to odd-numbered problems be provided at the back of the book; unfortunately, a roughly equal number adamantly oppose the idea. I have chosen to stick with the format of the first edition, providing occasional answers when this seemed appropriate, especially for problems at the end of each chapter. A complete solution manual is available (to instructors only) from the publisher.

There are two significant changes in the text itself: (1) The Dirac delta function is introduced in Chapter 1 and is used to simplify derivations in Chapters 2 and 5. I was gratified to find that virtually everyone I consulted favored this change. To the apprehensive student I offer my personal guarantee: The day or two you spend learning how to manipulate the delta function will ease and enrich your understanding not only of electrodynamics, but of many other subjects in physics and mathematics. (2) A brief section on wave guides has been added to Chapter 8. There are, of course, many other topics that might be included (plasmas, ac circuits, transmission lines, antennas, magnetic circuits, numerical methods, Lagrangians, etc.), and in a full-year course most instructors will supplement the text with applications of their own choosing. But I want the book itself to be as lean as possible, focusing exclusively on the essential theoretical development of the subject.

I have benefited from the comments of many readers—I cannot list them all here. But I would like to thank particularly the following people.

John Bjorkstam, *University of Washington*
Timothy Boyer, *City College of New York*
Burton Brody, *Bard College*
Jeffrey Dunham, *Middlebury College*
Stefan Estreichei, *Texas Tech University*
Mark Heald, *Swarthmore College*
Michael E. Michelson, *Denison University*
C. J. Mortoff, *Stanford University*
R. S. Rena, *College of the Holy Cross*
Larry Schecter, *Oregon State University*
Martin Tiersten, *City College of New York*
Nicholas Wheeler, *Reed College*

Finally, let me acknowledge now (as I should have in the first edition) that practically everything I know about electrodynamics—certainly about teaching electrodynamics—I owe to Edward Purcell.

David J. Griffiths

ADVERTISEMENT

What is electrodynamics, and where does it fit into the general scheme of physics?

FOUR REALMS OF MECHANICS

In the following diagram I have sketched out the four great realms of mechanics:

Higher Speeds ↓	Classical Mechanics (Newton)	Quantum Mechanics (Schrödinger, Heisenberg, Bohr *et al.*)
	Relativistic Mechanics (Einstein)	Quantum Field Theory (Dirac, Pauli, Schwinger, Feynman *et al.*)

→ **Smaller Distances**

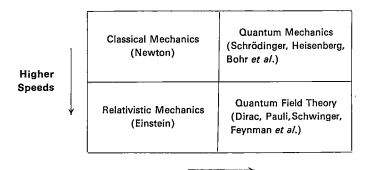

Newtonian mechanics was found to be inadequate in the early years of this century—it's all right in "everyday life," but for objects moving at high speeds (near the speed of light) it is incorrect, and must be replaced by special relativity (introduced by Einstein in 1905); for objects that are extremely small (near the size of atoms) it fails for different reasons, and is superseded by quantum mechanics (developed by Bohr, Schrödinger, Heisenberg, and many others in the twenties, mostly). For objects that are both very fast *and* very small (as is common in modern particle physics), a mechanics that combines relativity and quantum principles is in order: this relativistic quantum mechanics is known as quantum field theory—it was developed in the thirties and forties, primarily, but even today it cannot claim to be a completely satisfactory system. In this book, save for the last chapter, we shall work exclusively in the domain of classical mechanics, although the theory of electromagnetism extends with unique simplicity to the other three realms. (In fact, the theory in most respects *automatically* obeys special relativity, for which it was, historically, the main stimulus.)

FOUR KINDS OF FORCES

Mechanics tells us how a system will behave when subjected to a given *force*. There are just *four* basic forces known (presently) to physics: I list them in order of decreasing strength:

1. Strong
2. Electromagnetic
3. Weak
4. Gravitational

The brevity of this list may surprise you. Where is friction? Where is the "normal" force that keeps me from falling through the floor? Where are the chemical forces that bind molecules together? Where is the force of impact between two colliding billiard balls? The answer is that *all* these forces are *electromagnetic*. Indeed, it is scarcely an exaggeration to say that we live in an electromagnetic world—for virtually every force we experience in everyday life, with the exception of gravity, is electromagnetic in origin.

The "strong" forces, which hold protons and neutrons together in the atomic nucleus, have extremely short range, so we do not "feel" them, in spite of the fact that they are a hundred times stronger than electrical forces. The "weak" forces, which account for certain kinds of radioactive decay, are not only of short range; they are far less powerful than electromagnetic ones to begin with. As for gravity, it is so pitifully feeble (compared to all the others) that it is only in virtue of huge mass concentrations (like the earth and the sun) that we ever notice it at all. The electrical repulsion between two electrons is 10^{42} times as large as their gravitational attraction, and if atoms were held together by gravitational (instead of electrical) forces, a single hydrogen atom would be much larger than the known universe.

Not only are electromagnetic forces overwhelmingly the dominant ones in everyday life, they are also, at present, the *only* ones that are completely understood. There is, of course, a classical theory of gravity (Newton's law of universal gravitation) and a relativistic one (Einstein's general relativity), but no entirely satisfactory quantum mechanical theory of gravity has been constructed (though many people are working on it). At the present time there is a successful (if cumbersome) theory for the weak interactions, and a strikingly beautiful candidate (called **chromodynamics**) for the strong interactions. Both these theories draw their inspiration from electrodynamics; neither can claim conclusive experimental verification at this stage. So electrodynamics, a beautifully complete and successful theory, has become a kind of reference point for physicists: an ideal model that all other theories strive to emulate.

Classical electrodynamics was worked out in bits and pieces by Franklin, Coulomb, Ampère, Faraday, and others, but the person who put it all together and built it into the compact and consistent theory it is today, was James Clerk Maxwell. The theory is now a little over a hundred years old.

THE UNIFICATION OF PHYSICAL THEORIES

In the beginning, **electricity** and **magnetism** were entirely separate subjects. The one dealt with glass rods and cat's fur, pith balls, batteries, currents, electrolysis, and lightning; the other with bar magnets, iron filings, compass needles, and the North Pole. But in 1820 Oersted noticed that an *electric* current could deflect a *magnetic* compass needle. Soon afterward, Ampère correctly postulated that *all* magnetic phenomena are due to electric charges in motion. Then, in 1831, Faraday discovered that a moving *magnet* generates an *electric* current. By the time Maxwell and Lorentz put the finishing touches on the theory, electricity and magnetism were inextricably intertwined. They could no longer be regarded as separate subjects, but rather as two *aspects* of a *single* subject: **electromagnetism.**

Faraday had speculated that light, too, is electrical in nature. Maxwell's theory provided spectacular justification for this hypothesis, and soon **optics**—the study of lenses, mirrors, and prisms, and interference and diffraction—was incorporated into electromagnetism. Hertz, who presented the decisive experimental confirmation for Maxwell's theory in 1888, put it this way: "The connection between light and electricity is now established . . . . In every flame, in every luminous particle, we see an electrical process . . . . Thus, the domain of electricity extends over the whole of nature. It even affects ourselves intimately: we perceive that we possess . . . an electrical organ—the eye." By 1900, then, three great branches of physics, electricity, magnetism, and optics, had merged into a single unified theory. (And it was soon apparent that visible light represents only a tiny "window" in the vast spectrum of electromagnetic radiation, from radio through microwaves, infrared and ultraviolet, to X-rays and gamma rays.)

Einstein dreamed of a further unification, which would combine gravity and electrodynamics, in much the same way as electricity and magnetism had been combined a century earlier. His **unified field theory** was not particularly successful, but in recent years the same impulse has spawned a hierarchy of increasingly ambitious (and speculative) unification schemes, beginning in the 1960s with the **electroweak** theory of Glashow, Weinberg, and Salam (which joins the weak and electromagnetic forces), proceding in the 1970s to the **grand unified theories** (which stir in the strong forces), and culminating in the 1980s with the **superstring** theory (which, according to its proponents, incorporates all four forces in a single "theory of everything"). At each step in this hierarchy the mathematical difficulties mount, and the gap between inspired conjecture and experimental test widens; nevertheless, it is clear that the unification of forces initiated by electrodynamics has become a major theme in the development of physics.

THE FIELD FORMULATION OF ELECTRODYNAMICS

The fundamental problem a theory of electromagnetism hopes to solve is this: I hold up a bunch of electric charges *here* (and maybe shake them around)—what happens to some *other* charge, over *there*? The classical solution takes the form of a **field theory:** We say that the space around an electric charge is permeated by electric and magnetic **fields** (the electromagnetic "odor," as it were, of the charge). A second charge, in the presence of these fields, experiences a force; the fields, then, transmit the influence from one charge to the other—they "mediate" the interaction.

When a charge undergoes *acceleration,* a portion of the field "detaches" itself, in a sense, and travels off at the speed of light, carrying with it energy, momentum, and angular momentum. We call this **electromagnetic radiation.** Its existence invites (if not *compels*) us to regard the fields as independent dynamical entities in their own right, every bit as "real" as atoms or baseballs. Our interest accordingly shifts from the study of forces between charges to the theory of the fields themselves. But it takes a charge to *produce* an electromagnetic field, and it takes another charge to *detect* one, so we had best begin by considering the nature of electric charge.

PROPERTIES OF ELECTRIC CHARGE

1. *Charge comes in two varieties,* which we call "plus" and "minus," because their effects tend to *cancel* (if we have $+q$ and $-q$ at the same point, electrically it is the same as having no charge there at all). This may seem too obvious to warrant comment, but I encourage you to think sometime of the other possibilities: what if there were 8 or 10 different species of charge? (In chromodynamics there are, in fact, *three* quantities analogous to electric charge, each of which can be positive or negative.) Or what if the two kinds did not tend to cancel? The extraordinary fact is that plus and minus charges occur in *exactly* equal amounts, to fantastic precision, in bulk matter, so that their effects are almost completely neutralized. Were it not for this, we would be subjected to enormous forces: a potato would explode violently if the cancellation were imperfect by as little as one part in 10^{10}.

2. *Charge is conserved.* It cannot be created or destroyed; what there is now has always been. (A plus charge can "annihilate" an equal minus charge, but a plus charge cannot simply disappear by itself—*something* must account for that electric charge.) So the total charge of the universe is fixed for all time. This is called **global** conservation of charge. Actually, I can say something much stronger: global conservation would allow a charge to disappear in New York and instantly reappear in San Francisco (that wouldn't affect the *total* charge) and yet we know this doesn't happen. If the charge *was* in New York and it *went* to San Francisco, then it must have passed along some continuous path from one to the other. This is called **local** conservation of charge. Later on we'll see how to formulate a precise mathematical law expressing local conservation of charge—the **continuity equation.**

3. *Charge is quantized.* Although nothing in classical electromagnetic theory

requires that it be so, the *fact* is that electric charge comes only in discrete lumps—integer multiples of the basic unit of charge. If we call the charge on the proton $+e$, then the electron carries charge $-e$, the neutron charge zero, the pi mesons $+e$, 0, and $-e$, the carbon nucleus $+6e$, and so on (never $7.392e$, or even $1/2e$).[1] This fundamental unit of charge is extremely small, so for practical purposes it is usually appropriate to ignore quantization altogether. (Water, too, "really" consists of discrete lumps (molecules) yet, if we're dealing with reasonably large quantities of it we can treat it as a continuous fluid.) This is in fact much closer to Maxwell's own view; he knew nothing of electrons and protons—he must have pictured charge as a kind of "jelly" that could be divided up into portions of any size and smeared out at will.

These, then, are the three general properties of charge. Before we discuss the forces *between* charges, however, some mathematical tools are necessary; their introduction will occupy us in Chapter 1.

UNITS

The subject of electrodynamics is plagued by competing systems of units, which sometimes render it difficult for physicists to communicate with one another. The problem is far worse than in mechanics, where Neanderthals still speak of pounds and feet; for in mechanics at least all equations *look* the same, regardless of the units used to measure quantities. $\mathbf{F} = m\mathbf{a}$ remains $\mathbf{F} = m\mathbf{a}$, whether it is feet-pounds-seconds, kilograms-meters-seconds, or whatever. But this is not so in electromagnetism, where Coulomb's law may appear variously as

$$\mathbf{F} = \frac{q_1 q_2}{\imath^2}\hat{\boldsymbol{\imath}} \text{ (Gaussian)} \quad \text{or} \quad \mathbf{F} = \frac{1}{4\pi\epsilon_0}\frac{q_1 q_2}{\imath^2}\hat{\boldsymbol{\imath}} \text{ (SI)} \quad \text{or} \quad \mathbf{F} = \frac{1}{4\pi}\frac{q_1 q_2}{\imath^2}\hat{\boldsymbol{\imath}} \text{ (HL)}$$

Of the systems in common use, the two most popular are the Gaussian cgs system and the SI, or "rationalized" mks, system. (Elementary particle theorists favor yet a third system: Heaviside-Lorentz.) Although Gaussian units offer distinct theoretical advantages, most undergraduate instructors seem to prefer SI, I suppose because they incorporate the familiar household units (volts, amperes, and watts). In this book, therefore, I have used SI units. Appendix B provides a "dictionary" for converting the important results into Gaussian units.

[1]Actually, protons and neutrons are composed of three **quarks**, which carry fractional charges ($\pm\frac{2}{3}e$ and $\pm\frac{1}{3}e$). However, *free* quarks do not appear to exist in nature, and in any event this does not alter the fact that charge is quantized; it merely reduces the size of the basic unit.

1

VECTOR ANALYSIS

1.1 VECTOR ALGEBRA

1.1.1 Vector Operations

If you walk 4 mi due north and then 3 mi due east (Fig. 1.1), you have gone a total of 7 mi, but you're *not* 7 mi from where you set out—you're only 5. We need an algebra to describe quantities like this, which evidently do not add in the ordinary way. The reason they don't, of course, is that **displacements** (straight line segments going from one point to another) have *direction* as well as *magnitude* (length), and it is essential to take both into account when you combine them. Such objects are called **vectors:** velocity, acceleration, force, and momentum are other examples. By contrast, quantities that have magnitude but no direction are called **scalars:** Examples are mass, charge, density, and temperature. I shall use **boldface** type (**A, B,** and so on) for vectors and ordinary type for scalars. The magnitude of a vector **A** is written $|\mathbf{A}|$ or, more simply, A. In diagrams, vectors are denoted by arrows: the length of the arrow is proportional to the magnitude of the vector, and the arrowhead indicates its direction. *Minus* **A** ($-\mathbf{A}$) is a vector with the same magnitude as **A** but of opposite direction (Fig. 1.2). Note that vectors have magnitude and direction but *not location:* a displacement of 4 mi due north from Washington is represented by the same vector

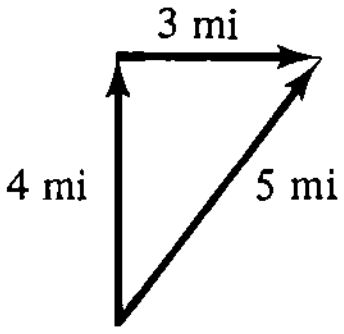

Figure 1.1

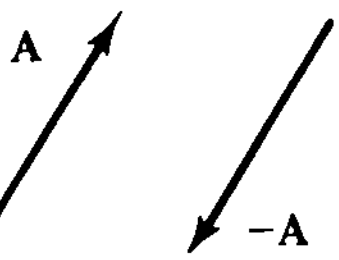

Figure 1.2

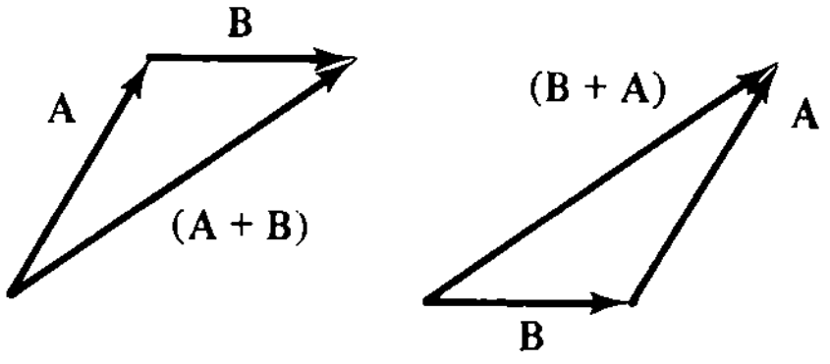

Figure 1.3

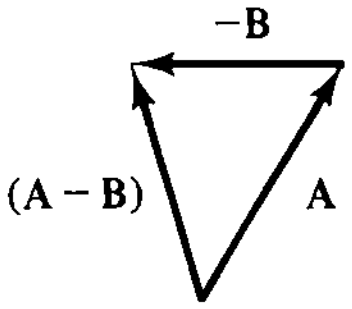

Figure 1.4

as a displacement 4 mi north from Baltimore (neglecting, of course, the curvature of the earth). On a diagram, therefore, you can slide the arrow around at will, as long as you don't change its length or direction.

We define four vector operations: addition and three kinds of multiplication.

(i) Addition of two vectors. Place the tail of **B** at the head of **A**; the sum, $\mathbf{A} + \mathbf{B}$, is the vector from the tail of **A** to the head of **B** (Fig. 1.3). (This rule generalizes the obvious procedure for combining two displacements.) Addition is *commutative:*

$$\mathbf{A} + \mathbf{B} = \mathbf{B} + \mathbf{A}$$

3 mi east followed by 4 mi north gets you to the same place as 4 mi north followed by 3 mi east. Addition is also *associative:*

$$(\mathbf{A} + \mathbf{B}) + \mathbf{C} = \mathbf{A} + (\mathbf{B} + \mathbf{C})$$

To subtract a vector (Fig. 1.4), add its opposite:

$$\mathbf{A} - \mathbf{B} = \mathbf{A} + (-\mathbf{B})$$

(ii) Multiplication by a scalar. Multiplication of a vector by a positive scalar a multiplies the *magnitude* but leaves the direction unchanged (Fig. 1.5). (If a is negative, the direction is reversed.) Scalar multiplication is *distributive:*

$$a(\mathbf{A} + \mathbf{B}) = a\mathbf{A} + a\mathbf{B}$$

(iii) Dot product of two vectors. The dot product of two vectors is defined by

$$\mathbf{A} \cdot \mathbf{B} = AB \cos\theta \tag{1.1}$$

where θ is the angle they form when placed tail-to-tail (Fig. 1.6). Note that $\mathbf{A} \cdot \mathbf{B}$ is itself a *scalar* (hence the alternative name **scalar product**). The dot product is *commutative,*

$$\mathbf{A} \cdot \mathbf{B} = \mathbf{B} \cdot \mathbf{A}$$

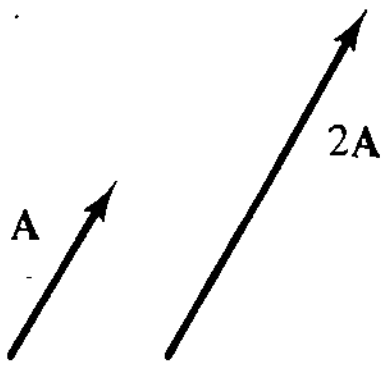

Figure 1.5

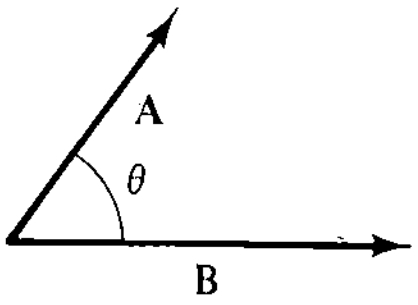

Figure 1.6

and *distributive,*

$$\mathbf{A}\cdot(\mathbf{B}+\mathbf{C}) = \mathbf{A}\cdot\mathbf{B}+\mathbf{A}\cdot\mathbf{C} \tag{1.2}$$

Geometrically, $\mathbf{A}\cdot\mathbf{B}$ is the product of A times the projection of $\mathbf{B}$ along $\mathbf{A}$ (or the product of B times the projection of $\mathbf{A}$ along $\mathbf{B}$). If the two vectors are parallel, then $\mathbf{A}\cdot\mathbf{B} = AB$. In particular, for any vector $\mathbf{A}$,

$$\mathbf{A}\cdot\mathbf{A} = A^2 \tag{1.3}$$

If $\mathbf{A}$ and $\mathbf{B}$ are perpendicular, then $\mathbf{A}\cdot\mathbf{B} = 0$.

Example 1

Let $\mathbf{C} = \mathbf{A} - \mathbf{B}$ (Fig. 1.7), and calculate the dot product of $\mathbf{C}$ with itself:

$$\mathbf{C}\cdot\mathbf{C} = (\mathbf{A}-\mathbf{B})\cdot(\mathbf{A}-\mathbf{B}) = \mathbf{A}\cdot\mathbf{A} - \mathbf{A}\cdot\mathbf{B} - \mathbf{B}\cdot\mathbf{A} + \mathbf{B}\cdot\mathbf{B}$$

or

$$C^2 = A^2 + B^2 - 2AB\cos\theta$$

This is called the **law of cosines.**

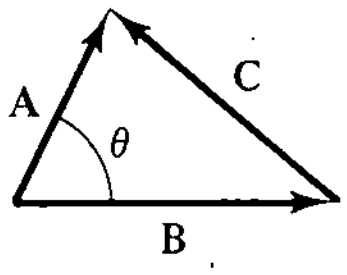

Figure 1.7

(iv) Cross product of two vectors. The cross product of two vectors is defined by

$$\mathbf{A}\times\mathbf{B} = AB\sin\theta\,\hat{n}, \tag{1.4}$$

where $\hat{n}$ is a unit vector (vector of length 1) pointing perpendicular to the plane of $\mathbf{A}$ and $\mathbf{B}$. (I shall use a hat (^) to designate unit vectors.) Of course, there are *two* directions perpendicular to any plane: "in" and "out." The ambiguity is resolved by the **right-hand rule:** let your fingers point in the direction of the first vector and curl around (via the smaller angle) toward the second; then your thumb indicates the direction of $\hat{n}$. (In the figure $\mathbf{A}\times\mathbf{B}$ points *into* the page, $\mathbf{B}\times\mathbf{A}$ points *out* of the page.) Note that $\mathbf{A}\times\mathbf{B}$ is itself a *vector* (hence the alternative name **vector product**). The cross product is *distributive,*

$$\mathbf{A}\times(\mathbf{B}+\mathbf{C}) = (\mathbf{A}\times\mathbf{B}) + (\mathbf{A}\times\mathbf{C}) \tag{1.5}$$

but *not commutative.* In fact,

$$(\mathbf{B}\times\mathbf{A}) = -(\mathbf{A}\times\mathbf{B}) \tag{1.6}$$

Geometrically, $|\mathbf{A}\times\mathbf{B}|$ is the area of the parallelogram generated by $\mathbf{A}$ and $\mathbf{B}$

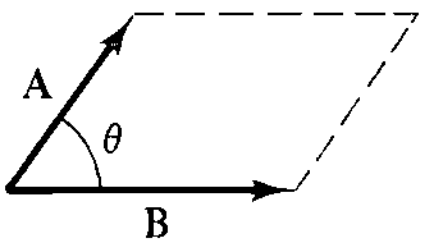

Figure 1.8

(Fig. 1.8). If two vectors are parallel, their cross product is zero. In particular,

$$\mathbf{A} \times \mathbf{A} = 0$$

for any vector **A**.

Problem 1.1 Using the definitions in (1.1) and (1.4) and appropriate diagrams, show that the dot product and cross product are distributive

(a) when the three vectors are coplanar;

! (b) in the general case.

Problem 1.2 Is the cross product associative?

$$(\mathbf{A} \times \mathbf{B}) \times \mathbf{C} \stackrel{?}{=} \mathbf{A} \times (\mathbf{B} \times \mathbf{C})$$

If so, *prove* it; if not, provide a counterexample.

1.1.2 Vector Algebra: Component Form

In the previous section I defined the four vector operations (addition, scalar multiplication, dot product, and cross product) in "abstract" form—that is, without reference to any particular coordinate system. In practice it is often easier to set up Cartesian coordinates x, y, z and work with vector "components." Let $\hat{i}$, $\hat{j}$, and $\hat{k}$ be unit vectors parallel to the x, y, and z axes, respectively (Fig. 1.9(a)). An arbitrary vector **A** can be expanded in terms of these **basis vectors** (Fig. 1.9(b)):

$$\mathbf{A} = A_x\hat{i} + A_y\hat{j} + A_z\hat{k}$$

The numbers A_x, A_y, and A_z are called **components** of **A**; geometrically, they are the projections of **A** along the three coordinate axes. We can now reformulate each of the four vector operations as a rule for manipulating components:

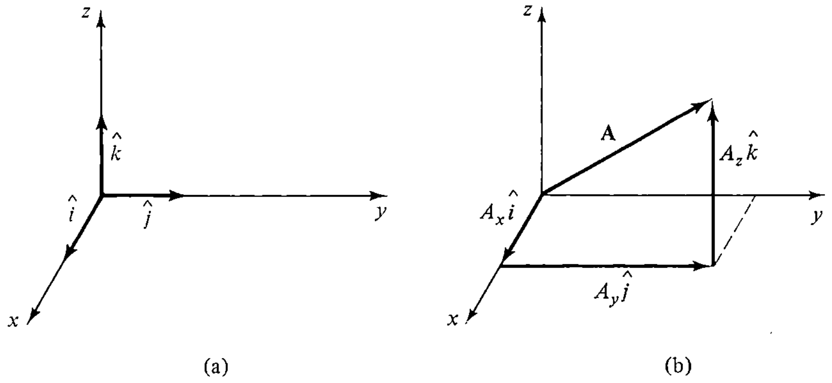

Figure 1.9

(i)

$$\begin{aligned}\mathbf{A}+\mathbf{B} &= (A_x\hat{\imath}+A_y\hat{\jmath}+A_z\hat{k})+(B_x\hat{\imath}+B_y\hat{\jmath}+B_z\hat{k}) \\ &= (A_x+B_x)\hat{\imath}+(A_y+B_y)\hat{\jmath}+(A_z+B_z)\hat{k}\end{aligned} \tag{1.7}$$

Rule: ***To add vectors, add like components.***

(ii)

$$a\mathbf{A} = a(A_x\hat{\imath}+A_y\hat{\jmath}+A_z\hat{k}) = (aA_x)\hat{\imath}+(aA_y)\hat{\jmath}+(aA_z)\hat{k} \tag{1.8}$$

Rule: ***To multiply by a scalar, multiply each component.***

(iii) Because $\hat{\imath}$, $\hat{\jmath}$, and $\hat{k}$ are mutually perpendicular unit vectors,

$$\hat{\imath}\cdot\hat{\imath}=\hat{\jmath}\cdot\hat{\jmath}=\hat{k}\cdot\hat{k}=1, \qquad \hat{\imath}\cdot\hat{\jmath}=\hat{\imath}\cdot\hat{k}=\hat{\jmath}\cdot\hat{k}=0 \tag{1.9}$$

Accordingly,

$$\begin{aligned}\mathbf{A}\cdot\mathbf{B} &= (A_x\hat{\imath}+A_y\hat{\jmath}+A_z\hat{k})\cdot(B_x\hat{\imath}+B_y\hat{\jmath}+B_z\hat{k}) \\ &= A_xB_x+A_yB_y+A_zB_z\end{aligned} \tag{1.10}$$

Rule: ***To calculate the dot product, multiply like components, and add.***

In particular,

$$\mathbf{A}\cdot\mathbf{A} = A_x^2+A_y^2+A_z^2$$

so

$$A = \sqrt{A_x^2+A_y^2+A_z^2} \tag{1.11}$$

(This is, if you like, the three-dimensional generalization of the Pythagorean theorem.) Note that the dot product of **A** with any *unit* vector is the component of **A** along that direction (thus $\mathbf{A}\cdot\hat{\imath}=A_x$, $\mathbf{A}\cdot\hat{\jmath}=A_y$, and $\mathbf{A}\cdot\hat{k}=A_z$).

(iv) Similarly,[1]

$$\begin{aligned}\hat{\imath}\times\hat{\imath}&=\hat{\jmath}\times\hat{\jmath}=\hat{k}\times\hat{k}=0, \\ \hat{\imath}\times\hat{\jmath}&=-\hat{\jmath}\times\hat{\imath}=\hat{k} \\ \hat{\jmath}\times\hat{k}&=-\hat{k}\times\hat{\jmath}=\hat{\imath} \\ \hat{k}\times\hat{\imath}&=-\hat{\imath}\times\hat{k}=\hat{\jmath}\end{aligned} \tag{1.12}$$

Therefore,

$$\begin{aligned}\mathbf{A}\times\mathbf{B} &= (A_x\hat{\imath}+A_y\hat{\jmath}+A_z\hat{k})\times(B_x\hat{\imath}+B_y\hat{\jmath}+B_z\hat{k}) \\ &= (A_yB_z-A_zB_y)\hat{\imath}+(A_zB_x-A_xB_z)\hat{\jmath}+(A_xB_y-A_yB_x)\hat{k}\end{aligned} \tag{1.13}$$

[1]These signs pertain to a *right-handed* coordinate system (x axis out of the page, y axis to the right, z axis up, or any rotated version thereof). In a *left-handed* system (z axis down) the signs are reversed: $\hat{\imath}\times\hat{\jmath}=-\hat{k}$, and so on. We shall use right-handed systems exclusively.

This cumbersome expression can be written more neatly as a determinant:

$$\mathbf{A} \times \mathbf{B} = \begin{vmatrix} \hat{\imath} & \hat{\jmath} & \hat{k} \\ A_x & A_y & A_z \\ B_x & B_y & B_z \end{vmatrix} \tag{1.14}$$

Rule: ***To calculate the cross product, form the determinant whose first row is*** $\hat{\imath}, \hat{\jmath}, \hat{k}$***, whose second row is* A *(in component form), and whose third row is* B.**

Example 2

Find the angle between the face diagonals of a cube.

Solution: We might as well use a cube of side 1, and place it as shown in Fig. 1.10, with one corner at the origin. The face diagonals **A** and **B** are

$$\mathbf{A} = 1\,\hat{\imath} + 0\,\hat{\jmath} + 1\,\hat{k}, \qquad \mathbf{B} = 0\,\hat{\imath} + 1\,\hat{\jmath} + 1\,\hat{k}$$

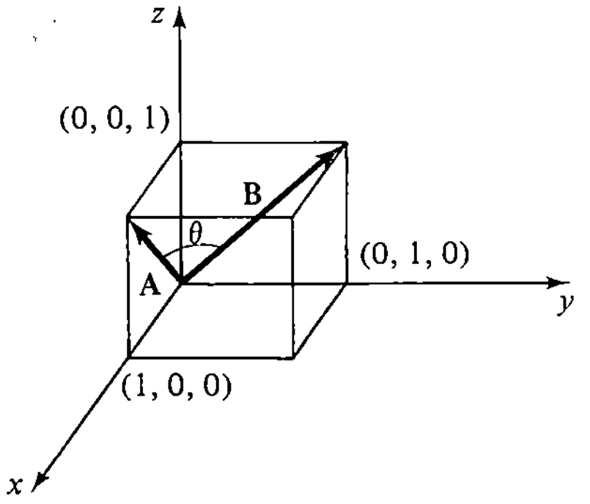

Figure 1.10

So, in component form,

$$\mathbf{A}\cdot\mathbf{B} = 1\cdot 0 + 0\cdot 1 + 1\cdot 1 = 1$$

On the other hand, in "abstract" form

$$\mathbf{A}\cdot\mathbf{B} = AB\cos\theta = \sqrt{2}\,\sqrt{2}\cos\theta = 2\cos\theta$$

Therefore,

$$\cos\theta = \tfrac{1}{2}, \quad \text{or} \quad \theta = 60^\circ$$

Of course, you can get the answer more easily by drawing in a diagonal across the top of the cube, completing an equilateral triangle. But in cases where the geometry is not so simple, this device of comparing the abstract and component forms of the dot product can be a very efficient means of finding angles.

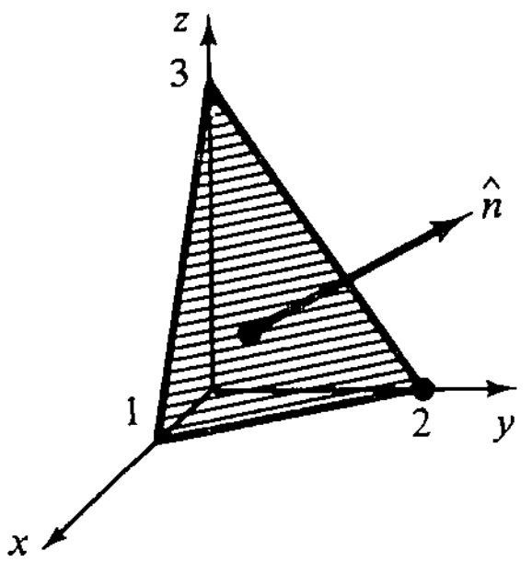

Figure 1.11

Problem 1.3

(a) Find the components of the displacement vector from the point (2, 8, 7) to the point (7, 5, 11).

(b) What are the components of the displacement vector from (x_0, y_0, z_0) to (x, y, z)?

Problem 1.4 Find the angle between the body diagonals of a cube.

Problem 1.5 Use the cross product to find the components of the unit vector $\hat{n}$ perpendicular to the plane shown in Fig. 1.11.

1.1.3 Triple Products

Since the cross product of two vectors is itself a vector, it can be dotted or crossed with a third vector to form a *triple* product.

(i) Scalar triple product: $\mathbf{A}\cdot(\mathbf{B}\times\mathbf{C})$. Geometrically, $|\mathbf{A}\cdot(\mathbf{B}\times\mathbf{C})|$ is the volume of the parallelepiped generated by **A**, **B**, and **C**, since $|\mathbf{B}\times\mathbf{C}|$ is the area of the base, and $|A\cos\theta|$ is the altitude (Fig. 1.12). Evidently,

$$\mathbf{A}\cdot(\mathbf{B}\times\mathbf{C}) = \mathbf{B}\cdot(\mathbf{C}\times\mathbf{A}) = \mathbf{C}\cdot(\mathbf{A}\times\mathbf{B}) \tag{1.15}$$

for they all correspond to the same figure. Note that "alphabetical" order is preserved—in view of equation (1.6), the "nonalphabetical" triple products

$$\mathbf{A}\cdot(\mathbf{C}\times\mathbf{B}) = \mathbf{B}\cdot(\mathbf{A}\times\mathbf{C}) = \mathbf{C}\cdot(\mathbf{B}\times\mathbf{A})$$

have the opposite sign. In component form

$$\mathbf{A}\cdot(\mathbf{B}\times\mathbf{C}) = \begin{vmatrix} A_x & A_y & A_z \\ B_x & B_y & B_z \\ C_x & C_y & C_z \end{vmatrix} \tag{1.16}$$

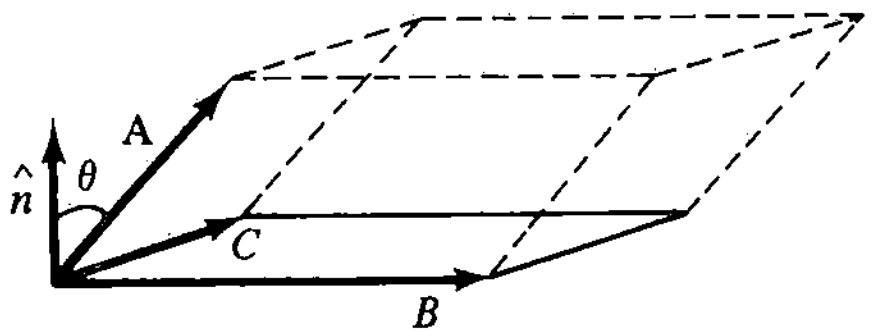

Figure 1.12

Note that the dot and cross can be interchanged:

$$\mathbf{A}\cdot(\mathbf{B}\times\mathbf{C}) = (\mathbf{A}\times\mathbf{B})\cdot\mathbf{C}$$

(this follows immediately from (1.15)); however, the placement of the parentheses is critical: $(\mathbf{A}\cdot\mathbf{B})\times\mathbf{C}$ is a nonsensical expression—you can't make a cross product from a *scalar* and a vector.

(ii) Vector triple product: $\mathbf{A}\times(\mathbf{B}\times\mathbf{C})$. The vector triple product can be simplified by the so-called **BAC-CAB** rule:

$$\mathbf{A}\times(\mathbf{B}\times\mathbf{C}) = \mathbf{B}(\mathbf{A}\cdot\mathbf{C}) - \mathbf{C}(\mathbf{A}\cdot\mathbf{B}) \tag{1.17}$$

Notice that

$$(\mathbf{A}\times\mathbf{B})\times\mathbf{C} = -\mathbf{C}\times(\mathbf{A}\times\mathbf{B}) = -\mathbf{A}(\mathbf{B}\cdot\mathbf{C}) + \mathbf{B}(\mathbf{A}\cdot\mathbf{C})$$

is an entirely different vector. Incidentally, all *higher* vector products can be similarly reduced, often by repeated application of (1.17), so it is never necessary for an expression to contain more than one cross product in any term. For instance,

$$\begin{aligned}(\mathbf{A}\times\mathbf{B})\cdot(\mathbf{C}\times\mathbf{D}) &= (\mathbf{A}\cdot\mathbf{C})(\mathbf{B}\cdot\mathbf{D}) - (\mathbf{A}\cdot\mathbf{D})(\mathbf{B}\cdot\mathbf{C})\\ \mathbf{A}\times(\mathbf{B}\times(\mathbf{C}\times\mathbf{D})) &= \mathbf{B}(\mathbf{A}\cdot(\mathbf{C}\times\mathbf{D})) - (\mathbf{A}\cdot\mathbf{B})(\mathbf{C}\times\mathbf{D})\end{aligned} \tag{1.18}$$

Problem 1.6 Prove the **BAC-CAB** rule by writing out both sides in component form.

Problem 1.7 Prove that

$$[\mathbf{A}\times(\mathbf{B}\times\mathbf{C})] + [\mathbf{B}\times(\mathbf{C}\times\mathbf{A})] + [\mathbf{C}\times(\mathbf{A}\times\mathbf{B})] = 0$$

Under what conditions does $\mathbf{A}\times(\mathbf{B}\times\mathbf{C}) = (\mathbf{A}\times\mathbf{B})\times\mathbf{C}$?

1.1.4 How Vectors Transform

The definition of a vector as "a quantity with magnitude and direction" is not altogether satisfactory: What precisely does "direction" mean? This may seem a pedantic question, but we shall shortly encounter a species of derivative that *looks* rather like a vector, and we'll want to know for sure whether it *is* one. You might be inclined to say that a vector is anything that has three components which combine properly under addition. Well, how about this: We have a barrel of fruit that contains N_x pears, N_y apples, and N_z bananas. Is $\mathbf{N} \equiv N_x\hat{i} + N_y\hat{j} + N_z\hat{k}$ a vector? It has three components, and when you add another barrel with M_x pears, M_y apples, and M_z bananas the result is $(N_x + M_x)$ pears, $(N_y + M_y)$ apples, and $(N_z + M_z)$ bananas. So it does *add* like a vector. Yet it's obviously *not* a vector, in the physicist's sense of the word, because it doesn't really have a direction. What exactly is wrong with it?

The answer is that $\mathbf{N}$ *does not transform properly when you change coordinates.* The coordinate frame we use to describe positions in space is of course entirely arbitrary, but there is a specific geometrical transformation law for converting vector components from one frame to another. Suppose, for instance, the x', y', z' system

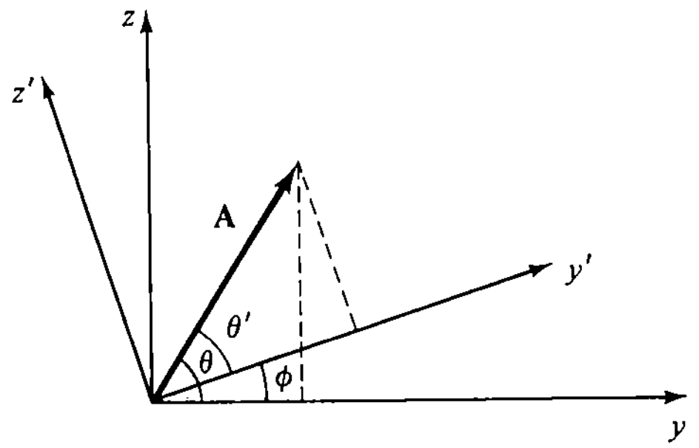

Figure 1.13

is rotated by angle ϕ, relative to x, y, z, about a common $x = x'$ axis. From the diagram (Fig. 1.13),

$$A_y = A\cos\theta, \qquad A_z = A\sin\theta$$

while

$$\begin{aligned} A'_y &= A\cos\theta' = A\cos(\theta-\phi) = A(\cos\theta\cos\phi + \sin\theta\sin\phi) \\ &= \cos\phi A_y + \sin\phi A_z \\ A'_z &= A\sin\theta' = A\sin(\theta-\phi) = A(\sin\theta\cos\phi - \cos\theta\sin\phi) \\ &= -\sin\phi A_y + \cos\phi A_z \end{aligned}$$

We might express this conclusion in matrix notation:

$$\begin{pmatrix} A'_y \\ A'_z \end{pmatrix} = \begin{pmatrix} \cos\phi & \sin\phi \\ -\sin\phi & \cos\phi \end{pmatrix} \begin{pmatrix} A_y \\ A_z \end{pmatrix} \tag{1.19}$$

More generally, for rotation about an arbitrary axis in three dimensions, the transformation law takes the form

$$\begin{pmatrix} A'_x \\ A'_y \\ A'_z \end{pmatrix} = \begin{pmatrix} R_{xx} & R_{xy} & R_{xz} \\ R_{yx} & R_{yy} & R_{yz} \\ R_{zx} & R_{zy} & R_{zz} \end{pmatrix} \begin{pmatrix} A_x \\ A_y \\ A_z \end{pmatrix} \tag{1.20}$$

or, more compactly,

$$A'_i = \sum_{j=1}^{3} R_{ij} A_j \tag{1.21}$$

where the index 1 stands for x, 2 for y, and 3 for z. The elements of the matrix R can be ascertained, for a given rotation, by the same sort of geometrical arguments as we used for a rotation about the x axis.

Now: *Do* the components of **N** transform in this way? Of *course* not—it doesn't matter what coordinates you use to represent positions in space, there is still the same number of apples in the barrel. You can't convert a pear into a banana by choosing a different set of axes, but you *can* turn $\hat{i}$ into $\hat{j}$. Formally, then, a *vector is any set of three components that transforms in the same manner as a displacement when you* *change coordinates.* As always, displacement is the model for the behavior of all vectors.

By the way, a (second rank) tensor is a quantity with *nine* components, T_{xx}, T_{xy}, T_{xz}, T_{yx}, . . . , T_{zz}, which transforms with *two* factors of R:

$$\begin{aligned} T'_{xx} = {} & R_{xx}(R_{xx}T_{xx} + R_{xy}T_{xy} + R_{xz}T_{xz}) \\ & + R_{xy}(R_{xx}T_{yx} + R_{xy}T_{yy} + R_{xz}T_{yz}) \\ & + R_{xz}(R_{xx}T_{zx} + R_{xy}T_{zy} + R_{xz}T_{zz}), \ldots \end{aligned}$$

or, more compactly,

$$T'_{ij} = \sum_{k=1}^{3} \sum_{l=1}^{3} R_{ik}R_{jl}T_{kl} \tag{1.22}$$

In general, an nth-rank tensor has n indices and 3^n components, and transforms with n factors of R. In this hierarchy, a vector is a tensor of rank 1, and a scalar is a tensor of rank zero.

Problem 1.8

(a) Prove that the two-dimensional rotation matrix (1.19) preserves the *length* of **A**. (That is, show that $(A'_y)^2 + (A'_z)^2 = A_y^2 + A_z^2$.)

(b) What constraints must the elements (R_{ij}) of the three-dimensional rotation matrix (1.20) satisfy in order to preserve the length of **A** (for *all* **A**)?

Problem 1.9 Find the transformation matrix R that describes a rotation by 120° about an axis from the origin through the point (1, 1, 1). The rotation is clockwise as you look down the axis toward the origin.

Problem 1.10

(a) How do the components of a vector transform under a *translation* of coordinates ($x' = x$, $y' = y - a$, $z' = z$, Fig. (1.14a))?[2]

(b) How do the components of a vector transform under an *inversion* of coordinates ($x' = -x$, $y' = -y$, $z' = -z$, Fig. 1.14b))?

(c) How does the cross product (1.13) of two vectors transform under inversion? (The

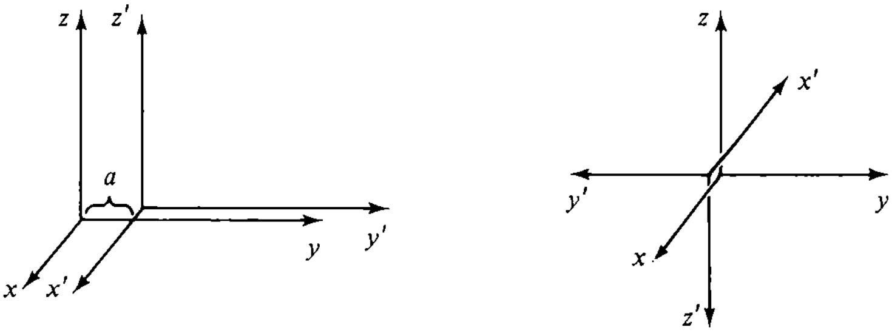

(a) (b) **Figure 1.14**

[2]There is a delicate issue here: Remember that a vector has no *location* (only magnitude and direction). The arrow from (0, 0, 0) to (0, 1, 2) is exactly the same *vector* as the arrow from (3, 8, 4) to (3, 9, 6). We sometimes speak casually of the "position" vector, going from the origin to a given point (x, y, z), but technically this is a bona fide vector only if by "origin" we mean a particular point in space, irrespective of the coordinate system used. Vectors are geometrical entities that exist "out there" entirely independent of our labels—though their *components do,* of course, depend on the specific coordinate system chosen.

cross-product of two vectors is properly called a **pseudovector** because of this anomalous behavior.) Is the cross product of two pseudovectors a vector, or a pseudovector? How about the cross product of a vector and a pseudovector? Is the vector triple product of three vectors a vector, or a pseudovector? Name two pseudovector quantities in classical mechanics.

(d) How does the scalar triple product of three vectors transform under inversions? (Such an object is called a **pseudoscalar.**)

1.2 DIFFERENTIAL CALCULUS

1.2.1 "Ordinary" Derivatives

Question. Suppose we have a function of one variable: $f(x)$. What does the derivative, df/dx, do for us? *Answer*: It tells us how rapidly the function $f(x)$ varies when we change the argument x by a tiny amount, dx:

$$df = \left(\frac{df}{dx}\right) dx \tag{1.23}$$

In words: If we change x by an amount dx, then f changes by an amount df; the derivative is the proportionality factor. For example, in Fig. 1.15(a), the function varies slowly with x, and the derivative is correspondingly small. In Fig. 1.15(b), f increases rapidly with x, and the derivative is large, as you get away from $x = 0$.

Geometrical Interpretation. The derivative df/dx is the *slope* of the curve $f(x)$.

1.2.2 Gradient

Suppose, next, that we have a function of *three* variables—say, the temperature $T(x, y, z)$ in a room. (Start out in one corner, and set up a system of axes; then for each point (x, y, z) in the room, T gives the temperature at that spot.) We want to generalize the notion of "derivative" to functions like T, which depend not on *one* but on *three* variables.

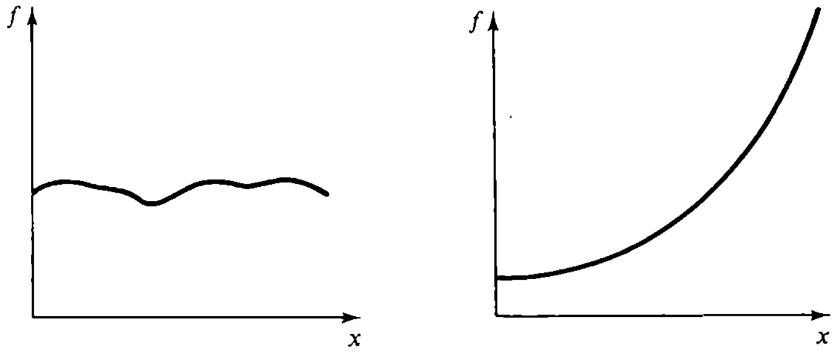

(a) (b) **Figure 1.15**

Now a derivative is supposed to tell us how fast the function varies, if we move a little distance. But this time the problem is more complicated, because obviously it depends on what *direction* we move: If we go straight up, then the temperature will probably increase fairly rapidly, but if we move horizontally, it may not change much at all; and if we look along some intermediate direction, the rate of change will presumably be somewhere in between. In short, the question "How fast does $T(x, y, z)$ vary?" has an infinite number of answers, one for each direction we might choose to explore. It appears, then, that we shall need infinitely many "derivatives" to describe the variation of $T(x, y, z)$.

Fortunately, the problem is not as bad as it looks. A theorem on partial derivatives states that

$$dT = \left(\frac{\partial T}{\partial x}\right) dx + \left(\frac{\partial T}{\partial y}\right) dy + \left(\frac{\partial T}{\partial z}\right) dz \tag{1.24}$$

This rule tells us how T changes when we alter all three variables by the infinitesimal amounts dx, dy, dz. Notice that we do *not* require an infinite number of derivatives—*three* will suffice: the *partial* derivatives along each of the three coordinate directions. What I'm going to do now is to rewrite this rule in a more convenient form and extract some of its implications.

Equation (1.24) is reminiscent of a dot product:

$$\begin{aligned} dT &= \left(\frac{\partial T}{\partial x}\hat{i} + \frac{\partial T}{\partial y}\hat{j} + \frac{\partial T}{\partial z}\hat{k}\right)\cdot(dx\hat{i} + dy\hat{j} + dz\hat{k}) \\ &= (\nabla T)\cdot(d\mathbf{l}) \end{aligned} \tag{1.25}$$

Here I have introduced the shorthand notation

$$d\mathbf{l} \equiv dx\hat{i} + dy\hat{j} + dz\hat{k} \tag{1.26}$$

(the **infinitesimal displacement vector**) and

$$\nabla T \equiv \frac{\partial T}{\partial x}\hat{i} + \frac{\partial T}{\partial y}\hat{j} + \frac{\partial T}{\partial z}\hat{k} \tag{1.27}$$

The latter is called the **gradient** of T; it is a *vector* quantity, with three components, and it is the generalization of the derivative for which we have been looking. Equation (1.25) is the three-dimensional version of (1.23).

Geometrical Interpretation of the Gradient. Like any vector, a gradient has *magnitude* and *direction.* To determine its geometrical meaning, let's rewrite the dot product in (1.25) in its abstract form:

$$dT = \nabla T\cdot d\mathbf{l} = |\nabla T||d\mathbf{l}|\cos\theta \tag{1.28}$$

where θ is the angle between ∇T and $d\mathbf{l}$. Now, if we *fix* the *magnitude* $|d\mathbf{l}|$ and search around in various *directions* (that is, vary θ), the *maximum* change in T evidently occurs when $\theta = 0$ (for then $\cos\theta = 1$). That is, for a fixed distance $|d\mathbf{l}|$, dT is greatest when I move in the *same direction* as ∇T. Thus,

> The gradient ∇T points in the direction of maximum increase of the function T.

Moreover,

> The magnitude $|\nabla T|$ gives the slope (rate of increase) along this maximal direction.

Imagine you are standing on a hillside. Look all around you, and find the direction of steepest ascent. That is the *direction* of the gradient. Now measure the *slope* in that direction (rise over run). That is the *magnitude* of the gradient. (Here the function we're talking about is the height of the hill, and the coordinates it depends on are positions—latitude and longitude, say. This function depends only on *two* variables, not *three*, but the geometrical meaning of the gradient is easier to grasp in two dimensions.) Notice from (1.28) that the direction of maximum *descent* is opposite to the direction of maximum *ascent*, while at right angles ($\theta = 90°$) the slope is zero (the gradient is perpendicular to the contour lines). You can conceive of surfaces that do not have these properties, but they always have "kinks" in them and correspond to nondifferentiable functions.

What would it mean for the gradient to vanish? If $\nabla T = 0$ at (x, y, z), then $dT = 0$ for small displacements about the point (x, y, z). This is, then, a **stationary point** of the function $T(x, y, z)$. It could be a maximum (a summit), a minimum (a valley), a "saddle point" (a pass), or a "shoulder." This is analogous to the situation for functions of *one* variable, where a vanishing derivative signals a maximum, a minimum, or an inflection. In particular, if you want to locate the extrema of a function of three variables, set its gradient equal to zero.

Example 3

Let $r(x, y, z)$ be the distance from the origin to the point (x, y, z); that is, $r = \sqrt{x^2 + y^2 + z^2}$ (Fig. 1.16). Let us figure out what ∇r is *before* calculating it, and then check that we were right. First, what's the *direction*? Clearly, r increases fastest as we move out *radially, away from the origin.* Moreover, if we go out radially a distance dl, then r increases by exactly that amount:

$$dr = dl$$

So the magnitude of ∇r is evidently 1. Hence, ∇r is a unit vector in the radial direction:

$$\nabla r = \hat{r}$$

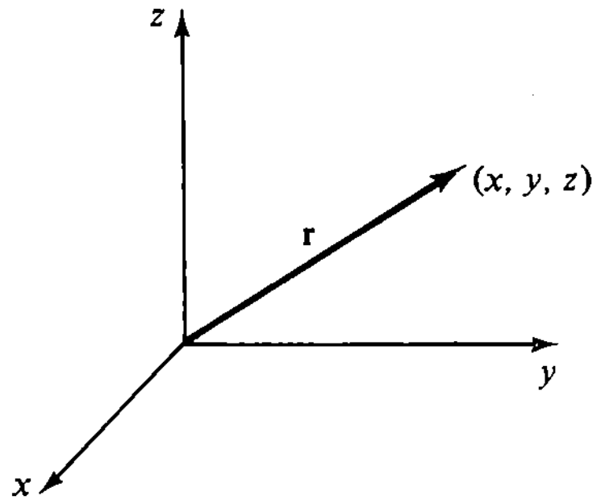

Figure 1.16

Now let's *check* it:

$$\nabla r = \frac{\partial r}{\partial x}\hat{\imath} + \frac{\partial r}{\partial y}\hat{\jmath} + \frac{\partial r}{\partial z}\hat{k} = \frac{1}{2}\frac{2x}{\sqrt{x^2+y^2+z^2}}\hat{\imath} + \frac{1}{2}\frac{2y}{\sqrt{x^2+y^2+z^2}}\hat{\jmath}$$

$$+ \frac{1}{2}\frac{2z}{\sqrt{x^2+y^2+z^2}}\hat{k} = \frac{x\hat{\imath} + y\hat{\jmath} + z\hat{k}}{\sqrt{x^2+y^2+z^2}} = \frac{\mathbf{r}}{r} = \hat{r}$$

Here $\mathbf{r} = x\hat{\imath} + y\hat{\jmath} + z\hat{k}$ is the vector from the origin to (x, y, z).

Problem 1.11 Find the gradients of the following functions:
(a) $f(x, y, z) = x^2 + y^3 + z^4$.
(b) $f(x, y, z) = x^2y^3z^4$.
(c) $f(x, y, z) = e^x \sin(y) \ln(z)$.

Problem 1.12 The height of a certain hill (in feet) is given by

$$h(x, y) = 10(2xy - 3x^2 - 4y^2 - 18x + 28y + 12)$$

where y is the distance (in miles) north, x the distance east of South Hadley.
(a) Where is the top of the hill located?
(b) How high is the hill?
(c) How steep is the slope (in feet per mile) at a point 1 mile north and one mile east of South Hadley? In what direction is the slope steepest, at that point?

• **Problem 1.13** Let $\boldsymbol{\imath}$ be the vector from some fixed point (x_0, y_0, z_0) to the point (x, y, z), and let $\imath$ be its length. Show that
(a) $\nabla(\imath^2) = 2\boldsymbol{\imath}$
(b) $\nabla\left(\frac{1}{\imath}\right) = -\frac{\hat{\boldsymbol{\imath}}}{\imath^2}$ ($\hat{\boldsymbol{\imath}}$ is a unit vector in the direction of $\boldsymbol{\imath}$).
(c) What is the *general* formula for $\nabla(\imath^n)$?

! **Problem 1.14** Suppose that f is a function of two variables (y and z) only. Show that the gradient $\nabla f = (\partial f/\partial y)\hat{\jmath} + (\partial f/\partial z)\hat{k}$ transforms as a vector under rotations (1.19). (*Hint*: $(\partial f/\partial y') = (\partial f/\partial y)(\partial y/\partial y') + (\partial f/\partial z)(\partial z/\partial y')$ (and an analogous formula for $\partial f/\partial z'$). We know that $y' = y\cos\theta + z\sin\phi$, $z' = -y\sin\phi + z\cos\phi$; "solve" these equations for y and z (as functions of y' and z') and compute the needed derivatives $\partial y/\partial y'$, $\partial z/\partial y'$, etc.)

1.2.3 The Operator ∇

The gradient has the formal appearance of a vector, ∇, "multiplying" a scalar T:

$$\nabla T = \left(\hat{\imath}\frac{\partial}{\partial x} + \hat{\jmath}\frac{\partial}{\partial y} + \hat{k}\frac{\partial}{\partial z}\right)T \tag{1.29}$$

(I write the unit vectors to the *left*, just so no one will think this means $\partial\hat{\imath}/\partial x$, and so on—which would be zero, since $\hat{\imath}$ is constant.) The term in parentheses is called "del":

$$\boxed{\nabla = \hat{\imath}\frac{\partial}{\partial x} + \hat{\jmath}\frac{\partial}{\partial y} + \hat{k}\frac{\partial}{\partial z}} \tag{1.30}$$

Of course, del is *not* a vector, in the usual sense. Indeed, it is without specific meaning until we provide it with a function to act upon. Furthermore, it does not "multiply" T; rather, it is an instruction to *differentiate* that function. To be precise, then, we should say that ∇ is a **vector operator** that *acts upon* T, not a vector that multiplies T.

With this qualification, though, ∇ mimics the behavior of an ordinary vector in virtually every way; almost anything that can be done with other vectors can also be done with ∇, if we merely translate "multiply" by "act upon." So by all means take the vector appearance of ∇ seriously: it is a marvellous piece of notational simplification, as you will appreciate if you ever consult Maxwell's original work on electromagnetism, written without the benefit of ∇.

Now an ordinary vector $\mathbf{A}$ can multiply in three ways:

1. Multiply a scalar a: $\mathbf{A}a$;
2. Multiply another vector $\mathbf{B}$, via the dot product: $\mathbf{A}\cdot\mathbf{B}$;
3. Multiply another vector via the cross product: $\mathbf{A}\times\mathbf{B}$.

Correspondingly, there are three ways the operator ∇ can act:

1. On a scalar function T: ∇T (the gradient);
2. On a vector function $\mathbf{v}$, via the dot product: $\nabla\cdot\mathbf{v}$ (the "divergence");
3. On a vector function, via the cross product: $\nabla\times\mathbf{v}$ (the "curl").

We have already discussed the gradient. In the following sections we examine the other two varieties of vector derivatives: divergence and curl.

1.2.4 The Divergence

From the definition of ∇ (1.30), we have

$$\begin{aligned}\nabla\cdot\mathbf{v} &= \left(\hat{\imath}\frac{\partial}{\partial x} + \hat{\jmath}\frac{\partial}{\partial y} + \hat{k}\frac{\partial}{\partial z}\right)\cdot(\hat{\imath}v_x + \hat{\jmath}v_y + \hat{k}v_z)\\ &= \frac{\partial v_x}{\partial x} + \frac{\partial v_y}{\partial y} + \frac{\partial v_z}{\partial z}.\end{aligned} \tag{1.31}$$

Observe that the divergence of a *vector* function $\mathbf{v}$ is itself a *scalar* $\nabla\cdot\mathbf{v}$. (You can't have the divergence of a scalar: that's meaningless.)

Geometrical Interpretation. The name **divergence** is well chosen, for $\nabla\cdot\mathbf{v}$ is a measure of how much the vector $\mathbf{v}$ spreads out (diverges) from the point in question. The vector function in Fig. 1.17 has a large (positive) divergence at the point P; it is spreading out. (If the arrows pointed *in*, it would be a large *negative* divergence.) On

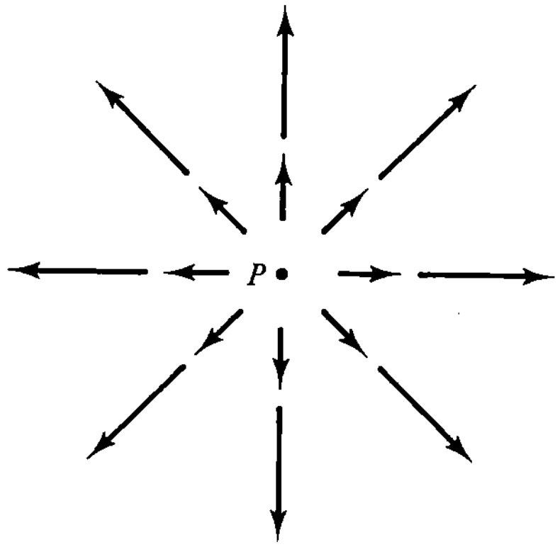

Figure 1.17

the other hand, the function in Fig. 1.18 has zero divergence at P; it is not spreading out at all. (Please understand that $\mathbf{v}$ here is a *vector-valued function*—in other words, there's a different vector associated with every point in space. In the diagrams, of course, I can only draw the arrows at a few representative locations.)

Example 4

(a) That first function looks about like $\mathbf{v}_1 = \mathbf{r}$, since each vector points radially out from the origin P and has a length equal to the distance from P:

$$\mathbf{v}_1 = x\hat{\imath} + y\hat{\jmath} + z\hat{k}$$

so

$$\nabla \cdot \mathbf{v}_1 = \frac{\partial}{\partial x}(x) + \frac{\partial}{\partial y}(y) + \frac{\partial}{\partial z}(z) = 1 + 1 + 1 = 3$$

As anticipated, this function has a positive divergence.

(b) The second function is $\mathbf{v}_2 = \hat{k}$ (say) they're all unit vectors pointing in the z-direction. So,

$$\nabla \cdot \mathbf{v}_2 = \frac{\partial}{\partial x}(0) + \frac{\partial}{\partial y}(0) + \frac{\partial}{\partial z}(1) = 0 + 0 + 0 = 0$$

as expected.

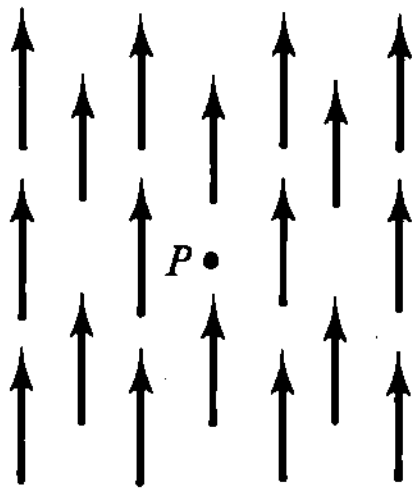

Figure 1.18

Problem 1.15 Calculate the divergence of the following vector functions:

(a) $\mathbf{v}_1 = x^2\hat{i} + 3xz^2\hat{j} - 2xz\hat{k}$

(b) $\mathbf{v}_2 = xy\hat{i} + 2yz\hat{j} + 3zx\hat{k}$

(c) $\mathbf{v}_3 = y^2\hat{i} + (2xy + z^2)\hat{j} + 2yz\hat{k}$

Problem 1.16 Show that $\nabla\cdot(t\mathbf{v}) = (\nabla t)\cdot\mathbf{v} + t(\nabla\cdot\mathbf{v})$, for a scalar function t and vector function $\mathbf{v}$.

• **Problem 1.17** Sketch the vector function

$$\mathbf{v} = \frac{\hat{\mathbf{r}}}{r^2}$$

(r is defined in Problem 1.13), and compute its divergence. The answer may surprise you . . . can you explain it?

! **Problem 1.18** In two dimensions, show that the divergence transforms as a scalar under rotations. (*Hint*: Use (1.19) to determine v_y' and v_z' and the method of Problem 1.14 to calculate the derivatives. Your aim is to show that

$$\partial v_y'/\partial y' + \partial v_z'/\partial z' = \partial v_y/\partial y + \partial v_z/\partial z.)$$

1.2.5 The Curl

From the definition of ∇ in (1.30), we have:

$$\nabla \times \mathbf{v} = \begin{vmatrix} \hat{i} & \hat{j} & \hat{k} \\ \frac{\partial}{\partial x} & \frac{\partial}{\partial y} & \frac{\partial}{\partial z} \\ v_x & v_y & v_z \end{vmatrix}$$

$$= \hat{i}\left(\frac{\partial v_z}{\partial y} - \frac{\partial v_y}{\partial z}\right) + \hat{j}\left(\frac{\partial v_x}{\partial z} - \frac{\partial v_z}{\partial x}\right) + \hat{k}\left(\frac{\partial v_y}{\partial x} - \frac{\partial v_x}{\partial y}\right) \quad (1.32)$$

Notice that the curl of a *vector* function $\mathbf{v}$ is, like any cross product, a *vector.* (You cannot have the curl of a scalar; that's meaningless.)

Geometrical Interpretation. The name **curl** is also well chosen, for $\nabla \times \mathbf{v}$ is a measure of how much the vector $\mathbf{v}$ curls around the point in question. Thus, the two functions $\mathbf{v}_1$ and $\mathbf{v}_2$ in Example 4 both have zero curl (as you can easily check for yourself), whereas the function in Fig. 1.19 has a large curl. In fact, this curl points in the z-direction, as the right-hand rule would suggest.

Imagine you are standing at the edge of a pond. Sprinkle some sawdust or pine needles or something on the surface. If the material spreads out, then you dropped it on a point of positive *divergence.* Now float a small paddlewheel (a cork with toothpicks pointing out radially would do). If it starts to turn, then you placed it at a point of nonzero *curl.* A point of large curl is a whirlpool; a point of large positive divergence is a source (or faucet); a point of large negative divergence is a sink (or drain).

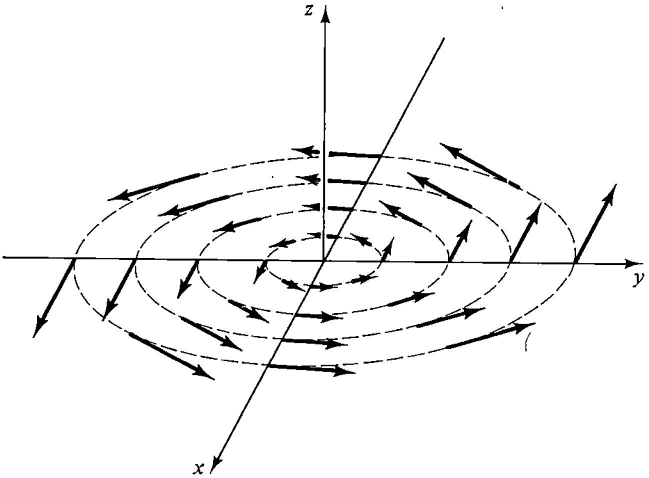

Figure 1.19

The vector function $\mathbf{v}$ in this model is the velocity of the surface water. As in the example for gradients, this function really depends only on *two* coordinates and has only *two* components—but I'm trying to give you a "feel" for what the divergence and curl are, and it's a bit easier to do this in two dimensions.

Example 5

The function sketched in Fig. 1.19 is

$$\mathbf{v}_3 = -y\hat{\imath} + x\hat{\jmath}$$

so

$$\nabla \times \mathbf{v}_3 = \begin{vmatrix} \hat{\imath} & \hat{\jmath} & \hat{k} \\ \dfrac{\partial}{\partial x} & \dfrac{\partial}{\partial y} & \dfrac{\partial}{\partial z} \\ -y & x & 0 \end{vmatrix} = 2\hat{k}$$

As expected, the curl of $\mathbf{v}_3$ points in the z-direction. (Incidentally, this function has zero divergence, as you might guess from the picture: nothing is "spreading out"—it just "curls around.")

Problem 1.19 Calculate the curls of the three vector functions of Problem 1.15.

Problem 1.20 Construct a vector function that has zero divergence and zero curl everywhere. (A *constant* will do the job, of course, but make it something a little more interesting than that!)

1.2.6 Product Rules

The calculation of ordinary derivatives is facilitated by a number of general rules, such as the sum rule,

$$\frac{d}{dx}(f+g)=\frac{df}{dx}+\frac{dg}{dx}$$

the rule for multiplication by a constant,

$$\frac{d}{dx}(kf)=k\frac{df}{dx}$$

the product rule,

$$\frac{d}{dx}(fg)=f\frac{dg}{dx}+g\frac{df}{dx}$$

and the quotient rule,

$$\frac{d}{dx}\left(\frac{f}{g}\right)=\frac{g\frac{df}{dx}-f\frac{dg}{dx}}{g^2}$$

Similar relations hold for the vector derivatives. Thus,

$$\nabla(f+g)=\nabla f+\nabla g, \qquad \nabla\cdot(\mathbf{A}+\mathbf{B})=(\nabla\cdot\mathbf{A})+(\nabla\cdot\mathbf{B}),$$
$$\nabla\times(\mathbf{A}+\mathbf{B})=(\nabla\times\mathbf{A})+(\nabla\times\mathbf{B})$$

and

$$\nabla(kf)=k\nabla f, \qquad \nabla\cdot(k\mathbf{A})=k(\nabla\cdot\mathbf{A}), \qquad \nabla\times(k\mathbf{A})=k(\nabla\times\mathbf{A})$$

as you can check for yourself. The product rules are not quite so simple. There are two ways to construct a scalar as the product of two functions:

fg (product of two scalar functions)

$\mathbf{A}\cdot\mathbf{B}$ (dot product of two vector functions)

and two ways to make a vector:

$f\mathbf{A}$ (scalar times vector)

$(\mathbf{A}\times\mathbf{B})$ (cross product of two vectors)

Accordingly, there are *six* product rules, two for gradients:

(i) $$\nabla(fg)=f\nabla g+g\nabla f$$

(ii) $$\nabla(\mathbf{A}\cdot\mathbf{B})=\mathbf{A}\times(\nabla\times\mathbf{B})+\mathbf{B}\times(\nabla\times\mathbf{A})+(\mathbf{A}\cdot\nabla)\mathbf{B}+(\mathbf{B}\cdot\nabla)\mathbf{A}$$

two for divergences:

(iii) $$\nabla\cdot(f\mathbf{A})=f(\nabla\cdot\mathbf{A})+\mathbf{A}\cdot(\nabla f)$$

(iv) $$\nabla\cdot(\mathbf{A}\times\mathbf{B}) = \mathbf{B}\cdot(\nabla\times\mathbf{A}) - \mathbf{A}\cdot(\nabla\times\mathbf{B})$$

and two for curls:

(v) $$\nabla\times(f\mathbf{A}) = f(\nabla\times\mathbf{A}) - \mathbf{A}\times(\nabla f)$$

(vi) $$\nabla\times(\mathbf{A}\times\mathbf{B}) = (\mathbf{B}\cdot\nabla)\mathbf{A} - (\mathbf{A}\cdot\nabla)\mathbf{B} + \mathbf{A}(\nabla\cdot\mathbf{B}) - \mathbf{B}(\nabla\cdot\mathbf{A})$$

You will be using these product rules so frequently that I have put them on the inside front cover for easy reference. The proofs come straight from the product rule for ordinary derivatives. For instance,

$$\begin{aligned}\nabla\cdot(f\mathbf{A}) &= \frac{\partial}{\partial x}(fA_x) + \frac{\partial}{\partial y}(fA_y) + \frac{\partial}{\partial z}(fA_z)\\ &= \left(\frac{\partial f}{\partial x}A_x + f\frac{\partial A_x}{\partial x}\right) + \left(\frac{\partial f}{\partial y}A_y + f\frac{\partial A_y}{\partial y}\right) + \left(\frac{\partial f}{\partial z}A_z + f\frac{\partial A_z}{\partial z}\right)\\ &= (\nabla f)\cdot\mathbf{A} + f(\nabla\cdot\mathbf{A})\end{aligned}$$

It is also possible to formulate three quotient rules:

$$\nabla\left(\frac{f}{g}\right) = \frac{g\nabla f - f\nabla g}{g^2}$$

$$\nabla\cdot\left(\frac{\mathbf{A}}{g}\right) = \frac{g(\nabla\cdot\mathbf{A}) - \mathbf{A}\cdot(\nabla g)}{g^2}$$

$$\nabla\times\left(\frac{\mathbf{A}}{g}\right) = \frac{g(\nabla\times\mathbf{A}) + \mathbf{A}\times(\nabla g)}{g^2}$$

However, since these can be obtained quickly from the corresponding product rules, I haven't bothered to put them on the inside front cover.

Problem 1.21 Prove product rules (i), (iv), and (v).

Problem 1.22 (a) If **A** and **B** are two vector functions, what does the expression $(\mathbf{A}\cdot\nabla)\mathbf{B}$ mean? (That is, what are its x-, y-, and z-components in terms of the Cartesian components of **A**, **B**, and ∇?)

(b) Compute $(\hat{r}\cdot\nabla)\hat{r}$, where $\hat{r}$ is the unit vector defined in Example 3.

(c) For the functions in Problem 1.15, evaluate $(\mathbf{v}_1\cdot\nabla)\mathbf{v}_2$.

Problem 1.23 (For masochists only.) Prove product rules (ii) and (vi). Refer to Problem 1.22 for the definition of $(\mathbf{A}\cdot\nabla)\mathbf{B}$.

Problem 1.24 Derive the three quotient rules.

Problem 1.25 (a) Check product rule (iv) (by calculating each term separately) for the functions

$$\mathbf{A} = x\hat{i} + 2y\hat{j} + 3z\hat{k};\qquad \mathbf{B} = 3y\hat{i} - 2x\hat{j}$$

(b) Do the same for product rule (ii).

(c) The same for rule (vi).

1.2.7 Second Derivatives

Gradient, divergence, and curl are all the first derivatives we can make with ∇; by applying ∇ *twice* we can construct five species of *second* derivatives. The gradient ∇T is a *vector*, so we can take the *divergence* and *curl* of it:

1. Divergence of gradient: $\nabla \cdot (\nabla T)$
2. Curl of gradient: $\nabla \times (\nabla T)$

The divergence $\nabla \cdot \mathbf{v}$ is a *scalar*—all we can do it take its *gradient:*

3. Gradient of divergence: $\nabla(\nabla \cdot \mathbf{v})$

The curl $\nabla \times \mathbf{v}$ is a *vector*, so we can take its *divergence* and *curl:*

4. Divergence of curl: $\nabla \cdot (\nabla \times \mathbf{v})$
5. Curl of curl: $\nabla \times (\nabla \times \mathbf{v})$

These five types of second derivative exhaust the possibilities and, in fact, not all of them give anything new. Let's consider them one at a time:

1. $$\nabla \cdot (\nabla T) = \left(\hat{i}\frac{\partial}{\partial x} + \hat{j}\frac{\partial}{\partial y} + \hat{k}\frac{\partial}{\partial z}\right) \cdot \left(\hat{i}\frac{\partial T}{\partial x} + \hat{j}\frac{\partial T}{\partial y} + \hat{k}\frac{\partial T}{\partial z}\right)$$
$$= \frac{\partial^2 T}{\partial x^2} + \frac{\partial^2 T}{\partial y^2} + \frac{\partial^2 T}{\partial z^2} \tag{1.33}$$

This object, which we write $\nabla^2 T$ for short, is called the **Laplacian** of T; we shall be studying it in great detail later on. Notice that the Laplacian of a *scalar* T is a *scalar*. Occasionally, we shall speak of the Laplacian of a *vector*, $\nabla^2 \mathbf{v}$. By this we mean a *vector* quantity whose x-component is the Laplacian of v_x, and so on:[3]

$$\nabla^2 \mathbf{v} \equiv (\nabla^2 v_x)\hat{i} + (\nabla^2 v_y)\hat{j} + (\nabla^2 v_z)\hat{k} \tag{1.34}$$

This is nothing more than a convenient *extension* of the meaning of ∇^2.

2. The curl of a gradient is always *zero:*

$$\nabla \times (\nabla T) = 0 \tag{1.35}$$

This is an important fact, which we shall use repeatedly; you can easily prove it from the definition of ∇, equation (1.30). *Beware*: You might think equation (1.35) is "obviously" true—isn't it just $(\nabla \times \nabla)T$, and isn't the cross product of any vector (in this case, ∇) with itself always zero? This reasoning is *suggestive* but not quite *conclusive*, since ∇ is an *operator* and does not "multiply" in the usual way. The proof of (1.35), in fact, hinges on the equality of cross derivatives:

[3] In curvilinear coordinates, where the unit vectors themselves depend on position, they too must be differentiated (see Section 1.4.1).

$$\frac{\partial}{\partial x}\left(\frac{\partial T}{\partial y}\right) = \frac{\partial}{\partial y}\left(\frac{\partial T}{\partial x}\right) \tag{1.36}$$

If you think I'm being fussy, test your intuition on this one:

$$(\nabla T) \times (\nabla S)$$

Is *that* always zero? (It *would* be, of course, if you replaced the ∇'s by an ordinary vector.)

3. $\nabla(\nabla \cdot \mathbf{v})$ for some reason seldom occurs in physical applications, and it has not been given any special name of its own—it's just **the gradient of the divergence.** Notice that $\nabla(\nabla \cdot \mathbf{v})$ is *not* the same as the Laplacian of a vector: $\nabla^2 \mathbf{v} = (\nabla \cdot \nabla)\mathbf{v} \neq \nabla(\nabla \cdot \mathbf{v})$.

4. The divergence of a curl, like the curl of a gradient, is *always zero:*

$$\nabla \cdot (\nabla \times \mathbf{v}) = 0 \tag{1.37}$$

You can prove this for yourself. (Again, there is a fraudulent short-cut proof, using the vector identity $\mathbf{A} \cdot (\mathbf{B} \times \mathbf{C}) = (\mathbf{A} \times \mathbf{B}) \cdot \mathbf{C}$.)

5. As you can check from the definition of ∇:

$$\nabla \times (\nabla \times \mathbf{v}) = \nabla(\nabla \cdot \mathbf{v}) - \nabla^2 \mathbf{v} \tag{1.38}$$

Thus, curl-of-curl gives us nothing new; the first term is just number (3) and the second is the Laplacian (of a vector). (In fact, (1.38) is often used to *define* the Laplacian of a vector, in preference to (1.34) which makes special reference to Cartesian coordinates.)

Really, then, there are just two kinds of second derivatives: the Laplacian (which is of fundamental importance) and the gradient-of-divergence (which we seldom encounter). We could go through a similar ritual to work out *third* derivates, but fortunately second derivatives suffice for practically all physical applications.

A final word on vector differential calculus: It *all* flows from the operator ∇, equation (1.30), and from taking seriously its vector character. Even if you remembered *only* the definition of ∇, you should be able, in principle, to reconstruct all the rest.

Problem 1.26 Calculate the Laplacian of the following functions:

(a) $T = x^2 + 2xy + 3z + 4$

(b) $T = \sin x \sin y \sin z$

(c) $T = e^{-5x} \sin 4y \cos 3z$

(d) $\mathbf{v} = x^2\hat{i} + 3xz^2\hat{j} - 2xz\hat{k}$

Problem 1.27 Prove that the divergence of a curl is always zero. *Check* it for function $\mathbf{v}_2$ in Problem 1.15.

Problem 1.28 Prove that the curl of a gradient is always zero. *Check* it for function (b) in Problem 1.11.

1.3 INTEGRAL CALCULUS

1.3.1 "Ordinary" Integration

Suppose $f(x)$ is a function of one variable. The **fundamental theorem of calculus** states:

$$\int_a^b \left(\frac{df}{dx}\right) dx = f(b) - f(a) \tag{1.39}$$

In case this doesn't look familiar, let's write it another way:

$$\int_a^b F(x)\, dx = f(b) - f(a)$$

where $df/dx = F(x)$. The fundamental theorem tells you how to integrate $F(x)$: you must think up a function $f(x)$ whose *derivative* is equal to F.

Geometrical Interpretation. According to equation (1.23), $df = (df/dx)\, dx$ is the infinitesimal change in f when you go from (x) to $(x + dx)$. The fundamental theorem (1.39) says that if you chop the interval from a to b (Fig. 1.20) into many tiny pieces dx, and add up the increments df from each little piece, the result is (not surprisingly) equal to the total change in f: $f(b) - f(a)$. In other words, there are two ways to determine the total change in the function: *either* subtract the values at the ends *or* go step-by-step, adding up all the tiny increments as you go. You'll get the same answer either way.

Notice the basic format of the fundamental theorem: *the integral of a derivative over some interval is given by the value of the function at the end points (or boundaries).* In vector calculus there are three species of derivative (gradient, divergence, and curl), and each has its own "fundamental theorem," with this same essential format. I shall not prove these theorems here; rather I shall try to explain what they *mean* and make them seem *plausible.* Proofs appear in Appendix A.

1.3.2 The Fundamental Theorem for Gradients

Suppose we have a scalar function of three variables $T(x, y, z)$. Starting at point $\mathbf{a} = (a_x, a_y, a_z)$, we move a small distance $d\mathbf{l}_1$ (Fig. 1.21)). According to equation (1.25), the function T will change by an amount

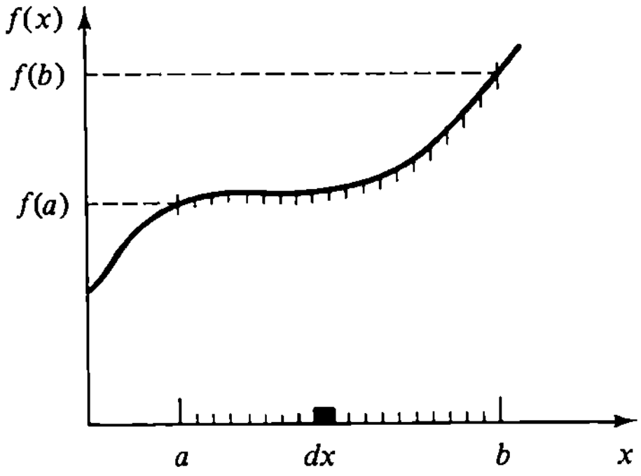

Figure 1.20

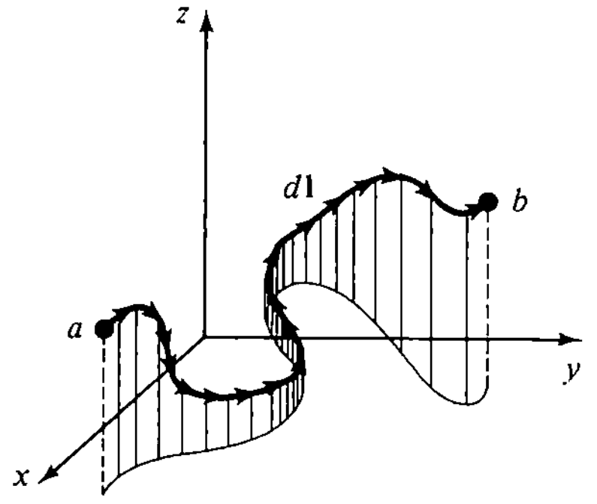

Figure 1.21

$$dT = (\nabla T) \cdot d\mathbf{l}_1$$

Now we move a little further, by an additional small distance $d\mathbf{l}_2$; the incremental change in T will be $(\nabla T) \cdot d\mathbf{l}_2$. In this manner, proceeding by infinitesimal steps, we make the journey to point $b = (b_x, b_y, b_z)$. At each step we compute the gradient of T (at that point) and dot it into the displacement $d\mathbf{l}$—this gives us the change in T. Evidently the *total* change in T in going from a to b *over the path selected* is

$$\boxed{\int_{\substack{a \\ \text{line}}}^{b} (\nabla T) \cdot d\mathbf{l} = T(b) - T(a)} \tag{1.40}$$

This is the **fundamental theorem for gradients;** like the "ordinary" fundamental theorem, it says that the integral (here a "line" or "path" integral along some selected curve) of a derivative (here the gradient) is given by the value of the function at the boundaries (a and b).

Incidentally, I have been careful to emphasize the words "over the selected path" because, off-hand, one would suppose that the integral on the left side of (1.40) depends critically on the path taken from a to b. But wait: the *right* side of this equation doesn't depend on the path *at all*! Only on the end points. Apparently, then:

Corollary 1: $\int_a^b (\nabla T) \cdot d\mathbf{l}$ is independent of path taken from a to b.

Corollary 2: $\oint (\nabla T) \cdot d\mathbf{l} = 0$. The circle on the integral sign indicates that the integration is to be taken around a closed loop. Since the beginning and end points are identical, $T(b) - T(a) = 0$.

Example 6

Let $T = xy^2$, and take point a to be the origin (0, 0, 0) and b the point (2, 1, 0). Check the fundamental theorem for gradients.

Solution: Although the integral is independent of path, we must *pick* a specific path in order to evaluate it. Let's go out along the x axis (step I) and then

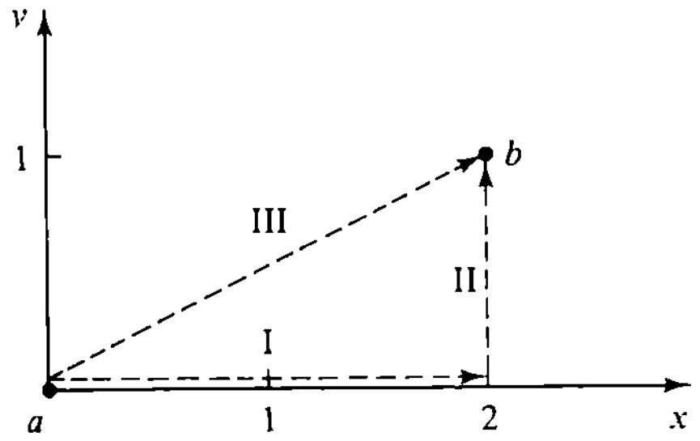

Figure 1.22

up (step II) (Fig. 1.22). As always, $d\mathbf{l} = dx\,\hat{\imath} + dy\,\hat{\jmath} + dz\,\hat{k}$; in this case $\nabla T = y^2\hat{\imath} + 2xy\hat{\jmath}$.

Step I: $x: 0 \to 2$; $y = 0$; $dy = dz = 0$. $\nabla T \cdot d\mathbf{l} = y^2\,dx = 0$ for this segment. So,

$$\int_{\mathrm{I}} \nabla T \cdot d\mathbf{l} = 0$$

Step II: $x = 2$; $y: 0 \to 1$; $dx = dz = 0$. $\nabla T \cdot d\mathbf{l} = 2xy\,dy = 4y\,dy$. So,

$$\int_{\mathrm{II}} \nabla T \cdot d\mathbf{l} = \int_0^1 4y\,dy = 2y^2\Big|_0^1 = 2$$

Thus, $\int_a^b \nabla T \cdot d\mathbf{l} = 2$. Is this consistent with the fundamental theorem? Yes, for $T(b) - T(a) = 2 - 0 = 2$.

Now, just to convince you that the answer is independent of path, let me calculate the same integral along path III (the straight line from a to b):

Path III: $x: 0 \to 2$; $y = \frac{1}{2}x$; $dy = \frac{1}{2}\,dx$. $\nabla T \cdot d\mathbf{l} = y^2\,dx + 2xy\,dy = \frac{3}{4}x^2\,dx$. So,

$$\int_{\mathrm{III}} \nabla T \cdot d\mathbf{l} = \int_0^2 \tfrac{3}{4}x^2\,dx = \tfrac{1}{4}x^3\Big|_0^2 = 2$$

(The strategy here is to get everything in terms of one variable; I could just as well have eliminated x in favor of y.)

Problem 1.29 Check the fundamental theorem for gradients, using the function $T = x^2 + 4xy + 2yz^3$ and the points $a = (0, 0, 0)$, $b = (1, 1, 1)$, and the following three paths (Fig. 1.23):

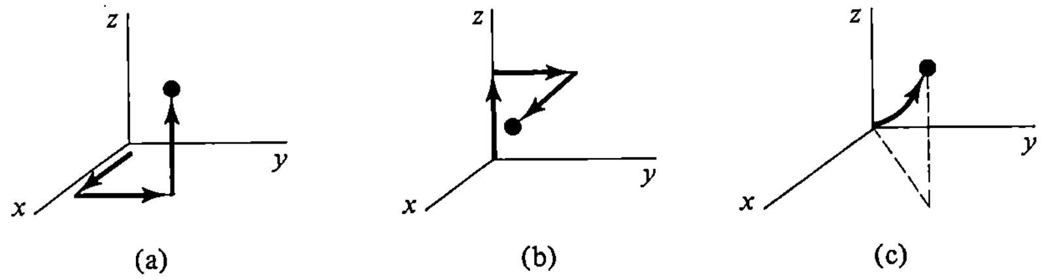

Figure 1.23

(a) $(0, 0, 0) \rightarrow (1, 0, 0) \rightarrow (1, 1, 0) \rightarrow (1, 1, 1)$
(b) $(0, 0, 0) \rightarrow (0, 0, 1) \rightarrow (0, 1, 1) \rightarrow (1, 1, 1)$
(c) The parabolic path $z = x^2$; $x = y$

Problem 1.30 Although the line integral of a *gradient* is independent of path (Corollary 1), the same is *not* true for *arbitrary* vector functions. Calculate

$$\int_a^b \mathbf{v} \cdot d\mathbf{l} \quad \text{from} \quad a = (1, 1, 0) \quad \text{to} \quad b = (2, 2, 0)$$

for the function $\mathbf{v} = y^2\hat{\imath} + 2x(y + 1)\hat{\jmath}$. Do it first by path (1), (Fig. 1.24); then by path

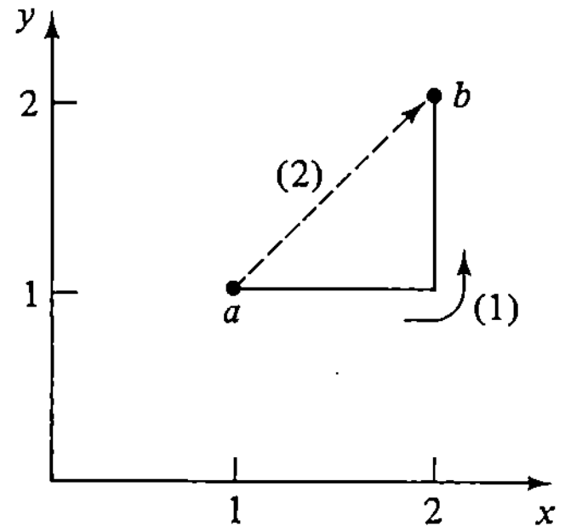

Figure 1.24

(2). What is $\oint \mathbf{v} \cdot d\mathbf{l}$ for the loop, from a to b and back, which goes *out* along (1) and *returns* along (2)?

1.3.3 The Fundamental Theorem for Divergences

The fundamental theorem for divergences states that:

$$\int_{\text{volume}} (\nabla \cdot \mathbf{v})\, d\tau = \oint_{\text{surface}} \mathbf{v} \cdot d\mathbf{a} \tag{1.41}$$

In honor, I suppose, of its great importance, this theorem has at least three special names: **Gauss's theorem, Green's theorem,** or, simply, the **divergence theorem.** Like the other "fundamental theorems," it says that the *integral* of a *derivative* (in this case, the *divergence*) over a *region* (in this case, a *volume*) is equal to the value of the function at the *boundary* (in this case, the *surface* which bounds the volume). $d\tau$ is an element of volume (in Cartesian coordinates, $d\tau = dx\, dy\, dz$), and the volume integration is really a *triple* integral. Instead of writing multiple integral signs, I indicate the integration region by a subscript.

Unlike (1.39) and (1.40), the boundary term in this case is *itself* an integral,

$$\oint_{\text{surface}} \mathbf{v} \cdot d\mathbf{a} \tag{1.42}$$

This is reasonable: the "boundary" of a *line* is just two end points, but the boundary of a *volume* is an entire continuous surface. (I put a circle on the integral to remind you that this is a *closed* surface, with no edges—just as a circle on a *line* integral indicates a closed loop, with no ends.) $d\mathbf{a}$ represents an infinitesimal element of area; it is a *vector*, whose magnitude is the *area* of the element and whose *direction* is perpendicular ("normal") to the surface, pointing *outward*. For instance, suppose the volume in question is the cube shown in Fig. 1.25. On the front face of the cube, a surface element is

$$d\mathbf{a}_1 = (dy\,dz)\hat{\imath}$$

on the right face, it would be

$$d\mathbf{a}_2 = (dx\,dz)\hat{\jmath}$$

whereas for the bottom it is

$$d\mathbf{a}_3 = (dx\,dy)(-\hat{k})$$

Geometrical Interpretation. An integral of the form (1.42) involving the *normal component* of a vector $\mathbf{v}$ integrated over some surface (the dot product picks out the normal component), is called the **flux** of $\mathbf{v}$ through the surface. If, as before, $\mathbf{v}$ represents the velocity of some fluid ($\mathbf{v}$ can, in general, vary from point to point), then the flux of $\mathbf{v}$ is simply the *total amount of fluid passing through the surface per unit time* (see Problem 1.32). Now, recall the geometrical interpretation of the divergence: it measures the "spreading out" of the vectors from a point; a place of high divergence is like a "faucet" pouring out liquid. If we have a lot of faucets in a region filled with incompressible fluid, an equal amount of liquid will be forced out through the boundaries of the region. In fact, there are *two* ways we could determine how much is being produced: (a) we could count up all the faucets, recording how much each puts out or (b) we could go around the boundary, measuring the flow at each point, and add it all up. Provided the fluid is incompressible (which is part of this geometrical picture of a divergence), we will get the same answer either way:

$$\int (\text{all faucets in the volume}) = \int (\text{flow out through the surface})$$

This, in essence, is what the divergence theorem states.

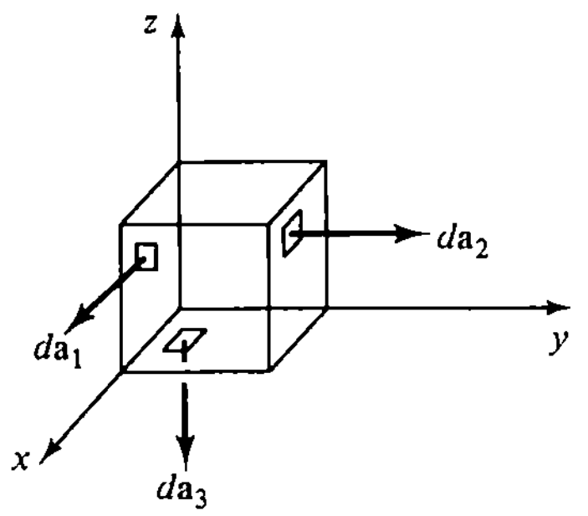

Figure 1.25

Example 7

Check the divergence theorem using the function

$$\mathbf{v} = y^2\hat{i} + (2xy + z^2)\hat{j} + (2yz)\hat{k}$$

and the unit cube situated at the origin (Fig. 1.26).

Solution: In this case

$$\nabla \cdot \mathbf{v} = 2(x + y)$$

and

$$\int_{\text{volume}} 2(x + y)\, d\tau = 2 \int_0^1 \int_0^1 \int_0^1 (x + y)\, dx\, dy\, dz$$

$$\int_0^1 (x + y)\, dx = \tfrac{1}{2} + y, \qquad \int_0^1 (\tfrac{1}{2} + y)\, dy = 1, \qquad \int_0^1 1\, dz = 1$$

So,

$$\int_{\text{volume}} (\nabla \cdot \mathbf{v})\, d\tau = 2$$

So much for the left side of the divergence theorem. To evaluate the surface integral we must consider separately the six sides of the cube:

(i) $\quad d\mathbf{a} = dy\, dz\, \hat{i}; \qquad x = 1; \qquad \mathbf{v} \cdot d\mathbf{a} = y^2\, dy\, dz$

$$\int \mathbf{v} \cdot d\mathbf{a} = \int_0^1 \int_0^1 y^2\, dy\, dz = \tfrac{1}{3}$$

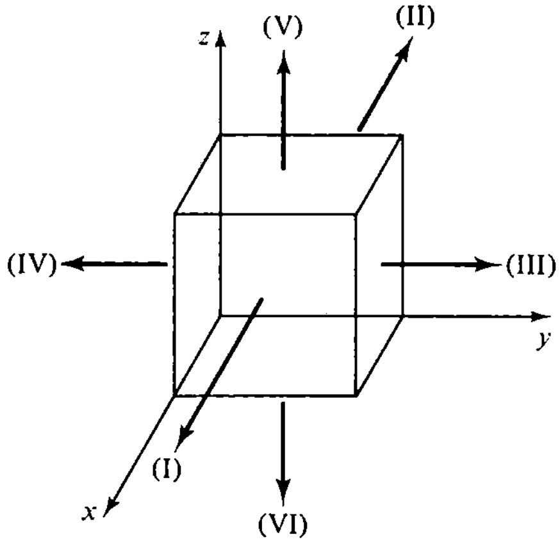

Figure 1.26

(ii) $d\mathbf{a} = -dy\,dz\,\hat{\imath}; \quad x = 0; \quad \mathbf{v}\cdot d\mathbf{a} = -y^2\,dy\,dz$

$$\int \mathbf{v}\cdot d\mathbf{a} = -\int_0^1\int_0^1 y^2\,dy\,dz = -\tfrac{1}{3}$$

(iii) $d\mathbf{a} = dx\,dz\,\hat{\jmath}; \quad y = 1; \quad \mathbf{v}\cdot d\mathbf{a} = (2x + z^2)\,dx\,dz$

$$\int \mathbf{v}\cdot d\mathbf{a} = \int_0^1\int_0^1 (2x + z^2)\,dx\,dz = \tfrac{4}{3}$$

(iv) $d\mathbf{a} = -dx\,dz\,\hat{\jmath}; \quad y = 0; \quad \mathbf{v}\cdot d\mathbf{a} = -z^2\,dx\,dz$

$$\int \mathbf{v}\cdot d\mathbf{a} = -\int_0^1\int_0^1 z^2\,dx\,dz = -\tfrac{1}{3}$$

(v) $d\mathbf{a} = dx\,dy\,\hat{k}; \quad z = 1; \quad \mathbf{v}\cdot d\mathbf{a} = 2y\,dx\,dy$

$$\int \mathbf{v}\cdot d\mathbf{a} = \int_0^1\int_0^1 2y\,dx\,dy = 1$$

(vi) $d\mathbf{a} = -dx\,dy\,\hat{k}; \quad z = 0; \quad \mathbf{v}\cdot d\mathbf{a} = 0$

$$\int \mathbf{v}\cdot d\mathbf{a} = 0$$

Total flux:

$$\oint_{\text{surface}} \mathbf{v}\cdot d\mathbf{a} = \tfrac{1}{3} - \tfrac{1}{3} + \tfrac{4}{3} - \tfrac{1}{3} + 1 + 0 = 2, \qquad \text{as expected}$$

Problem 1.31 Test the divergence theorem for the function $\mathbf{v} = (xy)\hat{\imath} + (2yz)\hat{\jmath} + (3zx)\hat{k}$. Take as your volume the cube shown in Fig. 1.27, with sides of length 2.

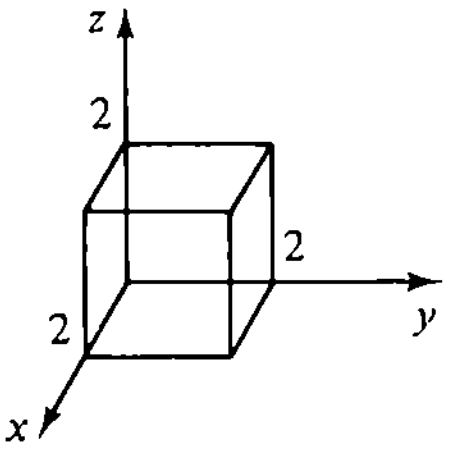

Figure 1.27

Problem 1.32 This problem is designed to develop the notion of flux. It is not supposed to be difficult, so don't look for tricks or booby traps.

(a) Water flows with speed v down a rectangular pipe of dimensions s and l (Fig. 1.28). What is the rate, Φ = volume per unit time, at which water accumulates in the bucket?

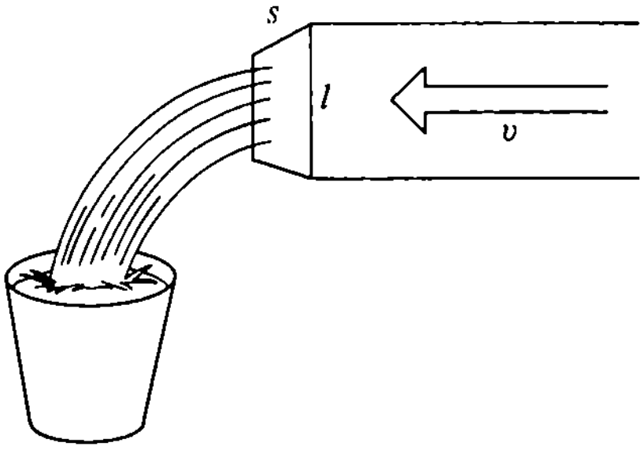

Figure 1.28

(b) Now we slice the end of the pipe off at some angle (Fig. 1.29). This doesn't change Φ, of course. Express your formula for Φ in terms of the dimensions s and l' of the end, and the angle of tilt θ.

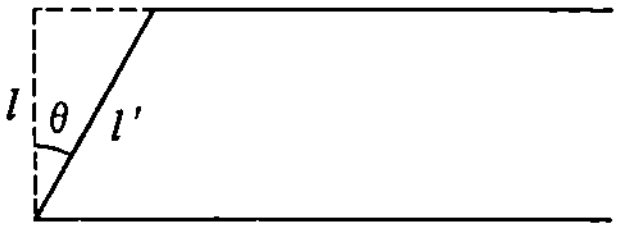

Figure 1.29

(c) Write the formula for Φ more compactly, in terms of the vector area **a** of the end and the vector velocity **v** in Fig. 1.30.

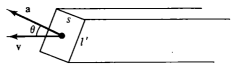

Figure 1.30

(d) What if the velocity varied (in magnitude and/or direction) from point to point, and the end of the pipe were cut off in some arbitrary way as in Fig. 1.31. How would you compute Φ in this general case?

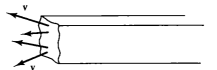

Figure 1.31

1.3.4 The Fundamental Theorem for Curls

The fundamental theorem for curls, which goes by the special name of **Stokes' theorem,** states that

$$\int_{\text{surface}} (\nabla \times \mathbf{v}) \cdot d\mathbf{a} = \oint_{\substack{\text{boundary}\\ \text{line}}} \mathbf{v} \cdot d\mathbf{l} \tag{1.43}$$

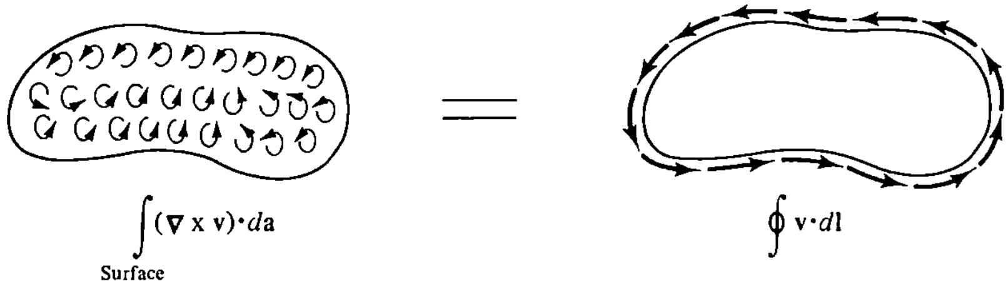

Figure 1.32

As always, the *integral* of a *derivative* (here, the curl) over a region (here, a patch of *surface*) is equal to the value of the function at the *boundary* (here, the perimeter of the patch). As in the case of the divergence theorem, the boundary term is itself an integral—in fact, a closed line integral.

Geometrical Interpretation. Remember that the curl measures the "twist" of the vectors **v**; a region of high curl is a whirlpool—if you put a paddle wheel there, it will rotate. Now the *integral* of the curl over some surface (or, more precisely, the *flux* of the curl *through* that surface) represents the "total amount of swirl," and we can determine that swirl just as well by going all around the *edge* and finding how much the flow is following the boundary (Fig. 1.32). You may find this a rather forced interpretation of Stokes' theorem, but it's a helpful mnemonic, if nothing else.

You might have noticed an apparent ambiguity in Stokes' theorem: concerning the boundary line integral, which *way* are we supposed to go around (clockwise or counterclockwise)? If we go the "wrong" way, we'll pick up an overall sign error. The answer is that it doesn't *matter* which way you go as long as you are *consistent*, for there is a compensating sign ambiguity in the surface integral: Which way does $d\mathbf{a}$ point? For a *closed* surface (as in the divergence theorem) $d\mathbf{a}$ points in the direction of the *outward* normal; but for an *open* surface, which way is "out"? Consistency in Stokes' theorem (as in all such matters) is given by the right-hand rule: if your fingers point in the direction of the line integral, then your thumb fixes the direction of $d\mathbf{a}$ (Fig. 1.33).

Now, there are plenty of surfaces (infinitely many) that share any given boundary line. Twist a paper clip into a loop and dip it into soapy water. The soap film constitutes a surface, with the wire loop as its boundary. If you blow on it, the soap film will expand, making a larger surface, with the same boundary. Ordinarily, a flux integral depends critically on what surface you integrate over,[4] but evidently this is *not* the case with curls. For Stokes' theorem says that $\int_{\text{surface}} (\nabla \times \mathbf{v}) \cdot d\mathbf{a}$ is equal to the line integral of **v** around the boundary, and the latter makes no reference to the specific surface you choose.

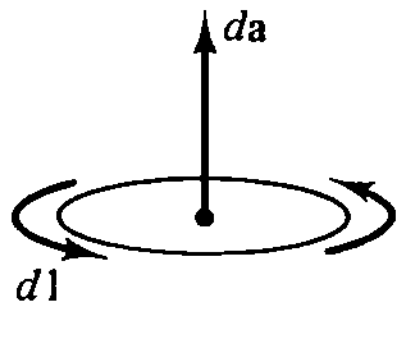

Figure 1.33

[4]For instance, in Example 7 the flux through the front face is $\frac{1}{3}$, but the total flux (in the "forward" direction) through the other five sides is $-\frac{5}{3}$.

Corollary 1: $\int_{\text{surface}} (\nabla \times \mathbf{v}) \cdot d\mathbf{a}$ depends only on the boundary line, not on the particular surface used.

Corollary 2: $\oint_{\text{surface}} (\nabla \times \mathbf{v}) \cdot d\mathbf{a} = 0$ for any *closed* surface, since the boundary line, like the mouth of a balloon, shrinks down to a point, and hence the right side of equation (1.43) vanishes.

These corollaries are analogous to those for the gradient theorem. We shall develop the parallel further in due course.

Example 8

Suppose $\mathbf{v} = (2xz + 3y^2)\hat{j} + (4yz^2)\hat{k}$. Let's check Stokes' theorem for the square surface shown in Fig. 1.34. Here

$$\nabla \times \mathbf{v} = (4z^2 - 2x)\hat{i} + 2z\hat{k} \quad \text{and} \quad d\mathbf{a} = dy\, dz\, \hat{i}$$

(In saying that $d\mathbf{a}$ points in the x direction, we are committing ourselves to a counterclockwise line integral. We could as well write $d\mathbf{a} = -dy\, dz\, \hat{i}$, but then we would be obliged to go clockwise.) Since $x = 0$ for this surface,

$$\int_{\text{surface}} (\nabla \times \mathbf{v}) \cdot d\mathbf{a} = \int_0^1 \int_0^1 4z^2\, dy\, dz = \tfrac{4}{3}$$

Now, what about the line integral? We must break this up into four segments:

(i) $x = 0;\ y: 0 \to 1;\ z = 0;\ \mathbf{v} \cdot d\mathbf{l} = 3y^2\, dy.\ \int \mathbf{v} \cdot d\mathbf{l} = \int_0^1 3y^2\, dy = 1$

(ii) $x = 0;\ y = 1;\ z: 0 \to 1;\ \mathbf{v} \cdot d\mathbf{l} = 4z^2\, dz.\ \int \mathbf{v} \cdot d\mathbf{l} = \int_0^1 4z^2\, dz = \tfrac{4}{3}$

(iii) $x = 0;\ y: 1 \to 0;\ z = 1;\ \mathbf{v} \cdot d\mathbf{l} = 3y^2\, dy.\ \int \mathbf{v} \cdot d\mathbf{l} = \int_1^0 3y^2\, dy = -1$

(iv) $x = 0;\ y = 0;\ z: 1 \to 0;\ \mathbf{v} \cdot d\mathbf{l} = 0.\ \int \mathbf{v} \cdot d\mathbf{l} = 0.$

$$\oint \mathbf{v} \cdot d\mathbf{l} = 1 + \tfrac{4}{3} - 1 + 0 = \tfrac{4}{3} \qquad \text{It checks.}$$

A point of strategy: notice how I handled step (iii). There is a temptation to write $d\mathbf{l} = -dy\, \hat{j}$ here, since the path goes to the left. You can get away with this, if you insist, by running the integral from $0 \to 1$. Personally, I prefer to say

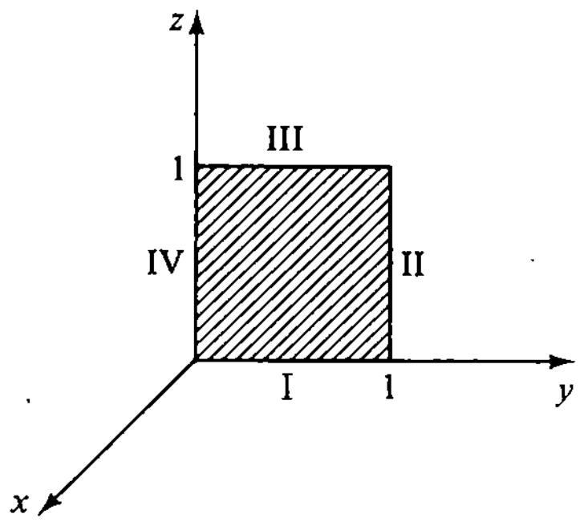

Figure 1.34

$d\mathbf{l} = dx\,\hat{\imath} + dy\,\hat{\jmath} + dz\,\hat{k}$ *always* (never any minus signs) and let the limits of the integral take care of the direction.)

Problem 1.33 Test Stokes' theorem for the function $\mathbf{v} = (xy)\hat{\imath} + (2yz)\hat{\jmath} + (3zx)\hat{k}$, using the triangular shaded area of Fig. 1.35.

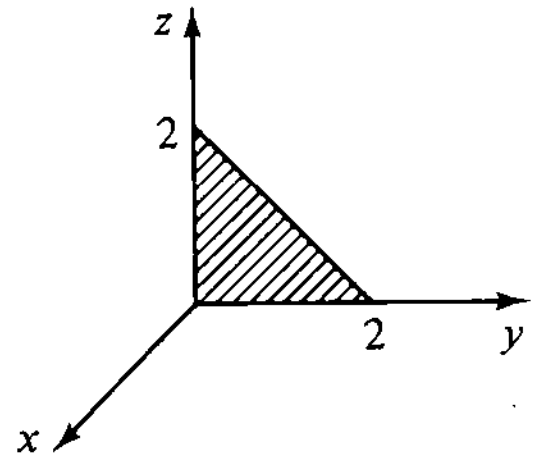

Figure 1.35

Problem 1.34 Check Corollary 1 by using the same function and boundary line as in Example 8 but integrating over the five sides of the cube in Fig. 1.36. The back of the cube is open.

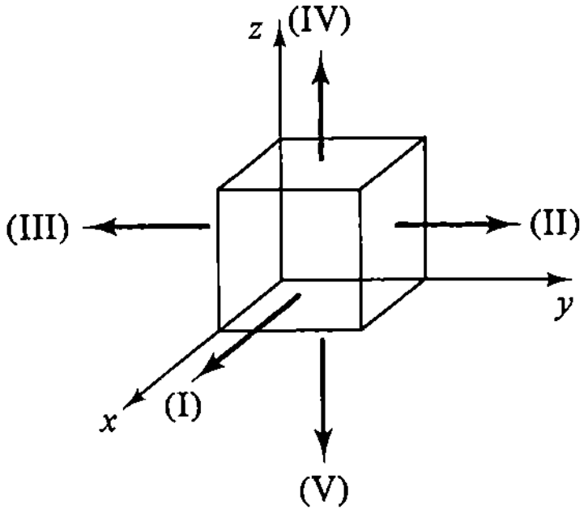

Figure 1.36

1.3.5 Relations Among the Fundamental Theorems

In this section we check the consistency of the three fundamental theorems:

1. Gradient theorem: $$\int_{a\ \text{line}}^{b} (\nabla T)\cdot d\mathbf{l} = T(b) - T(a)$$

2. Divergence theorem: $$\int_{\text{volume}} (\nabla\cdot\mathbf{v})\,d\tau = \oint_{\text{surface}} \mathbf{v}\cdot d\mathbf{a}$$

3. Curl theorem: $$\int_{\text{surface}} (\nabla\times\mathbf{v})\cdot d\mathbf{a} = \oint_{\text{line}} \mathbf{v}\cdot d\mathbf{l}$$

Notice that, of the six terms, *two* are line integrals and *two* are surface (more explicitly, *flux*) integrals. With this in mind, we can combine the theorems in two ways:

(i) **Combine (1) and (3).** Let points a and b coincide (Fig. 1.37); then (1) says

$$\oint_{\text{line}} (\nabla T)\cdot d\mathbf{l} = 0$$

(this is just Corollary 2, of the gradient theorem). But Theorem (3) says

$$\oint_{\text{line}} (\nabla T)\cdot d\mathbf{l} = \int_{\text{surface}} [\nabla \times (\nabla T)]\cdot d\mathbf{a}$$

so consistency of (1) and (3) requires that

$$\int_{\text{surface}} [\nabla \times (\nabla T)]\cdot d\mathbf{a} = 0$$

Moreover, this must hold for *any* surface (and for any scalar function T), so the integrand must vanish everywhere:

$$\nabla \times (\nabla T) = 0 \tag{1.44}$$

(Don't accept this last step without due thought. The fact that an integral is zero, by itself, proves nothing about the integrand. But if it is zero for *any* domain, then the integrand must vanish. For suppose the integrand were *not* zero in some region. Then, we could pick a surface where the integrand is, say, positive, and the integral would be nonzero.)

So consistency of (1) and (3) requires that $\nabla \times (\nabla T) = 0$. But *that's nothing new*! We noted this back in equation (1.35), and you proved it in Problem 1.28.

(ii) **Combine (2) and (3).** Applying Theorem (3) to a *closed* surface (Fig. 1.38), we have

$$\oint_{\text{surface}} (\nabla \times \mathbf{v})\cdot d\mathbf{a} = 0$$

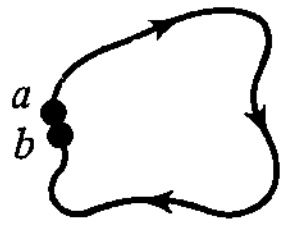

Figure 1.37

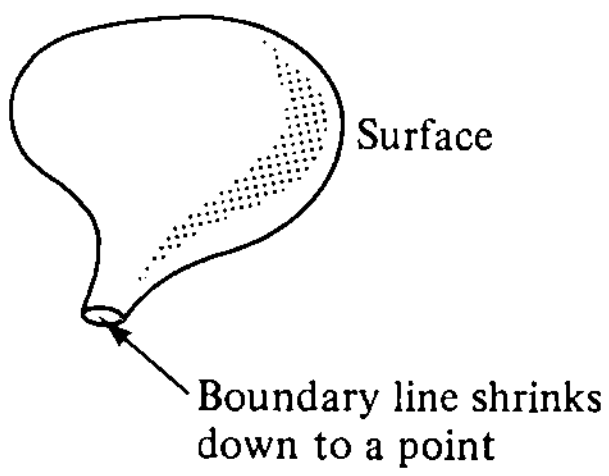

Figure 1.38

This is Corollary 2 to Stokes' theorem. But the divergence theorem gives

$$\oint_{\text{surface}} (\nabla \times \mathbf{v}) \cdot d\mathbf{a} = \int_{\text{volume}} [\nabla \cdot (\nabla \times \mathbf{v})] \, d\tau$$

So

$$\int_{\text{volume}} [\nabla \cdot (\nabla \times \mathbf{v})] \, d\tau = 0$$

Since this must hold for *any* volume, we conclude:

$$\nabla \cdot (\nabla \times \mathbf{v}) = 0$$

This, again, is nothing new; divergence of curl is the *other* vanishing derivative, (1.37), as you proved in Problem 1.27.

Thus, the three fundamental theorems are mutually consistent, and the consistency follows, in a delightfully simple way, from the properties of the second derivatives.

1.4 CURVILINEAR COORDINATES

1.4.1 Spherical Polar Coordinates

The spherical polar coordinates (r, θ, ϕ) of a point P are defined in Fig. 1.39; r is the distance from the origin, θ (the angle down from the z axis) is called the **polar angle,** and ϕ (the angle around from the x axis) is the **azimuthal angle.** Their relation to Cartesian coordinates (x, y, z) can be read from the figure:

$$x = r \sin\theta \cos\phi, \qquad y = r \sin\theta \sin\phi, \qquad z = r \cos\theta. \tag{1.45}$$

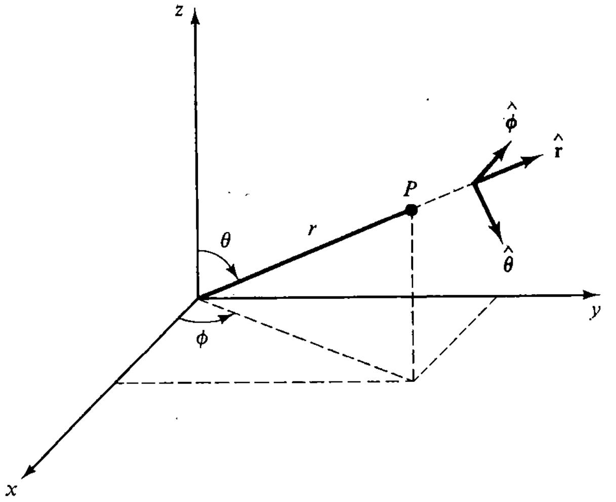

Figure 1.39

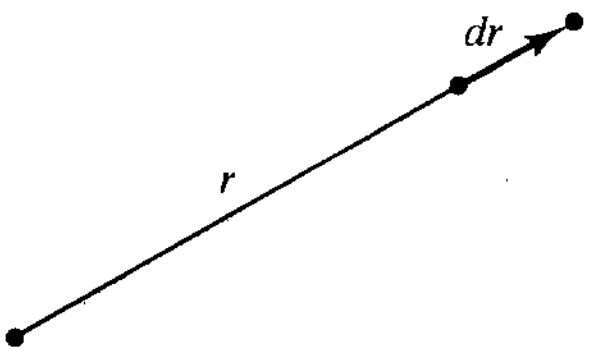

Figure 1.40

Figure 1.39 also shows three unit vectors, $\hat{r}$, $\hat{\theta}$, $\hat{\phi}$, pointing in the direction of increase of the corresponding coordinates. They constitute an orthogonal (mutually perpendicular) basis set, just like, $\hat{i}$, $\hat{j}$, $\hat{k}$, and any vector **A** can be expressed in terms of them in the usual way:

$$\mathbf{A} = A_r\hat{r} + A_\theta\hat{\theta} + A_\phi\hat{\phi}$$

(A_r, A_θ, and A_ϕ are the radial, polar, and azimuthal components of **A**). However, there is a very subtle and important thing to notice here: $\hat{r}$, $\hat{\theta}$, $\hat{\phi}$ are *associated with a particular point P*, and unlike $\hat{i}$, $\hat{j}$, and $\hat{k}$, they *change direction* as P moves around: $\hat{r}$ always points radially outward, for instance, but "radially outward" can be the x-direction, the y-direction, or any other direction, depending on where you are. This means, in particular, that you must be extremely careful in differentiating a vector which is expressed in spherical coordinates, since the unit vectors themselves have nonzero derivatives: $\partial\hat{r}/\partial\theta = \hat{\theta}$, for example (and the same goes for integration).

An infinitesimal element of length in the $\hat{r}$-direction is simply dr (Fig. 1.40), just as an infinitesimal element of length in the x-direction is dx:

$$dl_r = dr \tag{1.46}$$

On the other hand, an infinitesimal element of length in the $\hat{\theta}$-direction (Fig. 1.41) is *not* just $d\theta$ (that's an *angle*—it doesn't even have the right *units* for a length) but rather $r\,d\theta$:

$$dl_\theta = r\,d\theta \tag{1.47}$$

Similarly, an infinitesimal element of length in the $\hat{\phi}$-direction (Fig. 1.42) is $r \sin\theta\, d\phi$:

$$dl_\phi = r \sin\theta\, d\phi \tag{1.48}$$

Thus, the most general infinitesimal displacement $d\mathbf{l}$ is

$$d\mathbf{l} = dr\,\hat{r} + r\,d\theta\,\hat{\theta} + r\sin\theta\,d\phi\,\hat{\phi} \tag{1.49}$$

This plays the role (in line integrals, for example) that $d\mathbf{l} = dx\hat{i} + dy\hat{j} + dz\hat{k}$ played in Cartesian coordinates.

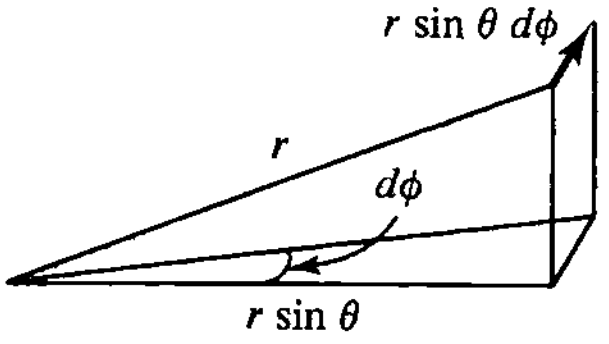

Figure 1.41

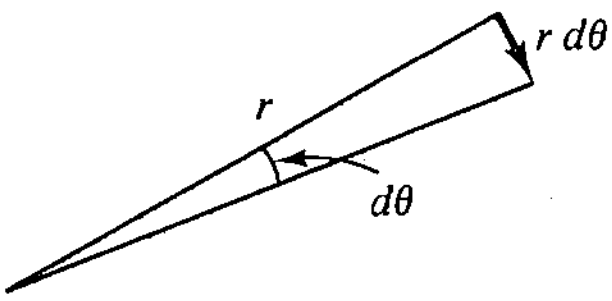

Figure 1.42

The infinitesimal volume element $d\tau$, in spherical coordinates, is the product of the three infinitesimal displacements:

$$d\tau = dl_r dl_\theta dl_\phi = r^2 \sin\theta \, dr \, d\theta \, d\phi \tag{1.50}$$

I cannot give you a general expression for *surface* elements $d\mathbf{a}$, since these depend entirely on the orientation of the surface. You simply have to analyse the geometry for any given case (this goes for Cartesian and curvilinear coordinates alike). If you are integrating over the surface of a sphere, for instance, then r is constant whereas θ and ϕ change (Fig. 1.43), so

$$d\mathbf{a}_1 = dl_\theta dl_\phi \hat{r} = r^2 \sin\theta \, d\theta \, d\phi \, \hat{r}$$

On the other hand, if the surface lies in the xy plane, say, so that θ is constant (to wit: $\pi/2$) while r and ϕ vary, then

$$d\mathbf{a}_2 = dl_r dl_\phi \hat{\theta} = r \sin\theta \, dr \, d\phi \, \hat{\theta} = r \, dr \, d\phi \, \hat{\theta}$$

Notice, finally, that r ranges from 0 to ∞, ϕ from 0 to 2π, and θ from 0 to π (*not* 2π—that would count every point twice).[5]

Example 9

Find the volume of a sphere of radius R.

Solution:

$$\text{volume} = \int_{\text{volume}} d\tau = \int_{r=0}^{R} \int_{\theta=0}^{\pi} \int_{\phi=0}^{2\pi} r^2 \sin\theta \, dr \, d\theta \, d\phi$$

$$= \left(\int_0^R r^2 \, dr\right)\left(\int_0^\pi \sin\theta \, d\theta\right)\left(\int_0^{2\pi} d\phi\right)$$

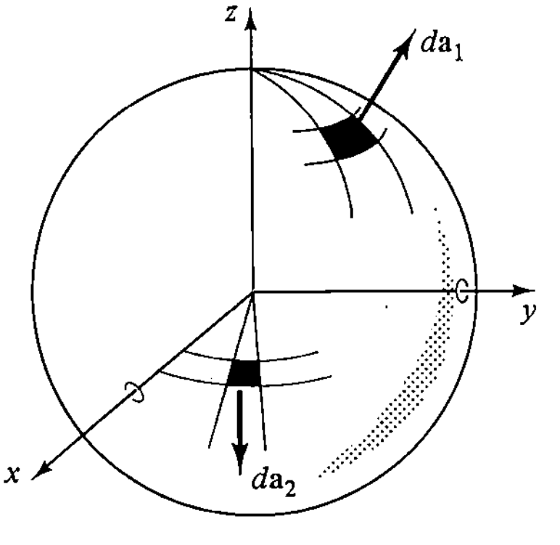

Figure 1.43

[5]Alternatively, you could run ϕ from 0 to π (the "eastern hemisphere") and cover the "western hemisphere" by extending θ from π up to 2π. But this is very bad notation, since, among other things, $\sin\theta$ will then run negative, and you'll have to put absolute value signs around that term in volume and surface elements (area and volume being intrinsically positive quantities).

$$= \left(\frac{R^3}{3}\right)(2)(2\pi) = \frac{4}{3}\pi R^3$$

This is scarcely a surprising answer.

So far we have talked only about the *geometry* of spherical coordinates. Now I would like to "translate" the vector derivatives (gradient, divergence, curl, and Laplacian) into r, θ, ϕ notation. In principle, this is entirely straightforward: in the case of the gradient,

$$\nabla T = \frac{\partial T}{\partial x}\hat{i} + \frac{\partial T}{\partial y}\hat{j} + \frac{\partial T}{\partial z}\hat{k}$$

for instance, we would first use the chain rule to reexpress the partials:

$$\frac{\partial T}{\partial x} = \frac{\partial T}{\partial r}\left(\frac{\partial r}{\partial x}\right) + \frac{\partial T}{\partial \theta}\left(\frac{\partial \theta}{\partial x}\right) + \frac{\partial T}{\partial \phi}\left(\frac{\partial \phi}{\partial x}\right)$$

The terms in parentheses we would work out from equations (1.45), or rather, the inverse equations of Problem 1.35. Then we'd do the same for $\partial T/\partial y$ and $\partial T/\partial z$. But *still* we would not be finished, for the unit vectors $\hat{i}$, $\hat{j}$, $\hat{k}$ would then have to be rewritten in terms of $\hat{r}$, $\hat{\theta}$, $\hat{\phi}$, using Problem 1.36. It would take an hour to determine the gradient in spherical coordinates by this brute-force method. I suppose this is how it was first done, but there is a much more efficient indirect approach, explained in Appendix A, which has the extra advantage of treating all coordinate systems in one format. I described the "straightforward" method only to show you that there is nothing subtle or mysterious about transforming to spherical coordinates: you're expressing the *same quantity* (gradient, divergence, or whatever) in different notation, that's all.

Here, then, are the vector derivatives in spherical coordinates:

Gradient:

$$\nabla T = \frac{\partial T}{\partial r}\hat{r} + \frac{1}{r}\frac{\partial T}{\partial \theta}\hat{\theta} + \frac{1}{r\sin\theta}\frac{\partial T}{\partial \phi}\hat{\phi} \tag{1.51}$$

Divergence:

$$\nabla \cdot \mathbf{v} = \frac{1}{r^2}\frac{\partial}{\partial r}(r^2 v_r) + \frac{1}{r\sin\theta}\frac{\partial}{\partial \theta}(\sin\theta\, v_\theta) + \frac{1}{r\sin\theta}\frac{\partial v_\phi}{\partial \phi} \tag{1.52}$$

Curl:

$$\nabla \times \mathbf{v} = \frac{1}{r\sin\theta}\left[\frac{\partial}{\partial \theta}(\sin\theta\, v_\phi) - \frac{\partial v_\theta}{\partial \phi}\right]\hat{r} + \frac{1}{r}\left[\frac{1}{\sin\theta}\frac{\partial v_r}{\partial \phi} - \frac{\partial}{\partial r}(r v_\phi)\right]\hat{\theta}$$

$$+ \frac{1}{r}\left[\frac{\partial}{\partial r}(r v_\theta) - \frac{\partial v_r}{\partial \theta}\right]\hat{\phi} \tag{1.53}$$

Laplacian:

$$\nabla^2 T = \frac{1}{r^2}\frac{\partial}{\partial r}\left(r^2 \frac{\partial T}{\partial r}\right) + \frac{1}{r^2 \sin\theta}\frac{\partial}{\partial\theta}\left(\sin\theta \frac{\partial T}{\partial\theta}\right) + \frac{1}{r^2 \sin^2\theta}\frac{\partial^2 T}{\partial\phi^2} \tag{1.54}$$

For reference, these formulas are listed inside the front cover.

Problem 1.35 Find the formulas for r, θ, ϕ in terms of x, y, z—the inverse, in other words, of equation (1.45).

Problem 1.36 Express the unit vectors $\hat{r}$, $\hat{\theta}$, $\hat{\phi}$, at the point r, θ, ϕ, in terms of $\hat{i}$, $\hat{j}$, $\hat{k}$. That is, find the Cartesian components of $\hat{r}$, $\hat{\theta}$, $\hat{\phi}$: $\hat{r} = (?)\hat{i} + (?)\hat{j} + (?)\hat{k}$, and so on. Check your answers several ways ($\hat{r}\cdot\hat{r} \stackrel{?}{=} 1$, $\hat{\theta}\cdot\hat{\phi} \stackrel{?}{=} 0$, $\hat{r}\times\hat{\theta} \stackrel{?}{=} \hat{\phi}$, . . .).

• **Problem 1.37**

(a) Check the divergence theorem for the function $\mathbf{v}_1 = r^2\hat{r}$, using as your volume the sphere of radius R, centered at the origin. (b) Do the same for $\mathbf{v}_2 = (1/r^2)\hat{r}$. (If the answer surprises you, look back at Problem 1.17.)

Problem 1.38 Compute the divergence of the function

$$\mathbf{v} = (r\cos\theta)\hat{r} + (r\sin\theta)\hat{\theta} + (r\sin\theta\cos\phi)\hat{\phi}$$

Check the divergence theorem for this function, using as your volume the inverted hemisphere of radius R, resting on the xy plane and centered at the origin (Fig. 1.44).

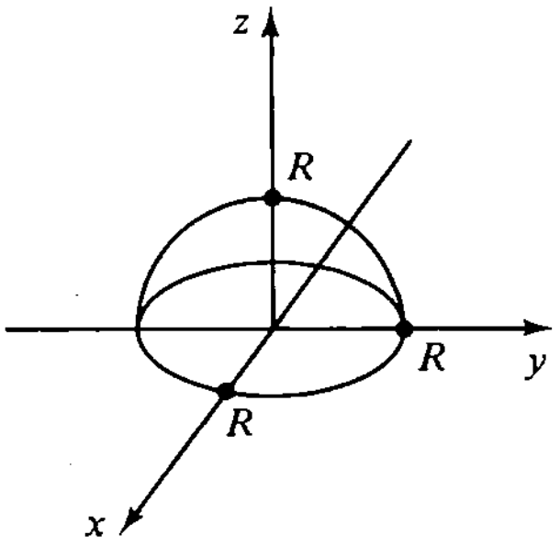

Figure 1.44

Problem 1.39 Compute the gradient and Laplacian of the function

$$T = r(\cos\theta + \sin\theta\cos\phi)$$

Check the Laplacian by converting T to Cartesian coordinates and using equation

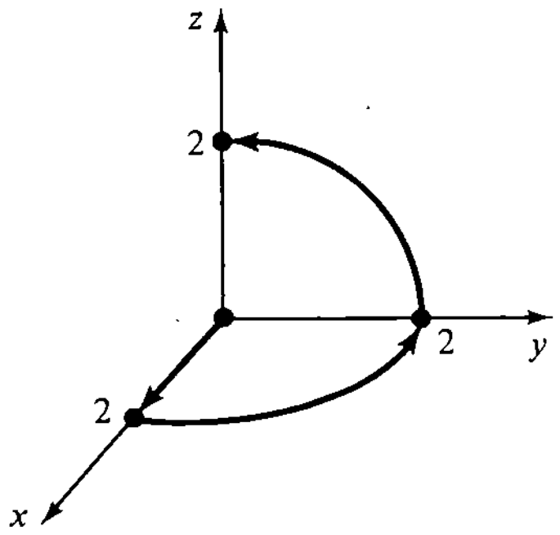

Figure 1.45

(1.33). Test the gradient theorem for this function, using the path shown in Fig. 1.45, from (0, 0, 0) to (0, 0, 2).

1.4.2 Cylindrical Coordinates

Cylindrical coordinates (r, ϕ, z) for a point P are defined by Fig. 1.46. Notice that whereas ϕ has the same meaning as in spherical polar coordinates (and z the same as in Cartesian), r is quite different (it is the distance to P *from the z axis*, not from the origin). For this reason some authors use an entirely different letter, usually ρ, for this coordinate; but we have a more important purpose in store for ρ. The relation to Cartesian coordinates is

$$x = r\cos\phi, \qquad y = r\sin\phi, \qquad z = z \tag{1.55}$$

The infinitesimal displacements are

$$dl_r = dr, \qquad dl_\phi = r\,d\phi, \qquad dl_z = dz \tag{1.56}$$

So that

$$d\mathbf{l} = dr\,\hat{\mathbf{r}} + r\,d\phi\,\hat{\boldsymbol{\phi}} + dz\,\hat{\mathbf{z}} \tag{1.57}$$

and

$$d\tau = r\,dr\,d\phi\,dz \tag{1.58}$$

The range of r is $0 \to \infty$, ϕ goes from $0 \to 2\pi$, and z from $-\infty \to +\infty$.

The vector derivatives in cylindrical coordinates are:

Gradient:

$$\nabla T = \frac{\partial T}{\partial r}\hat{\mathbf{r}} + \frac{1}{r}\frac{\partial T}{\partial \phi}\hat{\boldsymbol{\phi}} + \frac{\partial T}{\partial z}\hat{\mathbf{z}} \tag{1.59}$$

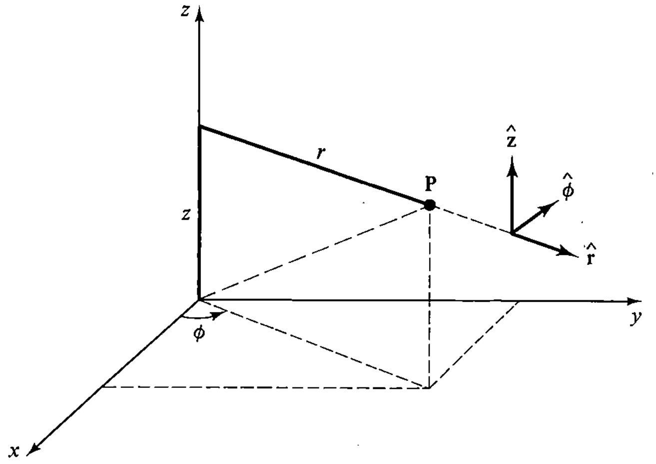

Figure 1.46

Divergence:

$$\nabla \cdot \mathbf{v} = \frac{1}{r}\frac{\partial}{\partial r}(r v_r) + \frac{1}{r}\frac{\partial v_\phi}{\partial \phi} + \frac{\partial v_z}{\partial z} \tag{1.60}$$

Curl:

$$\nabla \times \mathbf{v} = \left(\frac{1}{r}\frac{\partial v_z}{\partial \phi} - \frac{\partial v_\phi}{\partial z}\right)\hat{\mathbf{r}} + \left(\frac{\partial v_r}{\partial z} - \frac{\partial v_z}{\partial r}\right)\hat{\boldsymbol{\phi}} + \frac{1}{r}\left[\frac{\partial}{\partial r}(r v_\phi) - \frac{\partial v_r}{\partial \phi}\right]\hat{\mathbf{z}} \tag{1.61}$$

Laplacian:

$$\nabla^2 T = \frac{1}{r}\frac{\partial}{\partial r}\left(r\frac{\partial T}{\partial r}\right) + \frac{1}{r^2}\frac{\partial^2 T}{\partial \phi^2} + \frac{\partial^2 T}{\partial z^2} \tag{1.62}$$

These formulas are also listed inside the front cover.

Problem 1.40 Express the cylindrical unit vectors $\hat{\mathbf{r}}$, $\hat{\boldsymbol{\phi}}$, $\hat{\mathbf{z}}$ (at the point r, ϕ, z) in terms of $\hat{\mathbf{i}}$, $\hat{\mathbf{j}}$, $\hat{\mathbf{k}}$. "Invert" your formulas to get $\hat{\mathbf{i}}$, $\hat{\mathbf{j}}$, $\hat{\mathbf{k}}$ in terms of $\hat{\mathbf{r}}$, $\hat{\boldsymbol{\phi}}$, $\hat{\mathbf{z}}$.

Problem 1.41 (a) Find the divergence of the function

$$\mathbf{v} = r(2 + \sin^2\phi)\hat{\mathbf{r}} + r\sin\phi\cos\phi\,\hat{\boldsymbol{\phi}} + 3z^2\hat{\mathbf{z}}$$

(b) Test the divergence theorem for this function, using the quarter-cylinder (radius 2, height 5) shown in Figure 1.47.

(c) Find the curl of $\mathbf{v}$.

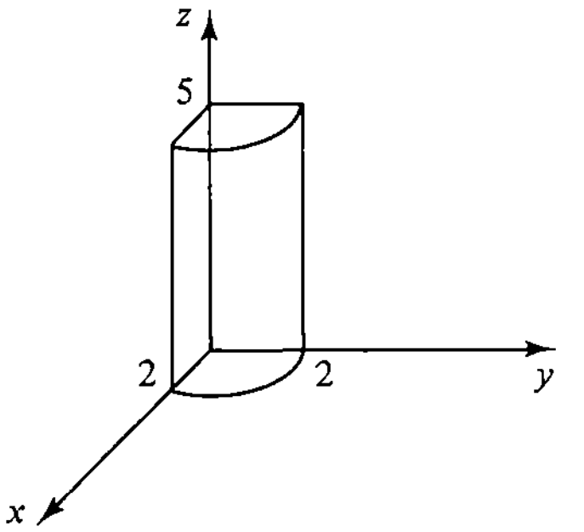

Figure 1.47

1.5 THE DIRAC DELTA FUNCTION

1.5.1 The Divergence of $\hat{\mathbf{r}}/r^2$

Consider the vector function

$$\mathbf{v} = \frac{1}{r^2}\hat{\mathbf{r}} \tag{1.63}$$

(I'm working in spherical coordinates, so r is the distance from the origin to the point in question). At every location, $\mathbf{v}$ is directed radially outward (Figure 1.48); if ever

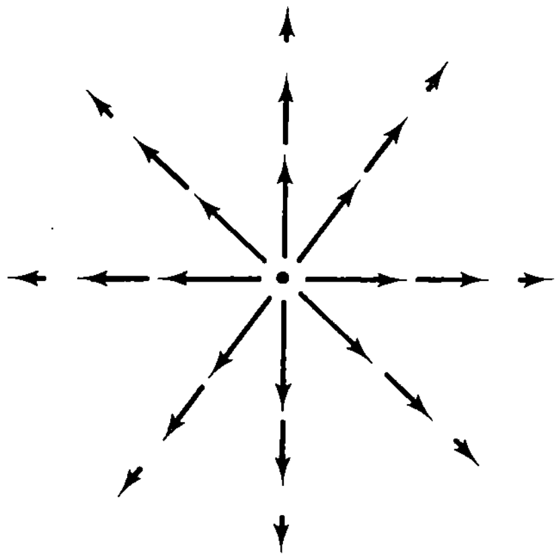

Figure 1.48

there was a function that ought to have a large positive divergence, this is it. And yet, when you actually *calculate* the divergence (using equation (1.52)), you get precisely *zero!*

$$\nabla \cdot \mathbf{v} = \frac{1}{r^2}\frac{\partial}{\partial r}\left(r^2 \frac{1}{r^2}\right) = \frac{1}{r^2}\frac{\partial}{\partial r}(1) = 0 \tag{1.64}$$

(You will have encountered this paradox, in a slightly different form, if you worked Problem 1.17.) The plot thickens when you apply the divergence theorem to this function. Suppose we integrate over a sphere of radius R, centered at the origin (Problem 1.37(b)); the surface integral is

$$\oint \mathbf{v} \cdot d\mathbf{a} = \int \left(\frac{1}{R^2}\hat{\mathbf{r}}\right) \cdot (R^2 \sin\theta \, d\theta \, d\phi \, \hat{\mathbf{r}}) = \left(\int_0^{\pi} \sin\theta \, d\theta\right)\left(\int_0^{2\pi} d\phi\right) = 4\pi \tag{1.65}$$

But the *volume* integral, $\int \nabla \cdot \mathbf{v} \, d\tau$, is *zero,* if we are really to believe equation (1.64). Does this mean that the divergence theorem is false? What's going on here?

The source of the problem is the point $r = 0$, where $\mathbf{v}$ blows up (and where, in equation (1.64), we have unwittingly divided by zero). It is quite true that $\nabla \cdot \mathbf{v} = 0$ everywhere *except* the origin, but right *at* the origin the situation is more complicated. Notice that the surface integral (1.65) is *independent* of R; if the divergence theorem is right (and it *is*), we should get $\int (\nabla \cdot \mathbf{v}) \, d\tau = 4\pi$ for *any* sphere centered at the origin, no matter how large or how small. Evidently the entire contribution must be coming from the point $r = 0$! Thus $\nabla \cdot \mathbf{v}$ has the bizarre property that it vanishes everywhere except at one point, and yet its *integral* (over any volume containing that point) is 4π. No ordinary function behaves like that. (On the other hand, a *physical* example *does* come to mind: the density (mass per unit volume) of a point particle. It's zero except at the exact location of the particle, and yet its *integral* is finite—namely, the mass of the particle.) What we have stumbled on is a mathematical object known to physicists as the **Dirac delta function.** It arises in many branches of theoretical physics. Moreover, the specific problem at hand (the divergence of the function $\hat{\mathbf{r}}/r^2$) is not just some arcane curiosity—it is, in fact, central to the whole theory of electrodynamics. So it is worthwhile to pause here and study the Dirac delta function with some care.

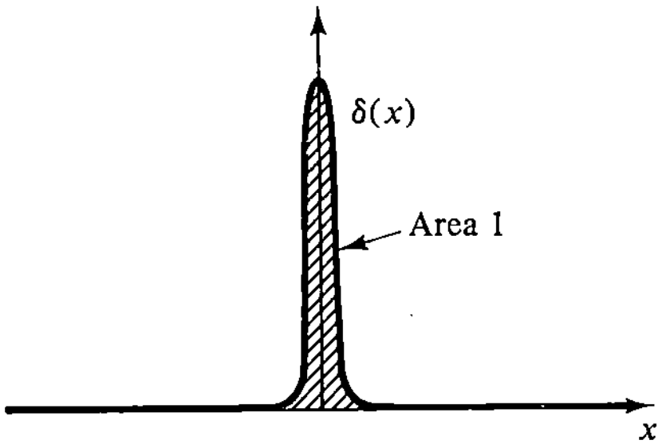

Figure 1.49

1.5.2 The One-Dimensional Dirac Delta Function

The one-dimensional Dirac delta function, $\delta(x)$, can be pictured as an infinitely high, infinitesimally narrow "spike", with area 1 (Fig. 1.49). That is to say:

$$\delta(x) = \begin{cases} 0, & \text{if } x \neq 0 \\ \infty, & \text{if } x = 0 \end{cases} \tag{1.66}$$

and

$$\int_{-\infty}^{\infty} \delta(x)\, dx = 1 \tag{1.67}$$

Technically, $\delta(x)$ is not a function at all, since its value is not finite at $x = 0$. In the mathematical literature it is known as a *generalized* function, or *distribution*. It is, if you like, the *limit* of a *sequence* of functions, such as rectangles $R_n(x)$, of height n and width $1/n$, or isosceles triangles $T_n(x)$, of height n and base $2/n$ (Fig. 1.50).

If $f(x)$ is some "ordinary" function (that is, *not* another delta function—in fact, just to be on the safe side let's say that $f(x)$ is *continuous*), then the *product* $f(x)\delta(x)$ is zero everywhere except at $x = 0$. It follows that

$$f(x)\delta(x) = f(0)\delta(x) \tag{1.68}$$

(This is the most important fact about the delta function, so make sure you understand why it is true. The point is that since the product is zero anyway *except* at $x = 0$, we may as well replace $f(x)$ by the value it assumes at the origin.) In particular

$$\int_{-\infty}^{\infty} f(x)\delta(x)\, dx = f(0) \int_{-\infty}^{\infty} \delta(x)\, dx = f(0) \tag{1.69}$$

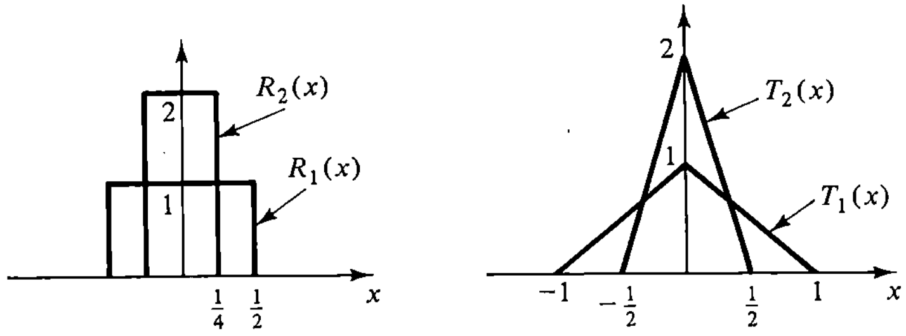

Figure 1.50

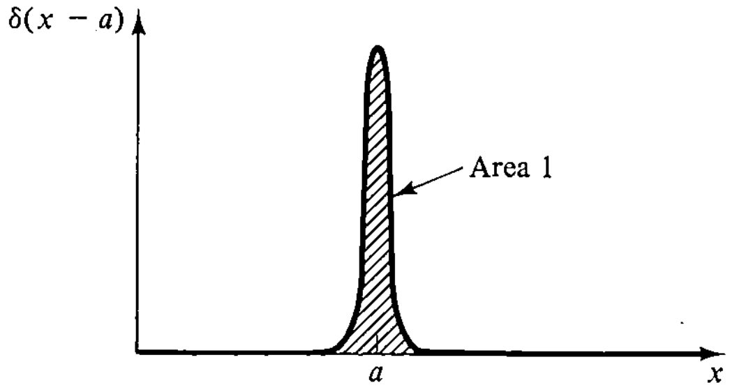

Figure 1.51

Under an integral, then, the delta function "picks out" the value of $f(x)$ at $x = 0$. (Here and below, the integral need not run from $-\infty$ to $+\infty$; it is sufficient that the domain extend across the delta function, and $-\epsilon$ to $+\epsilon$ would do just as well.)

Of course, we can shift the spike from $x = 0$ to some other point, $x = a$ (Fig. 1.51):

$$\delta(x-a) = \begin{cases} 0, & \text{if } x \neq a \\ \infty, & \text{if } x = a \end{cases} \quad \text{with} \quad \int_{-\infty}^{\infty} \delta(x-a)dx = 1 \tag{1.70}$$

Equation (1.68) becomes

$$f(x)\delta(x-a) = f(a)\delta(x-a) \tag{1.71}$$

and (1.69) generalizes to

$$\int_{-\infty}^{\infty} f(x)\delta(x-a)dx = f(a) \tag{1.72}$$

Example 10

$$\int_0^3 x^3\, \delta(x-2)\, dx = 2^3 = 8$$

Notice, however, that if the upper limit were 1 (instead of 3) the answer would be 0, because the spike is now outside the domain of integration.

Although δ itself is not a legitimate function, ***integrals*** over δ are perfectly acceptable. In fact, it's best to think of the delta function as something that is ***always intended for use under an integral sign.*** In particular, two expressions involving delta functions (say, $D_1(x)$ and $D_2(x)$) are considered equal if[6]

$$\int_{-\infty}^{+\infty} f(x)D_1(x)\, dx = \int_{-\infty}^{+\infty} f(x)D_2(x)\, dx \tag{1.73}$$

for all ("ordinary") functions $f(x)$.

[6]This is not as arbitrary as it may sound. The crucial point is that the integrals are equal for *any* $f(x)$. Suppose D_1 and D_2 actually *differed,* say, in the neighborhood of the point $x = 17$. Then we could pick a function $f(x)$ that was sharply peaked about $x = 17$, and the integrals would not be equal.

Example 11

Show that

$$\delta(kx) = \frac{1}{|k|}\delta(x) \tag{1.74}$$

where k is any (nonzero) constant.

Solution: For an arbitrary test function $f(x)$, consider the integral

$$\int_{-\infty}^{\infty} f(x)\delta(kx)\,dx$$

Changing variables, we let $y \equiv kx$, so that $x = y/k$, and $dx = 1/k\;dy$. If k is positive, the integration still runs from $-\infty$ to $+\infty$, but if k is *negative*, then $x = \infty$ implies $y = -\infty$, and vice versa, so the order of the limits is reversed. Restoring the "proper" order costs a minus sign. Thus

$$\int_{-\infty}^{\infty} f(x)\delta(kx)\,dx = \pm\int_{-\infty}^{\infty} f(y/k)\delta(y)\frac{dy}{k} = \pm\frac{1}{k}f(0) = \frac{1}{|k|}f(0)$$

(The lower signs apply when k is negative, and we can account for this neatly by putting absolute value bars around the final k, as indicated.) Under the integral sign, then, $\delta(kx)$ serves the same purpose as $(1/|k|)\delta(x)$:

$$\int_{-\infty}^{\infty} f(x)\delta(kx)\,dx = \int_{-\infty}^{\infty} f(x)\left[\frac{1}{|k|}\delta(x)\right]dx$$

According to criterion (1.73), therefore, $\delta(kx)$ and $(1/|k|)\delta(x)$ are equal.

Problem 1.42 Evaluate the following integrals:

(a) $\int_2^6 (3x^2 - 2x - 1)\delta(x - 3)\,dx$

(b) $\int_0^5 \cos x\;\delta(x - \pi)\,dx$

(c) $\int_0^3 x^3\delta(x + 1)\,dx$

(d) $\int_{-\infty}^{\infty} \ln(x + 3)\;\delta(x + 2)\,dx$

Problem 1.43 Evaluate the following integrals:

(a) $\int_{-2}^{2} (2x + 3)\delta(3x)\,dx$

(b) $\int_0^2 (x^3 + 3x + 2)\delta(1 - x)\,dx$

(c) $\int_{-1}^{1} 9x^2\delta(3x + 1)\,dx$

(d) $\int_{-\infty}^{a} \delta(x - b)\,dx$

Problem 1.44

(a) Show that

$$x\frac{d}{dx}(\delta(x)) = -\delta(x)$$

(*Hint*: Use integration by parts.)

(b) Let $\theta(x)$ be the "step function":

$$\theta(x) \equiv \begin{cases} 1, & \text{if } x > 0 \\ 0, & \text{if } x \le 0 \end{cases}$$

Show that $d\theta/dx = \delta(x)$.

1.5.3 The Three-Dimensional Delta Function

It is an easy matter to generalize the delta function to three dimensions:

$$\delta^3(\mathbf{r}) = \delta(x)\delta(y)\delta(z) \tag{1.75}$$

(as always, $\mathbf{r} \equiv x\hat{i} + y\hat{j} + z\hat{k}$ is the position vector, extending from the origin to the point (x, y, z)). This three-dimensional delta function is zero everywhere except at (0, 0, 0), where it blows up. Its volume integral is 1:

$$\int_{\text{all space}} \delta^3(\mathbf{r})\, d\tau = \int_{-\infty}^{\infty}\int_{-\infty}^{\infty}\int_{-\infty}^{\infty} \delta(x)\delta(y)\delta(z)\, dx\, dy\, dz = 1 \tag{1.76}$$

and, generalizing (1.72),

$$\int_{\text{all space}} f(\mathbf{r})\delta^3(\mathbf{r} - \mathbf{r}_0)\, d\tau = f(\mathbf{r}_0) \tag{1.77}$$

(where $\mathbf{r}_0$ is any fixed point). As in the one-dimensional case, integration with δ picks out the value of the function f at the location of the spike.

We are now in a position to resolve the paradox introduced in Section 1.5.1. As you will recall, we found that the divergence of $\hat{r}/r^2$ is zero everywhere except at the origin, and yet its *integral* over any volume containing the origin is a constant (to wit: 4π). These are precisely the defining conditions for the Dirac delta function; evidently

$$\nabla \cdot \left(\frac{\hat{r}}{r^2}\right) = 4\pi\delta^3(\mathbf{r}) \tag{1.78}$$

More generally,

$$\boxed{\nabla \cdot \left(\frac{\hat{\boldsymbol{\mathscr{r}}}}{\mathscr{r}^2}\right) = 4\pi\delta^3(\boldsymbol{\mathscr{r}})} \tag{1.79}$$

where, as earlier,

$$\boldsymbol{\mathscr{r}} \equiv \mathbf{r} - \mathbf{r}_0 \tag{1.80}$$

(I shall use the script $\mathscr{r}$ in this way throughout the book, to denote the displacement vector from a fixed point $\mathbf{r}_0$ to a variable point $\mathbf{r}$. Note that the differentiation here is with respect to $\mathbf{r}$, with $\mathbf{r}_0$ held constant.) Incidentally, since

$$\nabla\left(\frac{1}{\eta}\right) = -\frac{\hat{\boldsymbol{\eta}}}{\eta^2} \tag{1.81}$$

(see Problem 1.13b) it follows that

$$\nabla^2 \frac{1}{\eta} = -4\pi\delta^3(\boldsymbol{\eta}) \tag{1.82}$$

Example 12

Determine the value of the integral

$$J = \int_V (r^2 + 2)\nabla \cdot \left(\frac{\hat{\mathbf{r}}}{r^2}\right) d\tau$$

where V is a sphere of radius R centered at the origin.

Solution 1: Use equation (1.78) to rewrite the divergence and (1.77) to evaluate the integral:

$$J = \int_V (r^2 + 2)4\pi\delta^3(\mathbf{r})\, d\tau = 4\pi(0 + 2) = 8\pi$$

This one-line solution demonstrates something of the power and beauty of the delta function, but I would like to show you a second method, which is much more cumbersome but serves to illustrate a useful technique: integration by parts. In *one*-dimensional integration by parts we exploit the product rule,

$$\frac{d}{dx}(fg) = f\frac{dg}{dx} + g\frac{df}{dx}$$

to "transfer the derivative" from one term in the integrand to the other:

$$\int_a^b f\frac{dg}{dx}\, dx = -\int_a^b g\frac{df}{dx}\, dx + \int_a^b \frac{d}{dx}(fg)\, dx$$

This costs us a minus sign and a "boundary term":

$$\int_a^b \frac{d}{dx}(fg)\, dx = f(b)g(b) - f(a)g(a)$$

The vector product rules (Section 1.2.6) can be used in much the same way, to peel a ∇ off one term in the integrand and slap it onto the other. For example, product rule number (5), inside the front cover,

$$\nabla \cdot (f\mathbf{A}) = f(\nabla \cdot \mathbf{A}) + \mathbf{A} \cdot (\nabla f)$$

yields

$$\int_V f(\nabla \cdot \mathbf{A})\, d\tau = -\int_V \mathbf{A} \cdot (\nabla f)\, d\tau + \int_V \nabla \cdot (f\mathbf{A})\, d\tau$$

Applying the divergence theorem converts the last term into a surface integral, so

$$\int_V f(\nabla\cdot\mathbf{A})\,d\tau = -\int_V \mathbf{A}\cdot(\nabla f)\,d\tau + \oint_S f\mathbf{A}\cdot d\mathbf{a} \qquad (1.83)$$

where S is the surface bounding the volume V. As before, integration by parts has transferred the derivative under the integral sign from one function ($\mathbf{A}$) to the other (f)—at the cost of a minus sign and a boundary term (which in this case is itself an integral).

Solution 2: Using equation (1.83), we find

$$J = -\int_V \frac{\hat{\mathbf{r}}}{r^2}\cdot[\nabla(r^2+2)]\,d\tau + \oint_S (r^2+2)\frac{\hat{\mathbf{r}}}{r^2}\cdot d\mathbf{a}$$

From the gradient in spherical coordinates (1.51),

$$\nabla(r^2+2) = 2r\hat{\mathbf{r}},$$

so the volume integral is

$$\int_V \frac{2}{r}\,d\tau = \int_V \frac{2}{r}r^2\sin\theta\,dr\,d\theta\,d\phi = 8\pi\int_0^R r\,dr = 4\pi R^2$$

Meanwhile, for the boundary of the sphere (where $r = R$),

$$d\mathbf{a} = R^2\sin\theta\,d\theta\,d\phi\,\hat{\mathbf{r}}$$

so the surface integral becomes

$$\int_S (R^2+2)\sin\theta\,d\theta\,d\phi = 4\pi(R^2+2)$$

Putting it all together, then,

$$J = -4\pi R^2 + 4\pi(R^2+2) = 8\pi$$

as before.

Problem 1.45

(a) Write an expression representing the density (mass per unit volume) of a particle of mass m located at the position $\mathbf{r}_0$. Do the same for the electric charge density of a point charge q at $\mathbf{r}_0$.

(b) What is the charge density of an electric dipole, consisting of a point charge $-q$ at the origin and a point charge $+q$ at $\mathbf{r}_0$?

(c) What is the charge density of a uniform, infinitesimally thin spherical shell of radius R and total charge Q, centered at the origin. (Beware: the integral over all space must equal Q.)

Problem 1.46 Evaluate the following integrals:

(a) $\int_{\text{all space}} (r^2 + \mathbf{r}_1\cdot\mathbf{r}_0 + r_0^2)\delta^3(\mathbf{r}-\mathbf{r}_0)\,d\tau$, where $\mathbf{r}_0$ is a constant vector.

(b) $\int_V (\mathbf{r} - \mathbf{a})^2 \delta^3(5\mathbf{r})\, d\tau$, where V is a cube of side 2, centered on the origin, and $\mathbf{a} = 4\hat{y} + 3\hat{z}$.

(c) $\int_V (r^4 + r^2(\mathbf{r}\cdot\mathbf{c}) + c^4)\delta^3(\mathbf{r} - \mathbf{c})\, d\tau$, where V is a sphere of radius 6 about the origin, and $\mathbf{c} = 5\hat{x} + 3\hat{y} + 2\hat{z}$.

(d) $\int_V \mathbf{r}\cdot(\mathbf{a} - \mathbf{r})\delta^3(\mathbf{b} - \mathbf{r})\, d\tau$, where $\mathbf{a} = (1, 2, 3)$, $\mathbf{b} = (3, 2, 1)$, and V is a sphere of radius 1.5 centered at (2, 2, 2).

Problem 1.47 Show that

$$\int_V \mathbf{B}\cdot(\nabla \times \mathbf{A})\, d\tau = \int_V \mathbf{A}\cdot(\nabla \times \mathbf{B})\, d\tau + \int_S (\mathbf{A} \times \mathbf{B})\cdot d\mathbf{a}$$

where S is the surface bounding the volume V.

Problem 1.48 Evaluate the integral

$$J = \int_{\text{all space}} e^{-r}\left(\nabla\cdot\frac{\hat{r}}{r^2}\right) d\tau$$

by two different methods, as in Example 12.

1.6 THE THEORY OF VECTOR FIELDS

1.6.1 The Helmholtz Theorem

Ever since Faraday, the laws of electricity and magnetism have been expressed in terms of **electric** and **magnetic fields,** $\mathbf{E}$ and $\mathbf{B}$. Like many physical laws, these are most compactly expressed as differential equations. Since $\mathbf{E}$ and $\mathbf{B}$ are *vectors*, the differential equations naturally involve vector derivatives: divergence and curl. Indeed, Maxwell reduced the entire theory to four equations, specifying respectively the divergence and the curl of $\mathbf{E}$ and $\mathbf{B}$.[7] Maxwell's formulation raises an important mathematical question: To what extent is a vector function determined by its divergence and curl? In other words, suppose we are told that the divergence of $\mathbf{F}$ (which stands for $\mathbf{E}$ or $\mathbf{B}$, as the case may be) is a specified scalar function D:

$$\nabla\cdot\mathbf{F} = D \tag{1.84}$$

and the curl of $\mathbf{F}$ is a specified vector function $\mathbf{C}$:

$$\nabla \times \mathbf{F} = \mathbf{C} \tag{1.85}$$

(for consistency, $\mathbf{C}$ must be divergenceless,

$$\nabla\cdot\mathbf{C} = 0 \tag{1.86}$$

because the divergence of a curl is always zero); can we, on the basis of this information, find the function $\mathbf{F}$? If this is *insufficient* information, there may be more than one solution to the problem; if it is *too much* information, there may not be any solution at all. In either case the theory is no good. Before we devote too much energy

[7]Strictly speaking, this is only true in the static case; in general the divergence and curl are given in terms of time derivatives of the fields themselves.

to Maxwell's electrodynamics, then, we had better check that a solution to equations (1.84)-(1.86) exists and is unique. This is the burden of the **Helmholtz theorem.** (If you're content to take such things on faith, you're welcome to skip this section.)

Existence is proved by explicit construction of a solution:

$$\mathbf{F} = -\nabla U + \nabla \times \mathbf{W} \tag{1.87}$$

where

$$U(\mathbf{r}) \equiv \frac{1}{4\pi} \int_V \frac{D(\mathbf{r}')}{\imath} \, d\tau' \tag{1.88}$$

$$\mathbf{W}(\mathbf{r}) \equiv \frac{1}{4\pi} \int_V \frac{\mathbf{C}(\mathbf{r}')}{\imath} \, d\tau' \tag{1.89}$$

$$\imath \equiv \mathbf{r} - \mathbf{r}' \tag{1.90}$$

and V is all of space. For if $\mathbf{F}$ is given by (1.87), then its divergence (using 1.82) is

$$\nabla \cdot \mathbf{F} = -\nabla^2 U = -\frac{1}{4\pi} \int_V D \nabla^2 \left(\frac{1}{\imath}\right) d\tau' = \frac{1}{4\pi} \int_V D(\mathbf{r}') 4\pi \delta^3(\mathbf{r} - \mathbf{r}') \, d\tau' = D(\mathbf{r})$$

(Remember that the divergence of a curl is zero, so the $\mathbf{W}$ term drops out, and note that the differentiation is with respect to $\mathbf{r}$, which is contained in $\imath$.) So the divergence is right; how about the curl?

$$\nabla \times \mathbf{F} = \nabla \times (\nabla \times \mathbf{W}) = -\nabla^2 \mathbf{W} + \nabla(\nabla \cdot \mathbf{W}) \tag{1.91}$$

(Since the curl of a gradient is zero, the U term drops out.) Now

$$-\nabla^2 \mathbf{W} = -\frac{1}{4\pi} \int_V \mathbf{C} \nabla^2 \left(\frac{1}{\imath}\right) d\tau' = \int_V \mathbf{C}(\mathbf{r}') \delta^3(\mathbf{r} - \mathbf{r}') \, d\tau' = \mathbf{C}(\mathbf{r})$$

which is perfect—I'll be done if I can just persuade you that the *second* term on the right side of (1.91) vanishes. Using integration by parts (equation (1.83)) and noting that derivatives of $\imath$ with respect to *primed* coordinates will differ by a sign from those with respect to unprimed coordinates, we have

$$\begin{aligned} 4\pi \nabla \cdot \mathbf{W} &= \int_V \mathbf{C} \cdot \nabla \left(\frac{1}{\imath}\right) d\tau' = -\int_V \mathbf{C} \cdot \nabla' \left(\frac{1}{\imath}\right) d\tau' \\ &= \int_V \frac{1}{\imath} \nabla' \cdot \mathbf{C} \, d\tau' - \int_S \frac{1}{\imath} \mathbf{C} \cdot d\mathbf{a} \end{aligned}$$

But the divergence of $\mathbf{C}$ is zero (equation (1.86)), and the surface integral (way out at infinity, since V is all of space) will vanish, as long as $\mathbf{C}$ is localized. q.e.d.

Well . . . that's what's known as an "elegant" argument: fancy formal footwork that is nice to look at, but doesn't do much to illuminate the situation. Later on we'll work through these same steps in a concrete physical context, where their intuitive content will be more apparent. Of course, the proof tacitly assumes that the integrand (1.88) and (1.89) *converge*—otherwise U and $\mathbf{W}$ don't exist at all. There could be a problem at the upper limit on r', where $\mathscr{r} \approx r'$ and the integrals have the form

$$\int^{\infty} \frac{X(r')}{r'} r'^2 \, dr' = \int^{\infty} r' X(r') \, dr' \tag{1.92}$$

(X is D or $\mathbf{C}$, as the case may be). Obviously, X must go to zero at large r'—but that's not enough: If $X \sim 1/r'$, the integrand is constant, so the integral blows up, and even if $X \sim 1/r'^2$, the integral is a logarithm, which is still no good. Evidently the divergence and curl of $\mathbf{F}$ must go to zero at large distances *more rapidly than* $1/r^2$ for the proof to hold. (Incidentally, this is *more* than enough to ensure that the surface integral in $\nabla \cdot \mathbf{W}$ vanishes.)

Now, assuming these conditions on $D(\mathbf{r})$ and $\mathbf{C}(\mathbf{r})$ are met, is the solution (1.87) *unique?* The answer is clearly *no,* for we can add to $\mathbf{F}$ any vector function whose divergence and curl both vanish, and the result still has divergence D and curl $\mathbf{C}$. However, it so happens that there is *no* function which has zero divergence and zero curl everywhere *and* goes to zero at infinity (the proof of this will come more naturally in Chapter 3). So if we add the requirement that $\mathbf{F}(\mathbf{r})$ goes to zero as $r \to \infty$, then the solution (1.87) is unique. (Typically we *do* expect the electric and magnetic fields to go to zero at large distances from the charges and currents that produce them, so this is not an unreasonable stipulation. Occasionally we shall encounter artificial problems in which the charge distribution itself extends to infinity—along infinite wires, for instance, or over infinite planes. In such cases other means must be found to establish the existence and uniqueness of the solution.)

Now that all the cards are on the table, I can state the Helmholtz theorem more rigorously.

Helmholtz Theorem: If the divergence $D(\mathbf{r})$ and the curl $\mathbf{C}(\mathbf{r})$ of a vector function $\mathbf{F}(\mathbf{r})$ are specified, and if they both go to zero faster than $1/r^2$ as $r \to \infty$, and if $\mathbf{F}(\mathbf{r})$ goes to zero as $r \to \infty$, then $\mathbf{F}$ is given uniquely by equation (1.87).

There is an interesting corollary to this:

Corollary: Any (differentiable) vector function $\mathbf{F}(\mathbf{r})$, which goes to zero faster than $1/r$ as $r \to \infty$, can be expressed as the gradient of a scalar plus the curl of a vector:

$$\mathbf{F} = \nabla \left(\frac{-1}{4\pi} \int \frac{\nabla \cdot \mathbf{F}}{\mathscr{r}} \, d\tau \right) + \nabla \times \left(\frac{1}{4\pi} \int \frac{\nabla \times \mathbf{F}}{\mathscr{r}} \, d\tau \right) \tag{1.93}$$

Problem 1.49 Suppose $D(\mathbf{r}) = a\delta^3(\mathbf{r})$ and $\mathbf{C}(\mathbf{r}) = \mathbf{b}\delta^3(\mathbf{r})$, where a and $\mathbf{b}$ are constants. Determine $U(\mathbf{r})$, $\mathbf{W}(\mathbf{r})$, and $\mathbf{F}(\mathbf{r})$.

1.6.2 Scalar and Vector Potentials

For completeness, I conclude with two powerful theorems that are of fundamental importance in the development of electrodynamics, for they permit the introduction of electric and magnetic potentials.

Theorem 1: **Curl-less (or "irrotational") fields.** The following conditions are equivalent (that is, **F** satisfies one if and only if it satisfies all the others):

(a) $\nabla \times \mathbf{F} = 0$ everywhere.

(b) $\int_a^b \mathbf{F} \cdot d\mathbf{l}$ is independent of path, for any given end points.

(c) $\oint \mathbf{F} \cdot d\mathbf{l} = 0$ for any closed loop.

(d) **F** is the gradient of some scalar, $\mathbf{F} = -\nabla U$.

U is called the **scalar potential** for the field **F**; it is not unique—any constant can be added to U with impunity, since this will not affect the gradient.

Theorem 2: **Divergence-less (or "solenoidal") fields.** The following conditions are equivalent:

(a) $\nabla \cdot \mathbf{F} = 0$ everywhere.

(b) $\int_{\text{surface}} \mathbf{F} \cdot d\mathbf{a}$ is independent of surface, for any given boundary line.

(c) $\oint_{\text{surface}} \mathbf{F} \cdot d\mathbf{a} = 0$ for any closed surface.

(d) **F** is the curl of some vector, $\mathbf{F} = \nabla \times \mathbf{W}$.

W is called the **vector potential** for the field **F**; it is not unique—the gradient of any scalar function can be added to **W** without affecting the curl, since the curl of a gradient is zero (1.35).

You should be able to prove all the connections by now, save for the ones that say (a), (b), or (c) implies (d). Those are more subtle, and will come later. (They follow very simply from the corollary to the Helmholtz theorem, *provided* **F** goes to zero rapidly enough as $r \to \infty$. However, the two theorems here carry no such restriction, and we shall often use them in contexts where the Helmholtz theorem itself does not apply.)[8]

Problem 1.50

(a) Let $\mathbf{F}_1 = x^2\hat{k}$ and $\mathbf{F}_2 = x\hat{i} + y\hat{j} + z\hat{k}$. Calculate the divergence and curl of $\mathbf{F}_1$ and $\mathbf{F}_2$. Which one can be written as the gradient of a scalar? Find a scalar potential that does the job. Which one can be written as the curl of a vector? Find a suitable vector potential.

(b) Show that $\mathbf{F}_3 = yz\hat{i} + zx\hat{j} + xy\hat{k}$ can be written both as the gradient of a scalar and as the curl of a vector. Find scalar and vector potentials for this function.

Problem 1.51 For Theorem 1 show that (d) $\Rightarrow$ (a) $\Rightarrow$ (c) $\Rightarrow$ (b) and that (b) $\Rightarrow$ (c) $\Rightarrow$ (a).

Problem 1.52 For Theorem 2 show that (d) $\Rightarrow$ (a) $\Rightarrow$ (c) $\Rightarrow$ (b) and that (b) $\Rightarrow$ (c) $\Rightarrow$ (a).

Problem 1.53

(a) Which of the vectors in Problem 1.15 can be expressed as the gradient of a scalar? Find a scalar function that does the job.

(b) Which can be expressed as the curl of a vector? Find such a vector.

[8]The fact is, *any* vector function can be written as the gradient of a scalar plus the curl of a vector. But only if it goes to zero sufficiently rapidly at infinity are the scalar and vector potentials given by (1.93).

Further Problems on Chapter 1

Problem 1.54 Check the divergence theorem for the function

$$\mathbf{v} = r^2 \cos\theta\,\hat{r} + r^2 \cos\phi\,\hat{\theta} - r^2 \cos\theta \sin\phi\,\hat{\phi},$$

using as your volume one octant of the sphere of radius R (Fig. 1.52). Make sure you include the *entire* surface. (*Answer*: $\pi R^4/4$)

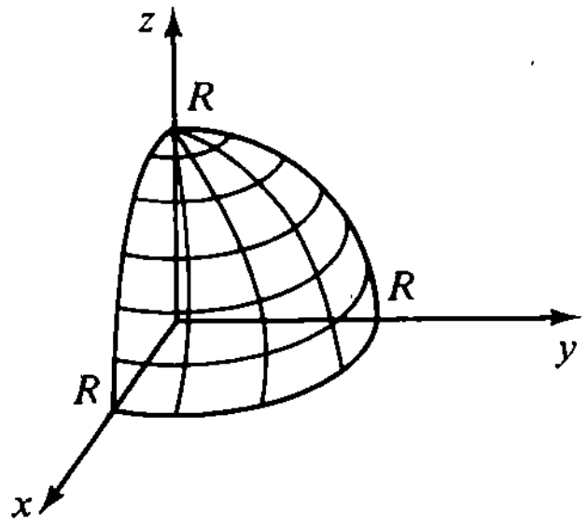

Figure 1.52

Problem 1.55 Check Stokes' theorem using the function

$$\mathbf{v} = ay\hat{\imath} + bx\hat{\jmath} \qquad (a \text{ and } b \text{ are constants})$$

and the circular path of radius ab, centered at the origin in the xy plane. (*Answer*: $\pi a^2 b^2 (b - a)$)

Problem 1.56 Compute the line integral of

$$\mathbf{v} = 6\hat{\imath} + yz^2\hat{\jmath} + (3y + z)\hat{k}$$

along the triangular path shown in Fig. 1.53. Check your answer using Stokes' theorem. (*Answer*: $\frac{8}{3}$)

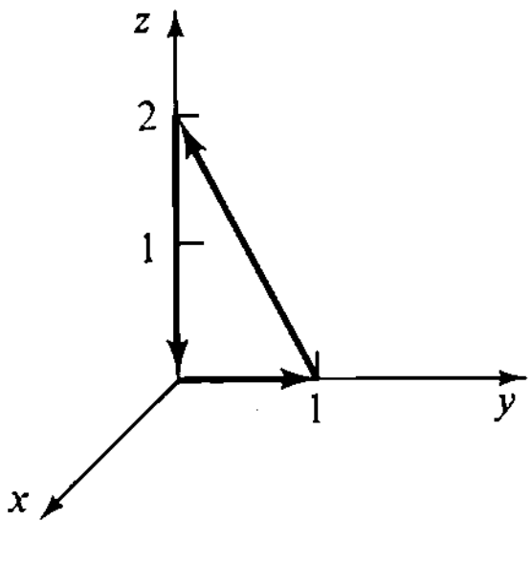

Figure 1.53

Problem 1.57 Compute the line integral of

$$\mathbf{v} = (r\cos^2\theta)\hat{r} - (r\cos\theta\sin\theta)\hat{\theta} + 3r\hat{\phi}$$

around the path shown in Fig. 1.54 (the points are labeled by their Cartesian coordinates). Do it either in cylindrical or in spherical coordinates. Check your answer, using Stokes' theorem. (*Answer*: $3\pi/2$)

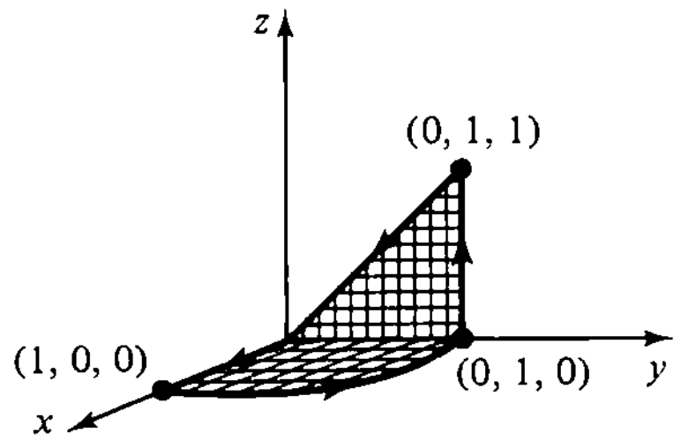

Figure 1.54

Problem 1.58 Check Stokes' theorem for the function

$$\mathbf{v} = y\hat{k}$$

using the triangular surface shown in Fig. 1.55 (*Answer*: a^2)

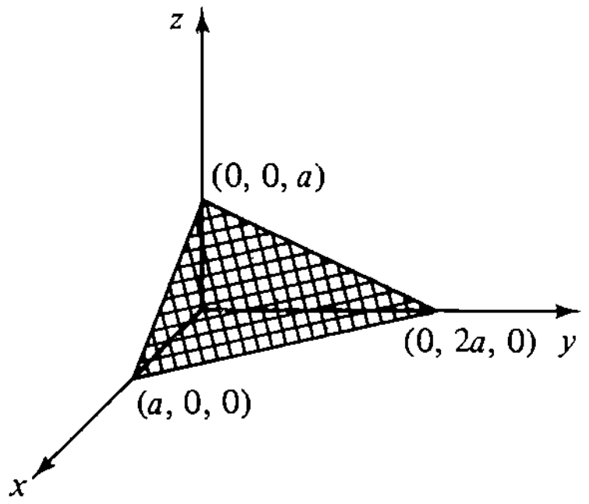

Figure 1.55

Problem 1.59 Check the divergence theorem for the function

$$\mathbf{v} = r^2 \sin\theta\, \hat{r} + 4r^2 \cos\theta\hat{\theta} + r \tan\theta\hat{\phi}$$

using the volume of the "ice-cream cone" shown in Fig. 1.56.

$\left(\textit{Answer}: \quad \frac{\pi R^4}{2}\left(\frac{\pi}{3} + \frac{\sqrt{3}}{2}\right)\right)$

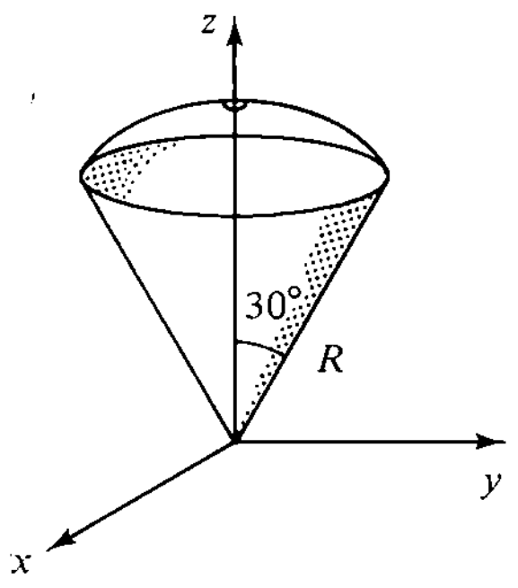

Figure 1.56

Problem 1.60 Show that

$$\int_S f(\nabla \times \mathbf{A})\cdot d\mathbf{a} = -\int_S (\mathbf{A} \times \nabla f)\cdot d\mathbf{a} + \oint_C f\mathbf{A}\cdot d\mathbf{l}$$

where S is an open surface and C is its perimeter.

• **Problem 1.61** Although the gradient, divergence, and curl theorems are the fundamental integral theorems of vector calculus, it is possible to derive a number of corollaries from them. Show that:

(a) $\int_{\text{volume}} (\nabla T)\, d\tau = \oint_{\text{surface}} T\, d\mathbf{a}$. (*Suggestion*: Let $\mathbf{v} = \mathbf{c}T$, where $\mathbf{c}$ is a constant, in the divergence theorem; use the product rules.)

(b) $\int_{\text{volume}} (\nabla \times \mathbf{v})\, d\tau = -\oint_{\text{surface}} \mathbf{v} \times d\mathbf{a}$. (Replace $\mathbf{v}$ by $(\mathbf{v} \times \mathbf{c})$ in the divergence theorem.)

(c) $\int_{\text{volume}} [T\nabla^2 U + (\nabla T)\cdot(\nabla U)]\, d\tau = \oint_{\text{surface}} (T\nabla U)\cdot d\mathbf{a}$. (Let $\mathbf{v} = T\nabla U$ in the divergence theorem.)

(d) $\int_{\text{volume}} (T\nabla^2 U - U\nabla^2 T)\, d\tau = \oint_{\text{surface}} (T\nabla U - U\nabla T)\cdot d\mathbf{a}$. (This is known as **Green's theorem**; it follows from (c), which is sometimes called **Green's identity.**)

(e) $\int_{\text{surface}} \nabla T \times d\mathbf{a} = -\oint T\, d\mathbf{l}$. (Let $\mathbf{v} = \mathbf{c}T$ in Stokes' theorem).

• **Problem 1.62** (a) What is the divergence of the function

$$\mathbf{v} = \frac{\hat{\mathbf{r}}}{r}$$

First compute it directly, as in equation (1.64). Test your result using the divergence theorem, as in equation (1.65). Is there a delta function at the origin, as there was for $\hat{\mathbf{r}}/r^2$? What is the general formula for the divergence of $r^n\hat{\mathbf{r}}$?

(b) Find the *curl* of $r^n\hat{\mathbf{r}}$. Test your conclusion using Problem 1.61(b).

(*Answer*: $\nabla\cdot(r^n\hat{\mathbf{r}}) = (n + 2)r^{n-1}$, unless $n = -2$, in which case it is $4\pi\delta^3(\mathbf{r})$; $\nabla \times (r^n\hat{\mathbf{r}}) = 0$)

2

ELECTROSTATICS

2.1 THE ELECTROSTATIC FIELD

2.1.1 Introduction

The fundamental problem which electromagnetic theory hopes to solve is this (Fig. 2.1): Suppose that we have some electric charges, $q_1, q_2, q_3, \ldots$ (call them **source charges**). *What force do they exert on another charge, Q* (call it the **test charge**)? (The positions of the source charges are *given,* as functions of time; the trajectory of the test charge is *to be calculated.*) In general, both the source charges and the test charge may be in motion.

The solution to this problem is facilitated by the **principle of superposition,** which states that the interaction between any two charges is completely unaffected by the presence of other charges. This means that to compute the force on Q, we can first compute the force $\mathbf{F}_1$, due to q_1 alone (ignoring all the others); then we compute the force $\mathbf{F}_2$, due to q_2 alone; and so on. Finally, we take the vector sum of all these individual forces: $\mathbf{F} = \mathbf{F}_1 + \mathbf{F}_2 + \mathbf{F}_3 + \ldots$. Thus, if we can find the force on Q due to a *single* source charge q, we are, in principle, done (the rest is just a question of repeating the same operation over and over, and adding it all up).[1]

Well, at first sight this sounds very easy: Why don't I just write down the formula for the force on Q due to q, and be done with it? I *could,* and in Chapter 9, I shall, but you would be shocked to see it at this stage, for not only does the force on Q depend on the separation distance $\mathscr{r}$ between the charges (Fig. 2.2), it also depends

[1]The principle of superposition may seem "obvious" to you, but it did not *have* to be so simple: if the electromagnetic force were proportional to the *square* of the total source charge, for instance, the principle of superposition would not hold, since $(q_1 + q_2)^2 \neq q_1^2 + q_2^2$ (there would be "cross terms" to consider).

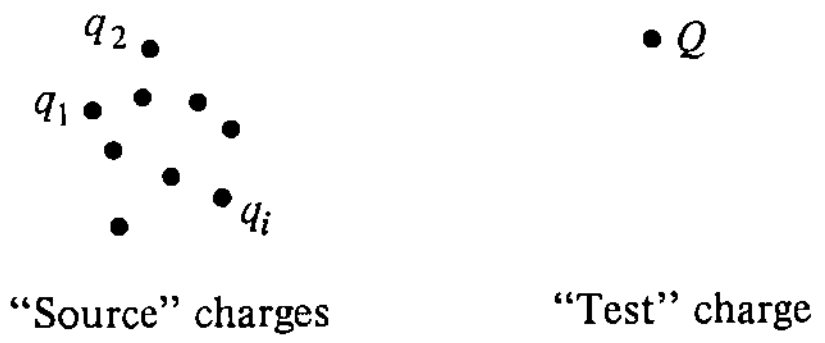

Figure 2.1

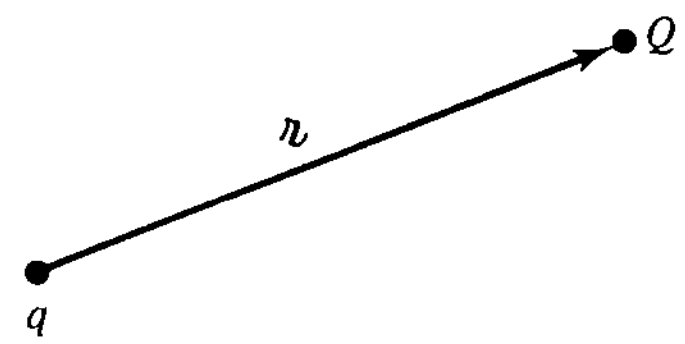

Figure 2.2

on *both* their velocities and on the acceleration of q. Moreover, it is not the position, velocity, and acceleration of q *right now* that matter: Electromagnetic "news" travels at the speed of light, so what concerns Q is the position, velocity, and acceleration q *had* at some earlier time, when the message left.

Thus, in spite of the fact that the basic question ("What is the force on Q due to q?") is easy to state, it does not pay to confront it head-on; rather, we shall go at it by stages. In the meantime, the theory we develop will permit the solution of more subtle electromagnetic problems that do not present themselves in quite this simple format. To begin with we shall consider the special case of **electrostatics** in which all the *source charges are stationary* (though the test charge may be moving).

2.1.2 Coulomb's Law

What is the force on a point charge Q due to a single point charge q which is at *rest* a distance $\mathscr{r}$ away? The answer is given by **Coulomb's law**:

$$\boxed{\mathbf{F} = \frac{1}{4\pi\epsilon_0}\frac{qQ}{\mathscr{r}^2}\hat{\mathscr{r}}} \tag{2.1}$$

The constant ϵ_0 is called the **permittivity of free space.** In mks units, where force is in newtons (N), distance is in meters (m), and charge is in coulombs (C)

$$\epsilon_0 = 8.85 \times 10^{-12}\frac{C^2}{N\cdot m^2}.$$

In words, the force on Q is proportional to the product of the charges and inversely proportional to the square of the separation distance. It points along the line from q to Q, and is repulsive if q and Q have the same sign, and attractive if their signs are opposite. (Throughout this book I shall reserve script $\mathscr{r}$ for situations such as Coulomb's law, in which we want to specify the vector displacement from one point to another; **r** will be saved for position vectors, and r for the cylindrical or spherical coordinate.) Coulomb's law and the principle of superposition constitute the physical input for electrostatics—the rest, except for some special properties of matter, is mathematical elaboration of these fundamental rules.

Problem 2.1

(a) Twelve equal charges, q, are situated at the corners of a regular 12-sided polygon (for instance, one on each numeral of a clock face). What is the net force on a test charge Q at the center?

(b) Suppose *one* of the 12 q's is removed (the one at "6 o'clock"). What is the force on Q? Explain your reasoning carefully.
(c) Now 13 equal charges, q, are placed at the corners of a regular 13-sided polygon. What is the force on a test charge Q at the center?
(d) If one of the 13 q's is removed, what is the force on Q? Explain your reasoning.

2.1.3 The Electric Field

If we have many point charges $q_1, q_2, \ldots, q_n$ at distances $\mathscr{r}_1, \mathscr{r}_2, \ldots, \mathscr{r}_n$ from Q, then according to the principle of superposition the total force on Q is

$$\mathbf{F} = \mathbf{F}_1 + \mathbf{F}_2 + \cdots = \frac{1}{4\pi\epsilon_0}\left(\frac{q_1 Q}{\mathscr{r}_1^2}\hat{\mathscr{r}}_1 + \frac{q_2 Q}{\mathscr{r}_2^2}\hat{\mathscr{r}}_2 + \cdots\right)$$
$$= \frac{Q}{4\pi\epsilon_0}\left(\frac{q_1\hat{\mathscr{r}}_1}{\mathscr{r}_1^2} + \frac{q_2\hat{\mathscr{r}}_2}{\mathscr{r}_2^2} + \frac{q_3\hat{\mathscr{r}}_3}{\mathscr{r}_3^2} + \cdots\right)$$

or

$$\boxed{\mathbf{F} = Q\mathbf{E}} \tag{2.2}$$

where

$$\mathbf{E}(P) = \frac{1}{4\pi\epsilon_0}\sum_{i=1}^{n}\frac{q_i}{\mathscr{r}_i^2}\hat{\mathscr{r}}_i \tag{2.3}$$

E is called the **electric field** of the source charges. Notice that it is a function of position $P = (x, y, z)$ because the vectors $\mathscr{r}_i$ depend on P (Fig. 2.3). But it makes no reference at all to the test charge Q. The electric field, then, is a vector quantity

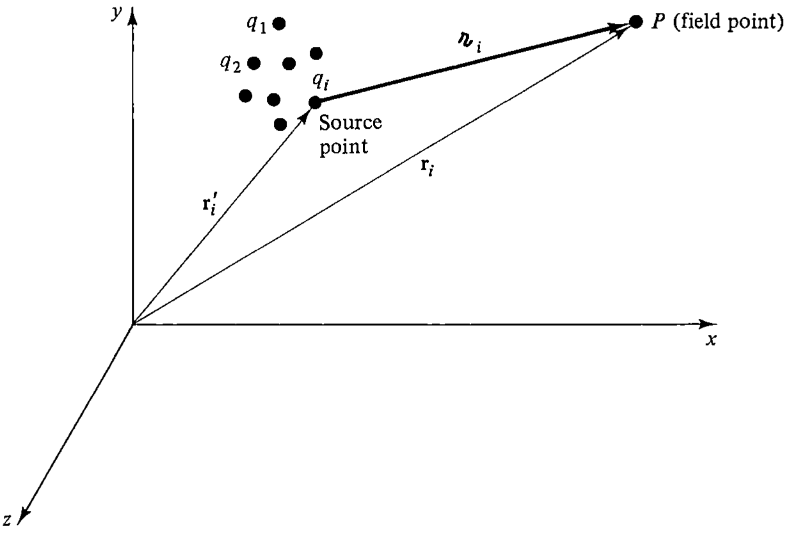

Figure 2.3

which varies from point to point and is determined by the configuration of source charges; physically, $\mathbf{E}(P)$ is the force per unit charge that would be exerted on a test charge placed at P.

What exactly *is* an electric field? I have deliberately begun with what you might call the "minimal" interpretation of $\mathbf{E}$, as an intermediate step in the calculation of electric forces. But I encourage you to think of the field as a "real" physical entity, filling the space in the neighborhood of any electric charge. Maxwell himself came to believe that electric and magnetic fields represented actual stresses and strains in an invisible primordial jellylike "ether." Special relativity has forced us to abandon the notion of ether, and with it Maxwell's mechanical interpretation of electromagnetic fields. It is even possible, though cumbersome, to formulate classical electrodynamics as an "action-at-a-distance" theory and dispense with the field concept altogether. I can't tell you, then, what a field "*is*"—but only how to calculate it and what it can do for you once you've got it.

Problem 2.2

(a) Find the electric field (magnitude and direction) a distance z above the midpoint between two equal charges q a distance d apart (Fig. 2.4). Check that your result is consistent with what you'd expect when $z \gg d$.

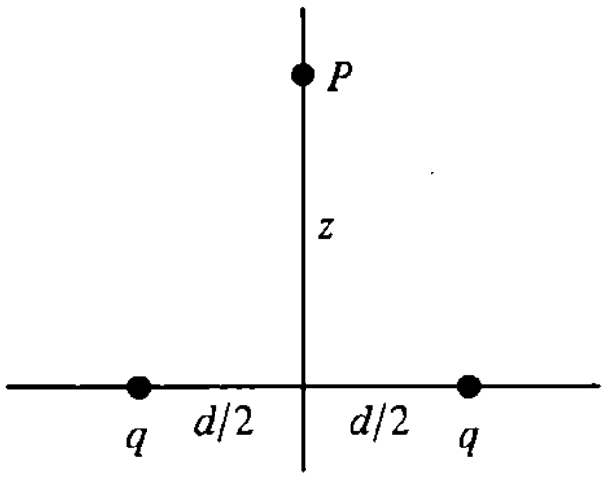

Figure 2.4

(b) Repeat part (a), only this time make the right-hand charge $-q$ instead of $+q$.

2.1.4 Continuous Charge Distributions

Our definition of the electric field, equation (2.3), assumes that the source of the field is a set of discrete point charges q_i. Often, we shall want to consider charges that are distributed continuously over some region. *Notation*: Charge spread out along a line is described by the letter λ (representing the charge per unit length); charge spread over a surface is described by σ (the charge per unit area); and charge spread throughout a volume is described by ρ (the charge per unit volume). It is no real problem to generalize equation (2.3) or any other equation pertaining to discrete point charges: We simply divide the charge up into infinitesimal pieces and treat each tiny bit as a point charge. Thus, q_i is replaced by $dq = \lambda\, dl$, $\sigma\, da$, or $\rho\, d\tau$, respectively, while the summation becomes an integral:

$$\sum_{i=1}^{n} (\quad) q_i \sim \int_{\text{line}} (\quad) \lambda\, dl \sim \int_{\text{surface}} (\quad) \sigma\, da \sim \int_{\text{volume}} (\quad) \rho\, d\tau$$

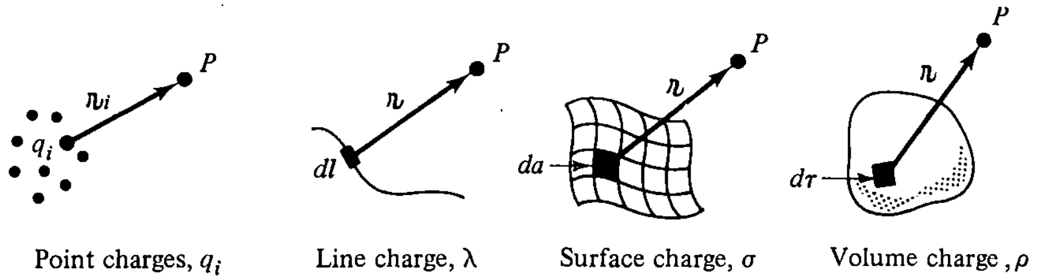

Figure 2.5

For instance, the electric field of a line charge is

$$\mathbf{E}(P) = \frac{1}{4\pi\epsilon_0} \int_{\text{line}} \frac{\hat{\mathscr{r}}}{\mathscr{r}^2} \lambda \, dl \tag{2.4}$$

for a surface charge,

$$\mathbf{E}(P) = \frac{1}{4\pi\epsilon_0} \int_{\text{surface}} \frac{\hat{\mathscr{r}}}{\mathscr{r}^2} \sigma \, da \tag{2.5}$$

and for a volume charge,

$$\boxed{\mathbf{E}(P) = \frac{1}{4\pi\epsilon_0} \int_{\text{volume}} \frac{\hat{\mathscr{r}}}{\mathscr{r}^2} \rho \, d\tau} \tag{2.6}$$

Equation (2.6) is often referred to as "Coulomb's Law", because it is such a short step from the original (2.1). Please note carefully the meaning of $\mathscr{r}$ in these equations. Originally, in equation (2.3), $\mathscr{r}_i$ represented the distance from the source charge q_i to P. Correspondingly, in equations (2.4)-(2.6), $\mathscr{r}$ is the distance from the element of charge (therefore, from dl, da, or $d\tau$) to the point P, as shown in Fig. 2.5. If the source point is (x', y', z') and $P = (x, y, z)$, then in Cartesian coordinates:[2]

$$\boldsymbol{\mathscr{r}} = (x - x')\hat{i} + (y - y')\hat{j} + (z - z')\hat{k},$$

and its length (by the Pythagorean theorem) is

$$\mathscr{r} = [(x - x')^2 + (y - y')^2 + (z - z')^2]^{1/2}$$

The unit vector $\hat{\mathscr{r}}$ is $\boldsymbol{\mathscr{r}}$ divided by its length:

$$\hat{\mathscr{r}} = \frac{\boldsymbol{\mathscr{r}}}{\mathscr{r}}$$

Thus, the field of a volume charge distribution, written out for once in painful detail, is

$$\mathbf{E}(x, y, z) = \frac{1}{4\pi\epsilon_0} \int_{\text{volume}} \frac{(x - x')\hat{i} + (y - y')\hat{j} + (z - z')\hat{k}}{[(x - x')^2 + (y - y')^2 + (z - z')^2]^{3/2}} \rho(x', y', z')dx'dy'dz' \tag{2.7}$$

[2]In terms of the *position* vectors of the source point ($\mathbf{r}'$) and the field point ($\mathbf{r}$), $\boldsymbol{\mathscr{r}} = \mathbf{r} - \mathbf{r}'$.

I hasten to say that one almost never actually *calculates* a field from equation (2.7); in the context of a particular problem, a simpler method usually suggests itself.[3]

Example 1

Find the electric field a distance z above the midpoint of a straight line segment of length $2L$, which carries a uniform line charge λ (Fig. 2.6).

It is advantageous to chop the line up into symmetrically placed pairs, for then the horizontal components of the two fields cancel, and the net field of the pair is

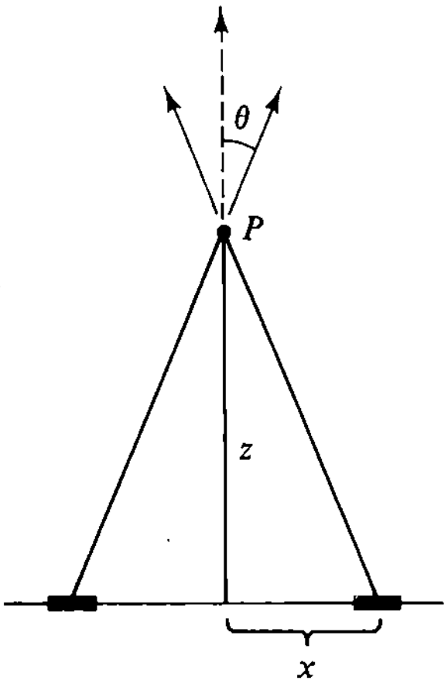

Figure 2.6

$$d\mathbf{E} = \frac{1}{4\pi\epsilon_0} 2\left(\frac{\lambda\, dx}{\imath^2}\right)\cos\theta\,\hat{k}$$

Here $\theta = z/\imath$, $\imath = \sqrt{z^2 + x^2}$, and the integral runs from 0 to L:

$$E = \frac{1}{4\pi\epsilon_0}\int_0^L \frac{2\lambda z}{(z^2+x^2)^{3/2}}\,dx = \frac{2\lambda z}{4\pi\epsilon_0}\left[\frac{x}{z^2\sqrt{z^2+x^2}}\right]\Bigg|_0^L = \frac{1}{4\pi\epsilon_0}\frac{2\lambda L}{z\sqrt{z^2+L^2}}$$

and it aims in the z-direction.

For points far from the line ($z \gg L$), the formula simplifies:

$$E \cong \frac{1}{4\pi\epsilon_0}\frac{2\lambda L}{z^2}$$

which makes sense: From this distance the line "looks" like a point charge $q = 2\lambda L$, so the field reduces to that of point charge $q/(4\pi\epsilon_0 z^2)$. In the limit $L \to \infty$, on the other hand, we get the field of an infinite straight wire:

$$E = \frac{1}{4\pi\epsilon_0}\frac{2\lambda}{z}$$

[3] *Warning*: The unit vector $\hat{\imath}$ is *not* constant; its *direction* depends on the source point (x', y', z'), and hence it *cannot* be taken outside the integral (2.6). In practice, you must work with *Cartesian* components ($\hat{i}$, $\hat{j}$, $\hat{k}$ *are* constant, and *do* come out)—even if you use spherical coordinates to perform the integration.

Problem 2.3 Find the electric field a distance z above one end of a straight line segment of length L (Fig. 2.7), which carries a uniform line charge λ. Check that your formula is consistent with what you would expect for the case $z \gg L$.

Problem 2.4 Find the electric field a distance z above the center of a square loop (side s) carrying uniform line charge λ (Fig. 2.8). (*Suggestion*: Use the result of Example 1.)

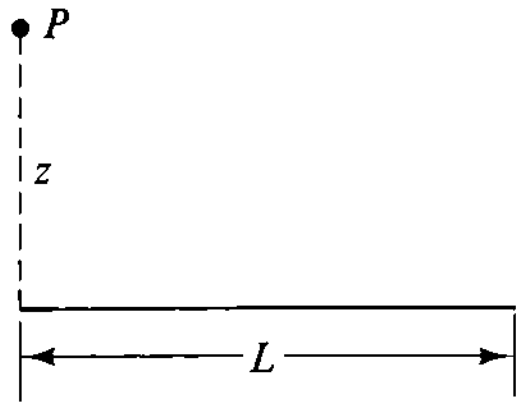

Figure 2.7

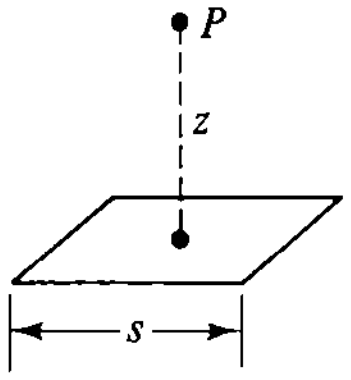

Figure 2.8

Problem 2.5 Find the electric field a distance z above the center of a circular loop of radius r (Fig. 2.9), which carries a uniform line charge λ.

Problem 2.6 Find the electric field a distance z above the center of a flat circular disc of radius R (Fig. 2.10), which carries a uniform surface charge σ. What does your formula give in the limit $R \to \infty$? Also check the case $z \gg R$.

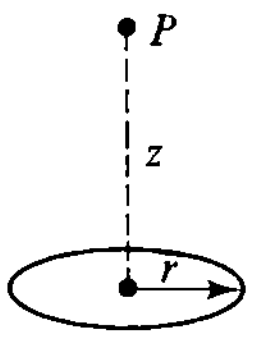

Figure 2.9

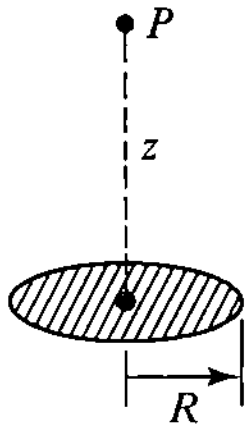

Figure 2.10

! **Problem 2.7** Find the electric field a distance z from the center of a spherical surface of radius R (Fig. 2.11), which carries a uniform charge density σ. Treat the case $z < R$ (inside) as well as $z > R$ (outside). Express your answers in terms of the total charge q on the sphere. (*Suggestion*: Use the law of cosines to write $\imath$ in terms of R and θ. Be sure to take the *positive* square root: $\sqrt{R^2 + z^2 - 2Rz} = (R - z)$ if $R > z$, but it's $(z - R)$ if $R < z$.)

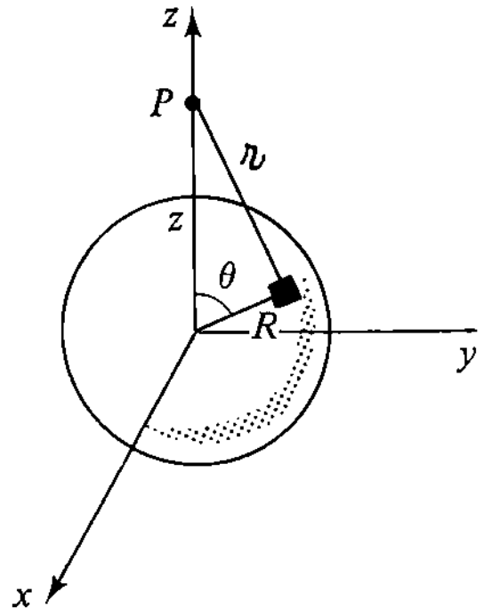

Figure 2.11

Problem 2.8 Use your result in Problem 2.7 to find the field inside and outside a sphere of radius R, which carries a uniform volume charge density ρ. Express your answers in terms of the total charge of the sphere.

2.2 DIVERGENCE AND CURL OF ELECTROSTATIC FIELDS

2.2.1 Field Lines and Gauss's Law

In *principle,* we are done with the subject of electrostatics. Equations (2.3)-(2.6) (or, more explicitly, (2.7)), tell us how to compute the field of a charge distribution, and equation (2.2) tells us what the force on any charge Q placed in this field will be. Unfortunately, as you may have discovered in working Problem 1.7, the integrals involved in computing **E** can be formidable, even for reasonably simple charge distributions. Much of the rest of electrostatics is devoted to assembling a bag of tools and tricks for *avoiding* these integrals. It all begins with the derivation of the divergence and curl of **E**. I shall calculate the divergence of **E** directly from equation (2.6) in Section 2.2.2, but first I want to show you a more qualitative, and perhaps more illuminating, heuristic approach.

Let's begin with the simplest possible case; a single point charge q, situated at the origin:

$$\mathbf{E}(\mathbf{r}) = \frac{1}{4\pi\epsilon_0}\frac{q}{r^2}\hat{r} \tag{2.8}$$

where **r** is the position in question. To get a "feel" for this field, I might sketch in a few representative vectors, here and there, as we have done before. Because the field falls off like $1/r^2$, the vectors get shorter as I go farther away from the origin (Fig. 2.12); they always point radially outward. But there is a nicer way to represent this field, and that's to connect up the arrows, to form **field lines** (Fig. 2.13). You might

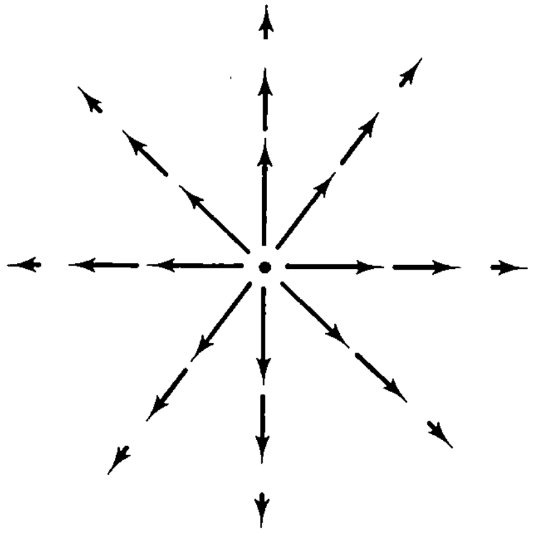

Figure 2.12

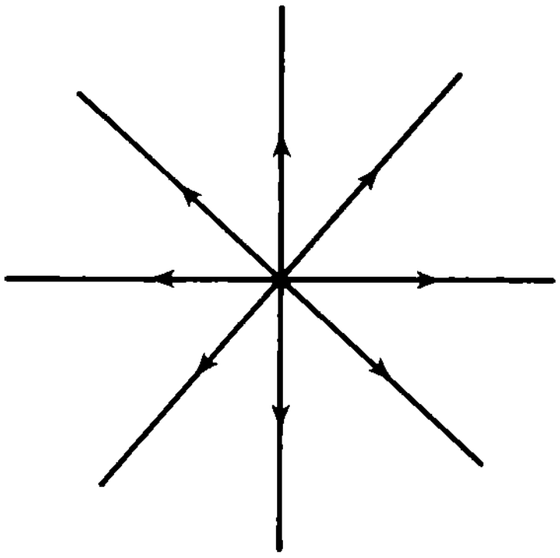

Figure 2.13

think that I have thereby thrown away information about the *strength* of the field, which was contained in the length of the arrows. But actually I have not. The magnitude of the field is indicated by the *density* of the field lines: it's strong near the center where the field lines are close together, and weak farther out, where they are far apart. [In truth, the field-line diagram is deceptive, when I draw it on a two-dimensional surface, for the density of lines passing through a circle of radius r is the total number divided by the circumference, $(n/2\pi r)$, which goes like $(1/r)$, not $(1/r^2)$. But if you imagine the model in three dimensions (a pincushion with needles sticking out in all directions) then the density of lines is the total number divided by the area of the sphere, $(n/4\pi r^2)$, which *does* go like $(1/r^2)$.]

Such diagrams are also convenient for representing more complicated fields. Of course, the number of lines you draw depends on how energetic you are (and how sharp your pencil is), though you ought to include enough to get an accurate sense of the field, and you must be consistent: If charge q gets 8 lines, then $2q$ deserves 16. And you must space them fairly—they emanate from a point charge symmetrically in all directions (Fig. 2.14). Field lines originate on positive charges and terminate on negative ones; they cannot simply stop in midair, though they may extend out to infinity. Moreover, field lines can never cross (at the intersection, the field would have two different directions at once!). With all this in mind, it is easy to sketch the field of any simple configuration of point charges: Begin by drawing the lines in the

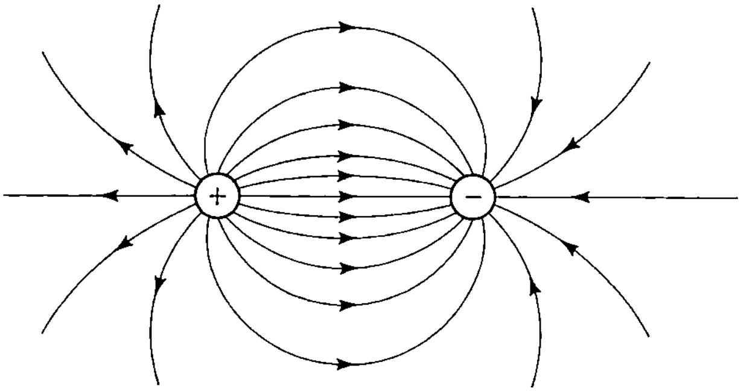

Equal but opposite charges

Figure 2.14

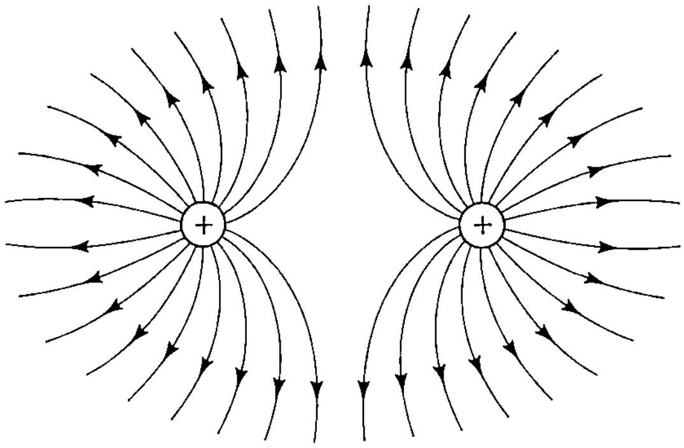

Equal charges **Figure 2.15**

neighborhood of each charge, and then connect them up or extend them to infinity (Fig. 2.15).

Since in this model the field strength is proportional to the number of lines per unit area (area perpendicular to the lines), the *flux* of **E** (i.e., $\int \mathbf{E} \cdot d\mathbf{a}$) through any surface is *proportional to the number of field lines passing through that surface* (the dot product in $\mathbf{E} \cdot d\mathbf{a}$ picks out the area perpendicular to the field lines; see Problem 1.32). Now, for the case of a point charge at the origin, the flux of **E** through a sphere of radius r is

$$\oint \mathbf{E} \cdot d\mathbf{a} = \int \frac{1}{4\pi\epsilon_0}\left(\frac{q}{r^2}\hat{r}\right) \cdot (r^2 \sin\theta \, d\theta \, d\phi \, \hat{r}) = \frac{1}{\epsilon_0} q \tag{2.9}$$

Notice that the radius of the sphere cancels out, for while the surface area goes *up* as r^2, the field goes *down* as $1/r^2$, and so the *product* is *constant*. In terms of the field-line picture, this makes good sense, since the same number of field lines passes through any sphere centered at the origin, regardless of its size. In fact, it didn't have to be a sphere—any old surface, whatever its shape, would trap the same number of field lines. So *the flux through any surface enclosing the charge is* q/ϵ_0.

Now suppose that instead of a single charge at the origin, we have a bunch of charges scattered about. According to the principle of superposition, the total field is simply the (vector) sum of all the individual fields:

$$\mathbf{E} = \sum_{i=1}^{n} \mathbf{E}_i$$

The flux through any surface that encloses them all, then, is

$$\oint \mathbf{E} \cdot d\mathbf{a} = \sum_{i=1}^{n} \left(\oint \mathbf{E}_i \cdot d\mathbf{a}\right) = \sum_{i=1}^{n} \left(\frac{1}{\epsilon_0} q_i\right)$$

On the other hand, a charge *outside* the surface would contribute *nothing* to the total flux, since its field lines go in one side and out the other (Fig. 2.16). It follows, then, that for any closed surface,

$$\oint_{\text{surface}} \mathbf{E} \cdot d\mathbf{a} = \frac{1}{\epsilon_0} Q_{\text{enc}} \tag{2.10}$$

where Q_{enc} is the total charge enclosed within the surface. This is **Gauss's law.** Although it contains no information that was not already present in Coulomb's law and the principle of superposition, it is of almost magical power, as you will see in Section (2.2.3). Notice that it all hinges on the $1/r^2$ character of Coulomb's law; without that the crucial cancellation in (2.9) would not take place, and the flux of **E** would depend on the surface chosen, not merely on the total charge enclosed. Other $1/r^2$ forces (I am thinking particularly of Newton's law of universal gravitation) will obey "Gauss's laws" of their own, and the applications we develop here carry over directly.

As it stands, Gauss's law is an *integral* equation, but we can readily turn it into a *differential* one, for continuous charge distributions, by applying the divergence theorem:

$$\oint_{\text{surface}} \mathbf{E} \cdot d\mathbf{a} = \int_{\text{volume}} (\nabla \cdot \mathbf{E})\, d\tau$$

Rewriting Q_{enc} in terms of the charge density ρ, we have

$$Q_{\text{enc}} = \int_{\text{volume}} \rho\, d\tau$$

So Gauss's law becomes

$$\int_{\text{volume}} (\nabla \cdot \mathbf{E})\, d\tau = \int_{\text{volume}} \left(\frac{1}{\epsilon_0} \rho\right) d\tau$$

Since this holds for *any* volume, the integrands must be equal:

$$\nabla \cdot \mathbf{E} = \frac{1}{\epsilon_0} \rho \tag{2.11}$$

Equation (2.11) carries the same message as equation (2.10); it is **Gauss's law in differential form.** The differential version is tidier, but the integral form has the advantage that it accommodates point, line, and surface charges more naturally.

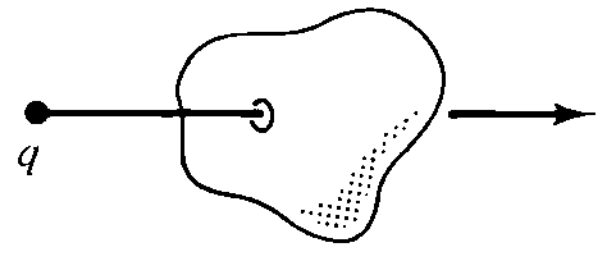

Figure 2.16

Problem 2.9 Suppose the electric field in some region is found to be $\mathbf{E} = kr^3\hat{\mathbf{r}}$, in spherical coordinates (k is some constant).

(a) Find the charge density ρ.

(b) Find the total charge contained in a sphere of radius R, centered at the origin. (Do it two different ways.)

Problem 2.10 A charge q sits at the back corner of a cube, as shown. What's the flux of $\mathbf{E}$ (that is, $\int \mathbf{E} \cdot d\mathbf{a}$) through the shaded side of Fig. 2.17?

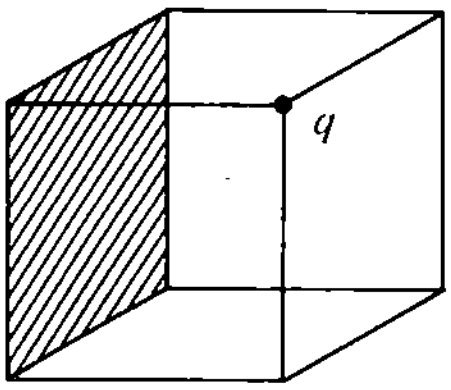

Figure 2.17

2.2.2 The Divergence of E

Let's go back now and calculate the divergence of $\mathbf{E}$ directly from equation (2.6), which I rewrite in more explicit notation:

$$\mathbf{E}(\mathbf{r}) = \frac{1}{4\pi\epsilon_0} \int_{\text{all space}} \frac{\hat{\mathbf{\mathscr{r}}}}{\mathscr{r}^2} \rho(\mathbf{r}')\, d\tau' \tag{2.12}$$

with

$$\boldsymbol{\mathscr{r}} \equiv \mathbf{r} - \mathbf{r}'$$

(Originally the integration was over the volume occupied by the charge distribution, but I may as well extend it to all space, since $\rho = 0$ in the exterior region anyway.) Noting that the $\mathbf{r}$-dependence is contained in $\boldsymbol{\mathscr{r}}$, we have

$$\nabla \cdot \mathbf{E} = \frac{1}{4\pi\epsilon_0} \int \nabla \cdot \left(\frac{\hat{\mathbf{\mathscr{r}}}}{\mathscr{r}^2}\right) \rho(\mathbf{r}')\, d\tau'$$

This is precisely the divergence we calculated in Chapter 1:

$$\nabla \cdot \left(\frac{\hat{\mathbf{\mathscr{r}}}}{\mathscr{r}^2}\right) = 4\pi\delta^3(\boldsymbol{\mathscr{r}}) \tag{1.79}$$

Thus

$$\nabla \cdot \mathbf{E} = \frac{1}{4\pi\epsilon_0} \int 4\pi\delta^3(\mathbf{r} - \mathbf{r}')\rho(\mathbf{r}')\, d\tau' = \frac{1}{\epsilon_0}\rho(\mathbf{r}) \tag{2.13}$$

which is Gauss's law in differential form (2.11). To recover the integral form (2.10), we run the argument preceding equation (2.11) in reverse—integrate over a volume and apply the divergence theorem:

$$\int_{\text{volume}} \nabla \cdot \mathbf{E}\, d\tau = \oint_{\text{surface}} \mathbf{E} \cdot d\mathbf{a} = \frac{1}{\epsilon_0} \int_{\text{volume}} \rho\, d\tau = \frac{1}{\epsilon_0} Q_{\text{enc}}$$

2.2.3 Applications of Gauss's Law

I must interrupt the theoretical development here to comment on the extraordinary power of Gauss's law, in its integral form. When symmetry permits, Gauss's law provides *by far* the quickest and easiest means for computing electric fields. An example will illustrate the method.

Example 2

Find the field outside a uniformly charged sphere of radius a.

Solution: Draw a spherical surface at radius r as in Fig. 2.18 (this is called a "Gaussian surface" in the trade). Gauss's law says that for this surface (as for any other)

$$\oint_{\text{surface}} \mathbf{E} \cdot d\mathbf{a} = \frac{1}{\epsilon_0} Q_{\text{enc}}$$

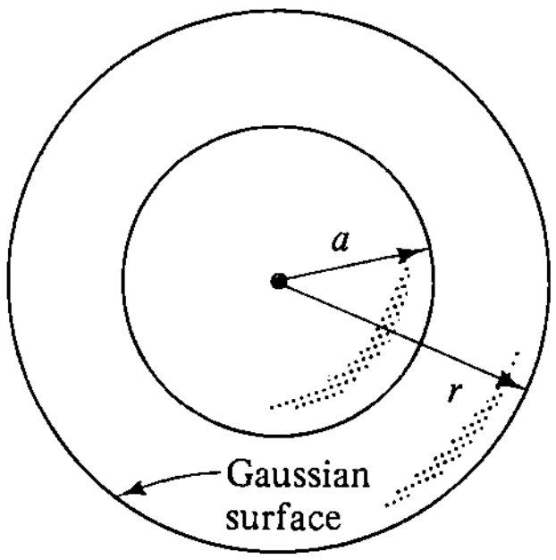

Figure 2.18

Here $Q_{\text{enc}} = q$, the total charge of the sphere. Normally speaking, this wouldn't get us very far because the quantity we want (namely, **E**) is buried inside that surface integral. However, in this particular case symmetry allows us to extract **E** from under the integral sign; for (a) **E** certainly points radially outward, as does $d\mathbf{a}$, so we can drop the dot product and deal only with magnitudes:

$$\int_{\text{surface}} \mathbf{E} \cdot d\mathbf{a} = \int_{\text{surface}} |\mathbf{E}|\, da$$

and (b) in magnitude, **E** is constant over the Gaussian surface, and hence comes outside the integral:

$$\int_{\text{surface}} |\mathbf{E}|\, da = |\mathbf{E}| \int_{\text{surface}} da = |\mathbf{E}|\, 4\pi r^2$$

Thus,

$$|\mathbf{E}|\, 4\pi r^2 = \frac{1}{\epsilon_0} q$$

or

$$\mathbf{E} = \frac{1}{4\pi\epsilon_0}\frac{q}{r^2}\hat{\mathbf{r}}$$

Notice a remarkable feature of this result: the field outside the sphere is exactly the *same as it would have been if all the charge had been concentrated at the center.*

Gauss's law is always *true,* but it is not always *useful.* If ρ had not been uniform (or, at any rate, not spherically symmetrical), or if I had chosen some nonspherical shape for my Gaussian surface, it would still be true that the flux of $\mathbf{E}$ is $(1/\epsilon_0)q$, but I would not have been certain that $\mathbf{E}$ was radially directed or constant in magnitude, and without this I cannot pull $\mathbf{E}$ out of the integral. *Symmetry is crucial* to this application of Gauss's law. As far as I know, there are only three kinds of symmetry which are sufficient:

1. *Spherical symmetry.* Make your Gaussian surface a concentric sphere.
2. *Cylindrical symmetry.* Make your Gaussian surface a coaxial cylinder (Fig. 2.19).
3. *Plane symmetry.* Use a Gaussian "pillbox," which straddles the surface (Fig. 2.20).

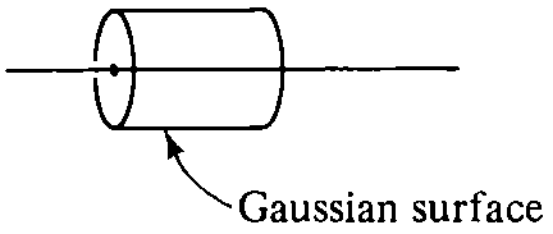

Figure 2.19

Figure 2.20

Although (2) and (3) technically require infinitely long cylinders, and planes extending to infinity in all directions, we shall often use them to get approximate answers for "long" cylinders or "large" plane surfaces, at points far from the edges.

Example 3

A long cylinder (Fig. 2.21) carries a charge density which is proportional to the distance from the axis: $\rho = kr$, for some constant k. Find the electric field inside this cylinder.

Solution: Draw a Gaussian cylinder of length l and radius r. For this surface, Gauss's law states:

$$\oint_{\text{surface}} \mathbf{E}\cdot d\mathbf{a} = \frac{1}{\epsilon_0} Q_{\text{enc}}.$$

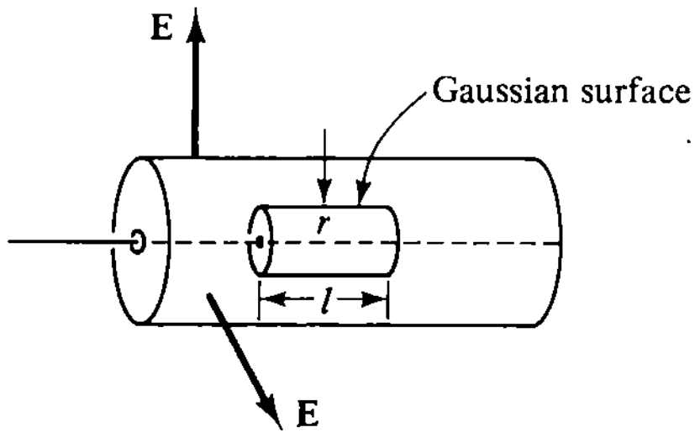

Figure 2.21

The enclosed charge is

$$Q_{\text{enc}} = \int \rho\, d\tau = \int (kr')(r'\, dr'\, d\phi\, dz) = 2\pi kl \int_0^r r'^2\, dr' = \tfrac{2}{3}\pi klr^3$$

in which we use the volume element appropriate to cylindrical coordinates, equation (1.58), and integrate ϕ from 0 to 2π, dz from 0 to l. I have put a prime on the integration variable, r', to distinguish it from the radius r of the Gaussian surface.

Now, symmetry dictates that **E** must point radially outward, so for the curved portion of the Gaussian cylinder we have:

$$\int \mathbf{E} \cdot d\mathbf{a} = \int |\mathbf{E}|\, da = |\mathbf{E}| \int da = |\mathbf{E}|\, 2\pi rl$$

whereas the two ends contribute nothing (here **E** is perpendicular to $d\mathbf{a}$). Thus,

$$|\mathbf{E}|\, 2\pi rl = \frac{1}{\epsilon_0} \frac{2}{3} \pi klr^3$$

or, finally,

$$\mathbf{E} = \frac{1}{3\epsilon_0} kr^2 \hat{\mathbf{r}}$$

Example 4

An infinite plane carries a uniform surface charge σ. Find its electric field.

Solution: Draw a "Gaussian pillbox," extending equal distances above and below the plane (Fig. 2.22). Apply Gauss's law to this surface:

$$\oint \mathbf{E} \cdot d\mathbf{a} = \frac{1}{\epsilon_0} Q_{\text{enc}}$$

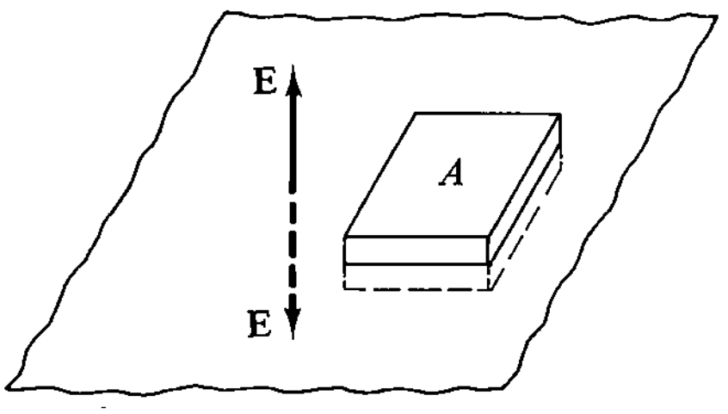

Figure 2.22

In this case, $Q_{enc} = \sigma A$, where A is the area of the lid of the pillbox. By symmetry, **E** points away from the plane (upward for points above, downward for points below). Thus, the top and bottom surfaces yield

$$\int \mathbf{E} \cdot d\mathbf{a} = 2A|\mathbf{E}|$$

whereas the sides contribute nothing. Thus,

$$2A|\mathbf{E}| = \frac{1}{\epsilon_0}\sigma A$$

or

$$\mathbf{E} = \frac{\sigma}{2\epsilon_0}\hat{n} \qquad (2.14)$$

where $\hat{n}$ is a unit vector pointing away from the surface. In Problem 2.6 you obtained this same result by a more laborious method.

It seems surprising, at first, that the field of an infinite plane is *independent of how far away you are.* What about the $1/r^2$ in Coulomb's law? Well, the point is that as you move farther and farther away from the plane, more and more charge comes into your "field of view," and this compensates for the diminishing influence of any particular piece. The field of a sphere falls off like $1/r^2$; the field of an infinite line falls off like $1/r$; and the field of an infinite plane does not fall off at all.

Although the direct use of Gauss's law to compute electric fields is limited to cases of spherical, cylindrical, and planar symmetry, we can put together *combinations* of objects possessing such symmetry, even though the arrangement as a whole is not symmetrical. Thus, using the principle of superposition, we could find the field in the vicinity of two uniformly charged parallel cylinders, say, or a sphere near an infinite charged plane.

Example 5

Two infinite parallel planes carry equal but opposite uniform charge densities $\pm\sigma$ (Fig. 2.23). Find the field in each of the three regions: (I) to the left of both, (II) between them, (III) to the right of both.

Solution: The left plate produces a field $(1/2\epsilon_0)\sigma$ which points away from it (Fig. 2.24)—to the left in region (I) and to the right in regions (II) and (III). The right plate, being negatively charged, produces a field $(1/2\epsilon_0)\sigma$ which points *toward* it—to the right in regions (I) and (II) and to the left in region (III). The two fields cancel in regions (I) and (III); they conspire in region (II). *Conclusion*: The field is $(1/\epsilon_0)\sigma$, and points to the right, between the planes; elsewhere it is zero.

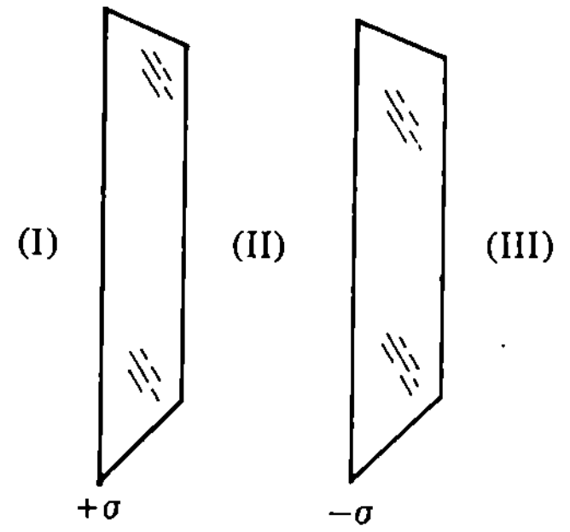

Figure 2.23

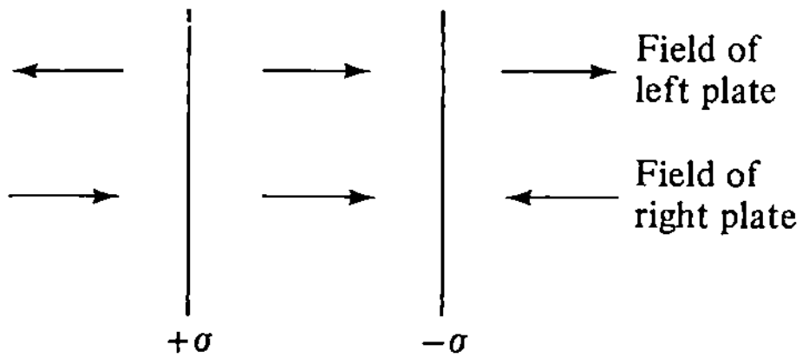

Figure 2.24

Problem 2.11 Use Gauss's law to find the electric field inside and outside a spherical shell of radius R, which carries a uniform surface charge density σ. Compare your answer to Problem 2.7.

Problem 2.12 Use Gauss's law to find the electric field inside a uniformly charged sphere (charge density ρ). Compare your answer to Problem 2.8.

Problem 2.13 Find the electric field a distance z from an infinitely long straight wire, which carries a uniform line charge λ. Compare the answer we obtained in Example 1.

Problem 2.14 Find the electric field inside a sphere which carries a charge density proportional to the distance from the origin, $\rho = kr$, for some constant k. (*Note*: This charge density is *not* uniform, and you must *integrate* to get the enclosed charge.)

Problem 2.15 A hollow spherical shell carries charge density

$$\rho = \frac{k}{r^2}$$

in the region $a \leq r \leq b$ (Fig. 2.25). Find the electric field in the three regions: (I) $r < a$, (II) $a < r < b$, (III) $r > b$. Plot $|\mathbf{E}|$ as a function of r.

Problem 2.16 A long coaxial cable (Fig. 2.26) carries a uniform *volume* charge density ρ on the inner cylinder (radius a), and a uniform *surface* charge density on the outer cylindrical shell (radius b). This surface charge is negative and of just the right magnitude so that the cable as a whole is electrically neutral. Find the electric field in each of the three regions: (I) inside the inner cylinder ($r < a$), (II) between the cylinders ($a < r < b$), (III) outside the cable ($r > b$). Plot $|\mathbf{E}|$ as a function of r.

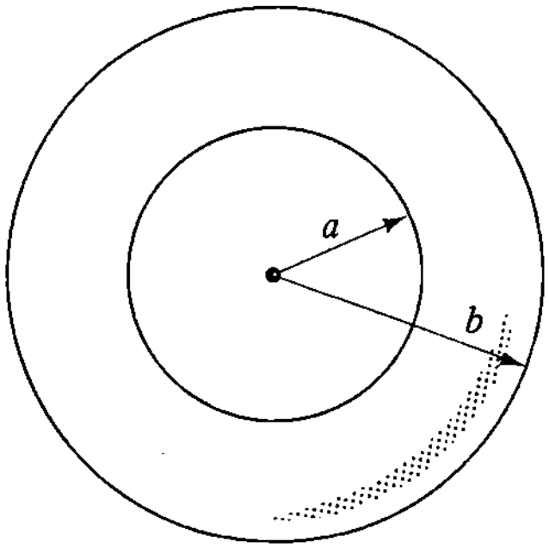

Figure 2.25

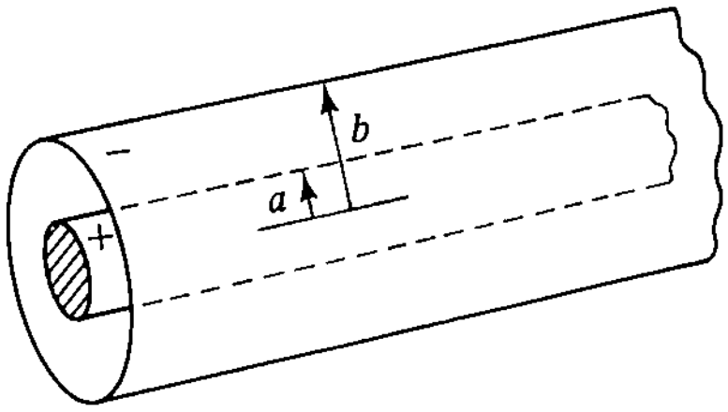

Figure 2.26

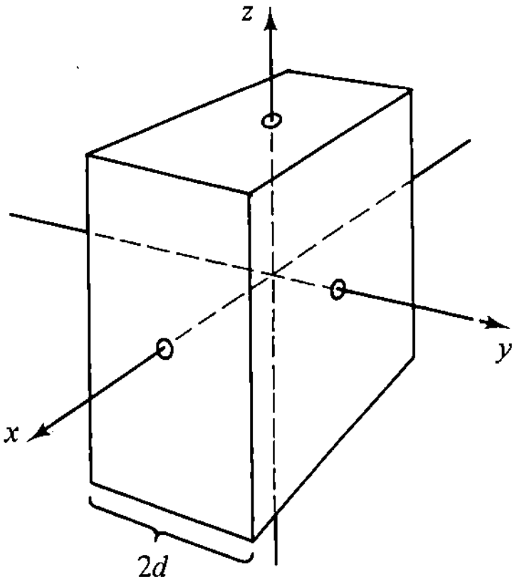

Figure 2.27

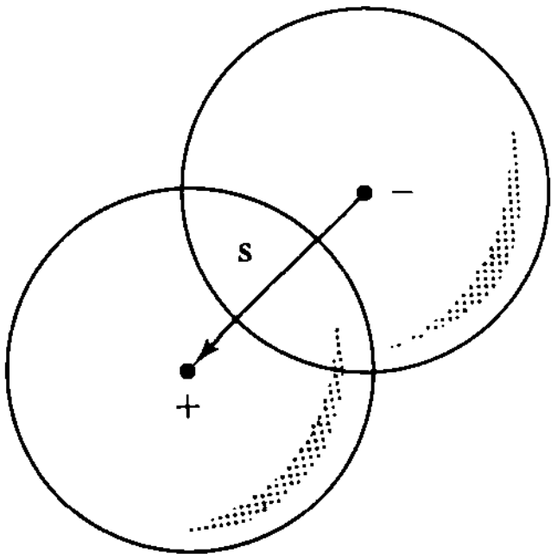

Figure 2.28

Problem 2.17 An infinite plane slab, of thickness $2d$, carries a uniform volume charge density ρ (Fig. 2.27). Find the electric field, as a function of y, the distance from the center. Plot E versus y, calling E positive when it points in the $+y$-direction and negative when it points in the $-y$-direction.

• **Problem 2.18** Two spheres, each of radius R and carrying uniform charge densities $+\rho$ and $-\rho$, respectively, are placed so that they partially overlap (Fig. 2.28). Call the vector from the negative center to the positive center $\mathbf{s}$. Show that the field in the region of overlap is constant, and find its value. (*Suggestion*: Use the answer to Problem 2.12.)

2.2.4 The Curl of E

We'll calculate the curl of $\mathbf{E}$, as we did the divergence in Section 2.2.1, by studying first the simplest possible configuration: a point charge at the origin. In this case

$$\mathbf{E} = \frac{1}{4\pi\epsilon_0}\frac{q}{r^2}\hat{\mathbf{r}}$$

Now, a glance at Fig. 2.12 should convince you that the curl of this field has to be zero, but I suppose we ought to come up with something a little more rigorous than a *coup d'oeil.* What if we calculate the line integral of this field from some point a to some other point b (Fig. 2.29):

$$\int_a^b \mathbf{E}\cdot d\mathbf{l}$$

In spherical coordinates, $d\mathbf{l} = dr\,\hat{\mathbf{r}} + r\,d\theta\,\hat{\boldsymbol{\theta}} + r\sin\theta\,d\phi\,\hat{\boldsymbol{\phi}}$, so

$$\mathbf{E}\cdot d\mathbf{l} = \frac{1}{4\pi\epsilon_0}\frac{q}{r^2}\,dr$$

Therefore,

$$\int_a^b \mathbf{E}\cdot d\mathbf{l} = \frac{1}{4\pi\epsilon_0}\int_a^b \frac{q}{r^2}\,dr = \frac{-1}{4\pi\epsilon_0}\frac{q}{r}\bigg|_{r_a}^{r_b} = \frac{1}{4\pi\epsilon_0}\left(\frac{q}{r_a} - \frac{q}{r_b}\right) \tag{2.15}$$

where r_a is the distance from the origin to point a and r_b is the distance to b.

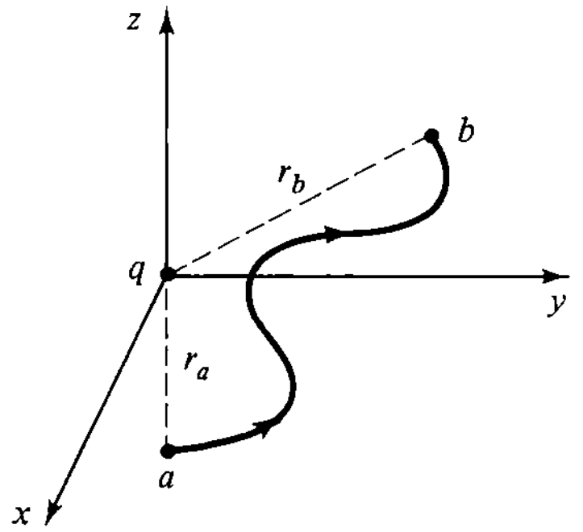

Figure 2.29

At the moment I'm not much interested in the actual *answer*; the important thing is that the line integral is plainly *independent of path.* All it depends on is the two end points; in fact, all that matters is how far a and b are from the charge. You see why this happened? Because **E** points in the radial direction, it costs me nothing to move around in the θ or ϕ directions: Any contribution from such a displacement is wiped out by the dot product $\mathbf{E} \cdot d\mathbf{l}$. The integral around a *closed* path is obviously zero (then $r_a = r_b$):

$$\boxed{\oint \mathbf{E} \cdot d\mathbf{l} = 0} \tag{2.16}$$

and hence, applying Stokes' theorem,

$$\boxed{\nabla \times \mathbf{E} = 0} \tag{2.17}$$

Now, I have proved (2.16) and (2.17) only for the field of a single point charge at the *origin,* but these results make no reference to what is, after all, a perfectly arbitrary choice of coordinates; they hold no matter *where* the charge is located. Moreover, if we have many charges, the principle of superposition states that the total field is the vector sum of their individual fields:

$$\mathbf{E} = \mathbf{E}_1 + \mathbf{E}_2 + \cdots$$

so

$$\nabla \times \mathbf{E} = \nabla \times (\mathbf{E}_1 + \mathbf{E}_2 + \cdots) = (\nabla \times \mathbf{E}_1) + (\nabla \times \mathbf{E}_2) + \cdots = 0$$

Consequently, (2.16) and (2.17) must hold for *any static charge distribution whatever.*

Problem 2.19 Calculate $\nabla \times \mathbf{E}$ directly from equation (2.6), by the method of Section (2.2.2). Refer to Problem 1.62 if you get stuck.

2.3 ELECTRIC POTENTIAL

2.3.1 Introduction to Potential

The electric field **E** is not just *any* old vector function; it is a very special *kind* of vector function: one whose curl is always zero. $\mathbf{E} = y\hat{\imath}$, for example, could not possibly be an electrostatic field; *no* set of charges, regardless of their sizes and positions, could ever produce such a field. In this section we're going to exploit this special property of electric fields to reduce a vector problem (finding **E**) down to a much simpler scalar problem. The first theorem in Section 1.6.2 asserts that any vector whose curl is zero is equal to the gradient of some scalar. What I'm going to do now amounts to a proof of that claim, in the context of electrostatics.

Because the line integral of **E** around any closed loop is zero, the integral from a to b is independent of path (we showed this explicitly for the field of a point charge at the origin, but it is equally true for any configuration of static charges, for if $\int_a^b \mathbf{E} \cdot d\mathbf{l}$ were *different* for paths (I) and (II), then we could go out along (I) and back along (II)—Fig. 2.30—and obtain $\oint \mathbf{E} \cdot d\mathbf{l} \neq 0$). *Because* $\oint_a^b \mathbf{E} \cdot d\mathbf{l}$ is independent of path, we can define a function

$$\boxed{V(P) \equiv -\int_{\mathcal{O}}^{P} \mathbf{E} \cdot d\mathbf{l}} \tag{2.18}$$

Here, $\mathcal{O}$ is some standard reference point on which we have agreed beforehand; V then depends only on the point P. V is called the **electric potential**. Evidently, the potential *difference* between two points a and b is

$$\begin{aligned} V(b) - V(a) &= -\int_{\mathcal{O}}^{b} \mathbf{E} \cdot d\mathbf{l} + \int_{\mathcal{O}}^{a} \mathbf{E} \cdot d\mathbf{l} \\ &= -\int_{\mathcal{O}}^{b} \mathbf{E} \cdot d\mathbf{l} - \int_{a}^{\mathcal{O}} \mathbf{E} \cdot d\mathbf{l} = -\int_{a}^{b} \mathbf{E} \cdot d\mathbf{l} \end{aligned} \tag{2.19}$$

Now, the fundamental theorem for gradients states that

$$V(b) - V(a) = \int_a^b (\nabla V) \cdot d\mathbf{l}$$

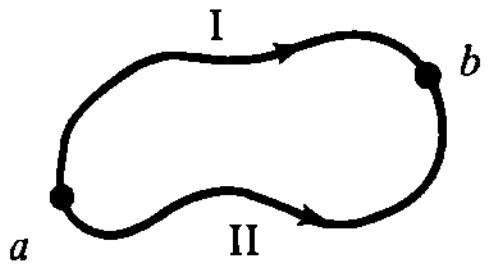

$$\oint \mathbf{E} \cdot d\mathbf{l} = \underset{\text{(I)}}{\int_a^b \mathbf{E} \cdot d\mathbf{l}} - \underset{\text{(II)}}{\int_a^b \mathbf{E} \cdot d\mathbf{l}}$$

Figure 2.30

so

$$\int_a^b (\nabla V)\cdot d\mathbf{l} = -\int_a^b \mathbf{E}\cdot d\mathbf{l}$$

Since, finally, this is true for *any* points a and b, the integrands must be equal:

$$\boxed{\mathbf{E} = -\nabla V} \tag{2.20}$$

Equation (2.20) is the differential version of (2.18); it says that the electric field is the gradient of a scalar potential, which is what we set out to prove.

Notice the crucial but subtle role played by path independence (or, equivalently, the fact that $\nabla \times \mathbf{E} = 0$) in this argument. If the line integral of $\mathbf{E}$ depended on the path taken, then the "definition" of V, equation (2.18), would be nonsense. It simply would not define a function, since changing the path would alter the value of $V(P)$. By the way, don't let the minus sign in equation (2.20) distract you; it carries over from (2.18) and is really just a matter of tradition.

Problem 2.20 One of these is an impossible electrostatic field. Which one?
(a) $\mathbf{E} = k[(xy)\hat{\imath} + (2yz)\hat{\jmath} + (3xz)\hat{k}]$
(b) $\mathbf{E} = k[(y^2)\hat{\imath} + (2xy + z^2)\hat{\jmath} + (2yz)\hat{k}]$.
(Here k is a constant with the appropriate units.) For the *possible* one, find the potential, using the origin as your reference point. Check your answer by computing ∇V. (*Warning*: You must select a specific path to integrate along. It doesn't matter *what* path you choose, since the answer is path-independent, but you simply cannot integrate unless you have a particular path in mind.)

2.3.2 Comments on Potential

(i) The name. The word "potential" is a hideous misnomer because it inevitably reminds you of potential *energy*. This is particularly confusing, because there *is* a connection between "potential" and "potential energy," as you will see in Section 2.4. I'm sorry that it is impossible to escape this word. The best I can do is to insist once and for all that "potential" and "potential energy" are completely different terms and should, by all rights, have completely different names. (Incidentally, a surface over which the potential is constant is called an **equipotential**.)

(ii) Advantage of the potential formulation. If you know V, you can easily get $\mathbf{E}$—just take the gradient: $\mathbf{E} = -\nabla V$. This is quite extraordinary when you stop to think about it, for $\mathbf{E}$ is a *vector* quantity (three components), but V is a *scalar* (one component). How can *one* function possibly contain all the information that *three* independent functions carry? The answer is that the three components of $\mathbf{E}$ are not really as independent as they look; in fact, they are explicitly interrelated by the very condition we started with, $\nabla \times \mathbf{E} = 0$. In terms of components,

$$\frac{\partial E_x}{\partial y} = \frac{\partial E_y}{\partial x}, \quad \frac{\partial E_z}{\partial y} = \frac{\partial E_y}{\partial z}, \quad \frac{\partial E_x}{\partial z} = \frac{\partial E_z}{\partial x}$$

This brings us back to my observation at the beginning of Section 2.3.1: **E** *is a very special kind of vector.* What the potential formulation does is to exploit this feature to maximum advantage, reducing, as I say, a vector problem down to a scalar one, in which there is no need to fuss with components or vector addition.

(iii) The reference point $\mathcal{O}$. There is an essential ambiguity in the definition of potential, since the choice of reference point $\mathcal{O}$ was arbitrary. Changing reference points amounts to adding a constant to the potential:

$$V'(P) = -\int_{\mathcal{O}'}^{P} \mathbf{E}\cdot d\mathbf{l} = -\int_{\mathcal{O}'}^{\mathcal{O}} \mathbf{E}\cdot d\mathbf{l} - \int_{\mathcal{O}}^{P} \mathbf{E}\cdot d\mathbf{l} = K + V(P)$$

where K is the line integral of **E** from the old reference point $\mathcal{O}$ to the new one $\mathcal{O}'$. Of course, adding a constant to V will not affect the potential *difference* between two points:

$$V'(b) - V'(a) = V(b) - V(a)$$

since the K's cancel out. (Actually, it is already clear from equation (2.19) that the potential difference is independent of $\mathcal{O}$ because it can be written as the line integral of **E** from a to b.) Nor does the ambiguity affect the gradient of V:

$$\nabla V' = \nabla V$$

since the derivative of a constant is zero. That's why all such V's, differing only in their choice of reference point, correspond to the same field **E**.

Evidently, potential as such carries no real physical significance, for at any given point we can adjust its value at will by a suitable relocation of $\mathcal{O}$. In this sense potential is rather like altitude: If I ask you how high Denver is, you will probably tell me its height above sea level, because that is a convenient and traditional reference point. But we could as well agree to measure altitude above Washington D.C., or Greenwich, or wherever. That would add (or, rather, subtract) a fixed amount from all our sea-level readings, but it wouldn't change anything about the real world. The only quantity of intrinsic significance is the *difference* in altitude between two points, and *that* is the same *whatever* your reference level.

Having said that, however, there *is* a "natural" spot to use for $\mathcal{O}$ in electrostatics—analogous to sea level for altitude—and that is a point infinitely far from the charge. Ordinarily, then, we "set the zero of potential at infinity." (Since $V(\mathcal{O}) = 0$, choosing a reference point is equivalent to selecting a place where V is to be zero.) But I must warn you that there is one special circumstance in which this convention fails: when the charge distribution itself extends to infinity. The symptom of trouble, in such cases, is that the potential blows up. For instance, the field of a uniformly charged plane is $(1/2\epsilon_0)\sigma\hat{\mathbf{n}}$, as we found in Example 4; if we naively put $\mathcal{O} = \infty$, then the potential at a height z above the plane becomes

$$V(z) = -\int_{\infty}^{z} \frac{1}{2\epsilon_0}\sigma\, dz = -\frac{1}{2\epsilon_0}\sigma(z - \infty)$$

The remedy is simply to choose some other reference point (in this problem you might use the point $z = 0$). Notice that the difficulty occurs only in phony textbook problems; in "real life" there is no such thing as a charge distribution that goes on forever, and we can *always* use infinity as our reference point.

(iv) Potential obeys the superposition principle. The original superposition principle of electrodynamics pertains to the force on a test charge Q. It says that the total force on Q is the vector sum of the forces attributable to the source charges individually:

$$\mathbf{F} = \mathbf{F}_1 + \mathbf{F}_2 + \cdots$$

Dividing through by Q, we find that the electric field, too, obeys the superposition principle:

$$\mathbf{E} = \mathbf{E}_1 + \mathbf{E}_2 + \cdots$$

Integrating from the common reference point to P, it follows that potential also satisfies such a principle:

$$V = V_1 + V_2 + \cdots$$

That is, the potential at any point is the sum of the potentials due to all the source charges separately. Only this time it is an *ordinary* sum, not a *vector* sum, which makes it a lot easier to work with.

(v) Units of potential. In our units, force is measured in newtons and charge in coulombs, so electric fields are in newtons per coulomb. Accordingly, potential is measured in newton-meters per coulomb or joules per coulomb. A joule per coulomb is called a **volt**.

Example 6

Find the potential inside and outside a spherical shell of radius R (Fig. 2.31), which carries a uniform surface charge. Set the reference point at infinity.

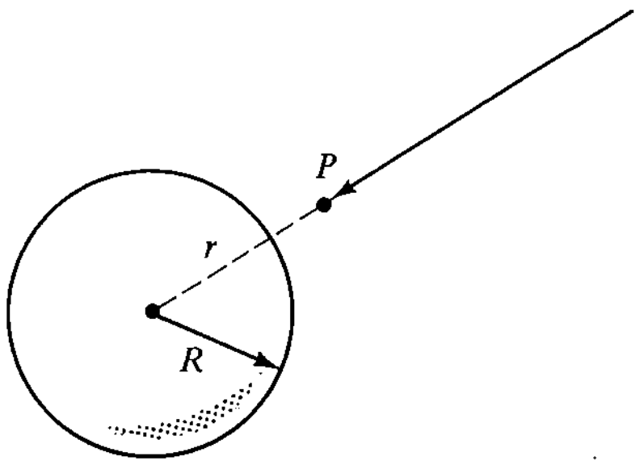

Figure 2.31

Solution: From Gauss's law, the field outside is

$$\mathbf{E} = \frac{1}{4\pi\epsilon_0}\frac{q}{r^2}\hat{r}$$

where q is the total charge on the sphere. The field inside is zero. For points outside the sphere, then,

$$V(r) = -\int_{\infty}^{r} \mathbf{E}\cdot d\mathbf{l} = \frac{-1}{4\pi\epsilon_0}\int_{\infty}^{r}\frac{q}{r'^2}\,dr' = \frac{1}{4\pi\epsilon_0}\frac{q}{r'}\bigg|_{\infty}^{r} = \frac{1}{4\pi\epsilon_0}\frac{q}{r}$$

$$(\text{for } r > R)$$

To find the potential inside the sphere ($r < R$), we must break the integral into two sections, using in each region the field that prevails there:

$$V(r) = -\int_{\infty}^{r} \mathbf{E}\cdot d\mathbf{l} = \frac{-1}{4\pi\epsilon_0}\int_{\infty}^{R}\frac{q}{r'^2}\,dr' - \int_{R}^{r}(0)\,dr' = \frac{1}{4\pi\epsilon_0}\frac{q}{r'}\bigg|_{\infty}^{R} + 0$$

$$= \frac{1}{4\pi\epsilon_0}\frac{q}{R} \qquad (\text{for } r < R)$$

Notice that the potential is *not* zero inside the shell, even though the field *is*. V is a *constant* in this region, to be sure, so that $\nabla V = 0$—that's what matters.

In problems of this type you must always *work your way in from the reference point*; that's where the potential is "nailed down." It is tempting to suppose that you could figure out the potential inside the sphere on the basis of the field there alone, but this is false: the potential inside the sphere is sensitive to what's going on outside the sphere as well. If I placed a second, charged spherical shell out at some radius $R' > R$, the potential inside R would change, even though the field would still be zero. Gauss's law guarantees that charge exterior to a given point (that is, at larger r) produces no net *field* at that point, provided it is spherically or cylindrically symmetric; but there is no such rule for *potential*, when infinity is used as the reference point.

Problem 2.21 Find the potential inside and outside a uniformly charged solid sphere whose radius is R and whose total charge is q. Use infinity as your reference point. Compute the gradient of V in each region, and check that it yields the correct field.

Problem 2.22 Find the potential a distance r from an infinitely long straight wire that carries a uniform line charge λ. Compute the gradient of your potential, and check that it yields the correct field.

Problem 2.23 For the charge configuration of Problem 2.15, find the potential at the center, using infinity as your reference point.

Problem 2.24 For the configuration of Problem 2.16, find the potential difference between a point on the axis and a point on the outer cylinder. Note that it is not necessary to commit yourself to a particular reference point if you use equation (2.19).

Problem 2.25 Show that any *central force* admits a potential. (A central force is one which depends only on the distance from the force center and is directed along the line from that center.)

2.3.3 Poisson's Equation and Laplace's Equation

We found in Section 2.3.1 that the electric field can be written as the gradient of a scalar potential,

$$\mathbf{E} = -\nabla V$$

The question arises: How do the fundamental equations for **E**,

$$\nabla \cdot \mathbf{E} = \frac{\rho}{\epsilon_0} \quad \text{and} \quad \nabla \times \mathbf{E} = 0$$

read, in terms of V? Well, $\nabla \cdot \mathbf{E} = \nabla \cdot (-\nabla V) = -\nabla^2 V$ so, apart from that persistent minus sign, the divergence of **E** is the Laplacian of V. Gauss's law then says

$$\boxed{\nabla^2 V = -\frac{\rho}{\epsilon_0}} \tag{2.21}$$

This is known as **Poisson's equation.** In regions where there is no charge, so that $\rho = 0$, Poisson's equation reduces to **Laplace's equation,**

$$\nabla^2 V = 0 \tag{2.22}$$

We'll explore these questions more fully in Chapter 3.

So much for Gauss's law. What about the curl law? This says that

$$\nabla \times \mathbf{E} = \nabla \times (-\nabla V)$$

must equal zero. But that's no condition on V—curl of gradient is *always* zero. Of course, we *used* the curl law to show that **E** could be expressed as the gradient of a scalar, so it's not really very surprising that this works out. $\nabla \times \mathbf{E} = 0$ *permits* $\mathbf{E} = -\nabla V$; in return, $\mathbf{E} = -\nabla V$ *guarantees* $\nabla \times \mathbf{E} = 0$. It takes only *one* differential equation (Poisson's) to determine V, because V is a scalar; for **E** we needed *two,* the divergence and the curl.

2.3.4 The Potential of a Localized Charge Distribution

We defined V in terms of **E**, in equation (2.18). Ordinarily, though, it's **E** that we're looking for (if we already knew **E** there wouldn't be much point in computing V). The idea is that it might be easier to get V first, and then calculate **E** by taking the gradient. Typically, then, we know where the charge is (that is, we know ρ), and we want to find V. Now, Poisson's equation relates V and ρ, but unfortunately it's "the wrong way around": It would give us ρ, if we knew V, whereas we want V, knowing ρ. What we must do, then, is to "invert" Poisson's equation. That's the program for this section, although we shall do it by roundabout means, beginning, as we have twice before, with a point charge at the origin.

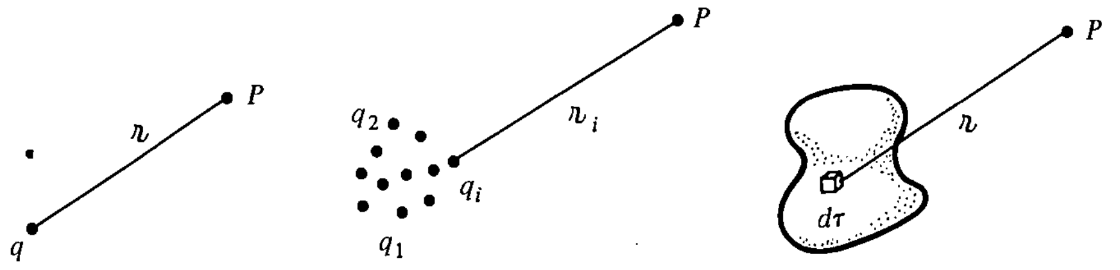

Figure 2.32

Setting the reference point at infinity, the potential of a point charge q at the origin is

$$V(r) = \frac{-1}{4\pi\epsilon_0} \int_{\infty}^{r} \frac{q}{r'^2}\, dr' = \frac{1}{4\pi\epsilon_0} \frac{q}{r'} \bigg|_{\infty}^{r} = \frac{1}{4\pi\epsilon_0} \frac{q}{r}$$

You see here the special virtue of infinity as reference point. Notice the sign of V; actually, the original minus sign in the definition of V (equation (2.18)) was chosen precisely in order to *make* the potential of a positive charge come out positive. It is useful to remember that regions of positive charge are potential "hills," regions of negative charge are potential "valleys," and the electric field points "downhill," from plus toward minus.

In general, the potential of a point charge q is

$$V(P) = \frac{1}{4\pi\epsilon_0} \frac{q}{\mathscr{r}} \tag{2.23}$$

where $\mathscr{r}$, as always, is the distance from q to P (Fig. 2.32). Invoking the superposition principle, then, the potential of a collection of charges is

$$V(P) = \frac{1}{4\pi\epsilon_0} \sum_{i=1}^{n} \frac{q_i}{\mathscr{r}_i}$$

or, for a continuous distribution,

$$\boxed{V(P) = \frac{1}{4\pi\epsilon_0} \int \frac{\rho}{\mathscr{r}}\, d\tau} \tag{2.24}$$

This is the equation we were looking for, telling us how to compute V when we know ρ; it is, if you like, the "solution" to Poisson's equation, for a localized charge distribution.[4] I invite you to compare equation (2.24) with the corresponding formula for the electric *field* in terms of ρ:

$$\mathbf{E}(P) = \frac{1}{4\pi\epsilon_0} \int \left(\frac{\hat{\mathscr{r}}}{\mathscr{r}^2}\right) \rho\, d\tau \tag{2.6}$$

The main point to notice is that the pesky unit vector $\hat{\mathscr{r}}$ is now missing, so there is no need to worry about components. Incidentally, the integrals analogous to (2.24) for line and surface charges are

[4] Equation (2.24) can be regarded as an application of the Helmholtz theorem (1.93), in the context of electrostatics, where the curl of $\mathbf{E}$ is zero and its divergence is ρ/ϵ_0.

$$\frac{1}{4\pi\epsilon_0}\int \frac{\lambda}{\imath}\,dl \quad \text{and} \quad \frac{1}{4\pi\epsilon_0}\int \frac{\sigma}{\imath}\,da$$

I should warn you that everything in this section is predicated on the assumption that the reference point is at infinity. This is hardly apparent in equation (2.24), but remember that we *got* (2.24) from the potential of a point charge at the origin, $(1/4\pi\epsilon_0)(q/r)$ and that is valid only when $\mathcal{O} = \infty$. If you try to apply (2.24) to one of those artificial problems in which the charge itself extends to infinity, the integral will diverge.

Example 7

Find the potential of a uniformly charged spherical shell of radius R (Fig. 2.33).

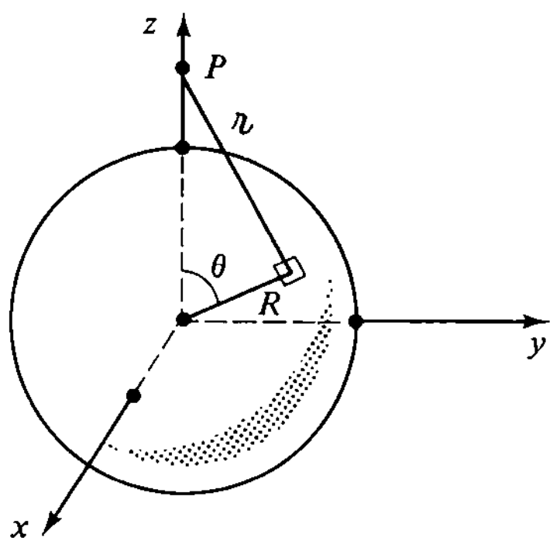

Figure 2.33

Solution: This is the same problem we solved in Example 6, but this time we shall do it using equation (2.24)—or, rather, the corresponding equation for surface charge distributions:

$$V(P) = \frac{1}{4\pi\epsilon_0}\int \frac{\sigma}{\imath}\,da$$

Let's set the point P on the z axis and use the law of cosines to express $\imath$ in terms of the polar angle θ:

$$\imath^2 = R^2 + z^2 - 2Rz\cos\theta$$

An element of surface area on this sphere is $R^2 \sin\theta\, d\theta\, d\phi$, so

$$\begin{aligned}
4\pi\epsilon_0 V(z) &= \sigma \int \frac{R^2 \sin\theta\, d\theta\, d\phi}{\sqrt{R^2 + z^2 - 2Rz\cos\theta}} \\
&= 2\pi R^2 \sigma \int_0^\pi \frac{\sin\theta}{\sqrt{R^2 + z^2 - 2Rz\cos\theta}}\,d\theta \\
&= 2\pi R^2 \sigma \left(\frac{1}{Rz}\sqrt{R^2 + z^2 - 2Rz\cos\theta} \right)\Bigg|_0^\pi
\end{aligned}$$

$$= \frac{2\pi R\sigma}{z}\left(\sqrt{R^2 + z^2 + 2Rz} - \sqrt{R^2 + z^2 - 2Rz}\right)$$

$$= \frac{2\pi R\sigma}{z}\left[\sqrt{(R + z)^2} - \sqrt{(R - z)^2}\right]$$

At this stage we must be very careful to take the *positive* root. For points *outside* the sphere, z is greater than R, and hence $\sqrt{(R - z)^2} = z - R$; for points *inside* the sphere, $\sqrt{(R - z)^2} = R - z$. Thus,

$$V(z) = \frac{R\sigma}{2\epsilon_0 z}[(R + z) - (z - R)] = \frac{R^2\sigma}{\epsilon_0 z}, \quad \text{outside}$$

$$V(z) = \frac{R\sigma}{2\epsilon_0 z}[(R + z) - (R - z)] = \frac{R\sigma}{\epsilon_0}, \quad \text{inside}$$

In terms of the total charge on the shell, $q = 4\pi R^2\sigma$, $V(z) = (1/4\pi\epsilon_0)(q/z)$ (or, in general, $V(r) = (1/4\pi\epsilon_0)(q/r)$) for points outside the sphere, and $V(r) = (1/4\pi\epsilon_0)(q/R)$ for points inside.

Of course, in this particular case, it was easier to get V by using (2.18) than (2.24) because Gauss's law gave us $\mathbf{E}$ with so little effort. But if you compare Example 7 with Problem 2.7, you will get a more accurate sense of the strength of the potential formulation.

Problem 2.26 Using equation (2.24) (or rather, the corresponding formula for point, line, and surface charges) find the potential at a distance z above the center of the charge distributions of Fig. 2.34. In each case, compute $\mathbf{E} = -\nabla V$, and compare your answers with Problem 2.2(a), Example 1, and Problem 2.6, respectively. Suppose that we changed the right-hand charge in Fig. 2.34(a) to $-q$. What then is the potential at P? What field does that suggest? Compare your answer to Problem 2.2(b), and explain carefully any discrepancy.

Problem 2.27 A conical surface (an empty ice-cream cone) carries a uniform surface charge σ. The height of the cone is a, as is the radius of the top. Find the potential difference between points P (the vertex) and Q (the center of the top).

Problem 2.28 Find the potential on the axis of a uniformly charged solid cylinder, a distance z from the center. The length of the cylinder is L, its radius is R, and the charge density is ρ. Use your result to calculate the electric field at this point. (Assume that $z > L/2$.)

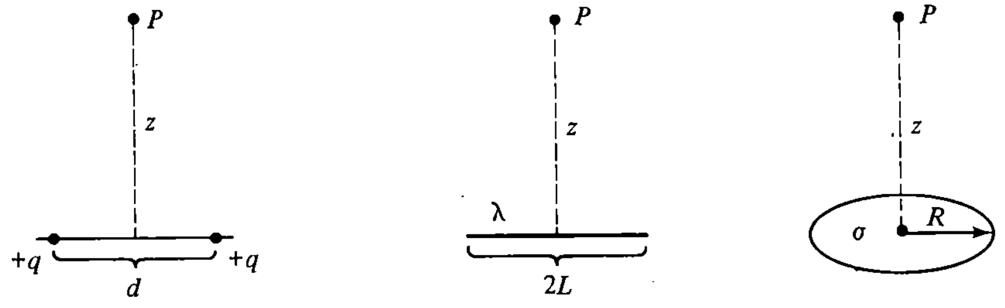

(a) Two point charges (b) Uniform line charge (c) Uniform surface charge **Figure 2.34**

Problem 2.29 Use equation (2.24) to calculate the potential inside a uniformly charged solid sphere of radius R and total charge q. Compare your answer to Problem 2.21.

Problem 2.30 Check that (2.24) satisfies Poisson's equation by applying the Laplacian and using equation (1.82).

2.3.5 Summary; Electrostatic Boundary Conditions

In the typical electrostatic problem you are given a source charge distribution ρ, and you want to find the electric field $\mathbf{E}$ it produces. Unless the symmetry of the problem admits a solution by Gauss's law, it is generally to your advantage to calculate the potential first, as an intermediate step. These, then, are the three fundamental quantities of electrostatics: ρ, $\mathbf{E}$, and V. We have, in the course of our discussion, derived all six formulas interrelating them. These equations are neatly summarized in Fig. 2.35.[5] We began with just two experimental observations: (1) the principle of superposition—a broad general rule applying to *all* electromagnetic forces, and (2) Coulomb's law—the fundamental law of electrostatics. From these, all else followed.

You may have noticed, in studying Examples 4 and 5 or working problems such as 2.7, 2.11, and 2.16, that the electric field always undergoes a discontinuity when you cross a surface charge σ. In fact, it is a simple matter to find the *amount* by which $\mathbf{E}$ changes at such a boundary. Suppose we draw a wafer-thin Gaussian pillbox, extending just barely over the edge in each direction (Fig. 2.36). Gauss's law states that

$$\oint_{\text{surface}} \mathbf{E}\cdot d\mathbf{a} = \frac{1}{\epsilon_0} Q_{\text{enc}} = \frac{1}{\epsilon_0}\sigma A$$

where A is the area of the pillbox lid. (If σ varies from point to point or the surface is curved, we must pick A to be extremely small.) Now, the *sides* of the pillbox contrib-

[5]This diagram is not original; I learned of it from E. M. Purcell, and it has apparently been "discovered" independently by a number of people, including T. W. Moore and M. A. Heald.

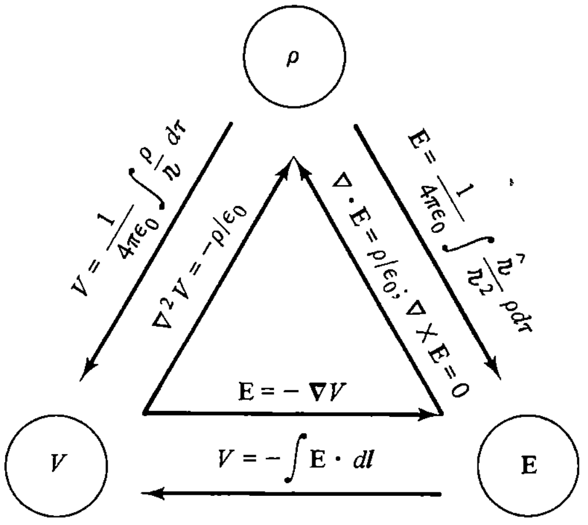

Figure 2.35

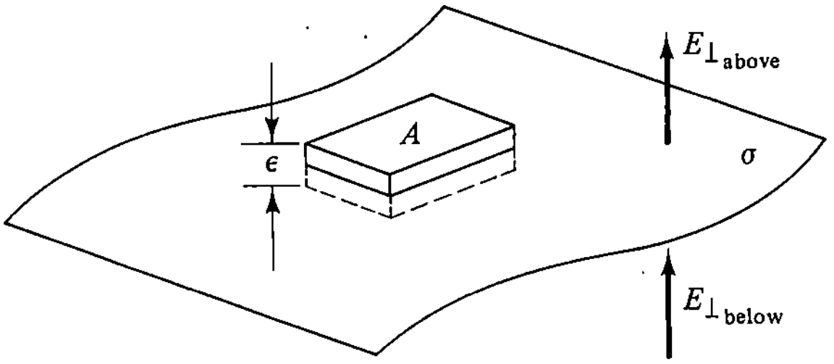

Figure 2.36

ute nothing to the flux, in the limit as the thickness ϵ goes to zero, so we are left with

$$E_{\perp\,\text{above}} - E_{\perp\,\text{below}} = \frac{1}{\epsilon_0}\sigma \tag{2.25}$$

where $E_{\perp\,\text{above}}$ denotes the component of **E** that is perpendicular to the surface immediately above, and $E_{\perp\,\text{below}}$ is the same, only just below the surface. For consistency we let "upward" be the positive direction for both. *Conclusion*: *The normal component of* **E** *is discontinuous by an amount* σ/ϵ_0 *at any boundary.* In particular, where there is *no* surface charge, $E_\perp$ is continuous, as for instance at the edge of a uniformly charged solid sphere.

The *tangential* component of **E**, by contrast, is *always* continuous. For if we apply equation (2.16),

$$\oint \mathbf{E} \cdot d\mathbf{l} = 0$$

to the thin rectangular loop of Fig. 2.37, the ends give nothing (as $\epsilon \to 0$), and the sides give $(E_{\parallel\,\text{above}}l - E_{\parallel\,\text{below}}l)$, so

$$E_{\parallel\,\text{above}} = E_{\parallel\,\text{below}} \tag{2.26}$$

where $E_\parallel$ represents the component of **E** which is *parallel* to the surface. The boundary conditions on **E** (equations (2.25) and (2.26)) can be combined into a single formula:

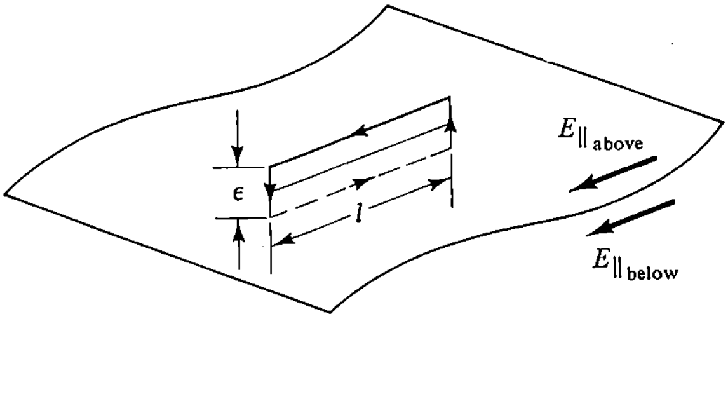

Figure 2.37

$$\boxed{\mathbf{E}_{\text{above}} - \mathbf{E}_{\text{below}} = \frac{\sigma}{\epsilon_0}\hat{n},} \tag{2.27}$$

where $\hat{n}$ is a unit vector perpendicular to the surface.[6]

The potential, meanwhile, is continuous across any boundary (Fig. 2.38), since

$$V_{\text{above}} - V_{\text{below}} = -\int_a^b \mathbf{E} \cdot d\mathbf{l}$$

As the path length shrinks to zero, so too does the integral:

$$\boxed{V_{\text{above}} = V_{\text{below}}} \tag{2.28}$$

However, the *gradient* of V inherits the discontinuity in $\mathbf{E}$. Since $\mathbf{E} = -\nabla V$, equation (2.27) implies that

$$\nabla V_{\text{above}} - \nabla V_{\text{below}} = -\frac{1}{\epsilon_0}\sigma\hat{n} \tag{2.29}$$

or, more conveniently,

$$\boxed{\frac{\partial V_{\text{above}}}{\partial n} - \frac{\partial V_{\text{below}}}{\partial n} = -\frac{1}{\epsilon_0}\sigma} \tag{2.30}$$

where

$$\frac{\partial V}{\partial n} = \nabla V \cdot \hat{n} \tag{2.31}$$

denotes the **normal derivative** of V (that is, the rate of change of V in the direction perpendicular to the surface). Please note that these boundary conditions on $\mathbf{E}$ and V apply at *any* surface, flat or curved, whether there happens to be charge there or not; they follow directly from Gauss's law and the fact that $\nabla \times \mathbf{E} = 0$.

[6]If you're only interested in the field *due to the* (essentially flat) *local patch* of *surface charge itself*, the answer is $(\sigma/2\epsilon_0)\hat{n}$ immediately above the surface, and $-(\sigma/2\epsilon_0)\hat{n}$ immediately below. This follows from Example 4, for if you are close enough to the patch it "looks" like an infinite plane. Evidently the entire *discontinuity* in $\mathbf{E}$ is attributable to this local patch of charge.

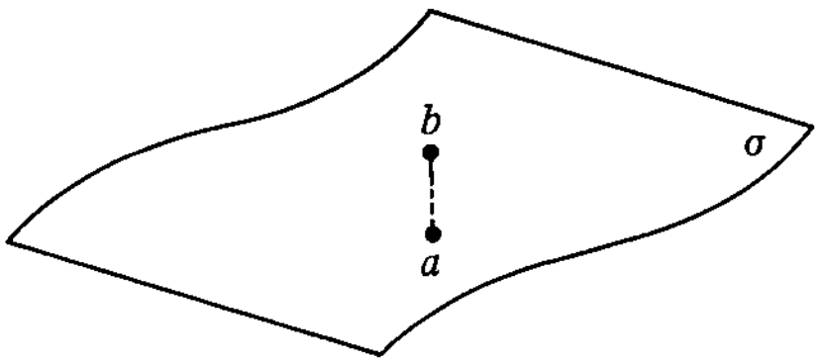

Figure 2.38

Problem 2.31

(a) Check that the results of Examples 4 and 5, and Problem 2.11, are consistent with (2.27).

(b) Use Gauss's law to find the field inside and outside a long hollow cylindrical tube, which carries a uniform surface charge σ. Check that your result is consistent with (2.27).

(c) Check that the result of Example 7 is consistent with boundary conditions (2.28) and (2.30).

2.4 WORK AND ENERGY IN ELECTROSTATICS

2.4.1 The Work Done in Moving a Charge

Suppose we have a stationary configuration of source charges, and we want to move a test charge Q from point a to point b, as in Fig. 2.39. *Question*: How much work will we have to do? At any point along the path, the electric force on Q is $\mathbf{F} = Q\mathbf{E}$. The force *we* must exert, in opposition to this electrical force, is $-Q\mathbf{E}$. (If the sign bothers you, think of it this way: when you lift a brick, gravity exerts a force *mg downward*, but *you* exert a force *mg upward*.) Thus, the total work we must do is

$$W = \int_a^b \mathbf{F}\cdot d\mathbf{l} = -Q\int_a^b \mathbf{E}\cdot d\mathbf{l} = Q[V(b) - V(a)]$$

Notice that the answer is independent of path; in mechanics, then, we would describe $\mathbf{F}$ as a "conservative" force. Dividing through by Q, we have

$$V(b) - V(a) = \frac{W}{Q} \tag{2.32}$$

In words, the *potential difference between points a and b is equal to the work per unit charge it takes to carry a particle from a to b.* In particular, if you want to bring the charge Q in from far away and stick it at point P, the work you must do is

$$W = Q[V(P) - V(\infty)]$$

so, if you have set the reference point at infinity,

$$W = QV(P) \tag{2.33}$$

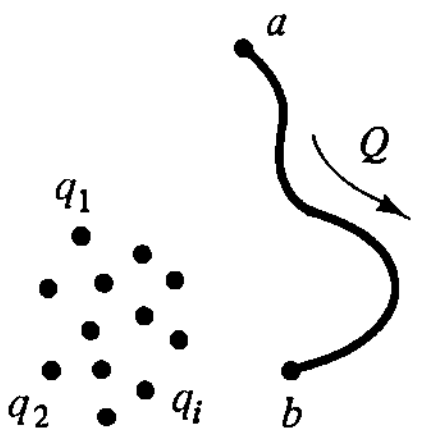

Figure 2.39

2.4.2 The Energy of a Point Charge Distribution

How much work is required to assemble a collection of point charges? Let's build it up, bringing the charges in, one by one, from far away (Fig. 2.40). It takes *no* work to bring in the first charge, q_1, since there is no field yet to fight against. Now, bring in q_2. According to equation (2.33), this will take an amount of work $q_2 V_1(P_2)$, where V_1 is the potential due to q_1, and P_2 is the place we're putting q_2:

$$W_2 = \frac{1}{4\pi\epsilon_0} q_2 \left(\frac{q_1}{\mathscr{r}_{12}}\right)$$

where $\mathscr{r}_{12}$ is the distance between q_1 and q_2 when they are in position. Now bring in q_3; this requires work $q_3 V_{1,2}(P_3)$, where $V_{1,2}$ is the potential due to charges q_1 and q_2, namely, $(q_1/\mathscr{r}_{13} + q_2/\mathscr{r}_{23})/4\pi\epsilon_0$. Thus

$$W_3 = \frac{1}{4\pi\epsilon_0} q_3 \left(\frac{q_1}{\mathscr{r}_{13}} + \frac{q_2}{\mathscr{r}_{23}}\right)$$

Similarly, the extra work to bring in q_4 will be

$$W_4 = \frac{1}{4\pi\epsilon_0} q_4 \left(\frac{q_1}{\mathscr{r}_{14}} + \frac{q_2}{\mathscr{r}_{24}} + \frac{q_3}{\mathscr{r}_{34}}\right)$$

The total work necessary to assemble the first four charges, then, is

$$W = \frac{1}{4\pi\epsilon_0} \left(\frac{q_1 q_2}{\mathscr{r}_{12}} + \frac{q_1 q_3}{\mathscr{r}_{13}} + \frac{q_1 q_4}{\mathscr{r}_{14}} + \frac{q_2 q_3}{\mathscr{r}_{23}} + \frac{q_2 q_4}{\mathscr{r}_{24}} + \frac{q_3 q_4}{\mathscr{r}_{34}}\right)$$

You see the general rule: Take the product of each pair of charges, divide by their separation distance, and add it all up:

$$W = \frac{1}{4\pi\epsilon_0} \sum_{i=1}^{n} \sum_{\substack{j=1 \\ j>i}}^{n} \frac{q_i q_j}{\mathscr{r}_{ij}} \tag{2.34}$$

The stipulation $j > i$ in equation (2.34) is to remind you not to count the same pair twice. A nicer way to accomplish the same purpose is *intentionally* to count each pair twice, and then divide by 2:

$$W = \frac{1}{8\pi\epsilon_0} \sum_{i=1}^{n} \sum_{\substack{j=1 \\ j\neq i}}^{n} \frac{q_i q_j}{\mathscr{r}_{ij}} \tag{2.35}$$

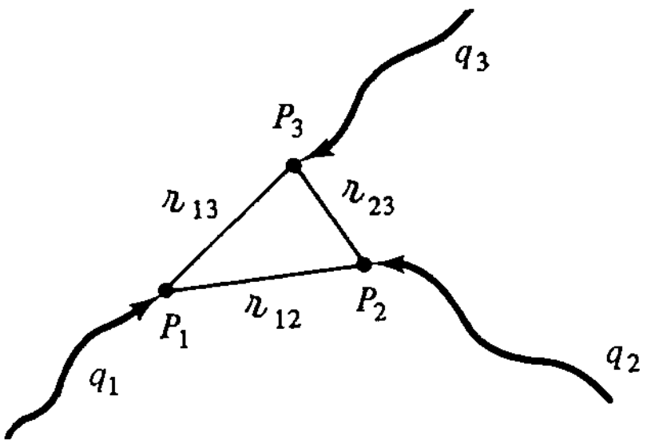

Figure 2.40

We must still avoid $i = j$, of course. Notice that in this form the answer plainly does not depend on the *order* in which you assemble the charges, since every pair occurs in the sum. Let me next pull out the factor q_i:

$$W = \frac{1}{2}\sum_{i=1}^{n} q_i\left(\sum_{\substack{j=1\\ j\neq i}}^{n} \frac{1}{4\pi\epsilon_0}\frac{q_j}{\mathscr{r}_{ij}}\right)$$

The term in parentheses is the potential at point P_i (the position of q_i) due to all the other charges—*all* of them, now, not just the ones that were present at some stage in the building-up process. Thus,

$$W = \frac{1}{2}\sum_{i=1}^{n} q_i V(P_i) \tag{2.36}$$

That's how much work is needed to put together a configuration of point charges; it's also the amount of work you'd get back out if you dismantled the system. In the meantime, it represents energy stored in the configuration ("potential" energy, if you like, though for obvious reasons I prefer to avoid that word in this context).

Problem 2.32

(a) Three charges are situated at the corners of a square (side s) as shown in Fig. 2.41. How much work does it take to bring in another charge, $+q$, from far away and place it at the fourth corner?

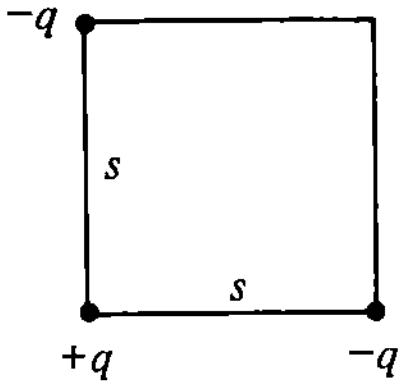

Figure 2.41

(b) How much work does it take to assemble the whole configuration of four charges?

2.4.3 The Energy of a Continuous Charge Distribution

For a volume charge density ρ, equation (2.36) becomes

$$W = \frac{1}{2}\int \rho V \, d\tau \tag{2.37}$$

The corresponding integrals for line and surface charges would be $\int \lambda V \, dl$ and $\int \sigma V \, da$, respectively.

There is a lovely way to rewrite equation (2.37), in which ρ and V are eliminated in favor of $\mathbf{E}$. First use Gauss's law (2.11) to express ρ in terms of $\mathbf{E}$:

$$\rho = \epsilon_0 \nabla\cdot\mathbf{E}, \quad \text{so} \quad W = \frac{\epsilon_0}{2}\int (\nabla\cdot\mathbf{E})V \, d\tau$$

Now product rule number 5 (front cover) says that

$$\nabla\cdot(\mathbf{E}V) = (\nabla\cdot\mathbf{E})V + \mathbf{E}\cdot(\nabla V)$$

On the other hand, $\nabla V = -\mathbf{E}$, so

$$W = \frac{\epsilon_0}{2}\left[\int \nabla\cdot(\mathbf{E}V)\,d\tau + \int E^2\,d\tau\right]$$

Finally, applying the divergence theorem to the first term,

$$W = \frac{\epsilon_0}{2}\left(\int_{\text{surface}} V\mathbf{E}\cdot d\mathbf{a} + \int_{\text{volume}} E^2\,d\tau\right) \tag{2.38}$$

But what volume is this we're integrating over, and what is its bounding surface? Let's go back to the formula we started with, equation (2.37). From its derivation, it is clear that we should integrate over the entire region where the charge is located. But actually any *larger* volume would do just as well; any "extra" territory we throw in will contribute nothing to the integral anyway, since $\rho = 0$ out there. With this in mind, we return to equation (2.38). What happens *here,* as we enlarge the volume beyond the minimum necessary to trap all the charge? Well, the integral of E^2 can only increase, the integrand being positive, so evidently the surface integral must decrease correspondingly to leave the sum intact. In fact, at large distances from the charge, E goes like $1/r^2$ and V like $1/r$, while the surface area grows like r^2. Roughly speaking, then, the surface integral goes down like $1/r$. Please understand that equation (2.38) gives you the correct energy W *whatever* volume you use (as long as it encloses all the charge), but the contribution from the volume integral goes up and that of the surface integral goes down as you pick larger and larger volumes. In particular, why not integrate over *all* space? Then the surface integral goes to zero, and we are left with

$$\boxed{W = \frac{\epsilon_0}{2}\int_{\text{all space}} E^2\,d\tau} \tag{2.39}$$

Example 8

Find the energy of a uniformly charged spherical shell of total charge q and radius R.

Solution 1: Use equation (2.37), in the version appropriate to surface charges:

$$W = \frac{1}{2}\int \sigma V\,da$$

Now the potential at the surface of this sphere is $(1/4\pi\epsilon_0)q/R$—a constant—so

$$W = \frac{1}{8\pi\epsilon_0}\frac{q}{R}\int \sigma\,da = \frac{1}{8\pi\epsilon_0}\frac{q^2}{R}$$

Solution 2: Use equation (2.39). Inside, the sphere $\mathbf{E} = 0$; outside,

$$\mathbf{E} = \frac{1}{4\pi\epsilon_0}\frac{q}{r^2}\,\hat{\mathbf{r}}, \qquad \text{so} \quad E^2 = \frac{q^2/r^4}{(4\pi\epsilon_0)^2}$$

Therefore,

$$W = \frac{\epsilon_0}{2(4\pi\epsilon_0)^2} \int_{\text{outside}} \left(\frac{q^2}{r^4}\right)(r^2 \sin\theta \, dr \, d\theta \, d\phi)$$

$$= \frac{1}{32\pi^2\epsilon_0} q^2 4\pi \int_R^\infty \frac{1}{r^2} dr = \frac{1}{8\pi\epsilon_0}\frac{q^2}{R}$$

Problem 2.33 Find the energy stored in a uniformly charged solid sphere of radius R and total charge q. Do it three different ways:

(a) Use equation (2.37). You found the potential in Problem 2.21.

(b) Use equation (2.39). Don't forget to integrate over *all space.*

(c) Use equation (2.38). Take a spherical volume of radius a. Notice what happens as $a \to \infty$.

Problem 2.34 Here is a fourth way of computing the energy of a uniformly charged sphere: Assemble the sphere layer by layer, each time bringing in an infinitesimal charge dq from far away and smearing it uniformly over the surface, thereby increasing the radius. How much work dW does it take to build up the radius by an amount dr? Integrate this to find the work necessary to put together the entire sphere of radius R and total charge q. Compare your answer to Problem 2.33.

2.4.4 Comments on Electrostatic Energy

(i) A perplexing "inconsistency." Equation (2.39) clearly implies that the energy of a stationary charge distribution is always *positive.* On the other hand, equation (2.36), from which (2.39) was in fact derived, can be positive or negative. For instance, according to (2.36)—or (2.34)—the energy of two equal but opposite charges a distance $\mathscr{r}$ apart would be $-(1/4\pi\epsilon_0)(q^2/\mathscr{r})$. What's gone wrong? Which equation is correct?

The answer is that *both* equations are correct, but they pertain to slightly different situations. For (2.36) does not take into account the work necessary to *make* the point charges in the first place; we *started* with point charges and simply found the work required to bring them together. This is wise policy, since (2.39) shows that the energy of a point charge itself is infinite:

$$W = \frac{\epsilon_0}{2(4\pi\epsilon_0)^2} \int \left(\frac{q^2}{r^4}\right)(r^2 \sin\theta \, dr \, d\theta \, d\phi) = \frac{q^2}{8\pi\epsilon_0}\int_0^\infty \frac{1}{r^2} dr = \infty$$

Equation (2.39) is more *complete*, in the sense that it tells you the *total* energy stored in a charge configuration, but (2.36) is more appropriate in the context of point charges, where we prefer (for good reason!) to leave out that portion of the total energy that is attributable to the fabrication of the point charges themselves. In practice, after all, the point charges (electrons, say) are *given* to us ready-made; all *we* do is move them around. Since we did not put them together, and we cannot take them apart, it's really an academic issue how much work the process would involve. Still, the infinite energy of point charges is a recurring source of embarrassment for electromagnetic theory, afflicting the quantum version as well as the classical. We shall return to the problem in Chapter 9.

Still, you may wonder where the inconsistency crept into an apparently watertight derivation. The "flaw" is between equations (2.36) and (2.37): In the former $V(P_i)$ represents the potential due to all the *other* charges *but not* q_i, whereas in the latter $V(P)$ is the *full* potential. For a continuous distribution there is no distinction, since the amount of charge *right at the point P* is vanishingly small, and its contribution to the total potential is zero.

(ii) Where is the energy stored? Equations (2.37) and (2.39) offer two different ways of calculating the same thing. The first is an integral over the charge distribution; the second is an integral over the field. These can involve completely different regions. For instance, in the case of the spherical shell (Example 8) the charge is confined to the surface, whereas the electric field is present everywhere *outside* this surface. Where *is* the energy, then? Is it stored in the field, as (2.39) seems to suggest, or is it stored in the charge, as (2.37) implies? At the present level, this is simply an unanswerable question: I can tell you what the total energy is, and I can provide you with several different ways to compute it, but it is idle to worry about *where* the energy is located. In the context of radiation theory (Chapter 9) it is useful (and in general relativity it is *essential*) to regard the energy as being stored in the field, at a density

$$\epsilon_0 \frac{E^2}{2} = \text{energy per unit volume} \tag{2.40}$$

But in electrostatics one could as well say it is stored in the charge, with a density $(\frac{1}{2}\rho V)$. The difference is purely a matter of bookkeeping.

(iii) The superposition principle. Because electrostatic energy is *quadratic* in the fields, it does *not* obey a superposition principle. The energy of a compound system is *not* the sum of the energies of its parts considered separately—there are also "cross terms":

$$W_{\text{tot}} = \frac{\epsilon_0}{2}\int E^2\,d\tau = \frac{\epsilon_0}{2}\int (\mathbf{E}_1 + \mathbf{E}_2)^2\,d\tau$$

$$= \frac{\epsilon_0}{2}\int (E_1^2 + E_2^2 + 2\mathbf{E}_1\cdot\mathbf{E}_2)\,d\tau = W_1 + W_2 + \epsilon_0\int \mathbf{E}_1\cdot\mathbf{E}_2\,d\tau$$

For example, if you double the charge everywhere, you *quadruple* the total energy.

2.5 CONDUCTORS

2.5.1 Basic Properties of Conductors

In an *insulator,* such as glass or rubber, each electron is attached to a particular atom. In a metallic *conductor,* by contrast, one or more electrons per atom are free to roam about at will through the material. (In liquid conductors such as salt water it is

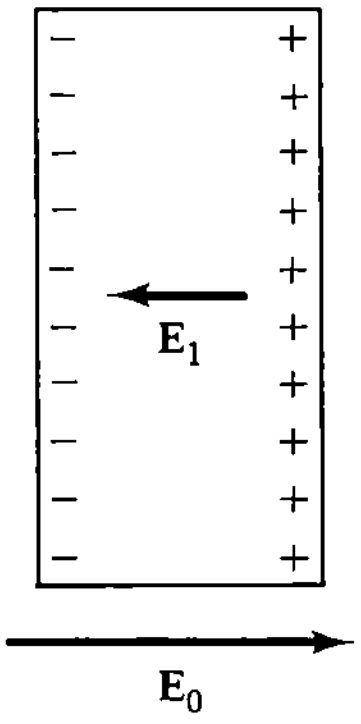

Figure 2.42

ions that do the moving.) A *perfect* conductor would be a material which contains an *unlimited* supply of completely free charges. In real life there are no perfect conductors, but many substances come amazingly close. From this definition the basic electrostatic properties of ideal conductors immediately follow:

(i) $\mathbf{E} = 0$ inside a conductor. Why? Because if there *were* any field, those free charges would move, and it wouldn't be electro*statics* anymore. Well . . . that's hardly a satisfactory explanation; maybe all it proves is that you can't have electrostatics when conductors are present. We had better examine what happens when you put a conductor into an external electric field $\mathbf{E}_0$ (Fig. 2.42). Initially, this will drive any free positive charges to the right, and negative ones to the left. (In practice, it's only the negative charges—electrons—that do the moving, but when they depart the right side is left with a net positive charge—the stationary protons—so it doesn't really matter which charges actually move, the effect is the same.) When they come to the edge of the material, the charges pile up: plus on the right side, minus on the left. Now, these **induced charges** produce a field of their own, $\mathbf{E}_1$, which, as you can see from the figure, is in the *opposite direction* to $\mathbf{E}_0$. That's the crucial point, for it means that the field of the induced charges *tends to cancel off the original field.* Charge will continue to flow until this cancellation is complete, and the resultant field inside the conductor is precisely zero.[7] The whole process is practically instantaneous.

(ii) $\rho = 0$ inside a conductor. This follows from Gauss's law: $\nabla \cdot \mathbf{E} = \rho/\epsilon_0$. If $\mathbf{E} = 0$, so also is ρ. There is still charge around, but exactly as much plus charge as minus, so the *net* charge density in the interior is zero.

(iii) Any net charge resides on the surface. That's the only other place it *can* be.

(iv) V is constant, throughout a conductor. For if a and b are any two points within (or at the surface of) a given conductor, $V(b) - V(a) = -\int_a^b \mathbf{E} \cdot d\mathbf{l} = 0$, and hence $V(a) = V(b)$. (Thus the surface of a conductor is always an equipotential.)

[7]*Outside* the conductor the field is *not* zero, for here $\mathbf{E}_0$ and $\mathbf{E}_1$ do not cancel.

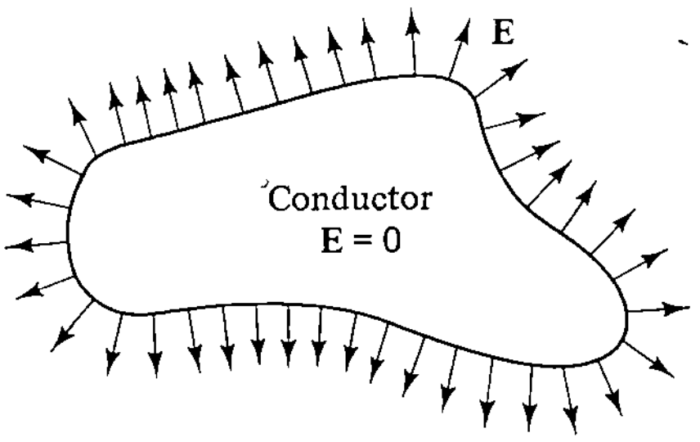

Figure 2.43

(v) E is perpendicular to the surface, just outside a conductor. Otherwise, as in (a), charge will immediately flow around the surface until it kills off the tangential component (Fig. 2.43). (*Perpendicular* to the surface, charge cannot flow, of course, since it is confined to the conducting object.)

It may seem strange to you that the charge on a conductor should flow to the surface. Suppose, for instance, that we charge up a metal sphere. At first glance it looks reasonable that all these charges, repelling one another, should push out to the surface in an effort to put as much distance as possible between them. But what a waste this is of all the space in the interior of the sphere! Surely we could do better, from the point of view of making each charge as far as possible from its neighbors, to sprinkle *some* of it throughout the volume . . . ? Well, it simply is not so. You do best to put *all* the charge on the surface, and this is true regardless of the size or shape of the conductor.

The problem can also be phrased in terms of energy. Like any other free dynamical system, the charge on a conductor will seek the configuration that minimizes its potential energy. What property (c) asserts is that the electrostatic energy for an object of specified shape and given total charge is a minimum when that charge is spread over the surface. For instance, the energy of a sphere is $(1/8\pi\epsilon_0)(q^2/R)$ if the charge is uniformly distributed over the surface, as we found in Example 8, but it is greater, $(3/20\pi\epsilon_0)(q^2/R)$, if the charge is uniformly distributed throughout the volume (see Problems 2.33 and 2.34).

Example 9

Suppose we place a point charge q at the center of a neutral spherical conducting shell (Fig. 2.44). It will attract negative charge to the inner surface of the conductor. How *much* induced charge will accumulate there?

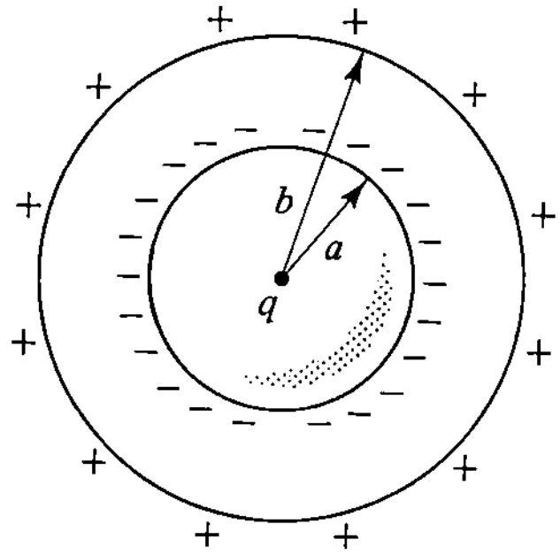

Figure 2.44

Solution: Charge will continue to flow until the net field in the conductor (due to q *and* the induced charge) is zero. Since the field of a uniform spherical surface charge is the same as if it were all concentrated at the center, perfect cancellation occurs when exactly $-q$ has piled up at a. Evidently, then, the induced surface charge is

$$\sigma_a = -\frac{q}{4\pi a^2}$$

Meanwhile, there is a leftover charge $+q$ on the conductor, which, feeling no force from q and σ_a (since their fields precisely cancel), distributes itself uniformly over the outer surface, with a density

$$\sigma_b = \frac{q}{4\pi b^2}$$

Problem 2.35 For the configuration in Example 9, find **E** and V as functions of r in the three regions $r < a$, $a < r < b$, and $r > b$. Graph $E(r)$ and $V(r)$.

2.5.2 Induced Charges

If you hold a charge $+q$ near an uncharged conductor (Fig. 2.45), the two will attract one another. The reason for this is that q will pull minus charges over to the near side and repel plus charges to the far side. (Another way to think of it is that the conductor moves charge around in such a way as to cancel off the field of q for points inside, where the total field must be zero.) Since the negative induced charge is closer to q, there is a net force of attraction. In Chapter 3 we shall calculate this force explicitly, for the case of a spherical conductor.

When I speak of the field, charge, or potential "inside" a conductor, I mean in the "meat" of the conductor; if there is some *cavity* in the conductor, and within that cavity there is some charge, then the field *in the cavity* will *not* be zero. But in a remarkable way the cavity and its contents are electrically isolated from the outside world by the surrounding conductor (Fig. 2.46). No external fields penetrate the conductor; they are canceled at the outer surface by the induced charge there. Similarly, the field due to charges within the cavity is killed off, for all exterior points, by the

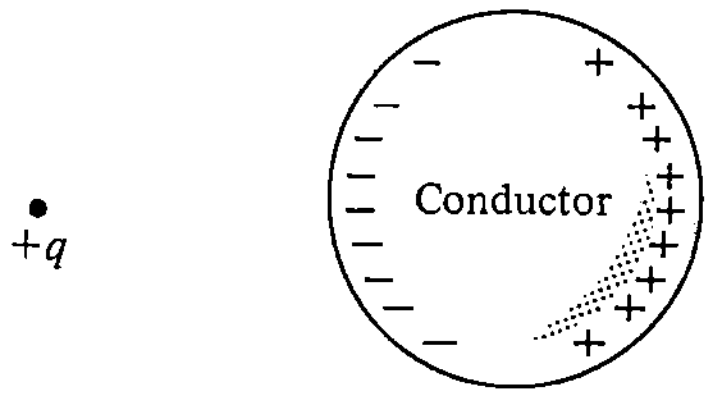

Figure 2.45

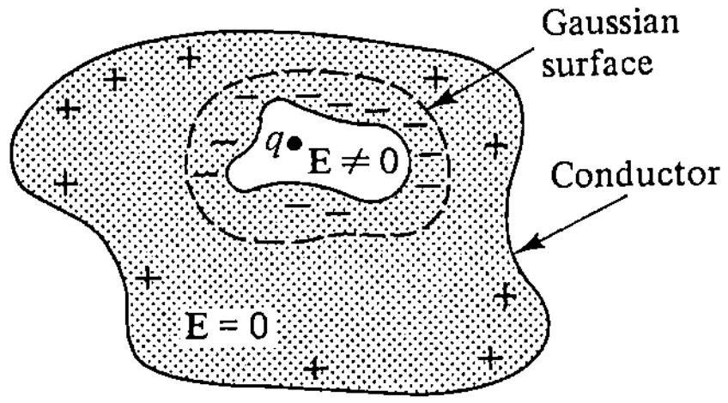

Figure 2.46

induced charge on the inner surface. (However, the compensating charge left over on the *outer* surface of the conductor effectively "communicates" the presence of q to the outside world, as we shall see in Example 10.) Incidentally, the total charge induced on the cavity wall is equal and opposite to the charge inside, for if we surround the cavity with a Gaussian surface, all points of which are in the conductor (Fig. 2.46), $\oint \mathbf{E} \cdot d\mathbf{a} = 0$, and hence (by Gauss's law) the net enclosed charge must be zero. But $Q_{\text{enc}} = q + q_{\text{induced}}$, so $q_{\text{induced}} = -q$.

Example 10

An uncharged spherical conductor centered at the origin has a cavity of some weird shape carved out of it (Fig. 2.47). Somewhere within the cavity is a charge $+q$. *Question*: What is the field outside the sphere?

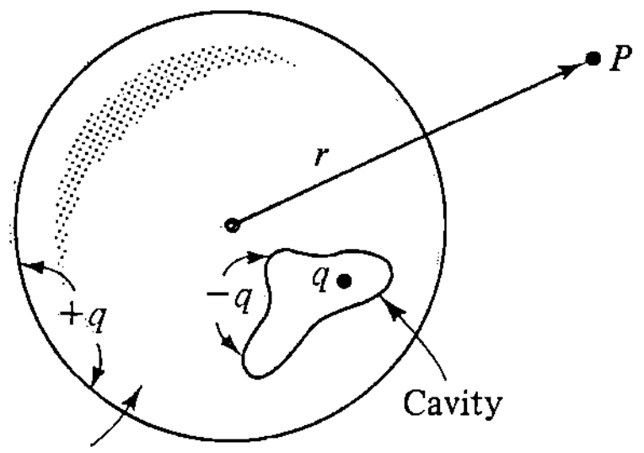

Figure 2.47

Solution: At first glance it would appear that the answer depends on the shape of the cavity and on the placement of the charge. But that's wrong: The answer is $(1/4\pi\epsilon_0)(q/r^2)\hat{\mathbf{r}}$, *regardless*. The conductor conceals from us all information concerning the nature of the cavity, revealing only the total charge it contains. How can this be? Well, the charge $+q$ induces an opposite charge $-q$ on the wall of the cavity, which distributes itself in such a way that its field cancels that of q, for all points exterior to the cavity. Since the conductor carries no net charge, this leaves $+q$ to distribute itself uniformly over the surface of the sphere. (It's *uniform* because the asymmetrical influence of the point charge $+q$ is negated by that of the induced charge $-q$ on the inner surface.) For points outside the sphere, then, the only thing that survives is the field of the leftover $+q$ on the outer surface, $(1/4\pi\epsilon_0)(q/r^2)\hat{\mathbf{r}}$, as advertised.

It may occur to you that in one respect the argument is open to challenge: There are actually *three* fields at work here, $\mathbf{E}_q$, $\mathbf{E}_{\text{induced}}$, and $\mathbf{E}_{\text{leftover}}$. All we know for certain is that the sum of the three is zero inside the conductor, yet I claimed that the first two *alone* cancel, while the third is separately zero there. Moreover, even if the first two cancel within the conductor, who is to say they still cancel for points outside? They do not after all cancel for points *inside* the cavity. I cannot give you a completely satisfactory answer, at the moment, but this much at least is true: There *exists* a way of distributing $-q$ over the inner surface so as to cancel the field of q at all exterior points. For that same cavity could have been carved out of a *huge* spherical conductor with a radius of 27 miles or light years or whatever. In that case the leftover $+q$ on the outer surface is simply too far away to produce a significant field, and the other two fields would *have* to accomplish the cancellation by themselves. So we know they *can* do it . . . but are we sure they *choose* to? Perhaps for small spheres nature prefers some complicated three-way cancellation. Nope: As we'll see in the uniqueness theorems of Chapter 3, electrostatics is very stingy with its options; there is always precisely one way—no more—of distributing the charge on a conductor so as to make the field inside zero. Having found a *possible* way, we are guaranteed that no alternative exists even in principle. Our conclusion is valid, therefore, $\mathbf{E} = (1/4\pi\epsilon_0)(q/r^2)\hat{r}$, though I was forced to invoke a later result prematurely to resolve all possible doubts.

If a cavity surrounded by conducting material is itself empty of charge, then the field within the cavity is zero. For any field line would have to begin and end on the cavity wall, going from a plus charge to a minus charge (Fig. 2.48). Letting that field line be part of a closed loop, the rest of which is entirely inside the conductor (where $\mathbf{E} = 0$), the integral $\oint \mathbf{E} \cdot d\mathbf{l}$ is distinctly *positive*, in violation of (2.16). It follows that $\mathbf{E} = 0$ within an *empty* cavity, and there is in fact *no* charge on the surface of the cavity. (This is why you are relatively safe inside a metal car during a thunderstorm—you may get *cooked*, if lightning strikes, but you will not be *electrocuted*. The same principle applies to the placement of sensitive apparatus inside a grounded **Faraday cage,** to shield out stray electric fields. In practice, the enclosure doesn't even have to be solid conductor—wire mesh will often do the job.)

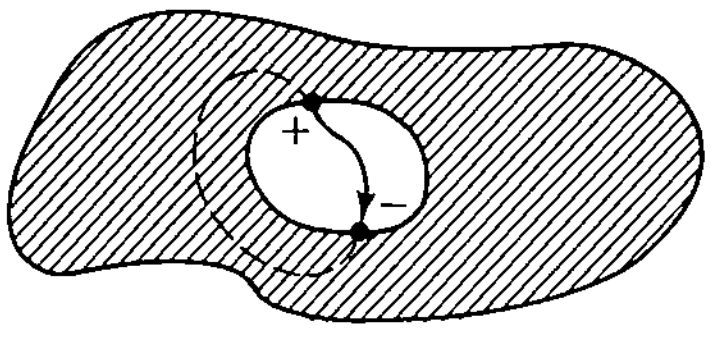

Figure 2.48

Problem 2.36 A metal sphere of radius R, carrying charge q, is surrounded by a thick concentric metal shell (inner radius a, outer radius b, as in Fig. 2.49). The shell carries no net charge.

(a) Find the surface charge density σ at R, at a, and at b.

(b) Find the potential at the center, using infinity as reference.

(c) Now the outer surface is touched to a grounding wire, which lowers its potential to zero (same as at infinity). How do your answers to (a) and (b) change?

Problem 2.37 Two spherical cavities, of radii a and b, are hollowed out from the interior of a (neutral) conducting sphere of radius R (Fig. 2.50). At the center of each cavity a point charge is placed; call these charges q_a and q_b.

(a) Find the surface charges σ_a, σ_b, and σ_R.

(b) What is the field outside the conductor?

(c) What is the field within each cavity?

(d) What is the force on q_a and q_b?

(e) Which of these answers would change if a third charge, q_c were brought near the conductor?

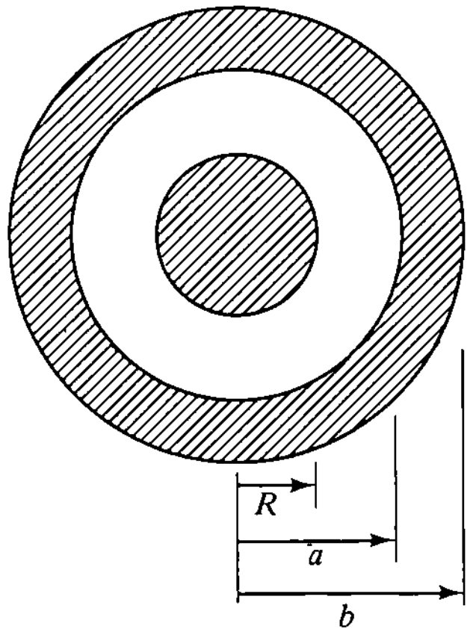

Figure 2.49

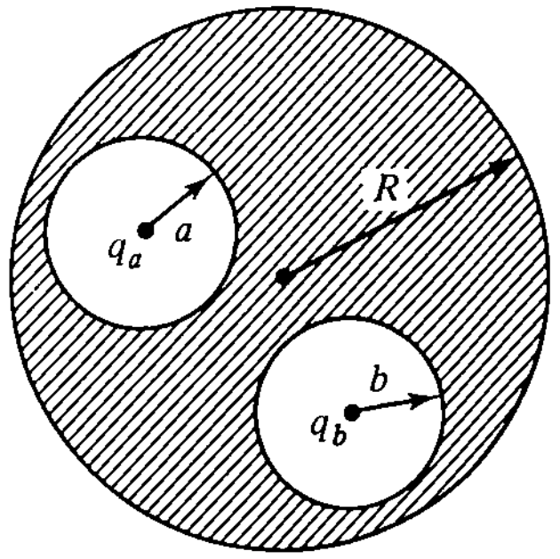

Figure 2.50

2.5.3 The Surface Charge on a Conductor; the Force on a Surface Charge

Because the field inside a conductor is zero, boundary condition (2.27) requires that the field immediately *outside* be

$$\mathbf{E} = \frac{\sigma}{\epsilon_0}\hat{n} \tag{2.41}$$

consistent with our earlier conclusion that the field is normal to the surface. In terms of potential, (2.30) yields

$$\boxed{\frac{\partial V}{\partial n} = -\frac{\sigma}{\epsilon_0}} \tag{2.42}$$

These equations enable you to calculate the surface charge on a conductor, if you can determine **E** or V; we shall use them frequently in the next chapter.

In the presence of an electric field, a surface charge will, naturally, experience a force. According to equation (2.2), the force per unit area, **f**, is $\sigma\mathbf{E}$. But there's a problem here, for the electric field is *discontinuous* at a surface charge, so which value are we supposed to use: $\mathbf{E}_{\text{above}}$, $\mathbf{E}_{\text{below}}$, or something in between? The answer is that we should use the *average* of the two:

$$\mathbf{f} = \sigma\mathbf{E}_{\text{average}} = \tfrac{1}{2}\sigma(\mathbf{E}_{\text{above}} + \mathbf{E}_{\text{below}}) \tag{2.43}$$

Why the average? The reason is very simple, though the telling makes it sound complicated. Let's focus our attention on a small patch of surface centered about the point in question (Fig. 2.51). Make it tiny enough so it is essentially flat, and the surface charge on it is essentially constant. The total field consists of two parts—that attributable to the patch itself, and that due to everything else (other regions of the surface, as well as any external sources that may be present):

$$\mathbf{E} = \mathbf{E}_{\text{patch}} + \mathbf{E}_{\text{other}}$$

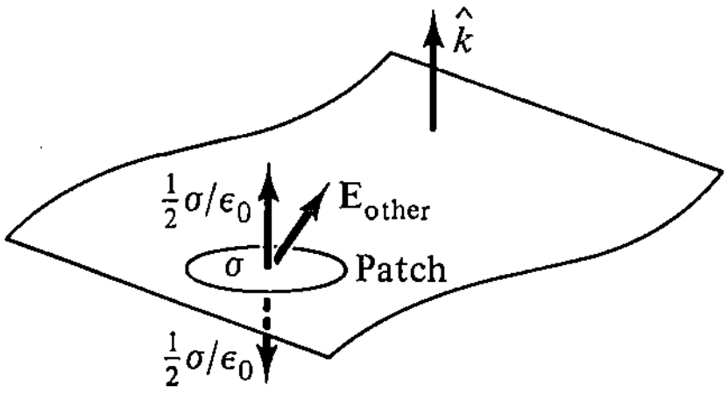

Figure 2.51

Now, the patch cannot exert a force on itself, any more than you can lift yourself by standing in a basket and pulling up on the handles. The force on the patch, then, is due exclusively to $\mathbf{E}_{\text{other}}$, and *this* suffers *no* discontinuity (if we removed the patch, the field in the "hole" would be perfectly smooth). The discontinuity is due entirely to the charge on the patch, which puts out a field $(\sigma/2\epsilon_0)$ on either side, pointing away from the surface (Fig. 2.51). Thus

$$\mathbf{E}_{\text{above}} = \mathbf{E}_{\text{other}} + \frac{\sigma}{2\epsilon_0}\hat{k}$$

$$\mathbf{E}_{\text{below}} = \mathbf{E}_{\text{other}} - \frac{\sigma}{2\epsilon_0}\hat{k}$$

and hence

$$\mathbf{E}_{\text{other}} = \tfrac{1}{2}(\mathbf{E}_{\text{above}} + \mathbf{E}_{\text{below}}) = \mathbf{E}_{\text{average}}$$

Averaging is really just a device for peeling off the contribution of the patch itself.

In the case of a conductor, the average field (equation 2.41) is $(1/2\epsilon_0)\sigma\hat{n}$, so the force per unit area is

$$\mathbf{f} = \frac{1}{2\epsilon_0}\sigma^2\hat{n} \tag{2.44}$$

This amounts to an outward **electrostatic pressure** on the surface, tending to draw the conductor into the field, regardless of the sign of σ. Expressing the pressure in terms of the field (2.41) just outside the surface,

$$P = \frac{\epsilon_0}{2}E^2 \tag{2.45}$$

Problem 2.38 Equation (2.43) is not restricted to infinitesimally thin surfaces. Consider a slab of thickness a, and assume for the sake of argument that the charge density ρ and the electric field $\mathbf{E}$ depend only on x (see Fig. 2.52). The normal force per unit area on this slab is then

$$f_x = \int_0^a \rho(x)E_x(x)\,dx$$

Use Gauss's law (in differential form) to rewrite ρ, and show that

$$f_x = \frac{1}{2}\epsilon_0[E_x(a)^2 - E_x(0)^2] = \sigma E_{x_{\text{ave}}}$$

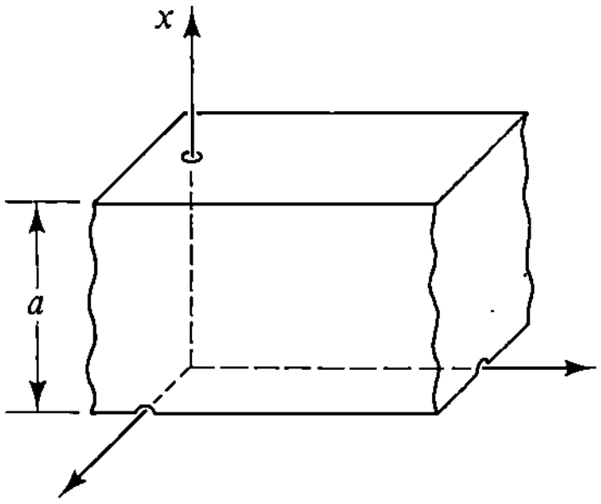

Figure 2.52

where

$$E_{x_{\text{ave}}} = \frac{1}{2}\left[E_x(a) + E_x(0)\right]$$

and σ is the total charge per unit area.

Problem 2.39 A metal sphere of radius R carries a total charge Q. What is the force of repulsion between the "northern" hemisphere and the "southern" hemisphere?

2.5.4 Capacitors

Suppose that we have two conductors, and we put charge $+Q$ on one and $-Q$ on the other (Fig. 2.53). Since V is constant over a conductor, we can speak unambiguously of the potential difference between them:

$$V = V_+ - V_- = -\int_{(-)}^{(+)} \mathbf{E}\cdot d\mathbf{l}$$

Now, we are not sure how the charge is distributed over the two conductors, and calculating the field would be a mess, if their shapes are complicated, but this much we *do* know: **E** is *proportional to* Q. For **E** is given by Coulomb's law:

$$\mathbf{E} = \frac{1}{4\pi\epsilon_0}\int\left(\frac{\hat{\boldsymbol{\mathscr{r}}}}{\mathscr{r}^2}\right)\rho\, d\tau \tag{2.6}$$

so if you double ρ, you double **E**. (Wait a minute! How do we know that doubling Q—and also $-Q$—simply doubles ρ? Maybe the charge *moves around* into a completely different configuration, quadrupling ρ in some places and halving it in others, just so the *total* charge on each conductor is doubled. The *fact* is that this concern is unwarranted—doubling Q *does* double ρ everywhere; it *doesn't* shift the charge around. The proof of this will come in Chapter 3; for now you'll just have to believe me.)

Figure 2.53

Since **E** is proportional to Q, so also is V. The constant of proportionality is called the **capacitance** of the arrangement:

$$C = \frac{Q}{V} \tag{2.46}$$

Capacitance is a purely geometrical quantity, determined by the sizes, shapes, and separation of the two conductors. In mks units, C is measured in *farads* (F): 1 farad is 1 coulomb per volt. Actually, this turns out to be inconveniently large—you'd need a forklift to carry around a 1F capacitor; more practical units are the microfarad (10^{-6} F) and the picofarad (10^{-12} F).

Notice that V is, by definition, the potential of the *positive* conductor less that of the negative one; likewise, Q is the charge of the *positive* conductor. Accordingly, capacitance is an intrinsically positive quantity. (By the way, you will occasionally hear someone speak of the capacitance of a *single* conductor. In this case the "second conductor," with the negative charge, is an imaginary spherical shell of infinite radius surrounding the one conductor. It contributes nothing to the field, so the capacitance is given by (2.46), where V is the potential with infinity as the reference point.)

Example 11

Find the capacitance of a "parallel-plate capacitor" consisting of two metal surfaces of area A held a distance d apart (Fig. 2.54).

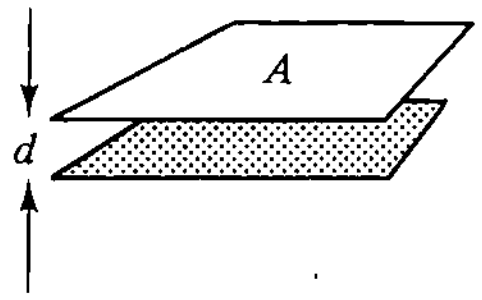

Figure 2.54

Solution: If we put $+Q$ on the top and $-Q$ on the bottom, they will spread out uniformly over the two surfaces, provided the area is reasonably large and the separation distance small. The surface charge density, then, is $\sigma = Q/A$ on the top plate, and so the field, according to Example 5, is $(1/\epsilon_0)\, Q/A$. The potential difference between the plates is therefore

$$V = \frac{Q}{A\epsilon_0} d$$

and, hence,

$$C = \frac{A\epsilon_0}{d} \tag{2.47}$$

If, for instance, the plates are square, with sides of 1 cm, and they are held 1 mm apart, then the capacitance is 9×10^{-13} F, or about 1 pF (picofarad).

Example 12

Find the capacitance of two concentric spherical metal shells, with radii a and b.

Solution: Place charge $+Q$ on the inner sphere, and $-Q$ on the outer one. The field between the spheres is

$$\mathbf{F} = \frac{1}{4\pi\epsilon_0}\frac{Q}{r^2}\hat{r}$$

so the potential difference between them is

$$V = -\int_b^a \mathbf{E}\cdot dl = -\frac{Q}{4\pi\epsilon_0}\int_b^a \frac{1}{r^2}\,dr = \frac{Q}{4\pi\epsilon_0}\left(\frac{1}{a} - \frac{1}{b}\right)$$

As promised, V is proportional to Q; the capacitance is

$$C = \frac{Q}{V} = 4\pi\epsilon_0\frac{ab}{(b-a)}$$

To "charge up" a capacitor, you have to remove electrons from the positive plate and carry them to the negative plate. In doing so you fight against the electric field, which is pulling them back toward the positive conductor and pushing them away from the negative one. How much work does it take, then, to charge the capacitor up to its final amount Q? Suppose that at some intermediate stage in the process the charge on the positive plate is q, so that the potential difference is q/C. According to equation (2.32), the work you must do to transport the next piece of charge, dq, is

$$dW = \left(\frac{q}{C}\right)dq$$

The total work necessary, then, to go from $q = 0$ to $q = Q$, is

$$W = \int_0^Q \left(\frac{q}{C}\right)dq = \frac{1}{2}\frac{Q^2}{C}$$

or, since $Q = CV$,

$$W = \frac{1}{2}CV^2 \tag{2.48}$$

where V is the final potential of the capacitor. Of course, one could also compute the energy stored in the capacitor using equation (2.37), (2.39), or even (2.38). You might try doing so for yourself.

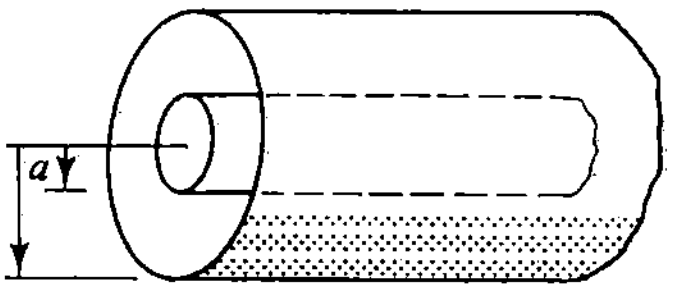

Figure 2.55

Problem 2.40 Find the capacitance per unit length of two coaxial metal cylindrical tubes, of radii a and b (Fig. 2.55).

Problem 2.41 Suppose the plates of a parallel-plate capacitor move closer together by an infinitesimal distance ϵ, as a result of their mutual attraction.

(a) Use equation (2.45) to express the amount of work done by electrostatic forces, in terms of the field E, and the area of the plates, A.

(b) Use equation (2.40) to express the energy lost by the field in this process.

(This problem is supposed to be easy, but it contains the embryo of an alternative derivation of equation (2.45), using conservation of energy.)

Further Problems on Chapter 2

Problem 2.42 Find the electric field at a height z above the center of a square sheet (side s) carrying a uniform surface charge σ. Check your result for the limiting cases $s \to \infty$ and $z \gg s$. (*Answer*: $E = (\sigma/2\epsilon_0)((4/\pi)\tan^{-1}\sqrt{1 + (s^2/2z^2)} - 1)$)

Problem 2.43 If the electric field in some region is given (in spherical coordinates) by the expression

$$\mathbf{E} = \frac{A\hat{\mathbf{r}} + B \sin\theta \cos\phi\, \hat{\boldsymbol{\phi}}}{r}$$

where A and B are constants, what is the charge density? (*Answer*: $(A - B \sin\phi)/r^2$)

Problem 2.44 Find the net force that the southern hemisphere of a uniformly charged sphere exerts on the northern hemisphere. Express your answer in terms of the radius R and the total charge Q. (*Answer*: $(1/4\pi\epsilon_0)(3Q^2/16R^2)$)

Problem 2.45 An inverted hemispherical bowl of radius R carries a uniform surface charge density σ. Find the potential difference between the "north pole" and the center. (*Answer*: $(R\sigma/2\epsilon_0)(\sqrt{2} - 1)$)

Problem 2.46 A sphere of radius R carries a charge density $\rho(r) = kr$ (where k is a constant). Find the energy of the configuration. Check your answer by calculating it in at least two different ways. (*Answer*: $\pi k^2 R^2/7\epsilon_0$)

Problem 2.47 Suppose the electric potential is given by the expression

$$V(\mathbf{r}) = A\,\frac{e^{-\lambda r}}{r},$$

for all r (A and λ are constants). Find the electric field $\mathbf{E}(\mathbf{r})$, the charge density $\rho(r)$, and the total charge Q. (*Answer*: $\rho = \epsilon_0 A(4\pi\delta^3(\mathbf{r}) - (\lambda^2 e^{-\lambda r}/r))$)

! **Problem 2.48** Two infinitely long wires running parallel to the x axis carry charge densities $+\lambda$ and $-\lambda$ (Fig. 2.56).

(a) Find the potential at any point (x, y, z), using the origin as your reference.

(b) Show that the equipotential surfaces are circular cylinders, and locate the axis and radius of the cylinder corresponding to a given potential V_0.

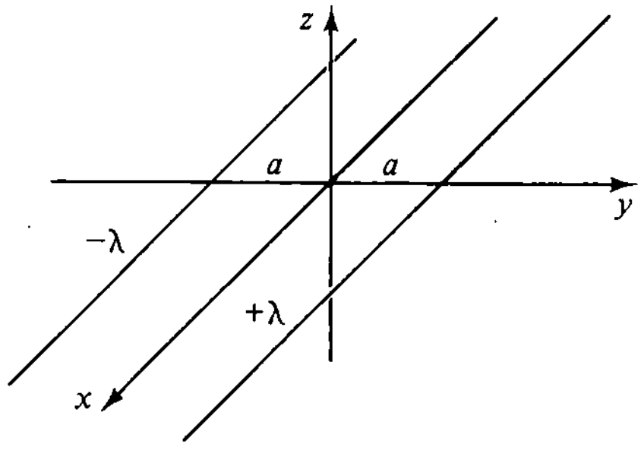

Figure 2.56

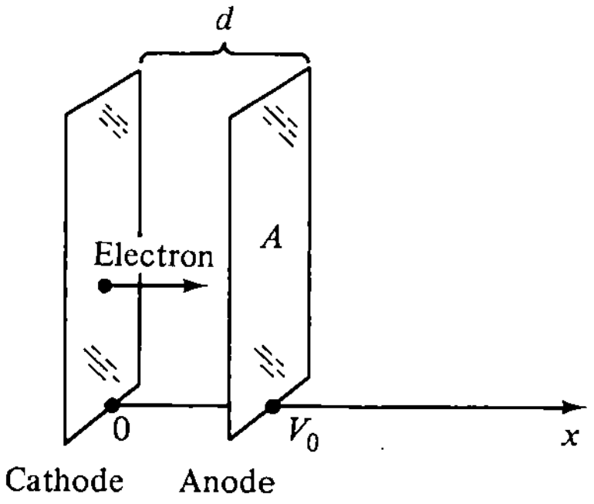

Figure 2.57

! **Problem 2.49** In a vacuum diode, electrons are "boiled" off a hot **cathode**, at potential zero, and accelerated across a gap to the **anode**, which is held at positive potential V_0. The cloud of moving electrons within the gap (called **space charge**) quickly builds up to the point where it reduces the field at the surface of the cathode to zero. From then on a steady current I flows between the plates.

Suppose the plates are large relative to the separation ($A \gg d^2$ in Fig. 2.57), so that edge effects can be neglected. Then V, ρ, and v (the speed of the electrons) are all functions of x alone.

(a) Write Poisson's equation for the region between the plates.

(b) Assuming the electrons start from rest at the cathode, what is their speed at point x where the potential is $V(x)$?

(c) In the steady state, I is independent of x. What, then, is the relation between ρ and v?

(d) Use these three equations to obtain a differential equation for V, by eliminating ρ and v.

(e) Solve this equation for V as a function of x. Plot $V(x)$, and compare the potential *without* space-charge. Also, find ρ and v as functions of x.

(f) Show that

$$I = KV_0^{3/2} \tag{2.49}$$

and find the constant K.

(Equation (2.49) is called the **Child-Langmuir law.** It holds for other geometries, such as cylindrical, where space-charge limits the current. Notice that the space-charge limited diode is *nonlinear*—it does not obey Ohm's law.)

! **Problem 2.50** Imagine that new and extraordinarily precise measurements have revealed an error in Coulomb's Law. The *actual* force of interaction between two point charges is found to be

$$\mathbf{F} = \frac{1}{4\pi\epsilon_0}\frac{q_1 q_2}{r^2}\left(1 + \frac{r}{\lambda}\right)e^{-r/\lambda}\,\hat{\mathbf{r}}$$

where λ is a new constant of nature (it has the dimensions of length, obviously, and is a huge number—about half the radius of the known universe—so that the correction is small, which is why no one ever noticed the discrepancy before). You are charged with the task of reformulating electrostatics to accommodate the new discovery. Assume the principle of superposition still holds.

(a) What is the electric field of a charge distribution ρ (replacing 2.6)?

(b) Does this electric field admit a scalar potential? Explain briefly how you reached your conclusion. (No formal proof necessary—just a persuasive argument.)
(c) Find the potential of a point charge q (analogous to equation (2.23)). (If your answer to (b) was "no," better go back and change it!) Use ∞ as your reference point.
(d) For a point charge q at the origin, show that

$$\oint_S \mathbf{E} \cdot d\mathbf{a} + \frac{1}{\lambda^2} \int_W V\, d\tau = \frac{1}{\epsilon_0} q$$

where S is the surface, and W the volume, of any sphere centered at q.
(e) Show that this result generalizes:

$$\oint_S \mathbf{E} \cdot d\mathbf{a} + \frac{1}{\lambda^2} \int_W V\, d\tau = \frac{1}{\epsilon_0} Q_{\text{enc}}$$

for *any* volume W with surface S, and *any* charge distribution. (This is the next best thing to Gauss's Law, in the new "electrostatics.")
(f) Draw the triangle diagram (Fig. 2.35) for this world, putting in all the appropriate formulas. (Think of Poisson's formula as an equation for ρ in terms of V, and Gauss's Law (differential form) as an equation for ρ in terms of $\mathbf{E}$.)

Problem 2.51 Suppose an electric field $\mathbf{E}(x, y, z)$ has the form

$$E_x = ax, \qquad E_y = 0, \qquad E_z = 0$$

where a is a constant. What is the charge density? How do you account for the fact that the field points in a particular direction, when the charge density is uniform? (This is a more subtle problem than it looks, and worthy of careful thought.)

3

SPECIAL TECHNIQUES FOR CALCULATING POTENTIALS

3.1 LAPLACE'S EQUATION

3.1.1 Introduction

The primary task of electrostatics is to find the electric field of a given stationary charge distribution. In principle, this purpose is accomplished by Coulomb's law, in the form (2.6):

$$\mathbf{E} = \frac{1}{4\pi\epsilon_0} \int \left(\frac{\hat{\mathscr{r}}}{\mathscr{r}^2} \right) \rho \, d\tau$$

Unfortunately, integrals of this type can be difficult to calculate for any but the simplest charge configurations. Occasionally we can get around this by exploiting symmetry and using Gauss's law, but ordinarily the best strategy is first to calculate the *potential*, V, which is given by the somewhat more tractable formula (2.24):

$$V = \frac{1}{4\pi\epsilon_0} \int \left(\frac{1}{\mathscr{r}} \right) \rho \, d\tau \qquad (3.1)$$

Still, even *this* integral is often too tough to handle analytically. Moreover, in problems involving conductors, ρ itself may not be known in advance: Since charge is free to move around, the only thing we control directly is the *total* charge (or perhaps the potential) of each conductor.

In such cases it is fruitful to recast the problem in differential form, using Poisson's equation (2.21),

$$\nabla^2 V = -\frac{1}{\epsilon_0} \rho \qquad (3.2)$$

which, together with appropriate boundary conditions, is equivalent to (3.1). Very often, in fact, we are interested in finding the potential in a region where $\rho = 0$. (If $\rho = 0$ *everywhere*, of course, then $V = 0$, and there is nothing further to say—that's not what I mean. There may be plenty of charge *elsewhere*, but we're confining our attention to those places where there is no charge.) In this case Poisson's equation reduces to Laplace's equation:

$$\nabla^2 V = 0 \tag{3.3}$$

or, written out in Cartesian coordinates,

$$\frac{\partial^2 V}{\partial x^2} + \frac{\partial^2 V}{\partial y^2} + \frac{\partial^2 V}{\partial z^2} = 0 \tag{3.4}$$

This formula is so fundamental to the subject that one might almost say electrostatics *is* the study of Laplace's equation. At the same time, it is a ubiquitous equation, appearing in such diverse branches of physics as gravitation and magnetism, the theory of heat, and the study of soap bubbles. In mathematics it plays a major role in analytic function theory. To get a feel for Laplace's equation and its solutions (which are called **harmonic functions**), we shall begin with the one- and two-dimensional versions, which are easier to picture, and illustrate all the essential properties of the three-dimensional case (though the one-dimensional example lacks the richness of the other two).

3.1.2 Laplace's Equation in One Dimension

Suppose V depends on only one variable, x. Then Laplace's equation becomes

$$\frac{d^2 V}{dx^2} = 0$$

The general solution to this equation is

$$V = mx + b \tag{3.5}$$

—the formula for a straight line. It contains two undetermined constants (m and b), as is appropriate for a second-order (ordinary) differential equation. They are to be fixed, in any particular case, by the boundary conditions of that problem. For instance, it might be specified that $V = 4$ at $x = 1$, and $V = 0$ at $x = 5$. In that case $m = -1$ and $b = 5$, so $V = -x + 5$ (see Fig. 3.1).

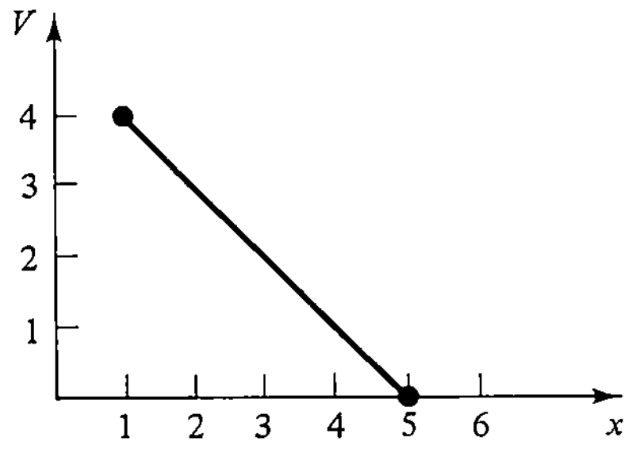

Figure 3.1

I want to mention two facts about the result (3.5), which may seem silly and obvious in one dimension, where I can write down the general solution explicitly. They are worth noting, however, because the analogs in two and three dimensions are powerful and by no means obvious:

1. $V(x)$ is the *average* of $V(x + R)$ and $V(x - R)$, for any R:

$$V(x) = \tfrac{1}{2}[V(x + R) + V(x - R)]$$

Laplace's equation is a kind of averaging instruction; it tells you to assign to the point x the average of the value to the left and to the right of x. Solutions to Laplace's equation are, in this sense, *as uninteresting as they could possibly be,* and yet fit the end points properly.

2. Laplace's equation tolerates *no local maxima or minima.* Extreme values of V must occur at the end points. Actually, this is a consequence of (1) for, if there *were* a local maximum, V at that point would be greater than on either side and therefore could not be the average. (Ordinarily, you expect the second derivative to be negative at a maximum and positive at a minimum. Since Laplace's equation requires on the contrary that the second derivative be zero, it seems appropriate that solutions should exhibit no extrema. However, this is not a *proof,* since there exist functions which have maxima and minima at points where the second derivative vanishes: x^4, for example, has such a minimum at the point $x = 0$.)

3.1.3 Laplace's Equation in Two Dimensions

If V depends on two variables, Laplace's equation becomes

$$\frac{\partial^2 V}{\partial x^2} + \frac{\partial^2 V}{\partial y^2} = 0$$

This is no longer an *ordinary* differential equation (that is, one involving ordinary derivatives only); it is a *partial* differential equation. As a consequence, some of the simple rules you may be familiar with do not apply. For instance, the general solution to this equation doesn't contain just two arbitrary constants—or for that matter *any* finite number—despite the fact that it's a second-order equation. Indeed, one cannot write down a "general solution" (at least, not in closed form like (3.5)). Nevertheless, it is possible to deduce certain general properties common to all solutions.

It may help to have a physical example in mind. Picture a thin rubber sheet (or a soap film) stretched over some support. For definiteness, suppose you take a cardboard box, cut a wavy line all the way around, and remove the top part (Fig. 3.2). Now glue a tightly stretched rubber membrane over the box, so that it fits like a drum head (it won't be a *flat* drumhead, of course, unless you chose to cut the edges off straight). Now, if you lay out coordinates (x, y) on the bottom of the box, the height $V(x, y)$ of the sheet above the point (x, y) will satisfy Laplace's equation.[1] (The one-

[1]Actually, the equation satisfied by a rubber sheet is

$$\frac{\partial}{\partial x}\left(g\frac{\partial V}{\partial x}\right) + \frac{\partial}{\partial y}\left(g\frac{\partial V}{\partial y}\right) = 0, \quad \text{where} \quad g = \left[1 + \left(\frac{\partial V}{\partial x}\right)^2 + \left(\frac{\partial V}{\partial y}\right)^2\right]^{-1/2}$$

It reduces (approximately) to Laplace's equation as long as the surface does not deviate too radically from a plane.

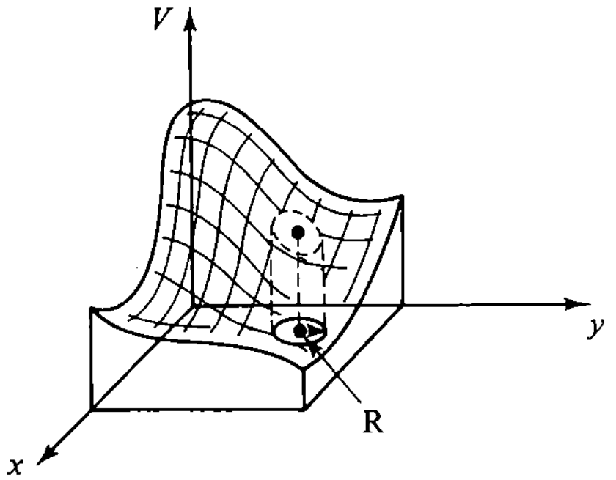

Figure 3.2

dimensional analog would be a rubber band stretched between two points. Of course, it would form a straight line.)

Harmonic functions in two dimensions have the same properties we noted in one dimension:

1. The value of V at a point (x, y) is the average of those *around* the point. More precisely, if you draw a circle of any radius R about the point (x, y), the average value of V on the circle is equal to the value at the center:

$$V(x, y) = \frac{1}{2\pi R} \oint_{\text{circle}} V\, dl$$

(This, incidentally, suggests the **method of relaxation** on which computer solutions to Laplace's equation are based: starting with specified values for V at the boundary, and reasonable guesses for V on a grid of interior points, the first pass reassigns to each point the average of its nearest neighbors. The second pass repeats the process, using the corrected values, and so on. After a few iterations, the numbers begin to settle down, so that subsequent passes produce negligible changes, and a numerical solution to Laplace's equation, with the given boundary values, has been achieved.)[2]

2. V has no local maxima or minima; all extrema occur at the boundaries. (As before, this follows from (1).) Again, Laplace's equation picks the most boring function possible, consistent with the boundary conditions: no hills, no valleys, just the smoothest surface available. For instance, if you put a Ping-Pong ball on the stretched rubber sheet of Fig. 3.2, it will roll over to one side and fall off—it will not find a "pocket" somewhere to roll into, for Laplace's equation allows no such dents in the surface. From a geometrical point of view, just as a straight line is the shortest distance between two points, so a harmonic function in two dimensions minimizes the surface area fitted to the given boundary line.

3.1.4 Laplace's Equation in Three Dimensions

In three dimensions I can neither provide you with an explicit solution (as in one dimension) nor offer a suggestive physical example to guide your intuition (as I did in

[2]See, for example, E. M. Purcell, *Electricity and Magnetism*, 2d ed. (New York: McGraw-Hill, 1985) problem 3.30 (p. 119).

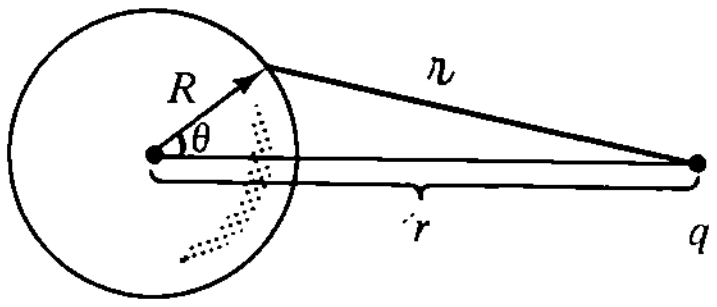

Figure 3.3

two dimensions). Nevertheless, the same two properties remain true, and this time I will sketch a proof.

1. The value of V at point P is the average value of V over a spherical surface of radius R centered at P:

$$V(P) = \frac{1}{4\pi R^2} \oint_{\text{sphere}} V \, da$$

2. As a consequence, V can have no local maxima or minima; the extreme values of V must occur at the boundaries. (For if V had a maximum at P, then by the very nature of a maximum I could draw a sphere around P over which all values of V (and *a fortiori* the average) would be less than at P.)

Proof of (1). Let us first calculate the average potential over a spherical surface of radius R, due to a single point charge q located outside the sphere. We may as well center the sphere at the origin and choose coordinates so that q lies on the z axis, a distance r from the center (Fig. 3.3). Then

$$V = \frac{1}{4\pi\epsilon_0} \frac{q}{\mathscr{r}}$$

and

$$\mathscr{r}^2 = r^2 + R^2 - 2rR\cos\theta$$

so that

$$\begin{aligned} V_{\text{ave}} &= \frac{1}{4\pi R^2} \frac{q}{4\pi\epsilon_0} \int [r^2 + R^2 - 2rR\cos\theta]^{-1/2} R^2 \sin\theta \, d\theta \, d\phi \\ &= \frac{q}{4\pi\epsilon_0} \frac{1}{2rR} \sqrt{r^2 + R^2 - 2rR\cos\theta} \, \Big|_0^{\pi} = \frac{q}{4\pi\epsilon_0} \frac{1}{2rR} [(r + R) - (r - R)] \\ &= \frac{1}{4\pi\epsilon_0} \frac{q}{r}. \end{aligned}$$

But this is precisely the potential due to q at the *center* of the sphere. By the superposition principle, the same goes for any *collection* of charges outside the sphere: Their average potential over the sphere is equal to the net potential they produce at the center. q.e.d.

Problem 3.1 Find the average potential over a spherical surface of radius R due to a point charge q located *inside* (same as above, in other words, only with $r < R$). (In this case, of course, Laplace's equation does not hold within the sphere.) Show that, in general,

$$V_{\text{ave}} = V_{\text{center}} + \frac{Q_{\text{enc}}}{4\pi\epsilon_0 R}$$

where V_{center} is the potential at the center due to all the *external* charges, and Q_{enc} is the total enclosed charge.

Problem 3.2 In one sentence, justify **Earnshaw's Theorem**: *a charged particle cannot be held in stable equilibrium by electrostatic forces alone.* As an example, consider the cubical lattice of fixed charges in Fig. 3.4. It *looks,* offhand, as though a positive charge at the center would be suspended in midair—repelled away from each corner. Where is the leak in this "electrostatic bottle"? (To harness nuclear fusion as a practical energy source it is necessary to heat a plasma (soup of charged particles) to fantastic temperatures—so hot that contact would vaporize any ordinary pot. Earnshaw's Theorem says that electrostatic containment is also out of the question. Fortunately, it *is* possible to confine a hot plasma *magnetically.*)

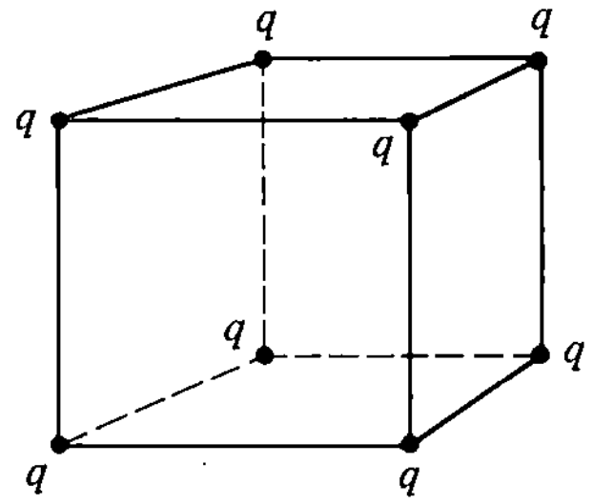

Figure 3.4

Problem 3.3 Find the general solution to Laplace's equation in spherical coordinates, for the case where V depends only on r. Then do the same for cylindrical coordinates.

3.1.5 Boundary Conditions and Uniqueness Theorems

Laplace's equation does not by itself determine V; in addition, a suitable set of boundary conditions must be supplied. This raises a delicate question: What are appropriate boundary conditions, sufficient to determine the answer and yet not so strong as to generate inconsistencies? The one-dimensional case is easy, for here the general solution $V = mx + b$ contains two arbitrary constants, and we therefore require two boundary conditions. We might, for instance, specify the value of the function at the two ends, or we might give the value of the function and its derivative at one end, or the value at one end and the derivative at the other, and so on. But we cannot get away with *just* the value or *just* the derivative at *one* end—this is insufficient information. (Nor could we specify the derivatives at both ends—they would in general be incompatible.)

In two or three dimensions we are confronted by a partial differential equation, and it is not so easy to see what constitutes acceptable boundary conditions. Is the shape of a taut rubber membrane, for instance, uniquely determined by the frame over which it is stretched, or, like a canning-jar lid, can it snap from one stable configuration to another? The answer, as I think your intuition would suggest, is that V *is* uniquely determined by its value at the boundary (canning jars evidently do not

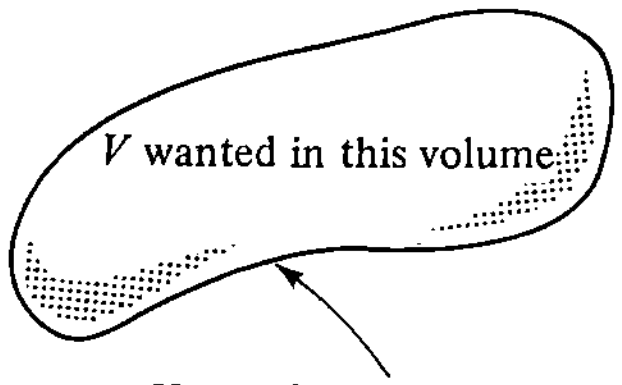

Figure 3.5

obey Laplace's equation). However, other boundary conditions can also be used (see Problem 3.4). The *proof* that a proposed set of boundary conditions will suffice is usually presented in the form of a **uniqueness theorem.** There are many such theorems for electrostatics, all sharing the same basic format—I'll show you the two most useful ones.[3]

First uniqueness theorem: The solution to Laplace's equation in some region is uniquely determined if the value of V is a specified function on all boundaries of the region.

Proof: In Fig. (3.5) I have drawn such a region and its boundary. There could also be "islands" inside, so long as V is specified on all their surfaces; also, the outer boundary could be at infinity, where V is ordinarily taken to be zero.

Suppose there are *two* solutions to Laplace's equation:

$$\nabla^2 V_1 = 0 \quad \text{and} \quad \nabla^2 V_2 = 0$$

both of which assume the given value on the surface. I want to prove that they must be equal. The trick is to look at their *difference:*

$$V_3 = V_1 - V_2$$

This obeys Laplace's equation,

$$\nabla^2 V_3 = \nabla^2 V_1 - \nabla^2 V_2 = 0$$

and it takes the value *zero* on all boundaries (since V_1 and V_2 are equal there). But Laplace's equation permits no local maxima or minima; the extrema occur on the boundaries. So the maximum and the minimum of V_3 are both zero. Therefore V_3 must be zero everywhere and hence,

$$V_1 = V_2 \quad \text{q.e.d.}$$

Example 1

Show that the potential is *constant* inside an enclosure completely surrounded by conducting material, provided there is no charge within the enclosure.

Solution: The potential on the cavity wall is some constant, V_0 (that's item

[3] I do not intend to prove the *existence* of solutions here—that's a much more difficult job. In context, the existence is generally clear on physical grounds.

(d), in Section 2.5.1), so the potential inside is a function which satisfies Laplace's equation and takes on the constant value V_0 at the boundary. With no work at all I can think of *one* solution to this problem: $V = V_0$ throughout the region. The uniqueness theorem guarantees that this is the *only* solution. (It follows that the *field* inside an empty cavity is zero—the same result we found in Section 2.5.2 on rather different grounds.)

The uniqueness theorem is a license to the imagination: It doesn't matter *how* you come by your solution; if (a) it satisfies Laplace's equation, and (b) it has the correct value on the boundaries, then it's *right,* and any complaints about your reasoning are moot. You'll see the power of this argument when we come to the method of images.

Incidentally, it is easy to improve on the first uniqueness theorem: I assumed there was no charge in the region in question, so that the potential obeyed Laplace's equation, but we may as well throw in some charge (in which case V obeys Poisson's equation). The argument is just the same, only

$$\nabla^2 V_1 = -\frac{1}{\epsilon_0}\rho, \qquad \nabla^2 V_2 = -\frac{1}{\epsilon_0}\rho$$

so

$$\nabla^2 V_3 = \nabla^2 V_1 - \nabla^2 V_2 = -\frac{1}{\epsilon_0}\rho + \frac{1}{\epsilon_0}\rho = 0$$

Again the *difference* $V_3 = V_1 - V_2$ satisfies Laplace's equation and has the value zero on all boundaries, so $V_3 = 0$ and hence $V_1 = V_2$, as before.

Corollary: The potential in some region is uniquely determined if (a) the charge density throughout the region, and (b) the value of V on all boundaries, are specified.

3.1.6 Conductors and the Second Uniqueness Theorem

The simplest way to set boundary conditions for an electrostatic problem is to specify the value of V on all surfaces surrounding the region of interest. And this situation often occurs in practice: In the laboratory, we have conductors connected to batteries, which maintain a given potential, or to **ground,** which is the experimentalist's word for $V = 0$. However, there are other circumstances in which we do not know the *potential* at the boundary, but rather the *charges* on various conducting surfaces. Suppose I put charge Q_1 on the first conductor, Q_2 on the second, and so on—I'm not telling you how the charge distributes itself over each conducting surface because, as soon as I put it on, it moves around in a way that I cannot directly control. And for good measure, let's say that there is some specified charge density ρ in the region between the conductors. Is the electric field now uniquely determined? Or are there perhaps a number of different ways the charges could arrange themselves on their respective conductors, each leading to a different field?

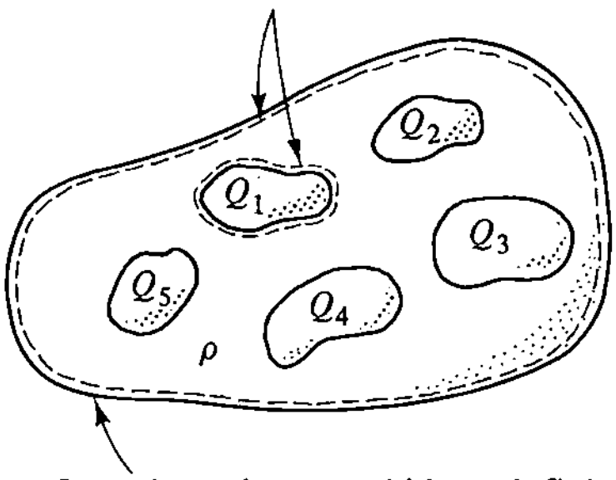

Figure 3.6

Second uniqueness theorem: In a region containing conductors and filled with a specified charge density ρ, the electric field is uniquely determined if the total charge on each conductor is given (Fig. 3.6). (The region as a whole can be bounded by another conductor, or else unbounded.)

Proof: Suppose there are *two* fields satisfying the conditions of the problem. Both obey Gauss's law in differential form in the space between the conductors,

$$\nabla \cdot \mathbf{E}_1 = \frac{1}{\epsilon_0}\rho, \qquad \nabla \cdot \mathbf{E}_2 = \frac{1}{\epsilon_0}\rho$$

and in integral form for a Gaussian surface enclosing each conductor.

$$\oint_{\substack{i\text{th conducting}\\ \text{surface}}} \mathbf{E}_1 \cdot d\mathbf{a} = \frac{1}{\epsilon_0} Q_i, \qquad \oint_{\substack{i\text{th conducting}\\ \text{surface}}} \mathbf{E}_2 \cdot d\mathbf{a} = \frac{1}{\epsilon_0} Q_i$$

Likewise, for the outer boundary (whether this is just inside an enclosing conductor or simply at infinity),

$$\oint_{\substack{\text{outer}\\ \text{boundary}}} \mathbf{E}_1 \cdot d\mathbf{a} = \frac{1}{\epsilon_0} Q_{\text{tot}}, \qquad \oint_{\substack{\text{outer}\\ \text{boundary}}} \mathbf{E}_2 \cdot d\mathbf{a} = \frac{1}{\epsilon_0} Q_{\text{tot}}$$

As before, we examine the difference

$$\mathbf{E}_3 = \mathbf{E}_1 - \mathbf{E}_2$$

which obeys

$$\nabla \cdot \mathbf{E}_3 = 0 \tag{3.6}$$

in the region between the conductors, and

$$\oint \mathbf{E}_3 \cdot d\mathbf{a} = 0 \tag{3.7}$$

over each boundary surface.

Now there is one final piece of information we must exploit: Although we do not know how the charge distributes itself, we *do* know that each conductor is an equipotential and, hence, V_3 is a constant (not necessarily the *same* constant) over each conducting surface. (It need not be *zero*, for the potentials V_1 and V_2 may not be equal—all we know for sure is that *both* are *constant* over any given conductor.) Next comes a trick. Invoking product rule number (5), we find that

$$\nabla \cdot (V_3 \mathbf{E}_3) = V_3(\nabla \cdot \mathbf{E}_3) + \mathbf{E}_3 \cdot (\nabla V_3) = -E_3^2$$

Here I have used equation (3.6), and $\mathbf{E}_3 = -\nabla V_3$. Integrating this over the entire region between the conductors and applying the divergence theorem to the left side:

$$\int_{\text{volume}} \nabla \cdot (V_3 \mathbf{E}_3)\, d\tau = \oint_{\text{surface}} V_3 \mathbf{E}_3 \cdot d\mathbf{a} = -\int_{\text{volume}} E_3^2\, d\tau$$

The surface integral covers all boundaries of the region in question, including the conductors and the outer boundary. Now V_3 is a constant over each surface, (if the outer boundary is infinity, $V_3 = 0$ there) so it comes outside the integral, and what is left is zero, according to equation (3.7). Therefore,

$$\int_{\text{volume}} E_3^2\, d\tau = 0$$

But this integrand is never negative; the only way the integral can vanish is if $E_3 = 0$ everywhere. Consequently, $\mathbf{E}_1 = \mathbf{E}_2$, and the theorem is proved.

This proof has not been a simple one, and there is a real danger that the theorem itself will seem more plausible to you than the proof. In case you think the second uniqueness theorem is obvious, consider this example of Purcell's:

Figure 3.7 is a comfortable electrostatic configuration, consisting of four conductors with charges $\pm Q$, situated so that the plusses are near the minuses. It looks very stable. Now, what happens if we join them in pairs, by tiny wires, as shown in Fig. 3.8? Since the positive charges are very near negative charges (which is where they *like* to be) you might well guess that *nothing* will happen—the configuration still looks stable.

Well, that is wrong. It takes no calculation at all to prove that the second configuration is *impossible*. For there are now effectively *two* conductors, and the total charge on each is zero. *One* possible way to distribute zero charge over these conductors is to have no accumulation of charge anywhere, and hence zero field everywhere (Fig. 3.9). By the second uniqueness theorem, this must be *the* solution: the charge will flow down the tiny wires, canceling itself off.

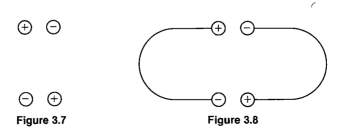

Figure 3.7

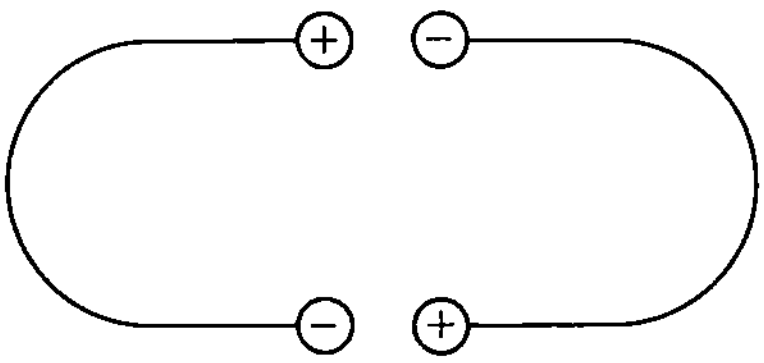

Figure 3.8

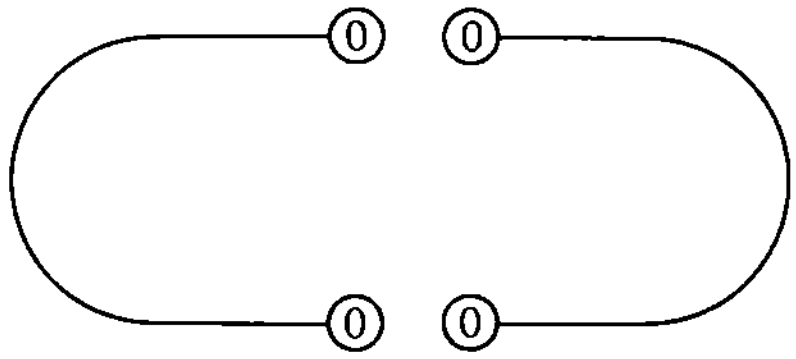

Figure 3.9

Problem 3.4 Prove that the field is uniquely determined when the charge density ρ is given and *either* V *or* the normal derivative $\partial V/\partial n$ is specified on each boundary surface.

Problem 3.5 A more elegant proof of the second uniqueness theorem uses Green's identity (Problem 1.61c), with $t = u = V_3$. Supply the details.

3.2 THE METHOD OF IMAGES

3.2.1 The Classic Image Problem

Suppose a point charge q is held a distance d above an infinite grounded conducting plane (Fig. 3.10). *Question*: What is the potential in the region above the plane? It's not just $(1/4\pi\epsilon_0)q/\mathscr{r}$, for q will induce a certain amount of negative charge on the nearby surface of the conductor; the total potential is due in part to q directly, and in part to this induced charge. But how can we possibly calculate the latter, when we don't know how much charge is induced or how it is distributed?

From a mathematical point of view our problem is to solve Poisson's equation in the region $z > 0$, with a single point charge q at $(0, 0, d)$, subject to the boundary conditions:

1. $V = 0$ when $z = 0$ (since the conducting plane is grounded), and
2. $V \to 0$ far from the charge (that is, for $x^2 + y^2 + z^2 \gg d^2$).

The first uniqueness theorem (actually, its corollary) guarantees that there is only one function which meets these requirements. If by trick or clever guess we can discover such a function, it's got to be the right answer.

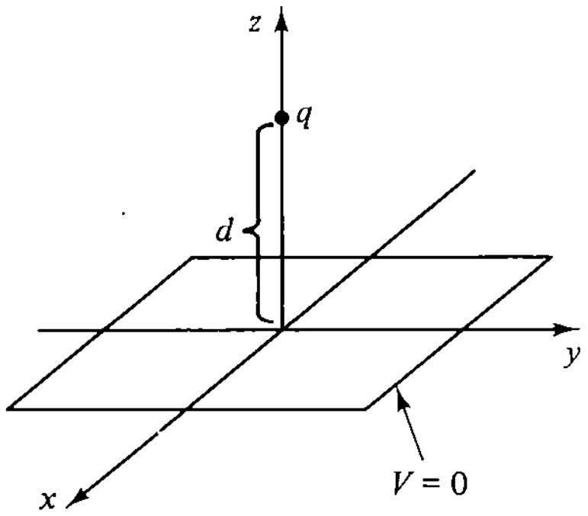

Figure 3.10

Trick: Forget about the actual problem; we're going to study a *completely different* situation for a moment. This new problem consists of *two* point charges, $+q$ at $(0, 0, d)$ and $-q$ at $(0, 0, -d)$, and *no* conducting plane (Fig. 3.11). For this configuration I can easily write down the potential:

$$V(x, y, z) = \frac{1}{4\pi\epsilon_0}\left[\frac{q}{\sqrt{x^2 + y^2 + (z - d)^2}} - \frac{q}{\sqrt{x^2 + y^2 + (z + d)^2}}\right] \quad (3.8)$$

(The denominators represent the distances from (x, y, z) to the charges $+q$ and $-q$, respectively.) Notice that

1. $V = 0$ when $z = 0$
2. $V \to 0$ for $x^2 + y^2 + z^2 \gg d^2$

and the only charge in the region $z > 0$ is the point charge $+q$ at $(0, 0, d)$. But these are precisely the conditions of our original problem! Evidently the "upper half" $(z \geq 0)$ of the second configuration happens to have exactly the same potential as the first configuration. (The "lower half," $z < 0$, is completely different, but who cares? The upper half is all we need, and it fits our specifications.) *Conclusion*: The potential of a point charge above an infinite grounded conductor is given by equation (3.8), with $z \geq 0$.

Notice the crucial role played by the uniqueness theorem in this argument: without it, no one would believe this solution, since it was, after all, obtained for a completely different charge distribution. Yet the uniqueness theorem assures us that if it satisfies Poisson's equation in the region of interest and assumes the correct value at the boundaries, then it must be right.

3.2.2 The Induced Surface Charge

Now that we know the potential, it is a straightforward matter to compute the surface charge σ induced on the conductor. According to equation (2.42),

$$\sigma = -\epsilon_0 \frac{\partial V}{\partial n}$$

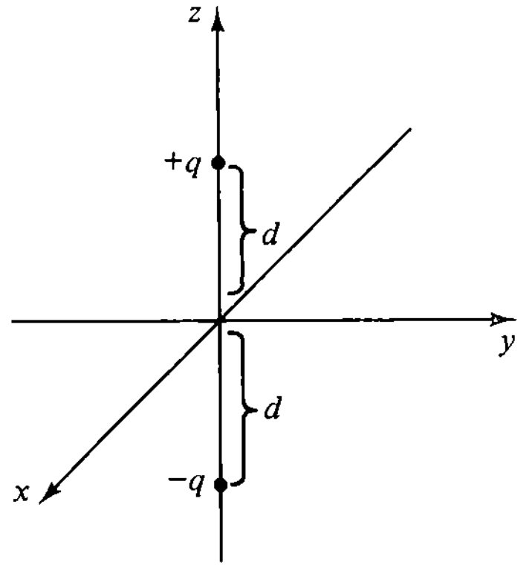

Figure 3.11

where $\partial V/\partial n$ is the normal derivative of V at the surface. In this instance, the normal direction is the z-direction, so

$$\sigma = -\epsilon_0 \frac{\partial V}{\partial z}\bigg|_{z=0}$$

From equation (3.8),

$$\frac{\partial V}{\partial z} = \frac{1}{4\pi\epsilon_0}\left\{\frac{-q(z-d)}{[x^2+y^2+(z-d)^2]^{3/2}} + \frac{q(z+d)}{[x^2+y^2+(z+d)^2]^{3/2}}\right\}$$

so

$$\sigma(x, y) = \frac{-qd}{2\pi(x^2+y^2+d^2)^{3/2}} \tag{3.9}$$

As expected, the induced charge is negative (assuming q is positive) and greatest at $x = y = 0$.

While we're at it, let's compute the *total* induced charge,

$$Q = \int \sigma\, da$$

This integral, over the xy plane, could be done in Cartesian coordinates, with $da = dx\, dy$, but it's a little easier to use plane polar coordinates (r, θ), with $r^2 = x^2 + y^2$ and $da = r\, dr\, d\theta$. Then,

$$\sigma(r) = \frac{-qd}{2\pi(r^2+d^2)^{3/2}}$$

and

$$Q = \int_0^{2\pi}\int_0^{\infty} \frac{-qd}{2\pi(r^2+d^2)^{3/2}}\, r\, dr\, d\theta = -qd\left(\frac{-1}{\sqrt{r^2+d^2}}\right)\bigg|_0^{\infty} = -q \tag{3.10}$$

So the total charge induced on the plane is $-q$ as, with benefit of hindsight, you can perhaps convince yourself it *had* to be.

3.2.3 Force and Energy

The charge q is attracted toward the plane, because of the negative induced charge. Let's calculate the force of attraction. Since the potential in the vicinity of q is the same as in the analog problem (the one with $+q$ and $-q$ but no conductor), so also is the field and, therefore, the force as well:

$$\mathbf{F} = -\frac{1}{4\pi\epsilon_0}\frac{q^2}{(2d)^2}\,\hat{\mathbf{z}} \tag{3.11}$$

Beware: It is easy to get careless and assume that *everything* is the same in the two problems; the energy, however, is *not* the same. With the two point charges and no conductor the energy is, according to equation (2.36),

$$W = -\frac{1}{4\pi\epsilon_0}\frac{q^2}{(2d)} \tag{3.12}$$

whereas for a single charge and conducting plane the energy is just *half* of this:

$$W = -\frac{1}{4\pi\epsilon_0}\frac{q^2}{(4d)} \tag{3.13}$$

Why half? Think of equation (2.39):

$$W = \frac{\epsilon_0}{2}\int E^2\,d\tau$$

In the first case both the upper region ($z > 0$) and the lower region ($z < 0$) contribute to this integral—and by symmetry they contribute equally. But in the second case only the upper region contains a nonzero field; accordingly, the energy is just half as much.

Of course, one can also compute the energy by integrating the force to find the work required to bring q in from $z = \infty$ to $z = d$. The force *I* exert, against (3.11), is $(1/4\pi\epsilon_0)(q^2/4z^2)\hat{k}$, so

$$\begin{aligned} W &= \int_\infty^d \mathbf{F}\cdot d\mathbf{l} = \frac{1}{4\pi\epsilon_0}\int_\infty^d \frac{q^2}{4z^2}\,dz \\ &= \frac{1}{4\pi\epsilon_0}\left(-\frac{q^2}{4z}\right)\Bigg|_\infty^d = -\frac{1}{4\pi\epsilon_0}\frac{q^2}{(4d)} \end{aligned}$$

As I bring q toward the conductor, I do work *only on q*. It is true that induced charge is moving in over the conductor, but this costs me nothing, since the whole conductor is at potential zero. By contrast, if I bring in two point charges (with no conductor), I do work on both of them, and the energy is twice as great.

3.2.4 Other Image Problems

The method just described is not limited to a single point charge near a grounded conducting plane. Any stationary distribution of charge near such a plane can be treated in the same way, by constructing its mirror image—hence the name **method of images.** (Remember that the image charges have the *opposite sign;* this is what guarantees that the xy plane will be at potential zero.) There are also a number of more exotic problems that can be handled in similar fashion, of which the nicest is the following.

Example 2

A point charge q is situated a distance s from the center of a grounded conducting sphere of radius R (Fig. 3.12). Find the potential everywhere. (Inside the sphere the potential is zero, so our only real concern is with the region outside.)

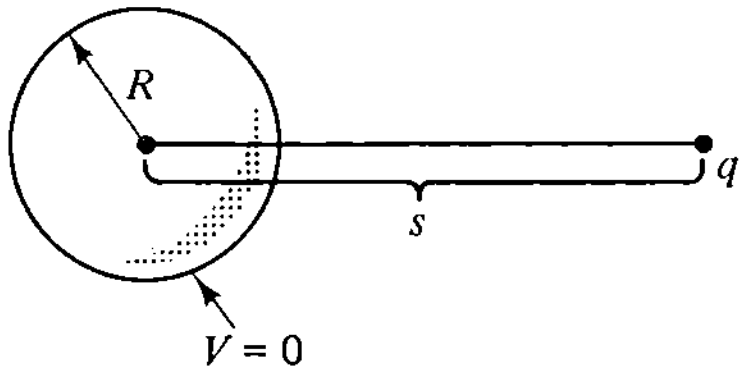

Figure 3.12

Solution: Examine the *completely different* configuration consisting of the point charge q, together with another point charge

$$q' = -\frac{R}{s} q \tag{3.14}$$

placed a distance

$$a = \frac{R^2}{s} \tag{3.15}$$

to the right of the center of the sphere (Fig. 3.13). No conductor, now—just the two point charges. The potential of this configuration is

$$V = \frac{1}{4\pi\epsilon_0}\left(\frac{q}{\mathscr{r}} + \frac{q'}{\mathscr{r}'}\right) \tag{3.16}$$

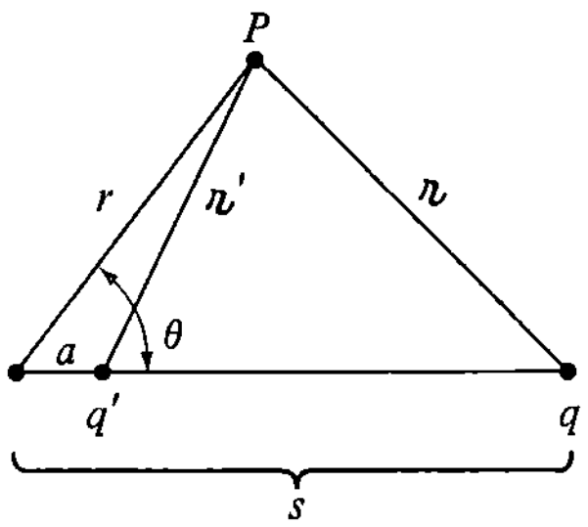

Figure 3.13

where $\mathscr{r}$ and $\mathscr{r}'$ are the distances from q and q', respectively. Now, it happens (see Problem 3.6) that this potential vanishes at all points on the sphere and therefore fits the boundary conditions for our original problem, in the exterior region.

Conclusion: Equation (3.16) represents the potential of the point charge and the conducting sphere, as well. (Notice that a is less than R, so the "image" charge q' is safely inside the sphere—*you cannot put image charges in the region where you're calculating* V, for if you do you have changed ρ, and you are solving Poisson's equation with the wrong source.) In particular, the force of attraction between the charge and the sphere is

$$F = \frac{1}{4\pi\epsilon_0}\frac{qq'}{(s-a)^2} = -\frac{1}{4\pi\epsilon_0}\frac{q^2Rs}{(s^2-R^2)^2} \tag{3.17}$$

The solution just presented is delightfully simple but also extraordinarily lucky. There's as much art as science in the method of images, for you must somehow think up the right "auxiliary problem" to look at. I doubt the first person who solved the problem of the point charge and the conducting sphere knew in advance what image charge q' to use or where to put it. Presumably, he (she?) started with an *arbitrary* charge at an *arbitrary* point inside the sphere, calculated the potential on the sphere, and then discovered that with q' and a just right, as in equations (3.14) and (3.15), the potential on the sphere would go to zero. But it is really a miracle that *any* choice does the job—had it been a cube instead of a sphere, for example, the method of images wouldn't have worked for *any* q'.

Problem 3.6

(a) Using the law of cosines, show that equation (3.16) can be written thus:

$$V(r, \theta) = \frac{1}{4\pi\epsilon_0}\left[\frac{q}{\sqrt{r^2 + s^2 - 2rs\cos\theta}} - \frac{q}{\sqrt{R^2 + (rs/R)^2 - 2rs\cos\theta}}\right] \tag{3.18}$$

where r and θ are the spherical polar coordinates, with the z axis set along the line through q, as in Fig. 3.13. In this form it is obvious that $V = 0$ on the sphere, $r = R$.

(b) Find the induced surface charge on the sphere, as a function of θ. Integrate this to get the total induced charge (what *should* it equal?).

(c) Calculate the electrostatic energy of this configuration.

Problem 3.7 In Example 2 we assumed that the conducting sphere was grounded ($V = 0$). But with the addition of a second image charge, the same basic model will handle the case of a sphere at *any* given potential V_0 (relative, of course, to zero at infinity). What charge should you use, and where should you put it? Find the force of attraction between a point charge q and a conducting sphere at potential V_0.

Problem 3.8 A uniform line charge λ is placed on an infinite straight wire, a distance d above a grounded conducting plane. (Let's say the wire runs parallel to the x axis and directly above it, and the conducting plane is the xy plane.)

(a) Find the potential in the region above the plane.

(b) Find the charge density σ induced on the conducting plane.

Problem 3.9 Two semi-infinite grounded conducting planes meet at right angles. In the region between them, there is a point charge q, situated as shown in Fig. 3.14. Set up the image configuration, and calculate the potential in this region—what charges do you need, and where should they be located? What is the force on q? How much work did it take to bring q in from infinity? Suppose the planes met at some angle other than 90°; would you still be able to solve the problem by the method of images? If not (in general), for what particular angles *does* the method work?

! **Problem 3.10** Two long, straight, copper pipes, each of radius R, are held a distance $2d$ apart. One is at potential V_0, the other at $-V_0$ (Fig. 3.15). Find the potential everywhere. (*Suggestion*: Exploit the result of Problem 2.48.) Describe how you could modify this method to handle the case of one cylinder at V_0 and the other at ground ($V = 0$).

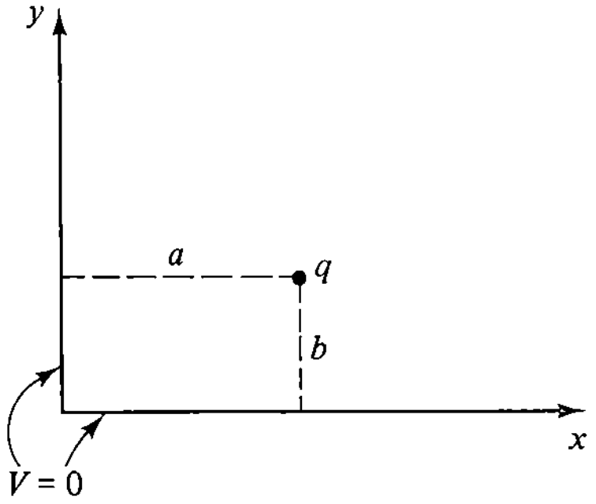

Figure 3.14

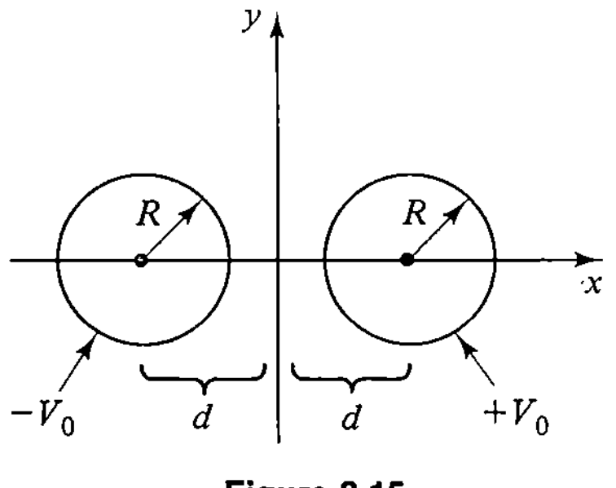

Figure 3.15

3.3 SEPARATION OF VARIABLES

In this section we shall attack Laplace's equation directly, using the method of **separation of variables,** which is the physicist's favorite tool for solving partial differential equations. The method is applicable in circumstances where the potential (V) or its normal derivative ($\partial V/\partial n$) is specified on the boundaries of some region, and we are asked to find the potential in the interior. The basic strategy is very simple: look for solutions that are *products* of functions, each of which depends on only *one* of the coordinates. The algebraic details, however, can become formidable, so I'm going to develop the method through a sequence of examples. We'll start out with *Cartesian* coordinates and then do *spherical* coordinates—I'll leave the *cylindrical* case for you to tackle on your own.

3.3.1 Cartesian Coordinates

Example 3

Two infinite, grounded, metal plates lie parallel to the xz plane, one at $y = 0$, the other at $y = \pi$ (Fig. 3.16).[4] The left end, at $x = 0$, is closed off with an infinite strip insulated from the two plates and maintained at a specified potential $V_0(y)$. Find the potential inside this "slot."

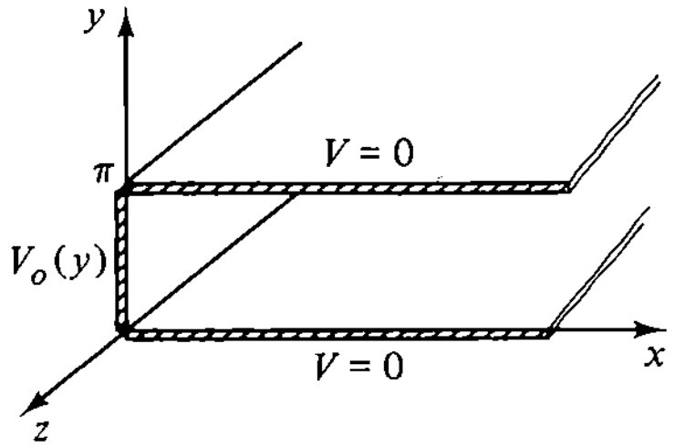

Figure 3.16

Solution: The configuration is independent of z, so this is really a *two*-dimensional problem. In mathematical terms, we must solve Laplace's equation,

$$\frac{\partial^2 V}{\partial x^2} + \frac{\partial^2 V}{\partial y^2} = 0 \tag{3.19}$$

subject to the boundary conditions

$$\left.\begin{array}{ll} \text{(i)} & V = 0 \text{ when } y = 0 \\ \text{(ii)} & V = 0 \text{ when } y = \pi \\ \text{(iii)} & V = V_0(y) \text{ when } x = 0 \\ \text{(iv)} & V \to 0 \text{ as } x \to \infty \end{array}\right\} \tag{3.20}$$

[4]This y dimension (π meters) may look strange, but it simplifies the notation considerably. It's easy to "scale down" the final result. See Problem 3.15.

(The latter, although not explicitly stated in the problem, is necessary on physical grounds: As you get farther and farther away from the "hot" strip at $x = 0$, the potential should drop off to zero.) Since the potential is specified on all boundaries, the answer is uniquely determined.

The first step is to look for solutions in the form of products:

$$V(x, y) = X(x)Y(y) \tag{3.21}$$

On the face of it, this is an absurd restriction—the overwhelming majority of solutions to Laplace's equation do *not* take such a form. For example, $V(x, y) = (5x + 6y)$ satisfies (3.19), but you can't express it as the product of a function x times a function of y. Obviously, we're only going to get a tiny subset of all possible solutions by this means, and it would be a *miracle* if one of them happened to fit the boundary conditions of our problem . . . but hang on, because the solutions we *do* get are very special, and it turns out that by pasting them together we can construct the general solution.

Anyway, putting (3.21) into (3.19), we have

$$Y\frac{d^2X}{dx^2} + X\frac{d^2Y}{dy^2} = 0$$

The next step is to "separate the variables" (that is, collect all the x-dependence into one term and all the y-dependence into another). Typically, this is accomplished by dividing through by V:

$$\frac{1}{X}\frac{d^2X}{dx^2} + \frac{1}{Y}\frac{d^2Y}{dy^2} = 0 \tag{3.22}$$

Here the first term depends only on x and the second only on y; in other words, we have an equation of the form

$$f(x) + g(y) = 0$$

Now, there's only one way such an equation could possibly be true: *f and g must both be constant.* For what if $f(x)$ *changed,* as you vary x—then if we held y fixed and fiddled with x, the sum $f(x) + g(y)$ would *change,* in violation of the condition that it always equals zero. (That's a simple but somehow rather elusive argument; don't accept it without due thought because the whole method rides on it.) It follows from (3.22), then, that

$$\frac{1}{X}\frac{d^2X}{dx^2} = C_1 \quad \text{and} \quad \frac{1}{Y}\frac{d^2Y}{dy^2} = C_2, \qquad \text{with} \quad C_1 + C_2 = 0$$

One of these constants is positive, the other negative (or perhaps both are zero). In general, one must investigate all the possibilities; however, in our particular problem we need C_1 positive and C_2 negative, for reasons that appear in a moment. Thus,

$$\frac{d^2X}{dx^2} = k^2X, \qquad \frac{d^2Y}{dy^2} = -k^2Y \tag{3.23}$$

Notice what has happened: A *partial* differential equation, (3.19), has been converted into two *ordinary* differential equations, (3.23). The advantage of this is obvious—ordinary differential equations are a lot easier to solve. In fact,

$$X(x) = Ae^{kx} + Be^{-kx}, \qquad Y(y) = C \sin ky + D \cos ky$$

so that

$$V(x, y) = (Ae^{kx} + Be^{-kx})(C \sin ky + D \cos ky) \tag{3.24}$$

We have found, now, the appropriate separable solutions to Laplace's equation—it remains to impose the boundary conditions. To begin at the end, condition (iv) requires that A equal zero.[5] Absorbing the constant B into C and D, we are left with

$$V(x, y) = e^{-kx}(C \sin ky + D \cos ky)$$

Condition (i) now demands that D equal zero, so that

$$V(x, y) = Ce^{-kx} \sin ky \tag{3.25}$$

while (ii) yields

$$\sin k\pi = 0$$

from which it follows that *k is an integer.* (At this point you can see why I chose C_1 positive and C_2 negative: If X were sinusoidal, we could never arrange for it to go to zero at infinity, and if Y were exponential we could not make it vanish at both zero and π.)

That's as far as we can go, using separable solutions and, unfortunately, unless $V_0(y)$ just happens to have the form $\sin ky$ for some integer k, we simply *can't fit* the final boundary condition at $x = 0$. But now comes the crucial step that redeems the method: Separation of variables has given us not *one* but an *infinite set* of solutions (one for each k) and whereas none of them *by itself* satisfies the final boundary condition, it is possible to combine them in a way that *does.* For Laplace's equation is *linear,* in the sense that if $V_1, V_2, V_3, \ldots$ satisfy it, so also does any linear combination of them, $V = \alpha_1 V_1 + \alpha_2 V_2 + \alpha_3 V_3 + \cdots$, where $\alpha_1, \alpha_2, \ldots$ are arbitrary constants:

$$\nabla^2 V = \alpha_1 \nabla^2 V_1 + \alpha_2 \nabla^2 V_2 + \cdots = 0\alpha_1 + 0\alpha_2 + \cdots = 0$$

Using this observation, we can patch together separable solutions (3.25) to construct a far more general solution:

$$V(x, y) = \sum_{k=1}^{\infty} C_k e^{-kx} \sin ky \tag{3.26}$$

Moreover, (3.26) still meets the first three boundary conditions (it was in anticipation of this that I fitted the $V = 0$ boundaries first). So the question is: Can we, by astute choice of the coefficients C_k, succeed in matching the last boundary condition,

$$V(0, y) = \sum_{k=1}^{\infty} C_k \sin ky = V_0(y) \tag{3.27}$$

[5] I'm assuming k is positive, but this involves no loss of generality—negative k gives the same solution, (3.24), only with the constants shuffled.

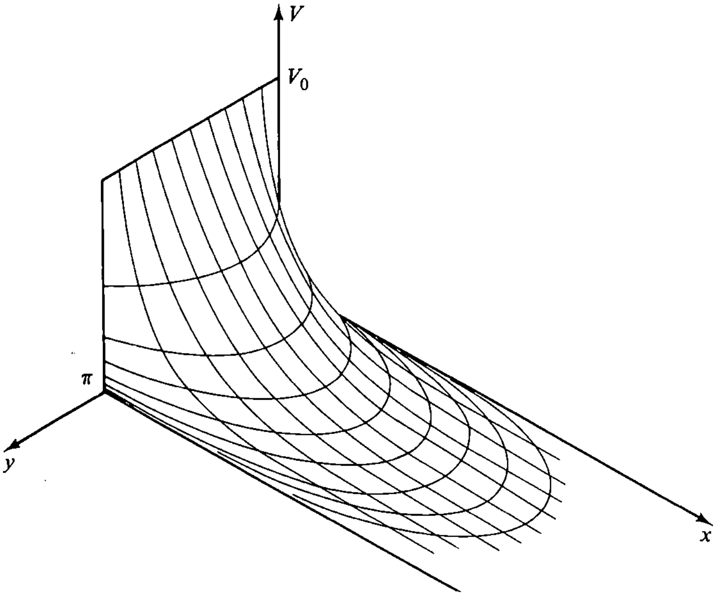

Figure 3.17

Well, you may recognize this sum—it's a Fourier sine series. And Dirichlet's theorem[6] guarantees that virtually *any* function $V_0(y)$—it can even have a finite number of discontinuities—can be expanded in such a series.

But how do we actually *evaluate* the coefficients C_k, buried as they are in the infinite sum? The device for accomplishing this is so lovely it deserves a name—I call it **Fourier's trick,** though it seems Euler had used essentially the same idea somewhat earlier. Here's how it goes: Multiply both sides of (3.27) by $\sin(ny)$, and integrate from 0 to π:

$$\sum_{k=1}^{\infty} C_k \int_0^\pi \sin ky \sin ny \, dy = \int_0^\pi V_0(y) \sin ny \, dy \tag{3.28}$$

Here n is some positive integer. You can do the integral on the left for yourself; the answer is

$$\int_0^\pi \sin ky \sin ny \, dy = \begin{cases} 0, & \text{if } k \neq n \\ \dfrac{\pi}{2}, & \text{if } k = n \end{cases} \tag{3.29}$$

Thus all the terms in the series drop out, save only the one where $k = n$, and the left side of (3.28) reduces to $(\pi/2)C_n$. *Conclusion*:

$$C_n = \frac{2}{\pi} \int_0^\pi V_0(y) \sin ny \, dy \tag{3.30}$$

That *does* it: (3.26) is the solution, with coefficients given by (3.30). As a concrete example, suppose the strip at $x = 0$ is a metal plate with constant

[6]Boas, M., *Mathematical Methods in the Physical Sciences*, 2d ed. (New York: John Wiley, 1983), p. 313.

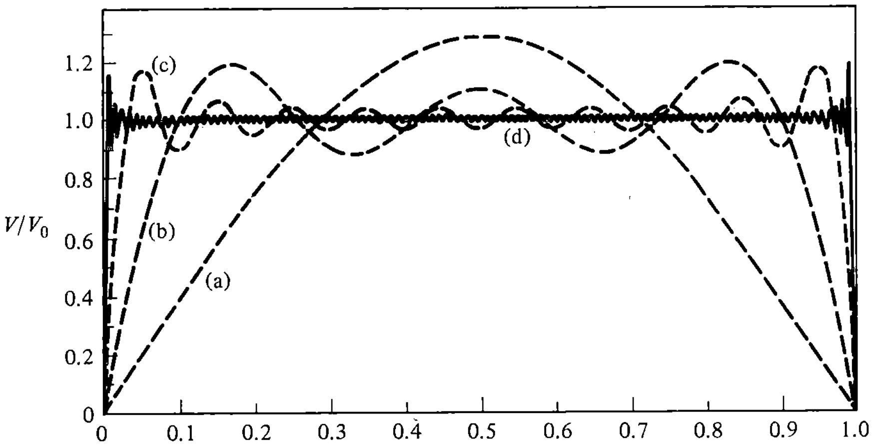

Figure 3.18

potential V_0 (naturally, it must be insulated from the grounded plates at $y = 0$ and $y = \pi$). Then,

$$C_n = \frac{2V_0}{\pi}\int_0^{\pi} \sin ny\, dy = \frac{2V_0}{n\pi}(1 - \cos n\pi) = \begin{cases} 0, & \text{if } n \text{ is even} \\ \dfrac{4V_0}{n\pi}, & \text{if } n \text{ is odd} \end{cases} \tag{3.31}$$

Consequently,

$$V(x, y) = \frac{4V_0}{\pi} \sum_{k=1,3,5\ldots} \frac{1}{k} e^{-kx} \sin ky \tag{3.32}$$

Figure 3.17[7] is a plot of this potential; Fig. 3.18 shows how the first few terms in the Fourier series combine to form a better and better approximation to the constant V_0 ((a) is $k = 1$ only, (b) includes k up to 5, and (d) is the sum of the first 100 terms).

Incidentally, the infinite series in equation (3.32) can be summed explicitly (try your hand at it, if you like); the result is

$$V(x, y) = \frac{2V_0}{\pi}\tan^{-1}\left(\frac{\sin y}{\sinh x}\right)$$

In this form it is easy to check that Laplace's equation is obeyed and the four boundary conditions (3.30) are satisfied.

The success of this method hinged on two extraordinary properties of the separable solutions (3.25): **completeness** and **orthogonality.** A set of functions $f_n(x)$ is *complete* if any other function $f(x)$ can be expressed as a linear combination of them:

$$f(x) = \sum_{n=1}^{\infty} C_n f_n(x) \tag{3.33}$$

[7]Figs. 3.17, 3.18, and 3.20 are from Paul Lorrain and Dale R. Carson, *Electromagnetic Fields and Waves*, 2d ed. (San Francisco: W. H. Freeman and Company, © 1970.)

For example, the functions $\sin(nx)$ are complete on the interval $0 \leq x \leq \pi$. It was this fact, guaranteed by Dirichlet's theorem, that assured us equation (3.27) could be satisfied, given the proper choice of the coefficients C_k. (The actual *proof* of completeness, for a particular set of functions, is an extremely difficult business, and I'm afraid physicists tend to *assume* it's true and leave the checking to others.) A set of functions is *orthogonal* if the integral of the product of any two different members of the set is zero:

$$\int_a^b f_n(x) f_m(x)\, dx = 0, \qquad \text{for} \quad n \neq m \tag{3.34}$$

For example, the sines are orthogonal for the interval $a = 0$ to $b = \pi$. This is the property on which Fourier's trick is based, allowing us to kill off all terms but one in the infinite series and thereby solve for the coefficients C_k. (Proof of orthogonality is generally quite simple; either by direct integration or by analysis of the differential equation from which the functions came.)

Example 4

Two infinitely long grounded metal plates, again at $y = 0$ and $y = \pi$, are connected at $x = \pm a$ by metal strips maintained at potential V_0, as shown in Fig. 3.19 (a thin sliver of insulation at each corner prevents them from shorting out). Find the potential inside the resulting rectangular pipe.

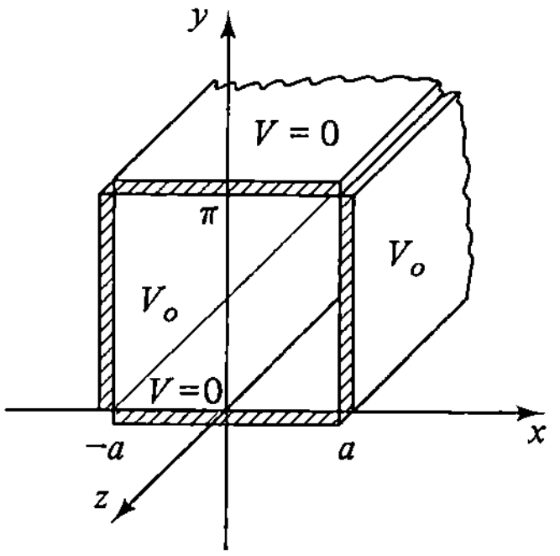

Figure 3.19

Solution: Once again, the configuration is independent of z. The problem is to solve Laplace's equation

$$\frac{\partial^2 V}{\partial x^2} + \frac{\partial^2 V}{\partial y^2} = 0$$

subject to the boundary conditions

$$\left.\begin{array}{ll} \text{(i)} & V = 0 \text{ when } y = 0 \\ \text{(ii)} & V = 0 \text{ when } y = \pi \\ \text{(iii)} & V = V_0 \text{ when } x = a \\ \text{(iv)} & V = V_0 \text{ when } x = -a \end{array}\right\} \tag{3.35}$$

The argument runs as before up to equation (3.24):

$$V(x, y) = (Ae^{kx} + Be^{-kx})(C \sin ky + D \cos ky)$$

At this point, however, we cannot set $A = 0$: the region in question does not extend to $x = \infty$, so e^{kx} is a perfectly acceptable contribution to the potential. On the other hand, the situation is obviously *even* with respect to x—that is, $V(-x, y) = V(x, y)$. It follows that $A = B$. Using

$$e^{kx} + e^{-kx} = 2 \cosh kx$$

and absorbing $2A$ into C and D, we have

$$V(x, y) = \cosh kx(C \sin ky + D \cos ky)$$

Boundary conditions (i) and (ii) require, as before, that $D = 0$ and k is a positive integer, leaving us with the separable solution

$$V(x, y) = C \cosh kx \sin ky \tag{3.36}$$

Because $V(x, y)$ is even in x, it will automatically meet condition (iv) if it fits (iii). It remains, therefore, to construct the general linear combination,

$$V(x, y) = \sum_{k=1}^{\infty} C_k \cosh kx \sin ky$$

and pick the coefficients C_k in such a way as to satisfy condition (iii):

$$V(a, y) = \sum_{k=1}^{\infty} C_k \cosh ka \sin ky = V_0$$

This is the same problem in Fourier analysis that we faced before: I quote the solution from equation (3.31):

$$C_n \cosh na = \begin{cases} 0, & \text{if } n \text{ is even} \\ \dfrac{4V_0}{n\pi}, & \text{if } n \text{ is odd} \end{cases}$$

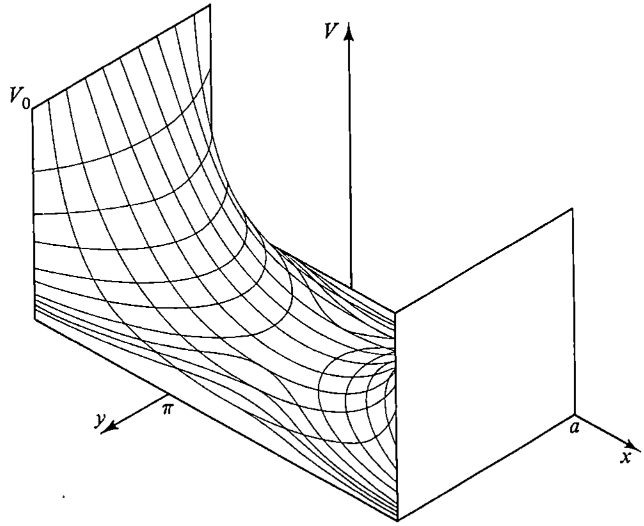

Figure 3.20

Conclusion: The potential in this case is given by

$$V(x, y) = \frac{4V_0}{\pi} \sum_{k=1,3,5\ldots} \frac{1}{k} \frac{\cosh kx}{\cosh ka} \sin ky \tag{3.37}$$

This function is shown in Fig. 3.20.

Example 5

An infinitely long square metal pipe (sides π) is grounded, but one end, at $x = 0$, is an insulator, carrying a specified surface charge $\sigma(y, z)$ (Fig. 3.21). Find the potential inside the pipe.

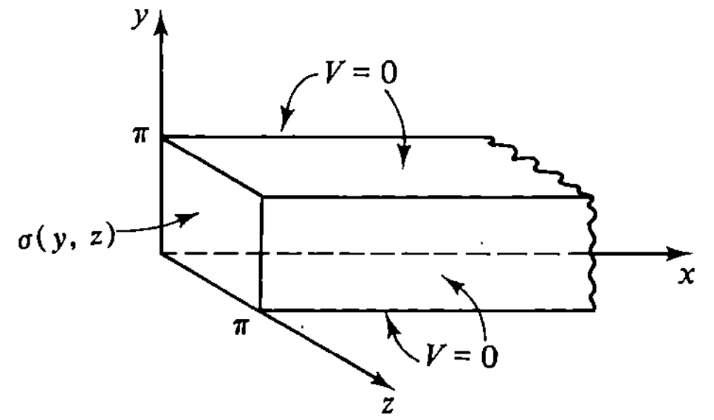

Figure 3.21

Solution: This is a genuinely three-dimensional problem,

$$\frac{\partial^2 V}{\partial x^2} + \frac{\partial^2 V}{\partial y^2} + \frac{\partial^2 V}{\partial z^2} = 0 \tag{3.38}$$

to be solved subject to the boundary conditions

$$\left.\begin{array}{ll} \text{(i)} & V = 0 \text{ when } y = 0 \\ \text{(ii)} & V = 0 \text{ when } y = \pi \\ \text{(iii)} & V = 0 \text{ when } z = 0 \\ \text{(iv)} & V = 0 \text{ when } z = \pi \\ \text{(v)} & V \to 0 \text{ as } x \to \infty \\ \text{(vi)} & \dfrac{\partial V}{\partial x} = -\dfrac{1}{\epsilon_0}\sigma(y, z) \text{ at } x = 0 \end{array}\right\} \tag{3.39}$$

(This last boundary condition follows from equation (2.42).) As always, we look for solutions that are products of functions of the separate coordinates:

$$V(x, y, z) = X(x)Y(y)Z(z) \tag{3.40}$$

Putting this into (3.38) and dividing by V, I find

$$\frac{1}{X}\frac{d^2X}{dx^2} + \frac{1}{Y}\frac{d^2Y}{dy^2} + \frac{1}{Z}\frac{d^2Z}{dz^2} = 0$$

It follows that

$$\frac{1}{X}\frac{d^2X}{dx^2} = C_1, \qquad \frac{1}{Y}\frac{d^2Y}{dy^2} = C_2, \qquad \frac{1}{Z}\frac{d^2Z}{dz^2} = C_3,$$

$$\text{with } C_1 + C_2 + C_3 = 0$$

Our previous experience with Example 3 suggests that C_1 must be positive, C_2 and C_3 negative. Setting $C_2 = -k^2$ and $C_3 = -l^2$, we have $C_1 = k^2 + l^2$, and hence

$$\frac{d^2X}{dx^2} = (k^2 + l^2)X, \qquad \frac{d^2Y}{dy^2} = -k^2Y, \qquad \frac{d^2Z}{dz^2} = -l^2Z \tag{3.41}$$

Once again, separation of variables has turned a *partial* differential equation into *ordinary* differential equations. The solutions are

$$X(x) = Ae^{\sqrt{k^2+l^2}\,x} + Be^{-\sqrt{k^2+l^2}\,x}$$

$$Y(y) = C\sin ky + D\cos ky$$

$$Z(z) = E\sin lz + F\cos lz$$

As before, boundary condition (v) implies $A = 0$, (i) implies $D = 0$, and (iii) implies $F = 0$, whereas (ii) and (iv) require that k and l be positive integers. Combining the remaining constants, we have

$$V(x, y, z) = Ce^{-\sqrt{k^2+l^2}\,x}\sin ky \sin lz \tag{3.42}$$

This separable solution meets all the boundary conditions except (vi). It contains *two* unspecified integers (k and l) and the most general linear combination is a *double* sum:

$$V(x, y, z) = \sum_{k=1}^{\infty}\sum_{l=1}^{\infty} C_{k,l}e^{-\sqrt{k^2+l^2}\,x}\sin ky \sin lz \tag{3.43}$$

We hope to fit the remaining boundary condition

$$-\left.\frac{\partial V}{\partial x}\right|_{x=0} = \sum_{k=1}^{\infty}\sum_{l=1}^{\infty}\sqrt{k^2+l^2}\,C_{k,l}\sin ky \sin lz = \frac{1}{\epsilon_0}\sigma(y, z) \tag{3.44}$$

by appropriate choice of the coefficients $C_{k,l}$. To evaluate these constants, multiply by $\sin ny \sin mz$, where n and m are arbitrary positive integers, and integrate:

$$\sum_{k=1}^{\infty}\sum_{l=1}^{\infty}\sqrt{k^2+l^2}\,C_{k,l}\int_0^{\pi}\sin ky \sin ny\, dy\int_0^{\pi}\sin lz \sin mz\, dz$$

$$= \frac{1}{\epsilon_0}\int_0^{\pi}\int_0^{\pi}\sigma(y, z)\sin ny \sin mz\, dy\, dz$$

Quoting (3.29), the left side is $(\pi/2)^2\sqrt{n^2+m^2}\,C_{n,m}$, so

$$C_{n,m} = \frac{1}{\epsilon_0}\left(\frac{2}{\pi}\right)^2\frac{1}{\sqrt{n^2+m^2}}\int_0^{\pi}\int_0^{\pi}\sigma(y, z)\sin ny \sin mz\, dy\, dz \tag{3.45}$$

Equation (3.43), with the coefficients given by (3.45), represents the solution to our problem.

For instance, if the end of the tube carries a *uniform* charge density σ_0, then

$$C_{n,m} = \frac{\sigma_0}{\epsilon_0}\left(\frac{2}{\pi}\right)^2 \frac{1}{\sqrt{n^2+m^2}} \int_0^\pi \sin ny\, dy \int_0^\pi \sin mz\, dz$$

$$= \begin{cases} 0, & \text{if } n \text{ or } m \text{ is even} \\ \dfrac{16\sigma_0}{\pi^2\epsilon_0 nm\sqrt{n^2+m^2}}, & \text{if } n \text{ and } m \text{ are odd} \end{cases} \tag{3.46}$$

and therefore,

$$V(x, y, z) = \frac{16\sigma_0}{\pi^2\epsilon_0} \sum_{n=1,3,5\ldots}^{\infty} \sum_{m=1,3,5\ldots}^{\infty} \frac{e^{-\sqrt{n^2+m^2}\,x} \sin ny \sin mz}{nm\sqrt{n^2+m^2}} \tag{3.47}$$

Notice that the successive terms decrease rapidly, as usual; a reasonable approximation would be obtained by keeping only the first few.

Problem 3.11 Find the potential in the infinite slot of Example 3 if the boundary at $x = 0$ consists of two metal strips: one, from $y = 0$ to $y = \pi/2$, is held at constant potential V_0, and the other, from $y = \pi/2$ to $y = \pi$ is at potential $-V_0$.

Problem 3.12 For the infinite slot (Example 3) determine the charge density $\sigma(y)$ on the strip at $x = 0$, both for the general case $V_0(y)$ and for the special case $V_0 =$ constant.

Problem 3.13 In the infinite slot problem (Example 3), suppose the charge density $\sigma(y)$ on the strip $x = 0$ is specified, instead of the potential $V_0(y)$. Work out the general solution for $V(x, y)$, and apply it to the case $\sigma = \sigma_0$ (a constant).

Problem 3.14 A long rectangular pipe, running parallel to the z axis, has three grounded metal sides, at $y = 0$, $y = \pi$, and $x = 0$. The fourth side, at $x = a$, is maintained at a specified potential $V_0(y)$.

(a) Develop a general formula for the potential within the pipe.

(b) Find the potential explicitly, for the case $V_0(y) = V_0$ (a constant).

Problem 3.15 In Examples 3, 4, and 5, I used rather unlikely dimensions (y running from 0 to π) in order to simplify the notation. Generalize equations (3.32), (3.37), and (3.47) to the case where y goes from 0 to b (also z goes $0 \to c$ in 3.47). (This takes some *thought* but not much *calculation*.)

Problem 3.16 A cubical box consists of five metal sides which are welded together and grounded (Fig. 3.22). The top is made of a separate sheet of metal, insulated from the

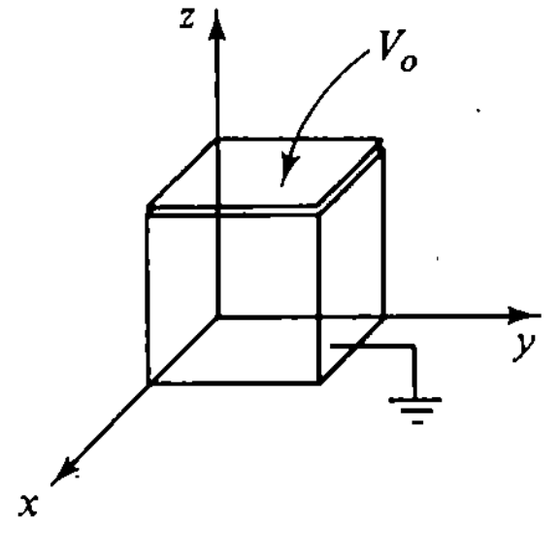

Figure 3.22

rest, and held at a constant potential V_0 by a battery. Find the potential inside the box. The sides are all of length π.

3.3.2 Spherical Coordinates

In the examples considered so far, Cartesian coordinates were clearly appropriate, since the boundaries were *planes*. But for round objects spherical coordinates are more natural. In the spherical system, Laplace's equation reads:

$$\nabla^2 V = \frac{1}{r^2}\frac{\partial}{\partial r}\left(r^2 \frac{\partial V}{\partial r}\right) + \frac{1}{r^2 \sin\theta}\frac{\partial}{\partial\theta}\left(\sin\theta\frac{\partial V}{\partial\theta}\right) + \frac{1}{r^2\sin^2\theta}\frac{\partial^2 V}{\partial\phi^2} = 0 \quad (3.48)$$

I shall assume that the problem has **azimuthal symmetry**, so that *V is independent of ϕ*. In that case (3.48) reduces to

$$\frac{\partial}{\partial r}\left(r^2 \frac{\partial V}{\partial r}\right) + \frac{1}{\sin\theta}\frac{\partial}{\partial\theta}\left(\sin\theta\frac{\partial V}{\partial\theta}\right) = 0 \quad (3.49)$$

As before, we look for solutions that are products:

$$V(r, \theta) = R(r)\Theta(\theta) \quad (3.50)$$

Putting this into (3.49) and dividing by V:

$$\frac{1}{R}\frac{d}{dr}\left(r^2\frac{dR}{dr}\right) + \frac{1}{\Theta\sin\theta}\frac{d}{d\theta}\left(\sin\theta\frac{d\Theta}{d\theta}\right) = 0 \quad (3.51)$$

Since the first term depends only on r and the second only on θ, it follows (as before) that each must be a constant:

$$\frac{1}{R}\frac{d}{dr}\left(r^2\frac{dR}{dr}\right) = l(l+1), \qquad \frac{1}{\Theta\sin\theta}\frac{d}{d\theta}\left(\sin\theta\frac{d\Theta}{d\theta}\right) = -l(l+1) \quad (3.52)$$

Here $l(l+1)$ is just a fancy way of writing the separation constant—you'll see in a minute why this is convenient.

As always, separation of variables has converted a *partial* differential equation (3.49) into *ordinary* differential equations (3.52). However, this time the ordinary differential equations are not quite so familiar. The radial equation,

$$\frac{d}{dr}\left(r^2\frac{dR}{dr}\right) = l(l+1)R \quad (3.53)$$

has the general solution

$$R(r) = Ar^l + \frac{B}{r^{l+1}} \quad (3.54)$$

as you can easily check. A and B are the two arbitrary constants to be expected in the solution of a second-order differential equation. As for the angular equation,

$$\frac{d}{d\theta}\left(\sin\theta\frac{d\Theta}{d\theta}\right) = -l(l+1)\sin\theta\,\Theta \quad (3.55)$$

the solutions are **Legendre polynomials:**

$$\Theta(\theta) = P_l(\cos\theta) \tag{3.56}$$

where $P_l(x)$ is defined by the **Rodrigues formula:**

$$P_l(x) = \frac{1}{2^l l!}\left(\frac{d}{dx}\right)^l (x^2 - 1)^l \tag{3.57}$$

The first few Legendre polynomials are listed in Table 3.1.

TABLE 3.1 LEGENDRE POLYNOMIALS

$P_0(x) = 1$
$P_1(x) = x$
$P_2(x) = \frac{1}{2}(3x^2 - 1)$
$P_3(x) = \frac{1}{2}(5x^3 - 3x)$
$P_4(x) = \frac{1}{8}(35x^4 - 30x^2 + 3)$
$P_5(x) = \frac{1}{8}(63x^5 - 70x^3 + 15x)$

Notice that $P_l(x)$ is (as the name suggests) an lth-order *polynomial* in x, containing only *even* powers, if l is even, and *odd* powers, if l is odd. The factor in front ($1/2^l l!$) was chosen in order that

$$P_l(1) = 1 \tag{3.58}$$

for all the Legendre polynomials. The Rodrigues formula obviously works only for nonnegative, integer values of l. Moreover, it provides us with only *one* solution—(3.55) is a *second*-order equation, and it should possess *two* independent solutions. However, the fact is that these "second solutions," and also the ones for noninteger l, are ordinarily unacceptable on physical grounds because they blow up at $\theta = 0$ and/or $\theta = \pi$. For instance, the second solution for $l = 0$ is

$$\Theta(\theta) = \ln\left(\tan\frac{\theta}{2}\right) \tag{3.59}$$

In the case of azimuthal symmetry, then, the most general *separable* solution to Laplace's equation consistent with minimal physical requirements is

$$V(r, \theta) = \left(Ar^l + \frac{B}{r^{l+1}}\right) P_l(\cos\theta)$$

(There was no need to include an overall constant in (3.56) because it can be absorbed into A and B at this stage.) As before, separation of variables leads to an infinite set of solutions, one for each l. The *general* solution is the linear combination of separable solutions:

$$\boxed{V(r, \theta) = \sum_{l=0}^{\infty}\left(A_l r^l + \frac{B_l}{r^{l+1}}\right) P_l(\cos\theta)} \tag{3.60}$$

The following examples illustrate the standard applications of this fundamental result.

Example 6

The potential $V_0(\theta)$ is specified on the surface of a hollow, empty sphere of radius R. Find the potential inside the sphere.

Solution: In this case $B_l = 0$ for all l—otherwise the potential would blow up at the origin. Thus,

$$V(r, \theta) = \sum_{l=0}^{\infty} A_l r^l P_l(\cos\theta) \tag{3.61}$$

At $r = R$ this must match the specified function $V_0(\theta)$:

$$V(R, \theta) = \sum_{l=0}^{\infty} A_l R^l P_l(\cos\theta) = V_0(\theta) \tag{3.62}$$

Can this equation be satisfied, for appropriate choice of the coefficients A_l? *Yes*, the Legendre polynomials (like the sines) constitute a *complete set* of functions, on the interval $-1 \le x \le 1$ $(0 \le \theta \le \pi)$. How do we evaluate the constants? Again, by Fourier's trick, for the Legendre polynomials (like the sines) are *orthogonal* functions:[8]

$$\int_{-1}^{1} P_l(x)P_m(x)\,dx = \int_0^{\pi} P_l(\cos\theta)P_m(\cos\theta)\sin\theta\,d\theta = \begin{cases} 0, & \text{if } l \ne m \\ \dfrac{2}{(2m+1)}, & \text{if } l = m \end{cases} \tag{3.63}$$

Thus, multiplying both sides of (3.62) by $P_m(\cos\theta)\sin\theta$ and integrating, we have

$$A_m R^m \frac{2}{(2m+1)} = \int_0^{\pi} V_0(\theta)P_m(\cos\theta)\sin\theta\,d\theta$$

or

$$A_m = \frac{(2m+1)}{2R^m}\int_0^{\pi} V_0(\theta)P_m(\cos\theta)\sin\theta\,d\theta \tag{3.64}$$

So (3.61) is the solution to our problem, with the coefficients given by (3.64).

In practice, it is frequently difficult, if not impossible, to evaluate integrals of the form (3.64) analytically, and in those cases where it *can* be done, it is often

[8] M. Boas, *Mathematical Methods in the Physical Sciences*, 2d ed. (New York: John Wiley, 1983), Section 12.7.

easier to solve equation (3.62) "by eyeball."[9] For instance, suppose we are told that the potential on the sphere is

$$V_0(\theta) = k \sin^2 \frac{\theta}{2} \tag{3.65}$$

where k is a constant. Using the half-angle formula, we can rewrite this as

$$V_0(\theta) = \frac{k}{2}\left(1 - \cos\theta\right) = \frac{k}{2}\left[P_0(\cos\theta) - P_1(\cos\theta)\right]$$

Putting this into (3.62), we read off immediately that $A_0 = k/2$ and $A_1 = -(k/2R)$, whereas all other A_l's are zero, and conclude that

$$V(r, \theta) = \frac{k}{2}\left[r^0 P_0(\cos\theta) - \frac{r^1}{R} P_1(\cos\theta)\right] = \frac{k}{2}\left(1 - \frac{r}{R}\cos\theta\right) \tag{3.66}$$

inside the sphere.

Example 7

The potential $V_0(\theta)$ is again specified on the surface of a sphere of radius R, but this time we are asked to find the potential *outside* the sphere, assuming there's no charge there,

Solution: In this case it's the A_l's that must be zero (else V would not go to zero at ∞), so

$$V(r, \theta) = \sum_{l=0}^{\infty} \frac{B_l}{r^{l+1}} P_l(\cos\theta) \tag{3.67}$$

At the surface of the sphere we require that

$$V(R, \theta) = \sum_{l=0}^{\infty} \frac{B_l}{R^{l+1}} P_l(\cos\theta) = V_0(\theta)$$

Multiplying by $P_m(\cos\theta)\sin\theta$, and integrating—exploiting, again, the orthogonality relation (3.63)—we have

$$\frac{B_m}{R^{m+1}} \frac{2}{(2m+1)} = \int_0^{\pi} V_0(\theta) P_m(\cos\theta) \sin\theta \, d\theta$$

or

$$B_m = \frac{(2m+1)}{2} R^{m+1} \int_0^{\pi} V_0(\theta) P_m(\cos\theta) \sin\theta \, d\theta \tag{3.68}$$

[9]This is certainly true whenever $V_0(\theta)$ can be expressed as a polynomial in $\cos\theta$. The degree of the polynomial tells us the highest l we require, and the leading coefficient determines the corresponding A_l. Subtracting off $A_l R^l P_l(\cos\theta)$ and repeating the process, we can systematically work our way down to A_0. Notice that if V_0 is an *even* function of $\cos\theta$, then only even terms will occur in the sum (and likewise for odd functions).

Equation (3.67), with the coefficients given by (3.68), represents the solution to our problem.

Example 8

An uncharged metal sphere of radius R is placed in an otherwise uniform electric field $\mathbf{E} = E_0\hat{k}$. The field will push positive charge to the "northern" surface of the sphere, leaving a negative charge on the "southern" surface (Fig. 3.23). This induced charge, in turn, distorts the field in the neighborhood of the sphere. Find the potential in the region outside the sphere.

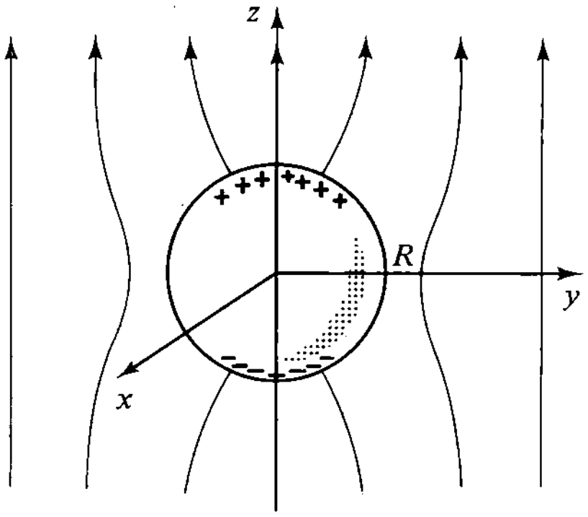

Figure 3.23

Solution: The sphere is an equipotential—we may as well call it potential zero. Then by symmetry the entire xy plane is at potential zero. This time, however, V does *not* go to zero at large z. In fact, far from the sphere the field is $E_0\hat{k}$, and hence

$$V \to -E_0 z + C$$

Since $V = 0$ in the equatorial plane, the constant C must be zero. Accordingly, the boundary conditions for this problem are

$$\left.\begin{array}{ll} \text{(i)} & V = 0 \text{ when } r = R \\ \text{(ii)} & V \to -E_0 r \cos\theta \text{ for } r \gg R \end{array}\right\} \tag{3.69}$$

We must fit these boundary conditions with a function of the form (3.60). The first condition yields

$$A_l R^l + \frac{B_l}{R^{l+1}} = 0$$

or

$$B_l = -A_l R^{2l+1} \tag{3.70}$$

so that

$$V(r, \theta) = \sum_{l=0}^{\infty} A_l\left(r^l - \frac{R^{2l+1}}{r^{l+1}}\right) P_l(\cos\theta)$$

For $r \gg R$, the second term in parentheses is negligible, and therefore condition (ii) requires that

$$\sum_{l=0}^{\infty} A_l r^l P_l(\cos\theta) = -E_0 r \cos\theta$$

Evidently only one term occurs: $l = 1$. In fact, since $P_1(\cos\theta) = \cos\theta$, we can read off immediately

$$A_1 = -E_0, \qquad \text{all other } A_l\text{'s zero.}$$

Conclusion:

$$V(r, \theta) = -E_0\left(r - \frac{R^3}{r^2}\right)\cos\theta \tag{3.71}$$

The first term ($-E_0 r \cos\theta$) is due to the external field; the contribution attributable to the induced charge is evidently

$$E_0 \frac{R^3}{r^2} \cos\theta$$

If you want to know the induced charge density, it can be calculated in the usual way:

$$\sigma(\theta) = -\epsilon_0 \left.\frac{\partial V}{\partial r}\right|_{r=R} = \epsilon_0 E_0\left(1 + 2\frac{R^3}{r^3}\right)\cos\theta\,\Big|_R = 3\epsilon_0 E_0 \cos\theta \tag{3.72}$$

As expected, it is positive in the "northern" hemisphere ($0 \leq \theta \leq \pi/2$) and negative in the "southern" ($\pi/2 \leq \theta \leq \pi$).

Example 9

A specified charge density $\sigma_0(\theta)$ is glued over the surface of a spherical shell of radius R. Find the resulting potential inside and outside the sphere.

Solution: You could, of course, do it by direct integration:

$$V = \frac{1}{4\pi\epsilon_0} \int \frac{\sigma_0}{\mathscr{r}}\, da$$

but separation of variables is frequently easier. For the interior region we have

$$V(r, \theta) = \sum_{l=0}^{\infty} A_l r^l P_l(\cos\theta) \qquad (r \leq R) \tag{3.73}$$

(no B_l terms—they blow up at the origin); for the exterior region

$$V(r, \theta) = \sum_{l=0}^{\infty} \frac{B_l}{r^{l+1}} P_l(\cos\theta) \qquad (r \geq R) \tag{3.74}$$

(no A_l terms—they don't go to zero at infinity). These two functions must be joined together by the appropriate boundary conditions at the surface itself. First, the potential is *continuous* at $r = R$ (equation 2.28):

$$\sum_{l=0}^{\infty} A_l R^l P_l(\cos\theta) = \sum_{l=0}^{\infty} \frac{B_l}{R^{l+1}} P_l(\cos\theta) \tag{3.75}$$

It follows that the coefficients of like Legendre polynomials are equal:

$$B_l = A_l R^{2l+1} \tag{3.76}$$

(To prove that formally, multiply both sides of (3.75) by $P_m(\cos\theta)\sin\theta$ and integrate from 0 to π, using the orthogonality relation (3.63).) Second, the radial derivative of V suffers a discontinuity at the surface (equation 2.30):

$$\left(\frac{\partial V_{\text{out}}}{\partial r} - \frac{\partial V_{\text{in}}}{\partial r}\right)\bigg|_{r=R} = -\frac{1}{\epsilon_0}\sigma_0(\theta) \tag{3.77}$$

Thus,

$$-\sum_{l=0}^{\infty} (l+1)\frac{B_l}{R^{l+2}} P_l(\cos\theta) - \sum_{l=0}^{\infty} l A_l R^{l-1} P_l(\cos\theta) = -\frac{1}{\epsilon_0}\sigma_0(\theta)$$

or, using (3.76):

$$\sum_{l=0}^{\infty} (2l+1) A_l R^{l-1} P_l(\cos\theta) = \frac{1}{\epsilon_0}\sigma_0(\theta) \tag{3.78}$$

From here, the coefficients can be evaluated by Fourier's trick:

$$A_m = \frac{1}{2\epsilon_0 R^{m-1}} \int_0^{\pi} \sigma_0(\theta) P_m(\cos\theta)\sin\theta\, d\theta \tag{3.79}$$

Equations (3.73) and (3.74) constitute the solution to our problem, with the coefficients given by (3.76) and (3.79).

For instance, if

$$\sigma_0(\theta) = k\cos\theta = kP_1(\cos\theta) \tag{3.80}$$

for some constant k, then all the A_m's are zero, except for $m = 1$, and

$$A_1 = \frac{k}{2\epsilon_0} \int_0^{\pi} [P_1(\cos\theta)]^2 \sin\theta\, d\theta = \frac{k}{3\epsilon_0}$$

The potential inside the sphere is therefore,

$$V(r, \theta) = \frac{k}{3\epsilon_0} r\cos\theta \qquad (r \leq R) \tag{3.81}$$

whereas outside the sphere

$$V(r, \theta) = \frac{kR^3}{3\epsilon_0} \frac{1}{r^2} \cos\theta \qquad (r \geq R) \tag{3.82}$$

In particular, if $\sigma_0(\theta)$ is the induced charge on a metal sphere in an external field $E_0\hat{k}$, so that $k = 3\epsilon_0 E_0$ (see (3.72)), then the potential inside is $E_0 r \cos\theta = E_0 z$, and the *field* is $-E_0\hat{k}$—exactly right to cancel off the external field, as of course it *should* be. Outside the sphere the potential due to this surface charge is

$$E_0 \frac{R^3}{r^2} \cos\theta$$

consistent with our conclusion in Example 8.

Problem 3.17 Derive $P_3(x)$ from the Rodrigues formula, and check that $P_3(\cos\theta)$ satisfies the angular equation (3.55) for $l = 3$. Check that P_3 and P_1 are orthogonal by explicit integration.

Problem 3.18 (a) Suppose the potential is a *constant* V_0 over the surface of a sphere. Use the results of Example 6 and Example 7 to find the potential inside and outside the sphere. (Of course, you know the answers in advance—this is just a consistency check on the method.)

(b) Find the potential inside and outside a spherical shell which carries a *uniform* surface charge σ_0, using the results of Example 9.

Problem 3.19 The potential at the surface of a sphere is given by

$$V_0(\theta) = k\cos(3\theta)$$

where k is some constant. Find the potential inside and outside the sphere, as well as the surface charge density $\sigma(\theta)$ on the sphere. (Assume there's no charge inside or outside the sphere.)

Problem 3.20 Suppose the potential $V_0(\theta)$ at the surface of a sphere is specified, and there is no charge inside or outside the sphere. Show that the charge density on the sphere is given by

$$\sigma(\theta) = \frac{\epsilon_0}{2R} \sum_{l=0}^{\infty} (2l+1)^2 C_l P_l(\cos\theta) \tag{3.83}$$

where

$$C_l = \int_0^{\pi} V_0(\theta) P_l(\cos\theta) \sin\theta \, d\theta \tag{3.84}$$

Problem 3.21 Find the potential outside a *charged* metal sphere (charge Q, radius R) placed in an otherwise uniform electric field $\mathbf{E}_0$. Explain clearly where you are setting the zero of potential.

Problem 3.22 In Problem 2.26, you found the potential on the axis of a uniformly charged disc:

$$V(r, 0) = \frac{\sigma}{2\epsilon_0}\left(\sqrt{r^2 + R^2} - r\right)$$

(a) Use this, together with the fact that $P_l(1) = 1$, to evaluate the first three terms in the expansion (3.67) for the potential of the disc at points *off* the axis, for $r > R$.
(b) Find the potential for $r < R$ by the same method, using (3.61) (*Note*: You must break the interior region up into two hemispheres, above and below the disc. Do *not* assume the coefficients A_l are the same in both hemispheres.)

Problem 3.23 A spherical shell of radius R carries a uniform surface charge σ_0 on the "northern" hemisphere and a uniform surface charge $-\sigma_0$ on the "southern" hemisphere. Find the potential inside and outside the sphere, calculating the coefficients explicitly up to A_6 and B_6.

• **Problem 3.24** Solve Laplace's equation by separation of variables in *cylindrical* coordinates, assuming there is no dependence on z (cylindrical symmetry). Make sure you find *all* solutions to the radial equation. Does your result accommodate the case of an infinite line charge?

Problem 3.25 Find the potential outside an infinitely long metal pipe, of radius R, placed at right angles to an otherwise uniform electric field $\mathbf{E}_0$. Find the surface charge induced on the pipe.

Problem 3.26 Charge density

$$\sigma = a \sin 5\phi$$

is glued over the surface of an infinite cylinder of radius R (Fig. 3.24). Find the potential inside and outside the cylinder.

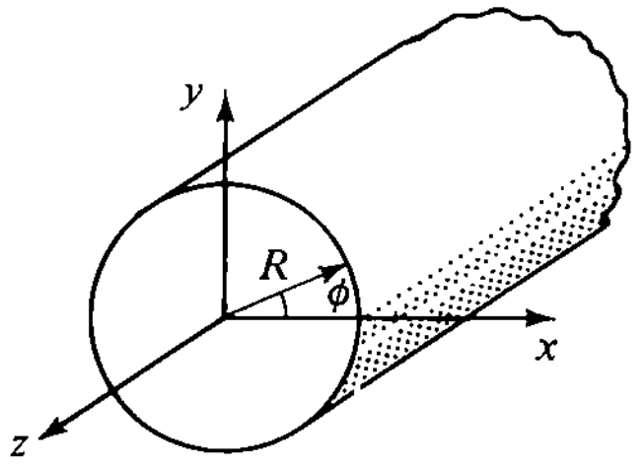

Figure 3.24

3.4 MULTIPOLE EXPANSION

3.4.1 Approximate Potentials at Large Distances

If you are very far away from a localized charge distribution, it "looks" like a point charge, and the potential is (to good approximation) $(1/4\pi\epsilon_0)\, Q/r$, where Q is the total charge. We have often used this observation as a check on formulas for V. But what if Q is *zero*? You might respond that the potential is then approximately zero, and of course, you're *right*, in a sense (indeed, the potential at large r is *pretty small* even if Q is *not* zero). But we're looking for something a bit more informative than that.

Example 10

A (physical) **electric dipole** consists of two equal and opposite charges ($\pm q$) separated by a distance s. Find the approximate potential at a point P far away.

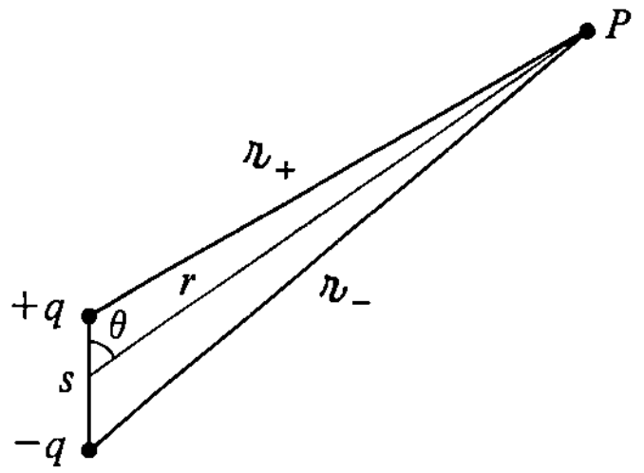

Figure 3.25

Solution: Let $\mathscr{r}_-$ be the distance from $-q$ and $\mathscr{r}_+$ the distance from $+q$ (Fig. 3.25). Then

$$V(P) = \frac{1}{4\pi\epsilon_0}\left(\frac{q}{\mathscr{r}_+} - \frac{q}{\mathscr{r}_-}\right)$$

and (from the law of cosines)

$$\mathscr{r}_\pm^2 = r^2 + \left(\frac{s}{2}\right)^2 \mp rs\cos\theta = r^2\left(1 \mp \frac{s}{r}\cos\theta + \frac{s^2}{4r^2}\right)$$

We're interested in the regime $r \gg s$ (the drawing is misleading, in this respect), so the third term is negligible, and the binomial expansion yields

$$\frac{1}{\mathscr{r}_\pm} \cong \frac{1}{r}\left(1 \mp \frac{s}{r}\cos\theta\right)^{-1/2} \cong \frac{1}{r}\left(1 \pm \frac{s}{2r}\cos\theta\right)$$

Thus

$$\left(\frac{1}{\mathscr{r}_+} - \frac{1}{\mathscr{r}_-}\right) \cong \frac{s}{r^2}\cos\theta$$

and hence

$$V(P) \cong \frac{1}{4\pi\epsilon_0}\frac{qs\cos\theta}{r^2} \qquad (3.85)$$

Evidently the potential of a dipole goes like $1/r^2$ at large r; as we might have anticipated, it falls off *more rapidly* than the potential of a distribution with nonzero total charge. Incidentally, if we put together a pair of equal and opposite *dipoles* to make a **quadrupole,** the potential goes like $1/r^3$; for back-to-back *quadrupoles* (an **octopole**), it goes like $1/r^4$; and so on. (Figure 3.26 summarizes this hierarchy; for completeness I have included the electric **monopole** (point charge), whose potential, of course, goes like $1/r$.)

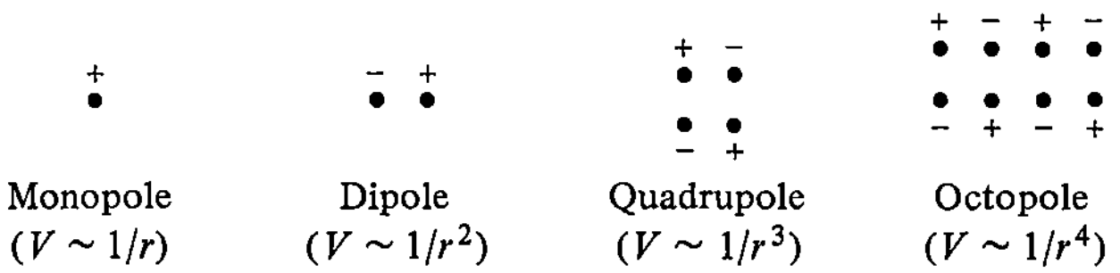

Figure 3.26

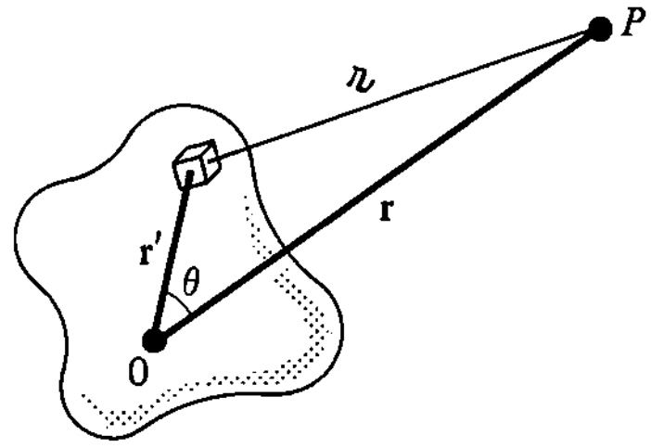

Figure 3.27

I propose now to develop a *systematic expansion for the potential of an arbitrary localized charge distribution, in powers of 1/r*. Figure 3.27 shows the configuration; the potential at P is given by

$$V(P) = \frac{1}{4\pi\epsilon_0}\int\left(\frac{1}{\imath}\right)\rho\,d\tau \tag{3.86}$$

where $\imath$ is the distance from $d\tau$ to P. Using the law of cosines, as before, we have

$$\imath^2 = r^2 + (r')^2 - 2rr'\cos\theta = r^2\left(1 + \left(\frac{r'}{r}\right)^2 - 2\left(\frac{r'}{r}\right)\cos\theta\right)$$

or

$$\imath = r\sqrt{1+\epsilon} \tag{3.87}$$

where

$$\epsilon = \left(\frac{r'}{r}\right)\left(\frac{r'}{r} - 2\cos\theta\right)$$

As long as P lies well outside the charge distribution, ϵ is always much less than 1, and this invites a binomial expansion:

$$\frac{1}{\imath} = \frac{1}{r}(1+\epsilon)^{-1/2} = \frac{1}{r}\left(1 - \frac{1}{2}\epsilon + \frac{3}{8}\epsilon^2 - \frac{5}{16}\epsilon^3 + \cdots\right) \tag{3.88}$$

or, in terms of r, θ, and r':

$$\begin{aligned}\frac{1}{\imath} &= \frac{1}{r}\left[1 - \frac{1}{2}\left(\frac{r'}{r}\right)\left(\frac{r'}{r} - 2\cos\theta\right) + \frac{3}{8}\left(\frac{r'}{r}\right)^2\left(\frac{r'}{r} - 2\cos\theta\right)^2\right.\\ &\qquad\left. - \frac{5}{16}\left(\frac{r'}{r}\right)^3\left(\frac{r'}{r} - 2\cos\theta\right)^3 + \cdots\right]\\ &= \frac{1}{r}\left[1 + \left(\frac{r'}{r}\right)(\cos\theta) + \left(\frac{r'}{r}\right)^2\left(\frac{3}{2}\cos^2\theta - \frac{1}{2}\right)\right.\\ &\qquad\left. + \left(\frac{r'}{r}\right)^3\left(\frac{5}{2}\cos^3\theta - \frac{3}{2}\cos\theta\right) + \cdots\right]\end{aligned}$$

In the last step I have collected together like powers of (r'/r); curiously enough, their coefficients (the terms in parentheses) are Legendre polynomials:

$$\frac{1}{\imath} = \frac{1}{r}\sum_{n=0}^{\infty}\left(\frac{r'}{r}\right)^n P_n(\cos\theta) \tag{3.89}$$

Putting this expression into equation (3.86), and noting that r is a constant, as far as the integration is concerned, I conclude that

$$V(P) = \frac{1}{4\pi\epsilon_0} \sum_{n=0}^{\infty} \frac{1}{r^{(n+1)}} \int (r')^n P_n(\cos\theta)\, \rho\, d\tau \tag{3.90}$$

or, more explicitly,

$$V(P) = \frac{1}{4\pi\epsilon_0}\left[\frac{1}{r}\int \rho\, d\tau + \frac{1}{r^2}\int r' \cos\theta\, \rho\, d\tau + \frac{1}{r^3}\int (r')^2\left(\frac{3}{2}\cos^2\theta - \frac{1}{2}\right)\rho\, d\tau + \cdots\right] \tag{3.91}$$

This is the desired result—the **multipole expansion** of V in powers of $1/r$. The first term ($n = 0$) is called the **monopole term** (it goes like $1/r$); the second ($n = 1$) is called the **dipole term** (it goes like $1/r^2$); the third is the **quadrupole term**; the fourth **octopole**; and so on. As it stands, equation (3.90) is *exact*, but it is *useful* primarily as an *approximation* scheme: The lowest nonzero term in the expansion provides the approximate potential at large r, and the successive terms tell us how to improve the approximation if greater precision is required.

Problem 3.27 A sphere of radius a centered at the origin carries charge density

$$\rho(r, \theta) = \rho_0 \frac{a}{r^2}(a - 2r)\sin\theta$$

where ρ_0 is a constant, and r, θ are the usual spherical coordinates. Find the approximate potential for points on the z axis, far from the sphere.

3.4.2 The Monopole and Dipole Terms

Ordinarily, the multipole expansion is dominated (at large r) by the monopole term:

$$V_{\text{mon}}(P) = \frac{1}{4\pi\epsilon_0}\frac{Q}{r} \tag{3.92}$$

where $Q = \int \rho\, d\tau$ is the total charge of the configuration. This is just what we expected for the approximate potential at large distances from the charge. Incidentally, for a *point* charge Q *at the origin*, V_{mon} represents the *exact* potential everywhere, not merely a first approximation at large r. Evidently in this case all the higher multipoles vanish.

If the total charge is zero, then the dominant term in the potential will be the dipole (unless, of course, it *also* vanishes):

$$V_{\text{dip}}(P) = \frac{1}{4\pi\epsilon_0}\frac{1}{r^2}\int (r'\cos\theta)\rho\, d\tau$$

Since θ is the angle between $\mathbf{r}'$ and $\mathbf{r}$,

$$r' \cos\theta = \hat{r}\cdot\mathbf{r}'$$

and the dipole potential can be written more succinctly as:

$$V_{\text{dip}}(P) = \frac{1}{4\pi\epsilon_0}\frac{1}{r^2}\hat{r}\cdot\int \mathbf{r}'\rho\,d\tau$$

This integral, which does not depend at all on the location of P, is called the **dipole moment** of the distribution:

$$\boxed{\mathbf{p} = \int \mathbf{r}'\rho\,d\tau} \tag{3.93}$$

With this notation, the dipole contribution to the potential reads

$$\boxed{V_{\text{dip}}(P) = \frac{\mathbf{p}\cdot\hat{r}}{4\pi\epsilon_0 r^2}} \tag{3.94}$$

The dipole moment is determined by the geometry (size, shape, and density) of the charge distribution. Equation (3.93) translates in the usual way (Section 2.1.4) for point, line, and surface charges. Thus, the dipole moment of a collection of *point* charges is

$$\mathbf{p} = \sum_{i=1}^{n} q_i\mathbf{r}_i' \tag{3.95}$$

For the physical dipole itself (equal and opposite charges, $\pm q$)

$$\mathbf{p} = q\mathbf{r}_+' - q\mathbf{r}_-' = q(\mathbf{r}_+' - \mathbf{r}_-') = q\mathbf{s} \tag{3.96}$$

where $\mathbf{s}$ is the vector from the negative charge to the positive one (Fig. 3.28).

If you put (3.96) into (3.94), you will recover (3.85). Notice, however, that this is only the *approximate* potential of the dipole—evidently there are higher multipole contributions. Of course, as you go farther and farther away, V_{dip} becomes a better and better approximation, since the higher terms die off more rapidly with increasing r. By the same token, at a fixed r the dipole approximation improves as you shrink

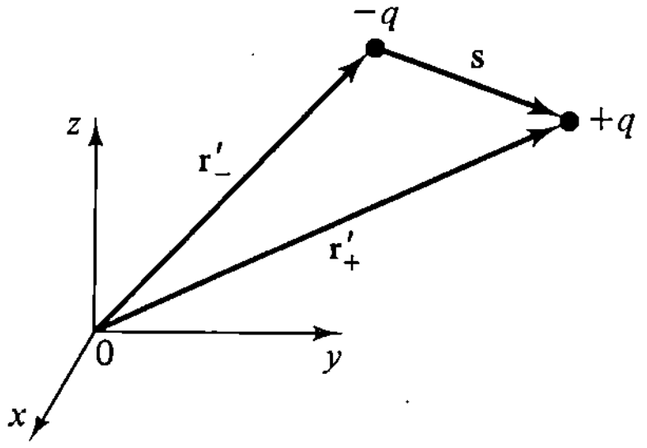

Figure 3.28

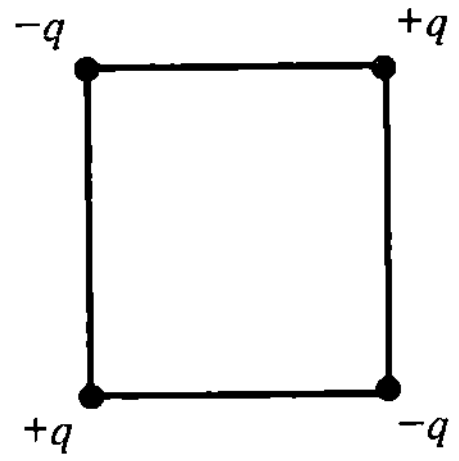

Figure 3.29

the separation (s). To construct a "pure" dipole whose potential is *exactly* (3.94), you'd have to let s approach zero. Unfortunately, you then lose the dipole term *too*, unless you simultaneously arrange for q to go to infinity! A *physical* dipole becomes a *pure* dipole, then, in the rather artificial limit $s \to 0$, $q \to \infty$, with the product $qs = p$ held fixed. (When someone uses the word "dipole," you can't always tell whether they mean a *physical* dipole (with finite separation between the charges) or a *pure* (point) dipole. If in doubt, assume that s is small enough (compared to r) that you can safely apply (3.94).)

Dipole moments are *vectors*, and they add accordingly: If you have two dipoles, $\mathbf{p}_1$ and $\mathbf{p}_2$, the total dipole moment is $\mathbf{p}_1 + \mathbf{p}_2$. For instance, with four charges at the corners of a square, as shown in Fig. 3.29, the net dipole moment is zero. You can see this by combining the charges in pairs (vertically, $\downarrow + \uparrow = 0$; or horizontally, $\rightarrow + \leftarrow = 0$) or by adding up the four contributions individually, using (3.95). (This is a *quadrupole*, as I indicated earlier, and its potential is dominated by the quadrupole term in the multipole expansion.)

Problem 3.28 Four particles (one of charge q, one of charge $3q$, and two of charge $-2q$) are placed as shown in Fig. 3.30, each a distance d from the origin. Find a simple approximate formula for the potential, valid at a point P far from the origin. (Express your answer in terms of the spherical coordinates (r, θ, ϕ) of P.)

Problem 3.29 In Example 9 we derived the exact potential of a spherical shell of radius R which carries a surface charge $\sigma = k \cos\theta$

(a) Calculate the dipole moment of this charge distribution.

(b) Find the approximate potential, at points far from the sphere, and compare the exact answer (3.82). What can you conclude about the higher multipoles?

Problem 3.30 For the dipole in Example 10, expand $1/\mathscr{r}_\pm$ to order $(s/r)^3$, and use this to determine the quadrupole and octopole terms in the potential.

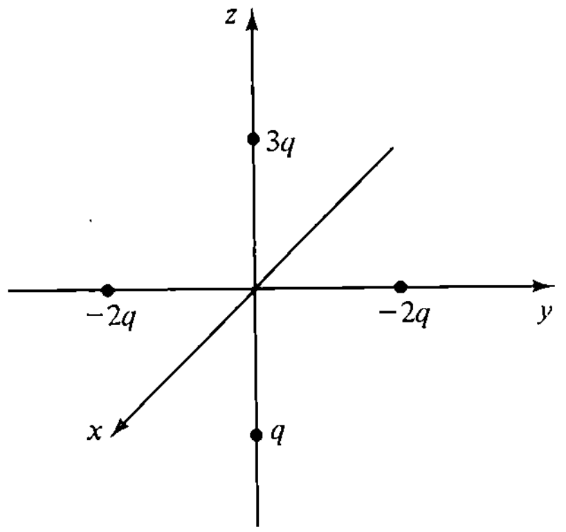

Figure 3.30

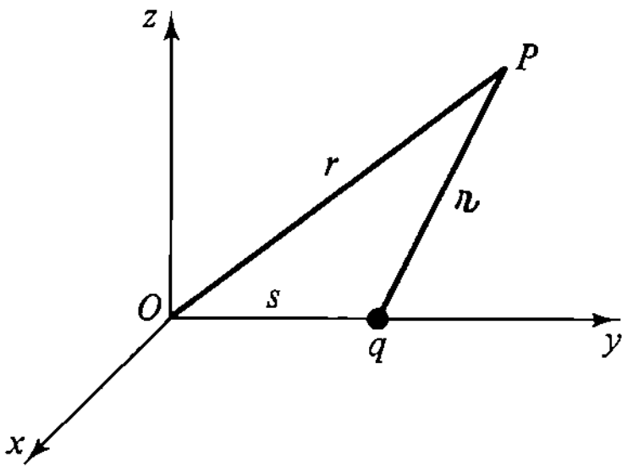

Figure 3.31

3.4.3 Origin of Coordinates in Multipole Expansions

I mentioned earlier that a point charge at the origin constitutes a "pure" monopole. If it is *not* at the origin, it's no longer a pure monopole. For instance, the charge in Fig. 3.31 has a dipole moment $\mathbf{p} = qs\hat{j}$ and a corresponding dipole term in its potential. The monopole potential $(1/4\pi\epsilon_0)\, q/r$ is not quite correct for this configuration; rather, the exact potential is $(1/4\pi\epsilon_0) q/\mathcal{R}$. The multipole expansion is, remember, a series in inverse powers of r (the distance to the *origin*), and when we expand $1/\mathcal{R}$ we get *all* powers, not just the first.

So moving the origin (or, what amounts to the same thing, moving the *charge*) can radically alter a multipole expansion. The monopole moment Q does not change, since the total charge is obviously independent of the coordinate system. (In Fig. 3.31 the monopole term was unaffected when we moved q away from the origin—it's just that it was no longer the whole story: a dipole term appeared as well.) Ordinarily, the dipole moment *does* change when you shift the origin, but there is an important exception: *If the total charge is zero, then the dipole moment is independent of the choice of origin.* For suppose we displace the origin by an amount $\mathbf{d}$ (Fig. 3.32). The new dipole moment is then

$$\bar{\mathbf{p}} = \int \bar{\mathbf{r}}' \rho \, d\tau = \int (\mathbf{r}' - \mathbf{d}) \rho \, d\tau$$

$$= \int \mathbf{r}' \rho \, d\tau - \mathbf{d} \int \rho \, d\tau = \mathbf{p} - \mathbf{d}Q$$

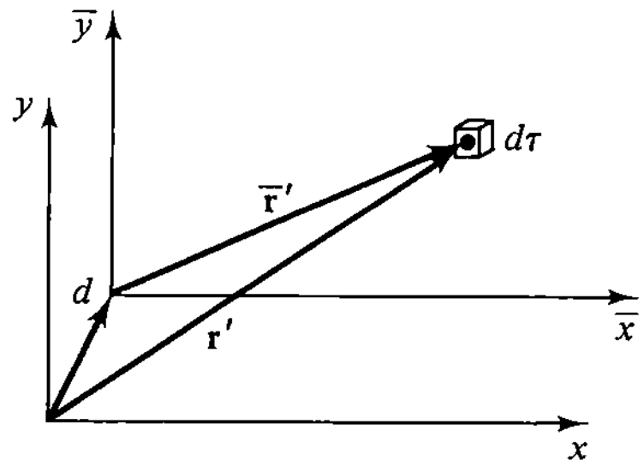

Figure 3.32

In particular, if $Q = 0$, then $\bar{\mathbf{p}} = \mathbf{p}$. So if someone asks for the dipole moment of

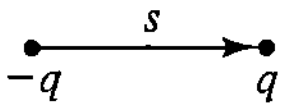

you can tell him it is $q\mathbf{s}$, but if he asks you for the dipole moment of

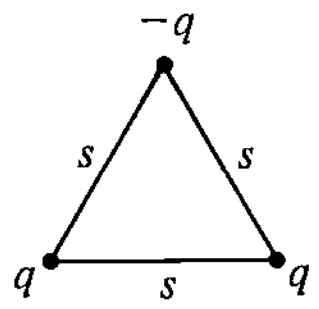

the appropriate response would be: "With respect to *what origin*?"

3.4.4 The Electric Field of a Dipole

So far we have worked only with *potentials*. Now I would like to calculate the electric *field* of a (pure) dipole. If we choose coordinates so that $\mathbf{p}$ lies at the origin and points in the z direction (Fig. 3.33), then the potential at r, θ is given by equation (3.94) as

$$V_{\text{dip}}(r, \theta) = \frac{\hat{r} \cdot \mathbf{p}}{4\pi\epsilon_0 r^2} = \frac{p \cos\theta}{4\pi\epsilon_0 r^2} \tag{3.97}$$

To get the field, we take the negative gradient of V:

$$E_r = -\frac{\partial V}{\partial r} = \frac{2p\cos\theta}{4\pi\epsilon_0 r^3}$$

$$E_\theta = -\frac{1}{r}\frac{\partial V}{\partial \theta} = \frac{p\sin\theta}{4\pi\epsilon_0 r^3}$$

$$E_\phi = -\frac{1}{r\sin\theta}\frac{\partial V}{\partial \phi} = 0$$

Thus,

$$\mathbf{E}_{\text{dip}}(r, \theta) = \frac{p}{4\pi\epsilon_0 r^3}(2\cos\theta\,\hat{r} + \sin\theta\,\hat{\theta}) \tag{3.98}$$

Notice that the dipole field falls off as the inverse *cube* of r; the *monopole* field $(Q/4\pi\epsilon_0 r^2)\hat{r}$ goes as the inverse *square*, of course. Quadrupole fields go like $1/r^4$,

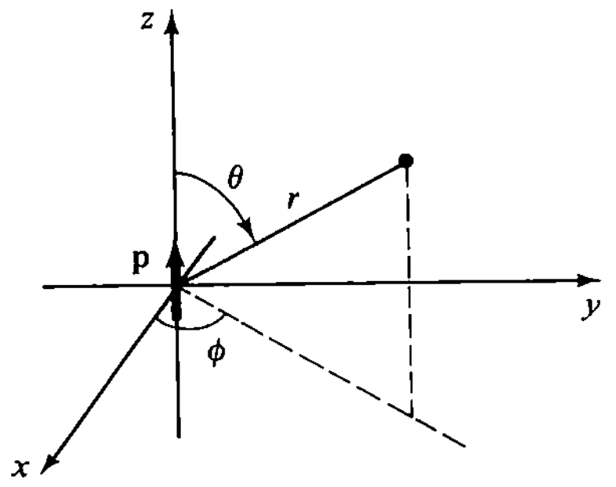

Figure 3.33

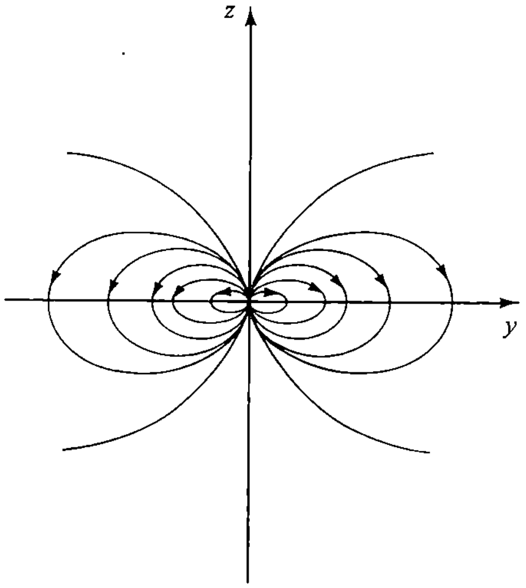

(a) Field of a "pure" dipole

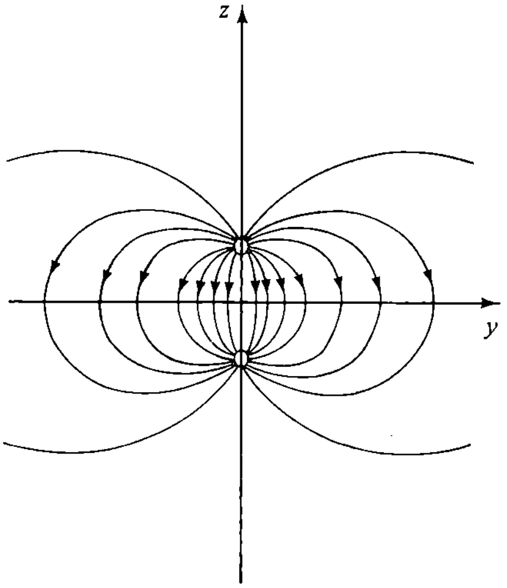

(b) Field of a "physical" dipole

Figure 3.34

octopole like $1/r^5$, and so on. (This merely reflects the fact that monopole *potentials* fall off like $1/r$, dipole like $1/r^2$, quadrupole like $1/r^3$, and so on—the gradient throws in another factor of $1/r$.)

Figure 3.34(a) shows the field lines of a "pure" dipole (equation (3.98)). For comparison, I have also sketched the field lines for a "physical" dipole, in Fig. 3.34(b). Notice how similar the two pictures become if you blot out the central region; up close, however, they are entirely different. Only for points $r \gg s$ does (3.98) represent a valid approximation to the field of a physical dipole. This regime can be reached either by going to large r or by squeezing the charges very close together.

Problem 3.31 A "pure" dipole p is situated at the origin, pointing in the z direction.

(a) What is the force on a point charge q at $(a, 0, 0)$ (Cartesian coordinates)?

(b) What is the force on q at $(0, 0, a)$?

(c) How much work does it take to move q from $(a, 0, 0)$ to $(0, 0, a)$?

Problem 3.32 Three point charges are situated as shown in Fig. 3.35, each a distance d from

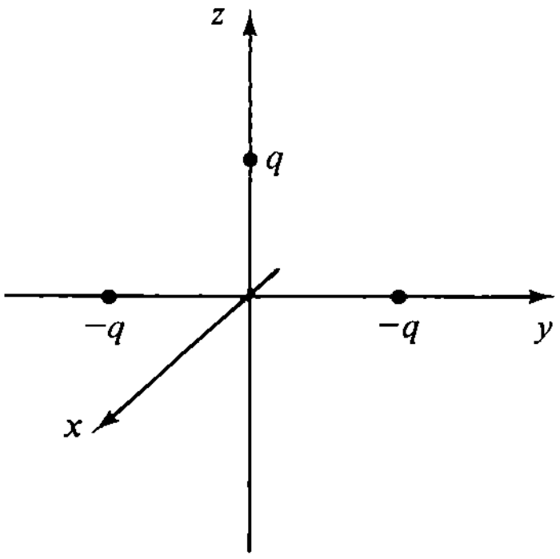

Figure 3.35

the origin. Find the approximate electric field at points far from the origin. Express your answer in spherical coordinates, and include the two lowest orders in the multipole expansion.

• **Problem 3.33** Show that the electric field of a ("pure") dipole can be written

$$\mathbf{E} = \frac{1}{4\pi\epsilon_0}\frac{1}{r^3}[3(\mathbf{p}\cdot\hat{r})\hat{r} - \mathbf{p}] \tag{3.99}$$

(This form has the advantage—over (3.98)—that it does not commit you to a particular coordinate system.)

Further Problems on Chapter 3

Problem 3.34 A point charge q of mass m is released from rest at a distance d from an infinite grounded conducting plane. How long will it take for the charge to hit the plane?

$$\left(\text{Answer: } \frac{\pi}{q}\sqrt{\frac{md^3}{2}}\right)$$

Problem 3.35 Two infinite parallel grounded conducting planes are held a distance a apart. A point charge q is placed in the region between them, a distance x from one plate. Find the force on q. Check that your answer is correct for the special cases $a \to \infty$ and $x = a/2$.

Problem 3.36 Two long straight wires, carrying opposite uniform line charges $\pm\lambda$, are situated on either side of a long conducting cylinder (Fig. 3.36). The cylinder (which carries no net charge) has radius R, and the wires are a distance s from the axis. Find the potential at point P.

$$\left(\text{Answer: } V(r,\theta) = \frac{\lambda}{4\pi\epsilon_0}\ln\left\{\frac{(r^2+s^2+2rs\cos\theta)[(rs/R)^2+R^2-2rs\cos\theta]}{(r^2+s^2-2rs\cos\theta)[(rs/R)^2+R^2+2rs\cos\theta]}\right\}\right)$$

Problem 3.37 A conducting sphere of radius a, at potential V_0, is surrounded by a thin concentric spherical shell of radius b, over which someone has glued a surface charge

$$\sigma(\theta) = \sigma_0 \cos\theta$$

(σ_0 is a constant, and θ is the usual spherical coordinate).

(a) Find the potential in each region:

(i) $r > b$

(ii) $a < r < b$

(b) Find the induced surface charge $\sigma_i(\theta)$ on the conductor.

(c) What is the total charge of this system: Check that your answer is consistent with the behavior of V at large r.

$$\left(\text{Answer: } V = \begin{cases} aV_0/r + (b^3 - a^3)\sigma_0\cos\theta/3r^2\epsilon_0, & r \geq b \\ aV_0/r + (r^3 - a^3)\sigma_0\cos\theta/3r^2\epsilon_0, & r \leq b \end{cases}\right)$$

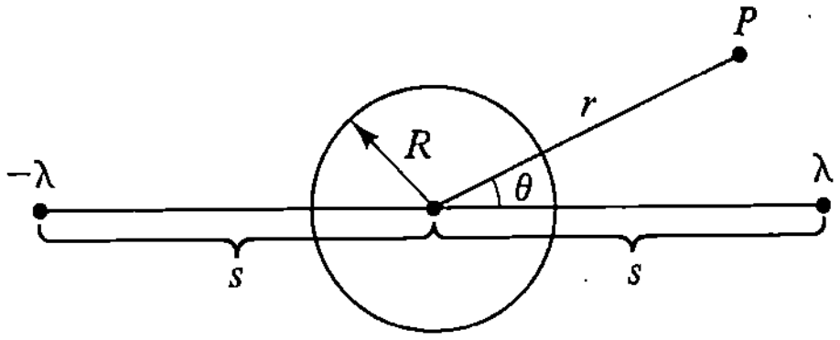

Figure 3.36

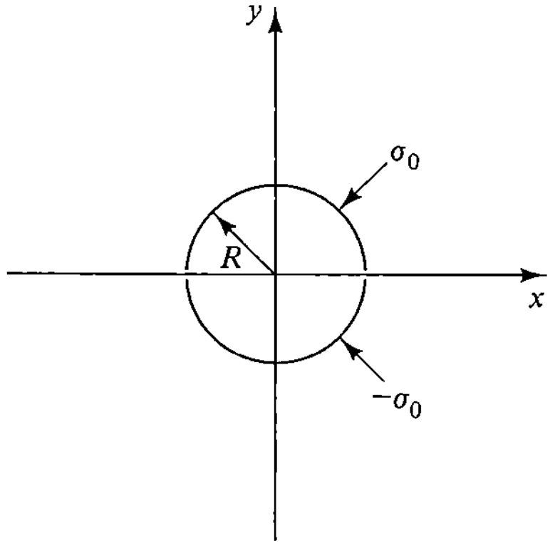

Figure 3.37

Problem 3.38 A charge $+Q$ is distributed uniformly along the z axis from $z = -a$ to $z = +a$. Show that the electric potential at a point (r, θ) is given by:

$$V(r, \theta) = \frac{Q}{4\pi\epsilon_0}\frac{1}{r}\left[1 + \frac{1}{3}\left(\frac{a}{r}\right)^2 P_2(\cos\theta) + \frac{1}{5}\left(\frac{a}{r}\right)^4 P_4(\cos\theta) + \cdots\right]$$

for $r > a$. (*Hint*: $\log((1 + x)/(1 - x)) = 2[x + (x^3/3) + (x^5/5) + (x^7/7) + \cdots]$, for $|x| < 1$.)

Problem 3.39 A long cylindrical shell of radius R carries a uniform surface charge σ_0 on the upper half and an opposite charge $-\sigma_0$ on the lower half (Fig. 3.37). Find the electric potential inside and outside the cylinder.

Problem 3.40 A thin insulating rod, running from $z = -a$ to $z = +a$, carries the indicated line charges. In each case, find the leading term in the multipole expansion of the potential. (a) $\lambda = \lambda_0 \cos(\pi z/2a)$; (b) $\lambda = \lambda_0 \sin(\pi z/a)$; (c) $\lambda = \lambda_0 \cos(\pi z/a)$ (λ_0 is a constant).

• **Problem 3.41** Show that the *average* field inside a sphere of radius R, due to all the charge within the sphere, is

$$E_{\text{ave}} = -\frac{1}{4\pi\epsilon_0}\frac{\mathbf{p}}{R^3} \tag{3.100}$$

where $\mathbf{p}$ is the total dipole moment. There are several ways to prove this delightfully simple result. Here's one method:[10]

(a) Show that the average field due to a single charge q at point P (inside the sphere) is the same as the field at P due to a uniformly charged sphere with $\rho = -q/(\frac{4}{3}\pi R^3)$, namely,

$$\frac{1}{4\pi\epsilon_0}\frac{1}{(\frac{4}{3}\pi R^3)}\int \frac{q}{\mathscr{r}^2}\hat{\mathscr{r}}\, d\tau$$

where $\mathscr{r}$ is the vector from P to $d\tau$.

(b) The latter may be found from Gauss's law (see Problem 2.12). Express the answer in terms of the dipole moment of q.

(c) Use the superposition principle to generalize to an arbitrary charge distribution.

(d) What is the average field, within a sphere, due to all charge *outside* the sphere?

Problem 3.42 (a) Using equation 3.98, calculate the average electric field of a dipole, over a sphere of radius R centered at the origin. Do the angular integrals first. (*Note*: You

[10] For a more direct approach see P. Lorrain and D. R. Corson, *Electromagnetic Fields and Waves*, 2d ed., San Francisco: (W. H. Freeman & Co., 1970) Sec. 2.13.

must express $\hat{r}$ and $\hat{\theta}$ in terms of $\hat{i}$, $\hat{j}$, and $\hat{k}$ (see back cover) before integrating. If you don't understand why, reread the comments on page (41) and footnote 3 in Chapter 2.) Compare your answer with the general theorem in Problem 3.41.

(*Comment*: The discrepancy here is related to the fact that the field of a dipole blows up at $r = 0$. The angular integral is zero, but the radial integral is infinite, so we really don't know *what* to make of the answer. To resolve the ambiguity, let's say that 3.98 applies *outside a tiny sphere of radius* ϵ*—its* contribution to E_{ave} is then *rigorously* zero, and the whole answer has to come from the field *inside* the ϵ-sphere.)

(b) What must the field *inside* the ϵ-sphere be, in order for the general theorem (3.100) to hold? (*Hint*: Since ϵ is arbitrarily small, we're talking about something that is infinite at $r = 0$ and whose integral over an infinitesimal volume is finite.)[11]

$$\left(\textit{Answer}: \frac{-\mathbf{p}}{3\epsilon_0}\,\delta^3(\mathbf{r})\right)$$

Problem 3.43

(a) Suppose a charge distribution ρ_1 produces a potential $V_1(\mathbf{r})$, and some other charge distribution ρ_2 produces a potential $V_2(\mathbf{r})$. (The two situations may have nothing in common, for all I care—perhaps number 1 is a uniformly charged sphere and number 2 is a parallel-plate capacitor. Please understand that ρ_1 and ρ_2 are not present *at the same time;* we are talking about two *different problems,* one in which only ρ_1 is present, and another in which only ρ_2 is present.) Prove **Green's reciprocity theorem:**

$$\int_{\text{all space}} \rho_1 V_2\, d\tau = \int_{\text{all space}} \rho_2 V_1\, d\tau$$

(*Hint*: Evaluate $\int \mathbf{E}_1 \cdot \mathbf{E}_2\, d\tau$ two ways, first writing $\mathbf{E}_1 = -\nabla V_1$ and using integration-by-parts to transfer the derivative to $\mathbf{E}_2$, then writing $\mathbf{E}_2 = -\nabla V_2$ and transferring the derivative to $\mathbf{E}_1$.)

(b) Suppose now that you have two separated conductors (Fig. 3.38). If you charge up conductor (a) by an amount Q (leaving (b) uncharged) the resulting potential of (b) is, say, V_b. On the other hand, if you put the same charge Q on conductor (b) (leaving (a) uncharged), the potential of (a) would be, say, V_a. Use Green's reciprocity theorem to show that $V_a = V_b$ (an astonishing result, since we assumed nothing about the shapes or placement of the conductors).

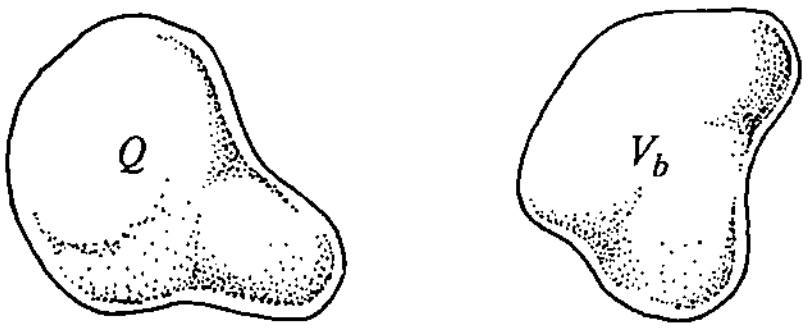

(a) (b) **Figure 3.38**

Problem 3.44 Use Green's reciprocity theorem (Problem 3.43) to solve the following two problems. (*Hint*: for distribution 1, use the actual situation; for distribution 2, remove q, and set one of the conductors at potential V_0.)

(a) Both plates of a parallel-plate capacitor are grounded, and a point charge q is placed between them at a distance x from plate 1. The plate separation is d. Find the induced charge on each plate.

[11]For details and applications, see D. J. Griffiths, *Am. J. Phys.* **50**, 698 (1982).

$\left(\textit{Answer}: Q_1 = q\left(\frac{x}{d} - 1\right); Q_2 = -q\,\frac{x}{d}\right)$

(b) Two concentric spherical conducting shells (radii a and b) are grounded, and a point charge q is placed between them (at radius r). Find the induced charge on each sphere.

Problem 3.45 (a) Show that the quadrupole term in the multipole expansion of V can be written

$$V_{\text{quad}} = \frac{1}{4\pi\epsilon_0}\,\frac{\left(\frac{1}{2}\sum\limits_{i,j=1}^{3} \hat{r}_i \hat{r}_j Q_{ij}\right)}{r^3}$$

where

$$Q_{ij} = \int (3r_i' r_j' - (r')^2 \delta_{ij})\rho\, d\tau$$

Here

$$\delta_{ij} = \begin{cases} 1 & \text{if } i = j \\ 0 & \text{if } i \neq j \end{cases}$$

is the **Kronecker delta.** Q_{ij} is the **quadrupole moment** of the charge distribution. Notice the hierarchy:

$$V_{\text{mono}} = \frac{1}{4\pi\epsilon_0}\frac{Q}{r}; \quad V_{\text{dip}} = \frac{1}{4\pi\epsilon_0}\frac{\sum \hat{r}_i p_i}{r^2}; \quad V_{\text{quad}} = \frac{1}{4\pi\epsilon_0}\frac{\frac{1}{2}\sum \hat{r}_i \hat{r}_j Q_{ij}}{r^3}; \ \ldots$$

The monopole moment (Q) is a scalar, the dipole moment ($\mathbf{p}$) is a vector, the quadrupole moment (Q_{ij}) is a second-rank tensor, and so on.

(b) Find all nine components of Q_{ij} for the configuration in Figure 3.29 (assume the square has side s and lies in the xy plane, centered at the origin).

(c) Show that the quadrupole moment is independent of origin if the monopole and dipole moments both vanish. (This works all the way up the hierarchy—the lowest nonzero multipole moment is always independent of origin.)

(d) How would you define the **octopole moment?** Express the octopole term in the multipole expansion in terms of the octopole moment.

Problem 3.46 In Example 8 we determined the electric field outside a spherical conductor (radius R) placed in a uniform external field $\mathbf{E}_0$. Solve the same problem now using the method of images, and check that your answer agrees with equation (3.71). (*Hint*: Use Example 2, but put another charge, $-q$, diametrically opposite q. Let $s \to \infty$, with $(1/4\pi\epsilon_0)(2q/s^2) = E_0$ held constant.)

4

ELECTROSTATIC FIELDS IN MATTER

4.1 POLARIZATION

4.1.1 Dielectrics

In this chapter we shall study the effects of electric fields on matter. Matter, of course, comes in many varieties—solids, liquids, gases, metals, woods, glasses—and these substances do not all respond in the same way to electrostatic fields. Nevertheless, *most* everyday objects belong (at least, in good approximation) to one of two large classes: **conductors** and **insulators** (or **dielectrics**). We have already talked about conductors; these are substances that contain an "unlimited" supply of charges that are free to move about through the material. In practice what this ordinarily means is that many of the electrons (one or two per atom in a typical metal) are not associated with any particular nucleus, but roam around at will. In dielectrics, by contrast, *all charges are attached to specific atoms or molecules*—they're on a tight leash, and all they can do is move a bit *within* the molecule. Such microscopic displacements are not as dramatic as the wholesale rearrangement of charge in a conductor, but their cumulative effects account for the characteristic behavior of dielectric materials. There are actually two principal mechanisms by which electric fields can distort the charge distribution of a dielectric atom or molecule: *stretching* and *rotating*. In the next two sections I'll discuss these processes.

4.1.2 Induced Dipoles

What happens to a neutral atom when it is placed in an electric field **E**? Your first guess might well be: Absolutely nothing—since the atom is not charged, the field has no effect on it. But that is incorrect. Although the atom as a whole is electrically

TABLE 4.1 ATOMIC POLARIZABILITIES

(The table lists $\frac{1}{4\pi\epsilon_0}\alpha$ in units of 10^{-30} m^3.)

Element:	H	He	Li	Be	C	Ne	Na	Ar	K
	0.66	0.21	12	9.3	1.5	0.4	27	1.6	34

Source: E. M. Purcell, *Electricity and Magnetism* (Berkeley Physics Course Vol. 2), 2d ed. (New York: McGraw-Hill, 1985), p. 363.

neutral, there *is* a positively charged core (the nucleus) and a negatively charged electron cloud surrounding it. These two regions of charge within the atom are influenced by the field: the nucleus is pushed in the direction of the field, and the electrons the opposite way. In principle, if the field is large enough, it can pull the atom apart completely, "ionizing" it (the substance then becomes a conductor). With less extreme fields, however, an equilibrium is soon established, for if the center of the electron cloud does not coincide with the nucleus, these positive and negative charges attract one another, and this holds the atom together. The two opposing forces—**E** pulling the electrons and nucleus apart, their mutual attraction drawing them together—reach a balance, leaving the atom **polarized,** with plus charge shifted slightly one way, and minus the other. The atom now has a tiny dipole moment **p** which points in the *same direction as* **E**. In fact, this induced dipole moment is approximately proportional to the field (as long as the latter is not too strong):

$$\mathbf{p} = \alpha \mathbf{E} \tag{4.1}$$

The constant of proportionality α is called the **atomic polarizability.** Its value depends on the detailed structure of the atom in question. Table 4.1 lists some experimentally determined atomic polarizabilities.

Example 1

A primitive model for an atom consists of a point nucleus $(+q)$ surrounded by a uniformly charged spherical cloud $(-q)$ of radius a (Fig. 4.1). Calculate the atomic polarizability of such an atom.

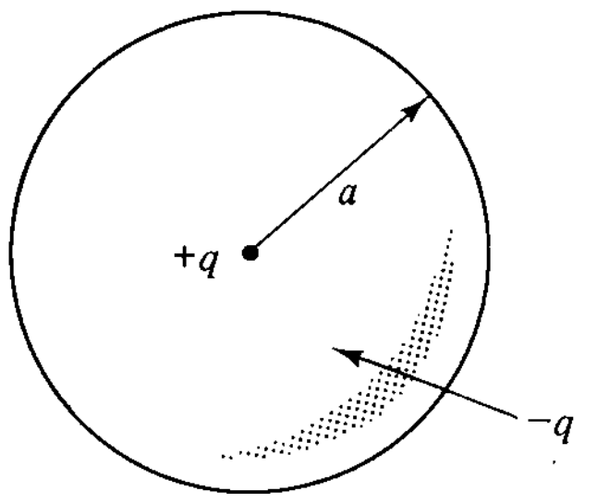

Figure 4.1

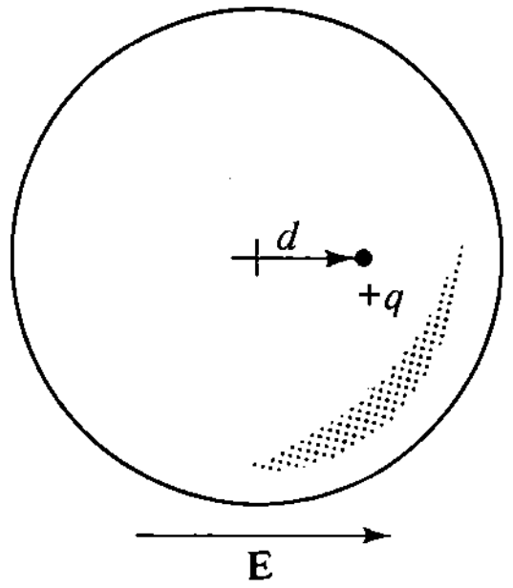

Figure 4.2

Solution: In the presence of an external field **E**, the nucleus will be shifted slightly to the right and the electron cloud to the left as shown in Fig. 4.2. (Because the actual displacements involved are extremely small, as you'll see in

Problem 4.1, it is reasonable to assume that the electron cloud retains its spherical shape as it moves.) Say that equilibrium occurs when the nucleus is displaced a distance d from the center of the sphere. At that point the external field pushing the nucleus to the right exactly balances the internal field pulling it to the left: $E = E_e$, where E_e is the field produced by the electron cloud. Now the field is at a distance d from the center of a uniformly charged sphere is

$$E_e = \frac{1}{4\pi\epsilon_0}\frac{qd}{a^3}$$

as you can easily check by using Gauss's law (see Problem 2.12). At equilibrium, then,

$$E = \frac{1}{4\pi\epsilon_0}\frac{qd}{a^3}, \quad \text{or} \quad p = qd = (4\pi\epsilon_0 a^3)E$$

The atomic polarizability is therefore

$$\alpha = 4\pi\epsilon_0 a^3 = 3\epsilon_0 v \tag{4.2}$$

where v is the volume of the atom. Although this atomic model is extremely crude, the result (4.2) is not too bad—it's accurate to within a factor of four or so for many simple atoms.

[For molecules the situation is not quite so simple, because frequently they polarize more readily in some directions than others. Carbon dioxide (Fig. 4.3), for instance, has a polarizability of $4.5 \times 10^{-40}\ C^2 \cdot m/N$ when you apply the field along the axis of the molecule, but only 2×10^{-40} for fields perpendicular to this direction. When the field is at some *angle* to the axis, you must resolve it into parallel and perpendicular components, and multiply each by the pertinent polarizability:

$$\mathbf{p} = \alpha_\perp \mathbf{E}_\perp + \alpha_\parallel \mathbf{E}_\parallel$$

In this case the induced dipole moment may not even be in the same *direction* as **E**. And CO_2 is relatively simple, as molecules go, since the atoms at least arrange themselves in a straight line; for a completely asymmetrical molecule equation (4.1) takes the form of the most general linear relation between **E** and **p**:

$$\left.\begin{aligned} p_x &= \alpha_{xx}E_x + \alpha_{xy}E_y + \alpha_{xz}E_z \\ p_y &= \alpha_{yx}E_x + \alpha_{yy}E_y + \alpha_{yz}E_z \\ p_z &= \alpha_{zx}E_x + \alpha_{zy}E_y + \alpha_{zz}E_z \end{aligned}\right\} \tag{4.3}$$

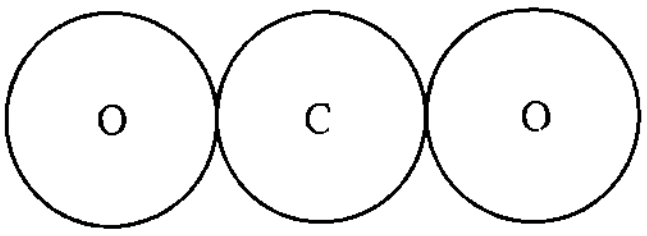

Figure 4.3

The set of nine constants α_{ij} constitute the **polarizability tensor** for the molecule. Their actual values depend on the orientation of the axes you chose, though it is possible to select axes such that all the "off-diagonal" terms (α_{xy}, α_{zx}, etc.) are zero, leaving just three "principal" polarizabilities: α_{xx}, α_{yy}, and α_{zz}.]

Problem 4.1 A hydrogen atom (with the Bohr radius of half an angstrom) is situated between two metal plates 1 mm apart, which are connected to opposite terminals of a 500-V battery. What fraction of the atomic radius does the separation distance d amount to, roughly? Estimate the voltage you would need with this apparatus to ionize the atom. (*Moral:* The displacements we're talking about are *minute,* even on an atomic scale.)

Problem 4.2 According to quantum mechanics, the electron cloud for a hydrogen atom in the ground state has a charge density

$$\rho = \frac{q}{\pi a^3} e^{-2r/a}$$

where q is the charge of the electron and a is the Bohr radius. Find the atomic polarizability of such an atom. (For a more sophisticated approach, see W. A. Bowers, Am. J. Phys. **54**, 347 (1986)).

Problem 4.3 According to equation (4.1) the induced dipole moment of an atom is approximately proportional to the external field. This is a "rule of thumb," not a fundamental law, and it is easy to concoct exceptions—in theory. Suppose, for example, the charge density of the electron cloud were proportional to the distance from the center, out to a radius R. To what power of E would p be proportional in that case? Find appropriate conditions on $\rho(r)$ such that equation (4.1) will hold in the weak-field limit.

Problem 4.4 A point charge q is situated a large distance r from a neutral atom of polarizability α. Find the force of attraction between them.

4.1.3 Alignment of Polar Molecules

The neutral atom discussed in Section 4.1.2 had no dipole moment to start with—**p** was *induced* by the applied field. Some molecules have built-in, permanent dipole moments. In the water molecule, for example, the electrons tend to cluster around the oxygen atom (Fig. 4.4), and since the molecule is bent at 105°, this leaves a negative charge at the vertex and a net positive charge at the opposite end. (The dipole moment of water is unusually large: 6.1×10^{-30} C · m; in fact, this is what accounts for its effectiveness as a solvent.) What happens when such molecules (called **polar molecules**) are placed in an electric field? If the field is uniform, the *force* on the

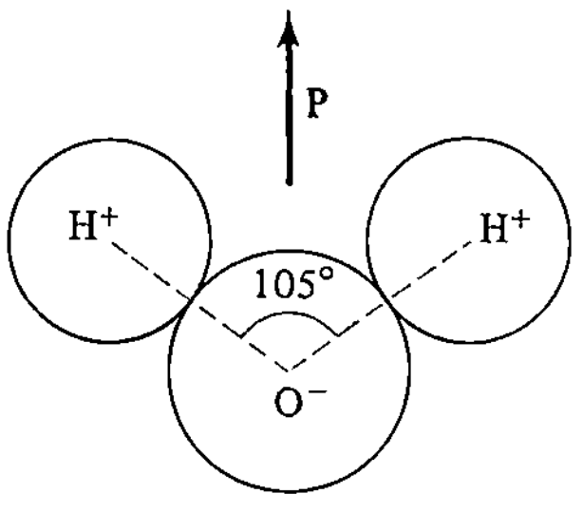

Figure 4.4

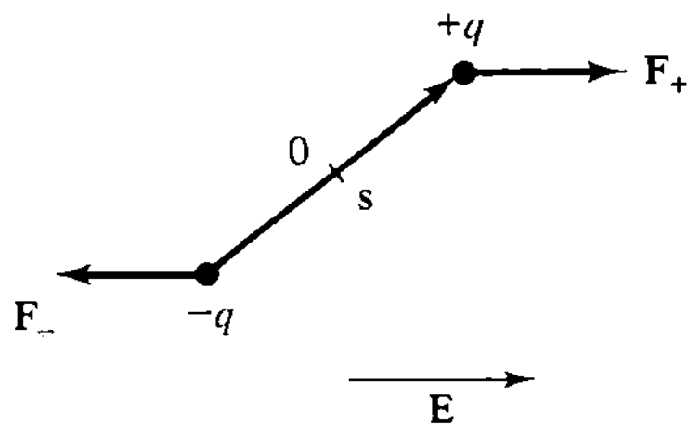

Figure 4.5

positive end, $\mathbf{F}_+ = q\mathbf{E}$, exactly cancels the force on the negative end, $\mathbf{F}_- = -q\mathbf{E}$ (Fig. 4.5). However, there will be a *torque:*

$$\mathbf{N} = (\mathbf{r}_+ \times \mathbf{F}_+) + (\mathbf{r}_- \times \mathbf{F}_-)$$
$$= \left[\left(\frac{\mathbf{s}}{2}\right) \times (q\mathbf{E})\right] + \left[\left(-\frac{\mathbf{s}}{2}\right) \times (-q\mathbf{E})\right] = q\mathbf{s} \times \mathbf{E}$$

Thus a dipole $\mathbf{p}$ in a uniform field $\mathbf{E}$ experiences a torque

$$\boxed{\mathbf{N} = \mathbf{p} \times \mathbf{E}} \tag{4.4}$$

Notice that $\mathbf{N}$ is in such a direction as to line $\mathbf{p}$ up *parallel* to $\mathbf{E}$; a polar molecule that is free to rotate will swing around until it points in the direction of the applied field.

If the field is *non*uniform, so that $\mathbf{F}_+$ does not exactly balance $\mathbf{F}_-$, there will be a net *force* on the dipole, in addition to the torque. Of course, $\mathbf{E}$ must change extremely rapidly for there to be significant variation in the space of one molecule, so this is not ordinarily a major consideration in discussing the behavior of dielectrics. Nevertheless, the formula for the force on a dipole in a nonuniform field is of some interest:

$$\mathbf{F} = \mathbf{F}_+ + \mathbf{F}_- = q(\mathbf{E}_+ - \mathbf{E}_-) = q(d\mathbf{E})$$

where $d\mathbf{E}$ represents the difference between the field at the plus end and the field at the minus end. Assuming the dipole is very short, we may use equation (1.25) to express the small change in E_x:

$$dE_x = (\nabla E_x) \cdot \mathbf{s}$$

with corresponding formulas for E_y and E_z. More compactly,

$$d\mathbf{E} = (\mathbf{s} \cdot \nabla)\mathbf{E}$$

and, therefore,

$$\boxed{\mathbf{F} = (\mathbf{p} \cdot \nabla)\mathbf{E}} \tag{4.5}$$

For a "perfect" dipole of infinitesimal length, (4.4) gives the exact torque even in a *non*uniform field, and (4.5) is the exact force. For "physical" dipoles these formulas are technically only approximate.[1]

Problem 4.5 In Fig. 4.6, $\mathbf{p}_1$ and $\mathbf{p}_2$ are (perfect) dipoles a distance r apart. What is the torque on $\mathbf{p}_1$ due to $\mathbf{p}_2$? What is the torque on $\mathbf{p}_2$ due to $\mathbf{p}_1$?

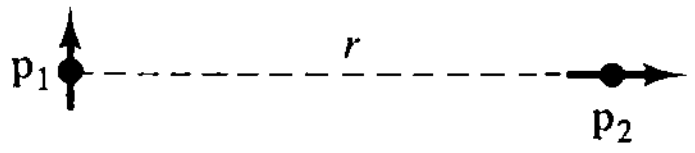

Figure 4.6

Problem 4.6 A dipole $\mathbf{p}$ is situated a distance d above an infinite grounded conducting plane (Fig. 4.7). The dipole makes an angle θ with the perpendicular to the plane. Find the torque on $\mathbf{p}$. If the dipole is free to rotate, in what orientation will it come to rest?

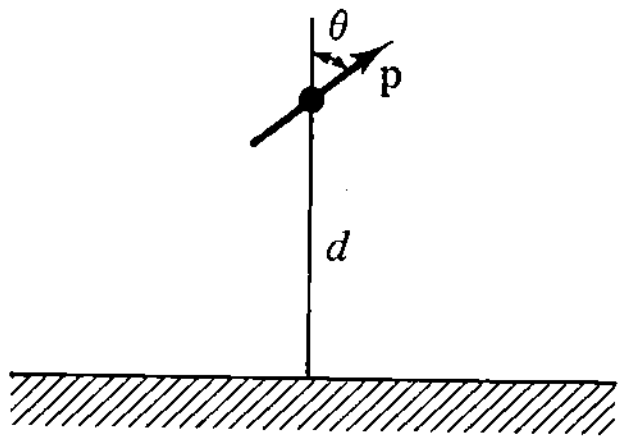

Figure 4.7

Problem 4.7 Show that the energy of a dipole in an electric field is given by

$$U = -\mathbf{p}\cdot\mathbf{E} \tag{4.6}$$

Problem 4.8 Show that the interaction energy of two dipoles separated by a displacement $\mathbf{r}$ is given by

$$U = \frac{1}{4\pi\epsilon_0}\frac{1}{r^3}[\mathbf{p}_1\cdot\mathbf{p}_2 - 3(\mathbf{p}_1\cdot\hat{r})(\mathbf{p}_2\cdot\hat{r})] \tag{4.7}$$

(*Hint*: Use Problem 4.7 and equation (3.99).)

Problem 4.9 Repeat Problem 4.5, only this time calculate the *force* on $\mathbf{p}_2$ due to $\mathbf{p}_1$. Also (this is a lot harder) calculate the force on $\mathbf{p}_1$ due to $\mathbf{p}_2$. Are the answers consistent with Newton's third law?

4.1.4 Polarization

In the previous two sections we have considered the effect of an external electric field on an individual atom or molecule. We are now in a position to answer (qualitatively) the original question: What happens to a piece of dielectric material when it is placed

[1]In the present context (4.5) could be written more conveniently as $\mathbf{F} = \nabla(\mathbf{p}\cdot\mathbf{E})$. However, it is safer to stick with (4.5), because we will be applying the formula to materials in which the dipole moment (per unit volume) is itself a function of position and this second expression would imply (incorrectly) that $\mathbf{p}$ *too* is to be differentiated.

in an electric field? If the substance consists of neutral atoms (or nonpolar molecules), the field will induce in each a tiny dipole moment, pointing in the same direction as the field.[2] If the material is made up of polar molecules, each permanent dipole will experience a torque, tending to line it up along the field direction. (Random thermal motions compete with this process: The molecules are continually undergoing collisions that destroy the alignment momentarily. For this reason the alignment is never complete, especially at higher temperatures, and disappears almost at once when the field is removed.)

Notice that these two mechanisms produce the same basic result: *a lot of little dipoles pointing along the direction of the field*—the dielectric becomes **polarized.** A convenient measure of this effect is

$$\mathbf{P} \equiv \textit{dipole moment per unit volume}$$

which is called the **polarization.** From now on we shall not worry much about how the polarization *got* there. Actually, the two mechanisms I described are not as clear-cut as I tried to pretend. Even in polar molecules there will be some polarization by displacement (though generally it's a lot easier to rotate a molecule than to stretch it, so the second mechanism predominates). It is even possible in some materials to "freeze in" polarization, so that it persists after the field is removed. But let's forget for a moment about the *cause* of the polarization and study the field which a chunk of polarized material *itself* produces. Then, in Section 4.3, we'll put it all together: the original field which was *responsible* for **P**, plus the new field which is *due* to **P**.

4.2 THE FIELD OF A POLARIZED OBJECT

4.2.1 Bound Charges

Suppose we have a piece of polarized material—that is, an object containing a lot of microscopic dipoles lined up. The dipole moment per unit volume, **P**, is given. *Question*: What is the field produced by this object (not the field which may have *caused* the polarization, but the field the polarization itself causes)? Well, we know what the field of an individual dipole looks like, so why not chop the material up into infinitesimal dipoles and integrate to get the total field? As usual it's easier to work with the potential. For a single dipole **p** we have (equation 3.94),

$$V = \frac{1}{4\pi\epsilon_0}\frac{\hat{\boldsymbol{\mathscr{r}}}\cdot\mathbf{p}}{\mathscr{r}^2} \tag{4.8}$$

where $\boldsymbol{\mathscr{r}}$ is the vector from the dipole to the point at which we are evaluating the potential (Fig. 4.8). In the present context we have a dipole moment $\mathbf{p} = \mathbf{P}\,d\tau$ in each volume element $d\tau$, so the total potential is

$$V = \frac{1}{4\pi\epsilon_0}\int_{\text{volume}} \frac{\mathbf{P}\cdot\hat{\boldsymbol{\mathscr{r}}}}{\mathscr{r}^2}\,d\tau \tag{4.9}$$

[2] In an asymmetrical molecule the induced dipole moment may not be parallel to the field, but if the molecules are randomly oriented, the perpendicular contributions will *average* to zero. Within a single crystal, the orientations are certainly *not* random, and we would have to treat this case separately.

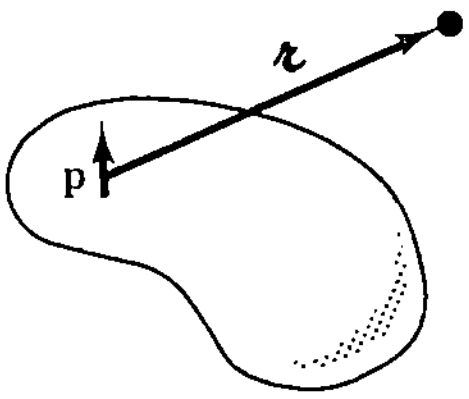

Figure 4.8

That *does* it, in principle. But a little sleight-of-hand casts this integral into a much more illuminating form. Observing that

$$\nabla\left(\frac{1}{\mathscr{r}}\right)=\frac{\hat{\mathscr{r}}}{\mathscr{r}^2}$$

where (unlike Problem 1.13) the differentiation is with respect to the *source* coordinates, we have

$$V=\frac{1}{4\pi\epsilon_0}\int_{\text{volume}}\mathbf{P}\cdot\nabla\left(\frac{1}{\mathscr{r}}\right)d\tau$$

Integrating by parts, using product rule number 5, gives

$$V=\frac{1}{4\pi\epsilon_0}\left[\int_{\text{volume}}\nabla\cdot\left(\frac{1}{\mathscr{r}}\mathbf{P}\right)d\tau-\int_{\text{volume}}\frac{1}{\mathscr{r}}(\nabla\cdot\mathbf{P})\,d\tau\right]$$

or, using the divergence theorem,

$$V=\frac{1}{4\pi\epsilon_0}\int_{\text{surface}}\frac{1}{\mathscr{r}}\mathbf{P}\cdot d\mathbf{a}-\frac{1}{4\pi\epsilon_0}\int_{\text{volume}}\frac{1}{\mathscr{r}}(\nabla\cdot\mathbf{P})\,d\tau \tag{4.10}$$

The first term looks just like the potential of a surface charge

$$\boxed{\sigma_b=\mathbf{P}\cdot\hat{n}} \tag{4.11}$$

(where $\hat{n}$ is the normal unit vector), while the second term looks like the potential of a volume charge

$$\boxed{\rho_b=-\nabla\cdot\mathbf{P}} \tag{4.12}$$

With these definitions (4.10) becomes

$$V=\frac{1}{4\pi\epsilon_0}\int_{\text{surface}}\frac{1}{\mathscr{r}}\sigma_b da+\frac{1}{4\pi\epsilon_0}\int_{\text{volume}}\frac{1}{\mathscr{r}}\rho_b\,d\tau \tag{4.13}$$

What this means is that the potential (and hence also the field) of a polarized object is the same as that produced by a volume charge density $\rho_b=-\nabla\cdot\mathbf{P}$ plus a surface charge density $\sigma_b=\mathbf{P}\cdot\hat{n}$. Instead of integrating the contributions of all the infinitesimal dipoles, as in equation (4.9), we can just find these **bound charges,** and then calculate the fields *they* produce, in the same way we calculate the field of any other volume and surface charges.

Example 2

Find the electric field produced by a uniformly polarized sphere of radius R.

Solution: We may as well choose the z axis to coincide with the direction of polarization (Fig. 4.9). The volume bound charge density ρ_b is zero, since $\mathbf{P}$ is constant, but

$$\sigma_b = \mathbf{P}\cdot\hat{n} = P\cos\theta$$

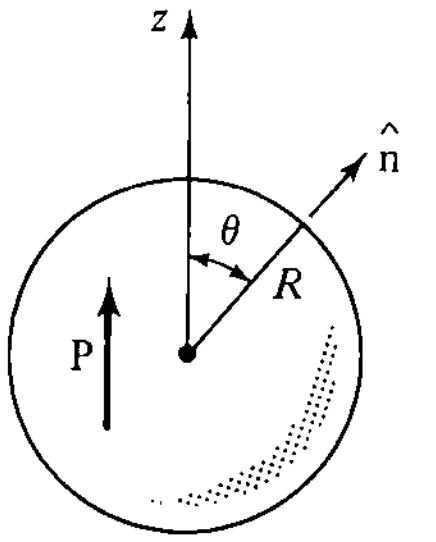

Figure 4.9

where θ is the usual spherical coordinate. What we want, then, is the field produced by a charge density $P\cos\theta$ plastered over the surface of a sphere. But we have already computed the potential of such a configuration in Example 9 of Chapter 3:

$$V(r,\theta) = \begin{cases} \dfrac{P}{3\epsilon_0} r\cos\theta, & \text{for } r \le R \\[2ex] \dfrac{P}{3\epsilon_0}\dfrac{R^3}{r^2}\cos\theta, & \text{for } r \ge R \end{cases}$$

Since $r\cos\theta = z$, the *field* inside the sphere is uniform,

$$\mathbf{E} = -\nabla V = -\frac{P}{3\epsilon_0}\hat{z} = -\frac{1}{3\epsilon_0}\mathbf{P}, \qquad \text{for } r < R \tag{4.14}$$

This result will be very useful in what follows. Outside the sphere the potential is identical to that of a perfect dipole at the origin,

$$V = \frac{1}{4\pi\epsilon_0}\frac{\mathbf{p}\cdot\hat{r}}{r^2}, \qquad \text{for } r \ge R \tag{4.15}$$

whose dipole moment is, not surprisingly, equal to the total dipole moment of the sphere:

$$\mathbf{p} = \tfrac{4}{3}\pi R^3\mathbf{P} \tag{4.16}$$

The field of the uniformly polarized sphere is shown in Fig. 4.10.

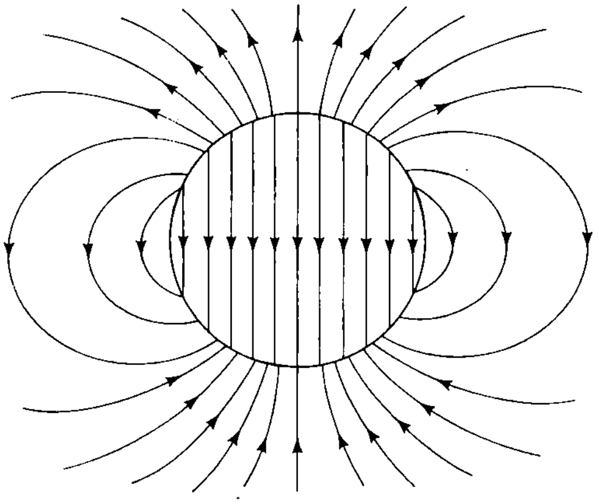

Figure 4.10

Problem 4.10 A sphere of radius R carries a polarization

$$\mathbf{P} = k\mathbf{r}$$

where k is a constant and $\mathbf{r}$ is the vector from the center.

(a) Calculate the bound charges σ_b and ρ_b.

(b) Find the field inside and outside the sphere.

Problem 4.11 A short circular cylinder, of radius R and length L, carries a "frozen-in" uniform polarization $\mathbf{P}$, parallel to its axis. Find the bound charge, and sketch the electric field of this cylinder. Make three sketches—one for $L \gg R$, one for $L \ll R$, and one for $L \approx R$. (This device is known as a **bar electret**; it is the electrical analog to the bar magnet. In practice, only very special materials (barium titanate is the most "familiar" example) will hold a permanent electric polarization. That's why you can't buy electrets at the toy store.)

Problem 4.12 Calculate the potential of a uniformly polarized sphere (Example 2) directly from equation (4.9).

4.2.2 Physical Interpretation of Bound Charge

In the last section we found that the field of a polarized object is identical to the field that would be produced by a certain distribution of "bound charges," σ_b and ρ_b. But this conclusion emerged in the course of abstract manipulations on the integral (4.9) and left us with no clue as to the physical meaning of these bound charges. Indeed, some authors give you the impression that bound charges are in some sense "fictitious"—mere bookkeeping devices used to facilitate the calculation of fields. Nothing could be farther from the truth; ρ_b and σ_b represent *perfectly genuine accumulations of charge*. In this section I'll show you how polarization leads to such accumulations of charge.

The basic idea is very simple: Suppose we have a long string of dipoles, as shown in Fig. 4.11. Along the line, the head of one effectively cancels the tail of its neighbor, but at the ends there are two charges left over: plus at the right end and

Figure 4.11

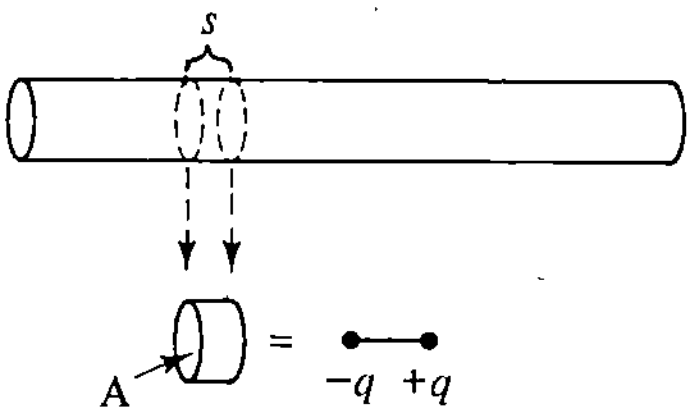

Figure 4.12

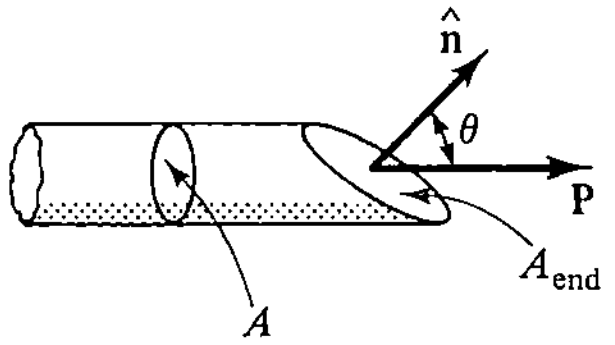

Figure 4.13

minus at the left. It is as if we had peeled off an electron at one end and carried it all the way down to the other end, though in fact no single electron made the whole trip—a lot of tiny displacements add up to one large one. We call the net charge at the ends *bound* charge to remind ourselves that it cannot be removed; in a dielectric every electron is permanently attached to a specific atom or molecule. But apart from that, bound charge is no different from any other kind.

To calculate the actual *amount* of bound charge resulting from a given uniform polarization, examine a "tube" of dielectric parallel to **P**. The dipole moment of the tiny chunk shown in Fig. 4.12 is $P(As)$, where A is the cross-sectional area of the tube and s is the length of the chunk. In terms of the charge at the end, this same dipole moment can be written qs. The bound charge that piles up at the right end of the tube is therefore

$$q = PA$$

If the ends have been sliced off perpendicularly, the surface charge density is

$$\sigma_b = \frac{q}{A} = P$$

For an oblique cut (Fig. 4.13), the *charge* is still the same, but $A = A_{\text{end}} \cos\theta$, so

$$\sigma_b = \frac{q}{A_{\text{end}}} = P\cos\theta = \mathbf{P}\cdot\hat{n}$$

The effect of uniform polarization, then, is to paint a bound charge $\sigma_b = \mathbf{P}\cdot\hat{n}$ over the surface of the material. This is exactly what we found by more rigorous means in Section 4.2.1. But now we know where the bound charge *comes* from.

If the polarization is *non*uniform, then we get accumulations of bound charge *within* the material, as well as on the surface. A glance at Fig. 4.14 suggests that a diverging **P** results in a pileup of negative charge. Indeed, the net bound charge $\int \rho_b\, d\tau$ in a given volume is equal and opposite to the amount that has been pushed out through the surface. The latter (by the same reasoning we used to treat uniform polarization) is $\mathbf{P}\cdot\hat{n}$ per unit area, so

$$\int_{\text{volume}} \rho_b\, d\tau = -\int_{\text{surface}} \mathbf{P}\cdot d\mathbf{a} = -\int_{\text{volume}} (\nabla\cdot\mathbf{P})\, d\tau$$

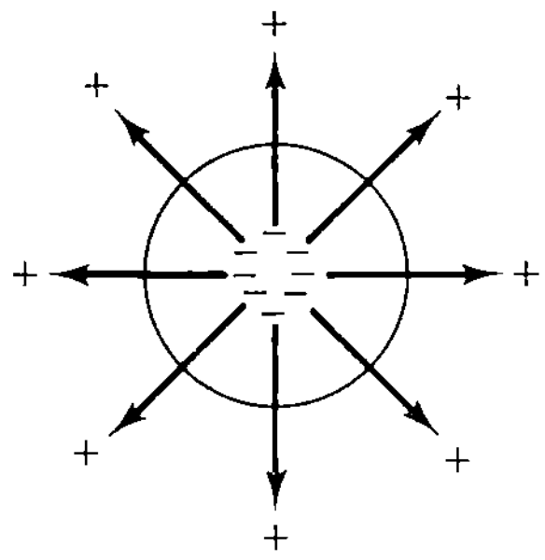

Figure 4.14

Since this is true for *any* volume, we have

$$\rho_b = -\nabla \cdot \mathbf{P}$$

reproducing, again, the more rigorous conclusion of Section 4.2.1.

Example 3

There is another way of analyzing the uniformly polarized sphere (Example 2), which nicely illustrates the idea of bound charge. What we have, really, is *two* spheres of charge: a positive sphere and a negative sphere. Without polarization the two are superimposed and cancel completely. But when the material is uniformly polarized, all the plus charges move slightly *upward* (the z direction), and all the minus charges move slightly *downward* (Fig. 4.15). The two spheres no longer overlap perfectly: at the top there's a "cap" of leftover positive charge and at the bottom a cap of negative charge. This "leftover" charge is, of course, the bound surface charge σ_b.

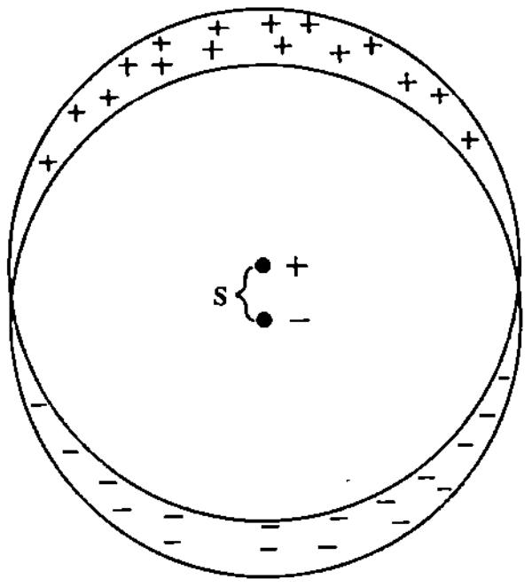

Figure 4.15

In Problem 2.18 you calculated the field in the region of overlap between two uniformly charged spheres; the answer was

$$\mathbf{E} = -\frac{1}{4\pi\epsilon_0}\frac{q\mathbf{s}}{R^3}$$

where q is the total charge of the positive sphere and $\mathbf{s}$ is the vector from the negative center to the positive center. We can now express this in terms of the polarization of the sphere, $\mathbf{p} = q\mathbf{s} = (\frac{4}{3}\pi R^3)\mathbf{P}$, as

$$\mathbf{E} = -\frac{1}{3\epsilon_0}\mathbf{P}$$

Meanwhile, for points *outside*, it is as though all the charge on each sphere were concentrated at the respective centers. We have, then, a dipole, with potential

$$V = \frac{1}{4\pi\epsilon_0}\frac{\mathbf{p}\cdot\hat{\mathbf{r}}}{r^2}$$

(Remember that $\mathbf{s}$ is some small fraction of an atomic radius; Fig. 4.15 is grossly exaggerated.) These answers agree, of course, with the results of Example 2.

Problem 4.13 A very long cylinder, of radius R, carries a uniform polarization $\mathbf{P}$ perpendicular to its axis. (Careful: I said *uniform*, not *radial*.) Find the electric field inside the cylinder. Show that the field *outside* the cylinder can be expressed in the form

$$\mathbf{E} = \frac{R^2}{2\epsilon_0 r^2}[2(\mathbf{P}\cdot\hat{\mathbf{r}})\hat{\mathbf{r}} - \mathbf{P}]$$

Problem 4.14 When you polarize a neutral dielectric, charge moves a bit, but the *total* remains zero. This fact should be reflected in the bound charges σ_b and ρ_b. Prove from equations (4.11) and (4.12) that the total bound charge vanishes.

4.2.3 The Field Inside a Dielectric

I have been sloppy about the distinction between "pure" dipoles and "physical" dipoles. In developing the theory of bound charges, I assumed we were working with the pure kind—indeed, I started with equation (4.8), the formula for the potential of a pure dipole. And yet an actual polarized dielectric consists of *physical* dipoles, albeit extremely tiny ones. What is more, I presumed to represent discrete molecular dipoles by a continuous density function $\mathbf{P}$. How can I justify this method? *Outside* the dielectric there is no real problem: Here we are far away from the molecules (r is many times greater than the separation distance between plus and minus charges), so the dipole potential dominates overwhelmingly and the detailed "graininess" of the source is blurred by distance. *Inside* the dielectric, however, we can hardly pretend to be far from all the dipoles, and the procedure I used in Section 4.2.1 is open to serious challenge.

In fact, when you stop to think about it, the actual electric field inside matter must be fantastically complicated, on the microscopic level. If you happen to be very near an electron, the field is gigantic, whereas a short distance away it may be small or point in a totally different direction. Moreover, an instant later, as the atoms move about, the field will have altered entirely. This true **microscopic** field would be utterly impossible to calculate, nor would it be of much interest if you could. Just as, for macroscopic purposes, we regard water as a continuous fluid, ignoring its molecular structure, so also we can ignore the microscopic bumps and wrinkles in the electric

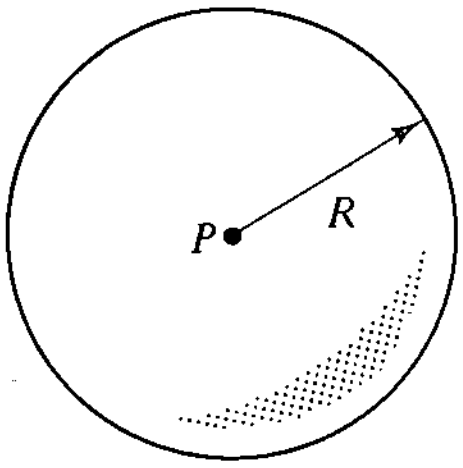

Figure 4.16

field inside matter and concentrate on the **macroscopic** field. This is defined as the *average* field over regions large enough to contain many thousands of atoms (so that the uninteresting microscopic fluctuations are smoothed over), and yet small enough to ensure that we do not wash out any significant large-scale variations in the field. (In practice, this means we must average over regions much smaller than the dimensions of the object itself). Ordinarily, the macroscopic field is what people *mean* when they speak of "the" field inside matter.[3]

It remains to show that the macroscopic field is what we actually obtain when we use the methods of Section 4.2.1. The argument is subtle, so hang on. Suppose I want to calculate the macroscopic field at some point P within a dielectric (Fig. 4.16). I know I must average the true, microscopic field over an appropriate volume, so let me draw a small sphere about P, of radius, say, a thousand times the size of a molecule. The macroscopic field at P, then, consists of two parts: the average field over the sphere due to all charges *outside* plus the average due to all charges *inside*:

$$\mathbf{E} = \mathbf{E}_{\text{out}} + \mathbf{E}_{\text{in}}$$

Now you proved in Problem 3.41 that the average field produced by charges outside a sphere is equal to the field they produce at the center. So $\mathbf{E}_{\text{out}}$ is the field at P due to the dipoles exterior to the sphere. These are far enough away that we can safely use equation (4.9):

$$V_{\text{out}} = \frac{1}{4\pi\epsilon_0}\int_{\text{outside}} \frac{\mathbf{P}\cdot\hat{\boldsymbol{\imath}}}{\imath^2}\,d\tau \tag{4.17}$$

The dipoles *inside* the sphere are too close to treat in this fashion. But fortunately all we need is their *average* field and that (again, by Problem 3.41) is

$$\mathbf{E}_{\text{in}} = -\frac{1}{4\pi\epsilon_0}\frac{\mathbf{p}}{R^3}$$

regardless of the details of the charge distribution within the sphere. The only relevant quantity is the total dipole moment, $\mathbf{p} = (\frac{4}{3}\pi R^3)\,\mathbf{P}$:

$$\mathbf{E}_{\text{in}} = -\frac{1}{3\epsilon_0}\mathbf{P} \tag{4.18}$$

Now, by assumption the sphere is small enough that $\mathbf{P}$ does not vary signifi-

[3]In case the introduction of the macroscopic field sounds suspicious to you, let me point out that you do *exactly* the same averaging whenever you speak of the *density* of a material.

cantly over its volume, so the term *left out* of the integral (4.17) corresponds to the field at the center of a *uniformly* polarized sphere, to wit: $-(1/3\epsilon_0)\,\mathbf{P}$ (equation (4.14)). But that is precisely what (4.18) puts back in! The macroscopic field, then, is given by the potential

$$V = \frac{1}{4\pi\epsilon_0}\int \frac{\mathbf{P}\cdot\hat{\boldsymbol{\mathscr{r}}}}{\mathscr{r}^2}\,d\tau \tag{4.19}$$

where the integral runs over the *entire* volume of the dielectric. This is, of course, the potential we used in Section 4.2.1; without realizing it, we were correctly calculating the averaged, macroscopic field for points inside the dielectric.

You may have to reread the last couple of paragraphs two or three times for the argument to sink in. Notice that it all revolves around the curious fact that the average field over *any* sphere (due to the charge inside) is the same as the field at the center of a *uniformly polarized* sphere with the same total dipole moment. This means that no matter how crazy the actual microscopic charge configuration, we can replace it by a nice smooth distribution of perfect dipoles, if all we want is the average, macroscopic field. Incidentally, while the argument ostensibly relies on the spherical shape I chose to average over, the macroscopic field is certainly independent of the geometry of the averaging region, and this is reflected in the final answer, (4.19). Presumably, one could reproduce the same argument for a cube or an ellipsoid or whatever—the calculation might be more difficult, but the conclusion would be the same.

4.3 THE ELECTRIC DISPLACEMENT

4.3.1 Gauss's Law in the Presence of Dielectrics

In Section 4.2 we found that the effect of polarization is to produce accumulations of bound charge $\rho_b = -\nabla\cdot\mathbf{P}$ within the dielectric and $\sigma_b = \mathbf{P}\cdot\hat{n}$ on the surface. The field due to polarization of the medium is just the field of this bound charge. We are now ready to put it all together: the field attributable to bound charge, plus the field due to everything *else* (which, for want of a better term, we call **free charge**). The free charge might consist of electrons on a conductor or ions embedded in the dielectric materials or whatever; any charge, in other words, that is *not* a result of polarization. Within the dielectric, then, the total charge density can be written:

$$\rho = \rho_b + \rho_f \tag{4.20}$$

and Gauss's law reads

$$\epsilon_0\nabla\cdot\mathbf{E} = \rho = \rho_b + \rho_f = -\nabla\cdot\mathbf{P} + \rho_f$$

where **E** is now the *total* field, not just that portion generated by polarization. It is convenient to combine the two divergence terms:

$$\nabla\cdot(\epsilon_0\mathbf{E} + \mathbf{P}) = \rho_f$$

The expression in parentheses, designated by the letter **D**

$$\boxed{\mathbf{D} \equiv \epsilon_0\mathbf{E} + \mathbf{P}} \tag{4.21}$$

is known as the **electric displacement**. In terms of **D**, Gauss's law reads

$$\boxed{\nabla \cdot \mathbf{D} = \rho_f} \tag{4.22}$$

or, in integral form,

$$\oint_{\text{surface}} \mathbf{D} \cdot d\mathbf{a} = Q_{f_{\text{enc}}} \tag{4.23}$$

where $Q_{f_{\text{enc}}}$ denotes the total free charge enclosed in the volume. This is a particularly useful way to express Gauss's law, in the context of dielectrics, because *it makes reference only to free charges,* and free charge is the stuff we control. Bound charge comes along for the ride: When we put the free charges in place, a certain polarization automatically ensues by the mechanisms of Section 4.1. In a typical problem, therefore, we know ρ_f but we do not initially know ρ_b; equation (4.23) lets us go right to work with the information at hand. In particular, whenever the requisite symmetry is present, we can immediately calculate **D** by the standard Gauss's law methods.

Example 4

A long straight wire, carrying uniform line charge λ, is surrounded by rubber insulation out to a radius R (Fig. 4.17). Find the electric displacement.

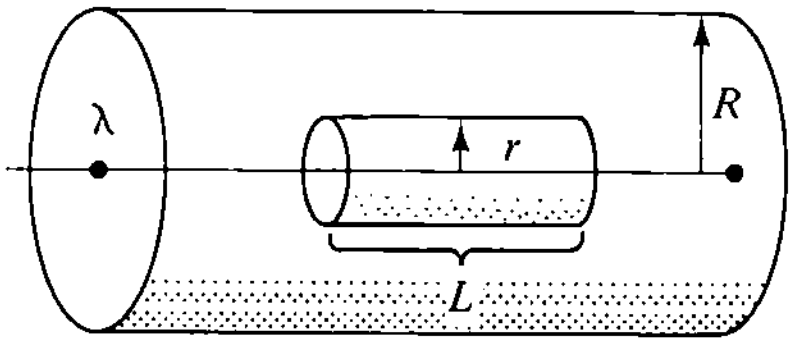

Figure 4.17

Solution: Drawing a cylindrical Gaussian surface, of radius r and length L, and applying (4.23), we find

$$D(2\pi r L) = \lambda L$$

Therefore,

$$\mathbf{D} = \frac{\lambda}{2\pi r}\,\hat{\mathbf{r}} \tag{4.24}$$

Notice that this formula applies within the insulation and outside it alike. In the latter region, $\mathbf{P} = 0$, so

$$\mathbf{E} = \frac{1}{\epsilon_0}\mathbf{D} = \frac{\lambda}{2\pi\epsilon_0 r}\,\hat{\mathbf{r}}, \qquad \text{for } r > R$$

Inside the rubber the electric field cannot be computed, since we have no way of knowing **P** at this stage.

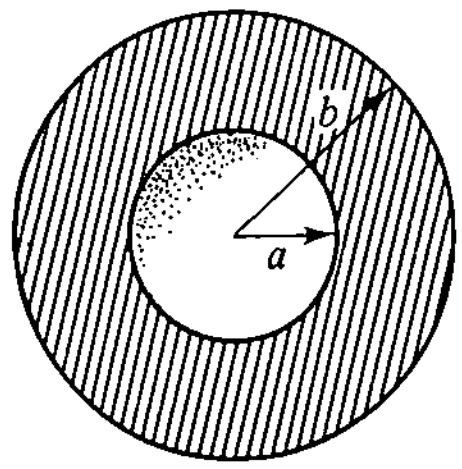

Figure 4.18

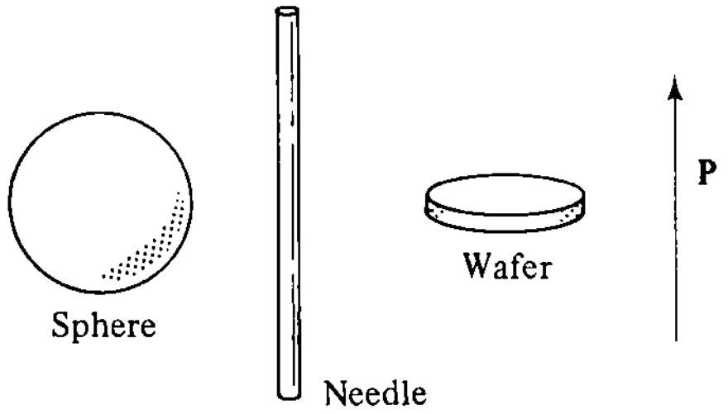

Figure 4.19

It may have appeared to you that I left out the surface bound charge σ_b in deriving equation (4.22), and in a sense that is true. We cannot apply Gauss's law precisely *at* the surface of a dielectric, for here ρ_b blows up, taking the divergence of **E** with it. But everywhere *else* the logic is sound, and in fact, if we picture the edge of the dielectric as having some finite thickness within which the polarization tapers off to zero (probably a more realistic model than an abrupt cut-off anyway), then there *is* no surface bound charge; ρ_b varies rapidly but smoothly within this "skin," and Gauss's law in differential form can be safely applied *everywhere*. At any rate, the integral form is free from this "defect."

Problem 4.15 A thick spherical shell (inner radius a, outer radius b) is filled with dielectric material with a "frozen-in" polarization

$$\mathbf{P} = \frac{k}{r}\hat{\mathbf{r}}$$

where k is a constant and r is the distance from the center (Fig. 4.18). (There is no *free* charge in the problem.) Find the electric field in all three regions by two different methods:

(a) As in Section 4.2, locate all the bound charge, and calculate the field it produces.

(b) Use Gauss's law in the form (4.23) to find **D**, and then get **E** from (4.21). Notice that the second method is much faster and avoids any explicit reference to the bound charges.

Problem 4.16 For the bar electret of Problem 4.11, make three careful sketches: one of **P**, one of **E**, and one of **D**. Assume L is about $2R$.

Problem 4.17 Suppose the field inside a large piece of dielectric is $\mathbf{E}_0$, so that the electric displacement is $\mathbf{D}_0 = \epsilon_0\mathbf{E}_0 + \mathbf{P}$.

(a) Now, a small spherical cavity (Fig. 4.19) is hollowed out of the material. Find the field at the center of the cavity in terms of $\mathbf{E}_0$ and **P**. Also find the displacement at the center of the cavity in terms of $\mathbf{D}_0$ and **P**.

(b) Do the same for a long needle-shaped cavity running parallel to **P**.

(c) Do the same for a thin wafer-shaped cavity perpendicular to **P**.

Assume the cavities are small enough that **P**, $\mathbf{E}_0$, and $\mathbf{D}_0$ are essentially constant. (*Hint*: Carving out a cavity is the same as superimposing an object of the same shape but opposite polarization.)

4.3.2 A Deceptive Parallel

Equation (4.22) looks just like Gauss's law, only the *total* charge density ρ is replaced by the *free* charge density ρ_f, and **D** is substituted for ϵ_0**E**. For this reason, you may be tempted to conclude that $\mathbf{D}$ is "just like" $\mathbf{E}$ (apart from the factor of ϵ_0), except that its source is ρ_f instead of ρ: "To solve problems involving dielectrics, you just forget all about the bound charge—calculate the field as you ordinarily would, only call the answer $\mathbf{D}$ instead of $\mathbf{E}$." This reasoning is seductive, but the conclusion is false; in particular, there is no "Coulomb's law" for $\mathbf{D}$:

$$\mathbf{D} \neq \frac{1}{4\pi}\int \rho_f \left(\frac{\hat{\boldsymbol{\imath}}}{\imath^2}\right) d\tau$$

The parallel between $\mathbf{E}$ and $\mathbf{D}$ is more subtle than that.

The reason is that the divergence alone is insufficient to determine a vector field; you need to know the curl as well. One tends to forget this in the case of electrostatic fields because the curl of $\mathbf{E}$ is always zero. But the curl of $\mathbf{D}$ is *not* always zero:

$$\nabla \times \mathbf{D} = \epsilon_0(\nabla \times \mathbf{E}) + (\nabla \times \mathbf{P}) = \nabla \times \mathbf{P} \tag{4.25}$$

and there is no reason, in general, to suppose that the curl of $\mathbf{P}$ vanishes. Sometimes it does, as in Example 4 and Problem 4.15, but more often it does not. The bar electret of Problem 4.16 is a case in point: here there is no free charge anywhere, so if you really believe that the only source of $\mathbf{D}$ is ρ_f, you will be forced to conclude that $\mathbf{D} = 0$ everywhere and hence that $\mathbf{E} = (-1/\epsilon_0)\,\mathbf{P}$ inside and $\mathbf{E} = 0$ outside the electret, which is quite wrong. (I leave it to you to find the place where $\nabla \times \mathbf{P} \neq 0$ in this problem.) Because $\nabla \times \mathbf{D} \neq 0$, moreover, $\mathbf{D}$ cannot be expressed as the gradient of a scalar—there is no "potential" for $\mathbf{D}$.

Advice: When you are asked to compute the electric displacement, first look for symmetry. If the problem exhibits spherical, infinite cylindrical, or planar symmetry, then you can get $\mathbf{D}$ directly from (4.23) by the usual Gauss's law methods. (Evidently in such cases $\nabla \times \mathbf{P}$ is automatically zero. Since symmetry alone dictates the answer, you're not really obliged to worry about the curl.) If the requisite symmetry is absent, you'll have to think of another approach and, in particular, you must *not* assume that $\mathbf{D}$ is determined exclusively by the free charge.

4.4 LINEAR DIELECTRICS

4.4.1 Susceptibility, Permittivity, Dielectric Constant

In Sections 4.2 and 4.3 we did not commit ourselves as to the *cause* of $\mathbf{P}$; we dealt only with the *effects* of polarization. From the qualitative discussion of Section 4.1, though, we know that the polarization of a dielectric ordinarily results from an electric field, which lines up the atomic or molecular dipoles. In many substances, in fact, the polarization is *proportional* to the field,

$$\boxed{\mathbf{P} = \epsilon_0 \chi_e \mathbf{E}} \tag{4.26}$$

provided $\mathbf{E}$ is not too strong. The constant of proportionality, χ_e, is called the **electric susceptibility** of the medium (a factor of ϵ_0 has been extracted to make χ_e dimensionless). The value of χ_e depends on the microscopic structure of the substance in question. I shall call materials that obey (4.26) **linear dielectrics.**

TABLE 4.2 DIELECTRIC CONSTANTS

Material	Dielectric Constant	Material	Dielectric Constant
Vacuum	1	Air (100 atm)	1.055
Helium	1.000068	Polyethylene	2.26
Neon	1.00013	Glass	4-7
Hydrogen	1.00025	Porcelain	6-8
Argon	1.00055	Methanol	33.6
Nitrogen	1.00058	Water	80.4
Air	1.00059	HCN (0°C)	158
Water vapor (110°C)	1.0126		

Source: Handbook of Chemistry and Physics, 67th ed. (Cleveland: CRC Press, Inc., 1986–87). Unless otherwise specified, values given are for 1 atm, 20°C.

Note that $\mathbf{E}$ in (4.26) is the *total* field; it may be due in part to free charges and in part to the polarization itself. If, for instance, we put a piece of dielectric into an external field $\mathbf{E}_0$, we cannot compute $\mathbf{P}$ directly from (4.26): The external field will polarize the material, and this polarization will produce its own field, which then contributes to the total field, and this in turn modifies the polarization, which . . . . Breaking out of this infinite regress is not always easy. You'll see some examples in a moment. The simplest approach is to begin with the *displacement*, at least in those cases where $\mathbf{D}$ can be deduced directly from the free charge distribution.

In linear media, we have

$$\mathbf{D} = \epsilon_0\mathbf{E} + \mathbf{P} = \epsilon_0\mathbf{E} + \epsilon_0\chi_e\mathbf{E} = \epsilon_0(1 + \chi_e)\mathbf{E} \tag{4.27}$$

Thus, not only is $\mathbf{P}$ proportional to $\mathbf{E}$; so also is $\mathbf{D}$:

$$\mathbf{D} = \epsilon\mathbf{E} \tag{4.28}$$

where

$$\epsilon = \epsilon_0(1 + \chi_e) \tag{4.29}$$

ϵ is called the **permittivity** of the material. In a vacuum, where there is no matter to polarize, the susceptibility is zero, and the permittivity is ϵ_0. That's why ϵ_0 is called the **permittivity of free space.** (I dislike the term, for it suggests that a vacuum is just a special kind of linear dielectric in which the permittivity happens to have the value $8.85 \times 10^{-12}\ \mathrm{C^2/N \cdot m^2}$.) If you remove a factor of ϵ_0, the remaining dimensionless quantity

$$K = 1 + \chi_e = \frac{\epsilon}{\epsilon_0} \tag{4.30}$$

is called the **dielectric constant.** Dielectric constants for some common materials are listed in Table 4.2. Of course, the permittivity and the dielectric constant do not convey any information that was not already available in the susceptibility, nor is there anything essentially new in (4.28). The *physics* of linear dielectrics is all contained in equation (4.26). Incidentally, the volume bound charge in a homogeneous linear dielectric is proportional to the density of free charge:[4]

$$\rho_b = -\nabla\cdot\mathbf{P} = -\nabla\cdot\left(\epsilon_0\frac{\chi_e}{\epsilon}\mathbf{D}\right) = -\left(\frac{\chi_e}{1+\chi_e}\right)\rho_f \tag{4.31}$$

[4]This does not apply to the surface charge (σ_b) because χ_e is *not* independent of position (obviously) at the boundary.

In particular, unless free charge is actually embedded in the medium, all bound charge on a linear dielectric is at the surface.

Example 5

A metal sphere of radius a carries a charge Q (Fig. 4.20). It is surrounded, out to radius b, by linear dielectric material of permittivity ϵ. Find the potential at the center (relative to infinity).

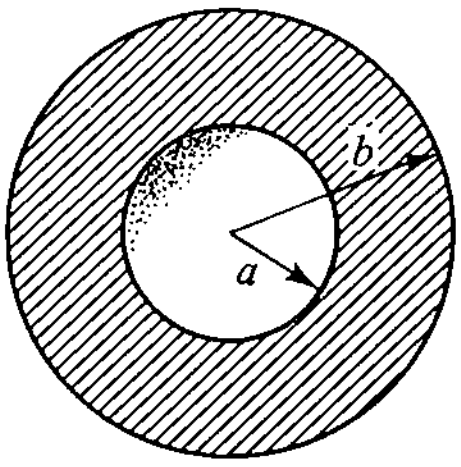

Figure 4.20

Solution: To compute V, we need to know $\mathbf{E}$; to find $\mathbf{E}$, we might try first to locate the bound charge. We could get the bound charge from $\mathbf{P}$, but we can't calculate $\mathbf{P}$ unless we already know $\mathbf{E}$ (equation (4.26)). What we *do* know is the *free* charge Q, and fortunately the arrangement is spherically symmetric, so let us begin by calculating $\mathbf{D}$, using (4.23):

$$\mathbf{D} = \frac{Q}{4\pi r^2}\hat{r}, \qquad \text{for all points } r > a$$

(Inside the metal sphere, of course, $\mathbf{E} = \mathbf{P} = \mathbf{D} = 0$.) Once we know $\mathbf{D}$, it is a trivial matter to obtain $\mathbf{E}$, using (4.28):

$$\mathbf{E} = \begin{cases} \dfrac{Q}{4\pi\epsilon r^2}\hat{r}, & \text{for } a < r < b \\[2ex] \dfrac{Q}{4\pi\epsilon_0 r^2}\hat{r}, & \text{for } r > b \end{cases}$$

The potential at the center is, therefore,

$$V = -\int_{\infty}^{0} \mathbf{E}\cdot d\mathbf{l} = -\int_{\infty}^{b}\left(\frac{Q}{4\pi\epsilon_0 r^2}\right)dr - \int_{b}^{a}\left(\frac{Q}{4\pi\epsilon r^2}\right)dr - \int_{a}^{0}(0)\,dr$$

$$= \frac{Q}{4\pi}\left(\frac{1}{\epsilon_0 b} + \frac{1}{\epsilon a} - \frac{1}{\epsilon b}\right)$$

As it turns out, it was not necessary for us to compute the polarization or the bound charge explicitly, though this can easily be done:

$$\mathbf{P} = \epsilon_0\chi_e\mathbf{E} = \frac{\epsilon_0\chi_e Q}{4\pi\epsilon r^2}\hat{r}$$

in the dielectric and, hence,

$$\rho_b = -\nabla\cdot\mathbf{P} = 0$$

(as we knew at any rate from (4.31)), while

$$\sigma_b = \mathbf{P}\cdot\hat{n} = \begin{cases} \dfrac{\epsilon_0\chi_e Q}{4\pi\epsilon b^2}, & \text{at the outer surface} \\ \dfrac{-\epsilon_0\chi_e Q}{4\pi\epsilon a^2}, & \text{at the inner surface} \end{cases}$$

Notice that the surface bound charge at a is *negative* ($\hat{n}$ points outward *with respect to the dielectric,* which is $+\hat{r}$ at b but $-\hat{r}$ at a). This is natural, since the charge on the metal sphere attracts its opposite in all the dielectric molecules. It is this layer of negative charge that cuts the field down, within the dielectric, from $1/4\pi\epsilon_0(Q/r^2)\hat{r}$ to $1/4\pi\epsilon(Q/r^2)\hat{r}$. In this respect a dielectric is rather like an imperfect conductor: On a *conducting* shell the induced surface charge would be such as to cancel the field of Q *completely* in the region $a < r < b$; the dielectric does the best it can, but the cancellation is only partial.

You might suppose that linear dielectrics would escape the defect in the parallel between **E** and **D**. Since **P** and **D** are now proportional to **E**, does it not follow that their curls, like **E**'s, must vanish? Unfortunately, it does *not,* for the line integral of **P** around a closed path which *straddles the boundary between one type of material and another* need not be zero, even though the integral of **E** around the same loop *must* be. The reason is that the proportionality factor $\epsilon_0\chi_e$ is different on the two sides. For instance, at the interface between a dielectric and a vacuum (Fig. 4.21), **P** is zero on one side but not the other. Around this loop $\oint \mathbf{P}\cdot d\mathbf{l} \neq 0$, and hence, by Stokes' theorem, the curl of **P** cannot vanish everywhere within the loop (in fact, it is *infinite* at the boundary).

Of course, if the space is *entirely* filled with linear dielectric of one uniform type, then this objection is void; in this rather special circumstance

$$\nabla\cdot\mathbf{D} = \rho_f \quad \text{and} \quad \nabla\times\mathbf{D} = 0$$

so **D** can be found from the free charge just as though the dielectric were not there:

$$\mathbf{D} = \epsilon_0\mathbf{E}_{\text{vac}}$$

where $\mathbf{E}_{\text{vac}}$ is the field the same free charge distribution would produce in the absence of any dielectric. According to (4.28) and (4.30), therefore,

$$\mathbf{E} = \frac{1}{\epsilon}\mathbf{D} = \frac{1}{K}\mathbf{E}_{\text{vac}} \tag{4.32}$$

When all space is filled with a homogeneous linear dielectric, the field everywhere is

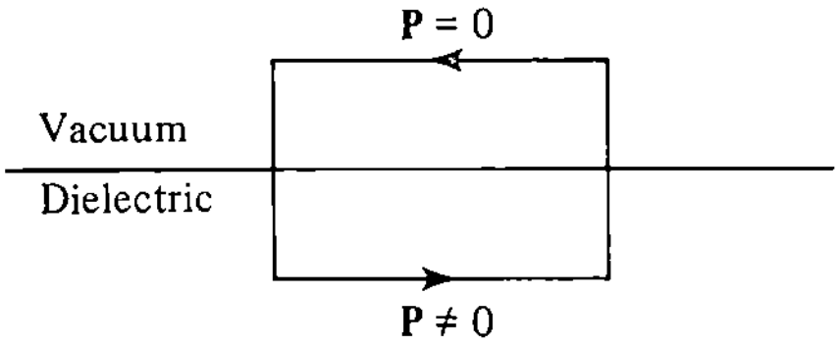

Figure 4.21

simply reduced by a factor of one over the dielectric constant. (Actually, it is not necessary for the dielectric to fill *all* space: In regions where the field is zero anyway, it can hardly matter whether we include dielectric or not, since there's no polarization in any event.)

Example 6

A parallel-plate capacitor (Fig. 4.22) is filled with insulating material of dielectric constant K. What effect does this have on its capacitance?

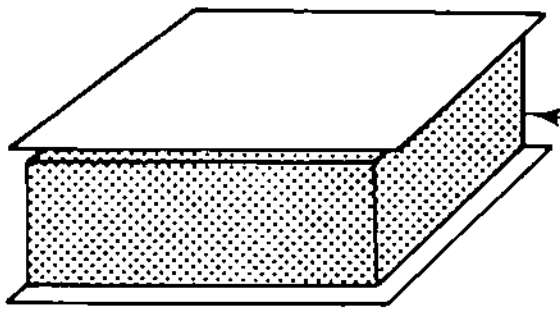

Figure 4.22

Solution: Since the field is essentially confined to the space between the plates, the dielectric will reduce **E** and hence also the potential difference V, by a factor $1/K$. Accordingly, the capacitance $C = Q/V$ is *increased by a factor of the dielectric constant,*

$$C = KC_{\text{vac}}$$

This is, in fact, a very common and efficient way to beef up a capacitor.

[A *crystal* is generally easier to polarize in some directions than others, and in this case (4.26) is replaced by the general linear relation

$$\left.\begin{aligned} P_x &= \epsilon_0(\chi_{e_{xx}}E_x + \chi_{e_{xy}}E_y + \chi_{e_{xz}}E_z) \\ P_y &= \epsilon_0(\chi_{e_{yx}}E_x + \chi_{e_{yy}}E_y + \chi_{e_{yz}}E_z) \\ P_z &= \epsilon_0(\chi_{e_{zx}}E_x + \chi_{e_{zy}}E_y + \chi_{e_{zz}}E_z) \end{aligned}\right\} \tag{4.33}$$

just as (4.1) was superseded by (4.3) for asymmetrical molecules. The nine coefficients, $\chi_{e_{xx}}$, $\chi_{e_{xy}}$, and so on, constitute the **susceptibility tensor.**]

Problem 4.18 The space between the plates of a parallel-plate capacitor (Fig. 4.23) is filled with two slabs of linear dielectric material. Each slab has thickness s, so the total distance between the plates is $2s$. Slab 1 has a dielectric constant of 2, and slab 2 has a dielectric constant of 1.5. The free charge density on the top plate is σ and on the bottom plate $-\sigma$.

(a) Find the electric displacement **D** in each slab.

(b) Find the electric field **E** in each slab.

(c) Find the polarization **P** in each slab.

(d) Find the potential difference between the plates.

(e) Find the location and amount of all bound charge.

(f) Now that you know all the charge (free and bound), recalculate the field in each slab, and compare with your answer to (b).

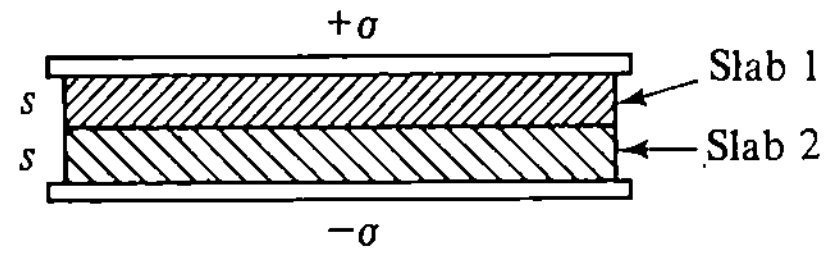

Figure 4.23

Problem 4.19 Suppose you have enough linear dielectric material, of dielectric constant K, to *half*-fill a parallel-plate capacitor (Fig. 4.24). By what fraction is the capacitance increased when you distribute the material as in Fig. 4.24(a)? How about Fig. 4.24(b)?

Problem 4.20 A sphere of linear dielectric material has embedded in it a uniform free charge density ρ. Find the potential at the center of the sphere, if its radius is R and its dielectric constant is K.

Problem 4.21 A certain coaxial cable consists of a copper wire, radius a, surrounded by a concentric copper tube of inner radius c (Fig. 4.25). The space between is partially filled (from b out to c) with material of dielectric constant K, as shown. Find the capacitance per unit length of this cable.

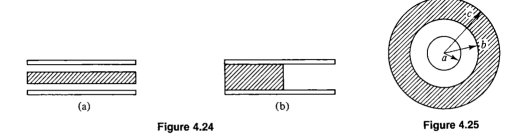

Figure 4.24

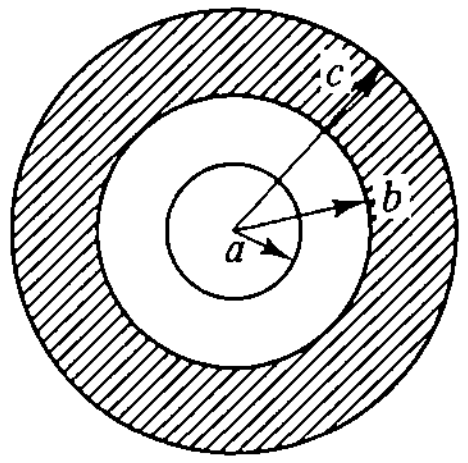

Figure 4.25

4.4.2 Special Problems Involving Linear Dielectrics

The examples I have treated so far in this chapter are either artificial, in the sense that the polarization is specified at the start, or else highly symmetrical, so that the displacement could be obtained directly from the free charge. When neither condition prevails, a more imaginative approach may be called for. In this section I'll show you two particularly nice problems of this kind.

Example 7

A sphere of linear dielectric material is placed in an originally uniform electric field $\mathbf{E}_0$ (Fig. 4.26). Find the new field inside the sphere.

Solution: This is reminiscent of Example 8 of Chapter 3, in which an uncharged *conducting* sphere is introduced into a uniform field. In that case the field of the induced charge completely canceled $\mathbf{E}_0$ within the sphere; in the *dielectric*, the cancellation resulting from the bound charge is only partial. We shall calculate $\mathbf{E}$ by a process of successive approximations. Imagine that the polarization takes place in a sequence of steps, as follows: When the external field $\mathbf{E}_0$ is first turned on, it creates a uniform polarization in the sphere,

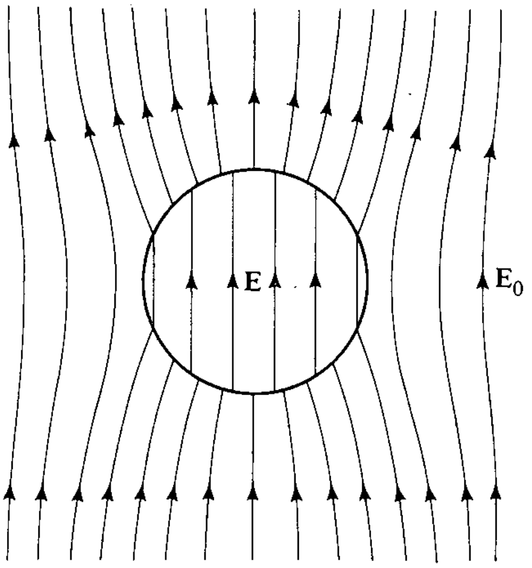

Figure 4.26

$$\mathbf{P}_0 = \epsilon_0 \chi_e \mathbf{E}_0$$

This is not the *final* polarization, since we have used $\mathbf{E}_0$ rather than $\mathbf{E}$ in applying equation (4.26). This polarization sets up a field of its own within the sphere, which according to (4.14) is uniform and has the value

$$\mathbf{E}_1 = -\frac{1}{3\epsilon_0}\mathbf{P}_0 = -\left(\frac{\chi_e}{3}\right)\mathbf{E}_0$$

The new field $\mathbf{E}_1$, in turn, will polarize the sphere some more:

$$\mathbf{P}_1 = \epsilon_0 \chi_e \mathbf{E}_1 = (\epsilon_0 \chi_e)\left(-\frac{\chi_e}{3}\right)\mathbf{E}_0$$

and this polarization will add to the field an amount

$$\mathbf{E}_2 = -\frac{1}{3\epsilon_0}\mathbf{P}_1 = \left(-\frac{\chi_e}{3}\right)^2\mathbf{E}_0$$

You can see that, as this process continues, the nth contribution will be

$$\mathbf{E}_n = \left(-\frac{\chi_e}{3}\right)^n\mathbf{E}_0$$

and the final field within the sphere is

$$\mathbf{E} = \mathbf{E}_0 + \mathbf{E}_1 + \mathbf{E}_2 + \cdots = \left[\sum_{n=0}^{\infty}\left(-\frac{\chi_e}{3}\right)^n\right]\mathbf{E}_0$$

The geometric series can be summed explicitly:

$$\mathbf{E} = \left(\frac{1}{1 + \chi_e/3}\right)\mathbf{E}_0 = \left(\frac{3}{2 + K}\right)\mathbf{E}_0 \tag{4.34}$$

Evidently the field, and thus also the polarization, is *uniform* within the sphere!

Of course, I do not pretend that the physical polarization of the sphere actually proceeds by this laborious sequence of discrete steps—it is just a conceptual device for "zeroing in" on the answer. If you have any doubts about the conclusion (4.34), *check* it: What polarization would the field **E** produce?

$$\mathbf{P} = \epsilon_0 \chi_e \mathbf{E} = \frac{\epsilon_0 \chi_e}{1 + \chi_e/3} \mathbf{E}_0$$

And what field would this polarization in turn generate?

$$\mathbf{E}_{\text{sphere}} = -\frac{1}{3\epsilon_0} \mathbf{P} = -\frac{\chi_e}{3 + \chi_e} \mathbf{E}_0$$

If you combine the field of the sphere with the background field $\mathbf{E}_0$, do you recover the field you started with?

$$\mathbf{E}_0 + \mathbf{E}_{\text{sphere}} = \mathbf{E}_0 - \frac{\chi_e}{3 + \chi_e} \mathbf{E}_0 = \frac{1}{1 + \chi_e/3} \mathbf{E}_0$$

Yes—the circle closes, and everything is consistent.

I must emphasize, however, that we were extraordinarily *lucky* to arrive at the solution by this means. Success depended critically on the fact that the field of a uniformly polarized sphere is *constant* inside the sphere. Without that, the "corrections" to the field, $\mathbf{E}_1$, $\mathbf{E}_2$, $\mathbf{E}_3$, . . . would have been increasingly complicated functions of position, and we could never have summed the series (though we might still have a useful *approximate* solution). As a matter of fact, it is only for *ellipsoids* (including the sphere and the infinite cylinder as special cases) that uniform polarization leads to a constant field within, and only for such shapes does a uniform $\mathbf{E}_0$ produce uniform polarization.[5]

Example 8

Suppose the entire region below the plane $z = 0$ in Fig. 4.27 is filled with uniform linear dielectric material of susceptibility χ_e. Calculate the force on a point charge q situated a distance d above the origin.

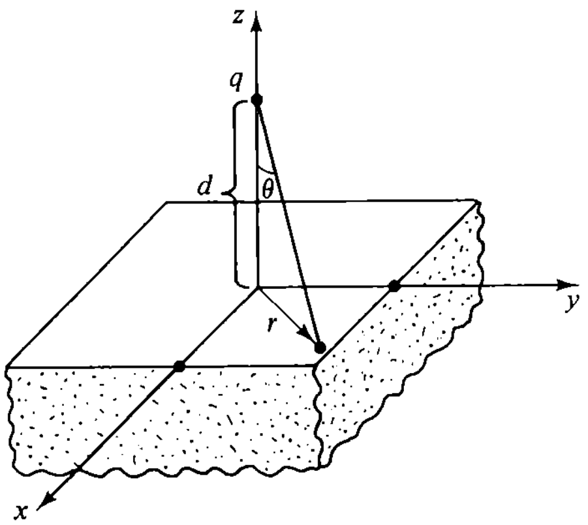

Figure 4.27

Solution: The surface bound charge on the xy plane is of opposite sign to q, so

[5]See H. S. Zapolsky, *Am. J. Phys.* **55**, 77 (1987).

the force will be attractive. (In view of equation (4.31), there is no volume bound charge.) Let us first calculate σ_b, using (4.11) and (4.26):

$$\sigma_b = \mathbf{P}\cdot\hat{n} = P_z = \epsilon_0\chi_e E_z$$

where E_z is the z-component of the total field just inside the dielectric, at $z = 0$. This field is due in part to q and in part to the bound charge itself. From Coulomb's law, the former contribution is

$$-\frac{1}{4\pi\epsilon_0}\frac{q}{(r^2 + d^2)}\cos\theta = -\frac{1}{4\pi\epsilon_0}\frac{qd}{(r^2 + d^2)^{3/2}}$$

where $r = \sqrt{x^2 + y^2}$ is the distance from the origin. The z-component of the field of the bound charge, meanwhile, is $-\sigma_b/2\epsilon_0$ (see footnote 6, p. 91). Thus,

$$\sigma_b = \epsilon_0\chi_e\left[-\frac{1}{4\pi\epsilon_0}\frac{qd}{(r^2 + d^2)^{3/2}} - \frac{\sigma_b}{2\epsilon_0}\right]$$

which we can solve for σ_b:

$$\sigma_b = -\frac{1}{2\pi}\left(\frac{\chi_e}{\chi_e + 2}\right)\frac{qd}{(r^2 + d^2)^{3/2}} \tag{4.35}$$

Apart from the factor $\chi_e/(\chi_e + 2)$, this is exactly the same as the induced charge on an infinite *conducting* plane under similar circumstances (equation (3.9)).[6] Evidently the *total* bound charge is

$$q_b = -\left(\frac{\chi_e}{\chi_e + 2}\right)q \tag{4.36}$$

We could, of course, obtain the field of σ_b by direct integration:

$$\mathbf{E} = \frac{1}{4\pi\epsilon_0}\int\left(\frac{\hat{\imath}}{\imath^2}\right)\sigma_b\, da$$

But as in the case of the conducting plane, there is a nicer solution by the method of images. Indeed, if we replace the dielectric by a single point charge q_b at the image position $(0, 0, -d)$, we have

$$V = \frac{1}{4\pi\epsilon_0}\left[\frac{q}{\sqrt{x^2 + y^2 + (z - d)^2}} + \frac{q_b}{\sqrt{x^2 + y^2 + (z + d)^2}}\right] \tag{4.37}$$

in the region $z > 0$. Meanwhile, a charge $(q + q_b)$ at $(0, 0, d)$ yields the potential

$$V = \frac{1}{4\pi\epsilon_0}\left[\frac{q + q_b}{\sqrt{x^2 + y^2 + (z - d)^2}}\right] \tag{4.38}$$

for the region $z < 0$. Taken together, (4.37) and (4.38) constitute a function which satisfies Poisson's equation with a point charge q at $(0, 0, d)$, which goes

[6]For some purposes a conductor can be regarded as the limiting case of a linear dielectric, with $\chi_e \to \infty$. This is often a useful check—you might apply it to Examples 5, 6, and 7 as well.

to zero at infinity, which is continuous at the boundary $z = 0$, and whose normal derivative exhibits the discontinuity appropriate to a surface charge σ_b at $z = 0$:

$$-\epsilon_0 \left(\frac{\partial V}{\partial z}\bigg|_{z=0^+} - \frac{\partial V}{\partial z}\bigg|_{z=0^-} \right) = -\frac{1}{2\pi}\left(\frac{\chi_e}{\chi_e + 2}\right)\frac{qd}{(x^2 + y^2 + d^2)^{3/2}}$$

Accordingly, this is the correct potential for our problem. In particular, the force on q is

$$\mathbf{F} = \frac{1}{4\pi\epsilon_0}\frac{qq_b}{(2d)^2}\hat{k} = -\frac{1}{4\pi\epsilon_0}\left(\frac{\chi_e}{\chi_e + 2}\right)\frac{q^2}{4d^2}\hat{k} \tag{4.39}$$

I do not offer any *motivation* for (4.37) and (4.38)—like all image solutions, this one owes its justification to the fact that it *works:* It solves Poisson's equation, and it meets the boundary conditions. Still, discovering an image solution is not entirely a matter of guesswork. There are at least two "rules of the game": (1) You must never put an image charge into the region where you're computing the potential. (Thus, (4.37) gives the potential for $z > 0$, but this image charge q_b is at $z = -d$. When we turn to the potential for $z < 0$ (4.38), the image charge $(q + q_b)$ is at $z = +d$.) (2) The image charges must add up to the correct total in each region. (That's how I knew to use q_b to account for the charge in the region $z \leq 0$, and $(q + q_b)$ to cover the region $z \geq 0$.)

Problem 4.22 A very long cylinder of linear dielectric material is placed in an otherwise uniform electric field $\mathbf{E}_0$. Find the resulting field within the cylinder. (The radius is R, the susceptibility χ_e, and the axis is perpendicular to $\mathbf{E}_0$.) You may want to refer back to Problem 4.13.

Problem 4.23 Solve for the field inside a sphere of linear dielectric in an otherwise uniform electric field $\mathbf{E}_0$ (Example 7) by the method of separation of variables. (Refer to Example 8 of Chapter 3. Note that: (a) V is continuous at R; (b) the discontinuity in the derivative of V is equal to $-\sigma_b/\epsilon_0$; (c) because the dielectric is linear, $\sigma_b = \epsilon_0\chi_e(\mathbf{E} \cdot \hat{r})$.)

Problem 4.24 Suppose the region *above* the xy plane in Example 8 is *also* filled with linear dielectric but of a different susceptibility χ_e'. Find the force on q in this case.

Problem 4.25 Prove the following uniqueness theorem: A region $\mathcal{R}$ contains a specified free charge distribution ρ_f and various pieces of linear dielectric material, with the susceptibility of each one given. If the potential is specified on the boundaries of $\mathcal{R}$ ($V = 0$ at infinity would be suitable) then the potential throughout $\mathcal{R}$ is uniquely determined. (*Suggestion*: Integrate $\nabla \cdot (V_3 \mathbf{D}_3)$ over $\mathcal{R}$.)

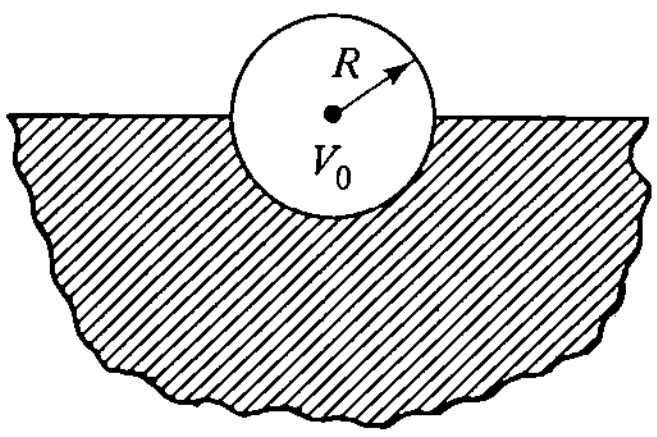

Figure 4.28

Problem 4.26 A conducting sphere (Fig. 4.28) at potential V_0 is half embedded in linear dielectric material of susceptibility χ_e, which occupies the region $z < 0$. *Claim*: the potential everywhere is exactly the same as it would have been in the absence of the dielectric! Check this claim, as follows:

(a) Write down the formula for the suggested potential $V(r)$, in terms of V_0, R, and r. Use it to determine the field, the polarization, the bound charge, and the free charge distribution on the sphere.

(b) Show that the total charge configuration would indeed produce the potential $V(r)$.

(c) Appeal to the uniqueness theorem in Problem 4.25 to complete the argument.

Could you solve the configurations in Fig. 4.29 with the same potential? If not, explain *why*.

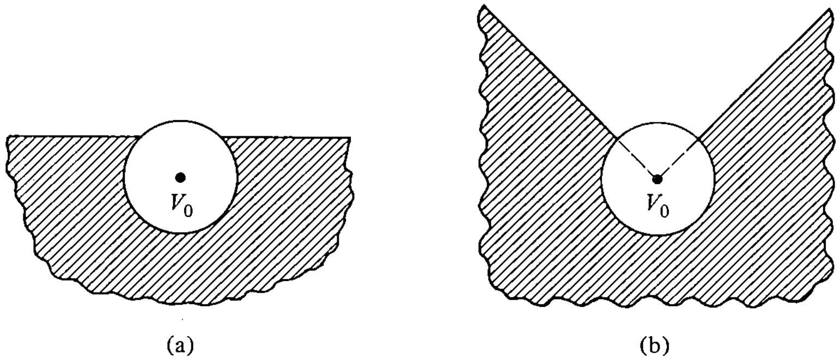

Figure 4.29

4.4.3 Energy in Dielectric Systems

It takes an amount of work

$$W = \tfrac{1}{2} CV^2$$

to charge a capacitor up to potential V (equation (2.48)). If the capacitor is filled with linear dielectric, its capacitance exceeds the vacuum value by a factor of the dielectric constant,

$$C = KC_{\text{vac}}$$

as we found in Example 6. Accordingly, the work necessary to charge a dielectric-filled capacitor is increased by the same factor. The reason is simple: You have to pump more (free) charge on to achieve a given potential because part of the field is cancelled off by the bound charges.

In Chapter 2, we derived a general formula for the energy stored in any electrostatic system (equation (2.39)):

$$W = \frac{\epsilon_0}{2} \int E^2 \, d\tau \tag{4.40}$$

The case of the dielectric-filled capacitor suggests that this should be changed to

$$W = \frac{\epsilon_0}{2} \int KE^2 \, d\tau = \frac{1}{2} \int \mathbf{D} \cdot \mathbf{E} \, d\tau \tag{4.41}$$

in the presence of linear dielectrics. To *prove* it, suppose the dielectric material is fixed in position, and we bring in the free charge, a bit at a time. As ρ_f is increased by an amount $\Delta\rho_f$, the polarization will change and with it the bound charge distribution—but we're interested only in the work done on the incremental *free* charge:

$$\Delta W = \int (\Delta\rho_f) V \, d\tau \tag{4.42}$$

Since $\nabla \cdot \mathbf{D} = \rho_f$, $\Delta\rho_f = \nabla \cdot (\Delta\mathbf{D})$, and

$$\Delta W = \int (\nabla \cdot (\Delta\mathbf{D})) V \, d\tau$$

Now,

$$\nabla \cdot (\Delta\mathbf{D} V) = [\nabla \cdot (\Delta\mathbf{D})] V + \Delta\mathbf{D} \cdot (\nabla V)$$

so

$$\Delta W = \int \nabla \cdot (\Delta\mathbf{D} V) \, d\tau + \int (\Delta\mathbf{D}) \cdot \mathbf{E} \, d\tau$$

The divergence theorem turns the first term into a surface integral, which vanishes if we integrate over all of space. Therefore, the work done is equal to

$$\Delta W = \int (\Delta\mathbf{D}) \cdot \mathbf{E} \, d\tau \tag{4.43}$$

So far this applies to *any* material. Now, if the medium is a *linear* dielectric, then $\mathbf{D} = \epsilon\mathbf{E}$, so that

$$\tfrac{1}{2}\Delta(\mathbf{D} \cdot \mathbf{E}) = \tfrac{1}{2}\Delta(\epsilon E^2) = \epsilon(\Delta\mathbf{E}) \cdot \mathbf{E} = (\Delta\mathbf{D}) \cdot \mathbf{E}$$

(for infinitesimal increments). Thus,

$$\Delta W = \Delta\left(\frac{1}{2} \int \mathbf{D} \cdot \mathbf{E} \, d\tau\right)$$

The total work done, then, as we build the free charge up from zero to the final configuration, is

$$W = \frac{1}{2} \int \mathbf{D} \cdot \mathbf{E} \, d\tau$$

confirming (4.41).[7]

It may puzzle you that (4.40), which we derived quite generally in Chapter 2, does not seem to apply in the context of dielectrics, where it is replaced by (4.41). The point is not that one or the other of these equations is *wrong*, but rather that they speak to somewhat different issues. The distinction is subtle, so let's go right back to

[7]In case you wonder why I did not derive (4.41) more easily by the method of Chapter 2 (Section 2.4.3), starting with $W = \frac{1}{2}\int \rho_f V \, d\tau$, the reason is that *this* formula is untrue, in general. Study the derivation of (2.36) and you will see that it applies only to the *total* charge. For *linear* dielectrics it happens to hold for the free charge alone, but this is scarcely obvious a priori and, in fact, is most easily confirmed by working backward from (4.41).

the beginning: what do we mean by "the energy of a system"? *Answer*: It is the work required to assemble the system. All right—but in the presence of dielectrics there are two very different ways we might construe this process: (1) We bring in all the charges (free *and* bound), one by one, with tweezers, and glue each one down in its proper final location. If *this* is what you mean by "assemble the system," then (4.40) is your formula for the energy stored. Notice, however, that this will *not* include the work involved in stretching and twisting the dielectric molecules (if we picture the positive and negative charges as held together by tiny springs, it does not include the spring energy, $\frac{1}{2}kx^2$, associated with polarizing each molecule).[8] (2) With the unpolarized dielectric in place, we bring in the *free* charges, one by one, allowing the dielectric to respond as it sees fit. If *this* is what you mean by "assemble the system" (and ordinarily it *is*, since free charge is what we actually push around), then (4.41) is the formula you want. In this case the "spring" energy *is* included, albeit indirectly, because the force you must apply to the *free* charge depends on the disposition of the *bound* charge, and that in turn depends on nature of the binding force. To put it another way, in method (2) the total energy of the system consists of three parts: the electrostatic energy of the free charge, the electrostatic energy of the bound charge, and the "spring" energy:

$$W_{\text{tot}} = W_{\text{free}} + W_{\text{bound}} + W_{\text{spring}}$$

The last two are equal and opposite (in procedure (2) the bound charges are always in equilibrium, and hence the *net* work done on them is zero); thus method (2), in calculating W_{free}, actually delivers W_{tot}, whereas method (1), by calculating $W_{\text{free}} + W_{\text{bound}}$, leaves out W_{spring}.

Incidentally, it is sometimes alleged that (4.41) represents the energy even for *non*linear dielectrics, but this is false: To proceed beyond (4.43) one must assume linearity. In fact, for many nonlinear systems the whole notion of "stored energy" loses its meaning, because the work done depends not only on the final configuration but on *how it got there*. If the molecular "springs" are allowed to have some *friction*, for instance, then W_{spring} can be made as large as you like, by assembling the charges in such a way that the spring is obliged to expand and contract many times before reaching its final state. In particular, you can get nonsensical results if you try to apply (4.41) to electrets, with frozen-in polarization (see Problem 4.28).

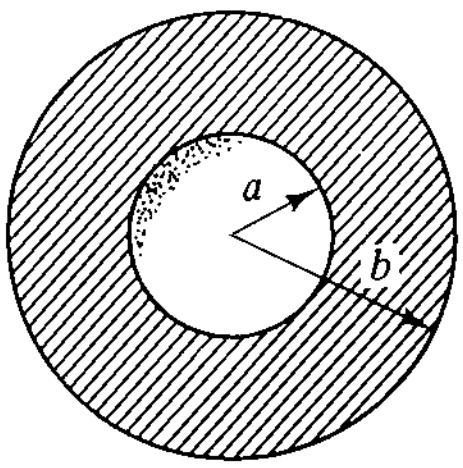

Figure 4.30

Problem 4.27 A spherical conductor, of radius a, carries a charge Q (Fig. 4.30). It is surrounded by linear dielectric material of susceptibility χ_e, out to a radius b. Find the energy of this configuration.

[8] The "spring" itself may be electrical in nature, but it is still not included in (4.40), where **E** is the *macroscopic* field.

Problem 4.28 Calculate W in (4.40) and (4.41) for a sphere of radius R with frozen-in uniform polarization $\mathbf{P}$. Comment on the discrepancy. Which (if either) is the "true" energy of this system?

4.4.4 Forces on Dielectrics

Just as a conductor is attracted into an electric field (equation (2.44)), so too is a dielectric—and for essentially the same reason: The bound charge tends to accumulate near to free charge of the opposite sign. The actual calculation of forces on dielectrics can be surprisingly tricky. Consider, for example, the case of a slab of linear dielectric material, partially inserted between the plates of a parallel-plate capacitor (Fig. 4.31). We have always pretended that the field is uniform inside a parallel-plate capacitor, and zero outside. Were this entirely correct, there would be no force on the dielectric at all. However, there is in reality a **fringing field** around the edges, which for most purposes can be ignored but in this case is responsible for the whole effect. (Indeed, the field *could* not terminate abruptly at the edge of the capacitor, for if it did the line integral of $\mathbf{E}$ around the closed loop shown in Fig. 4.32 would not be zero.) It is this nonuniform fringing field that pulls the dielectric into the capacitor.

It would be very tough to determine the fringing field; luckily, we can avoid this altogether, by the following ingenious method. Let W be the energy of the system—it depends, of course, on the overlap distance s. If we pull the dielectric out an infinitesimal distance ds, the energy is changed by an amount equal to the work done:

$$dW = F_{us}\, ds \tag{4.44}$$

where F_{us} is force we exert. Note that we must pull *just* hard enough to overcome the electrical force F on the dielectric: $F_{us} = -F$. Otherwise the slab will *accelerate,* and we'll have to include the increase in *kinetic* energy. Thus, the electrical force on the slab is

$$F = -\frac{dW}{ds} \tag{4.45}$$

So, if we can figure out how the energy of the system varies with s, we can calculate F *without knowing a thing about the fringing fields that are ultimately responsible for it!* (Of course, it's built into the whole structure of electrostatics that $\nabla \times \mathbf{E} = 0$, and

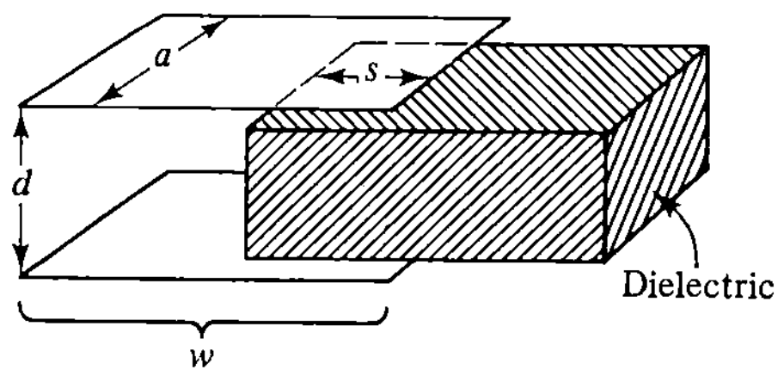

Figure 4.31

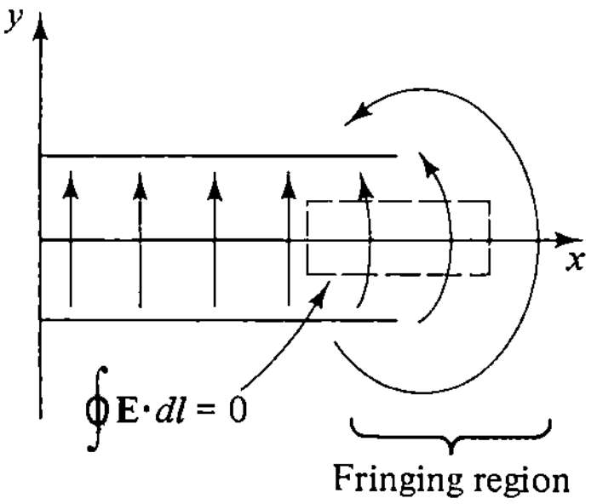

Figure 4.32

hence that the fringing fields must be present; we're not really getting something for nothing here—just cleverly exploiting the internal consistency of the theory.)

Now, the energy stored in the capacitor is

$$W = \tfrac{1}{2}CV^2 \tag{4.46}$$

and the capacitance in this case is

$$C = \frac{\epsilon_0 a}{d}(w + \chi_e s) \tag{4.47}$$

where w is the length of the plates (see Fig. 4.30). As the dielectric moves, the potential will change; what stays constant is the total *charge* on the plates, $Q = CV$. In terms of Q,

$$W = \frac{1}{2}\frac{Q^2}{C} \tag{4.48}$$

so that

$$F = -\frac{dW}{ds} = \frac{1}{2}\frac{Q^2}{C^2}\frac{dC}{ds} = \frac{1}{2}V^2\frac{dC}{ds} \tag{4.49}$$

But

$$\frac{dC}{ds} = \frac{\epsilon_0\chi_e a}{d}$$

and hence

$$F = \frac{1}{2}\epsilon_0\chi_e\frac{a}{d}V^2 \tag{4.50}$$

It is a common error to use (4.46) (with V constant) rather than (4.48) (with Q constant) in computing the force. One then obtains

$$F = -\frac{1}{2}V^2\frac{dC}{ds}$$

which is off by a sign. It is, of course, *possible* to maintain the capacitor at a fixed potential, by connecting it up to a battery. But in that case the *battery also does work* as the dielectric moves; instead of (4.44), we now have

$$dW = F_{us}\,ds + V\,dQ \tag{4.51}$$

where $V\,dQ$ is the work done by the battery. It follows that

$$F = -\frac{dW}{ds} + V\frac{dQ}{ds} = -\frac{1}{2}V^2\frac{dC}{ds} + V^2\frac{dC}{ds} = \frac{1}{2}V^2\frac{dC}{ds} \tag{4.52}$$

the same as before (4.49), with the *correct* sign. (Please understand: The force on the dielectric cannot possible depend on whether you plan to hold Q constant or V—it is determined entirely by the distribution of charge (free and bound). It's simpler to *calculate* the force assuming constant Q, because then you don't have to worry about work done by the battery, but if you insist, it can be done correctly either way.)

Problem 4.29 Two long coaxial cylindrical metal tubes (inner radius a, outer radius b) stand vertically in a tank of dielectric oil (susceptibility χ_e, mass density ρ). The inner one is maintained at potential V, and the outer one is grounded (Fig. 4.33). To what height (h) does the oil rise in the space between the tubes?

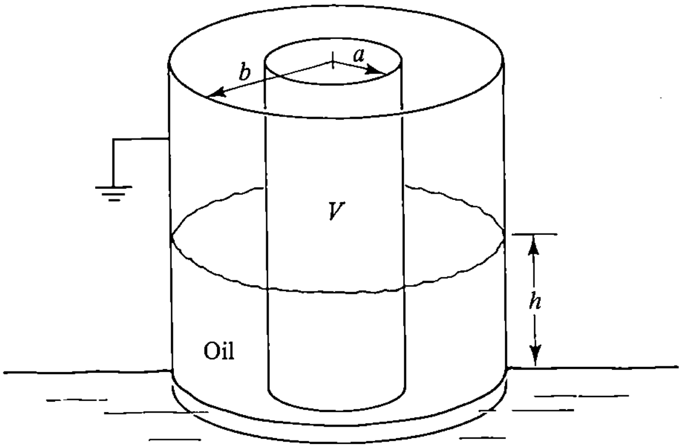

Figure 4.33

Problem 4.30 According to equation (4.5), the force on a single dipole is $(\mathbf{p} \cdot \nabla)\,\mathbf{E}$, so the total force on a piece of dielectric material must be

$$\mathbf{F} = \int (\mathbf{P} \cdot \nabla)\,\mathbf{E}_{\text{ext}}\,d\tau \tag{4.53}$$

[Here $\mathbf{E}_{\text{ext}}$ is the field of everything *except* the dielectric. You might assume—as others have been known to[9]—that it wouldn't matter if you used the *total* field; after all, the dielectric can't exert a force on *itself*. However, because the field of the dielectric is discontinuous at the location of any bound surface charge, the derivative introduces a spurious delta function, and you must either add a compensating surface term, or (better) stick with $\mathbf{E}_{\text{ext}}$, which suffers no such discontinuity.] Use equation (4.53) to determine the force on a tiny sphere of radius a, composed of linear dielectric material of susceptibility χ_e, which is situated a distance r from a fine wire carrying a uniform line charge λ.

4.4.5 Polarizability and Susceptibility

In a linear dielectric, the polarization is proportional to the field (equation (4.26)):

$$\mathbf{P} = \epsilon_0 \chi_e \mathbf{E} \tag{4.54}$$

If the material consists of nonpolar atoms or molecules, the induced dipole moment of each one is likewise proportional to the field (equation (4.1)):

$$\mathbf{p} = \alpha \mathbf{E} \tag{4.55}$$

[9]I thank Prof. M. Tiersten for pointing out this error in the first edition.

Question: What is the connection between the atomic polarizability α and the susceptibility χ_e? Since **P** (the dipole moment per unit volume) is **p** (the dipole moment per atom) times N (the number of atoms per unit volume),

$$\mathbf{P} = N\mathbf{p} = N\alpha\mathbf{E} \tag{4.56}$$

one's first inclination is to say that

$$\chi_e = \frac{N\alpha}{\epsilon_0} \tag{4.57}$$

And in fact this is not far off, if the density of atoms is small. But closer inspection reveals a subtle problem, for the field in (4.54) is the *total macroscopic* field in the medium, whereas the field in (4.55) is the *microscopic* field due to *everything except the particular atom* under consideration (atomic polarizability was defined for an isolated atom subject to a given external field). Now, the microscopic field, as always, would be quite impossible to calculate exactly, but we can get a pretty good estimate, as follows.

Imagine that the space allotted to each atom is a sphere of radius R, with the atom at its center. The density of atoms is then

$$N = \frac{1}{\frac{4}{3}\pi R^3} \tag{4.58}$$

The macroscopic field **E** can be written

$$\mathbf{E} = \mathbf{E}_{\text{self}} + \mathbf{E}_{\text{else}} \tag{4.59}$$

where $\mathbf{E}_{\text{self}}$ is the average over the sphere of the field due to the atom itself, and $\mathbf{E}_{\text{else}}$ is the average field due to everything outside the sphere. Equations (4.55) and (4.56) should properly be written, in this context, in terms of $\mathbf{E}_{\text{else}}$:

$$\mathbf{p} = \alpha\mathbf{E}_{\text{else}}, \qquad \mathbf{P} = N\alpha\mathbf{E}_{\text{else}} \tag{4.60}$$

(Actually, it is the field at the center, not the average over the sphere, that belongs here; but the two are equal, as you found in Problem 3.41). Now,

$$\mathbf{E}_{\text{self}} = -\frac{1}{4\pi\epsilon_0}\frac{\mathbf{p}}{R^3}$$

(from Problem 3.41, again) so (4.59) becomes

$$\mathbf{E} = -\frac{1}{4\pi\epsilon_0}\frac{\alpha}{R^3}\mathbf{E}_{\text{else}} + \mathbf{E}_{\text{else}} = \left(1 - \frac{\alpha}{4\pi\epsilon_0 R^3}\right)\mathbf{E}_{\text{else}}$$

Or, using (4.58) to eliminate R in favor of N:

$$\mathbf{E} = \left(1 - \frac{N\alpha}{3\epsilon_0}\right)\mathbf{E}_{\text{else}} \tag{4.61}$$

Thus,

$$\mathbf{P} = \frac{N\alpha}{(1 - N\alpha/3\epsilon_0)}\mathbf{E}$$

and therefore

$$\chi_e = \frac{N\alpha/\epsilon_0}{(1 - N\alpha/3\epsilon_0)} \tag{4.62}$$

This is a more accurate formula for the susceptibility than the naive version (4.57). To get a "feel" for this equation, let's put in the expression for α we obtained from the primitive model in Example 1: $\alpha = 4\pi\epsilon_0 a^3$, where a is the radius of the atom (equation (4.2)):

$$\chi_e = \frac{N(4\pi a^3)}{1 - N(\frac{4}{3}\pi a^3)} = \frac{3f}{1 - f} \tag{4.63}$$

where $f = (\frac{4}{3}\pi a^3)/(\frac{4}{3}\pi R^3)$ is the fraction of the total volume which is occupied by the atoms themselves. In a gas, where the space between atoms is large, f is correspondingly small and (4.57) is a suitable approximation. In a liquid or solid, f is greater (though it never reaches 1, obviously). Actually, (4.62) is not particularly accurate for dense materials anyway. The assignment of each atom to a spherical cell is a bit farfetched. For one thing, spheres don't "pack" properly to fill the space. The argument can be reformulated using cubical cells, and the result is the same, but even so, (4.62) cannot pretend to be a rigorous and exact law.

Nevertheless, it is a remarkable thing that a macroscopic quantity χ_e should be so simply related to an atomic parameter α. In fact, (4.62) is usually written as a formula for the polarizability in terms of the dielectric constant:

$$\alpha = \frac{3\epsilon_0}{N}\left(\frac{K - 1}{K + 2}\right) \tag{4.64}$$

Thus K, which is simple to measure by recording the change in C when you fill a capacitor with the substance, gives you important quantitative information about the structure of the atoms themselves. Equation (4.64) is known as the **Clausius-Mossotti** formula, or, in its application to optics, the **Lorentz-Lorenz** equation.[10]

In the case of polar molecules, the microscopic theory of susceptibility is not an exclusively electrostatic problem. Indeed, were it not for thermal agitation, the alignment of molecular dipoles would be total, no matter *how* weak the field, and the material would not be linear at all. The hotter the substance, the more frequently molecular alignments are upset by collisions; in general, therefore, we expect the susceptibility to decline with increasing temperature. But the details relate to statistical mechanics, and I shall not go into them here. (See Problem 4.33.)

[10]For a more complete discussion of the Clausius-Mossotti relation, see R. A. Levy, *Principles of Solid State Physics* (New York: Academic Press, 1968), Chapter 5.

Problem 4.31 Check the Clausius-Mossotti equation for the gases listed in Table 4.1. (Dielectric constants are given in Table 4.2.) (The densities here are so small that equations (4.64) and (4.57) are indistinguishable. For experimental data that confirm the Clausius-Mossotti correction term I refer you to the first edition of Purcell's *Electricity and Magnetism,* Problem 9.28.)[11]

! **Problem 4.32** In the last section we assigned each atom to a spherical cell of radius R. Carry through the same argument, but using *cubical* cells of side s. (You will need to calculate the average field over a cube, due to a dipole $\mathbf{p}$ at its center. To accomplish this, first find the average field of a *point* charge q a small distance d above the center (Fig. 4.34). By symmetry, the field averages to zero, save for the shaded slab at the bottom. Assume that the vertical component of $\mathbf{E}$ does not vary much across the thickness ($2d$) of this slab, so it may be replaced by its value on the median plane, and use Gauss's law to evaluate $\int \mathbf{E} \cdot d\mathbf{a}$ over this surface. See Section 9.13 in the first edition of Purcell's *Electricity and Magnetism.*[11])

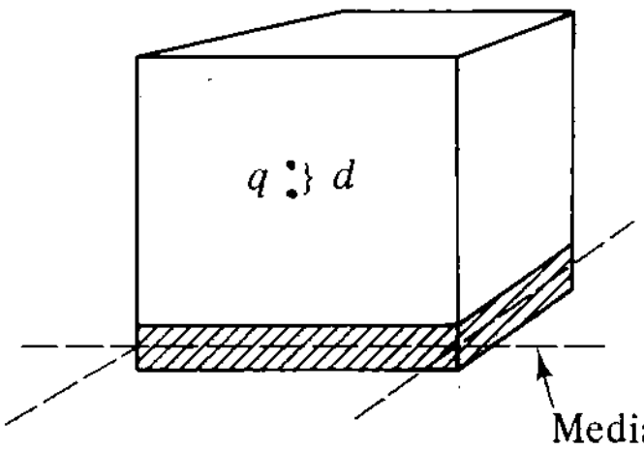

Figure 4.34

Further Problems on Chapter 4

Problem 4.33 An electric dipole $\mathbf{p}$, pointing in the y direction, is placed midway between two large conducting plates, as shown in Fig. 4.35. Each plate makes a small angle θ with respect to the x axis, and they are maintained at potentials $\pm V$. What is the direction of the net force on $\mathbf{p}$? Explain.

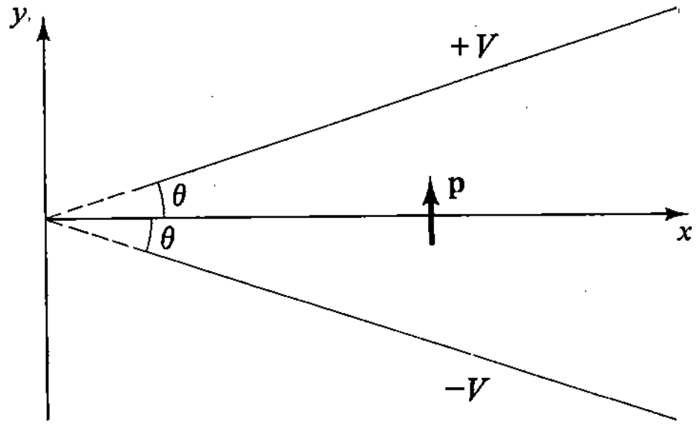

Figure 4.35

Problem 4.34 A dielectric cube of side s, centered at the origin, carries a "frozen-in" polarization $\mathbf{p} = k\mathbf{r}$, where k is a constant. Find all the bound charges, and check that they add up to zero.

[11]E. M. Purcell, *Electricity and Magnetism* (Berkeley Physics Course Vol. 2), (New York: McGraw-Hill, 1963).

Problem 4.35 A point charge q is imbedded at the center of a sphere of linear dielectric material (with susceptibility χ_e and radius R). Find the electric field, the polarization, and the bound charge, at a point $r < R$. What is the total bound charge at the surface? Where is the compensating negative bound charge located?

Problem 4.36 At the interface between one linear dielectric and another the electric field lines bend (see Fig. 4.36). Show that $\tan\theta_2/\tan\theta_1 = \epsilon_1/\epsilon_2$, assuming there is no *free* charge at the boundary.

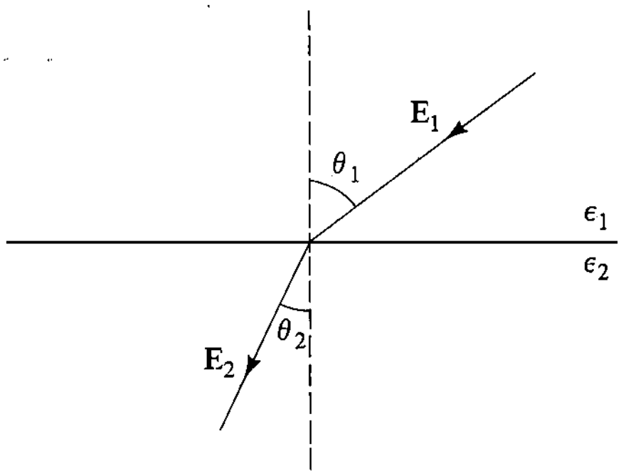

Figure 4.36

! **Problem 4.37** A point dipole $\mathbf{p}$ is imbedded at the center of a sphere of linear dielectric material (with radius R and dielectric constant K). Find the electric potential inside and outside the sphere.

$$\left(\textit{Answer}: \left(\frac{p\cos\theta}{4\pi\epsilon_0 r^2}\right)\frac{1}{K}\left[1 + 2\frac{r^3}{R^3}\frac{(K-1)}{(K+2)}\right], \qquad (r \le R);\right.$$

$$\left.\left(\frac{p\cos\theta}{4\pi\epsilon_0 r^2}\right)\left(\frac{3}{K+2}\right), \qquad (r \ge R).\right)$$

! **Problem 4.38** The Clausius-Mossotti equation, in the form (4.62), tells you how to calculate the susceptibility of a *nonpolar* substance, in terms of the atomic polarizability α. The Langevin equation tells you how to calculate the susceptibility of a *polar* substance, in terms of the permanent molecular dipole moment p. Here's how it goes:

(a) The energy of a dipole in an external field $\mathbf{E}$ is $u = -\mathbf{p}\cdot\mathbf{E}$—it ranges from $-pE$ to $+pE$, depending on the orientation. Statistical mechanics says that for a material in equilibrium at absolute temperature T, the probability of a given molecule having energy u is proportional to the Boltzmann factor,

$$\exp(-u/kT)$$

The average energy of the dipoles is therefore

$$\langle u\rangle = \frac{\int ue^{-(u/kT)}\,du}{\int e^{-(u/kT)}\,du}$$

where the integrals run from $-pE$ to $+pE$. Use this to show that the polarization of a substance containing N molecules per unit volume is

$$P = Np[\coth(pE/kT) - (kT/pE)] \tag{4.65}$$

That's the **Langevin formula.** Sketch P/Np as a function of pE/kT.

(b) Notice that for large fields/low temperatures virtually *all* the molecules are lined up, and the material is *non*linear. Ordinarily, however, kT is much greater than pE. Show that in this régime the material *is* linear, and calculate its susceptibility, in terms of N, p, T, and k. Compute the susceptibility of water at 20° C, and compare the experimental value—Table 4.2. (The dipole moment of water is 6.1×10^{-30} C · m.) This is rather far off, because we have again neglected the distinction between $\mathbf{E}$ and $\mathbf{E}_{\text{else}}$. The agreement is better in low-density gases, for which the difference between $\mathbf{E}$ and $\mathbf{E}_{\text{else}}$ is negligible. Try it for water vapor at 110° and 1 atm.

5

MAGNETOSTATICS

5.1 THE LORENTZ FORCE LAW

5.1.1 Magnetic Fields

Remember the original problem of classical electrodynamics: We have a collection of charges $q_1, q_2, q_3, \ldots$ (the "source" charges), and we want to calculate the force they exert on another charge Q (the "test" charge). (See Fig. 5.1.) According to the principle of superposition, it is sufficient to find the force of a *single* source charge—the total is then the vector sum of all the individual forces. Up to the present we have confined our attention to the simplest case, *electrostatics*, in which the source charge is *at rest* (though Q need not be). Now the time has come to consider forces between charges *in motion*.

To give you some sense of what is in store, imagine that I set up the following demonstration: Two wires hang from the ceiling, a few inches apart. When I turn on a current, so that it passes up one wire and back down the other, the wires jump apart—they plainly repel one another (Fig. 5.2(a)). How do you explain this? Well, you might suppose that the battery (or whatever drives the current) is actually charging up the wire, so naturally the different sections repel. But this "explanation" is incorrect. I could hold up a test charge near these wires and there would be no force on it, indicating that the wires are in fact electrically neutral. (It's true that electrons are flowing down the line—that's what a current *is*—but there are still just as many plus as minus charges on any given segment.) Moreover, I could hook up my demonstration so as to make the current flow up *both* wires (Fig. 5.2(b)); in this case the wires are found to *attract!*

Evidently, whatever force accounts for the attraction of parallel currents and the repulsion of antiparallel ones is *not* electrostatic in nature. It is our first encoun-

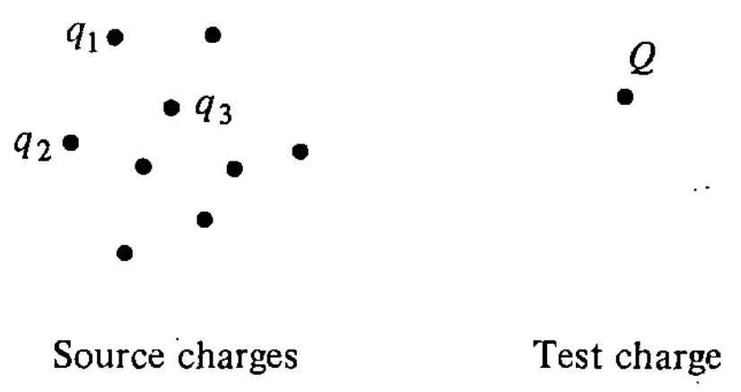

Figure 5.1

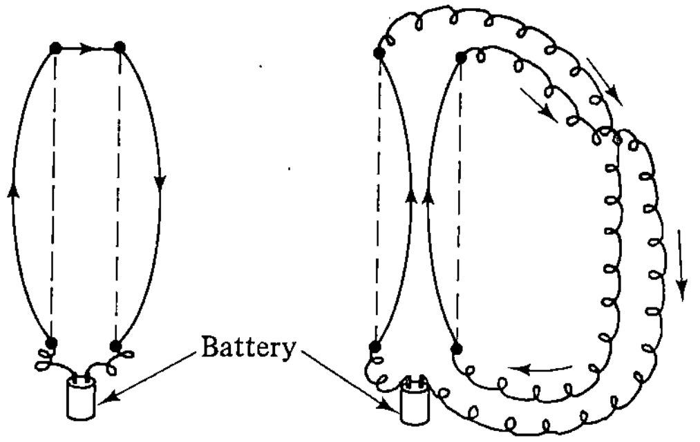

(a) Currents in opposite directions repel.

(b) Currents in same direction attract.

Figure 5.2

ter with a *magnetic* force. Whereas a *stationary* charge produces only an electric field **E** in the space around it, a *moving* charge generates, in addition, a magnetic field **B**. In fact, magnetic fields are a lot easier to detect, in practice—all you need is a boy scout compass. How these devices work is irrelevant at the moment—it is enough to know that the needle points in the direction of the local magnetic field. Ordinarily, this means *north*, in response to the earth's magnetic field, but in the laboratory, where typical fields may be hundreds of times stronger than that, the compass indicates the direction of whatever magnetic field is present. Now, if you hold up a tiny compass in the vicinity of a current-carrying wire, you quickly discover a peculiar thing: The field does not point *toward* the wire, nor *away* from it, but rather it *circles around the wire*. In fact, if you grab the wire with your right hand—thumb in the direction of the current—your fingers curl around in the direction of the magnetic field (Fig. 5.3). How can such a field lead to a force of *attraction* on a nearby parallel current? At the second wire the magnetic field points *into the page* (Fig. 5.4), the velocity of the charges is *upward*, and yet the resulting force is *to the left*. It's going to take a strange law to account for these directions; we'll introduce this law in the next section. Later on, in Section 5.2, we'll return to what is logically the prior question: How do you calculate the magnetic field of the first wire?

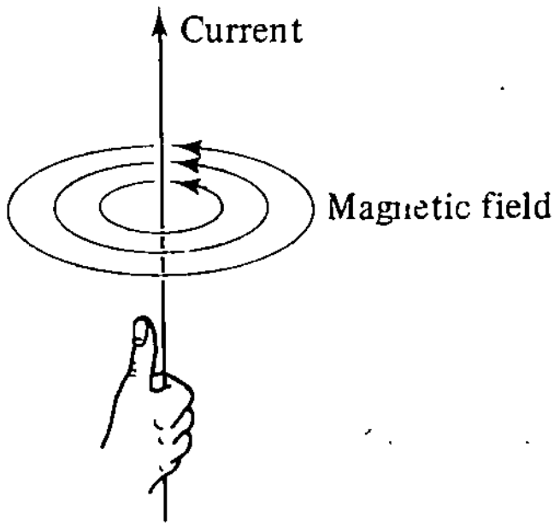

Figure 5.3

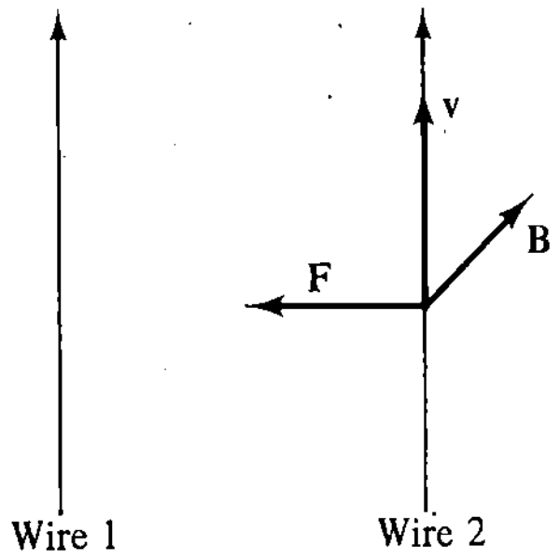

Figure 5.4

5.1.2 Magnetic Forces

It may have occurred to you that the combination of directions in Fig. 5.4 is just right for a cross product. In fact, the magnetic force on a charge Q, moving with velocity $\mathbf{v}$ in a magnetic field $\mathbf{B}$ is[1]

$$\boxed{\mathbf{F}_{\text{mag}} = Q(\mathbf{v} \times \mathbf{B})} \tag{5.1}$$

If we combine this with the electric force, $\mathbf{F}_{\text{elec}} = Q\mathbf{E}$, we obtain the **Lorentz force law**,

$$\mathbf{F} = Q[\mathbf{E} + (\mathbf{v} \times \mathbf{B})] \tag{5.2}$$

for the total electromagnetic force on Q. I do not pretend to have *derived* (5.1), of course. Like Coulomb's law, it is an axiom of the theory, whose justification is to be found in experiments such as the one I described in Section 5.1.1. Our main job from now on is to calculate the magnetic field $\mathbf{B}$ (and for that matter the electric field $\mathbf{E}$ as well, for the rules are more complicated when the source charges are in motion). But before we proceed, it is worthwhile to take a closer look at the Lorentz force law itself, for the magnetic term (5.1) is an extraordinary thing, and it leads to some very peculiar particle trajectories.

Example 1 Cyclotron motion

The archetypical motion of a charged particle in a magnetic field is circular, with the magnetic force providing the centripetal acceleration. In Fig. 5.5, a

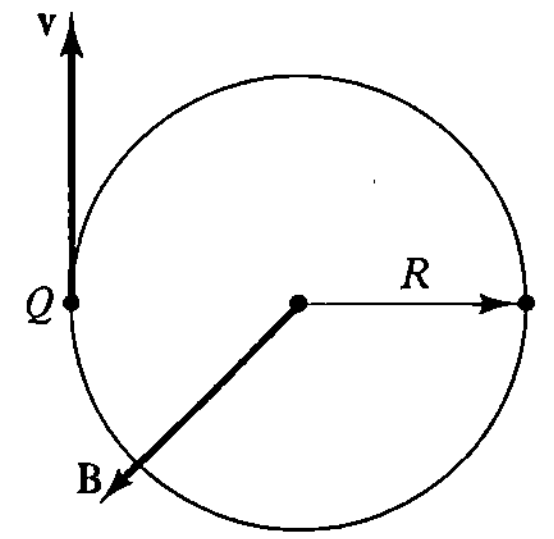

Figure 5.5

uniform magnetic field points out of the page; if the charge Q moves clockwise, with speed v, around a circle of radius R, the magnetic force (5.1) points inward and has a fixed magnitude (QvB)—just right to sustain uniform circular motion:

$$QvB = m\frac{v^2}{R}, \qquad \text{or} \quad p = QBR \tag{5.3}$$

where m is the particle's mass and p is its momentum. Equation (5.3) is known

[1] Since $\mathbf{F}$ and $\mathbf{v}$ are vectors, $\mathbf{B}$ must actually be a *pseudo*vector.

as the **cyclotron formula** because it describes the motion of a particle in a cyclotron—the first of the modern particle accelerators. It also suggests a simple experimental technique for finding the momentum of a particle: Shoot it into a region of known magnetic field, and measure the radius of its circular trajectory. This is in fact the standard means for determining the momentum of elementary particles. Incidentally, I assumed that the charge moves in a plane perpendicular to **B**. If it starts out with some additional speed $v_{\parallel}$ *parallel* to **B**, this component of the motion is unaffected by the magnetic force (5.1), and the particle moves in a helix (Fig. 5.6). The radius is still given by (5.3), only the velocity in question is the component perpendicular to **B**, $v_{\perp}$.

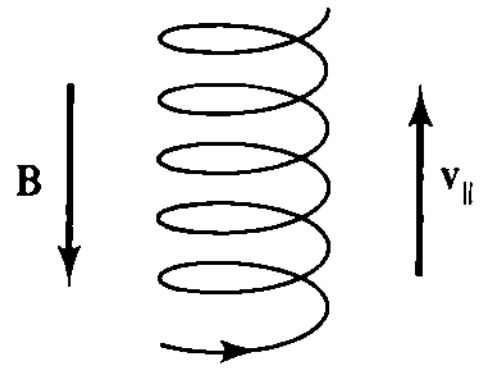

Figure 5.6

Example 2 Cycloid Motion

A more exotic trajectory occurs if we include a uniform electric field, at right angles to the magnetic one. Suppose, for instance, that **B** points in the x-direction, and **E** in the z-direction, as shown in Fig. 5.7. A particle is released from the origin: What path will it follow?

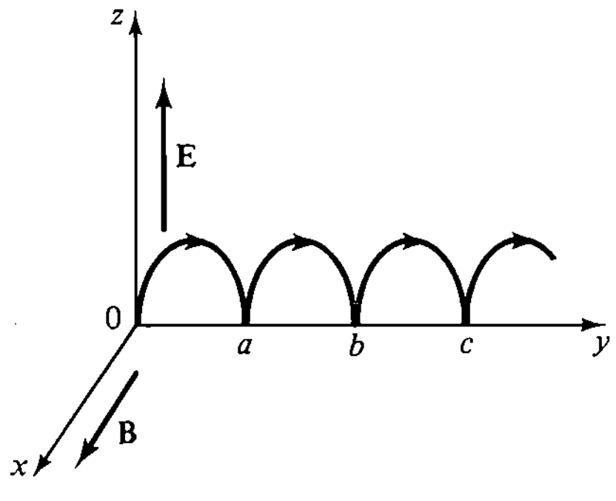

Figure 5.7

Solution: Let's piece it out qualitatively, first. Initially, the particle is at rest, so the magnetic force is zero, and the electric field accelerates the charge in the z-direction. As it picks up speed, a magnetic force develops which, according to (5.1), pulls the charge around to the right. The faster it goes, the stronger F_{mag} becomes; eventually, it curves the particle back around toward the y axis. At this point the charge is moving *against* the electrical force, so it begins to slow down—the magnetic force then decreases, and the electrical force takes over, bringing the charge to rest at point a, in Fig. 5.7. There the entire process commences anew, carrying the particle over to point b, and so on.

Now let's do it quantitatively. There being no force in the x-direction, the position of the particle at any time t can be described by the vector $(0, y(t), z(t))$; the velocity is therefore

$$\mathbf{v} = (0, \dot{y}, \dot{z})$$

where dots indicate time derivatives. Thus,

$$\mathbf{v} \times \mathbf{B} = \begin{vmatrix} \hat{i} & \hat{j} & \hat{k} \\ 0 & \dot{y} & \dot{z} \\ B & 0 & 0 \end{vmatrix} = (B\dot{z})\hat{j} - (B\dot{y})\hat{k}$$

and hence, applying Newton's law,

$$\mathbf{F} = Q(\mathbf{E} + \mathbf{v} \times \mathbf{B}) = Q[(E)\hat{k} + (B\dot{z})\hat{j} - (B\dot{y})\hat{k}] = m\mathbf{a} = m[\ddot{y}\hat{j} + \ddot{z}\hat{k}]$$

Or, treating the $\hat{j}$ and $\hat{k}$ components separately,

$$QB\dot{z} = m\ddot{y}, \qquad QE - QB\dot{y} = m\ddot{z}$$

For convenience, let

$$\frac{QB}{m} = \omega \tag{5.4}$$

(This is the **cyclotron frequency**, at which the particle would revolve in the absence of any electric field.) Then the equations of motion take the form

$$\ddot{y} = \omega\dot{z}, \qquad \ddot{z} = \omega\left(\frac{E}{B} - \dot{y}\right) \tag{5.5}$$

Their general solution[2] is

$$y(t) = C_1 \cos \omega t + C_2 \sin \omega t + \frac{E}{B} t + C_3 \tag{5.6}$$

$$z(t) = C_2 \cos \omega t - C_1 \sin \omega t + C_4$$

In this instance the particle started from rest ($\dot{y}(0) = \dot{z}(0) = 0$) at the origin ($y(0) = z(0) = 0$). These four conditions determine the constants C_1, C_2, C_3, and C_4:

$$y(t) = \frac{E}{\omega B}(\omega t - \sin \omega t), \qquad z(t) = \frac{E}{\omega B}(1 - \cos \omega t) \tag{5.7}$$

In this form the answer is not terribly enlightening, but if we let

$$R = \frac{E}{\omega B} \tag{5.8}$$

[2] As coupled differential equations, (5.5) are easily solved by differentiating the first and then using the second to eliminate $\ddot{z}$.

and eliminate the sines and cosines by exploiting the trigonometric identity $\sin^2 \omega t + \cos^2 \omega t = 1$, we find that

$$(y - R\omega t)^2 + (z - R)^2 = R^2 \tag{5.9}$$

This is the formula for a *circle* of radius R, whose center $(0, R\omega t, R)$ travels in the y-direction at a constant speed,

$$v = \omega R = \frac{E}{B} \tag{5.10}$$

The particle moves as though it were a spot on the rim of a wheel of radius R rolling down the y axis at speed v. The curve generated in this way is called a **cycloid**. Notice that the overall motion is not in the direction of **E**, as you might suppose, but perpendicular to it.

One feature of the magnetic force law (5.1) warrants special attention:

Magnetic forces do no work.

For if Q moves an amount $d\mathbf{l} = \mathbf{v}\,dt$, the work done is

$$dW_{\text{mag}} = \mathbf{F}_{\text{mag}} \cdot d\mathbf{l} = Q(\mathbf{v} \times \mathbf{B}) \cdot \mathbf{v}\,dt = 0 \tag{5.11}$$

[$(\mathbf{v} \times \mathbf{B})$ is perpendicular to $\mathbf{v}$, so $(\mathbf{v} \times \mathbf{B}) \cdot \mathbf{v} = 0$.] Magnetic forces may alter the *direction* in which a particle moves, but they cannot speed it up or slow it down. The fact that magnetic forces do no work is an elementary and direct consequence of the Lorentz force law, but there are many situations in which it *appears* so manifestly false that one's confidence is bound to waver. When a magnetic crane lifts the carcass of a junked car, for instance, *something* is obviously doing work, and it seems perverse to deny that the magnetic force is responsible. Well, perverse or not, deny it we must, and it can be a very subtle matter to figure out what agency *does* deserve the credit in such circumstances.

Problem 5.1 A particle of charge q enters a region of uniform magnetic field **B** (pointing *into* the page). The field deflects the particle a distance d above the original line of flight, as shown in Fig. 5.8. Is the charge positive, or negative? In terms of a, d, B, and q, find the momentum of the particle.

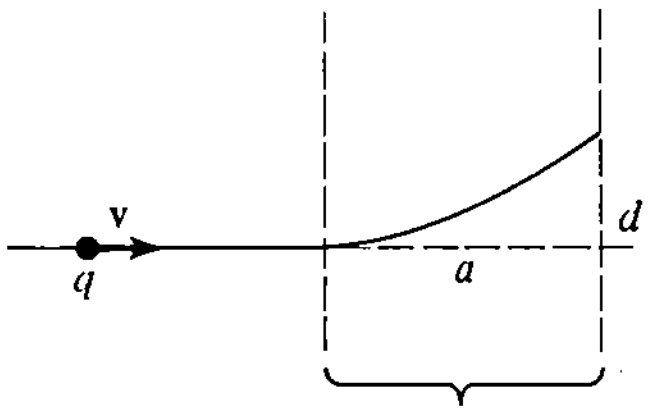

Figure 5.8

Problem 5.2 Find and sketch the trajectory of the particle in Example 2 if it starts at the origin with velocity

(a) $\mathbf{v}(0) = \frac{E}{B}\hat{y}$

(b) $\mathbf{v}(0) = \frac{E}{2B}\hat{y}$

(c) $\mathbf{v}(0) = \frac{E}{B}(\hat{y} + \hat{z})$

5.1.3 Currents

The **current** in a wire is the *charge per unit time* passing a given point. By definition, negative charges moving to the left count the same as positive ones to the right. This conveniently reflects the *physical* fact that almost all phenomena involving moving charges depend on the *product* of charge times velocity—if you change the sign of q *and* $\mathbf{v}$, you get the same answer, so it doesn't really matter which you have. (The Lorentz force law is a case in point; the Hall effect (Problem 5.51) is a notorious exception.) In practice, it is ordinarily the negatively charged electrons that do the moving—in the direction *opposite* to the electric current. To avoid the petty complications this entails, I shall often pretend it's the positive charges that move, as in fact everyone assumed they did for a century or so after Benjamin Franklin established his unfortunate convention.[3] Current is measured in coulombs per second, or **amperes** (A):

$$1\ \mathrm{A} = 1\ \mathrm{C/s} \tag{5.12}$$

A line charge λ traveling down a wire at speed v (Fig. 5.9) constitutes a current

$$I = \lambda v \tag{5.13}$$

because a segment of length $v\Delta t$, carrying charge $\lambda v \Delta t$, passes by point A in a time interval Δt. The current at each point is actually a *vector*:

$$\mathbf{I} = \lambda \mathbf{v} \tag{5.14}$$

However, since the path of the flow is dictated by the shape of the wire, most people don't bother to display the vectorial character of $\mathbf{I}$ explicitly. A neutral wire, of course, contains as many stationary positive charges as mobile negative ones. The former do not contribute to the current—the charge density λ in (5.13) refers only to the *moving* charges (if *both* charges move, then $\mathbf{I} = \lambda_+\mathbf{v}_+ + \lambda_-\mathbf{v}_-$).

[3]Had we from the start called the electron plus and the proton minus, the problem would never arise. In the context of Franklin's experiments with cats' furs and glass rods, the choice was completely arbitrary.

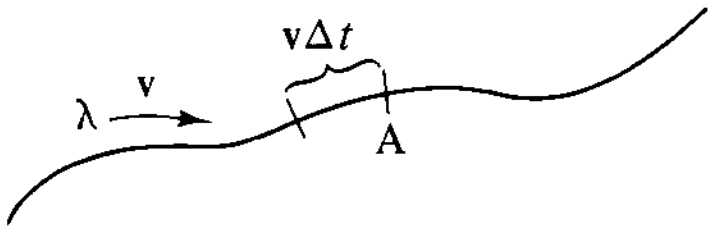

Figure 5.9

The magnetic force on a segment of current-carrying wire is evidently

$$\mathbf{F}_{\text{mag}} = \int (\lambda\, dl)(\mathbf{v} \times \mathbf{B}) = \int (\mathbf{I} \times \mathbf{B})\, dl \tag{5.15}$$

Inasmuch as **I** and $d\mathbf{l}$ both point in the same direction, we can equally well write this as

$$\boxed{\mathbf{F}_{\text{mag}} = \int I(d\mathbf{l} \times \mathbf{B})} \tag{5.16}$$

Typically, the current is constant (in magnitude) along the wire, and in that case I comes outside the integral:

$$\mathbf{F}_{\text{mag}} = I \int (d\mathbf{l} \times \mathbf{B}) \tag{5.17}$$

Example 3

A rectangular loop of wire, supporting a mass m, hangs vertically with one end in a uniform magnetic field **B**, which points into the page in the shaded region of Fig. 5.10. For what current I, in the loop, would the magnetic force upward exactly balance the gravitational force downward?

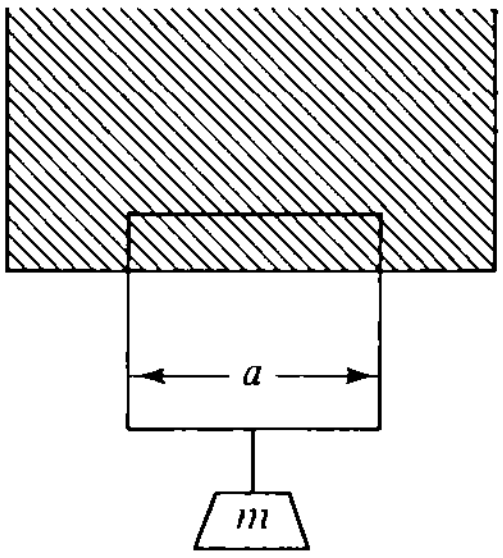

Figure 5.10

Solution: First of all, the current must circulate clockwise in order for $(\mathbf{I} \times \mathbf{B})$ in the horizontal segment to point upward. Then

$$F_{\text{mag}} = IBa$$

where a is the width of the loop. (The magnetic forces on the two vertical segments cancel.) For F_{mag} to balance the weight, we must therefore have

$$I = \frac{mg}{Ba}$$

When charge flows over a *surface,* we describe it by the **surface current density**, **K**, defined as follows: Consider a "ribbon" of infinitesimal width $dl_\perp$, running paral-

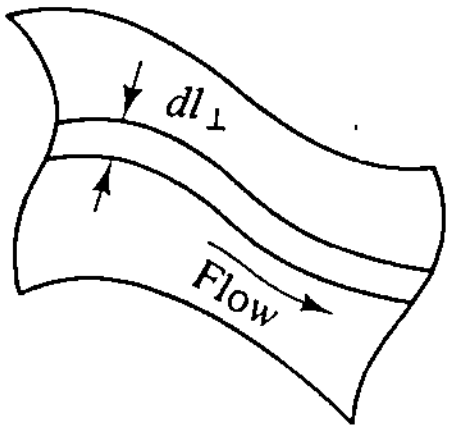

Figure 5.11

lel to the flow (Fig. 5.11). If the current in this ribbon is $d\mathbf{I}$, the surface current density is

$$\mathbf{K} = \frac{d\mathbf{I}}{dl_\perp} \tag{5.18}$$

In words, K is the *current per unit length-perpendicular-to-flow.* In particular, if the (mobile) surface charge density is σ and its velocity is $\mathbf{v}$, then

$$\mathbf{K} = \sigma\mathbf{v} \tag{5.19}$$

since the net line charge on the ribbon is $\sigma\, dl_\perp$, and hence $d\mathbf{I} = (\sigma\, dl_\perp)\mathbf{v}$. In general, $\mathbf{K}$ will vary from point to point over the surface in response to variations in σ and/or $\mathbf{v}$. The magnetic force on a surface current is

$$\mathbf{F}_{\text{mag}} = \int (\sigma\, da)(\mathbf{v} \times \mathbf{B}) = \int (\mathbf{K} \times \mathbf{B})\, da \tag{5.20}$$

Caveat: Just as $\mathbf{E}$ suffers a discontinuity at a surface *charge,* so $\mathbf{B}$ is discontinuous at a surface *current.* In equation (5.20), you must be careful to use the *average* field, just as we did in Section 2.5.3.

When the flow of charge is distributed throughout a three-dimensional region, we describe it by the **volume current density, J**, defined thus: Consider a "tube" of infinitesimal cross section $da_\perp$, running parallel to the flow (Fig. 5.12). If the current in this tube is $d\mathbf{I}$, the volume current density is

$$\mathbf{J} = \frac{d\mathbf{I}}{da_\perp} \tag{5.21}$$

In words, J is the *current per unit area-perpendicular-to-flow.* If the (mobile) volume charge density is ρ and the velocity is $\mathbf{v}$, then

$$\mathbf{J} = \rho\mathbf{v} \tag{5.22}$$

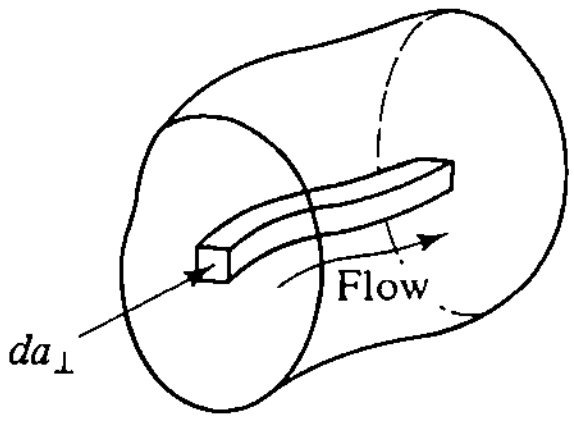

Figure 5.12

The magnetic force on a volume current is

$$\mathbf{F}_{\text{mag}} = \int (\rho \, d\tau)(\mathbf{v} \times \mathbf{B}) = \int (\mathbf{J} \times \mathbf{B}) \, d\tau \tag{5.23}$$

Example 4

(a) A current I is uniformly distributed over a wire of circular cross section, with radius R (Fig. 5.13). Find the volume current density J.

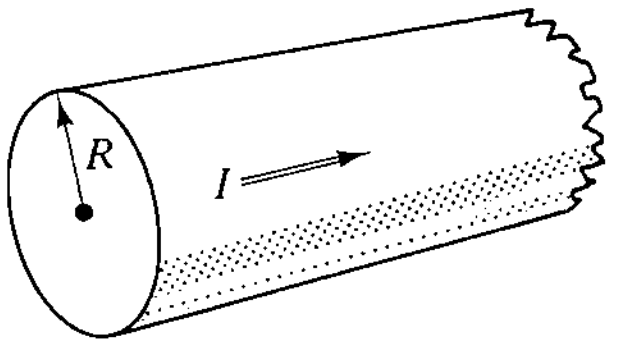

Figure 5.13

Solution: The area-perpendicular-to-flow is πR^2, so

$$J = \frac{I}{\pi R^2}$$

This was trivial because the current density was uniform.

(b) Suppose the current density in this wire is proportional to the distance from the axis,

$$J = kr$$

(for some constant k). Find the total current in the wire.

Solution: Because J varies with r, we must *integrate* equation (5.21). The current in the shaded tube (Fig. 5.14) is $J \, da_\perp$ and $da_\perp = r \, dr \, d\theta$. So,

$$I = \int (kr)(r \, dr \, d\theta)$$

$$= 2\pi k \int_0^R r^2 \, dr = \frac{2\pi k R^3}{3}$$

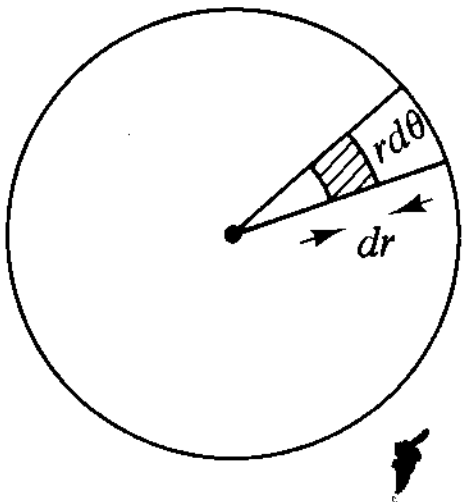

Figure 5.14

According to (5.21), the current crossing a surface S can be written as

$$I = \int_S J\, da_\perp = \int_S \mathbf{J} \cdot d\mathbf{a} \tag{5.24}$$

The dot product serves nicely to pick out the appropriate component of $d\mathbf{a}$. In particular, the total charge per unit time leaving a volume V is

$$\oint_S \mathbf{J} \cdot d\mathbf{a} = \int_V (\nabla \cdot \mathbf{J})\, d\tau$$

Because charge is conserved, whatever flows out through the surface must come at the expense of that remaining inside:

$$\int_V (\nabla \cdot \mathbf{J})\, d\tau = -\frac{d}{dt}\int_V \rho\, d\tau = -\int_V \left(\frac{\partial \rho}{\partial t}\right) d\tau$$

(The minus sign reflects the fact that an *outward* flow *decreases* the charge left in V.) Since this applies to *any* volume, we conclude that

$$\boxed{\nabla \cdot \mathbf{J} = -\frac{\partial \rho}{\partial t}.} \tag{5.25}$$

This is the precise mathematical statement of local charge conservation; it is called the **continuity equation.**

For future reference, let me summarize the "dictionary" we have implicitly developed for translating equations into the forms appropriate to point, line, surface, and volume currents:

$$\sum_{i=1}^{n} (\quad) q_i \mathbf{v}_i \sim \int_{\text{line}} (\quad) \mathbf{I}\, dl \sim \int_{\text{surface}} (\quad) \mathbf{K}\, da \sim \int_{\text{volume}} (\quad) \mathbf{J}\, d\tau \tag{5.26}$$

This correspondence, which is analogous to $q \sim \lambda\, dl \sim \sigma\, da \sim \rho\, d\tau$ for the various charge distributions, generates (5.15), (5.20), and (5.23) from the original magnetic force law (5.1).

Problem 5.3 Suppose the magnetic field in some region has the form

$$\mathbf{B} = kz\hat{\imath} \qquad (k \text{ is some constant})$$

Find the force on a square loop of side s, lying in the yz plane, centered at the origin, which carries a current I.

Problem 5.4 A current I flows down a wire of radius R.

(a) If it is uniformly distributed over the surface, what is the surface current density K?

(b) If it is distributed in such a way that the volume current density is inversely proportional to the distance from the axis, what is J?

Problem 5.5
(a) A phonograph record carries a uniform density of "static electricity" σ. If it rotates at angular velocity ω, what is the surface current density K at a distance r from the center?
(b) A uniformly charged sphere of radius R and total charge Q is centered at the origin and spinning at a constant angular velocity ω about the z axis. Find the current density $\mathbf{J}$ at any point (r, θ, ϕ) within the sphere.

Problem 5.6 In Example 3, what happens if the current is *greater* than mg/Ba, and is held constant? (A complete analysis of this problem is surprisingly subtle. Draw a diagram carefully, showing the actual motion of a charge in the upper segment of the loop and the direction of the force on this charge. Notice that a nonmagnetic horizontal force is now necessary in order to sustain the current. What agency might provide this extra force? Show that the work it does, as the loop rises a distance y, is $(IBay)$, and confirm that this equals the energy gained by the loop. How much work did the magnetic force do?)

Problem 5.7 For a configuration of charges and currents confined within a volume V, show that $\int_V \mathbf{J}\, d\tau = d\mathbf{p}/dt$, where $\mathbf{p}$ is the total dipole moment. (*Hint*: Evaluate $\int_V \nabla \cdot (x\mathbf{J})\, d\tau$.)

5.2 THE BIOT-SAVART LAW

5.2.1 Steady Currents

Stationary charges produce electric fields that are constant in time; hence the term **electrostatics.**[4] *Steady currents* produce magnetic fields that are constant in time. For this reason the theory of steady currents is called **magnetostatics**:

$$\begin{cases}\text{stationary charges} \Rightarrow \text{constant electric field: electrostatics}\\ \text{steady currents} \Rightarrow \text{constant magnetic field: magnetostatics}\end{cases}$$

By **steady current** I mean a flow of charge that has been going on forever—never increasing, never decreasing, and never changing course. In any region of space, the charge per unit time passing through right now is the same as it was 10 seconds ago or 10 years ago. (Some people call these "stationary currents"—to me, that's a contradiction in terms.) Of course, there's no such thing in practice as a *truly* steady current, any more than there is a *truly* stationary charge. In this sense both electrostatics and magnetostatics describe artificial worlds that exist only in textbooks. However, they represent suitable *approximations* as long as the actual fluctuations are reasonably slow; in fact, for most purposes magnetostatics applies very well to household currents, which alternate 60 times a second!

Notice that a moving *point* charge cannot possibly constitute a steady current. If it's here one instant, it's gone the next. This may seem like a minor thing to *you*, but it's a major headache for *me*. You see, I developed each topic in electrostatics by starting out with the simple case of a point charge at rest; then I generalized to an

[4] Actually, it is not necessary that all charges be stationary, but only that the charge *density* at each point be constant. For example, the sphere in Problem 5.5 produces an electrostatic field $1/4\pi\epsilon_0\,(Q/r^2)\hat{\mathbf{r}}$, even though it is rotating, because ρ does not depend on t.

arbitrary charge distribution by invoking the superposition principle. This line of attack is not open to us in magnetostatics because the point charge does not yield a static field in the first place. We are *forced* to deal with extended current distributions right from the start, and as a result the arguments are bound to be more cumbersome.

When a steady current flows in a wire, its magnitude I must be the same all along the line; otherwise, charge would be piling up somewhere, and the current could not be maintained indefinitely. By the same token, $\partial\rho/\partial t = 0$ in magnetostatics, and hence the continuity equation (5.25) becomes

$$\nabla \cdot \mathbf{J} = 0 \tag{5.27}$$

5.2.2 The Magnetic Field of a Steady Current

The magnetic field of a steady line current is given by the **Biot-Savart law**:

$$\boxed{\mathbf{B}(P) = \frac{\mu_0}{4\pi}\int \frac{\mathbf{I} \times \hat{\boldsymbol{\mathscr{r}}}}{\mathscr{r}^2}\,dl = \frac{\mu_0}{4\pi} I \int \frac{d\mathbf{l} \times \hat{\boldsymbol{\mathscr{r}}}}{\mathscr{r}^2}} \tag{5.28}$$

The integration is along the current path, in the direction of flow; $d\mathbf{l}$ is an element of length along the wire, and $\boldsymbol{\mathscr{r}}$, as always, is the vector from the source to the point P (Fig. 5.15). The constant μ_0 is called the **permeability of free space**; its value is[5]

$$\mu_0 = 4\pi \times 10^{-7}\ \mathrm{N/A^2} \tag{5.29}$$

These units are such that $\mathbf{B}$ itself comes out in newtons/ampere-meter, as required by the magnetic force law (5.1)—or, in the official terminology, **teslas** (T):

$$1\ \text{tesla} = 1\ \mathrm{N/(A \cdot m)} \tag{5.30}$$

(For some reason, in this one case the cgs unit, the **gauss**, is more commonly used than the mks unit: 1 tesla $= 10^4$ gauss. The earth's magnetic field is about half a gauss; a strong laboratory magnetic field is, say, 10,000 gauss.) As the fundamental starting equation of magnetostatics, the Biot-Savart law plays a role analogous to Coulomb's law in electrostatics. The $1/\mathscr{r}^2$ factor is in fact somewhat reminiscent of Coulomb's law.

[5] This is an exact number, not an empirical constant. It serves (via equation 5.33) to define the ampere, and the ampere in turn defines the coulomb.

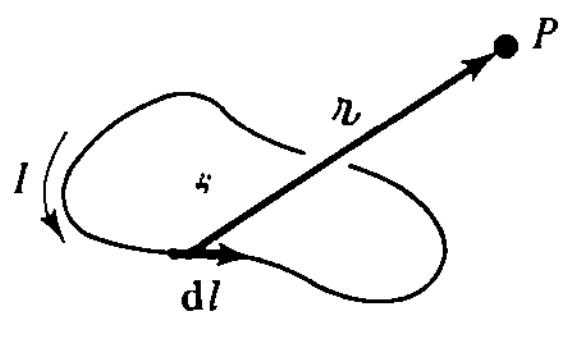

Figure 5.15

Example 5

Find the magnetic field a distance z above a long straight wire carrying a steady current I (Fig. 5.16).

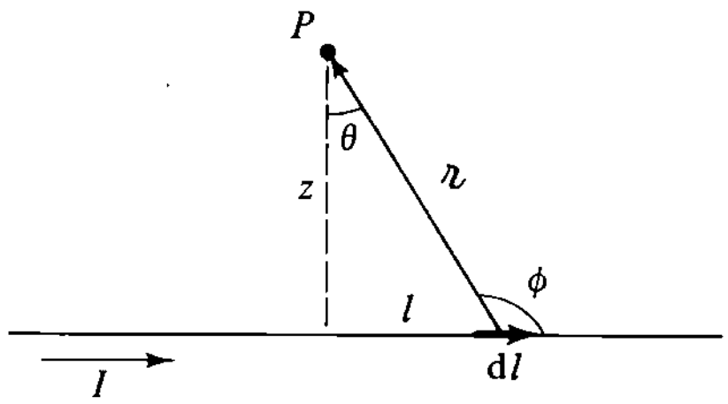

Figure 5.16

Solution: In the diagram $(dl \times \hat{\mathscr{r}})$ points *out* of the page and has the magnitude

$$dl \sin \phi = dl \cos \theta$$

Also, $l = z \tan \theta$, so

$$dl = \frac{z}{\cos^2 \theta}\, d\theta, \text{ and } z/\mathscr{r} = \cos \theta, \text{ so } \frac{1}{\mathscr{r}^2} = \frac{\cos^2 \theta}{z^2}$$

Thus,

$$B = \frac{\mu_0 I}{4\pi} \int_{\theta_1}^{\theta_2} \left(\frac{\cos^2 \theta}{z^2}\right)\left(\frac{z}{\cos^2 \theta}\right) \cos \theta \, d\theta$$

$$= \frac{\mu_0 I}{4\pi z} \int_{\theta_1}^{\theta_2} \cos \theta \, d\theta = \frac{\mu_0 I}{4\pi z} (\sin \theta_2 - \sin \theta_1) \tag{5.31}$$

Equation (5.31) gives the field of any straight segment of wire, in terms of the initial and final angles θ_1 and θ_2 (Fig. 5.17). Of course, a finite segment by itself could never support a steady current (where would the charge go when it got to the end?), but it might be a *portion* of some closed circuit, and (5.31) would then represent its contribution to the total field. In the case of an infinite wire, $\theta_1 = -\pi/2$ and $\theta_2 = \pi/2$, so we obtain

$$B = \frac{\mu_0 I}{2\pi z} \tag{5.32}$$

pointing out of the page. Notice that the field is inversely proportional to the distance from the wire—just like the electric field of an infinite line charge. Had

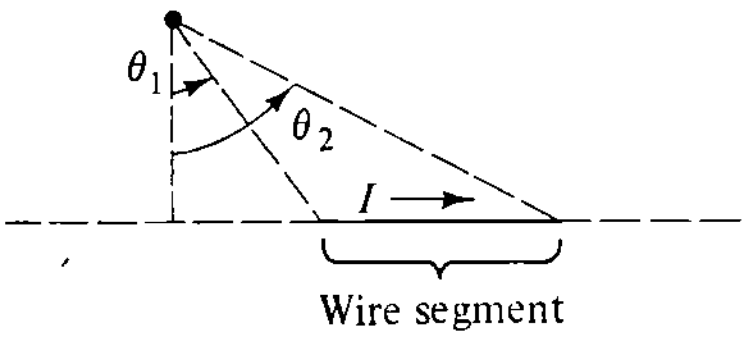

Figure 5.17

I chosen a point P below the wire, $\mathbf{B}$ would aim *into* the page. In general, it "circles around" the wire, in consonance with the right-hand rule of Section 5.1.1.

As an application, let's find the force of attraction between two long, parallel wires a distance d apart, carrying currents I_1 and I_2 (Fig. 5.18). According to (5.32) the field at (2) due to (1) is

$$B = \frac{\mu_0 I_1}{2\pi d}$$

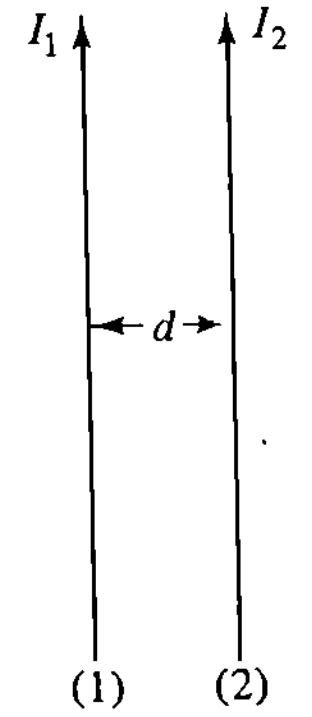

Figure 5.18

and it points into the page. The magnetic force law (5.16) then predicts a force directed toward (1), of magnitude

$$F = I_2\left(\frac{\mu_0 I_1}{2\pi d}\right)\int dl$$

The *total* force, not surprisingly, is infinite, but the force per unit length is

$$f = \frac{\mu_0}{2\pi}\left(\frac{I_1 I_2}{d}\right) \tag{5.33}$$

If the currents are antiparallel (one up, one down), the force is repulsive—consistent again with the qualitative observations of Section 5.1.1.

Example 6

Find the magnetic field a distance z above the center of a circular loop of radius R, which carries a steady current I (Fig. 5.19).

Solution: The field $d\mathbf{B}$ attributable to the segment dl points as shown. As we integrate dl around the loop, $d\mathbf{B}$ sweeps out a cone. The horizontal components cancel, and the vertical components combine to give

$$B = \frac{\mu_0}{4\pi} I \int \frac{dl}{\mathscr{r}^2} \cos\theta$$

(Notice that dl and $\hat{\mathscr{r}}$ are perpendicular, in this case; the factor of $\cos\theta$ projects

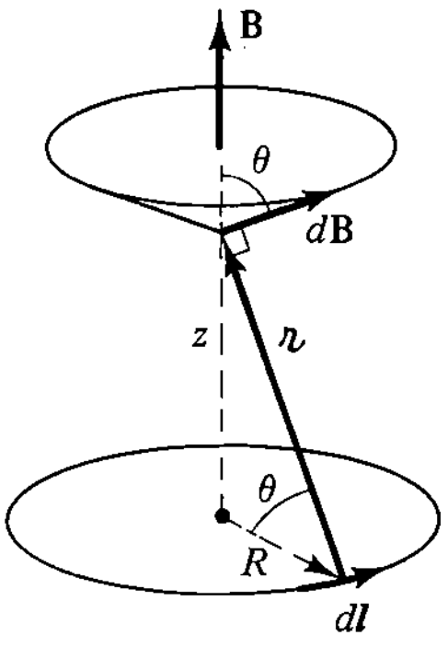

Figure 5.19

out the vertical component.) Now, $\cos\theta$ and $\mathscr{r}^2$ are constants, and $\int dl$ is simply the circumference, $2\pi R$, so

$$B = \frac{\mu_0 I}{4\pi}\left(\frac{\cos\theta}{\mathscr{r}^2}\right)2\pi R = \frac{\mu_0 I}{2}\frac{R^2}{(R^2+z^2)^{3/2}} \tag{5.34}$$

For surface and volume currents the Biot-Savart law becomes

$$\mathbf{B} = \frac{\mu_0}{4\pi}\int \frac{\mathbf{K}\times\hat{\mathscr{r}}}{\mathscr{r}^2}\,da \quad \text{and} \quad \mathbf{B} = \frac{\mu_0}{4\pi}\int \frac{\mathbf{J}\times\hat{\mathscr{r}}}{\mathscr{r}^2}\,d\tau \tag{5.35}$$

respectively. You might be tempted to write down the corresponding formula for a moving point charge, using the "dictionary" (5.26):

$$\mathbf{B} = \frac{\mu_0}{4\pi}q\frac{\mathbf{v}\times\hat{\mathscr{r}}}{\mathscr{r}^2} \tag{5.36}$$

but this is simply *wrong.*[6] As I mentioned earlier, a point charge does not constitute a steady current, and the Biot-Savart law, which only applies to steady currents, does not correctly determine its field.

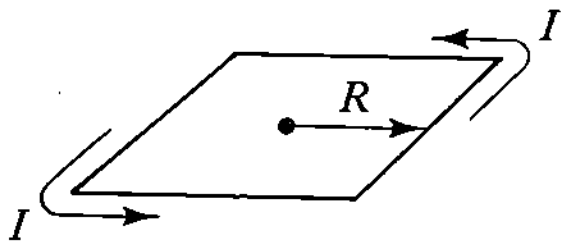

Figure 5.20

Problem 5.8 (a) Find the magnetic field at the center of a square loop, which carries a steady current I. Let R be the distance from center to side (Fig. 5.20).

(b) Find the field at the center of a regular n-sided polygon, carrying a steady current I. Again, let R be the distance from the center to any side.

(c) Check that your formula reduces to the field at the center of a circular loop in the limit $n \to \infty$.

[6] I say this loud and clear to emphasize the point of principle; actually, (5.36) is *approximately* correct for nonrelativistic charges ($v \ll c$), under conditions where retardation can be neglected. (See Chapter 9, Example 13.)

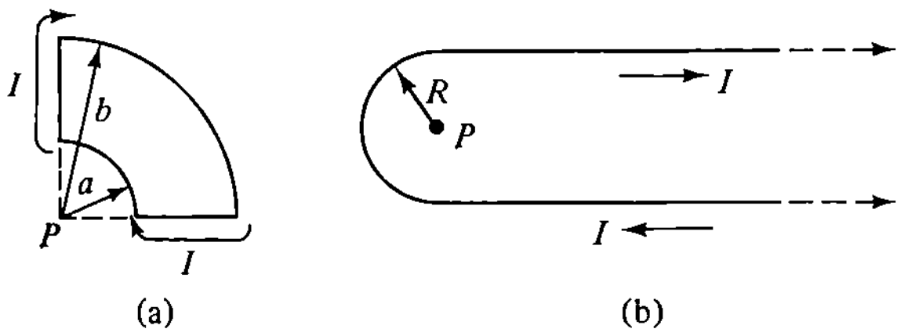

Figure 5.21

Problem 5.9 Find the magnetic field at point P for each of the steady current configurations shown in Fig. 5.21.

Problem 5.10 (a) Find the force on a square loop placed as shown in Fig. 5.22, near an infinite straight wire. Both the loop and the wire carry a steady current I.
(b) Find the force on the triangular loop.

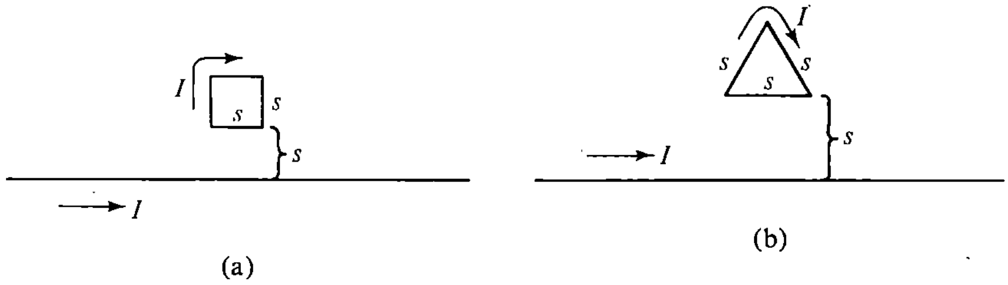

Figure 5.22

Problem 5.11 Find the magnetic field on the axis of a tightly wound solenoid (helical coil) consisting of N turns per unit length wrapped around a cylindrical tube of radius R and carrying current I (Fig. 5.23). Express your answer in terms of θ_1 and θ_2 (it's easiest that way). Consider the turns to be essentially circular, and use the result of Example 6. What is the field on the axis of an *infinite* solenoid (infinite in both directions)?

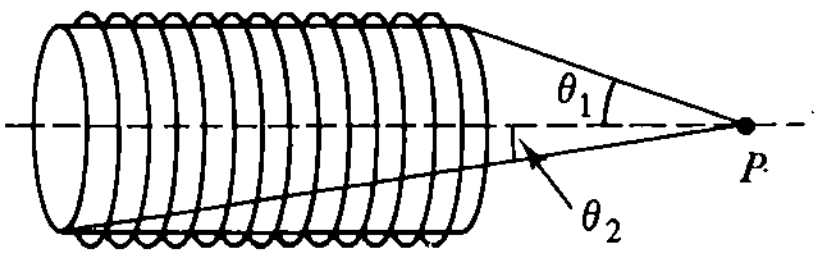

Figure 5.23

Problem 5.12 Suppose you have two infinite straight line charges λ, a distance d apart, moving along at a constant speed v (Fig. 5.24). How fast would v have to be in order for the magnetic attraction to balance the electrical repulsion? Work out the actual number . . . is this a reasonable sort of speed?[7]

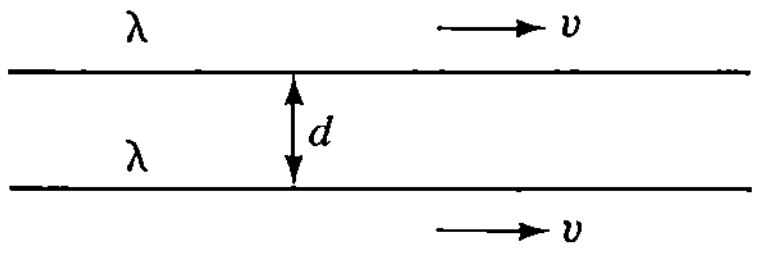

Figure 5.24

Problem 5.13 Find the magnetic field at a point $z > R$ on the axis of (a) the rotating disc and (b) the rotating sphere, in Problem 5.5.

[7]If you've studied special relativity, you may be tempted to look for complexities in this problem that are not really there—λ and v are both measured in the laboratory frame, and this is *ordinary electrostatics* (see footnote 4).

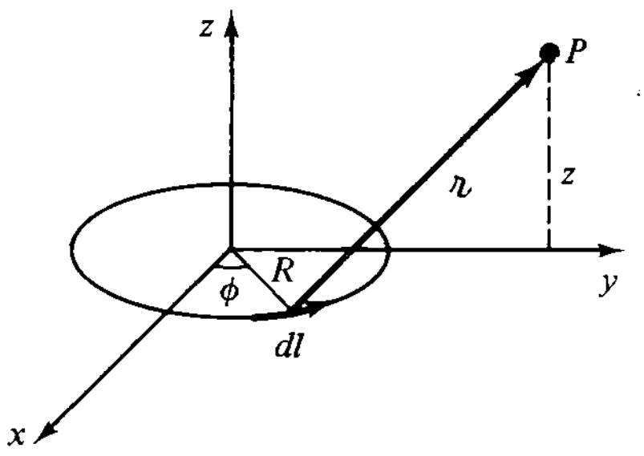

Figure 5.25

Problem 5.14 Suppose you wanted to find the field of the circular loop of Example 6 at a point P which is *not* directly above the center (Fig. 5.25). You might as well choose your axes so that P lies in the yz plane at $(0, y, z)$. The source point is $(R \cos \phi, R \sin \phi, 0)$, and ϕ runs from 0 to 2π. Set up the integrals from which you could calculate B_x, B_y, and B_z, and evaluate B_x explicitly.

5.3 THE DIVERGENCE AND CURL OF B

5.3.1 Straight-Line Currents

The magnetic field of an infinite straight wire can be represented by the field-line diagram of Fig. 5.26. As before, the density of lines serves to indicate the strength of the field. Here **B** is inversely proportional to the distance from the wire, so the lines are progressively farther and farther apart as you go out. At a glance, it is clear that this field has a curl but no divergence. Of course, we're talking about a very special case, but let's see how far we can pursue it.

According to equation (5.32), the integral of **B** around a circular path of radius R, centered at the wire, would be

$$\oint \mathbf{B} \cdot d\mathbf{l} = \oint \left(\frac{\mu_0 I}{2\pi R}\right) dl = \left(\frac{\mu_0 I}{2\pi R}\right) \oint dl = \mu_0 I$$

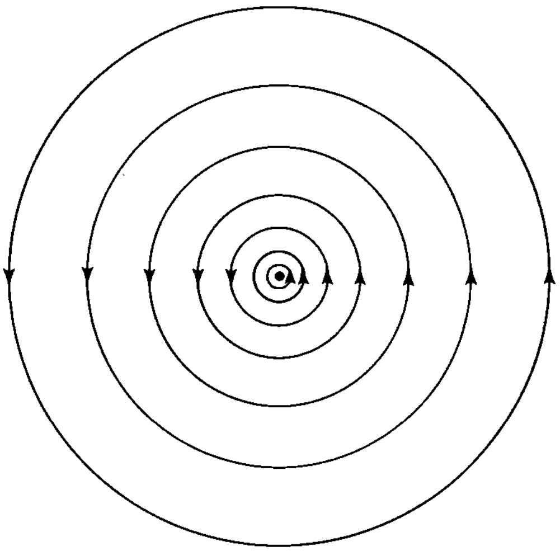

Figure 5.26

Notice that the answer is independent of R; that's because B *de*creases at the same rate as the circumference *in*creases. In fact, it doesn't have to be a circle: *Any* old loop that encloses the wire would give the same answer. For if we use cylindrical coordinates (r, ϕ, z), with the current flowing along the z axis,

$$\mathbf{B} = \left(\frac{\mu_0 I}{2\pi r}\right)\hat{\boldsymbol{\phi}} \quad \text{and} \quad d\mathbf{l} = dr\,\hat{\mathbf{r}} + r\,d\phi\,\hat{\boldsymbol{\phi}} + dz\,\hat{\mathbf{z}}$$

so that

$$\oint \mathbf{B}\cdot d\mathbf{l} = \frac{\mu_0 I}{2\pi}\oint\left(\frac{1}{r}\right) r\,d\phi = \frac{\mu_0 I}{2\pi}\int_0^{2\pi} d\phi = \mu_0 I$$

This assumes the loop circles the wire exactly once; if it went around twice, then ϕ would run from 0 to 4π, and if it didn't enclose the wire at all, then ϕ would go from ϕ_1 to ϕ_2 and back again, with $\int d\phi = 0$ (Fig. 5.27).

Now, suppose we had a *bundle* of straight wires. Each wire that passes through our loop would contribute $\mu_0 I$, and those outside would contribute nothing (Fig. 5.28). The line integral then would be

$$\oint \mathbf{B}\cdot d\mathbf{l} = \mu_0 I_{\text{enc}} \tag{5.37}$$

where I_{enc} stands for the total current enclosed by the integration path. If the flow of charge is represented by a volume current density $\mathbf{J}$, the enclosed current is

$$I_{\text{enc}} = \int \mathbf{J}\cdot d\mathbf{a} \tag{5.38}$$

with the integral taken over the surface bounded by the loop. Applying Stokes' theorem to equation (5.36), then,

$$\int (\nabla \times \mathbf{B})\cdot d\mathbf{a} = \mu_0 \int \mathbf{J}\cdot d\mathbf{a}$$

and hence

$$\nabla \times \mathbf{B} = \mu_0 \mathbf{J} \tag{5.39}$$

With minimal labor we have obtained the general formula for the curl of $\mathbf{B}$ (the divergence is zero, just as it was for a single wire). But our derivation is seriously

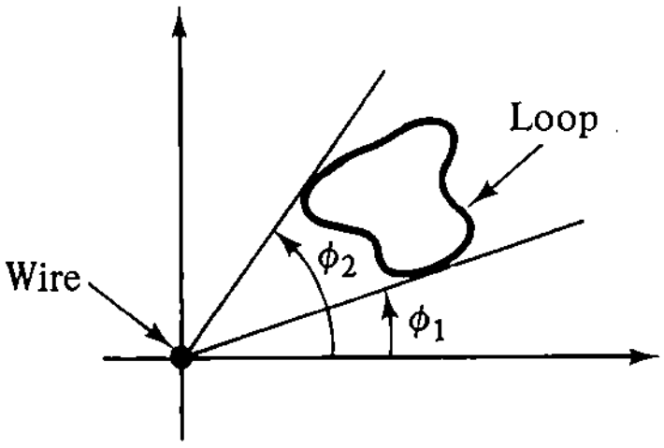

Figure 5.27

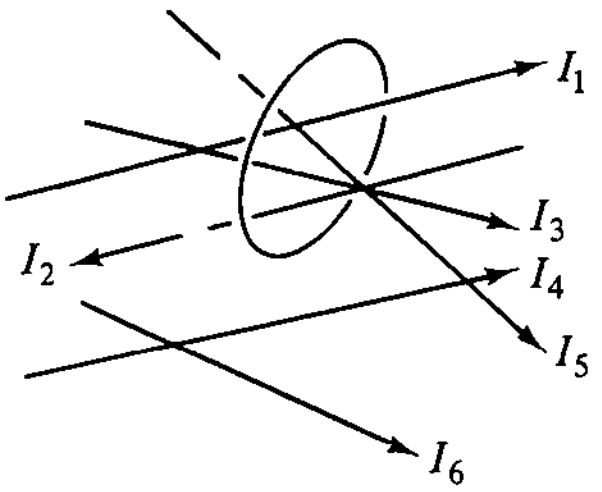

Figure 5.28

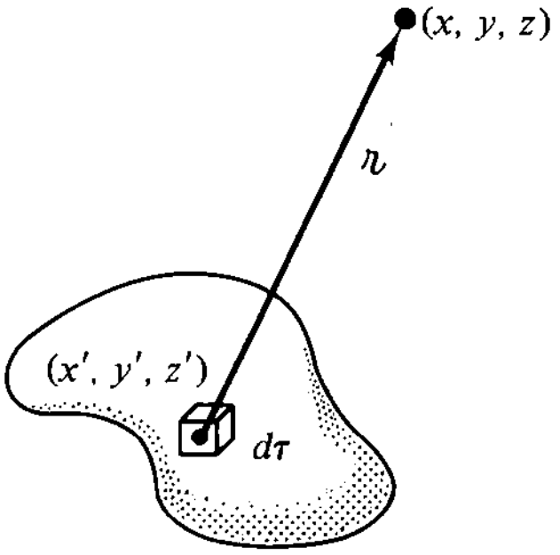

Figure 5.29

flawed by the restriction to infinite straight line currents and combinations thereof. There are plenty of current configurations which cannot be constructed out of infinite straight wires, and we have no right to assume that (5.39) applies to them. So the next section is devoted to the formal derivation of the divergence and curl of **B**, starting from the Biot-Savart law itself.

5.3.2 The Divergence and Curl of B

The Biot-Savart law for the general case of volume current reads

$$\mathbf{B} = \frac{\mu_0}{4\pi} \int \frac{\mathbf{J} \times \hat{\mathbf{\imath}}}{\imath^2} \, d\tau \tag{5.40}$$

This formula gives the magnetic field at a point $P = (x, y, z)$ in terms of an integral over the current distribution $\mathbf{J}(x', y', z')$ (Fig. 5.29). It is best to be absolutely explicit at this stage:

$$\mathbf{B} \text{ is a function of } (x, y, z)$$

$$\mathbf{J} \text{ is a function of } (x', y', z')$$

$$\imath = (x - x')\hat{i} + (y - y')\hat{j} + (z - z')\hat{k}$$

$$d\tau = dx'\, dy'\, dz'$$

The integration is over the primed coordinates; the divergence and the curl are to be taken with respect to the *unprimed* coordinates.

Applying the divergence to equation (5.40), we obtain:

$$\nabla \cdot \mathbf{B} = \frac{\mu_0}{4\pi} \int \nabla \cdot \left(\mathbf{J} \times \frac{\hat{\imath}}{\imath^2} \right) d\tau \tag{5.41}$$

Invoking product rule number (6),

$$\nabla \cdot \left(\mathbf{J} \times \frac{\hat{\imath}}{\imath^2} \right) = \frac{\hat{\imath}}{\imath^2} \cdot (\nabla \times \mathbf{J}) - \mathbf{J} \cdot \left(\nabla \times \frac{\hat{\imath}}{\imath^2} \right) \tag{5.42}$$

But $\nabla \times \mathbf{J} = 0$, because $\mathbf{J}$ doesn't depend on the unprimed variables (x, y, z), and $\nabla \times (\hat{\imath}/\imath^2) = 0$ (Problem 1.62), so

$$\boxed{\nabla \cdot \mathbf{B} = 0} \tag{5.43}$$

Applying the curl to (5.40), we obtain:

$$\nabla \times \mathbf{B} = \frac{\mu_0}{4\pi} \int \nabla \times \left(\mathbf{J} \times \frac{\hat{\boldsymbol{\mathscr{r}}}}{\mathscr{r}^2} \right) d\tau. \tag{5.44}$$

Again, our strategy is to expand out the integrand, using the appropriate product rule—in this case number 8:

$$\nabla \times \left(\mathbf{J} \times \frac{\hat{\boldsymbol{\mathscr{r}}}}{\mathscr{r}^2} \right) = \mathbf{J} \left(\nabla \cdot \frac{\hat{\boldsymbol{\mathscr{r}}}}{\mathscr{r}^2} \right) - (\mathbf{J} \cdot \nabla) \frac{\hat{\boldsymbol{\mathscr{r}}}}{\mathscr{r}^2} \tag{5.45}$$

(I have dropped terms involving derivatives of $\mathbf{J}$, as before, because $\mathbf{J}$ does not depend on x, y, z.) The second term integrates to zero, as we'll see in the next paragraph. The first term involves the divergence we were at pains to calculate in Chapter 1:

$$\nabla \cdot \frac{\hat{\boldsymbol{\mathscr{r}}}}{\mathscr{r}^2} = 4\pi\, \delta^3(\boldsymbol{\mathscr{r}}) \tag{1.79}$$

Thus

$$\nabla \times \mathbf{B} = \frac{\mu_0}{4\pi} \int \mathbf{J}(\mathbf{r}')\, 4\pi\, \delta^3(\mathbf{r} - \mathbf{r}')\, d\tau' = \mu_0 \mathbf{J}(\mathbf{r})$$

which confirms that equation (5.39) is not restricted to straight-line currents, but holds quite generally in magnetostatics.

To complete the argument, however, we must check that the second term in (5.45) integrates to zero. Because the derivative acts only on $(\hat{\boldsymbol{\mathscr{r}}}/\mathscr{r}^2)$, we can switch from ∇ to ∇' at the cost of a minus sign:

$$-(\mathbf{J} \cdot \nabla) \frac{\hat{\boldsymbol{\mathscr{r}}}}{\mathscr{r}^2} = (\mathbf{J} \cdot \nabla') \frac{\hat{\boldsymbol{\mathscr{r}}}}{\mathscr{r}^2} \tag{5.46}$$

The x-component, in particular, is

$$(\mathbf{J} \cdot \nabla') \left(\frac{x - x'}{\mathscr{r}^3} \right) = \nabla' \cdot \left[\frac{(x - x')}{\mathscr{r}^3} \mathbf{J} \right] - \left(\frac{x - x'}{\mathscr{r}^3} \right) (\nabla' \cdot \mathbf{J})$$

(using product rule 5). Now, for *steady* currents the divergence of $\mathbf{J}$ is zero (5.27), so

$$\left[-(\mathbf{J} \cdot \nabla) \frac{\hat{\boldsymbol{\mathscr{r}}}}{\mathscr{r}^2} \right]_x = \nabla' \cdot \left[\frac{(x - x')}{\mathscr{r}^3} \mathbf{J} \right]$$

and therefore this contribution to the integral (5.44) can be written

$$\int_{\text{volume}} \nabla' \cdot \left[\frac{(x - x')}{\mathscr{r}^3} \mathbf{J} \right] d\tau = \oint_{\text{surface}} \frac{(x - x')}{\mathscr{r}^3} \mathbf{J} \cdot d\mathbf{a} \tag{5.47}$$

(The reason for switching from ∇ to ∇' was precisely to permit this application of the divergence theorem. What I have done amounts to an integration by parts, in which I took the derivative off $\hat{\boldsymbol{\mathscr{r}}}/\mathscr{r}^2$ and put it onto $\mathbf{J}$; we are left now with the boundary term.) But what region are we integrating over? Well, it's the volume which appears in the Biot-Savart law (5.40)—large enough, that is, to include all the current. You can make it *bigger* than that, if you like; $\mathbf{J} = 0$ out there anyway, so it will add nothing to the integral. The essential point is that *on the boundary* the current is *zero* (all current is safely *inside*) and hence the surface integral (5.47) vanishes.[8]

5.3.3 Applications of Ampère's Law

The equation for the curl of **B**

$$\boxed{\nabla \times \mathbf{B} = \mu_0 \mathbf{J}} \tag{5.48}$$

is called **Ampère's law** (in differential form). It can be converted to integral form by the usual device of applying one of the fundamental theorems—in this case Stokes' theorem:

$$\int (\nabla \times \mathbf{B}) \cdot d\mathbf{a} = \oint \mathbf{B} \cdot d\mathbf{l} = \mu_0 \int \mathbf{J} \cdot d\mathbf{a}$$

Now, $\int \mathbf{J} \cdot d\mathbf{a}$ is the total current passing through the surface (Fig. 5.30), which we call I_{enc} ("enclosed" by the line integral). Thus

$$\boxed{\oint \mathbf{B} \cdot d\mathbf{l} = \mu_0 I_{\text{enc}}} \tag{5.49}$$

This is the integral version of Ampère's law; it generalizes (5.36) to all of magnetostatics. Notice that (5.49) inherits the sign ambiguity of Stokes' theorem (Section 1.3.4): Which *way* around the loop am I supposed to go? And which *direction* through the surface corresponds to a "positive" current? The resolution, as always, is the right-hand rule: If the fingers of your right hand indicate the direction of integration around the boundary, then your thumb defines the direction of a positive current.

[8] If **J** itself extends to infinity (as in the case of the an infinite straight wire), the surface integral is still typically zero, though the analysis calls for greater care.

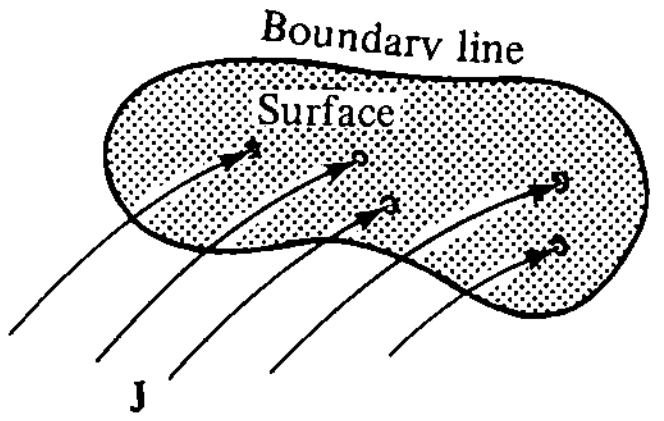

Figure 5.30

Just as the Biot-Savart law plays the role in magnetostatics that Coulomb's law assumed in electrostatics, so Ampère's plays the role of Gauss's:

Electrostatics: Coulomb → Gauss

Magnetostatics: Biot-Savart → Ampère

In particular, for currents with special symmetry, Ampère's law in integral form offers a lovely and extraordinarily efficient means for finding the magnetic field.

Example 7

Find the magnetic field a distance r from a long straight wire (Fig. 5.31), carrying a steady current I (the same problem we solved in Example 5, using the Biot-Savart law).

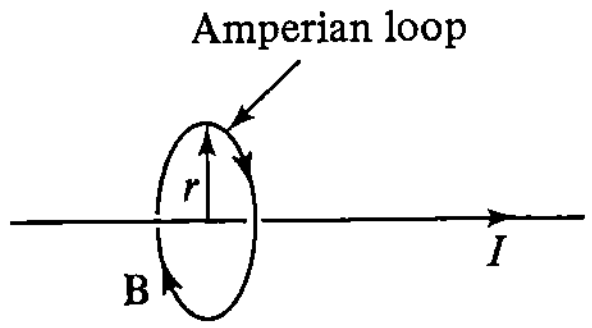

Figure 5.31

Solution: We know the direction of **B** is "circumferential," circling around the wire as indicated. By symmetry, the magnitude of **B** is constant around an "amperian loop" of radius r, centered on the wire. So Ampère's law gives

$$\oint \mathbf{B} \cdot d\mathbf{l} = B \oint dl = B 2\pi r = \mu_0 I_{\text{enc}} = \mu_0 I$$

or

$$B = \frac{\mu_0 I}{2\pi r}$$

This is the same answer we got before (eq. 5.32), but it was obtained this time with far less effort.

Example 8

Find the magnetic field of an infinite uniform surface current $\mathbf{K} = K\hat{\imath}$ covering the xy plane (Fig. 5.32).

Solution: First of all, what is the *direction* of **B**? Could it have any x-component? *No*: A glance at the Biot-Savart law (5.35) reveals that **B** is *perpendicular* to **K**. Could it have a z-component? *No again.* You could confirm this by noting that any vertical contribution from a filament at $+y$ is canceled by the corresponding filament at $-y$. But there is a nicer argument: Suppose the field pointed *away* from the plane. By reversing the direction of the current, I could make it point *toward* the plane (in the Biot-Savart law changing the sign of the

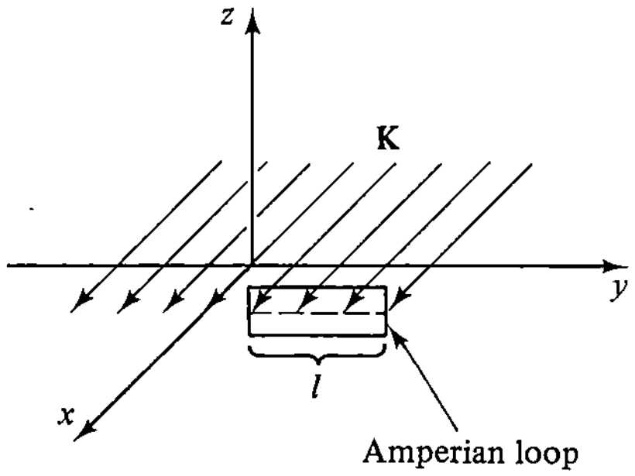

Figure 5.32

current switches the sign of the field). But the z-component of **B** cannot depend on the direction of the current in the xy plane. (Think about it!) So **B** can only have a y-component, and a quick check with your right hand should convince you that it points to the *left* above the plane and to the *right* below it.

With this in mind we draw a rectangular amperian loop as shown in Fig. 5.32, parallel to the yz plane and extending an equal distance above and below the surface $z = 0$. Applying Ampère's law, we have

$$\oint \mathbf{B} \cdot d\mathbf{l} = 2Bl = \mu_0 I_{\text{enc}} = \mu_0 K l$$

so that $B = (\mu_0/2)K$, or, more precisely,

$$\mathbf{B} = \begin{cases} \dfrac{\mu_0}{2} K\hat{\mathbf{j}}, & \text{for} \quad z < 0 \\[2ex] -\dfrac{\mu_0}{2} K\hat{\mathbf{j}}, & \text{for} \quad z > 0 \end{cases} \tag{5.50}$$

Notice that the field is independent of the distance from the plane, just like the *electric* field of a uniform surface *charge* (Example 4 of Chapter 2).

Example 9

Find the magnetic field of a very long solenoid, consisting of N closely wound turns per unit length on a cylinder of radius R and carrying a steady current I (Fig. 5.33).

(The point of making the windings so close is that one can then pretend each turn is circular. If this troubles you (after all, there is a net current I in the direction of the solenoid's axis, no matter *how* tight the winding) picture instead a sheet of tinfoil wrapped around the cylinder, carrying the equivalent uniform surface current $K = NI$ (Fig. 5.34). Or make a double winding, going up to one end and then—always in the same sense—back down again, thereby eliminating the net longitudinal current. But, in truth, this is all unnecessary fastidiousness, for the field inside a solenoid is huge (relatively speaking), and the field of the longitudinal current is a tiny refinement.)

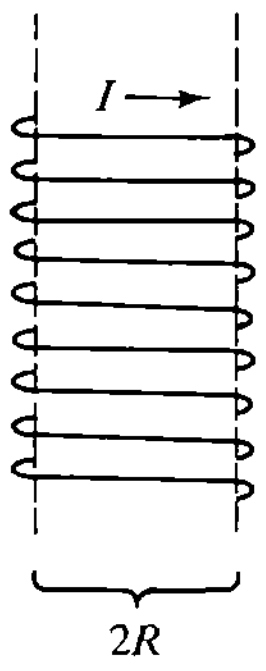

Figure 5.33

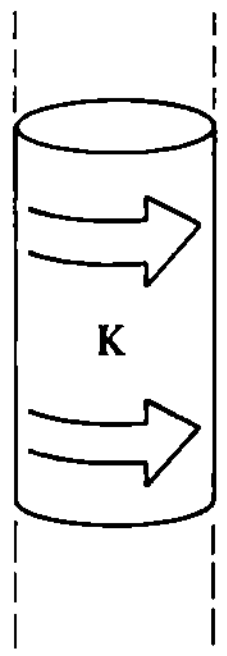

Figure 5.34

Solution: First of all, what is the *direction* of **B**? Could it have a radial component? *No.* For suppose B_r were *positive*; if we reversed the direction of the current, B_r would then be *negative.* But switching I is physically equivalent to turning the solenoid upside down, and that certainly should not alter the radial field. How about a "circumferential" component? *No.* For B_ϕ would be constant around an amperian loop concentric with the solenoid (Fig. 5.35), and hence

$$\oint \mathbf{B} \cdot d\mathbf{l} = B_\phi(2\pi r) = \mu_0 I_{enc} = 0$$

since the loop encloses no current (or at most I, if the loop is exterior to the solenoid, but for the reason given above, we shall ignore such contributions).

So the magnetic field of an infinite, closely wound solenoid points *parallel to the axis.* From the right-hand rule, we expect that it points upward inside the solenoid and downward outside. Moreover, it certainly approaches zero as you go very far away. With this in mind, let's apply Ampère's law to the two rectangular loops in Fig. 5.36. Loop 1 lies entirely outside the solenoid, with its sides at distances a and b from the axis:

$$\oint \mathbf{B} \cdot d\mathbf{l} = [B(a) - B(b)]L = \mu_0 I_{enc} = 0$$

so that

$$B(a) = B(b)$$

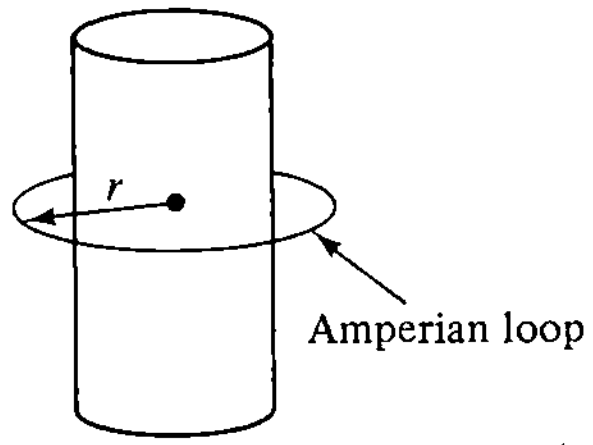

Figure 5.35

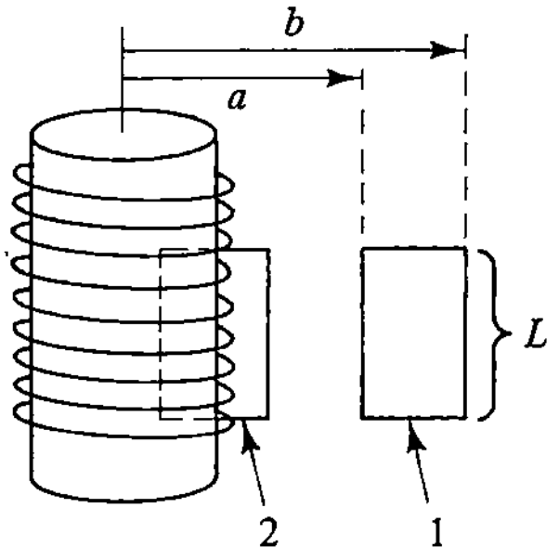

Figure 5.36

Evidently the *field outside does not depend on the distance from the axis.* But we know it goes to *zero* for large r. It must therefore be zero *everywhere*! (This surprising result can also be derived from the Biot-Savart law, of course, but it's much more difficult. See Problem 5.44.) As for loop 2, which is half inside and half outside, Ampère's law gives

$$\oint \mathbf{B}\cdot d\mathbf{l} = BL = \mu_0 I_{\text{enc}} = \mu_0 NIL$$

where B is the field inside the solenoid. The right side of the loop contributes nothing, since $B = 0$ outside. In conclusion then,

$$\mathbf{B} = \begin{cases} \mu_0 NI\hat{\mathbf{z}}, & \text{inside the solenoid} \\ 0, & \text{outside the solenoid} \end{cases} \tag{5.51}$$

Notice that the field inside is *uniform*; in this sense the solenoid is to magnetostatics what the parallel-plate capacitor is to electrostatics: a simple device for producing uniform fields.

Like Gauss's law, Ampère's law is always *true* (for steady currents), but it is not always *useful.* Only when the symmetry of the problem enables you to pull B outside the integral $\oint \mathbf{B}\cdot d\mathbf{l}$ can you calculate the magnetic field from Ampère's law. When it *does* work, it's by far the fastest method; when it doesn't, you have to fall back on the Biot-Savart law. The standard current configurations which can be handled by Ampère's law are

1. Infinite straight lines (prototype in Example 7);
2. Infinite planes (prototype in Example 8);
3. Infinite solenoids (prototype in Example 9);
4. Toroids

The latter is a surprising and elegant application of Ampère's law; it is treated in the following example.

Example 10 Magnetic field of a toroidal coil

A toroidal coil consists of a circular ring, or "donut," around which a long wire is wrapped (Fig. 5.37). The winding is uniform and tight enough so that each turn can be considered a closed loop with negligible "circumferential" component. The cross-sectional shape of the coil is immaterial. I made it rectangular in Fig. 5.37 for the sake of simplicity, but it could just as well be circular or even some weird asymmetrical form, as in Fig. 5.38, just as long as the shape remains the same all the way around the ring. In that case it follows that the *magnetic field of the toroid is circumferential at all points, both inside and outside the coil.*

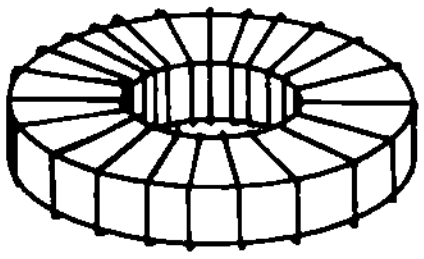

Figure 5.37

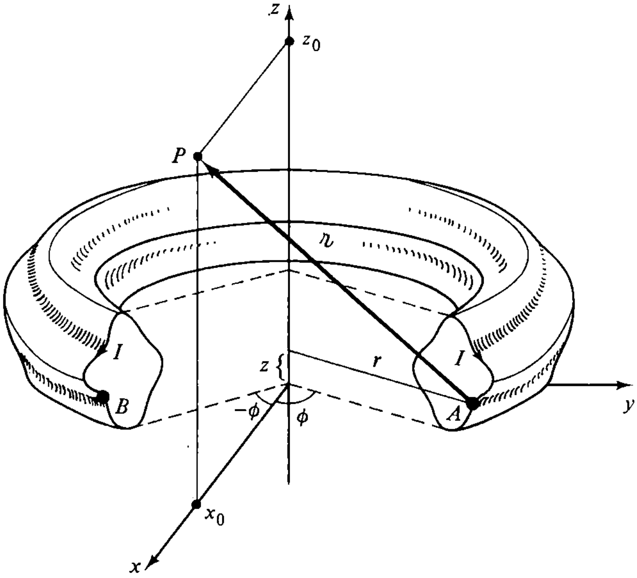

Figure 5.38

Proof: According to the Biot-Savart law, the field at P due to the current element at A is

$$d\mathbf{B} = \frac{\mu_0}{4\pi} \frac{\mathbf{I} \times \boldsymbol{\mathscr{r}}}{\mathscr{r}^3} \, dl$$

The Cartesian coordinates of P are (Fig. 5.38)

$$(x_0, 0, z_0)$$

whereas those of A are

$$(r \cos \phi, r \sin \phi, z)$$

Therefore,

$$\boldsymbol{\mathscr{r}} = (x_0 - r \cos \phi, -r \sin \phi, z_0 - z)$$

Since the current has no ϕ-component,

$$\mathbf{I} = I_r \hat{r} + I_z \hat{z}$$

or

$$\mathbf{I} = (I_r \cos\phi, I_r \sin\phi, I_z)$$

Accordingly,

$$\mathbf{I} \times \boldsymbol{\mathscr{r}} = \begin{vmatrix} \hat{i} & \hat{j} & \hat{k} \\ I_r \cos\phi & I_r \sin\phi & I_z \\ (x_0 - r\cos\phi) & (-r\sin\phi) & (z_0 - z) \end{vmatrix}$$

$$= \hat{i}[\sin\phi(I_r(z_0 - z) + rI_z)]$$

$$+ \hat{j}[I_z(x_0 - r\cos\phi) - I_r\cos\phi(z_0 - z)] + \hat{k}[-I_r x_0 \sin\phi]$$

Because the coil is very tightly wound, there is a symmetrically situated current element at B with the same r, same $\mathscr{r}$, same dl, same I_r, same I_z *but negative* ϕ. Because $\sin\phi$ changes sign, the $\hat{i}$ and $\hat{k}$ contributions from A and B cancel, leaving only a $\hat{j}$ term. Thus, the field at P is in the $\hat{j}$-direction, and hence, in general, the field at any point is in the $\hat{\phi}$ direction. q.e.d.

Now that we know the field is circumferential, determining its magnitude is ridiculously easy. Just apply Ampère's law to a circle of radius r about the axis of the toroid:

$$B 2\pi r = \mu_0 I_{\text{enc}}$$

and hence

$$\mathbf{B}(r) = \begin{cases} \dfrac{\mu_0 n I}{2\pi r}\hat{\phi}, & \text{for points inside the coil} \\ 0, & \text{for points outside the coil} \end{cases} \tag{5.52}$$

where n is the total number of turns.

Problem 5.16 A steady current I flows down a long cylindrical wire of radius R (Fig. 5.39). Find the magnetic field, both inside and outside the wire, if

(a) The current is uniformly distributed over the outside surface of the wire;

(b) The current is distributed in such a way that J is proportional to r, the distance from the axis.

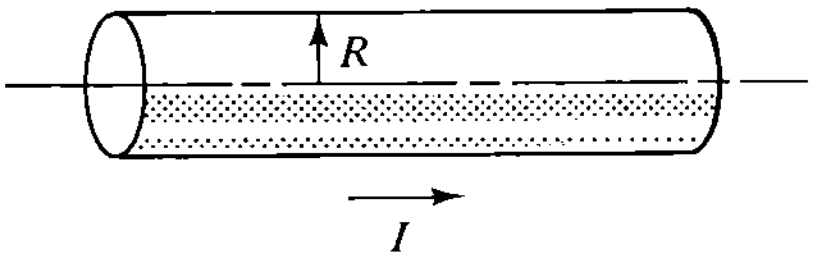

Figure 5.39

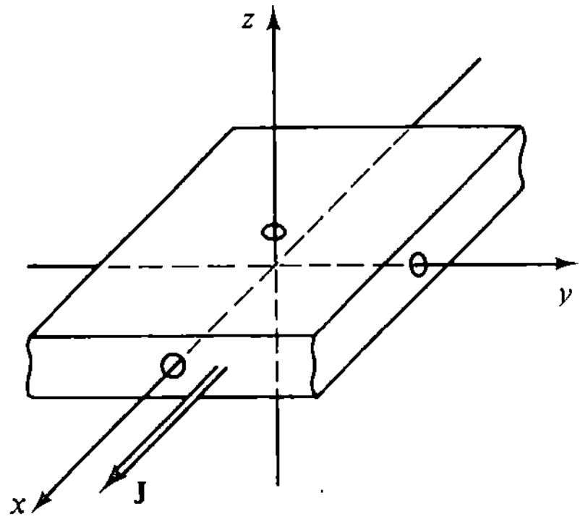

Figure 5.40

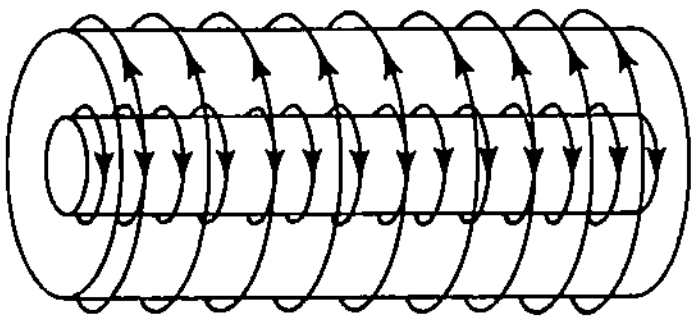

Figure 5.41

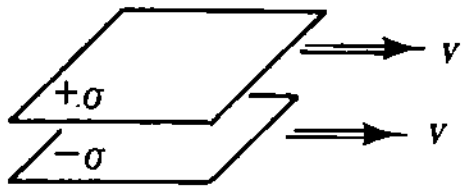

Figure 5.42

Problem 5.17 A thick slab extending from $z = -a$ to $z = +a$ carries a uniform volume current $\mathbf{J} = J\hat{\imath}$ (Fig. 5.40). Find the magnetic field both inside and outside the slab.

Problem 5.18 Two long coaxial solenoids each carry current I, but in opposite directions, as shown in Fig. 5.41. The inner solenoid (radius a) has N_1 turns per unit length, the outer (radius b) has N_2. Find **B** in each of the three regions: (a) inside the inner solenoid, (b) between them, and (c) outside both.

Problem 5.19 A large parallel-plate capacitor with uniform surface charge σ on the upper plate and $-\sigma$ on the lower is moving with constant speed v as shown in Fig. 5.42.
(a) Find the magnetic field between the plates and also above and below them.
(b) Find the magnetic force per unit area on the upper plate, including its direction.
(c) At what speed v would the magnetic force balance the electrical force?[9]

! **Problem 5.20** Show that the magnetic field of an infinite solenoid runs parallel to the axis, regardless of the cross-sectional shape of the coil, as long as that shape is constant along the length of the solenoid. What is the magnitude of the field, inside and outside such a coil? Show that the toroid field (5.52) reduces to the solenoid field, when the radius of the donut is so large that a segment can be considered essentially straight.

Problem 5.21 In calculating the current enclosed by an amperian loop, one must, in general, evaluate an integral of the form

$$I_{\text{enc}} = \int_{\text{surface}} \mathbf{J} \cdot d\mathbf{a}$$

The trouble is, there are an infinite number of surfaces which share the same boundary line. Which one are we supposed to use?

5.3.4 Comparison of Magnetostatics and Electrostatics

The divergence and curl of the *electrostatic* field are

$$\nabla \cdot \mathbf{E} = \frac{1}{\epsilon_0}\rho \qquad \text{(Gauss's law)}$$

$$\nabla \times \mathbf{E} = 0 \qquad \text{(no name)}$$

[9]See footnote 7.

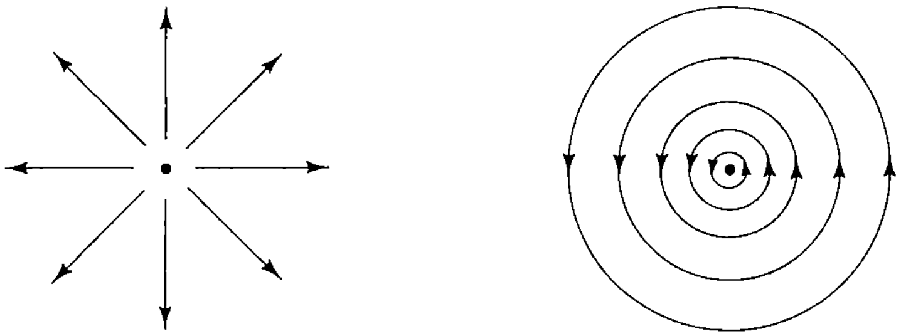

(a) Electrostatic field of point charge. (b) Magnetostatic field of long wire. **Figure 5.43**

These are **Maxwell's equations** for electrostatics. Together with the boundary condition $\mathbf{E} \to 0$ far from all charges, Maxwell's equations determine the field, if the source charge density ρ is given; they contain essentially the same information as Coulomb's law plus the principle of superposition. The divergence and curl of the *magnetostatic* field are

$$\nabla \cdot \mathbf{B} = 0 \qquad \text{(no name)}$$

$$\nabla \times \mathbf{B} = \mu_0 \mathbf{J} \qquad \text{(Ampère's law)}$$

These are Maxwell's equations for magnetostatics. Again, together with the boundary condition $\mathbf{B} \to 0$ far from all currents, Maxwell's equations determine the magnetic field; they contain no more and no less information than the Biot-Savart law. Maxwell's equations and the Lorentz force law,

$$\mathbf{F} = Q(\mathbf{E} + \mathbf{v} \times \mathbf{B})$$

constitute the most elegant formulation of the fundamental laws of electrostatics and magnetostatics.

The electric field *diverges away from* a (positive) charge; the magnetic field line *curls around* a current (Fig. 5.43). Electric field lines originate on positive charges and terminate on negative ones; magnetic field lines do not begin or end anywhere—to do so would imply a nonzero divergence. They either form closed loops or extend out to infinity. To put it another way, *there are no point sources for* **B**, as there are for **E**—there exists no magnetic analog to electric charge. This is the physical content of the statement $\nabla \cdot \mathbf{B} = 0$. Coulomb and others believed that magnetism was produced by magnetic "charges" (**magnetic monopoles**, we call them today), and in some older books you will still find references to a magnetic version of Coulomb's law, giving the force of attraction or repulsion between them. It was Ampère who first speculated that all magnetic effects are attributable to *electric* charges *in motion* (currents). As far as we know, Ampère was right; nevertheless, it remains an open experimental question whether magnetic monopoles exist in nature (they are obviously pretty *rare*, or somebody would have found one[10]), and in fact some recent elementary particle theories *require* them. For our purposes, though, $\nabla \cdot \mathbf{B} = 0$: there are no magnetic monopoles.

[10]An apparent detection (B. Cabrera, *Phys. Rev. Lett.* **48**, 1378 (1982)) has never been reproduced—and not for want of trying. For a delightful brief history of ideas about magnetism, see Chapter 1 in D. C. Mattis, *The Theory of Magnetism* (New York: Harper and Row, 1965).

So it takes a *moving* electric charge to *produce* a magnetic field (that's why we never encountered magnetic fields in electrostatics), and it takes another moving charge to "feel" a magnetic field. If the sources are stationary, the Lorentz force law reduces to $\mathbf{F} = Q\mathbf{E}$ (because $\mathbf{B} = 0$); if the test charge is at rest, again $\mathbf{F} = Q\mathbf{E}$ (because $\mathbf{v} = 0$)—though in the latter case $\mathbf{E}$ is not an electro*static* field, if the sources are moving. Only when *both* source and "recipient" are in motion do we get any magnetic force.

Typically, electric forces are enormously larger than magnetic ones. That's not something you can tell from the theory as such; it has to do with the sizes of the fundamental constants ϵ_0 and μ_0. In general, it is only when both the source charges and the test charge are moving at velocities comparable to the speed of light that the magnetic force approaches the electric force in strength. (Problems 5.12 and 5.19 illustrate this rule; later on you will see why it is so.) How is it, then, that we ever notice magnetic effects at all? The answer is that both in the production of a magnetic field (Biot-Savart) and in its detection (Lorentz) it is the *current* (charge times velocity) that enters, and we can compensate for a smallish velocity by pouring huge quantities of charge down the wire. Ordinarily, this charge would simultaneously generate so large an *electric* force as to swamp the magnetic one. But if we arrange to keep the wire *neutral,* by embedding in it an equal amount of opposite charge at rest, the electric field will cancel out, and the magnetic field will stand alone. It sounds very elaborate, but of course this is precisely what happens in an ordinary current-carrying wire.

Problem 5.22

(a) Find the density ρ of mobile charges in a piece of copper, assuming each atom contributes two free electrons.

(b) Calculate the average electron velocity in a copper wire 1 mm in diameter which carries a current of 1 A. (*Note*: this is literally a *snail's* pace. How, then, can you carry on a long distance telephone conversation?)

(c) What is the force of attraction between two such wires, 1 cm apart?

(d) If you could somehow remove the stationary positive ions, what would the electrical repulsion force be? How many times greater than the magnetic force is it?

Problem 5.23 Is Ampère's law consistent with the general rule (equation (1.37)) that divergence of curl is zero? Show that Ampère's law *cannot* be valid, in general, outside magnetostatics. Is there any such "defect" in the other three of Maxwell's equations?

Problem 5.24 Suppose there *did* exist magnetic monopoles. How would you modify Maxwell's equations (for electro/magnetostatics) and the Lorentz force law to accommodate them? If you think there are several plausible options, list them, and suggest how you might determine experimentally which one is right.

5.4 MAGNETIC VECTOR POTENTIAL

5.4.1 The Vector Potential

Just as $\nabla \times \mathbf{E} = 0$ permitted us to introduce a scalar potential (V) in electrostatics,

$$\mathbf{E} = -\nabla V$$

so $\nabla \cdot \mathbf{B} = 0$ invites the introduction of a *vector* potential $\mathbf{A}$ in magnetostatics:

$$\mathbf{B} = \nabla \times \mathbf{A} \tag{5.53}$$

The former is authorized by Theorem 1 (of Section 1.6.2), the latter by Theorem 2 (the proof of Theorem 2 is developed in Problem 5.33). The potential formulation (5.53) automatically takes care of $\nabla \cdot \mathbf{B} = 0$ (since the divergence of a curl is *always* zero); there remains Ampère's law:

$$\nabla \times \mathbf{B} = \nabla \times (\nabla \times \mathbf{A}) = \nabla(\nabla \cdot \mathbf{A}) - \nabla^2 \mathbf{A} = \mu_0 \mathbf{J} \tag{5.54}$$

Now, the electric potential had a built-in ambiguity: You can add to V any function whose gradient is zero (which is to say, any *constant*) without altering the *physical* quantity $\mathbf{E}$. Likewise, you can add to the magnetic potential any function whose *curl* vanishes (which is to say, the *gradient* of any scalar), with no effect on $\mathbf{B}$. We can exploit this freedom to eliminate the divergence of $\mathbf{A}$:

$$\boxed{\nabla \cdot \mathbf{A} = 0} \tag{5.55}$$

For suppose our original potential, $\mathbf{A}'$, is *not* divergence-less. If we add to it the gradient of λ, so that $\mathbf{A} = \mathbf{A}' + \nabla\lambda$, the new divergence is

$$\nabla \cdot \mathbf{A} = \nabla \cdot \mathbf{A}' + \nabla^2 \lambda$$

We can accommodate (5.55), provided a function λ can be found that satisfies

$$\nabla^2 \lambda = -(\nabla \cdot \mathbf{A}')$$

But this is *mathematically* identical to Poisson's equation (2.21),

$$\nabla^2 V = -\frac{\rho}{\epsilon_0}$$

with $\nabla \cdot \mathbf{A}'$ in place of ρ/ϵ_0 as the "source." And we *know* how to solve Poisson's equation—that's what electrostatics is all about ("given the charge distribution, find the potential"). In particular, if ρ goes to zero at infinity, then (equation (2.24))

$$V = \frac{1}{4\pi\epsilon_0} \int \frac{\rho}{\mathscr{r}} \, d\tau$$

and by the same token, if $\nabla \cdot \mathbf{A}'$ goes to zero at infinity, then

$$\lambda = \frac{1}{4\pi} \int \frac{\nabla \cdot \mathbf{A}'}{\mathscr{r}} \, d\tau$$

If $\nabla \cdot \mathbf{A}'$ does *not* go to zero at infinity, then we'll have to use other means to discover the appropriate λ—just as we get the potential by other means when the charge distribution extends to infinity. But the *essential* point remains: *It is always possible to make the vector potential divergence-less.* To put it the other way around: the definition $\mathbf{B} = \nabla \times \mathbf{A}$ specifies the *curl* of $\mathbf{A}$, but it doesn't say anything about the *divergence*—we are at liberty to pick that as we see fit, and zero is evidently the simplest choice.

With this condition on **A**, Ampère's law (5.54) becomes

$$\boxed{\nabla^2 \mathbf{A} = -\mu_0 \mathbf{J}} \tag{5.56}$$

This *again* is nothing but Poisson's equation—or rather, it is *three* Poisson's equations (one for each Cartesian[11] component); this time $\mu_0 \mathbf{J}$ is the source. Assuming **J** goes to zero at infinity, we can read off the solution:[12]

$$\boxed{\mathbf{A} = \frac{\mu_0}{4\pi} \int \frac{\mathbf{J}}{\mathscr{r}} \, d\tau} \tag{5.57}$$

For line and surface currents,

$$\mathbf{A} = \frac{\mu_0}{4\pi} \int \frac{\mathbf{I}}{\mathscr{r}} \, dl = \frac{\mu_0 I}{4\pi} \int \frac{1}{\mathscr{r}} \, d\mathbf{l}; \qquad \mathbf{A} = \frac{\mu_0}{4\pi} \int \frac{\mathbf{K}}{\mathscr{r}} \, da \tag{5.58}$$

If the current does *not* go to zero at infinity, we have to find other ways to get **A**; some of these are explored in Example 12 and in the Problems at the end of the section.[13]

It must be said that **A** is not as *useful* as V. For one thing, it's still a *vector*, and although (5.57) and (5.58) are somewhat easier to work with than the Biot-Savart law, you still have to fuss with components. It would be nice if we could get away with a *scalar* potential,

$$\mathbf{B} = -\nabla U \tag{5.59}$$

but this is inconsistent with Ampère's law, since the curl of a gradient is always zero. (A magnetostatic scalar potential *can* be used, if you stick scrupulously to simply-connected, current-free regions, but as a theoretical tool it is of limited interest. See Problem 5.31.) Moreover, since magnetic forces do no work, **A** does not admit a simple physical interpretation in terms of the work done per unit charge. Nevertheless, the vector potential has substantial theoretical importance, as you will see in Chapter 9.

[11] In Cartesian coordinates,

$$\nabla^2 \mathbf{A} = (\nabla^2 A_x)\hat{x} + (\nabla^2 A_y)\hat{y} + (\nabla^2 A_z)\hat{z}$$

but in curvilinear coordinates the unit vectors *themselves* must be differentiated. In spherical coordinates, for instance,

$$\nabla^2 \mathbf{A} \neq (\nabla^2 A_r)\hat{r} + (\nabla^2 A_\theta)\hat{\theta} + (\nabla^2 A_\phi)\hat{\phi}$$

To determine the Laplacian of a *vector* in terms of its curvilinear components, it is safest to use

$$\nabla^2 \mathbf{A} = \nabla(\nabla \cdot \mathbf{A}) - \nabla \times (\nabla \times \mathbf{A})$$

[12] *Beware*: Even if you *calculate* the integral in (5.57) using curvilinear coordinates, you must first express **J** in terms of its *Cartesian* components (see footnote 3, p. 66).

[13] Equation (5.57) can be regarded as an application of the Helmholtz theorem (1.93), in the context of magnetostatics, where the divergence of **B** is zero and its curl is $\mu_0 \mathbf{J}$.

Example 11

A spherical shell of radius R, carrying a uniform surface charge σ, is set spinning at angular velocity ω. Find the vector potential it produces at point P in Fig. 5.44(a).

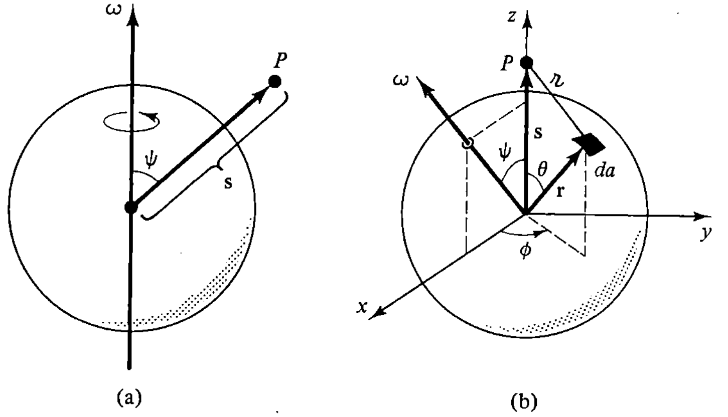

Figure 5.44

Solution: It might seem natural to align the polar axis along the axis of rotation, but in fact the integration is easier if we let P lie on the z axis, so that ω is tilted at an angle ψ. We may as well orient the x axis so that ω lies in the xz plane, as in Fig. 5.44(b). According to (5.58),

$$\mathbf{A}(P) = \frac{\mu_0}{4\pi}\int \frac{\mathbf{K}}{\imath}\,da$$

where $\mathbf{K} = \sigma\mathbf{v}$, $\imath = \sqrt{R^2 + s^2 - 2Rs\cos\theta}$, and $da = R^2 \sin\theta\, d\theta\, d\phi$. Now the velocity of a point $\mathbf{r}$ in a rotating rigid body is given by $(\boldsymbol{\omega} \times \mathbf{r})$; in this case,

$$\mathbf{v} = \boldsymbol{\omega} \times \mathbf{r} = \begin{vmatrix} \hat{\imath} & \hat{\jmath} & \hat{k} \\ \omega\sin\psi & 0 & \omega\cos\psi \\ R\sin\theta\cos\phi & R\sin\theta\sin\phi & R\cos\theta \end{vmatrix}$$

$$= R\omega[-(\cos\psi\sin\theta\sin\phi)\hat{\imath} + (\cos\psi\sin\theta\cos\phi - \sin\psi\cos\theta)\hat{\jmath} + (\sin\psi\sin\theta\sin\phi)\hat{k}]$$

Notice that each of these terms, save one, involves either $\sin\phi$ or $\cos\phi$. Since

$$\int_0^{2\pi} \sin\phi\, d\phi = \int_0^{2\pi} \cos\phi\, d\phi = 0$$

such terms integrate to zero. There remains

$$\mathbf{A}(P) = -\left(\frac{\mu_0 R^3 \sigma\omega\sin\psi}{2}\right)\left(\int_0^{\pi} \frac{\cos\theta\sin\theta}{\sqrt{R^2 + s^2 - 2Rs\cos\theta}}\,d\theta\right)\hat{\jmath}.$$

Letting $u = \cos\theta$, the integral becomes

$$\begin{aligned}\int_{-1}^{+1} \frac{u}{\sqrt{R^2+s^2-2Rsu}}\,du &= -\frac{(R^2+s^2+Rsu)}{3R^2s^2}\sqrt{R^2+s^2-2Rsu}\,\Bigg|_{-1}^{+1} \\ &= -\frac{1}{3R^2s^2}\left[(R^2+s^2+Rs)|R-s| - (R^2+s^2-Rs)(R+s)\right]\end{aligned}$$

If the point P lies *inside* the sphere, then $R > s$ and this expression reduces to $(2s/3R^2)$. If P lies *outside* the sphere, so that $R < s$, it reduces to $(2R/3s^2)$. Noting that $(\boldsymbol{\omega} \times \mathbf{s}) = -\omega s \sin\psi\,\hat{\jmath}$, we have, finally,

$$\mathbf{A}(P) = \begin{cases} \dfrac{\mu_0 R\sigma}{3}(\boldsymbol{\omega}\times\mathbf{s}), & \text{for points } \textit{inside} \text{ the sphere,} \\[2ex] \dfrac{\mu_0 R^4\sigma}{3s^3}(\boldsymbol{\omega}\times\mathbf{s}), & \text{for points } \textit{outside} \text{ the sphere.} \end{cases} \tag{5.60}$$

Having evaluated the integral, I revert to the "natural" coordinates of Fig. 5.44(a), in which $\boldsymbol{\omega}$ coincides with the z axis and the point P is at $r(=s)$, $\theta(=\psi)$, ϕ:

$$\mathbf{A}(r,\theta,\phi) = \begin{cases} \dfrac{\mu_0 R\omega\sigma}{3} r\sin\theta\,\hat{\phi}, & (r \le R) \\[2ex] \dfrac{\mu_0 R^4\omega\sigma}{3}\dfrac{\sin\theta}{r^2}\,\hat{\phi}, & (r \ge R) \end{cases} \tag{5.61}$$

Curiously enough, the field inside this spherical shell is *uniform*:

$$\mathbf{B} = \nabla\times\mathbf{A} = \frac{2\mu_0 R\omega\sigma}{3}(\cos\theta\,\hat{r} - \sin\theta\,\hat{\theta}) = \frac{2}{3}\mu_0\sigma R\omega\hat{z} = \frac{2}{3}\mu_0\sigma R\boldsymbol{\omega} \tag{5.62}$$

Example 12

Find the vector potential of an infinite solenoid with N turns per unit length, radius R, and current I.

Solution: This time we cannot use (5.58), since the current itself extends to infinity. But here's a cute method that does the job. Notice that

$$\oint \mathbf{A}\cdot d\mathbf{l} = \int(\nabla\times\mathbf{A})\cdot d\mathbf{a} = \int \mathbf{B}\cdot d\mathbf{a} = \Phi \tag{5.63}$$

where Φ is the flux of $\mathbf{B}$ through the loop in question. This is reminiscent of Ampère's law in integral form (5.49),

$$\oint \mathbf{B}\cdot d\mathbf{l} = \mu_0 I_{\text{enc}}$$

In fact, it's the same equation, with $\mathbf{B} \to \mathbf{A}$ and $\mu_0 I_{enc} \to \Phi$. If symmetry permits, we can determine $\mathbf{A}$ from Φ in the same way we got $\mathbf{B}$ from I_{enc} in Section 5.3.3. The present problem (with a uniform longitudinal magnetic field $\mu_0 NI$ inside the solenoid and no field outside) is analogous to the Ampère's law problem of a fat wire carrying a uniformly distributed current. The vector potential is "circumferential" (it mimics the magnetic field of the wire); using a circular "amperian loop" at radius r *inside* the solenoid, we have

$$\oint \mathbf{A} \cdot d\mathbf{l} = A(2\pi r) = \int \mathbf{B} \cdot d\mathbf{a} = \mu_0 NI(\pi r^2)$$

so

$$\mathbf{A} = \frac{\mu_0 NI}{2} r \hat{\boldsymbol{\phi}}, \qquad \text{for } r < R \tag{5.64}$$

For an amperian loop *outside* the solenoid, the flux is

$$\int \mathbf{B} \cdot d\mathbf{a} = \mu_0 NI(\pi R^2)$$

since the field only extends out to R. Thus,

$$\mathbf{A} = \frac{\mu_0 NI}{2} \frac{R^2}{r} \hat{\boldsymbol{\phi}}, \qquad \text{for } r > R \tag{5.65}$$

If you have any doubts about this answer, *check* it: Does $\nabla \times \mathbf{A} = \mathbf{B}$? Does $\nabla \cdot \mathbf{A} = 0$? If so, we're done.

Problem 5.25 Find the magnetic vector potential of a finite segment of straight wire carrying a current I. Check that your answer is consistent with (5.31).

Problem 5.26 What current density would produce a constant azimuthal potential, $A_\phi = k$, in cylindrical coordinates?

Problem 5.27 If $\mathbf{B}$ is *uniform*, show that $\mathbf{A} = -\frac{1}{2}(\mathbf{r} \times \mathbf{B})$, where $\mathbf{r}$ is a vector from the origin to the point in question. That is, check that $\nabla \cdot \mathbf{A} = 0$ and $\nabla \times \mathbf{A} = \mathbf{B}$.

Problem 5.28

(a) By whatever means you can think of (short of looking it up), find the vector potential a distance r from an infinite straight wire carrying a current I. Check that $\nabla \cdot \mathbf{A} = 0$ and $\nabla \times \mathbf{A} = \mathbf{B}$.

(b) Find the magnetic potential *inside* the wire if it has radius R and the current is uniformly distributed.

Problem 5.29 Find the vector potential above and below the plane surface current of Example 8.

Problem 5.30

(a) Check that Equation (5.57) is consistent with Equation (5.55) by applying the *divergence.*

(b) Check that (5.57) is consistent with (5.40) by applying the *curl.*

(c) Check that (5.57) is consistent with (5.56) by applying the *Laplacian.*

Problem 5.31 Suppose you want to define a magnetic scalar potential U (eq. 5.59) in the vicinity of a current-carrying wire. First of all, you must stay away from the wire itself

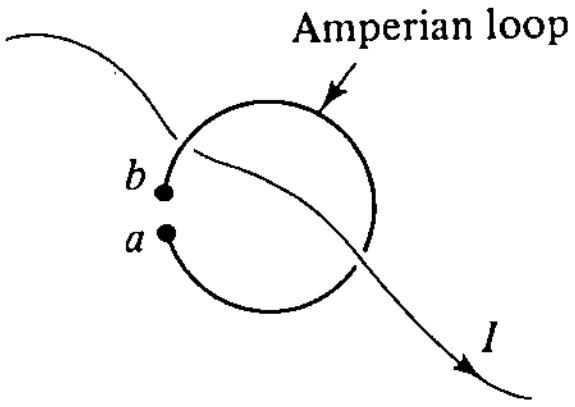

Figure 5.45

(there $\nabla \times \mathbf{B} \neq 0$), but that's not enough. Show, by applying Ampère's law to a path that starts at a and circles the wire, returning to b (Fig. 5.45), that the scalar potential cannot be single-valued (that is, $U(a) \neq U(b)$, even if they represent the same physical point). As an example, find the scalar potential for an infinite straight wire. (To avoid a multivalued potential, you must restrict yourself to simply-connected regions which remain on one side or the other of every wire but never allow you to go all the way around.)

Problem 5.32 Use the results of Example 11 to find the field inside a uniformly charged sphere, of total charge Q and radius R, which is rotating at a constant angular velocity ω.

Problem 5.33

(a) Complete the proof of Theorem 2, Section 1.6.2. That is, show that any divergence-less vector ($\mathbf{F}$) can be written as the curl of a vector ($\mathbf{W}$)—even if $\nabla \times \mathbf{F}$ does *not* go to zero at infinity (so that the Helmholtz theorem cannot be used.)

(*Hint*: What you have to do is find W_x, W_y, and W_z such that: (1) $\partial W_y/\partial z - \partial W_z/\partial y = F_x$; (2) $\partial W_z/\partial x - \partial W_x/\partial z = F_y$; and (3) $\partial W_x/\partial y - \partial W_y/\partial x = F_z$. Pick $W_x = 0$, and solve (2) and (3) for W_y and W_z. Note that the "constants of integration" here are themselves functions of y and z—they're constant only with respect to x. Now plug these expressions into (1), and use the fact that $\nabla \cdot \mathbf{F} = 0$ to obtain

$$W_y = \int_0^x F_z(x', y, z)\, dx'; \qquad W_z = \int_0^y F_x(0, y', z)\, dy' - \int_0^x F_y(x', y, z)\, dx'$$

(b) By direct differentiation, show that the $\mathbf{W}$ you obtained in part (a) satisfies $\nabla \times \mathbf{W} = \mathbf{F}$. Is $\mathbf{W}$ divergence-less? (This was a very asymmetrical construction, and it would be surprising if $\mathbf{W}$ *were* divergenceless—although we know that there *exists* a vector whose curl is $\mathbf{F}$ *and* whose divergence is zero.)

(c) As an example, let $\mathbf{F} = y\hat{i} + z\hat{j} + x\hat{k}$; calculate $\mathbf{W}$, and confirm that $\nabla \times \mathbf{W} = \mathbf{F}$. (For further discussion see Problem 5.60.)

5.4.2 Summary: Magnetostatic Boundary Conditions

In Chapter 2 I drew a triangular diagram to summarize the relations among the three fundamental quantities of electrostatics: the charge density ρ, the electric field $\mathbf{E}$, and the potential V. A similar diagram can be constructed for magnetostatics (Fig. 5.46), relating the current density $\mathbf{J}$, the field $\mathbf{B}$, and the potential $\mathbf{A}$. Filling in the arrows, we discover that one formula is missing: We have never derived an equation for $\mathbf{A}$ in terms of $\mathbf{B}$. It's unlikely that you would ever want such a formula, but just for the record, let's work it out. You may have noticed a striking parallel between the divergence and curl of $\mathbf{A}$, on the one hand:

$$\nabla \cdot \mathbf{A} = 0, \qquad \nabla \times \mathbf{A} = \mathbf{B}$$

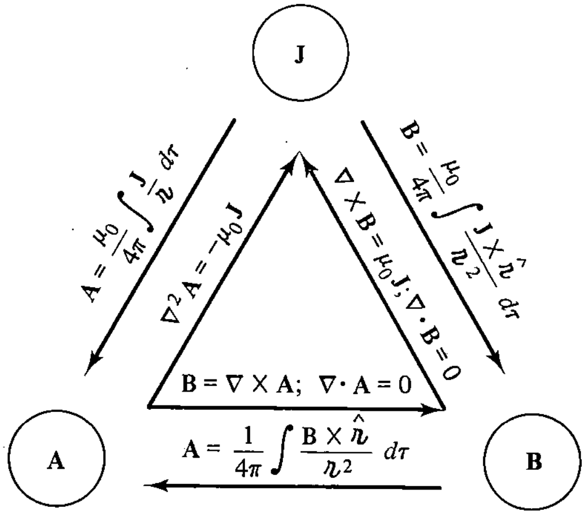

Figure 5.46

and the divergence and curl of **B**, on the other:

$$\nabla \cdot \mathbf{B} = 0, \qquad \nabla \times \mathbf{B} = \mu_0 \mathbf{J}$$

Evidently, **A** depends on **B** in exactly the same way that **B** depends on $\mu_0 \mathbf{J}$.[14] Since

$$\mathbf{B} = \frac{\mu_0}{4\pi}\int \frac{\mathbf{J} \times \hat{\mathscr{r}}}{\mathscr{r}^2} d\tau$$

(the Biot-Savart law), it must be the case that

$$\mathbf{A} = \frac{1}{4\pi}\int \frac{\mathbf{B} \times \hat{\mathscr{r}}}{\mathscr{r}^2} d\tau \tag{5.66}$$

(See Problem 5.60 for commentary on this formula.)

Just as the electric field suffers a discontinuity at a surface *charge*, so the magnetic field is discontinuous at a surface *current*. Only this time it is the *tangential* component that changes. For if we apply (5.43), in the integral form

$$\int \mathbf{B} \cdot d\mathbf{a} = 0$$

to a wafer-thin pillbox straddling the surface (Fig. 5.47), we have

$$B_{\perp\,\text{above}} = B_{\perp\,\text{below}} \tag{5.67}$$

As for the tangential components, an amperian loop running perpendicular to the current (Fig. 5.48) yields

$$\oint \mathbf{B} \cdot d\mathbf{l} = (B_{\parallel\,\text{above}} - B_{\parallel\,\text{below}})l = \mu_0 I_{\text{enc}} = \mu_0 K l$$

or

$$B_{\parallel\,\text{above}} - B_{\parallel\,\text{below}} = \mu_0 K \tag{5.68}$$

[14]This assumes that the boundary conditions for **A** and **B** are the same. In practice, as long as $\mathbf{J} \to 0$ at infinity, so also does **B** and (by choice) **A**. Therefore, except for those artificial problems in which (5.57) itself fails, (5.66) is valid.

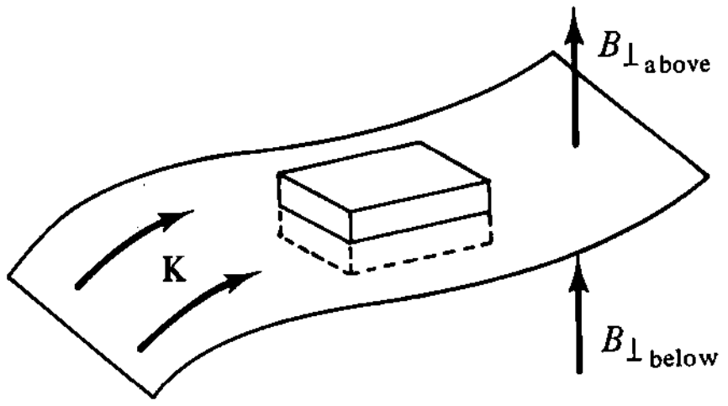

Figure 5.47

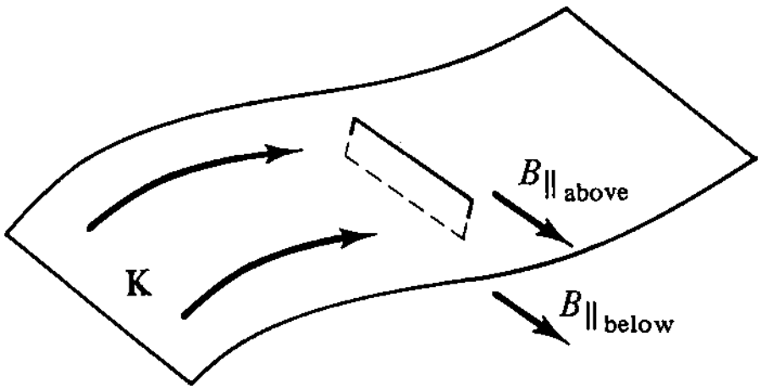

Figure 5.48

Thus, the component of **B** which is parallel to the surface but perpendicular to the current is discontinuous in the amount $\mu_0 K$. A similar amperian loop running *parallel* to the current serves to show that the component parallel is *continuous*. These results can be summarized in a single formula:

$$\mathbf{B}_{\text{above}} - \mathbf{B}_{\text{below}} = \mu_0(\mathbf{K} \times \hat{n}) \tag{5.69}$$

where $\hat{n}$ is a unit vector perpendicular to the surface, pointing "upward."

Like the scalar potential in electrostatics, the vector potential is continuous across any boundary:

$$\mathbf{A}_{\text{above}} = \mathbf{A}_{\text{below}} \tag{5.70}$$

for $\nabla \cdot \mathbf{A} = 0$ guarantees that the *normal* component is continuous, and $\nabla \times \mathbf{A} = \mathbf{B}$ in the form

$$\oint \mathbf{A} \cdot d\mathbf{l} = \int \mathbf{B} \cdot d\mathbf{a} = \Phi$$

means that the tangential components are continuous—the flux through an amperian loop of vanishing thickness is zero. But the *derivative* of **A** inherits the discontinuity of **B**:

$$\frac{\partial \mathbf{A}_{\text{above}}}{\partial n} - \frac{\partial \mathbf{A}_{\text{below}}}{\partial n} = -\mu_0 \mathbf{K} \tag{5.71}$$

Problem 5.34

(a) Check equation (5.69) for the configuration of Example 9.

(b) Check equations (5.70) and (5.71) for the configuration of Example 11.

Problem 5.35 Prove equation (5.71) from (5.69) and (5.70). (*Suggestion*: I'd set up Cartesian coordinates at the surface, with z perpendicular to the surface and y parallel to the current.)

5.4.3 Multipole Expansion of the Vector Potential

If you want an approximate formula for the vector potential of a localized current distribution, valid at distant points, a multipole expansion is in order. Remember: the idea of a multipole expansion is to write the potential in the form of a power series

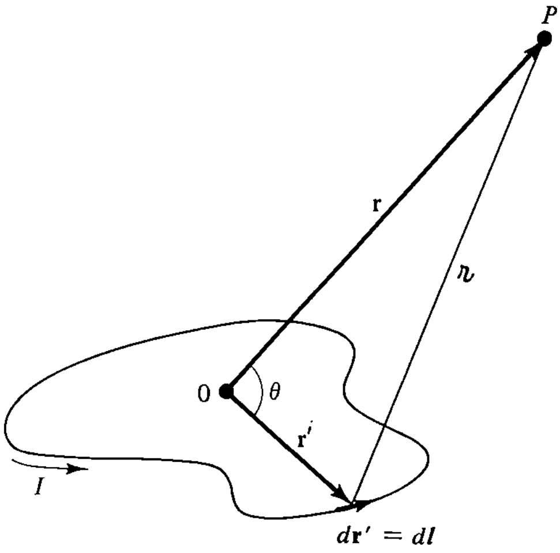

Figure 5.49

in $1/r$, where r is the distance to the point in question (Fig. 5.49); if r is sufficiently large, the series will be dominated by the lowest nonvanishing contribution, and the higher terms can be ignored. As we found in Section 3.4.1 (equation (3.89)),

$$\frac{1}{\mathscr{r}} = \frac{1}{\sqrt{r^2 + (r')^2 - 2rr'\cos\theta}} = \frac{1}{r}\sum_{n=0}^{\infty}\left(\frac{r'}{r}\right)^n P_n(\cos\theta) \tag{5.72}$$

Accordingly, the vector potential of a current loop can be written

$$\mathbf{A} = \frac{\mu_0 I}{4\pi}\oint \frac{1}{\mathscr{r}}\, d\mathbf{l} = \frac{\mu_0 I}{4\pi}\sum_{n=0}^{\infty}\frac{1}{r^{n+1}}\oint (r')^n P_n(\cos\theta)\, d\mathbf{l} \tag{5.73}$$

or, more explicitly:

$$\mathbf{A} = \frac{\mu_0 I}{4\pi}\left[\frac{1}{r}\oint d\mathbf{l} + \frac{1}{r^2}\oint r'\cos\theta\, d\mathbf{l} + \frac{1}{r^3}\oint (r')^2\left(\frac{3}{2}\cos^2\theta - \frac{1}{2}\right) d\mathbf{l} + \cdots\right] \tag{5.74}$$

As in the multipole expansion of V, we call the first term (which goes like $1/r$) the **monopole** term; the second (which goes like $1/r^2$) is the **dipole**, the third **quadrupole**, and so on.

Now, it happens that the *magnetic monopole term is always zero,* for the integral represents the total vector displacement around a closed loop:

$$\oint d\mathbf{l} = 0 \tag{5.75}$$

This reflects the fact that there are (apparently) no magnetic monopoles in nature (an assumption contained in Maxwell's equation $\nabla \cdot \mathbf{B} = 0$, on which the entire theory of vector potential is predicated).

In the absence of any monopole contribution, the dominant term is the dipole—except in the rare case where it, too, vanishes.

$$\mathbf{A}_{\text{dip}}(\mathbf{r}) = \frac{\mu_0 I}{4\pi r^2} \oint r' \cos\theta \, dl = \frac{\mu_0 I}{4\pi r^2} \oint (\hat{r} \cdot \mathbf{r}') \, dl \tag{5.76}$$

This integral can be cast in a more useful form by the following unlovely manipulations: First note that

$$d[(\hat{r} \cdot \mathbf{r}')\mathbf{r}'] = (\hat{r} \cdot d\mathbf{r}')\mathbf{r}' + (\hat{r} \cdot \mathbf{r}')d\mathbf{r}'$$

so that

$$\oint [(\hat{r} \cdot d\mathbf{r}')\mathbf{r}' + (\hat{r} \cdot \mathbf{r}')d\mathbf{r}'] = \oint d[(\hat{r} \cdot \mathbf{r}')\mathbf{r}'] = 0$$

(the total change in $[(\hat{r} \cdot \mathbf{r}')\mathbf{r}']$ around a *closed* loop is zero), and hence

$$\oint (\hat{r} \cdot d\mathbf{r}')\mathbf{r}' = -\oint (\hat{r} \cdot \mathbf{r}') \, d\mathbf{r}' \tag{5.77}$$

Therefore, using the BAC-CAB rule,

$$\hat{r} \times \oint (\mathbf{r}' \times d\mathbf{r}') = \oint [\mathbf{r}'(\hat{r} \cdot d\mathbf{r}') - d\mathbf{r}'(\hat{r} \cdot \mathbf{r}')] = -2 \oint (\hat{r} \cdot \mathbf{r}') \, d\mathbf{r}'$$

Now $d\mathbf{r}'$ is just the infinitesimal displacement vector we have always called dl (Fig. 5.49); so

$$\oint (\hat{r} \cdot \mathbf{r}') \, dl = -\tfrac{1}{2}\hat{r} \times \oint (\mathbf{r}' \times dl) \tag{5.78}$$

With this the dipole potential (5.76) becomes

$$\mathbf{A}_{\text{dip}}(\mathbf{r}) = \frac{\mu_0 I}{4\pi r^2} \left[-\frac{1}{2} \hat{r} \times \oint (\mathbf{r}' \times dl) \right]$$

or, more compactly,

$$\boxed{\mathbf{A}_{\text{dip}} = \frac{\mu_0}{4\pi} \frac{\mathbf{m} \times \hat{r}}{r^2}} \tag{5.79}$$

where $\mathbf{m}$, the **magnetic dipole moment** of the loop, is defined by

$$\boxed{\mathbf{m} = \tfrac{1}{2} I \oint (\mathbf{r}' \times dl)} \tag{5.80}$$

In the special case of a *plane* loop, the integral in (5.80) admits a nice geometrical interpretation: $\frac{1}{2}(\mathbf{r}' \times dl)$ is the area of the shaded triangle, in Fig. 5.50, so the integral is the area of the whole loop:

$$\tfrac{1}{2} \oint (\mathbf{r}' \times dl) = \mathbf{a} \tag{5.81}$$

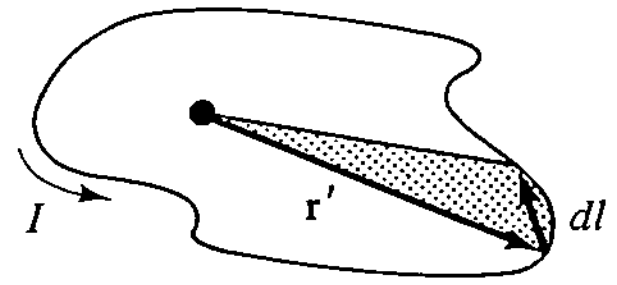

Figure 5.50

As a vector, it points perpendicular to the plane, in the direction given by the usual right-hand rule. For *flat* current loops, then, the dipole moment is simply the product of the current and the area:

$$\boxed{\mathbf{m} = I\mathbf{a}} \tag{5.82}$$

Since this is, in practice, by far the most common situation, some people regard (5.82) as the *definition* of the magnetic dipole moment; arbitrary currents can then be built up out of infinitesimal, essentially flat loops whose dipole moments combine vectorially.

Example 13

Find the magnetic dipole moment of the "bookend-shaped" loop shown in Fig. 5.51. All sides have length s, and it carries a current I.

Solution: This wire could be considered the superposition of two plane square loops (Fig. 5.52). The "extra" sides (AB) cancel when the two are put together,

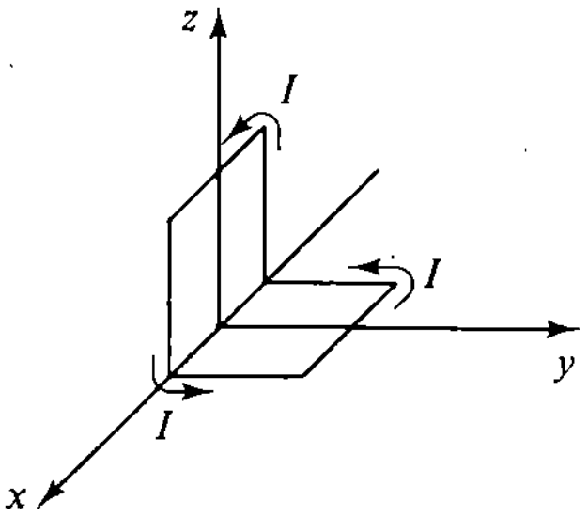

Figure 5.51

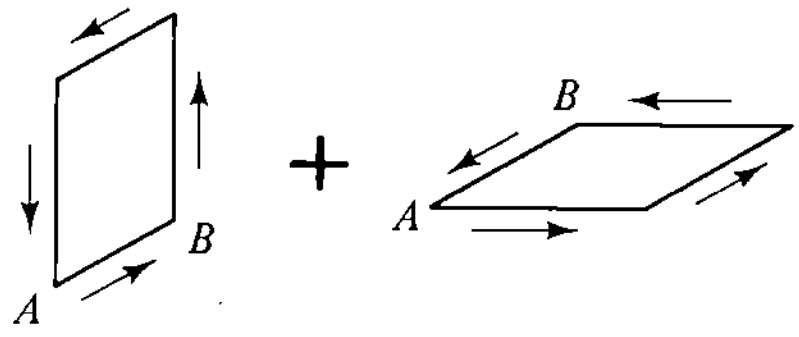

Figure 5.52

since the currents flow in opposite directions. The net magnetic dipole moment is therefore

$$\mathbf{m} = Is^2\hat{\jmath} + Is^2\hat{k}$$

its magnitude is $\sqrt{2}Is^2$, and it points along the 45° line $z = y$.

It is clear from (5.82) that the magnetic dipole moment of a flat current loop is independent of the choice of origin, and you can readily prove this for the general case from (5.80). You may remember that *electric* dipole moments are independent of origin only when the total charge is zero (Section 3.4.3)—that is, only when the

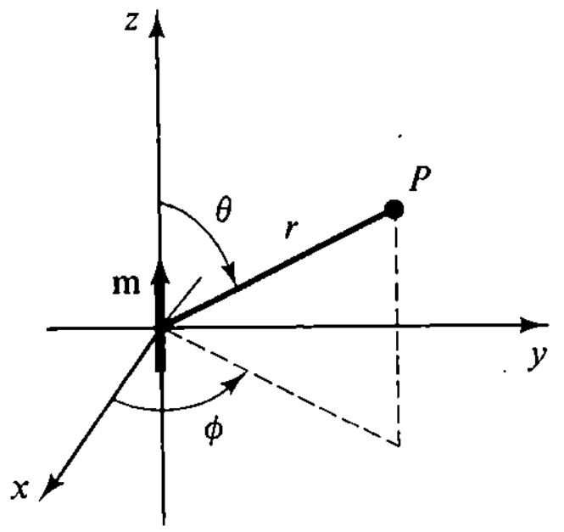

Figure 5.53

electric monopole term is zero. Since the *magnetic* monopole term is *always* zero, it is not really so surprising that the magnetic dipole moment is always independent of origin.

Although the dipole term *dominates* the multipole expansion (unless $\mathbf{m} = 0$), and thus offers a good approximation to the true potential, it is not ordinarily the *exact* potential; there will be quadrupole, octopole, and higher contributions. You might ask, is it possible to devise a current distribution whose potential is "pure" dipole—for which (5.79) is the *exact* potential? Well, yes and no: Like the electrical analog, it can be done in principle, but the model is a bit contrived. To begin with, you must take an *infinitesimally small* loop at the origin, but then, in order to keep the dipole moment finite, you have to crank the current up to infinity, with the product $m = Ia$ held fixed. In practice, the dipole potential is a suitable approximation whenever the distance r greatly exceeds the size of the loop.

The magnetic *field* of a (pure) dipole is easiest to calculate if we put $\mathbf{m}$ at the origin and let it point in the z-direction (Fig. 5.53). According to equation (5.79) the potential at point (r, θ, ϕ) is

$$\mathbf{A} = \frac{\mu_0}{4\pi}\frac{m \sin\theta}{r^2}\hat{\phi} \tag{5.83}$$

and hence

$$\mathbf{B} = \nabla \times \mathbf{A} = \frac{\mu_0}{4\pi}\frac{m}{r^3}(2\cos\theta\,\hat{r} + \sin\theta\,\hat{\theta}) \tag{5.84}$$

Surprisingly, this is *identical* in structure to the field of an *electric* dipole (3.98). (Up close, however, the field of a *physical* magnetic dipole—a small current loop—looks quite different from the field of a physical electric dipole—plus and minus charges a short distance apart. Compare Fig. 5.54 with Fig. 3.34.)

Problem 5.36 Show that the magnetic field of a dipole can be written in coordinate-free form thus:

$$\mathbf{B} = \frac{\mu_0}{4\pi}\frac{1}{r^3}[3(\mathbf{m}\cdot\hat{r})\hat{r} - \mathbf{m}] \tag{5.85}$$

Problem 5.37 A circular loop of wire, with radius R, lies in the xy plane, centered at the origin, and carries a current I running counterclockwise as viewed from the positive z axis.

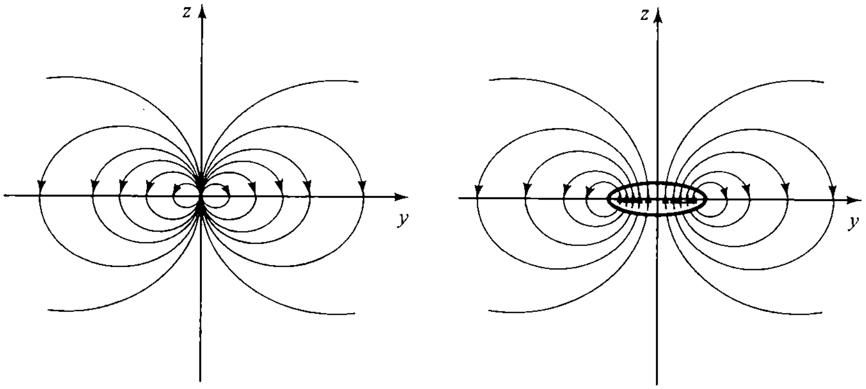

(a) Field of a "pure" magnetic dipole (b) Field of a "physical" magnetic dipole **Figure 5.54**

(a) What is its magnetic dipole moment?

(b) What is its (approximate) magnetic field at points far from the origin?

(c) Show that, for points on the z axis, your answer is consistent with the *exact* field as calculated in Example 6.

Problem 5.38 A phonograph record of radius R, carrying a uniform surface charge σ, is rotating at constant angular velocity ω. Find its magnetic dipole moment.

Problem 5.39 Find the magnetic dipole moment of the spinning spherical shell in Example 11. Show that for points $r > R$ the potential is that of a perfect dipole.

Problem 5.40 Find the exact magnetic field a distance z above the center of a square loop of side s, carrying a current I. Verify that it reduces to the field of a dipole, with the appropriate dipole moment, when $z \gg s$.

Problem 5.41 Show that the magnetic dipole moment of an arbitrary localized current loop (5.80) is independent of the location of the reference point.

Problem 5.42 A thin, uniform donut, carrying charge Q and mass M, rotates about its axis as shown in Fig. 5.55.

(a) Find the ratio of its magnetic dipole moment to its angular momentum. This is called the **gyromagnetic ratio** (or, occasionally, the "magnetomechanical ratio").

(b) What is the gyromagnetic ratio for a uniform spinning sphere? (This requires no new calculation; simply decompose the sphere into infinitesimal rings, and apply the result of part (a).

(c) According to quantum mechanics, the angular momentum of a spinning electron is $\frac{1}{2}\hbar$, where $\hbar$ is Planck's constant. What, then, is the electron's magnetic dipole moment, in $\mathrm{A \cdot m^2}$? (This semiclassical value is actually off by a factor of 2. The precise calculation of the electron's magnetic dipole moment remains the finest achievement of relativistic quantum electrodynamics. The quantity $(e\hbar/2m)$, where e is the charge of the electron and m is its mass, is called the **Bohr magneton.**)

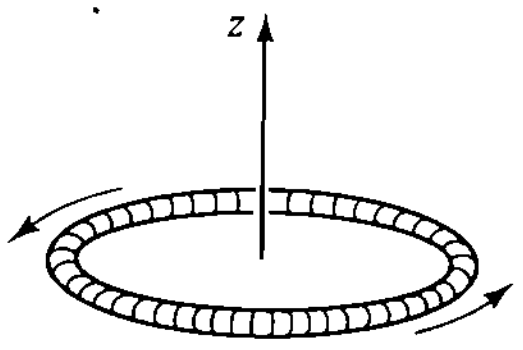

Figure 5.55

Further Problems on Chapter 5

Problem 5.43 It may have occurred to you that since parallel currents attract, the current within a single wire should contract into a tiny concentrated stream along the axis.[15] Yet in practice the current typically distributes itself quite uniformly over the wire. How do you explain this? What integral formula relating the charge and current densities within the wire does your explanation entail? If the current density *is* uniform, can the wire be electrically neutral?

! **Problem 5.44** Use the Biot-Savart law (most conveniently in the form (5.35) appropriate to surface currents) to find the field inside and outside an infinitely long solenoid of radius R, with N turns per unit length, carrying a steady current I.

Problem 5.45 A magnetic dipole $\mathbf{m} = -m_0\hat{k}$ is situated at the origin, in an otherwise uniform magnetic field $\mathbf{B} = B_0\hat{k}$. Show that there exists a sphere, centered at the origin, through which no magnetic field lines pass. Find the radius of this sphere, and sketch the field lines, inside and outside.

Problem 5.46 A circularly symmetrical magnetic field ($\mathbf{B}$ depends only on r), pointing perpendicular to the page, occupies the shaded region in Fig. 5.56. If the total flux ($\int \mathbf{B} \cdot d\mathbf{a}$) is zero, show that a charged particle which starts out at the center will emerge from the field region on a *radial* path (provided it escapes at all—if the initial velocity is too great, it may simply circulate around forever.) By the same token, a particle fired at the center from outside will hit its target, though it may follow a weird trajectory getting there. (*Suggestion*: Calculate the total angular momentum acquired by the particle, using the magnetic force law.)

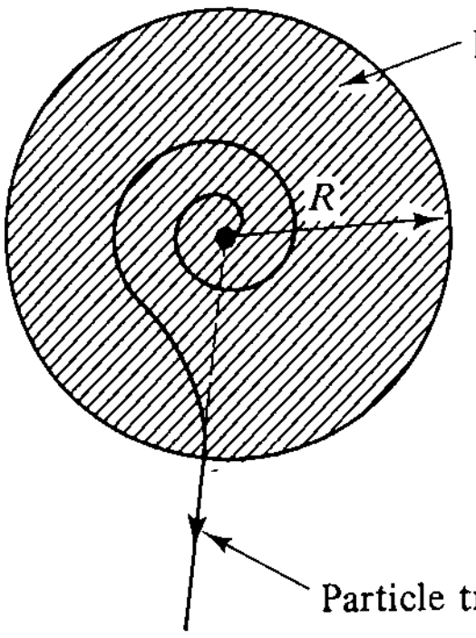

Figure 5.56

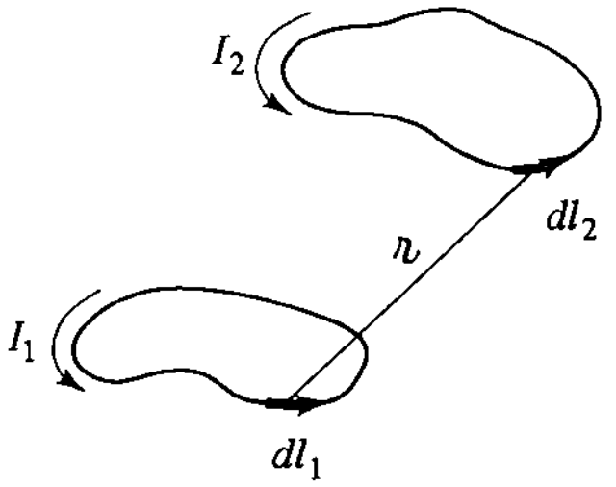

Figure 5.57

Problem 5.47 Magnetostatics treats the "source current" (the one that sets up the field) and the "recipient current" (the one that experiences the force) so asymmetrically that it is by no means obvious that the magnetic force between two current loops is consistent with Newton's third law. Show, starting with the Biot-Savart law (5.28) and the Lorentz force law (5.16), that the force on loop 2 due to loop 1 (Fig. 5.57) can be written as

$$\mathbf{F}_2 = -\frac{\mu_0}{4\pi} I_1 I_2 \oint\oint \frac{\hat{\imath}}{\imath^2}\, d\mathbf{l}_1 \cdot d\mathbf{l}_2 \tag{5.86}$$

In this form it is clear that $\mathbf{F}_2 = -\mathbf{F}_1$, since $\hat{\imath}$ changes direction when the roles of 1 and 2 are interchanged. (If you seem to be getting an "extra" term, it will help to note that $d\mathbf{l}_2 \cdot \hat{\imath} = d\imath$.)

[15] In a plasma this phenomenon is known as the "pinch effect."

• **Problem 5.48**

(a) Prove that the average magnetic field over a sphere of radius R, due to steady currents within the sphere, is

$$\mathbf{B}_{\text{ave}} = \frac{\mu_0}{4\pi}\frac{2\mathbf{m}}{R^3} \tag{5.87}$$

where $\mathbf{m}$ is the total dipole moment of the sphere. Contrast the electrostatic result, (3.100). [This is tough, so I'll give you a start:

$$\mathbf{B}_{\text{ave}} = \frac{1}{\frac{4}{3}\pi R^3}\int \mathbf{B}\, d\tau$$

Write $\mathbf{B}$ as $(\nabla \times \mathbf{A})$, and apply Problem 1.61(b). Now put in (5.57), and do the surface integral first, showing that

$$\int \frac{1}{\imath}\, d\mathbf{a} = \frac{4}{3}\pi \mathbf{r}$$

(see Fig. 5.58).]

(b) Show that the average magnetic field due to steady currents *outside* the sphere is the same as the field they produce at the center.

Problem 5.49 I worked out the multipole expansion for the vector potential of a *line* current because that's the most common type, and in some respects the easiest to work with. For a *volume* current $\mathbf{J}$:

(a) Write down the multipole expansion, analogous to equation (5.73).

(b) Find the monopole potential, and prove that it vanishes.

(c) Find the dipole moment, by appropriate modification of (5.80).

Problem 5.50 A uniformly charged solid sphere of radius R carries total charge Q, and is set spinning with angular velocity ω about the z axis.

(a) What is the magnetic dipole moment of the sphere?

(b) Find the average magnetic field within the sphere (see Problem 5.48).

(c) Find the approximate vector potential at a point (r, θ) where $r \gg R$.

(d) Find the *exact* potential at a point (r, θ) outside the sphere, and check that it is consistent with (c). [*Hint*: refer to Example 11.]

(e) Find the magnetic field at a point (r, θ) *inside* the sphere, and check that it is consistent with (b).

Problem 5.51 (The **Hall effect.**) A current I flows to the right through a rectangular bar of conducting material, in the presence of a uniform magnetic field **B** pointing out of the page (Fig. 5.59).

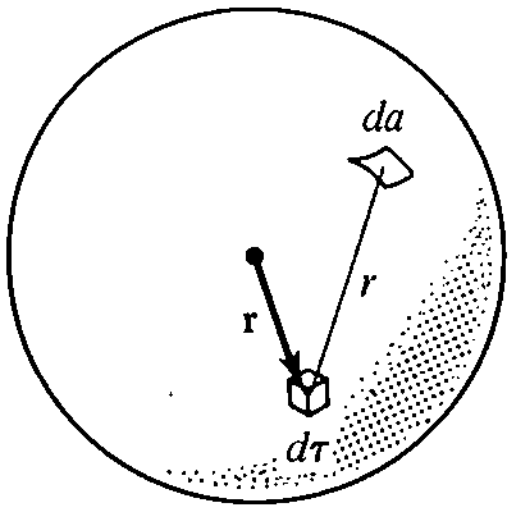

Figure 5.58

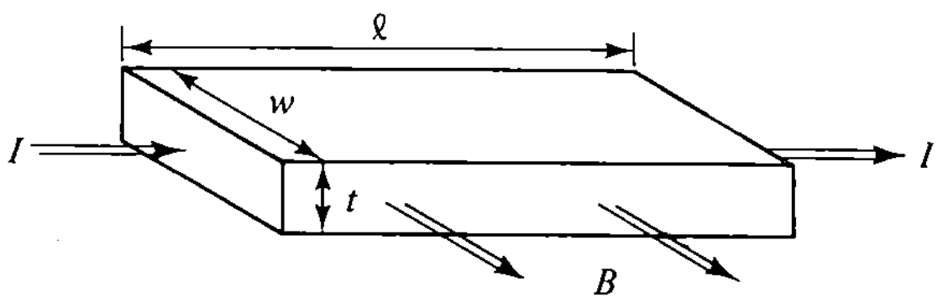

Figure 5.59

(a) If the moving charges are *positive,* in which direction are they deflected by the magnetic field?

This deflection results in an accumulation of charge on the upper and lower surfaces of the bar, which in turn produces an electrical force to counteract the magnetic one. Equilibrium occurs when the two exactly cancel.

(b) Find the resulting potential difference (the "Hall voltage") between the top and bottom of the bar, in terms of B, v (the speed of the charges), and the relevant dimensions of the bar.

(c) How would your analysis change if the moving charges were *negative*? [The Hall effect is the classic way of determining the sign of the mobile charge carriers in a material.]

Problem 5.52 Calculate the magnetic force of attraction between the northern and southern hemispheres of a spinning charged spherical shell (Example 11).

$$\left(\textit{Answer}: \frac{\pi}{4}\,\mu_0\sigma^2\omega^2R^4.\right)$$

Problem 5.53 A semicircular wire carries a steady current I (it must be hooked up to some *other* wires to complete the circuit, but we're not concerned with them here). Find the magnetic field at a point P on the other semicircle (Fig. 5.60).

$$\left(\textit{Answer}: \frac{\mu_0 I}{4\pi R}\ln\left[\frac{\tan\left(\frac{\theta+\pi}{4}\right)}{\tan\left(\frac{\theta}{4}\right)}\right]\right)$$

Problem 5.54 Just as $\nabla\cdot\mathbf{B} = 0$ allows us to express $\mathbf{B}$ as the curl of a vector potential ($\mathbf{B} = \nabla\times\mathbf{A}$), so $\nabla\cdot\mathbf{A} = 0$ permits us to write $\mathbf{A}$ itself as the curl of a "higher" potential: $\mathbf{A} = \nabla\times\mathbf{W}$. (And this hierarchy can be extended ad infinitum.)

(a) Find the general formula for $\mathbf{W}$ (as an integral over $\mathbf{B}$) which holds when $\mathbf{B}\to 0$ at ∞.

(b) Determine $\mathbf{W}$ for the case of a *uniform* magnetic field $\mathbf{B}$. (*Hint*: See Problem 5.27.)

(c) Find $\mathbf{W}$ inside and outside an infinite solenoid. (*Hint*: See Example 12.)

Problem 5.55 Using equation 5.84, calculate the average magnetic field of a dipole over a sphere of radius R centered at the origin. Do the angular integrals first. Compare your answer with the general theorem in Problem 5.48. Explain the discrepancy, and show how equation 5.84 can be corrected to resolve the ambiguity at $r = 0$. (If you get stuck, refer to Problem 3.42.)

$$\left(\textit{Answer}: \text{add } \frac{2\mu_0}{3}\,\mathbf{m}\,\delta^3(\mathbf{r}).\right)$$

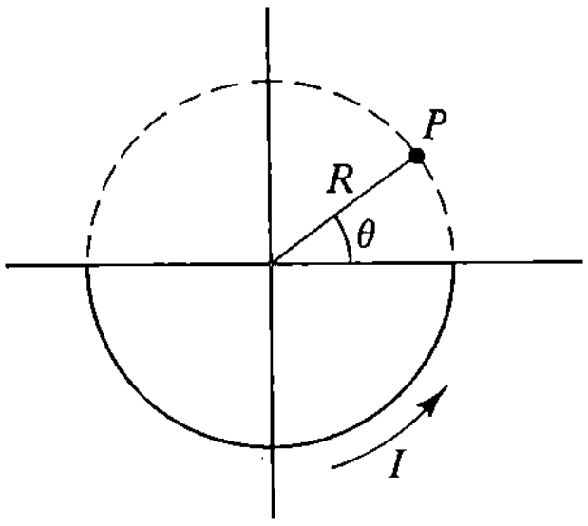

Figure 5.60

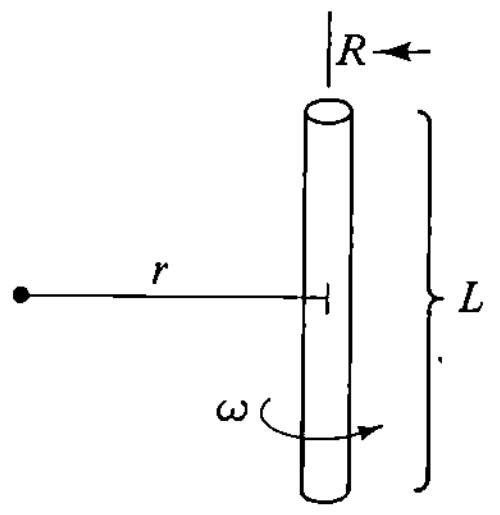

Figure 5.61

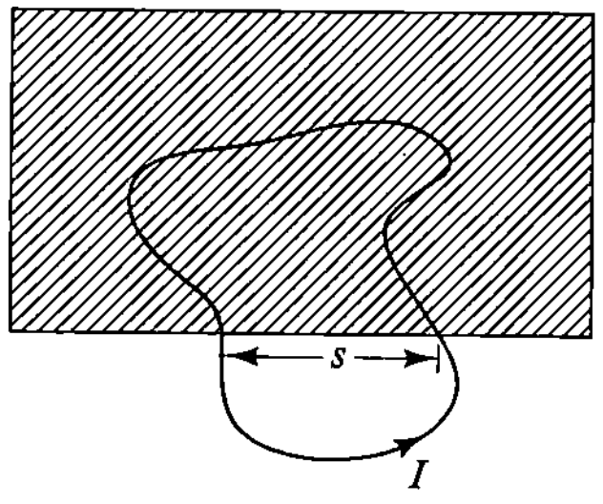

Figure 5.62

Problem 5.56 A thin glass rod of radius R and length L carries a uniform surface charge σ. It is set spinning about its axis, at an angular velocity ω. Find the magnetic field at a distance $r \gg R$ from the center of the rod (Fig. 5.61). (*Hint*: Treat it as a stack of magnetic dipoles.)

$$\left(\textit{Answer}: \frac{\mu_0 \omega \sigma L R^3}{4[r^2 + (L/2)^2]^{3/2}}\right)$$

Problem 5.57 Prove the following uniqueness theorem: If the current density **J** is specified throughout a volume V, and *either* the potential **A** *or* the magnetic field **B** is specified on the surface S bounding V, then the vector potential is determined to within the gradient of a scalar function throughout V (and hence the magnetic field itself is uniquely determined in V). Do *not* assume $\nabla \cdot \mathbf{A} = 0$. (*Hint*: First use the divergence theorem to show that

$$\int \{(\nabla \times \mathbf{U}) \cdot (\nabla \times \mathbf{V}) - \mathbf{U} \cdot [\nabla \times (\nabla \times \mathbf{V})]\}\, d\tau = \oint (\mathbf{U} \times (\nabla \times \mathbf{V})) \cdot d\mathbf{a}$$

for arbitrary vector functions **U** and **V**.)

Problem 5.58 A plane wire loop of irregular shape is situated so that part of it is in a uniform magnetic field **B** (in Figure 5.62 the field occupies the shaded region, and points perpendicular to the plane of the loop). The loop carries a current I. Show that the net magnetic force on the loop is $F = IBs$, where s is the chord subtended. Generalize this result to the case where the magnetic field region itself has an irregular shape. What is the direction of the force?

Problem 5.59 The magnetic field on the axis of a circular current loop (equation 5.34) is far from uniform (it falls off sharply with increasing z). You can produce a more nearly uniform field by using *two* such loops a distance s apart (Fig. 5.63).

(a) Find the field (B) as a function of z, and show that $\partial B/\partial z$ is zero at the point midway between them ($z = 0$).

If you pick s just right the *second* derivative of B will *also* vanish at the midpoint. This arrangement is known as a **Helmholtz coil**; it's a convenient way of producing relatively uniform magnetic fields.

(b) Determine s such that $\partial^2 B/\partial z^2 = 0$ at the midpoint, and find the resulting magnetic field at the center.

(*Answer*: $B = 8\mu_0 I/(5^{3/2} R)$)

! **Problem 5.60** In Section 5.4.2 I derived a formula for **A** in terms of **B**, but the result (5.66) is strikingly different in structure from the standard formula for V in terms of **E** (2.18). Actually, there exists an electrostatic formula analogous to 5.66:

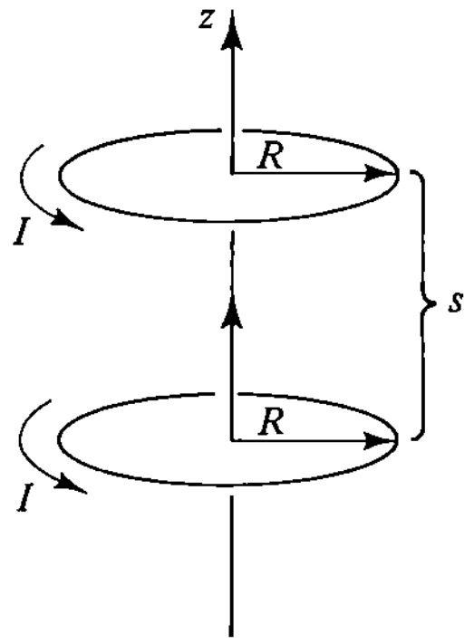

Figure 5.63

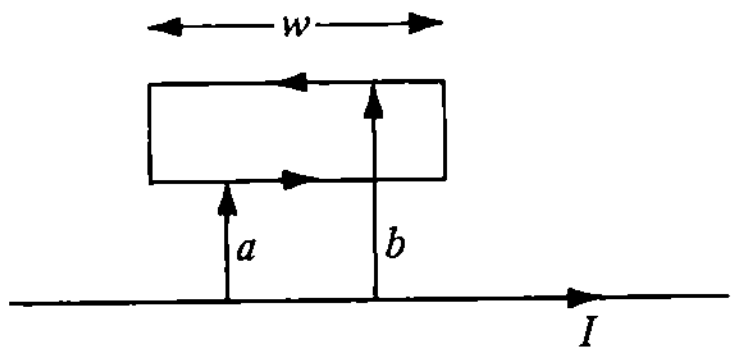

Figure 5.64

(a) Show that

$$V = -\frac{1}{4\pi}\int \frac{\mathbf{E}\cdot\hat{\boldsymbol{\imath}}}{\imath^2}\, d\tau.$$

Does there exist a magnetostatic analog to (2.18)? The obvious candidate would be

$$\mathbf{A}(P) = \int_{\mathcal{O}}^{P} (\mathbf{B} \times d\mathbf{l})$$

(b) Check this formula for the simplest possible case—uniform **B** (use the origin as your reference point, and let **B** point in the z direction). Is the result consistent with Problem 5.27? You could cure this problem by throwing in a factor of 1/2, but the flaws in this equation run deeper:

(c) Show that $\int (\mathbf{B} \times d\mathbf{l})$ is *not* independent of path, by calculating $\oint (\mathbf{B} \times d\mathbf{l})$ around the rectangular loop shown in Fig. (5.64).

As far as I know[16] the best one can do along these lines (using the origin as reference point) is the pair of equations

(i) $V(\mathbf{r}) = -\mathbf{r}\cdot\int_0^1 \mathbf{E}(\lambda\mathbf{r})d\lambda$

(ii) $\mathbf{A}(\mathbf{r}) = -\mathbf{r} \times \int_0^1 \lambda\mathbf{B}(\lambda\mathbf{r})d\lambda$

Equation (i) amounts to selecting a *radial* path for the integral in (2.18).

(d) Use (ii) to find the vector potential for *uniform* **B**.

(e) Use (ii) to find the vector potential of an infinite straight wire carrying a steady current I. Does (ii) automatically satisfy $\nabla\cdot\mathbf{A} = 0$?

$$\left(\textit{Answer}\text{: } \mathbf{A}(r, \phi, z) = \frac{\mu_0 I}{2\pi r}(z\hat{r} - r\hat{z}).\right)$$

---

[16] R. L. Bishop and S. I. Goldberg, *Tensor Analysis on Manifolds* (New York: Macmillan, 1968), Section 4.5. Equation (ii) constitutes a more "symmetrical" solution to problem 5.33.

6

MAGNETOSTATIC FIELDS IN MATTER

6.1 MAGNETIZATION

6.1.1 Diamagnets, Paramagnets, Ferromagnets

If you ask the average person what "magnetism" is, you will probably be told about horseshoe magnets, compass needles, and the North Pole—none of which has any obvious connection with moving charges or current-carrying wires. Yet all magnetic fields are due to electric charges in motion, and in fact, if you could examine a piece of magnetic material on an atomic scale you *would* find tiny currents: electrons orbiting around nuclei and electrons spinning on their axes. For macroscopic purposes, these current loops are so small that we may treat them as magnetic dipoles. Ordinarily, they cancel each other out because of the random orientation of the atoms. But when a magnetic field is applied, a net alignment of these magnetic dipoles occurs, and the medium becomes magnetically polarized, or **magnetized.**

Unlike electric polarization, which is almost always in the same direction as **E**, some materials acquire a magnetization *parallel* to **B** (*para*magnets) and some *opposite* to **B** (*dia*magnets). A few substances (called *ferro*magnets, in deference to the most common example, iron) retain a substantial magnetization indefinitely after the external field has been removed—for these the magnetization is not determined by the *present* field but by the whole magnetic "history" of the object. Permanent magnets made of iron are the most familiar examples of magnetism, though from a theoretical point of view they are the most complex; I shall save ferromagnetism for the end of the chapter, and begin now with qualitative models for paramagnetism and diamagnetism.

6.1.2 Torques and Forces on Magnetic Dipoles

A magnetic dipole experiences a torque in a magnetic field, just as an electric dipole does in an electric field. Let's calculate the torque on a rectangular current loop in a uniform field **B**. (Since any current loop could be built up from infinitesimal rectangles, with all the "internal" sides canceling, as indicated in Fig. 6.1, there is no real loss of generality in using this shape; but if you prefer to start from scratch with an arbitrary shape, see Problem 6.2.) Center the loop at the origin, and tilt it an angle θ from the z axis toward the y axis (Fig. 6.2). Let **B** point in the z-direction. The forces on the two sloping sides cancel—they tend to *stretch* the loop, but they don't *rotate* it. The forces on the "horizontal" sides are likewise equal and opposite (so the net *force* on the loop is zero), but they do generate a torque:

$$\mathbf{N} = s_1 F \sin\theta\,\hat{i}$$

The magnitude of the force on each of these segments is

$$F = I s_2 B$$

and therefore

$$\mathbf{N} = (I s_1 s_2) B \sin\theta\,\hat{i} = mB \sin\theta\,\hat{i}$$

or

$$\boxed{\mathbf{N} = \mathbf{m} \times \mathbf{B}} \tag{6.1}$$

where $m = I s_1 s_2$ is the magnetic dipole moment of the loop. Equation (6.1) gives the exact torque on any localized current distribution, in the presence of a *uniform* field—in a *nonuniform* field it is exact only for a "perfect" dipole of infinitesimal size.

Notice that equation (6.1) is identical in form to the electrical analog, equation (4.4): $\mathbf{N} = \mathbf{p} \times \mathbf{E}$. In particular, the torque is again in such a direction as to line the dipole up *parallel* to the field. It is this torque which accounts for **paramagnetism.** Since every electron constitutes a magnetic dipole (picture it, if you wish, as a tiny spinning sphere of charge), you might expect paramagnetism to be a universal phenomenon. Actually, the laws of quantum mechanics (specifically, the Pauli exclusion principle) dictate that the electrons within a given atom lock together in pairs with opposing spins, and this effectively neutralizes the torque on the combination. As a result, paramagnetism normally occurs in atoms or molecules with an odd number of

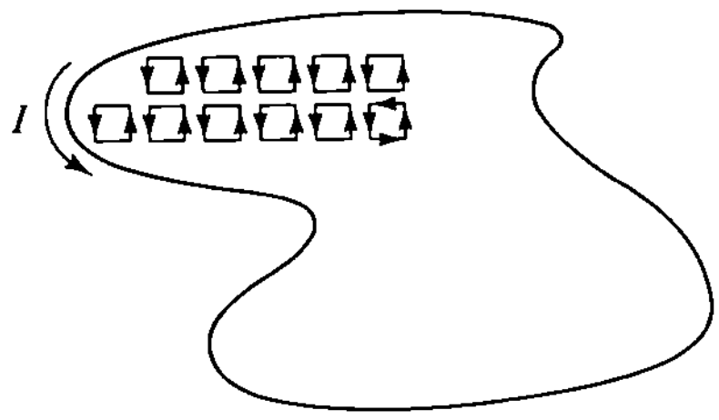

Figure 6.1

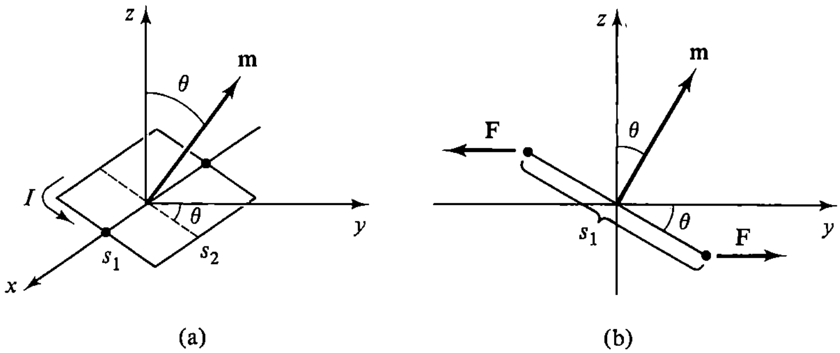

Figure 6.2

electrons, where the "extra," unpaired member is subject to the magnetic torque. Even here the alignment is far from complete, since random thermal collisions tend to destroy the order.

In a *uniform* field, the net *force* on a current loop is zero:

$$\mathbf{F} = I \oint (d\mathbf{l} \times \mathbf{B}) = I\left(\oint d\mathbf{l}\right) \times \mathbf{B} = 0$$

the constant **B** comes outside the integral, and the net displacement $\oint d\mathbf{l}$ around a closed loop vanishes. In a *nonuniform* field this is no longer the case. For example, suppose a circular wire of radius R, carrying a current I is suspended above a short solenoid in the "fringing" region (Fig. 6.3). Here **B** has a radial component, and there is a net downward force on the loop (Fig. 6.4):

$$F = 2\pi IRB \cos\theta \tag{6.2}$$

For an *infinitesimal* loop of dipole moment **m** in a field **B** the force is

$$\boxed{\mathbf{F} = \nabla(\mathbf{m}\cdot\mathbf{B})} \tag{6.3}$$

(see Problem 6.4). Using product rule 4, we have

$$\mathbf{F} = \mathbf{m} \times (\nabla \times \mathbf{B}) + (\mathbf{m}\cdot\nabla)\,\mathbf{B} \tag{6.4}$$

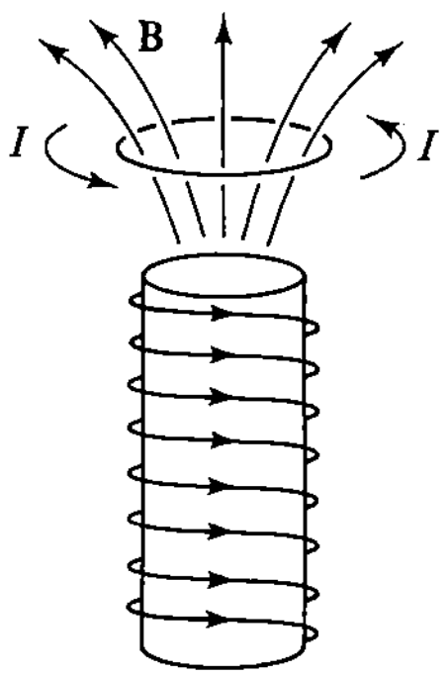

Figure 6.3

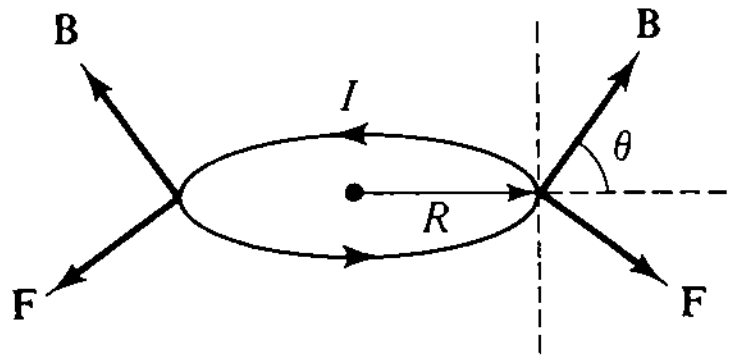

Figure 6.4

B here is the *external* field, of course (the dipole doesn't exert a force on *itself*); since $\nabla \times \mathbf{B} = \mu_0 \mathbf{J}$, it follows that

$$\mathbf{F} = (\mathbf{m} \cdot \nabla) \mathbf{B} \tag{6.5}$$

provided there is no external current at the actual location of the dipole. (By contrast, the *electrical* analogs to (6.3) and (6.5) are entirely equivalent—that is, $\nabla(\mathbf{p} \cdot \mathbf{E}) = (\mathbf{p} \cdot \nabla)\mathbf{E}$—because the "extra" term $\nabla \times \mathbf{E}$ is *always* zero, in electrostatics.) Once again the magnetic formula is identical to its electrical "twin," if we agree to write the latter in the form $\mathbf{F} = \nabla(\mathbf{p} \cdot \mathbf{E})$.[1]

If you're starting to get a sense of *déjà vu,* perhaps you will have more respect for those early physicists who thought magnetic dipoles consisted of positive and negative magnetic "charges" (north and south "poles," they called them), separated by a small distance, just like electric dipoles (Fig. 6.5(a)). They wrote down a "Coulomb's law" for the attraction and repulsion of these poles and developed the whole of magnetostatics in exact analogy to electrostatics. It's not a bad model, for many purposes—it gives the correct field of a dipole (at least, away from the origin), the right torque on a dipole (at least, on a *stationary* dipole), and the proper force on a dipole (at least, in the absence of external currents). But it's bad physics, because *there's no such thing* as a single magnetic north pole or south pole. If you break a bar magnet in half, you don't get a north pole in one hand and a south pole in the other; you get two complete magnets. Magnetism is *not* due to "magnetic charges" but rather to *moving electric* charges; magnetic dipoles are tiny current loops (Fig. 6.5(c)), and it's an extraordinary thing, really, that the formulas involving **m** bear any resemblance at all to formulas involving **p**.

Sometimes it is easier to think in terms of the "Gilbert" model of a magnetic dipole (separated monopoles) instead of the physically correct "Ampère" model (current loop). Indeed, this picture occasionally offers a quick and clever solution to an otherwise cumbersome problem (you just copy the corresponding result from electrostatics, changing **p** to **m**, $1/\epsilon_0$ to μ_0, and **E** to **B**). But whenever the *close-up* features of the dipole come into play, the two models can yield strikingly different answers. In particular, the Gilbert model gives (6.5) for the force on a dipole, whereas the Ampère model gives (6.3).[2] Moreover, the *field* of a dipole at $r \cong 0$ carries a term

[1] For reasons that are irrelevant here (see footnote 1, p. 163, $(\mathbf{p} \cdot \nabla)\mathbf{E}$ is a slightly preferable way to *write* the formula. In the same sense, (6.4) is slightly preferable to (6.3).

[2] T. H. Boyer, *Am. J. Phys.* **56**, 688 (1988). Experimentally, the force on a neutron (measured in scattering processes) fits the Ampère model.

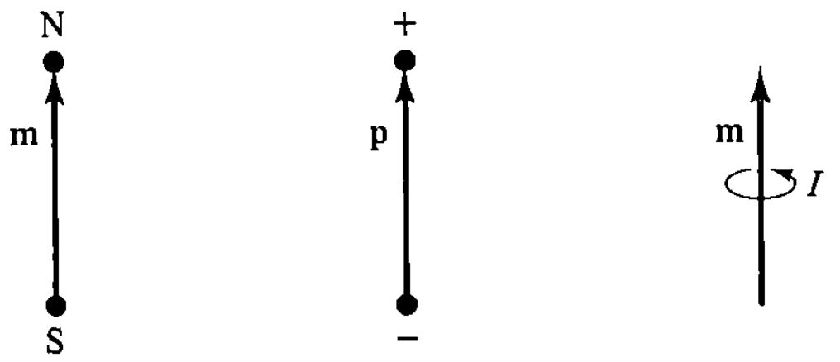

(a) Magnetic dipole (Gilbert model). (b) Electric dipole. (c) Magnetic dipole (Ampère model). **Figure 6.5**

$-(\frac{1}{3})\mu_0 \mathbf{m}\delta^3(\mathbf{r})$ in the Gilbert model (Problem 3.41) but $+(\frac{2}{3})\mu_0 \mathbf{m}\delta^3(\mathbf{r})$ in the Ampère model (Problem 5.48).[3] My advice is to use the Gilbert model, if you like, to get an intuitive "feel" for a problem, but never rely on it for quantitative results.

Problem 6.1 Calculate the torque exerted on the square loop shown in Fig. 6.6 due to the circular loop (assume r is much larger than a or s). If the square loop is free to rotate, what will its equilibrium orientation be?

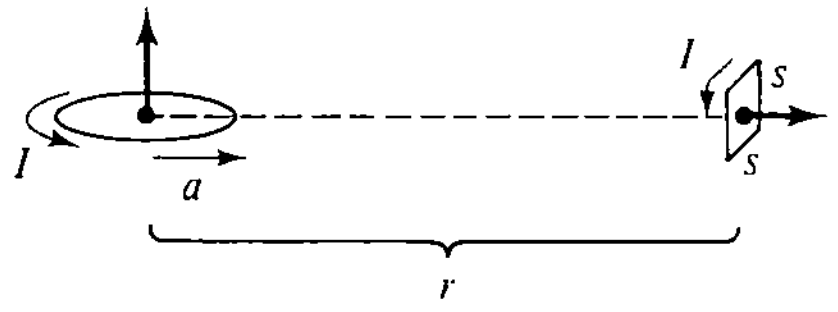

Figure 6.6

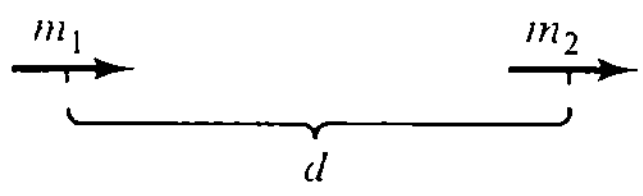

Figure 6.7

Problem 6.2 Show that the torque on *any* steady current distribution (not just a square loop) in a uniform field $\mathbf{B}$ is $(\mathbf{m} \times \mathbf{B})$.

Problem 6.3 Find the force of attraction between two magnetic dipoles, $\mathbf{m}_1$ and $\mathbf{m}_2$, oriented as shown in Fig. 6.7, a distance d apart, (a) using equation (6.2) and (b) using equation (6.3).

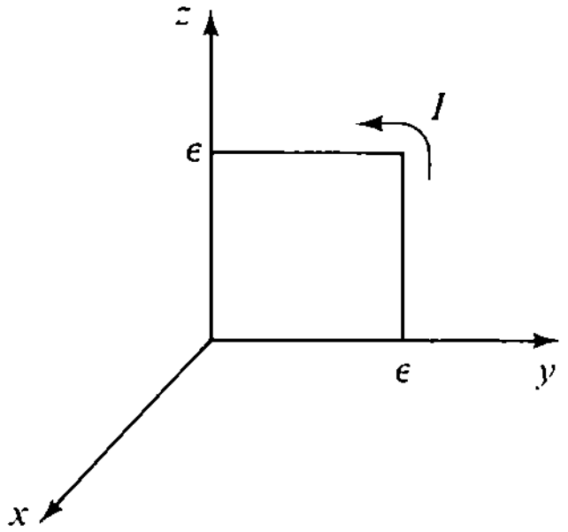

Figure 6.8

Problem 6.4 Derive equation (6.3). Here's one way to do it: Assume the dipole is an infinitesimal square, of side ϵ (if it's not, chop it up into squares, and apply the argument to each one). Choose axes as shown in Fig. 6.8, and calculate $\mathbf{F} = I \int (d\mathbf{l} \times \mathbf{B})$ along each of the four sides. Expand $\mathbf{B}$ in a Taylor series—on the right side, for instance,

$$\mathbf{B} = \mathbf{B}(0, \epsilon, z) \cong \mathbf{B}(0, \epsilon, 0) + z \left.\frac{\partial \mathbf{B}}{\partial z}\right|_{(0,\epsilon,0)}$$

Problem 6.5 A uniform current density $\mathbf{J} = J_0 \hat{k}$ fills a slab straddling the yz plane, from $x = -a$ to $x = +a$. A magnetic dipole $\mathbf{m} = m_0 \hat{i}$ is situated at the origin.

(a) Find the force on the dipole (i) using equation (6.3), and (ii) using equation (6.5). Show that (i) is correct for an "Ampère" dipole (take the current loop to be a square, and find the force on each side), but that (ii) would be correct for a "Gilbert" dipole (find the force on each pole).

(b) Do the same for a dipole pointing in the y-direction: $\mathbf{m} = m_0 \hat{j}$.

---

[3] D. J. Griffiths, *Am. J. Phys.* **50**, 698 (1982). Experimentally, the field of a proton (measured in hyperfine splitting) fits the Ampère model.

6.1.3 Effect of a Magnetic Field on Atomic Orbits

Electrons not only *spin;* they also *revolve* about the nucleus—for simplicity, let's assume the orbit is a circle of radius r (Fig. 6.9). Although technically this orbital motion does not constitute a steady current, in practice the period $T = 2\pi r/v$ is so short that unless you blink awfully fast, it's going to *look* like a steady current,

$$I = \frac{e}{T} = \frac{ev}{2\pi r}$$

Accordingly, the orbital dipole moment is

$$\mathbf{m} = -\tfrac{1}{2}evr\hat{k} \tag{6.4}$$

(The minus sign accounts for the negative charge of the electron.) Like any other magnetic dipole, this one is subject to a torque ($\mathbf{m} \times \mathbf{B}$) when the atom is placed in a magnetic field. But it's a lot harder to tilt the entire orbit than it is the spin, so the orbital contribution to paramagnetism is small. There is, however, a more significant effect on the orbital motion: The electron *speeds up or slows down,* depending on the orientation of $\mathbf{B}$. For whereas the centripetal acceleration v^2/r is ordinarily sustained by electrical forces alone,[4]

$$\frac{1}{4\pi\epsilon_0}\frac{e^2}{r^2} = m_e\frac{v^2}{r} \tag{6.5}$$

in the presence of a magnetic field there is an additional term, $-e(\mathbf{v} \times \mathbf{B})$. For the sake of argument, let's say that $\mathbf{B}$ is perpendicular to the plane of the orbit, as shown in Fig. 6.10; then

$$\frac{1}{4\pi\epsilon_0}\frac{e^2}{r^2} + ev'B = m_e\frac{v'^2}{r} \tag{6.6}$$

Under these conditions, the new speed v' is *greater* than v:

$$ev'B = \frac{m_e}{r}(v'^2 - v^2) = \frac{m_e}{r}(v' + v)(v' - v)$$

[4]To avoid confusion with the magnetic dipole moment m, I'll write the electron mass with a subscript: m_e.

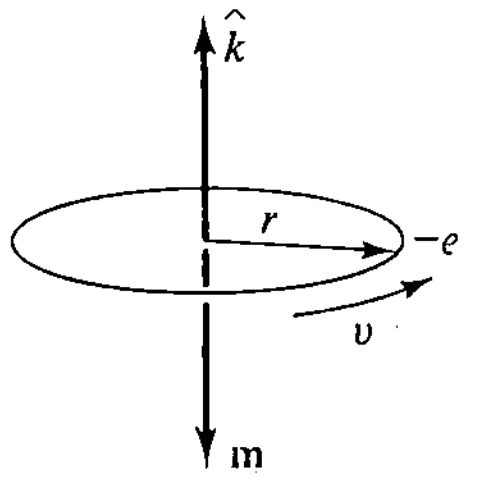

Figure 6.9

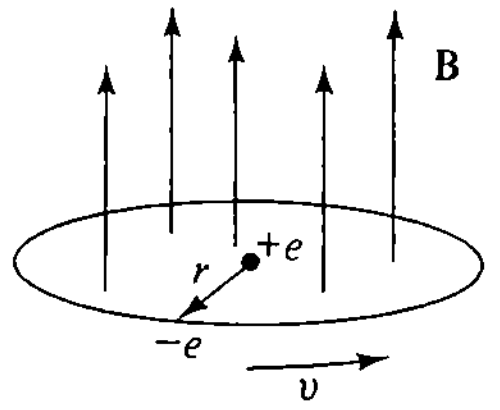

Figure 6.10

or, assuming the change $\Delta v = v' - v$ is small,

$$\Delta v = \frac{erB}{2m_e} \tag{6.7}$$

When **B** is turned on, then, the electron speeds up.

A change in orbital speed means a change in the dipole moment (6.4):

$$\Delta \mathbf{m} = -\frac{1}{2}e(\Delta v)r\hat{k} = -\frac{e^2r^2}{4m_e}\mathbf{B} \tag{6.8}$$

Notice that *the change in* **m** *is opposite to the direction of* **B**. An electron circling the other way would have a dipole moment pointing *upward,* but such an orbit would be *slowed down* by the field, so the *change* is still opposite to **B**. In ordinary, unmagnetized matter, the electron orbits are randomly oriented, and the orbital dipole moments cancel out. But in the presence of a magnetic field, each atom picks up a little "extra" dipole moment, and these increments are all *antiparallel* to the field. This is the mechanism responsible for diamagnetism. It is a universal phenomenon, affecting all atoms. However, it is typically much weaker than paramagnetism, and is therefore observed mainly in atoms with *even* numbers of electrons, where paramagnetism is absent.

In deriving equation (6.8) I assumed that the orbit remains circular, with its original radius r. I cannot offer a justification for this at the present stage. If the atom is stationary while the field is turned on, then my assumption can be proved—this is not magneto*statics,* however, and the details will have to await Chapter 7 (see Problem 7.17). If the atom is moved into the field, the situation is enormously more complicated. But never mind—I'm only trying to give you a qualitative account of diamagnetism. Assume, if you prefer, that the velocity remains the same while the *radius* changes—the formula (6.8) is altered (by a factor of 2), but the *conclusion* is unaffected. The truth is that this classical model is fundamentally flawed (diamagnetism is really a *quantum* phenomenon), so there's not much point in refining the details.[5] What *is* important is the *empirical* fact that in diamagnetic materials the induced dipole moments point *opposite* to the magnetic field.

6.1.4 Magnetization

In the presence of a magnetic field, matter becomes *magnetized*; that is, upon microscopic examination it will be found to contain many tiny dipoles, with a net alignment along some direction. We have discussed two mechanisms which account for this magnetic polarization: (1) paramagnetism (the dipoles associated with the spins of unpaired electrons experience a torque tending to line them up parallel to the field) and (2) diamagnetism (the orbital speed of the electrons is altered in such a way as to change the orbital dipole moment in a direction opposite to the field). Whatever

[5] S. L. O'Dell and R. K. P. Zia, *Am. J. Phys.* **54**, 32 (1986); R. Peierls, *Surprises in Theoretical Physics* (Princeton, N.J.: Princeton University Press, 1979); Section 4.3; R. P. Feynman, R. B. Leighton, and M. Sands, *The Feynman Lectures on Physics* (New York: Addison-Wesley, 1966), vol. 2, sec. 34-36.

the *cause,* we describe the state of magnetic polarization of a material by the vector quantity

$$\mathbf{M} \equiv \textit{magnetic dipole moment per unit volume} \tag{6.9}$$

M is called the **magnetization;** it plays a role analogous to the polarization **P** in electrostatics. In the following section, we will not worry about how the magnetization *got* there—it could be paramagnetism, diamagnetism, or even ferromagnetism—we shall take **M** as *given,* and calculate the field this magnetization itself produces.

Incidentally, it may have surprised you to learn that materials other than the famous ferromagnetic trio (iron, nickel, and cobalt) are affected by a magnetic field *at all.* You cannot, of course, pick up a piece of wood or aluminum with a magnet. The reason is that diamagnetism and paramagnetism are extremely weak: it takes a delicate experiment and a powerful magnet to detect them at all. If you were to suspend a piece of paramagnetic material above a solenoid, as in Fig. 6.3, the induced magnetization would be upward, and hence the force downward. By contrast, the magnetization of a diamagnetic object would be downward and the force upward. In general, when a sample is placed in a region of nonuniform field, the *paramagnet is attracted into the field,* whereas the *diamagnet is repelled away.* But the actual forces are pitifully weak—in a typical experimental arrangement the force on a comparable sample of iron would be 10^4 or 10^5 times as great. That's why it was reasonable for us to calculate the field inside a piece of copper wire, say, in Chapter 5, without worrying about the effects of magnetization.

Problem 6.6 Of the following materials, which would you expect to be paramagnetic and which diamagnetic? Aluminum, copper, copper chloride ($CuCl_2$), carbon, lead, nitrogen (N_2), salt (NaCl), sodium, sulfur, water. (Actually, copper is slightly *dia*magnetic; otherwise they're all what you'd expect.)

6.2 THE FIELD OF A MAGNETIZED OBJECT

6.2.1 Bound Currents

Suppose we have a piece of magnetized material; the magnetic dipole moment per unit volume, **M**, is given. What field does this object produce? Well, the vector potential of a single dipole **m** is given by equation (5.78):

$$\mathbf{A} = \frac{\mu_0}{4\pi} \frac{\mathbf{m} \times \hat{\mathbf{\imath}}}{\imath^2} \tag{6.10}$$

In the magnetized object, each volume element $d\tau$ carries a dipole moment $\mathbf{M}\,d\tau$, so the total vector potential is (Fig. 6.11)

$$\mathbf{A} = \frac{\mu_0}{4\pi} \int \frac{\mathbf{M} \times \hat{\mathbf{\imath}}}{\imath^2}\, d\tau \tag{6.11}$$

That *does* it, in principle. But as in the electrical case (Section 4.2.1), the integral can be cast into a more useful form by using the identity

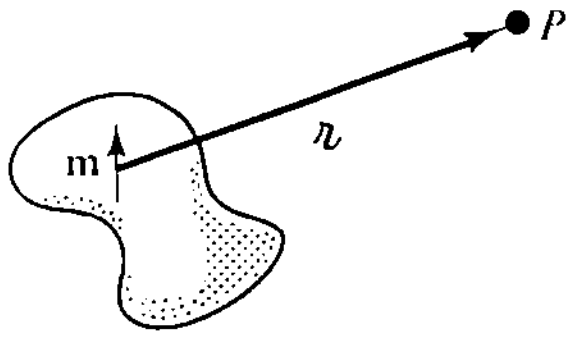

Figure 6.11

$$\nabla \frac{1}{\mathscr{r}} = \frac{\hat{\mathscr{r}}}{\mathscr{r}^2}$$

(The differentiation is with respect to *source* coordinates.) With this,

$$\mathbf{A} = \frac{\mu_0}{4\pi} \int \left(\mathbf{M} \times \nabla \frac{1}{\mathscr{r}} \right) d\tau$$

Integrating by parts, using product rule 7, gives

$$\mathbf{A} = \frac{\mu_0}{4\pi} \left[\int \frac{1}{\mathscr{r}} (\nabla \times \mathbf{M})\, d\tau - \int \nabla \times \left(\frac{1}{\mathscr{r}} \mathbf{M} \right) d\tau \right]$$

Problem 1.61(b) permits us to express the latter as a surface integral,

$$\mathbf{A} = \frac{\mu_0}{4\pi} \int \frac{1}{\mathscr{r}} (\nabla \times \mathbf{M})\, d\tau + \frac{\mu_0}{4\pi} \oint \frac{1}{\mathscr{r}} (\mathbf{M} \times d\mathbf{a}) \tag{6.12}$$

The first term looks just like the potential of a *volume* current,

$$\boxed{\mathbf{J}_b = \nabla \times \mathbf{M}} \tag{6.13}$$

while the second term looks like the potential of a surface current,

$$\boxed{\mathbf{K}_b = \mathbf{M} \times \hat{n}} \tag{6.14}$$

where $\hat{n}$ is the normal unit vector. With these definitions,

$$\mathbf{A} = \frac{\mu_0}{4\pi} \int_{\text{volume}} \frac{1}{\mathscr{r}} \mathbf{J}_b\, d\tau + \frac{\mu_0}{4\pi} \oint_{\text{surface}} \frac{1}{\mathscr{r}} \mathbf{K}_b\, da \tag{6.15}$$

What this means is that the potential (and hence also the field) of a magnetized object is the same as would be produced by a volume current density $\mathbf{J}_b = \nabla \times \mathbf{M}$ throughout the material, plus a surface current $\mathbf{K}_b = \mathbf{M} \times \hat{n}$, on the boundary. Instead of integrating the contributions of all the infinitesimal dipoles, as in equation (6.11), we can first evaluate these **bound currents**, and then calculate the field *they* produce, in the same way that we calculate the field of any other volume and surface currents. Notice the striking parallel with the electrical case: There the field of a polarized object was the same as that of a bound volume charge $\rho_b = -\nabla \cdot \mathbf{P}$ plus a bound surface charge $\sigma_b = \mathbf{P} \cdot \hat{n}$.

Example 1

Find the magnetic field of a uniformly magnetized sphere.

Solution: Choosing the z axis along the direction of **M** (Fig. 6.12), we have

$$\mathbf{J}_b = \nabla \times \mathbf{M} = 0, \qquad \mathbf{K}_b = \mathbf{M} \times \hat{n} = M \sin\theta \hat{\phi}$$

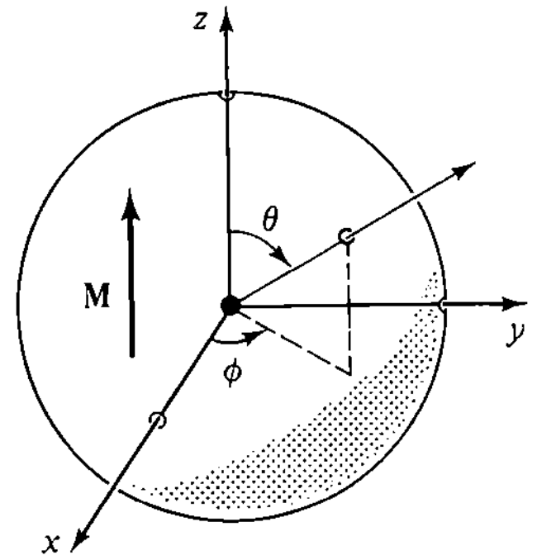

Figure 6.12

Now, a rotating spherical shell, of uniform surface charge σ, carries a surface current density

$$\mathbf{K} = \sigma \mathbf{v} = \sigma\omega R \sin\theta \hat{\phi}$$

It follows, therefore, that the field of a uniformly magnetized sphere is identical to the field of a spinning spherical shell, with the identification $\sigma R\omega \rightarrow \mathbf{M}$. Referring back to Example 11 of Chapter 5, I conclude that

$$\mathbf{B} = \tfrac{2}{3}\mu_0 \mathbf{M} \tag{6.16}$$

inside the sphere, whereas the field outside is the same as that of a pure dipole,

$$\mathbf{m} = \tfrac{4}{3}\pi R^3 \mathbf{M}$$

Notice that the internal field is *uniform,* like the *electric* field inside a uniformly *polarized* sphere (equation (4.14)), although the actual *formulas* for the two cases are curiously different ($\frac{2}{3}$ in place of $-\frac{1}{3}$). The external fields are also analogous: pure dipole in both instances.

Problem 6.7 An infinitely long circular cylinder carries a uniform magnetization **M** parallel to its axis. Find the magnetic field (due to **M**) inside and outside the cylinder.

Problem 6.8 A long circular cylinder of radius R carries a magnetization $\mathbf{M} = kr^2\hat{\phi}$, where k is a constant, r is the distance from the axis, and $\hat{\phi}$ is the "circumferential" unit vector (Fig. 6.13). Find the magnetic field due to **M**, for points inside and outside the cylinder.

Problem 6.9 A short circular cylinder of radius R and length L carries a "frozen-in" uniform magnetization **M** parallel to its axis. Find the bound current, and sketch the magnetic field of the cylinder. (Make three sketches: one for $L \gg R$, one for $L \ll R$, and one for $L \approx R$.) Compare this "bar magnet" with the "bar electret" of Problem 4.11.

Problem 6.10 An iron rod of length L and square cross section (side s), is given a uniform

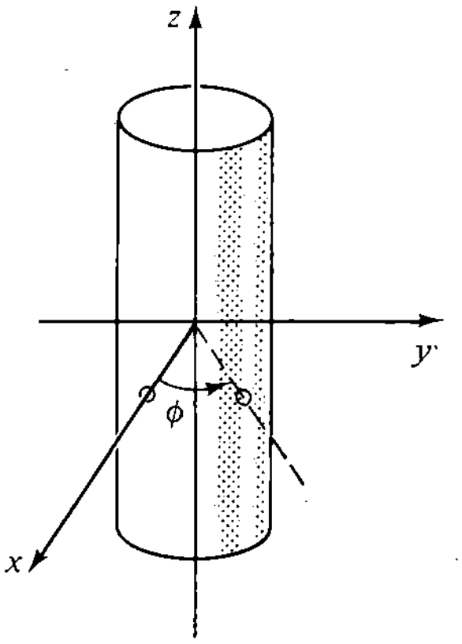

Figure 6.13

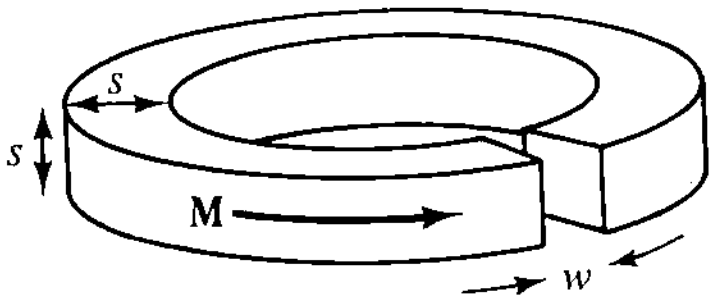

Figure 6.14

longitudinal magnetization M and then bent around into a circle with a narrow gap (width w) as shown in Fig. 6.14. Find the magnetic field in the center of the gap, assuming $w \ll s \ll L$. (*Hint*: Treat it as the superposition of a *complete* torus plus a square loop with reversed current.)

6.2.2 Physical Interpretation of Bound Currents

In the last section we found that the field of a magnetized object is identical to the field that would be produced by a certain distribution of "bound" currents $\mathbf{J}_b$ and $\mathbf{K}_b$. I want to show you how these bound currents arise physically. This will be a *heuristic* argument—the *rigorous* derivation has already been given. Figure 6.15 represents a thin slab of uniformly magnetized material, with the dipoles indicated by tiny current loops. Notice that all the "internal" currents cancel: every time there is one going to the right, a contiguous one is going to the left. However, at the edge there is *no adjacent loop to do the canceling*. The whole thing, then, is equivalent to a single ribbon of current I flowing around the boundary (Fig. 6.16).

What *is* this current, in terms of $\mathbf{M}$? Say that one of the tiny loops has area a

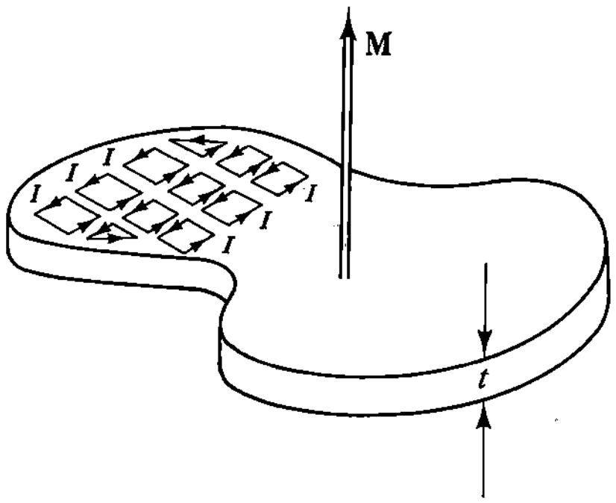

Figure 6.15

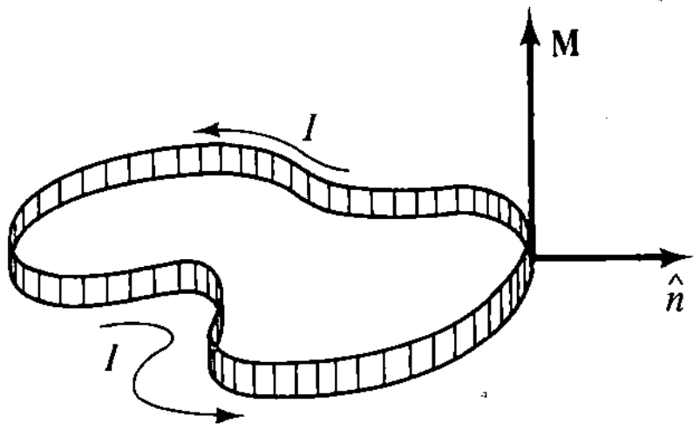

Figure 6.16

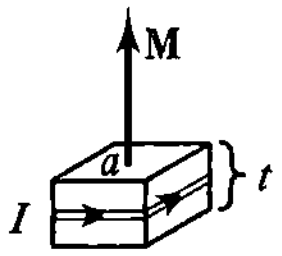

Figure 6.17

and thickness t (Fig. 6.17). In terms of the magnetization M, its dipole moment is

$$m = Mat$$

In terms of the circulating current I, however,

$$m = Ia$$

Therefore, $I = Mt$, so the surface current is $K_b = I/t = M$. Using the outward-drawn unit vector $\hat{n}$ (Fig. 6.16), the direction of $\mathbf{K}_b$ is conveniently indicated by the cross product

$$\mathbf{K}_b = \mathbf{M} \times \hat{n}$$

(This expression also records the fact that there is *no* current on the top or bottom surface of the slab; here $\mathbf{M}$ is parallel to $\hat{n}$, so the cross product vanishes.)

This bound surface current is exactly what we obtained in Section 6.2.1. It is a peculiar *kind* of current, in the sense that no single charge makes the whole trip—on the contrary, each charge moves only in a tiny little loop within a single atom. Nevertheless, the net effect is a macroscopic current flowing over the surface of the magnetized object. We call it a "bound" current to remind ourselves that every charge is attached to a particular atom, but it's a perfectly genuine current, and it produces a magnetic field in the same way any other current does.

When the magnetization is *non*uniform, the internal currents no longer cancel. Figure 6.18(a) shows two adjacent chunks of magnetized material, with a larger arrow on the one to the right, suggesting greater magnetization at that point. On the surface where they join there is a net current in the x-direction, given by

$$I_x = [M_z(y + dy) - M_z(y)]\, dz = \frac{\partial M_z}{\partial y}\, dy\, dz$$

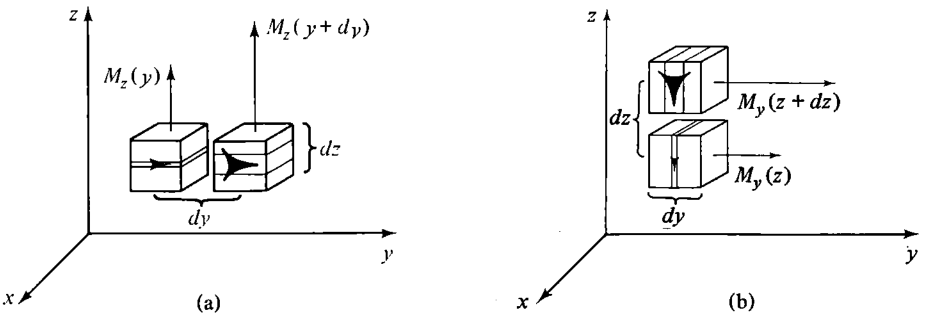

Figure 6.18

The corresponding volume current density is therefore

$$(J_b)_x = \frac{\partial M_z}{\partial y}$$

By the same token, a nonuniform magnetization in the y-direction would contribute an amount $-\partial M_y/\partial z$ (Fig. 6.18(b)), so that

$$(J_b)_x = \frac{\partial M_z}{\partial y} - \frac{\partial M_y}{\partial z}$$

In general, then,

$$\mathbf{J}_b = \nabla \times \mathbf{M}$$

consistent, again, with the result of Section 6.2.1. Incidentally, like any other steady current, $\mathbf{J}_b$ should obey the conservation law (5.27),

$$\nabla \cdot \mathbf{J} = 0$$

Does it? *Yes,* for the divergence of a curl is *always* zero.

6.2.3 The Magnetic Field Inside Matter

Like the electric field, the actual *microscopic* magnetic field inside matter fluctuates wildly from point to point and instant to instant. When we speak of "the" magnetic field in matter, we mean the *macroscopic* field: the average over regions large enough to contain many atoms. (The magnetization **M** is "smoothed out" in the same sense.) It is this macroscopic field that one obtains when the methods of Section 6.2.1 are applied to points inside magnetized material, as you can prove for yourself in the following problem.

Problem 6.11 In Section 6.2.1 we began with the potential of a *perfect* dipole (6.10), whereas in *fact* we are dealing with *physical* dipoles. Show, by the method of Section 4.2.3, that we nevertheless get the correct macroscopic field.

6.3 THE AUXILIARY FIELD H

6.3.1 Ampère's Law in Magnetized Materials

In Section 6.2 we found that the effect of magnetization is to establish bound currents $\mathbf{J}_b = \nabla \times \mathbf{M}$ within the material and $\mathbf{K}_b = \mathbf{M} \times \hat{n}$ on the surface. The field due to magnetization of the medium is just the field produced by these bound currents. We are now ready to put everything together: the field attributable to bound currents, plus the field due to everything else—which we shall call "free" currents. The free current might flow through wires imbedded in the magnetized substance or, if the latter is a conductor, through the material itself. In any event, the total current can be written as

$$\mathbf{J} = \mathbf{J}_b + \mathbf{J}_f \tag{6.17}$$

There is no new physics in equation (6.17); it is simply a *convenience* to separate the current into these two parts because they *got* there by quite different means: The free current is there because someone hooked up a wire to a battery—it involves actual transport of charge; the bound current is there because of magnetization—it results from the conspiracy of many aligned atomic dipoles.

In view of (6.17) and (6.13), Ampère's law can be written

$$\frac{1}{\mu_0}(\nabla \times \mathbf{B}) = \mathbf{J} = \mathbf{J}_f + \mathbf{J}_b = \mathbf{J}_f + (\nabla \times \mathbf{M})$$

or, collecting together the two curls:

$$\nabla \times \left(\frac{1}{\mu_0}\mathbf{B} - \mathbf{M}\right) = \mathbf{J}_f$$

The quantity in parentheses is designated by the letter **H**:

$$\boxed{\mathbf{H} \equiv \frac{1}{\mu_0}\mathbf{B} - \mathbf{M}} \tag{6.18}$$

In terms of **H**, then, Ampère's law reads

$$\boxed{\nabla \times \mathbf{H} = \mathbf{J}_f} \tag{6.19}$$

or, in integral form,

$$\oint \mathbf{H} \cdot d\mathbf{l} = I_{f_{\text{enc}}} \tag{6.20}$$

where $I_{f_{\text{enc}}}$ is the total *free* current passing through the Amperian loop.

H plays a role in magnetostatics analogous to **D** in electrostatics: Just as **D** allowed us to write *Gauss's* law in terms of the free *charge* alone, **H** permits us to express *Ampère's* law in terms of the free *current* alone—and free current is what we control directly. Bound current, like bound charge, comes along for the ride—the material gets magnetized, and this results in bound currents; we cannot turn them on or off independently, as we can free currents. In applying (6.20) all we need to worry about is the *free* current, which we know about because we *put* it there. In particular, when symmetry permits, we can calculate H immediately from (6.20) by the usual Ampère's law methods.

Example 2

A long copper rod of radius R carries a uniformly distributed (free) current I (Fig. 6.19). Find H inside and outside the rod.

Solution: Copper is weakly diamagnetic, so the dipoles will line up *opposite* to the field. This results in a bound current running *antiparallel* to I within the

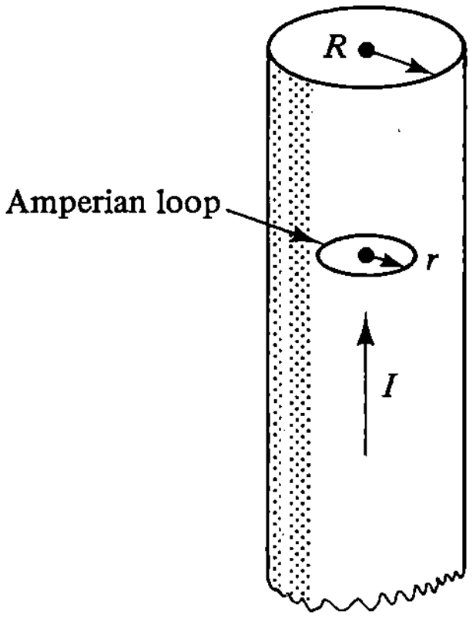

Figure 6.19

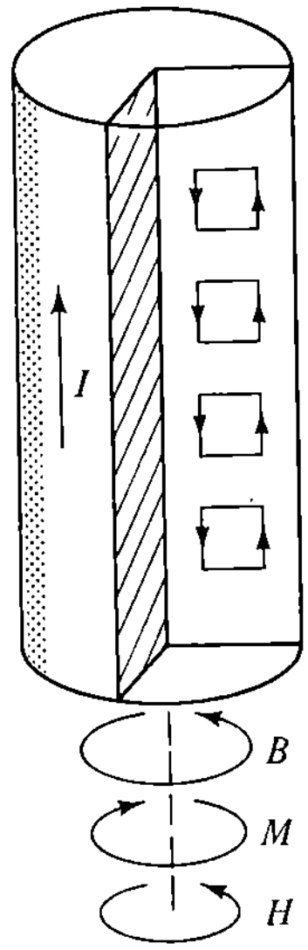

Figure 6.20

wire and *parallel* to I along the surface (see Fig. 6.20). Just how *great* these bound currents will be we are not yet in a position to say—but in order to calculate **H** it is sufficient to realize that all the currents are longitudinal, whereas **B**, **M**, and therefore also **H** are circumferential. Applying (6.20) to an Amperian loop of radius $r < R$,

$$H(2\pi r) = I_{f_{\text{enc}}} = I\left(\frac{\pi r^2}{\pi R^2}\right)$$

so that

$$\mathbf{H} = \frac{I}{2\pi R^2} r\hat{\phi} \qquad (r \leqslant R) \tag{6.21}$$

within the wire. Meanwhile, outside the wire

$$\mathbf{H} = \frac{I}{2\pi r}\hat{\phi} \qquad (r \geqslant R) \tag{6.22}$$

In the latter region (as always, in empty space) $\mathbf{M} = 0$; so

$$\mathbf{B} = \mu_0\mathbf{H} = \frac{\mu_0}{2\pi}\frac{I}{r}\hat{\phi} \quad (r \geqslant R)$$

the same as for a *non*magnetized wire (Example 7, Chapter 5). *Inside* the wire **B** cannot be determined at this stage; since we have no way of knowing **M** (though in practice the magnetization in copper is so slight that for most purposes we can ignore it altogether).

As it turns out, **H** is a more useful quantity than **D**. In the laboratory you will frequently hear people talking about **H** (more often even than **B**), but you will never hear anyone speak of **D** (only **E**). The reason is this: To build an electromagnet you run a certain (free) current through a coil. The *current* is the thing you read on the dial, and this determines **H** (or, at any rate, the line integral of **H**). **B** depends on the specific materials you used and even, if iron is present, on the history of your magnet. On the other hand, if you want to set up an *electric* field, you do *not* plaster a known free charge on the plates of a parallel-plate capacitor; rather, you connect them to a battery of known *voltage*. It's the *potential difference* you read on your dial, and that determines **E** (or, at any rate, the line integral of **E**). **D** depends on the details of the dielectric you're using. If it were easy to measure charge, and hard to measure potential, then you'd find experimentalists talking about **D** instead of **E**. So the relative familiarity of **H**, as contrasted with **D**, derives from purely practical considerations; theoretically, they're on an equal footing.

Many authors call **H**, not **B**, the "magnetic field." Then they have to invent a new word for **B**: the "flux density," or magnetic "induction" (an absurd choice, since that term already has at least two other meanings in electrodynamics). Anyway, **B** is indisputably the fundamental quantity, so I shall continue to call it the "magnetic field," as everyone does in the spoken language. **H** has no sensible name: just call it "**H**".[6]

Problem 6.12 An infinitely long cylinder, of radius R, carries a "frozen-in" magnetization, parallel to the axis,

$$M = kr$$

where k is a constant and r is the distance from the axis (there is no free current anywhere). Find the magnetic field inside and outside the cylinder by two different methods:

(a) As in Section 6.2, locate all the bound currents, and calculate the field they produce.

(b) Use Ampère's law, in the form (6.20), to find **H**, and then get **B** from (6.18). (Notice that the second method is much faster, and avoids any explicit reference to the bound currents.)

Problem 6.13 For the bar magnet of Problem 6.9, make careful sketches of **M**, **B**, and **H**, assuming L is about $2R$. Compare Problem 4.16.

Problem 6.14 Suppose the field inside a large piece of magnetic material is $\mathbf{B}_0$, so that $\mathbf{H}_0 = (1/\mu_0)\mathbf{B}_0 - \mathbf{M}$.

(a) Now a small spherical cavity is hollowed out of the material (Fig. 6.21). Find the field at the center of the cavity, in terms of $\mathbf{B}_0$ and **M**. Also find **H** at the center of the cavity, in terms of $\mathbf{H}_0$ and **M**.

(b) Do the same for a long needle-shaped cavity running parallel to **M**.

(c) The same for a thin wafer-shaped cavity perpendicular to **M**.

Assume the cavities are small enough so that **M**, $\mathbf{B}_0$, and $\mathbf{H}_0$ are essentially constant.

[6]For those who disagree, I quote A. Sommerfeld's *Electrodynamics* (New York: Academic Press, 1952) p. 45: "The unhappy term 'magnetic field' for **H** should be avoided as far as possible. It seems to us that this term has led into error none less than Maxwell himself. . .".

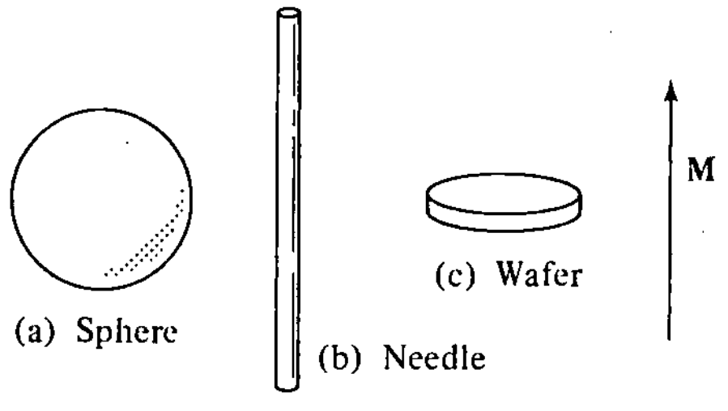

Figure 6.21

Compare Problem 4.17. (*Hint*: Carving out a cavity is the same as superimposing an object of the same shape but opposite magnetization.)

6.3.2 A Deceptive Parallel

Equation (6.19) looks just like Ampère's original law (5.48), only the *total* current is replaced by the *free* current, and **B** is replaced by $\mu_0\mathbf{H}$. As in the case of **D**, however, I must warn you against reading too much into this correspondence. It does *not* say that $\mu_0\mathbf{H}$ is "just like **B**, only its source is $\mathbf{J}_f$ instead of **J**." For the curl alone does not determine a vector field—you must know the divergence as well. And whereas $\nabla\cdot\mathbf{B} = 0$, the divergence of **H** is *not*, in general, zero. In fact, from (6.18)

$$\nabla\cdot\mathbf{H} = -\nabla\cdot\mathbf{M} \tag{6.23}$$

Only when the divergence of **M** vanishes is the parallel between **B** and $\mu_0\mathbf{H}$ faithful.

If you think I'm being pedantic, consider for example the bar magnet: a short cylinder of iron that carries a permanent uniform magnetization **M** parallel to its axis. (See Problems 6.9 and 6.13). In this case there is no free current anywhere, and a naïve application of (6.20) might lead you to suppose that $\mathbf{H} = 0$, and hence that $\mathbf{B} = \mu_0\mathbf{M}$ inside the magnet and $\mathbf{B} = 0$ outside, which is nonsense. It is quite true that the *curl* of **H** vanishes everywhere, but the divergence does not (can you see where $\nabla\cdot\mathbf{M} \neq 0$?) *Advice*: When you are asked to find **B** or **H** in a problem involving magnetic materials, first look for symmetry. If the problem exhibits cylindrical, plane, solenoidal, or toroidal symmetry, then you can get **H** directly from (6.20) by the usual Ampère's law methods. (Apparently, in such cases $\nabla\cdot\mathbf{M}$ is automatically zero, since the free current alone determines the answer.) If the requisite symmetry is absent, you'll have to think of another approach, and in particular you must *not* assume that **H** is zero just because you see no free current.

6.4 LINEAR AND NONLINEAR MEDIA

6.4.1 Magnetic Susceptibility and Permeability

In paramagnetic and diamagnetic materials, the magnetization is maintained by the field; when **B** is removed, **M** disappears. In fact, for most substances the magnetization is *proportional* to the field, provided the field is not too great. For notational

TABLE 6.1 MAGNETIC SUSCEPTIBILITIES

Material	Magnetic Susceptibility
Diamagnetic:	
Bismuth	-16.5×10^{-5}
Gold	-3.0×10^{-5}
Silver	-2.4×10^{-5}
Copper	-0.96×10^{-5}
Water	-0.90×10^{-5}
Carbon Dioxide	-1.2×10^{-8}
Hydrogen	-0.22×10^{-8}
Paramagnetic:	
Oxygen	190×10^{-8}
Sodium	0.85×10^{-5}
Aluminum	2.1×10^{-5}
Tungsten	7.8×10^{-5}
Gadolinium	$48{,}000 \times 10^{-5}$

Source: Handbook of Chemistry and Physics, 67*th ed.* (Cleveland: CRC Press, Inc., 1986-87.) All figures are for atmospheric pressure and room temperature.

consistency with the electrical case (equation 4.26), I *should* express the proportionality thus:

$$\mathbf{M} = \frac{1}{\mu_0}\chi_m \mathbf{B} \qquad (incorrect) \tag{6.24}$$

But custom dictates that it be written in terms of **H** instead of **B**:

$$\boxed{\mathbf{M} = \chi_m \mathbf{H}} \tag{6.25}$$

The constant of proportionality χ_m is called the **magnetic susceptibility**; it is a dimensionless quantity that varies from one substance to another—positive for paramagnets and negative for diamagnets. Typical values are around 10^{-5} (see Table 6.1).

A material that obeys (6.25) is called a **linear medium.** In view of (6.18),

$$\mathbf{B} = \mu_0(\mathbf{H} + \mathbf{M}) = \mu_0(1 + \chi_m)\mathbf{H} \tag{6.26}$$

for linear media. Thus, **B** is also proportional to **H**:[7]

$$\mathbf{B} = \mu \mathbf{H} \tag{6.27}$$

where

$$\mu = \mu_0(1 + \chi_m) \tag{6.28}$$

[7]Physically, therefore, (6.24) would say exactly the same as (6.25), only the constant χ_m would have a different value. Equation (6.25) is a little more convenient because the experimentalist finds it handier to work with **H** than **B**.

μ is called the **permeability** of the material. In a vacuum, where there is no matter to magnetize, the susceptibility χ_m vanishes, and the permeability is μ_0. That's why μ_0 is called the **permeability of free space.**

Example 3

An infinite solenoid (N turns per unit length, current I) is filled with linear material of susceptibility χ_m. Find the magnetic field inside the solenoid.

Solution: Since **B** is due in part to bound currents (which we don't yet know), we cannot compute it directly. However, this is one of those symmetrical cases in which we can get **H** from the free current alone, using Ampère's law in the form (6.20):

$$\mathbf{H} = NI\hat{k}$$

($\hat{k}$ is a unit vector along the axis—Fig. 6.22). According to (6.26), then,

$$\mathbf{B} = \mu_0(1 + \chi_m)NI\hat{k}$$

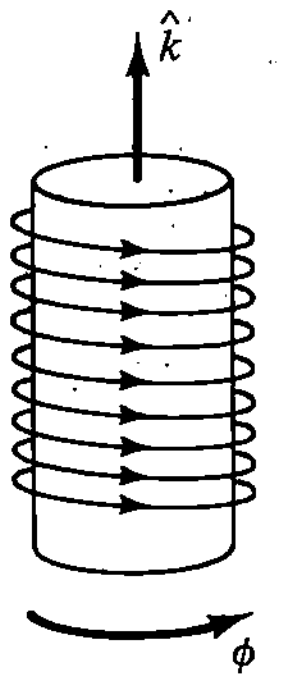

Figure 6.22

If the medium is paramagnetic, the field is slightly enhanced; if it's diamagnetic, the field is somewhat reduced. This reflects the fact that the bound surface current

$$\mathbf{K}_b = \mathbf{M} \times \hat{n} = \chi_m(\mathbf{H} \times \hat{n}) = \chi_m NI\hat{\phi}$$

is in the same direction as I, in the former case ($\chi_m > 0$), and opposite in the latter ($\chi_m < 0$).

You might suppose that linear media avoid the defect in the parallel between **B** and **H**: Since **M** and **H** are now proportional to **B**, does it not follow that their divergence, like **B**'s, must always vanish? Unfortunately, it does *not*; at the *boundary* between two materials of different permeability the divergence of **M** can actually be infinite. For instance, at the end of a cylinder of linear paramagnetic material, **M** is zero on one side but not on the other. For the "Gaussian pillbox" shown in Fig. 6.23, $\oint \mathbf{M} \cdot d\mathbf{a} \neq 0$, and hence, by the divergence theorem, $\nabla \cdot \mathbf{M}$ cannot vanish everywhere within.

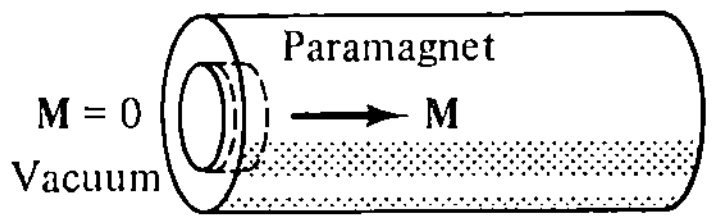

Figure 6.23

Incidentally, the volume bound current density in a homogeneous linear medium is proportional to the *free* current density:

$$\mathbf{J}_b = \nabla \times \mathbf{M} = \nabla \times (\chi_m \mathbf{H}) = \chi_m \mathbf{J}_f \tag{6.29}$$

In particular, unless free current actually flows *through* the material, all bound current will be at the surface.

Problem 6.15 A coaxial cable consists of two very long cylindrical tubes, separated by linear insulating material of magnetic susceptibility χ_m. A current I flows down the inner conductor and returns along the outer one; in each case the current distributes itself uniformly over the surface (Fig. 6.24). Find the magnetic field in the region between the tubes. As a check, calculate the magnetization and bound currents and confirm that (together, of course, with the free currents) they generate the correct field.

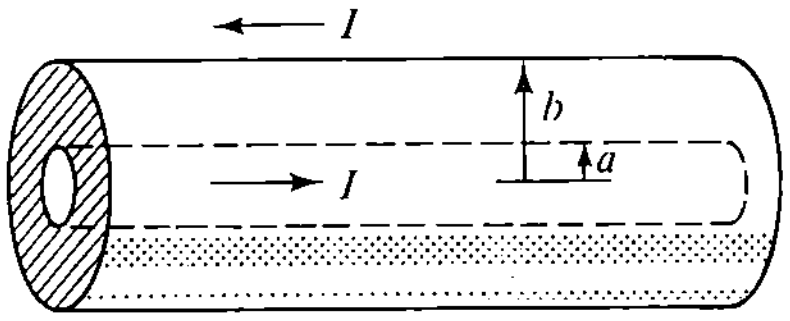

Figure 6.24

Problem 6.16 A current I flows down a long straight wire of radius R. If the wire is made of linear material (copper, say, or aluminum) with susceptibility χ_m, and the current is distributed uniformly, what is the magnetic field a distance r from the center? Find all the bound currents. What is the net bound current flowing down the wire?

! **Problem 6.17** A sphere of linear magnetic material is placed in an originally uniform magnetic field $\mathbf{B}_0$. Find the new field inside the sphere.

Problem 6.18 On the basis of the naïve model presented in Section 6.1.3, estimate the magnetic susceptibility of a diamagnetic metal such as copper. Compare your answer with the empirical value in Table 6.1, and comment on any discrepancy.

6.4.2 Ferromagnetism

In a linear medium the alignment of atomic dipoles is maintained by a magnetic field imposed from the outside. Ferromagnets—which are emphatically *not* linear[8]—require no external fields to sustain the magnetization: The alignment is frozen in. Like paramagnetism, ferromagnetism involves the magnetic dipoles associated with the spins of unpaired electrons. The new feature, which makes ferromagnetism so different from paramagnetism, is the interaction between nearby dipoles: In a ferromag-

[8]In this sense it is misleading to speak of the susceptibility or permeability of a ferromagnet. The terms *are* used for such materials, but they refer to the proportionality factor between a *differential* increase in H and the resulting *differential* change in M (or B); moreover, they are not *constants*, but functions of H.

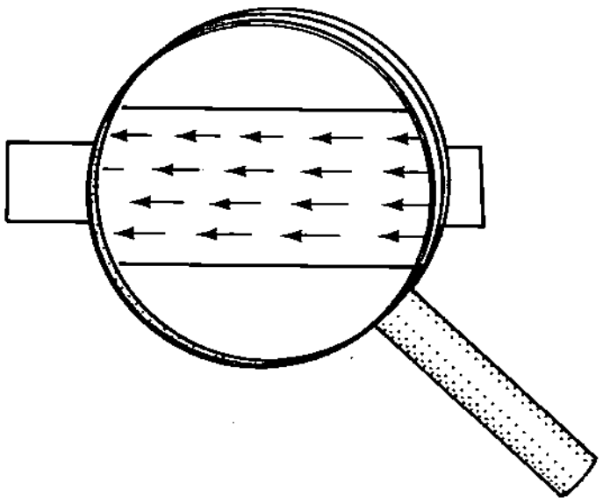

Figure 6.25

net, *each dipole "likes" to point in the same direction as its neighbor.* The *reason* for this preference is essentially quantum mechanical, and I shall not endeavor to explain it here; it is enough to know that the correlation is so strong as to align virtually 100% of the unpaired electron spins. If you could somehow magnify a piece of iron and "see" the individual dipoles as tiny arrows, it would look something like Fig. 6.25, with all the spins pointing the same way.

But if that is true, why isn't every wrench and nail a powerful magnet? The answer is that the alignment occurs in relatively small patches, called **domains.** Each domain contains billions of dipoles, all lined up (these domains are actually *visible* under a microscope, using suitable etching techniques—see Fig. 6.26), but the domains *themselves* are randomly oriented. The household wrench contains an enormous number of domains, and their magnetic fields cancel, so the wrench as a whole is not magnetized. (Actually, the orientation of domains is not *completely* random: Within a given crystal there may be some preferential alignment along the crystal

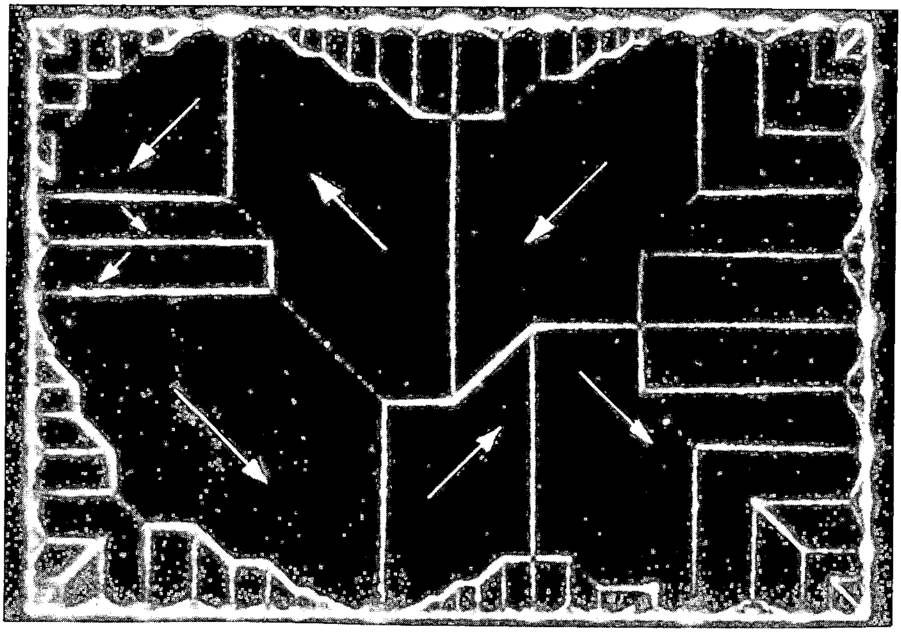

Figure 6.26 Ferromagnetic domains. (Photo courtesy of R. W. DeBlois)

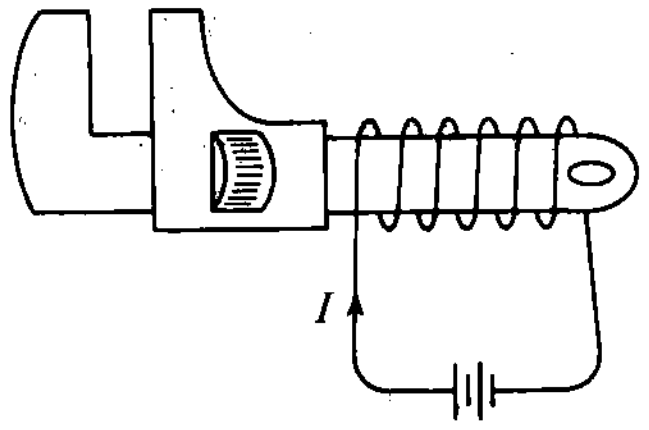

Figure 6.27

axes. But there will be just as many domains pointing one way along each axis as the other, so there is still no large-scale magnetization. Moreover, the crystals themselves are randomly oriented within any sizable chunk of metal.)

How, then, would you produce a "permanent magnet," such as they sell in toy stores? If you put a piece of iron into a strong magnetic field, the torque $\mathbf{N} = \mathbf{m} \times \mathbf{B}$ will try to rotate the dipoles parallel to the field. Since they like to stay parallel to their neighbors, most of the dipoles will resist this torque. However, at the *boundary* between two domains, there are *competing* neighbors, and the torque will throw its weight on the side of the domain most nearly parallel to the field; this domain will win over some converts, at the expense of the less favorably oriented one. The net effect of the magnetic field, then, is to *move the domain boundaries.* Domains parallel to the field grow, and the others shrink. If the field is strong enough, one domain takes over entirely, and the iron is said to be "saturated." Now, this process (the shifting of domain boundaries in response to an external field) is not entirely reversible: When the field is switched off, there will be *some* return to randomly oriented domains, but it is far from complete—there remains a preponderance of domains in the original direction. The object is now a permanent magnet.

A simple way to accomplish this, in practice, is to wrap a coil of wire around the object to be magnetized (Fig. 6.27). Run a current I through the coil; this provides the external magnetic field (pointing to the left in the diagram). As you increase the current, the field increases, the domain boundaries move, and the magnetization grows. Eventually, you reach the saturation point, with all dipoles aligned, and a further increase in the current has no effect on M (Fig. 6.28, point b).

Now suppose you *reduce* the current. Instead of retracing the path back to $M = 0$, there is only a *partial* return to randomly oriented domains. M decreases,

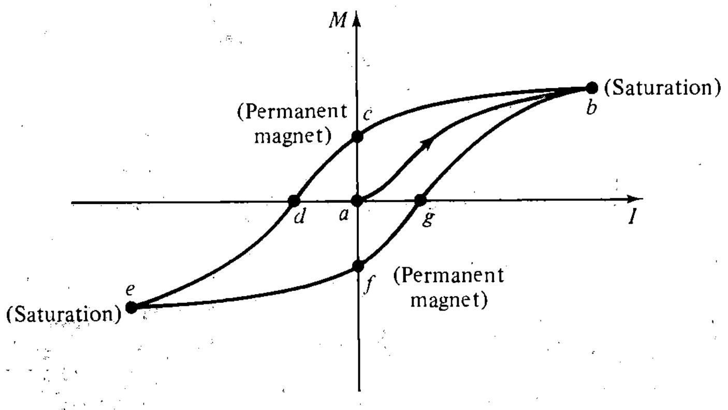

Figure 6.28

but even with the current off there is some residual magnetization (point c). The wrench is now a permanent magnet. If you want to eliminate the remaining magnetization, you'll have to run a current backward through the coil (a negative I). Now the external field points to the right, and as you increase I (negatively), M drops down to zero (point d). If you turn I still higher, you soon reach saturation in the other direction—all the dipoles now pointing to the *right* (e). At this stage switching off the current will leave the wrench with a permanent magnetization to the right (point f). To complete the story, turn I on again in the positive sense: M returns to zero (point g), and eventually to the forward saturation point (b).

The path we have traced out is called a **hysteresis loop.** Notice that the magnetization of the wrench depends not only on the applied field (that is, on I) but also on its previous magnetic "history."[9] For instance, at three different times in our experiment the current was zero (a, c, and f), yet the magnetization was different for each of them. Actually, it is customary to draw hysteresis loops as plots of B against H, rather than M against I. (If our coil is approximated by a long solenoid, with N turns per unit length, then $H = NI$, so H and I are proportional. Meanwhile, $\mathbf{B} = \mu_0(\mathbf{H} + \mathbf{M})$, but in practice M is huge compared to H, so that to all intents and purposes $\mathbf{B}$ is proportional to $\mathbf{M}$.)

To make the units consistent, I have plotted $(\mu_0 H)$ horizontally (Fig. 6.29); notice, however, that the vertical scale is 10^4 times greater than the horizontal one. $\mu_0\mathbf{H}$ is the field our coil *would* have produced in the absence of any iron; $\mathbf{B}$ is what we *actually* got, and compared to $\mu_0\mathbf{H}$ it is gigantic. A little current goes a long way when you have ferromagnetic materials around. That's why anyone who wants to make a powerful electromagnet will wrap the coil around an iron core. It doesn't take much of an external field to move the domain boundaries, and as soon as you've done that, you have all the dipoles in the iron working with you.

One final point concerning ferromagnetism: It all follows, remember, from the fact that the dipoles within a given domain line up parallel to one another. Random thermal motions compete with this ordering, but as long as the temperature doesn't get *too* high, they cannot budge the dipoles out of line. It's not surprising, though,

[9]Etymologically, the word *hysteresis* has nothing to do with the word *history*—nor with the word *hysteria*. It derives from a Greek verb meaning "to lag behind."

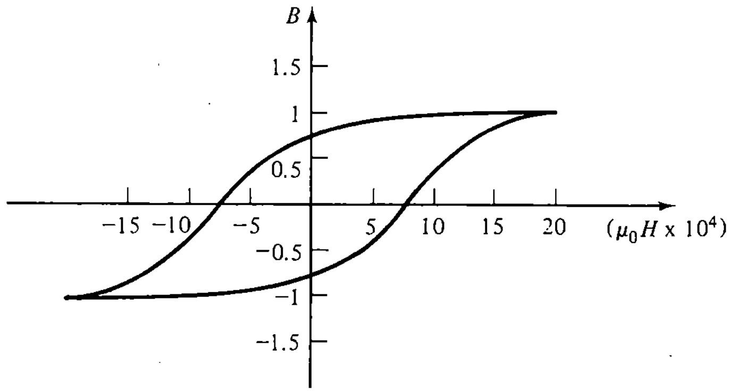

Figure 6.29

that *very* high temperatures do destroy the alignment. What *is*, perhaps, surprising is that this occurs at a precise temperature (770°C, for iron). Below this temperature (called the **Curie point**), iron is ferromagnetic; above, it is paramagnetic. The Curie point is rather like the boiling point or the freezing point in that there is no *gradual* transition from ferro- to para-magnetic behavior, any more than there is between water and ice. These abrupt changes in the properties of a substance, occurring at sharply defined temperatures, are known in statistical mechanics as **phase transitions.**

Problem 6.19 How would you go about *de*magnetizing a permanent magnet (such as the wrench we have been discussing, at point *c* in the hysteresis loop)? That is, how could you restore it to its original state, with $M = 0$ at $I = 0$?

Problem 6.20

(a) Show that the energy of a magnetic dipole in a magnetic field $\mathbf{B}$ is given by

$$U = -\mathbf{m}\cdot\mathbf{B} \tag{6.30}$$

Compare equation (4.6).

(b) Show that the interaction energy of two magnetic dipoles separated by a displacement $\mathbf{r}$ is given by

$$U = \frac{\mu_0}{4\pi}\frac{1}{r^3}[\mathbf{m}_1\cdot\mathbf{m}_2 - 3(\mathbf{m}_1\cdot\hat{r})(\mathbf{m}_2\cdot\hat{r})] \tag{6.31}$$

Compare equation (4.7).

(c) Express your answer to (b) in terms of the angles θ_1 and θ_2 in Fig. 6.30, and use the result to find the stable configuration two dipoles would adopt if held a fixed distance apart, but left free to rotate.

(d) Suppose you had a large collection of compass needles, mounted on pins at regular intervals along a straight line. How do you think they would point, ignoring the earth's magnetic field? (A rectangular array of compass needles also aligns itself spontaneously, and this is sometimes used as a demonstration of "ferromagnetic" behavior on a large scale. It's a bit of a fraud, however, since the mechanism here is purely classical, whereas true ferromagnetism is due to quantum mechanical exchange forces.)

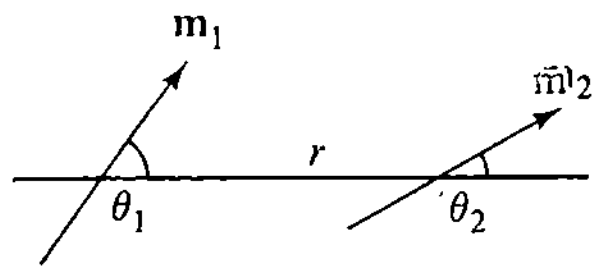

Figure 6.30

Further Problems on Chapter 6

Problem 6.21 Notice the following parallel:

$$\begin{cases} \nabla\cdot\mathbf{D} = 0, & \nabla\times\mathbf{E} = 0, & \epsilon_0\mathbf{E} = \mathbf{D} - \mathbf{P}, & \text{(no free charge)} \\ \nabla\cdot\mathbf{B} = 0, & \nabla\times\mathbf{H} = 0, & \mu_0\mathbf{H} = \mathbf{B} - \mu_0\mathbf{M}, & \text{(no free current)} \end{cases}$$

Thus, the transcription $D \to B$, $E \to H$, $P \to \mu_0 M$, $\epsilon_0 \to \mu_0$ turns an electrostatic problem into an analogous magnetostatic one. Use this observation, together with your knowledge of the electrostatic results, to rederive

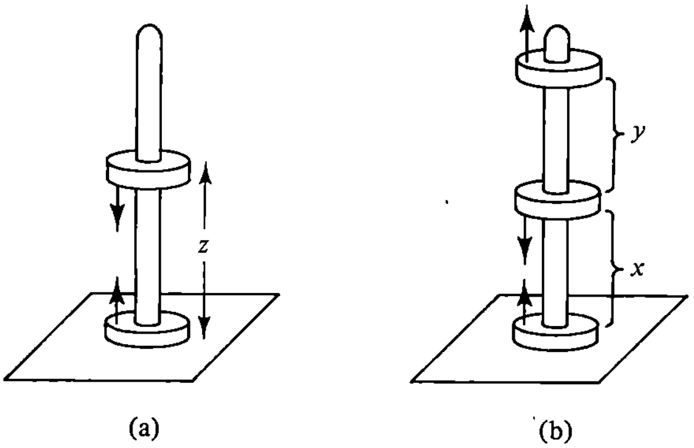

Figure 6.31

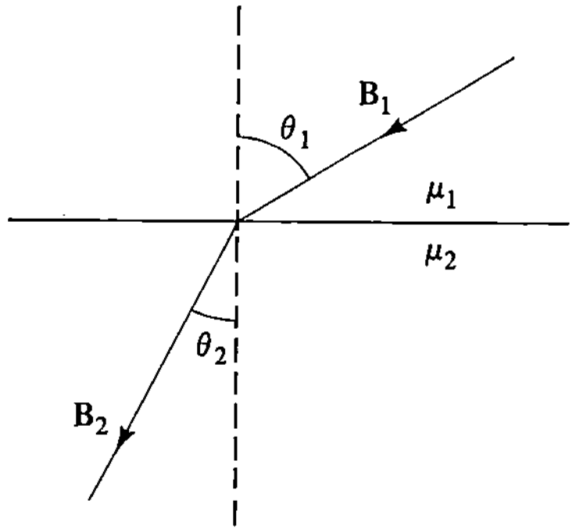

Figure 6.32

(a) the magnetic field of a uniformly magnetized sphere (6.16),

(b) the magnetic field inside a sphere of linear magnetic material in an otherwise uniform magnetic field (Problem 6.17),

(c) the average magnetic field over a sphere, due to steady currents within the sphere (equation (5.87)).

Problem 6.22 A familiar toy consists of donut-shaped permanent magnets (magnetization parallel to the axis), which slide frictionlessly on a vertical rod (Fig. 6.31). Treat the magnets as dipoles, with mass M and dipole moment $\mathbf{m}$.

(a) If you put two back-to-back magnets on the rod, the upper one will "float"—the magnetic force upward balancing the gravitational force downward. At what height (z) does it float? (*Hint*: Use (6.31).)

(b) If you now add a *third* magnet (parallel to the bottom one), what is the *ratio* of the two heights? (Determine the actual number, to three significant digits.)

(*Answer:* (a) $z = [3\mu_0 m^2/2\pi M g]^{1/4}$; (b) $x/y = 0.8501$)

Problem 6.23 At the interface between one linear magnetic material and another the magnetic field lines bend (see Fig. 6.32). Show that $\tan\theta_2/\tan\theta_1 = \mu_1/\mu_2$, assuming there is no free current at the boundary.

Problem 6.24 A magnetic dipole $\mathbf{m}$ is imbedded at the center of a sphere (radius R) of linear magnetic material (permeability μ). Show that the magnetic field inside the sphere ($0 < r \leqslant R$) is

$$\frac{\mu}{4\pi}\left\{\frac{1}{r^3}[3(\mathbf{m}\cdot\hat{\mathbf{r}})\hat{\mathbf{r}} - \mathbf{m}] + \frac{\mathbf{m}}{R^3(1 - 3\mu/2\mu_0)}\right\}$$

What is the field *outside* the sphere?

7

ELECTRODYNAMICS

7.1 ELECTROMOTIVE FORCE

7.1.1 Ohm's Law

When I first introduced electrostatics, I said that it applies whenever the source charges are *at rest.* Actually, it's more general than that: *electrostatics and magnetostatics apply whenever ρ and $\mathbf{J}$ are independent of time.* A rotating uniformly charged sphere, for example, produces exactly the same electric field as it would when stopped. There is one exception—one "law" of electrostatics that holds only when the charges are actually stationary. I have in mind the rule that $\mathbf{E} = 0$ inside a conductor. Remember the argument: If the field were *not* zero, then the free charge in the conductor would flow, so if the charge is, *in fact,* at rest, then $\mathbf{E}$ must be zero. This argument breaks down if (steady) currents are permitted—in this case $\mathbf{E}$ need *not* be zero inside a conductor, though it's still an electrostatic configuration.

To make charges move in a conductor, you have to *push* them. How *fast* they will move, in response to a given push, depends on the nature of the material. For most substances, the current density $\mathbf{J}$ is proportional to the *force per unit charge,* $\mathbf{f}$:

$$\mathbf{J} = \sigma \mathbf{f} \tag{7.1}$$

The proportionality factor σ is an empirical constant that varies from one material to another; it's called the **conductivity** of the medium. (Actually, the handbooks usually list the *reciprocal* of σ, called the **resistivity:** $\rho = 1/\sigma$. Some typical values are shown in Table 7.1. Notice that even *insulators* conduct slightly, though the conductivity of a metal is astronomically greater—by a factor of 10^{22} or so. In fact, for most purposes metals can be regarded as **perfect conductors**, with $\sigma = \infty$.)

In principle, the force that drives the charges to produce the current could be anything—chemical, gravitational, or trained ants with tiny harnesses. For *our* pur-

TABLE 7.1 RESISTIVITIES

Material	Resistivity (ohm-meters)
Conductors:	
Silver	1.6×10^{-8}
Copper	1.7×10^{-8}
Gold	2.3×10^{-8}
Aluminum	2.8×10^{-8}
Nichrome	100×10^{-8}
Semiconductors:	
Silicon	0.03–0.04 (depending on purity)
Salt water (saturated)	0.044
Germanium	0.46
Insulators:	
Water (pure)	2.5×10^{5}
Wood	10^{8}–10^{11}
Glass	10^{10}–10^{14}
Quartz	10^{13}
Sulfur	2×10^{15}
Rubber	10^{13}–10^{16}

Source: Handbook of Chemistry and Physics, 67th ed. (Cleveland: CRC Press, Inc., 1986–87.) Figures are for room temperature.

poses, though, it's always an electromagnetic force that does the job. In this case (7.1) becomes

$$\mathbf{J} = \sigma(\mathbf{E} + \mathbf{v} \times \mathbf{B}) \tag{7.2}$$

Ordinarily, the velocity of the charges is sufficiently small that the second term can be ignored:

$$\boxed{\mathbf{J} = \sigma\mathbf{E}} \tag{7.3}$$

(However, in plasmas, for instance, the magnetic contribution to $\mathbf{f}$ can be of crucial importance.) Equation (7.3) is known as **Ohm's law,** though the physics behind it is really contained in (7.1), of which (7.3) is just a special case.

Example 1

A cylindrical wire of cross-sectional area A and length L is made of material with conductivity σ. (See Fig. 7.1; note that the cross section need not be circu-

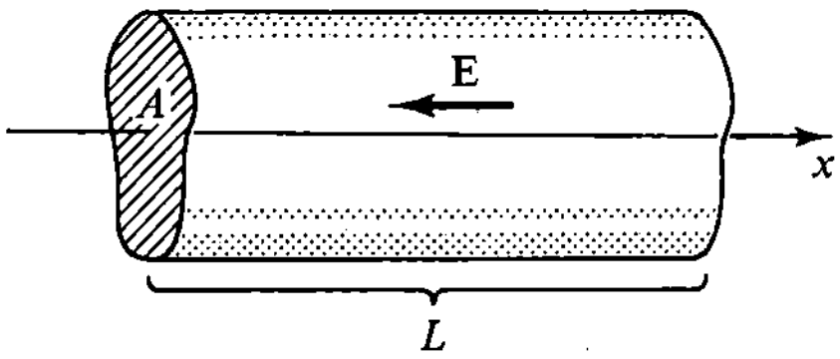

Figure 7.1

lar, but it must be *uniform.*) If the potential difference between the ends is V, what current flows? Assume the potential is constant over each end.

Solution: Assume that the electric field is *uniform* within the wire (I'll *prove* this in a moment). It follows from (7.3) that the current density is also uniform, so

$$I = JA = \sigma EA = \sigma A \frac{V}{L}$$

Example 2

Two long cylinders (radii a and b) are separated by material of conductivity σ (Fig. 7.2). If they are maintained at a potential difference V, what current flows from one to the other, in a length L?

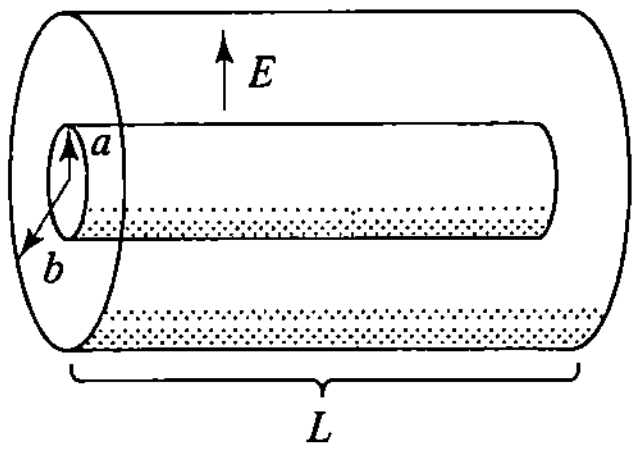

Figure 7.2

Solution: The field between the cylinders is

$$\mathbf{E} = \frac{\lambda}{2\pi\epsilon_0 r}\hat{\mathbf{r}}$$

where λ is the charge per unit length on the inner cylinder. The current is therefore

$$I = \int \mathbf{J}\cdot d\mathbf{a} = \sigma \int \mathbf{E}\cdot d\mathbf{a} = \sigma \frac{1}{\epsilon_0}\lambda L$$

(The integral is over any surface enclosing the inner cylinder, and I used Gauss's law in the last step). Meanwhile, the potential difference between the cylinders is

$$V = -\int_b^a \mathbf{E}\cdot d\mathbf{l} = \frac{\lambda}{2\pi\epsilon_0}\ln\left(\frac{b}{a}\right)$$

so

$$I = \frac{2\pi\sigma L}{\ln(b/a)} V$$

As these examples illustrate, the total current flowing from one electrode to another is proportional to the potential difference between them, though for some reason the relation is traditionally written the other way around:

$$V = IR \tag{7.4}$$

This, of course, is the familiar version of Ohm's law. The constant of proportionality R is called the **resistance;** it's a function of the geometry of the arrangement and the conductivity of the medium connecting the electrodes. (In Example 1, $R = (L/\sigma A)$; in Example 2, $R = \ln(b/a)/2\pi\sigma L$.) Resistance is measured in **ohms** (Ω): an ohm is a volt per ampere. Notice that the proportionality between V and I is a consequence of (7.3): If you want to double V, you double the charge everywhere—but that doubles $\mathbf{E}$, which doubles $\mathbf{J}$, which doubles I.

Within a material of uniform conductivity,

$$\nabla \cdot \mathbf{E} = \frac{1}{\sigma} \nabla \cdot \mathbf{J} = 0$$

for steady currents (equation (5.27)), and therefore the charge density is zero. Any unbalanced charge resides on the *surface.* (We proved this long ago, for the case of *stationary* charges, using the fact that $\mathbf{E} = 0$; evidently it is still true when the charges are allowed to move, though for rather different reasons.) It follows, in particular, that Laplace's equation holds within a conductor, so that all the tools and tricks of Chapter 3 are available for computing the potential, and from it $\mathbf{E}$ and $\mathbf{J}$.

Example 3

I asserted that the field in the wire of Example 1 was *uniform*. Here's the proof.

Within the cylinder V obeys Laplace's equation. What are the boundary conditions? At the left end the potential is constant—we may as well set it equal to zero. At the right end the potential is likewise constant—call it V_0. On the cylindrical surface, $\mathbf{J} \cdot \hat{n} = 0$, else charge would be leaking out into the surrounding space (which we assume to be nonconducting). Therefore $\mathbf{E} \cdot \hat{n} = 0$, and hence $\partial V/\partial n = 0$. With V or its normal derivative specified on all surfaces, the potential is uniquely determined (Problem 3.4). But it's *easy* to guess *one* potential that obeys Laplace's equation and fits these boundary conditions:

$$V(x) = \frac{V_0 z}{L}$$

where z is measured along the axis. The uniqueness theorem guarantees that this is *the* solution. The corresponding field is

$$\mathbf{E} = -\nabla V = -\frac{V_0 \hat{k}}{L}$$

which is uniform, as advertised. (Incidentally, since $\mathbf{E}_{\parallel}$ is continuous across any boundary, $\mathbf{E}$ has this same value immediately *outside* the wire as well.)

You might contrast the *enormously* more difficult problem that arises if the conducting material is removed, leaving only a metal plate at either end

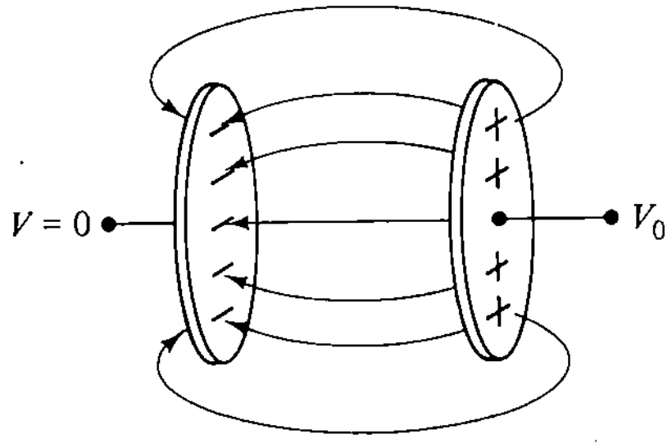

Figure 7.3

(Fig. 7.3). Evidently, in the present case charge arranges itself over the surface of the wire in just such a way as to produce a nice uniform field within (see Problem 7.54).

I don't suppose there is any formula in physics more widely known than Ohm's law, and yet it's not really a true law, in the sense of Gauss's law or Ampère's law; rather, it is a rule of thumb, which applies astonishingly well to many substances. In fact, when you stop to think about it, it's a little surprising that Ohm's law should *ever* hold. After all, a given field **E** produces a force $q\mathbf{E}$ (on a charge q), and according to Newton's second law the charge will accelerate. But if the charges are *accelerating,* why doesn't the current *increase* with time, growing larger and larger the longer you leave the field on? Ohm's law implies, on the contrary, that a constant field produces a constant *velocity,* and hence a constant current. Isn't that a contradiction of Newton's law?

No, for we are forgetting the frequent collisions electrons make as they pass down the wire. It's rather like this: Suppose you're driving down a street with a stop sign at every intersection, so that, although you accelerate constantly in between, you are obliged to start all over again with each new block. Your *average* speed is then a constant, in spite of the fact that (save for the periodic abrupt stops) you are always accelerating. If the length of a block is λ and the acceleration a, the time it takes to go a block is

$$t = \sqrt{\frac{2\lambda}{a}}$$

and hence the average velocity is

$$v_{\text{ave}} = \frac{1}{2}at = \sqrt{\frac{\lambda a}{2}}$$

But wait! That's no good *either*! It says that the velocity is proportional to the *square root* of the acceleration, and therefore, for electrons in a conductor, that the current is proportional to the square root of the field! There's another twist to the story: The charges in practice are already moving extremely fast because of their thermal energy. But the thermal velocities have random directions, and average to zero. The net "drift" velocity we're concerned with is a tiny extra bit. So the time between collisions is actually much shorter than we supposed; in fact,

$$t = \frac{\lambda}{v_{\text{thermal}}}$$

and therefore

$$v_{\text{ave}} = \frac{1}{2}at = \frac{a\lambda}{2v_{\text{thermal}}}$$

If there are N molecules per unit volume and f free electrons per molecule, each with charge q and mass m, the current density is

$$\mathbf{J} = Nfq\mathbf{v}_{\text{ave}} = \frac{fNq\lambda}{2v_{\text{thermal}}}\frac{\mathbf{F}}{m} = \left(\frac{fN\lambda q^2}{2mv_{\text{thermal}}}\right)\mathbf{E} \tag{7.5}$$

I don't claim that the term in parentheses is an accurate formula for the conductivity,[1] but it does indicate the basic ingredients, and it correctly predicts that conductivity is proportional to the density of the moving charges and (ordinarily) decreases with increasing temperature.

Incidentally, the conductivity of metal is so much greater than that of other materials that for most purposes we can consider it to be *infinite.* In this approximation $\mathbf{E} = 0$ inside metal, and any metal object is an equipotential. In analyzing electrical circuits, for instance, we ignore the (minute) voltage drop along each wire; if resistance is *wanted* in a circuit, we connect in a special element (a **resistor**) made of *weakly* conducting material.

As a result of all the collisions, the work done by the electrical force is converted into heat in the resistor. Since the work done per unit charge is V and the charge flowing per unit time is I, the power delivered is

$$\boxed{P = VI = I^2R} \tag{7.6}$$

This is the **Joule heating law.** With I in amperes and R in ohms, P comes out in watts (joules per second). In the more general case of equation (7.3), the power can be computed as force times velocity:

$$P = \int \rho\mathbf{E}\cdot\mathbf{v}\,d\tau = \int \mathbf{E}\cdot\mathbf{J}\,d\tau = \sigma\int E^2\,d\tau \tag{7.7}$$

Problem 7.1 Two concentric metal spherical shells, of radius a and b, respectively, are separated by weakly conducting material of conductivity σ (Fig. 7.4(a)).

(a) If they are maintained at a potential difference V, what current flows from one to the other?

(b) What is the resistance between the shells?

(c) Notice that the outer radius b is irrelevant, if $b \gg a$. In view of this, what current would flow between two metal spheres, each of radius a, immersed deep in the sea and

[1] This classical model bears little resemblance to the modern quantum theory of conductivity. See, for instance, D. Park's *Introduction to the Quantum Theory,* 2d ed. (New York: McGraw-Hill, 1974), chap. 13.

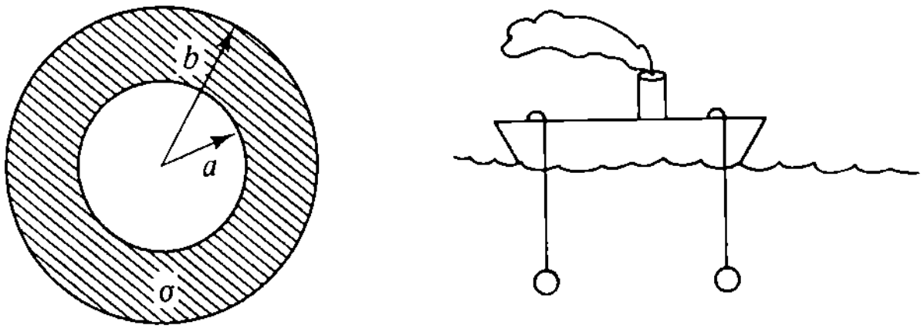

(a) (b) **Figure 7.4**

held quite far apart (Fig. 7.4(b)), if the potential difference between them is V? (This arrangement can be used to measure the conductivity of sea water.)

Problem 7.2

(a) Two metal objects are embedded in weakly conducting material of conductivity σ (Fig. 7.5). Show that the resistance between them is related to the capacitance of the arrangement by

$$R = \frac{\epsilon_0}{\sigma C}$$

(b) Suppose you connected a battery between 1 and 2 and charged them up to a potential difference V_0. If you then disconnect the battery, the charge will gradually leak off. Show that $V(t) = V_0 e^{-t/\tau}$, and find the "time constant" τ in terms of ϵ_0 and σ

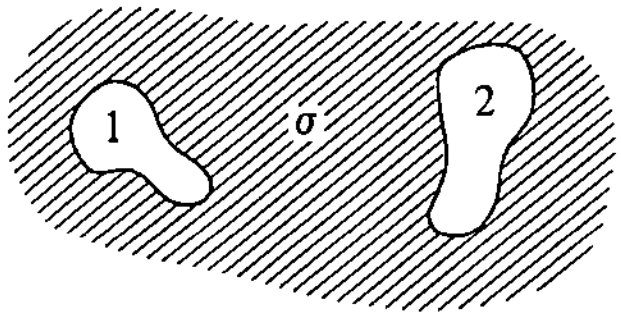

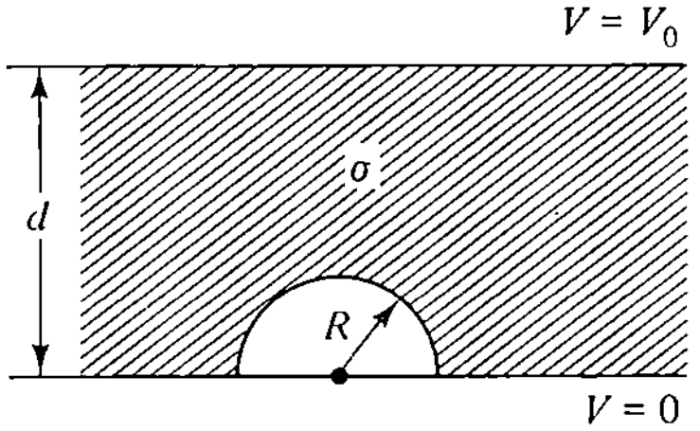

Figure 7.5 **Figure 7.6**

Problem 7.3 Two very large metal plates are held a distance d apart, one at potential zero, the other at potential V_0 (Fig. 7.6). A metal sphere of radius R ($R \ll d$) is sliced in two, and one hemisphere placed on the grounded plate, so that its potential is likewise zero. If the region between the plates is filled with weakly conducting material of conductivity σ, what current flows to the hemisphere? (*Answer*: $I = (3\pi R^2 \sigma / d) V_0$. *Hint*: Study Example 8 in Chapter 3.)

! **Problem 7.4** Two long, straight, copper pipes, each of radius R, are held a distance $2d$ apart (see Fig. 7.7). One is at potential V_0, the other at $-V_0$. The space surrounding the pipes is filled with weakly conducting material of conductivity σ. Find the current, per unit length, which flows from one pipe to the other. (*Hint*: Refer to Problem 3.10.)

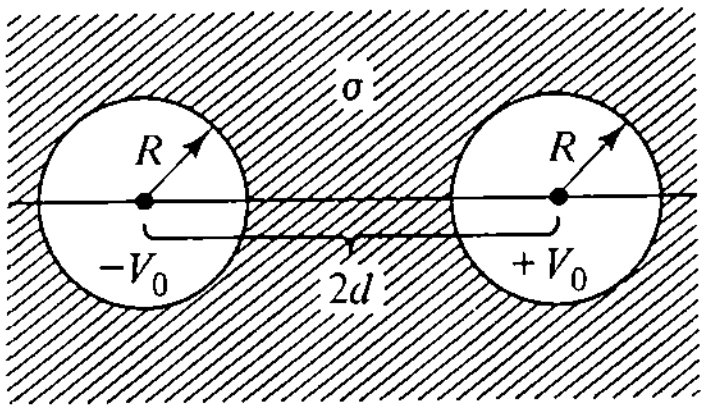

Figure 7.7

Figure 7.8

7.1.2 Electromotive Force

We have seen that it takes some *force* to drive charge through a conducting medium; in its most general form Ohm's law states

$$\mathbf{J} = \sigma \mathbf{f}$$

where $\mathbf{f}$ is the force exerted per unit charge. If you consider a typical electric circuit (Fig. 7.8), with a battery hooked up to a light bulb, say, there arises a curious question: In practice, the *current is the same all the way around the loop,* at any given instant of time—it's not 2.3 A in the battery and 0.7 A in the light bulb. But *why* is this the case, when the only obvious driving force is inside the battery (where some kind of chemical process pushes the charge along)? Offhand, you might expect this to produce a large current in the battery and none at all in the lamp. Who's doing the pushing in the rest of the circuit, and how does it happen that this push is exactly right to produce the same current in each segment? What's more, given that the charges in a typical wire move (literally) at a *snail's* pace (see Problem 5.22), why doesn't it take half an hour for the news to reach the light bulb? How do all the charges know to start moving at the same instant?

Answer: If the current is *not* the same all the way around (for instance, during the first split second after the switch is closed) then charge is piling up somewhere, and—here's the crucial point—the electric field of this accumulating charge is in such a direction as to even out the flow. Suppose, for instance, that the current *into* the bend in Fig. 7.9 is greater than the current *out.* Then charge piles up at the "knee," and this produces a field aiming *away* from the kink. This field *opposes* the current flowing in (slowing it down) and *promotes* the current flowing out (speeding it up) until these currents are equal, at which point there is no further accumulation of charge, and equilibrium is established. It's a beautiful system, automatically self-correcting to keep the current uniform, and it does it all so quickly that, in practice, you can safely assume the current is the same all around the circuit even in systems that oscillate at radio frequencies.

The upshot of all this is that there are really *two* forces involved in driving current around a circuit: the *source,* $\mathbf{f}_s$, which is ordinarily confined to one portion of the

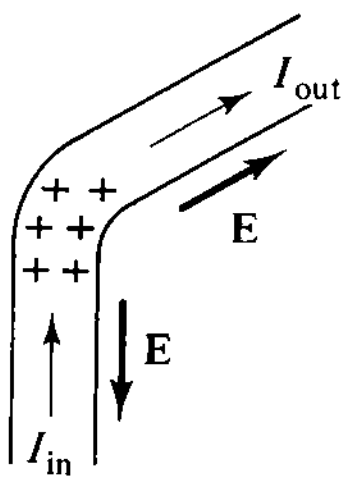

Figure 7.9

loop (a battery, say), and the *electrostatic* force whose function it is to smooth out the flow and communicate the influence of the source to distant parts of the circuit:[2]

$$\mathbf{f} = \mathbf{f}_s + \mathbf{E} \tag{7.8}$$

The physical agency responsible for $\mathbf{f}_s$ could be any one of many different things: In a battery it's a chemical force; in a piezoelectric crystal mechanical pressure is converted into an electrical impulse; in a thermocouple it's a temperature gradient that does the job; in a photoelectric cell it's light; and in a Van de Graaff generator the electrons are literally loaded onto a conveyer belt and swept along. Whatever the *mechanism,* its net effect is determined by the line integral of $\mathbf{f}$ around the circuit:

$$\oint \mathbf{f} \cdot d\mathbf{l} = \oint \mathbf{f}_s \cdot d\mathbf{l} \equiv \mathcal{E} \tag{7.9}$$

$\mathcal{E}$ is called the **electromotive force,** or the **emf,** of the circuit. (It's a poor word, since it's not a *force* at all—it's an *integral* of a *force per unit charge.* Some people prefer the term **electromotance,** but emf is so ingrained that we'd better stick with it.) Because $\oint \mathbf{E} \cdot d\mathbf{l} = 0$ for electrostatic fields, it doesn't matter whether you include $\mathbf{E}$ or not, in calculating $\mathcal{E}$.

It's the electromotive force that determines how much current will flow in a given circuit:

$$\mathcal{E} = \oint \mathbf{f} \cdot d\mathbf{l} = \oint \frac{\mathbf{J}}{\sigma} \cdot d\mathbf{l} = \oint \frac{I}{a\sigma}\, dl = I \oint \frac{1}{a\sigma}\, dl = IR, \tag{7.10}$$

where R is the total resistance of the loop.[3] As the line integral of $\mathbf{f}_s$, $\mathcal{E}$ can also be interpreted as the work done, per unit charge, by the source—indeed, in some treatments electromotive force is *defined* this way. However, as you will see in the next section, there is some subtlety involved in this interpretation, so I prefer (7.9).

Problem 7.5

(a) Show that electrostatic forces alone cannot be used to drive current around a circuit. (In fact, *within the source* current flows in the direction *opposite* to $\mathbf{E}$.)

(b) A rectangular loop of wire is situated so that one end is between the plates of a parallel-plate capacitor (Fig. 7.10), oriented parallel to the field $E = (\sigma/\epsilon_0)$. The other end is way outside, where the field is essentially zero. If the width of the loop is h and its total resistance is R, what current flows? Explain.

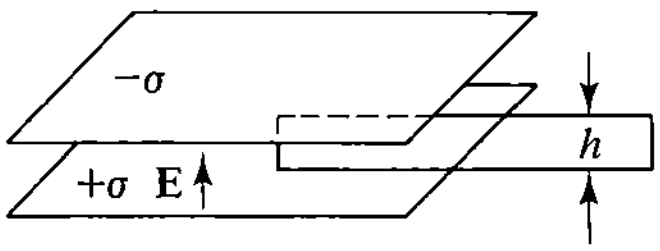

Figure 7.10

[2]For further discussion, see M. A. Heald, *Am. J. Phys.* **52,** 522 (1984).

[3]I am assuming the current is uniformly distributed over the cross section a at each point in the circuit. If it's *not,* I would (conceptually) divide the current up into tiny tubes of flow, small enough that the current *is* uniform over each one. For the ith tube, $I_i R_i = \mathcal{E}$, so the total current $I_1 + I_2 + \cdots = \mathcal{E}[(1/R_1) + (1/R_2) + \cdots]$, or $\mathcal{E} = I_{\text{tot}} R_{\text{tot}}$, with the resistances combined in parallel.

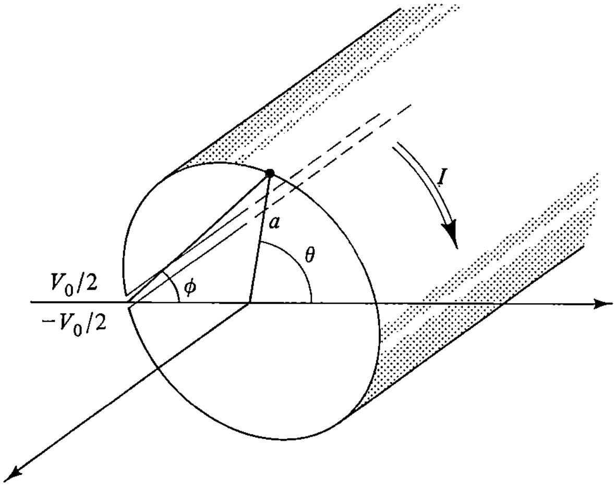

Figure 7.11

! Problem 7.6 A rare case in which the electrostatic field **E** for a circuit can actually be *calculated* is the following (from the article by Heald cited in footnote 2). Imagine an infinitely long cylindrical sheet, of uniform resistivity and radius a. A slot (corresponding to the battery) is maintained at $\pm V_0/2$, at $\theta = \pm\pi$, and a steady current flows over the surface, as indicated in Fig. 7.11. According to Ohm's law, then,

$$V(a, \theta) = \frac{V_0\theta}{2\pi} \qquad (-\pi < \theta < +\pi)$$

(a) Use separation of variables in cylindrical coordinates to determine $V(r, \theta)$ inside and outside the cylinder. (*Answer*: $V_0\phi/\pi$ $(r < a)$; $(V_0/\pi)\tan^{-1}[a \sin\theta/(r + a \cos\theta)]$ $(r > a)$)

(b) Find the surface charge density on the cylinder. (*Answer*: $(\epsilon_0 V_0/\pi a)\tan\phi$)

7.1.3 Motional emf

In the last section I mentioned several possible sources of electromotive force in a circuit, the battery being the most familiar example. Now I want to discuss another common source of emf: the *generator.* Generators typically exploit *motional emf's,* that is, emf's which arise when you *move a wire through a magnetic field.* Figure 7.12 shows a primitive model for a generator. In the shaded region there is a uniform magnetic field **B**, pointing into the page, and the resistor R represents whatever it is (maybe a light bulb or a toaster) we're trying to drive current through. If the entire loop is pulled to the right with a speed v, the charges in segment ab experience a

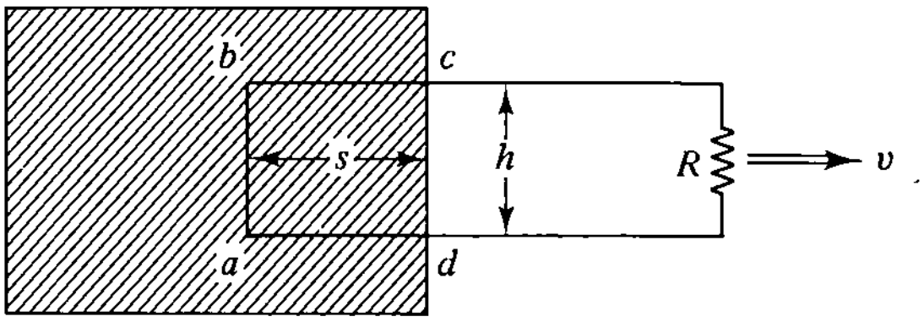

Figure 7.12

magnetic force whose vertical component qvB drives current around the loop in the clockwise direction. The emf is

$$\mathcal{E} = \oint \mathbf{f}_{\text{mag}} \cdot d\mathbf{l} = vBh \tag{7.11}$$

where h is the width of the loop. (The horizontal segments bc and ad contribute nothing, since the force here is perpendicular to the wire.)

Simple as this arrangement may seem, it contains a curious paradox, for although the magnetic force is responsible for establishing the emf, it is certainly *not* responsible for doing the work—magnetic forces *never* do work. Who, then, *is* doing the work? *Answer*: The person who's pulling on the loop! With the current flowing, charges in segment ab have a vertical velocity (call it $\mathbf{u}$) in addition to the horizontal velocity $\mathbf{v}$. Accordingly, the magnetic force has a component quB to the left. To counteract this, the person pulling on the wire must exert a force per unit charge

$$f_{\text{pull}} = uB$$

to the *right* (see Fig. 7.13(a)—this force is transmitted to the charge by the structure of the wire). Meanwhile, the particle is actually moving in the direction of the resultant velocity $\mathbf{w}$, and the distance it goes is $(h/\cos\theta)$. The work done per unit charge is therefore

$$\int \mathbf{f}_{\text{pull}} \cdot d\mathbf{l} = (uB)\left(\frac{h}{\cos\theta}\right)\sin\theta = vBh$$

($\sin\theta = \cos(90° - \theta)$ coming from the dot product). As it turns out, then, the *work done per unit charge is exactly equal to the emf,* though the integrals are taken along quite different paths (Fig. 7.13(b) and (c)) and completely different forces are involved: $\mathbf{f}_{\text{pull}}$ contributes nothing to the emf, because it is perpendicular to the wire;

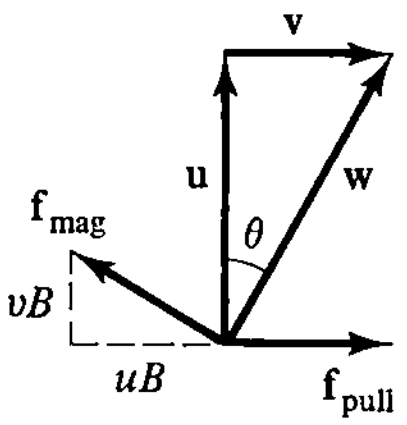

(a) Forces on a charge in the wire.

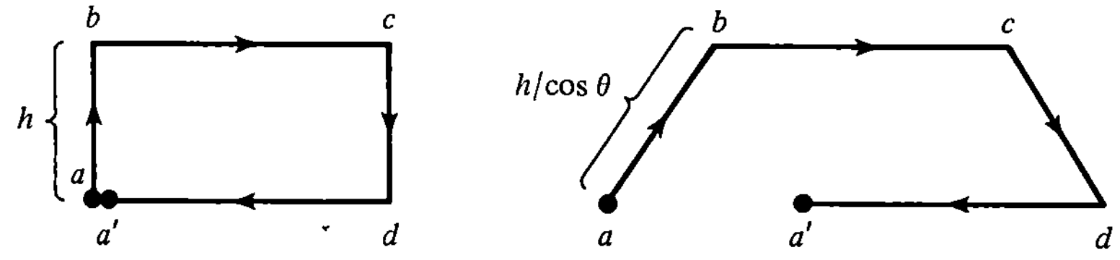

(b) Integration path for computing $\mathcal{E}$ (follow the wire at one instant of time). Only segment $a\,b$ contributes.

(c) Integration path for calculating work done (follow the charge around the loop). Only segment $a\,b$ contributes.

Figure 7.13

$\mathbf{f}_{mag}$ contributes nothing to the work because it is perpendicular to the motion of the charge.[4]

There is a particularly nice way of expressing the emf generated in a moving loop. Let Φ be the flux of $\mathbf{B}$ through the loop:

$$\Phi = \int \mathbf{B} \cdot d\mathbf{a} \tag{7.12}$$

For the rectangular loop in Fig. 7.12,

$$\Phi = Bhs$$

As the loop moves, the flux decreases:

$$\frac{d\Phi}{dt} = Bh\frac{ds}{dt} = -Bhv$$

(The minus sign accounts for the fact that ds/dt is negative.) But this is precisely the emf (equation (7.11)); evidently, the emf generated in the loop is minus the rate of change of flux through the loop:

$$\boxed{\mathcal{E} = -\frac{d\Phi}{dt}} \tag{7.13}$$

This is the **flux rule** for motional emf. Apart from its delightful simplicity, it has the virtue of applying to *non*rectangular loops moving in *arbitrary* directions through *non*uniform magnetic fields; in fact, the loop need not even maintain a fixed shape.

Proof of the flux rule for any loop moving in a static magnetic field. Figure 7.14 shows a loop of wire at time t and also at a short time dt later. Suppose we compute the flux at time t, using surface S, and the flux at time $(t + dt)$, using the

[4]For further discussion, see E. P. Mosca, *Am. J. Phys.* **42**, 295 (1974).

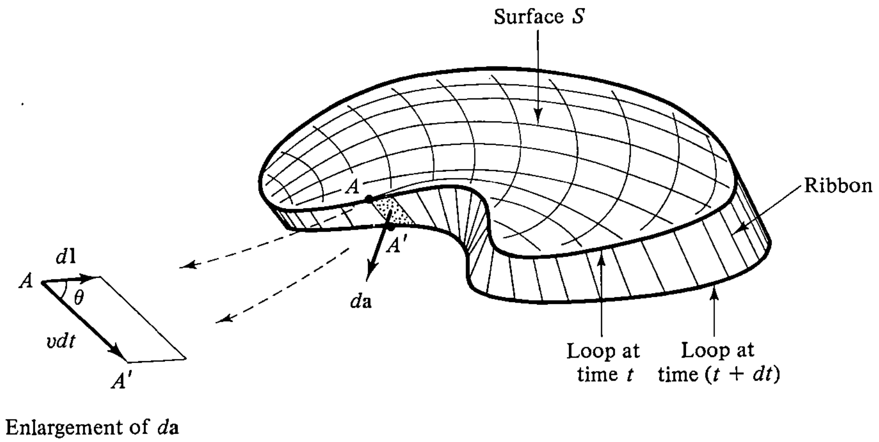

Figure 7.14

surface consisting of S plus the "ribbon" that connects the new position of the loop to the old. The *change* in flux, then, is

$$d\Phi = \Phi(t + dt) - \Phi(t) = \Phi_{\text{ribbon}} = \int_{\text{ribbon}} \mathbf{B} \cdot d\mathbf{a}$$

Focus attention on point A of the loop: In time dt it moves to A'. Let $\mathbf{v}$ be the velocity of the *wire* at A and $\mathbf{u}$ be the velocity of a charge *down* the wire at A; $\mathbf{w} = \mathbf{v} + \mathbf{u}$ is the resultant velocity of a charge at A. The infinitesimal element of area on the ribbon can be written as

$$d\mathbf{a} = (\mathbf{v} \times d\mathbf{l})\, dt$$

(see inset in Fig. 7.14). Therefore,

$$\frac{d\Phi}{dt} = \oint \mathbf{B} \cdot (\mathbf{v} \times d\mathbf{l})$$

Since $\mathbf{w} = (\mathbf{v} + \mathbf{u})$ and $\mathbf{u}$ is parallel to $d\mathbf{l}$, we can also write this as

$$\frac{d\Phi}{dt} = \oint \mathbf{B} \cdot (\mathbf{w} \times d\mathbf{l})$$

Now the scalar triple-product can be rewritten:

$$\mathbf{B} \cdot (\mathbf{w} \times d\mathbf{l}) = -(\mathbf{w} \times \mathbf{B}) \cdot d\mathbf{l}$$

so

$$\frac{d\Phi}{dt} = -\oint (\mathbf{w} \times \mathbf{B}) \cdot d\mathbf{l}$$

But $(\mathbf{w} \times \mathbf{B})$ is the magnetic force per unit charge, $\mathbf{f}_{\text{mag}}$, so

$$\frac{d\Phi}{dt} = -\oint \mathbf{f}_{\text{mag}} \cdot d\mathbf{l}$$

and the integral of $\mathbf{f}_{\text{mag}}$ is the emf

$$\mathcal{E} = -\frac{d\Phi}{dt} \qquad \text{q.e.d.}$$

There is a sign ambiguity in the definition of emf (equation 7.9)): Which *way* around the loop are you supposed to integrate? There is a compensatory ambiguity in the definition of *flux* (7.12): Which is the positive direction for $d\mathbf{a}$? In applying the flux rule, sign consistency is governed (as always) by your right hand: If the fingers define the positive direction around the loop, then the thumb indicates the direction of $d\mathbf{a}$. Should the emf then come out negative, it means the current will flow in the negative direction around the circuit.

The flux rule is a nifty short-cut for calculating motional emf's. It does *not* contain any new physics—just the Lorentz force law. Occasionally you will run across motional emf's which cannot be handled by the flux rule. Here's an example.

Example 4

A metal disc rotates about a vertical axis, through a uniform field **B**, pointing up. A circuit is made by connecting one end of a resistor to the axle and the other end to a sliding contact which touches the outer edge of the disc (Fig. 7.15). Current will flow in the direction indicated, yet the flux through the cir-

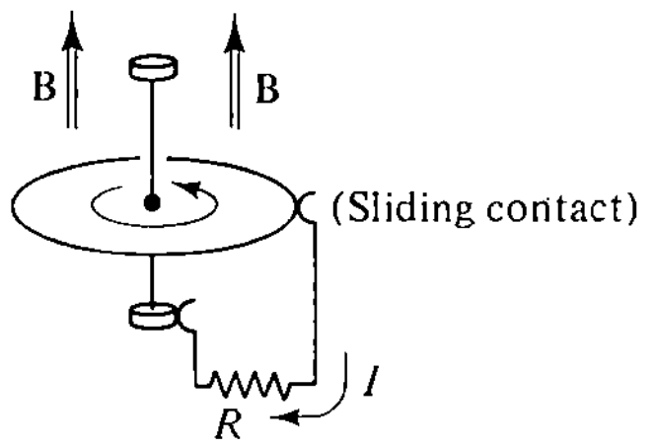

Figure 7.15

cuit does not seem to be changing. I'll let you calculate the emf (Problem 7.9), using the Lorentz force law. (The trouble with the flux rule is that it assumes the current flows along a well-defined path, whereas in this example the current spreads out over the whole disc. When you stop to think about it, it's not even clear what the "flux through the circuit" *means* in this context.)

Problem 7.7 A metal bar of mass m slides frictionlessly on two parallel conducting rails a distance l apart (Fig. 7.16). A resistor R is connected across the rails and a uniform magnetic field **B**, pointing into the page, fills the entire region.

(a) If the bar moves to the right at speed v, what is the current in the resistor? In what direction does it flow?

(b) What is the magnetic force on the bar? In what direction?

(c) If the bar starts out with speed v_0 at time $t = 0$, and is left to slide, what is its speed at a later time t?

(d) The initial kinetic energy of the bar was, of course, $\frac{1}{2}mv_0^2$. Where does this energy go? Prove that energy is conserved in this process by showing that the energy gained elsewhere is exactly $\frac{1}{2}mv_0^2$.

Problem 7.8 A square loop of wire (side s) lies on a table near a very long straight wire which carries a current I, as shown in Fig. 7.17.

(a) Find the flux of **B** through the loop.

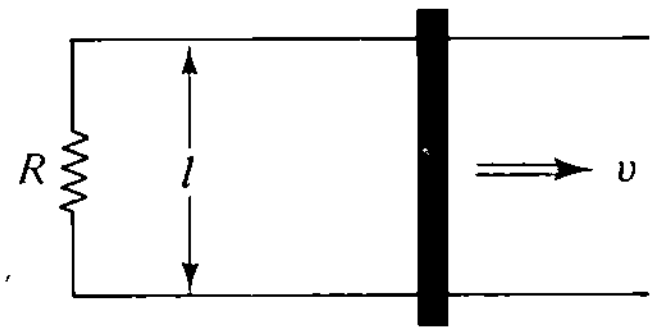

Figure 7.16

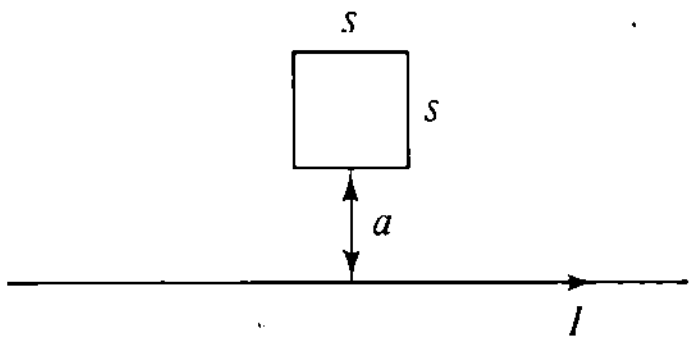

Figure 7.17

(b) If someone now pulls the loop directly away from the wire, at speed v, what emf is generated? In what direction (clockwise or counterclockwise) does the current flow?

(c) What if the loop is pulled to the *right* at speed v, instead of away?

Problem 7.9 Suppose the disc in Fig. 7.15 is rotating at an angular velocity ω, and its radius is a. What is the emf of the circuit, and what current flows through the resistor?

Problem 7.10 An infinite number of different surfaces can be fitted to a given boundary line, and yet, in defining the magnetic flux through a loop, $\Phi = \int \mathbf{B} \cdot d\mathbf{a}$, I never specified the particular surface to be used. Justify this apparent oversight.

Problem 7.11 A square loop (side s) is mounted on a vertical shaft and rotated at angular velocity ω (Fig. 7.18). A uniform magnetic field $\mathbf{B}$ points to the right. Find the emf of this alternating current generator (it will, of course, be a function of the time t).

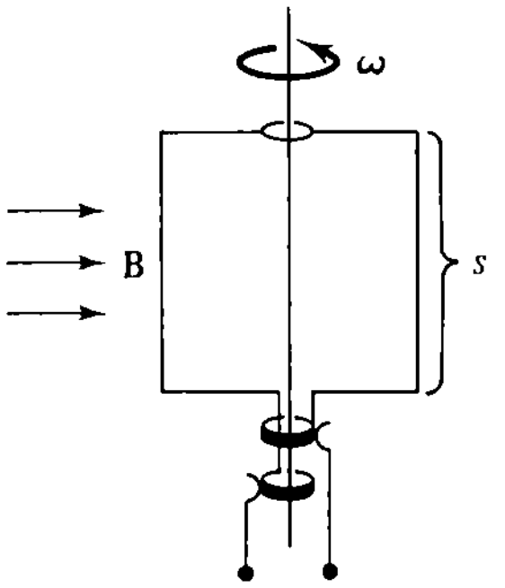

Figure 7.18

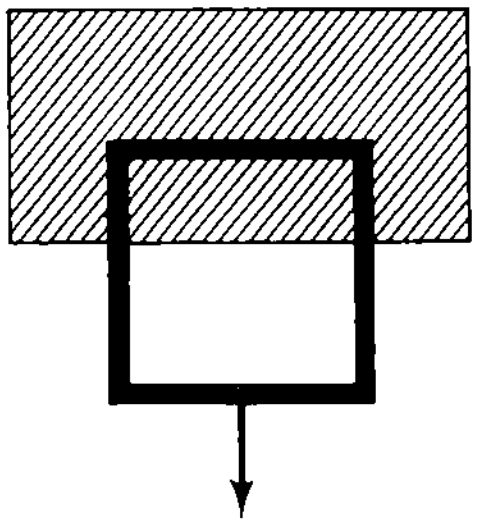

Figure 7.19

Problem 7.12 A square loop is cut out of a thick sheet of aluminum. It is then placed so that the top portion is in a uniform magnetic field $\mathbf{B}$, and allowed to fall under gravity (Fig. 7.19). (In the diagram, shading indicates the field region; $\mathbf{B}$ points into the page.) If the magnetic field is 1 T (a pretty standard laboratory field), find the terminal velocity of the loop (in m/s). Find the velocity of the loop as a function of time. How long does it take (in seconds) to reach, say, 90% of the terminal velocity? What would happen if you cut a tiny slit in the ring, breaking the circuit? (*Note:* The dimensions of the loop cancel out; determine the actual *numbers*, in the units indicated.)

7.2 FARADAY'S LAW

7.2.1 Electromagnetic Induction

In the last section we studied the case of a loop of wire moving to the right in the presence of a fixed magnetic field; the emf generated in the loop is given by the flux rule:

$$\mathcal{E} = -\frac{d\Phi}{dt}$$

What if, instead, we held the loop still and moved the *magnet* to the *left*? I don't think it will surprise you to learn that the same emf arises either way—all that really matters is the *relative* motion of the magnet and the loop. Indeed, in the light of special relativity it *has* to be so. But Faraday, who made this discovery empirically, knew nothing of relatively, and in the context of classical electrodynamics this simple reciprocity has remarkable implications. For if the *loop* moves, it's a *magnetic* force that sets up the emf, but if the loop is *stationary,* the force *cannot* be magnetic—stationary charges experience no magnetic forces. In that case, what *is* the driving force? What sort of field exerts a force on charges at rest? *Answer:* An *electric* field![5] Not an electro*static* field, to be sure, for we know that electrostatic fields cannot generate emf's, but an entirely new *kind* of electric field whose presence is associated with the fact that the magnet is moving, so that the magnetic field in the neighborhood of the loop is changing. Evidently a *changing magnetic field induces an electric field.*

Indeed, since the emf is the same as for the stationary magnet and moving loop, we have

$$\oint \mathbf{E}\cdot d\mathbf{l} = \mathcal{E} = -\frac{d\Phi}{dt} \tag{7.14}$$

This is **Faraday's law** in integral form. We can easily turn it into a differential law by applying Stokes' theorem:

$$\oint \mathbf{E}\cdot d\mathbf{l} = \int (\nabla\times\mathbf{E})\cdot d\mathbf{a} = -\frac{d}{dt}\int \mathbf{B}\cdot d\mathbf{a} = -\int \frac{\partial \mathbf{B}}{\partial t}\cdot d\mathbf{a}$$

so that

$$\boxed{\nabla\times\mathbf{E} = -\frac{\partial \mathbf{B}}{\partial t}} \tag{7.15}$$

Faraday's law reduces to the old rule $\oint \mathbf{E}\cdot d\mathbf{l} = 0$ (or, in differential form, $\nabla\times\mathbf{E} = 0$) in the static case (constant **B**) as, of course, it should.

Although the two experiments I described *look* reasonably similar and are described by the same formula ($\mathcal{E} = -d\Phi/dt$) their physical explanations could hardly be more different: When you move the loop, it's just the Lorentz force law in the guise of the flux rule; the emf is due to *magnetic* forces. But when you move the magnet, it's an *electrical* force that does the job—the changing magnetic field induces an electric field, in accordance with Faraday's law. Viewed in this light, it is quite astonishing that the two processes yield identical emf's. In fact, it was precisely this "coincidence" that led Einstein to the special theory of relativity—he sought a deeper explanation of what is, in classical electrodynamics, a peculiar accident. But that's a story for Chapter 10.

To exploit the parallel between motional emf's and Faraday emf's, I assumed that the changing flux in the latter case was due to motion of the magnet. But Fara-

[5]You could, I suppose, introduce an entirely new word to denote the field generated by a changing **B**. Electrodynamics would then involve *three* fields: E-fields, produced by electric charges [$\nabla\cdot\mathbf{E} = (1/\epsilon_0)\rho$, $\nabla\times\mathbf{E} = 0$], **B**-fields, produced by electric currents [$\nabla\cdot\mathbf{B} = 0$, $\nabla\times\mathbf{B} = \mu_0\mathbf{J}$]; and **G**-fields, produced by changing magnetic fields [$\nabla\cdot\mathbf{G} = 0$, $\nabla\times\mathbf{G} = -\partial\mathbf{B}/\partial t$]. Because **E** and **G** exert *forces* in the same way [$\mathbf{F} = q(\mathbf{E}+\mathbf{G})$], it is tidier to regard their sum as a *single* entity and call the whole thing "the electric field."

day's law applies *whatever* the reason for the change in **B**: It could be because the source is moving or because the *strength* of the source is changing (perhaps someone is fiddling with the current in a solenoid). The wire loop, in fact, has no way of knowing *why* the field changed. All that matters is that the magnetic field in some region is varying: Whenever that happens, an induced electric field will accompany the change in **B**.

Keeping track of the *signs* in Faraday's law can be a headache, but there's a special rule that often facilitates matters; it's called **Lenz's law:**[6]

> If a current flows, it will be in such a direction that the magnetic field *it* produces tends to counteract the change in flux that induced the emf.

If the flux is decreasing, the current will flow so that its field adds to the original flux; if the flux is increasing, the current will flow the opposite way. Notice carefully that it's the *change* in flux the induced current opposes—not the flux itself. Faraday induction is a sort of "inertial" phenomenon: A conducting loop "likes" to keep a constant flux through it. If you try to change the flux, the loop responds by sending a current around in such a direction as to counter your efforts. (It doesn't *succeed* completely; in fact, the flux produced by the induced current may be only a tiny fraction of the original; all Lenz's law tells you is the *direction* of the flow.)

Example 5 (The "jumping ring" demonstration)

If you wind a solenoidal coil around an iron core (the iron is there to beef up the magnetic field), place a metal ring on top, and plug it in, the ring will jump several feet in the air (Fig. 7.20). Why?

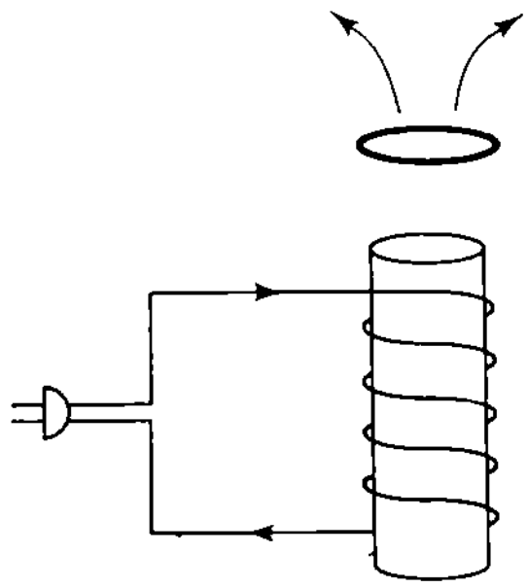

Figure 7.20

Solution: *Before* you turned on the current, the flux through the ring was *zero*. *Afterward* a flux appeared (upward in the diagram), and the emf generated in the ring led to a current (in the ring) which, according to Lenz's law, was in such a direction that *its* field tended to cancel this new flux. This means that the current in the loop is *opposite* to the current in the solenoid. And opposite currents repel, so the ring flies off.

---

[6]Lenz's law applies to *motional* emf's, too, but for them it is usually easier to get the direction of the current from the Lorentz force law.

The induced electric field associated with a given changing magnetic field can be calculated by exploiting the analogy between Faraday's law,

$$\nabla \times \mathbf{E} = -\frac{\partial \mathbf{B}}{\partial t}$$

and Ampère's law,

$$\nabla \times \mathbf{B} = \mu_0 \mathbf{J}$$

Of course, the curl alone is not enough to determine a field—you must also specify the divergence. But as long as $\mathbf{E}$ is a *pure* Faraday field, due exclusively to a changing $\mathbf{B}$, Gauss's law states that

$$\nabla \cdot \mathbf{E} = 0$$

while for magnetic fields,

$$\nabla \cdot \mathbf{B} = 0$$

as always. So the parallel is complete, and I conclude that *Faraday-induced electric fields are determined by* $-(\partial \mathbf{B}/\partial t)$ *in exactly the same way as magnetostatic fields are determined by* $\mu_0 \mathbf{J}$. Thus, the Biot-Savart law,

$$\mathbf{B} = \frac{\mu_0}{4\pi} \int \frac{\mathbf{J} \times \hat{\boldsymbol{\mathscr{r}}}}{\mathscr{r}^2}\, d\tau$$

translates into

$$\mathbf{E} = -\frac{1}{4\pi} \int \frac{(\partial \mathbf{B}/\partial t) \times \hat{\boldsymbol{\mathscr{r}}}}{\mathscr{r}^2}\, d\tau = -\frac{d}{dt}\left\{\frac{1}{4\pi} \int \frac{\mathbf{B} \times \hat{\boldsymbol{\mathscr{r}}}}{\mathscr{r}^2}\, d\tau\right\} \tag{7.16}$$

or, in terms of the vector potential (equation 5.66):

$$\mathbf{E} = -\frac{\partial \mathbf{A}}{\partial t} \tag{7.17}$$

This result is easily checked by applying the curl:

$$\nabla \times \mathbf{E} = -\frac{\partial}{\partial t}(\nabla \times \mathbf{A}) = -\frac{\partial \mathbf{B}}{\partial t}$$

If symmetry permits, we can use all the tricks associated with Ampère's law in integral form,

$$\oint \mathbf{B} \cdot d\mathbf{l} = \mu_0 I_{\text{enc}}$$

only this time it's *Faraday's* law in integral form:

$$\oint \mathbf{E} \cdot d\mathbf{l} = -\frac{d\Phi}{dt}$$

The rate of change of flux through the Amperian loop plays the role formerly assigned to $\mu_0 I_{\text{enc}}$.

Example 6

A uniform magnetic field $\mathbf{B}_0(t)$, pointing straight up, fills the shaded circular region of Fig. 7.21. If it is changing with time, what is the induced electric field?

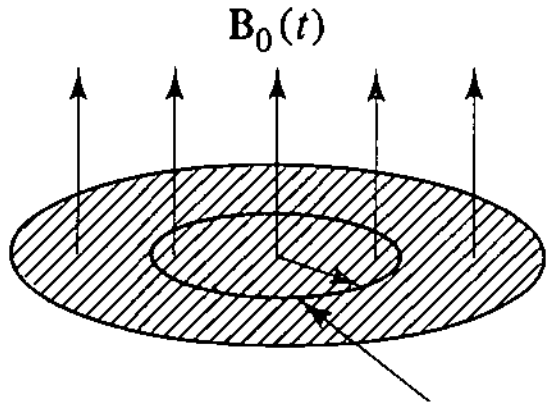

Figure 7.21

Solution: Draw an Amperian loop of radius r, and apply Faraday's law:

$$\oint \mathbf{E}\cdot d\mathbf{l} = E(2\pi r) = -\frac{d\Phi}{dt} = -\frac{d}{dt}\left(\pi r^2 B_0(t)\right) = -\pi r^2 \frac{dB_0}{dt}$$

Therefore,

$$E = -\frac{r}{2}\frac{dB_0}{dt}$$

$\mathbf{E}$ points in the circumferential direction, just like the *magnetic* field inside a long straight wire carrying a uniform *current* density. If B_0 is *increasing*, $\mathbf{E}$ runs *clockwise*, as viewed from above.

Example 7

A line charge λ is glued onto the rim of a wheel of radius R, which is then suspended horizontally, as in Fig. 7.22, so that it is free to rotate (the spokes are

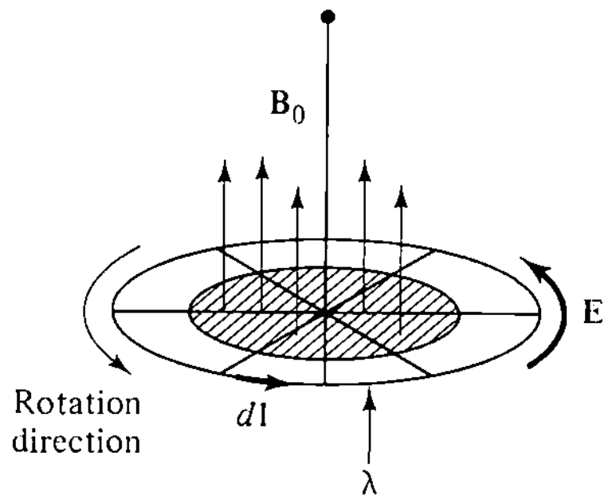

Figure 7.22

made of some nonconducting material—wood, maybe). In the central region, out to radius a, there is a uniform magnetic field $\mathbf{B}_0$, pointing up. Now someone turns the field off. What happens?

Solution: The changing magnetic field will induce an electric field, curling around the axis of the wheel. This electric field exerts a force on the charges on the rim, driving them around: The wheel will rotate. According to Lenz's law, it will rotate in such a direction that its field tends to restore the upward flux. The motion, then, is in the direction indicated—counterclockwise, as viewed from above.

Quantitatively, Faraday's law states that

$$\oint \mathbf{E} \cdot dl = -\frac{d\Phi}{dt} = -\pi a^2 \frac{dB}{dt}$$

Now, the torque on a segment of length dl is $(\mathbf{R} \times \mathbf{F})$, or $(R\lambda E)\, dl$. The total torque on the wheel is therefore

$$N = R\lambda \oint E\, dl = -R\lambda\pi a^2 \frac{dB}{dt}$$

and the total angular momentum imparted to the wheel is

$$\int N\, dt = -R\lambda\pi a^2 \int_{B_0}^{0} dB = R\lambda\pi a^2 B_0$$

As it turns out, it doesn't matter how fast or slow you turn off the field, the ultimate angular velocity of the wheel is the same regardless. (If you find yourself wondering where the angular momentum *came* from, you're getting ahead of the game! Wait for Example 17.)

A final word on this example: It's the *electric* field that did the rotating. To convince you of this I deliberately set things up so that the *magnetic* field is always *zero* at the location of the charge (on the rim). The experimenter may tell you he never put in any electric fields—all he did was switch off the magnetic field. But when he did that, an electric field automatically appeared, and it's this electric field that turned the wheel.

I must warn you, now, of a small fraud that tarnishes many applications of Faraday's law: Faraday's law, of course, pertains to *changing* magnetic fields, and yet we would like to use the apparatus of magneto*statics* (Ampère's law, the Biot-Savart law, and the rest) to *calculate* those fields. Technically, any result derived in this way is only approximately correct. But in practice the error is usually negligible unless the field fluctuates *extremely* rapidly. Even the case of a wire snipped by a pair of scissors (Problem 7.15) is *static enough* for Ampère's law to apply. This regime, in which magnetostatic rules can be used in conjunction with Faraday's law, is called **quasistatic.** Generally speaking, it is only when we come to electromagnetic waves and radiation that we must worry seriously about the breakdown of magnetostatics itself.

Example 8

A spherical shell of radius R carries a uniform surface charge σ. It spins about a fixed axis at angular velocity $\omega(t)$, which changes slowly with time. Find the electric field inside and outside the sphere.

Solution: First of all there is the ordinary *Coulomb* field:

$$\mathbf{E}_c = \frac{1}{4\pi\epsilon_0}\frac{Q}{r^2}\hat{\mathbf{r}} = \frac{R^2\sigma}{\epsilon_0 r^2}\hat{\mathbf{r}}$$

outside, and zero inside. In *addition,* however, there is the *Faraday* field induced by the changing magnetic field. The vector potential of a spinning spherical shell was derived in Example 11 of Chapter 5. Of course, that was magneto*statics,* and the result is *exact* only for *constant* ω. But as long as ω changes slowly, it shouldn't be too far off. Plugging (5.61) into (7.17), we find

$$\mathbf{E}_f = \begin{cases} \dfrac{-\mu_0 R\sigma}{3}\dot{\omega} r\sin\theta\,\hat{\boldsymbol{\phi}}, & (r \le R) \\[2ex] \dfrac{-\mu_0 R^4\sigma}{3}\dot{\omega}\dfrac{\sin\theta}{r^2}\hat{\boldsymbol{\phi}}, & (r \ge R) \end{cases}$$

(Here and below, a dot denotes the time derivative: $\dot{\omega} = d\omega/dt$.)

Example 9

An infinitely long straight wire carries a slowly varying current $I(t)$. Determine the induced electric field.[7]

Solution: In the quasistatic approximation, the magnetic field is $(\mu_0 I/2\pi r)$, and it circles around the wire. Like the **B**-field of a solenoid, **E** here runs parallel to the axis. For the rectangular "Amperian loop" in Fig. 7.23, Faraday's law gives:

$$\oint \mathbf{E}\cdot d\mathbf{l} = E(r_0)l - E(r)l = -\frac{d}{dt}\int \mathbf{B}\cdot d\mathbf{a}$$

$$= -\frac{\mu_0 \dot{I} l}{2\pi}\int_{r_0}^{r}\frac{1}{r'}\,dr' = -\frac{\mu_0 \dot{I} l}{2\pi}(\ln r - \ln r_0)$$

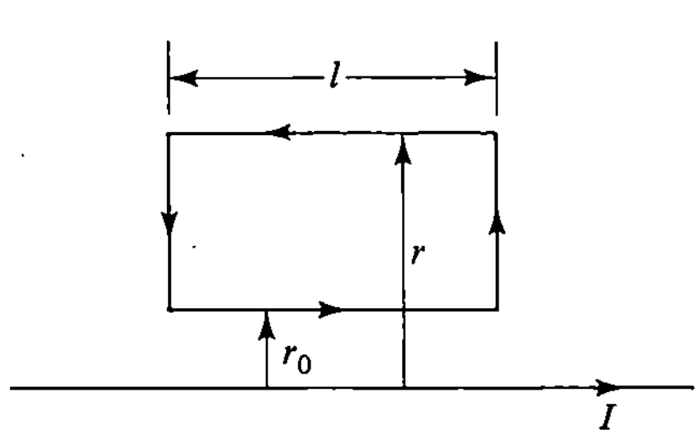

Figure 7.23

[7]This example is artificial, and not just in the usual sense of involving infinite wires, but in a more subtle respect. It assumes that the current is the same (at any given instant) all the way down the line. This is a safe assumption for the *short* wires in typical electric circuits, but not (in practice) for *long* wires (transmission lines), unless you supply a distributed and synchronized driving mechanism. But never mind—the problem doesn't ask how you would *produce* such a current, but only what *fields* would result if you *did.* (Variations on this problem are discussed in M. A. Heald, *Am. J. Phys.* **54,** 1142 (1986) and references cited therein.)

Thus

$$E(r) = \frac{\mu_0 \dot{I}}{2\pi} \ln r + K, \tag{7.19}$$

where $K = E(r_0) - (\mu_0 \dot{I}/2\pi) \ln r_0$ is a constant (that is to say, it is independent of r—it may well depend on t). The actual *value* of K depends on the whole history of the function $I(t)$; we'll see some examples in Chapter 9.

Equation (7.19) has the peculiar implication that E blows up as r goes to infinity. *That* can't be right . . . what's gone wrong? *Answer*: We have overstepped the limits of the quasistatic approximation. As we shall see in Chapter 9, electromagnetic "news" travels at the speed of light, and at large distances **B** depends not on the current *now*, but on the current *as it was* at some earlier time (indeed, a whole *range* of earlier times, since different points on the wire are different distances away). If τ is the time it takes I to change substantially, then the quasistatic approximation should hold only for

$$r \ll c\tau \tag{7.20}$$

and hence equation (7.19) does not apply, at extremely large r. (This didn't raise any obvious difficulties in Example 8, because E went to zero at large r anyway, but even there the result is strictly valid only within a sphere defined by (7.20).)

Problem 7.13 An alternating current $I = I_0 \cos(\omega t)$ flows down a long straight wire and back along a coaxial conducting cylinder of radius R. (See footnote 7.)
(a) In what *direction* does the induced electric field point (radial, circumferential, or longitudinal)?
(b) Find **E** as a function of r (the distance from the axis).

Problem 7.14 A long solenoid of radius a, carrying N turns per unit length, is looped by a wire with resistance R, as shown in Fig. 7.24.
(a) If the current in the solenoid is increasing,

$$\frac{dI}{dt} = k \quad \text{(a constant)}$$

what current flows in the loop, and which way (left or right) does it pass through the resistor?
(b) If the current I in the solenoid is constant but the solenoid is pulled out of the loop and reinserted in the opposite direction, what total charge passes through the resistor?

Problem 7.15 A square loop, side s, resistance R, lies a distance s from an infinite straight wire which carries current I (Fig. 7.25). Now someone cuts the wire, so that I drops to

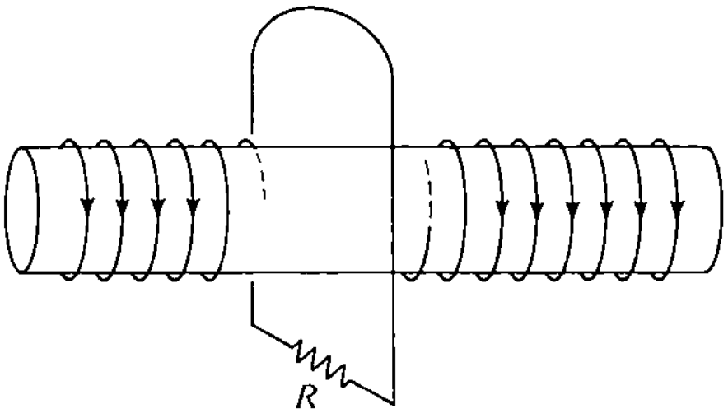

Figure 7.24

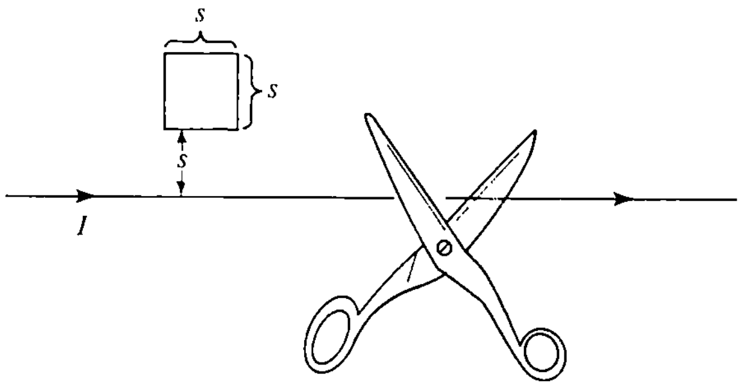

Figure 7.25

zero. In what direction does the induced current in the square loop flow, and what total charge passes a given point in the loop during the time this current flows?

If you don't like the scissors model, turn the current down *gradually*:

$$I(t) = \begin{cases} (1 - \alpha t)I, & \text{for } 0 \le t \le 1/\alpha \\ 0, & \text{for } t > 1/\alpha \end{cases}$$

Problem 7.16 Electrons undergoing cyclotron motion can be speeded up by increasing the magnetic field; the accompanying electric field will impart tangential acceleration. This is the principle behind the **betatron**. One would like to keep the radius of the orbit constant during the process. Show that this can be achieved by designing a magnet such that the average field over the area of the orbit is twice the field at the circumference (Fig. 7.26). Assume the electrons start from rest in zero field, and that the apparatus is symmetric about the center of the orbit. (Assume also that the electron velocity remains well below the speed of light, so that nonrelativistic mechanics applies.)

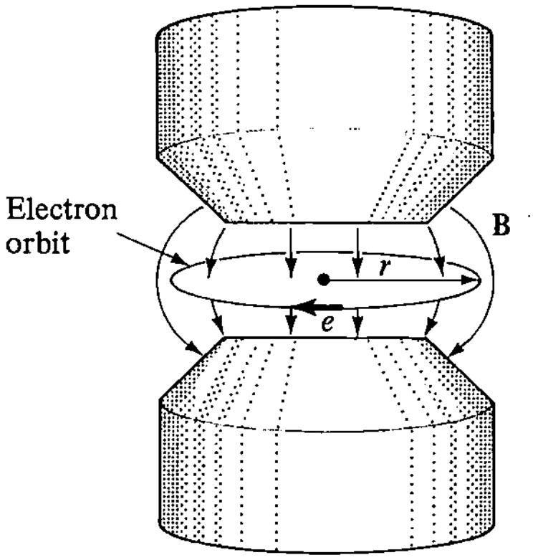

Figure 7.26

Problem 7.17 An atomic electron (charge q) circles about the nucleus (charge Q) in an orbit of radius r; the centripetal acceleration is provided, of course, by the Coulomb attraction of opposite charges. Now a small magnetic field dB is slowly turned on, perpendicular to the plane of the orbit. Show that the increase in kinetic energy, dT, imparted by the induced electric field, is just right to sustain circular motion *at the same radius r*. (That's why, in my discussion of diamagnetism, I assumed the radius is fixed. See Section 6.1.3 and the references cited there.)

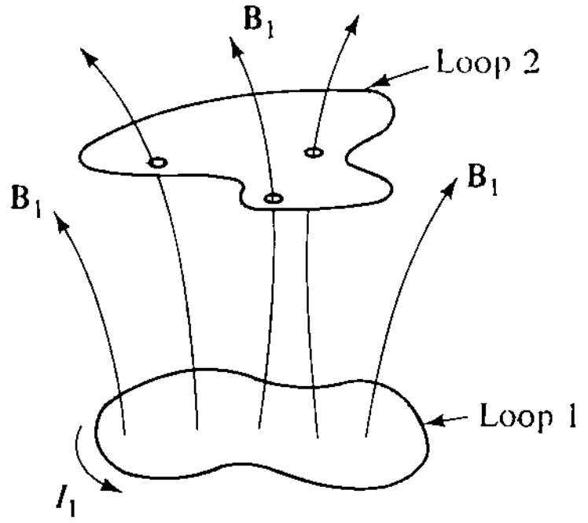

Figure 7.27

7.2.2 Inductance

Suppose we have two loops of wire at rest. If we run a steady current I_1 around loop 1, it will produce a magnetic field $\mathbf{B}_1$ (Fig. 7.27). Some of the field lines pass through loop 2; let Φ_2 be the flux of $\mathbf{B}_1$ through 2. Unless the shape of loop 1 is particularly simple, we might have a tough time actually *calculating* $\mathbf{B}_1$, but a glance at the Biot-Savart law,

$$\mathbf{B}_1 = \frac{\mu_0}{4\pi} I_1 \oint \frac{d\mathbf{l}_1 \times \hat{\boldsymbol{\mathscr{r}}}}{\mathscr{r}^2}$$

reveals one significant fact about this field: *It is proportional to the current* I_1. Therefore, so also is the flux through loop 2, since

$$\Phi_2 = \int \mathbf{B}_1 \cdot d\mathbf{a}_2$$

Thus

$$\Phi_2 = M_{21} I_1 \tag{7.21}$$

where M_{21} is the constant of proportionality; it is known as the **mutual inductance** of the two loops.

Although the mutual inductance is notoriously difficult to calculate in most practical cases, I can offer a simple-looking formula for it by expressing the flux in terms of the vector potential:

$$\Phi_2 = \int \mathbf{B}_1 \cdot d\mathbf{a}_2 = \int (\nabla \times \mathbf{A}_1) \cdot d\mathbf{a}_2 = \oint \mathbf{A}_1 \cdot d\mathbf{l}_2$$

Now, according to equation (5.58),

$$\mathbf{A}_1 = \frac{\mu_0 I_1}{4\pi} \oint \frac{d\mathbf{l}_1}{\mathscr{r}}$$

and hence

$$\Phi_2 = \frac{\mu_0 I_1}{4\pi} \oint \left(\oint \frac{d\mathbf{l}_1}{\mathscr{r}} \right) \cdot d\mathbf{l}_2$$

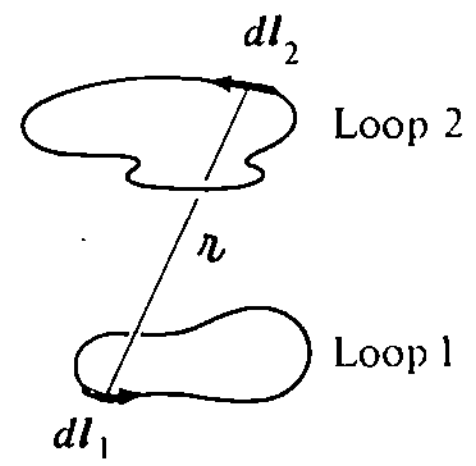

Figure 7.28

Therefore, comparing (7.21),

$$M_{21} = \frac{\mu_0}{4\pi} \oint \oint \frac{d\mathbf{l}_1 \cdot d\mathbf{l}_2}{\mathcal{R}} \tag{7.22}$$

This is the **Neumann formula;** it involves a double line integral, one integration around loop 1, the other around loop 2 (see Fig. 7.28). It's not an easy formula to work with, but it does illuminate two important things about mutual inductance:

1. M_{21} is a purely geometrical quantity, having to do with the sizes, shapes, and relative positions of the two loops.
2. The integral in (7.22) is unchanged if we switch the roles of loops 1 and 2; evidently

$$M_{21} = M_{12} \tag{7.23}$$

We may as well drop the subscripts and call them both M. This is an astonishing conclusion: *Whatever the shapes and positions of the loops, the flux through* 2 *when we run a current I around* 1 *is exactly the same as the flux through* 1 *when we send the same current I around* 2.

Example 10

A short solenoid (length l and radius R, N_1 turns per unit length) lies on the axis of a very long solenoid (N_2 turns per unit length) as shown in Fig. 7.29.

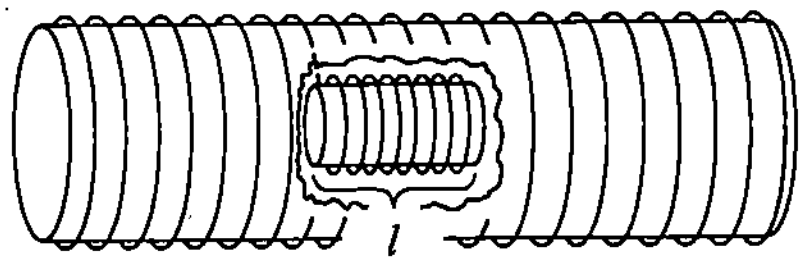

Figure 7.29

Current I flows in the short solenoid. What is the flux through the long solenoid?

Solution: Since the inner solenoid is short, it has a very complicated field; moreover, it puts a different amount of flux through each turn of the outer solenoid. It would be a *miserable* task to compute the flux this way. However, if we exploit the equality of the mutual inductances, the problem becomes very easy. Let's look at the reverse situation: run the current I through the *outer* solenoid, and calculate the flux through the *inner* one. The field inside the long solenoid is constant:

$$B = \mu_0 N_2 I$$

(equation (5.51)), so the flux through a single loop of the short solenoid is

$$B\pi R^2 = \mu_0 N_2 I \pi R^2$$

There are $(N_1 l)$ turns in all, so the total flux through the inner solenoid is

$$\Phi = (\mu_0 \pi R^2 N_1 N_2 l) I$$

This is also the flux a current I in the *short* solenoid would put through the *long* one, which is what we set out to find. Incidentally, the mutual inductance, in this case, is

$$M = \mu_0 \pi R^2 N_1 N_2 l$$

Suppose now that we *vary* the current in loop 1. The flux through loop 2 will vary accordingly, and Faraday's law says this changing flux will induce an emf in loop 2:

$$\mathcal{E}_2 = -\frac{d\Phi_2}{dt} = -M\frac{dI_1}{dt} \tag{7.24}$$

(In quoting (7.21)—which was based on the Biot-Savart law—I tacitly assume that the currents change slowly enough for the configuration to be considered quasi-static.) What a remarkable thing! Every time we *change* the current in loop 1, an induced current flows in loop 2, even though there are no wires connecting them.

Come to think of it, a changing current not only induces an emf in any nearby loops, it also induces an emf in the source loop *itself* (Fig. 7.30). As before, the field, and therefore the flux, is proportional to the current:

$$\Phi = LI \tag{7.25}$$

The constant of proportionality L is called the **self-inductance** (or simply the **inductance**) of the loop. As with M, it depends on the geometry (size and shape) of the loop. If the current changes, the emf induced in the loop is given by Faraday's law:

$$\mathcal{E} = -L\frac{dI}{dt} \tag{7.26}$$

Inductance is measured in **henries** (H); 1 henry is 1 volt-second per ampere.

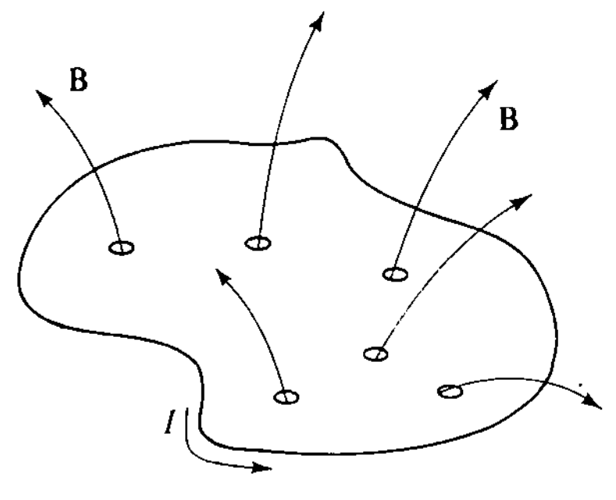

Figure 7.30

Example 11

Find the self-inductance of a toroidal coil with rectangular cross section (inner radius a, outer radius b, height h), which carries a total of n turns.

Solution: The magnetic field inside the toroid is (equation (5.52))

$$B = \frac{\mu_0 n I}{2\pi r}$$

The flux through a single turn (Fig. 7.31) is

$$\int \mathbf{B} \cdot d\mathbf{a} = \frac{\mu_0 n I}{2\pi} h \int_a^b \frac{1}{r}\, dr = \frac{\mu_0 n I}{2\pi} h \ln\left(\frac{b}{a}\right)$$

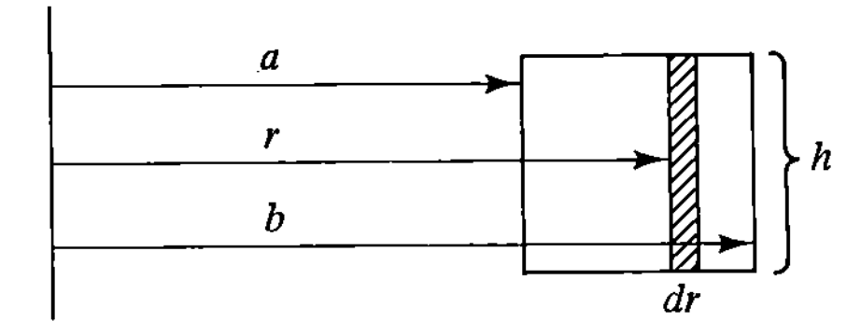

Figure 7.31

The *total* flux is n times this, so the self-inductance (equation (7.25)) is

$$L = \frac{\mu_0 n^2 h}{2\pi} \ln\left(\frac{b}{a}\right) \tag{7.27}$$

Inductance (like capacitance) is an intrinsically *positive* quantity. Lenz's law, which is enforced by the minus sign in (7.26), dictates that the emf is in such a direction as to *oppose* any *change in current.* For this reason, it is called a **back emf.** Whenever you try to alter the current in a wire, you must fight against this back emf. Thus inductance plays somewhat the same role in electric circuits that *mass* plays in mechanical systems: The greater L is, the harder it is to change the current, just as the larger the mass, the harder it is to change an object's velocity.

Example 12

Suppose a current I is flowing around a loop, when suddenly someone cuts the wire. The current drops "instantaneously" to zero. That generates a whopping back emf, for although I may be small, dI/dt is enormous. That's why you often draw a spark when you unplug an iron or a toaster—electromagnetic induction is desperately trying to keep the current going, even if it has to jump the gap in the circuit.

Nothing so dramatic occurs when you plug *in* a toaster or iron. In this case induction opposes the sudden *increase* in current, prescribing instead a smooth and continuous buildup. Suppose, for instance, that a battery (which supplies a constant emf $\mathcal{E}_0$) is connected to a circuit of resistance R and inductance L (Fig. 7.32). What current flows?

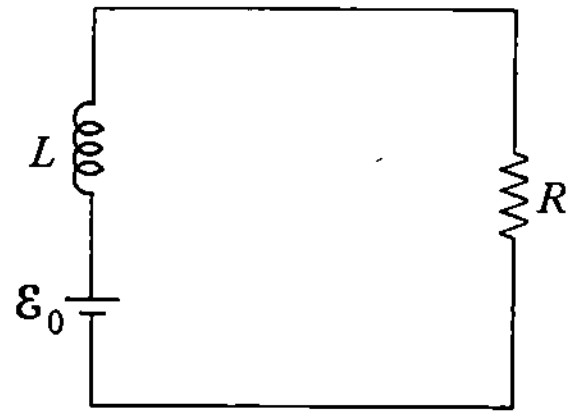

Figure 7.32

Solution: The total emf in this circuit is that provided by the battery plus that resulting from the self-inductance. Ohm's law gives, therefore,

$$\mathcal{E}_0 - L\frac{dI}{dt} = IR$$

This is a first-order differential equation for I as a function of time. The general solution, as you can easily derive for yourself, is

$$I(t) = \frac{\mathcal{E}_0}{R} + ke^{-(R/L)t}$$

where k is a constant to be determined by the boundary conditions of the problem. In particular, if the circuit is "plugged in" at time $t = 0$ (so that $I(0) = 0$), then k has the value $-\mathcal{E}_0/R$, and

$$I(t) = \frac{\mathcal{E}_0}{R}(1 - e^{-(R/L)t}) \tag{7.28}$$

This function is plotted in Fig. 7.33. Had there been no inductance in the circuit, the current would have jumped immediately to $\mathcal{E}_0/R$. In practice, *every*

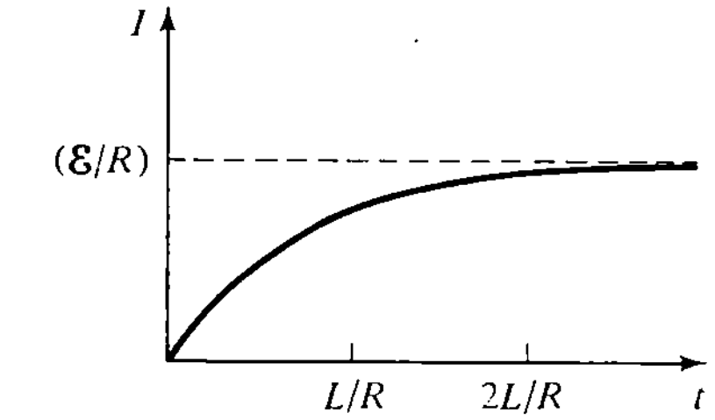

Figure 7.33

circuit has *some* self-inductance, and the current approaches $\mathcal{E}_0/R$ asymptotically. L/R is called the "time constant": it tells you how long the current takes to reach a substantial fraction (roughly two-thirds) of its final value.

Problem 7.18 A small loop of wire (radius a) lies a distance z above the center of a large loop (radius b), as in Fig. 7.34. The planes of the two loops are parallel, and perpendicular to a common axis.

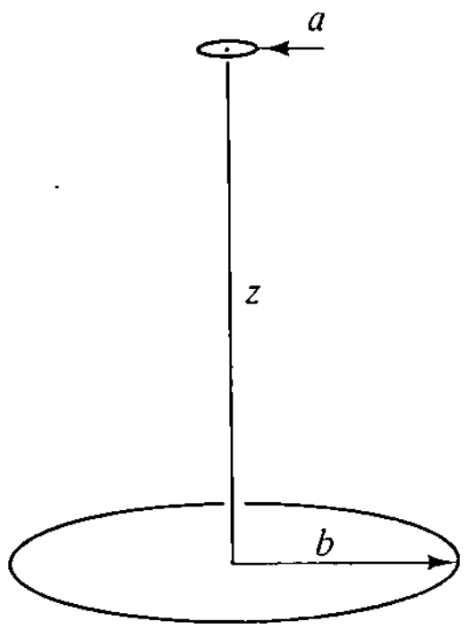

Figure 7.34

(a) Suppose current I flows in the big loop. Find the flux through the little loop. (The little loop is so small that you may consider the field of the big loop to be essentially constant.)

(b) Suppose current I flows in the little loop. Find the flux through the big loop. (The little loop is so small that you may treat it as a magnetic dipole.)

(c) Find the mutual inductances, and confirm that $M_{12} = M_{21}$.

Problem 7.19 A square loop of wire, of side s, lies midway between two long wires, $3s$ apart, and in the same plane. (Actually, the long wires are sides of a large rectangular loop, but the short ends are so far away that they can be neglected.) A clockwise current I in the square loop is gradually increasing: $dI/dt = k$ (a constant). Find the emf induced in the big loop. Which way will the induced current flow?

Problem 7.20 Find the self-inductance per unit length of a long solenoid, of radius R, carrying N turns per unit length.

Problem 7.21 Try to compute the self-inductance of the "hairpin" loop shown in Fig. 7.35. (Neglect the contribution from the ends; most of the flux comes from the long straight section.) You'll run into a snag that is characteristic of many self-inductance calculations. To get a definite answer, assume the wire has a tiny radius ϵ, and ignore any flux through the wire itself.

Problem 7.22 Two solenoidal coils are wrapped around a cylinder in such a way that the same amount of flux passes through every turn. (In practice this is achieved by inserting an iron core through the cylinder, which has the effect of concentrating the flux.) The "primary" coil has N_1 turns and the "secondary" N_2 (Fig. 7.36). If the current I in the primary is changing, show that the emf induced in the secondary is given by

$$\frac{\mathcal{E}_2}{\mathcal{E}_1} = \frac{N_2}{N_1} \tag{7.29}$$

where $\mathcal{E}_1$ is the (back) emf in the primary. (What we have here is a primitive **transformer**—a device for raising or lowering the emf of an alternating current source. By

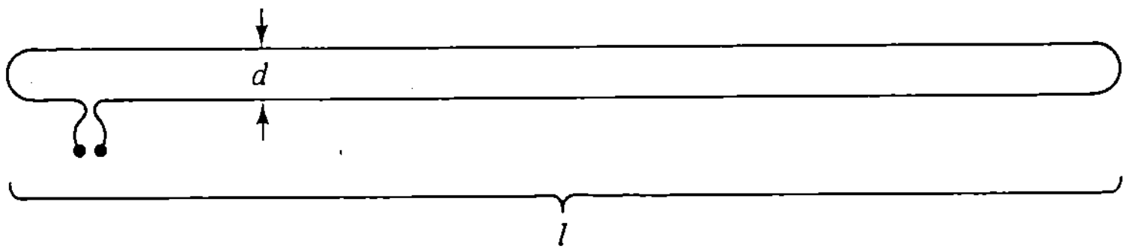

Figure 7.35

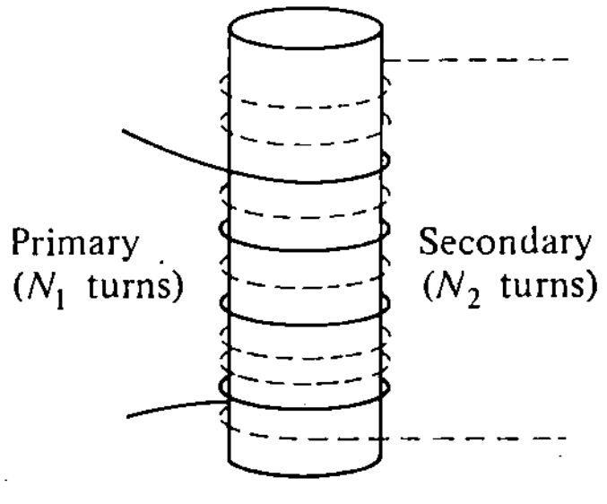

Figure 7.36

choosing the appropriate number of turns, any desired secondary emf can be obtained. If you think this violates conservation of energy, see Problem 7.58.)

Problem 7.23 A capacitor C is charged up to potential V and connected to an inductor L, as shown schematically in Fig. 7.37. At time $t = 0$ the switch S is closed. Find the current in the circuit as a function of time. How does your answer change if a resistor R is included in series with C and L?

Problem 7.24 By what ratio does the inductance change when an inductor is immersed in linear material of permeability μ?

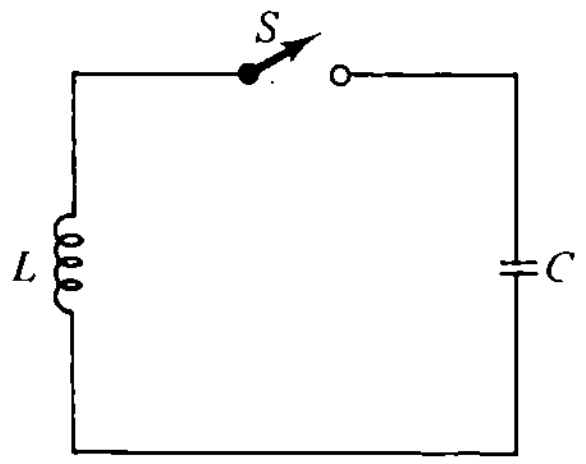

Figure 7.37

7.2.3 Energy in Magnetic Fields

It takes a certain amount of *energy* to start a current flowing in a circuit. I'm not talking about the energy delivered to the resistors and converted into heat—that is irretrievably lost as far as the circuit is concerned and can be large or small, depending on how long you let the current run. What I am concerned with, rather, is the work you must do *against the back emf* to get the current going. This is a *fixed* amount, and it is *recoverable:* you get it back when the current is turned off. In the meantime it represents a kind of potential energy of the circuit; as we'll see in a moment, it can be regarded as energy stored in the magnetic field.

The work done on a unit charge, against the back emf, in one trip around the circuit is $-\mathcal{E}$ (the minus sign records the fact that this is the work done *by you against* the emf, not the work done by the emf). The amount of charge per unit time passing down the wire is I. So the total work done per unit time is

$$\frac{dW}{dt} = -\mathcal{E}I = LI\frac{dI}{dt}$$

If we start with no current and build it up to final value I, the work done (integrating the last equation over time) is

$$\boxed{W = \frac{1}{2}LI^2} \tag{7.30}$$

It does not depend on how *long* we take to crank up the current, only on the geometry of the loop (in the form of L) and the final current I.

There is a nicer way to write W, which has the advantage that it is readily generalized to surface and volume currents. Remember that the flux Φ through the loop is equal to LI (7.25). On the other hand,

$$\Phi = \int_S \mathbf{B}\cdot d\mathbf{a} = \int_S (\nabla \times \mathbf{A})\cdot d\mathbf{a} = \oint_C \mathbf{A}\cdot d\mathbf{l}$$

where C is the perimeter of the loop and S is (any) surface bounded by C. Thus,

$$LI = \oint_C \mathbf{A}\cdot d\mathbf{l}$$

and therefore

$$W = \frac{1}{2} I \oint_C \mathbf{A}\cdot d\mathbf{l}$$

The vector sign might as well go on the I:

$$W = \frac{1}{2}\oint_C (\mathbf{A}\cdot\mathbf{I})\, dl \tag{7.31}$$

In this form the generalization to volume currents is obvious:

$$W = \frac{1}{2}\int_{\text{volume}} (\mathbf{A}\cdot\mathbf{J})\, d\tau \tag{7.32}$$

But we can do even better, and express W entirely in terms of the magnetic field: Ampère's law, $\nabla \times \mathbf{B} = \mu_0\mathbf{J}$, lets us eliminate $\mathbf{J}$:

$$W = \frac{1}{2\mu_0}\int_{\text{volume}} \mathbf{A}\cdot(\nabla \times \mathbf{B})\, d\tau \tag{7.33}$$

Integration by parts enables us to move the derivative from $\mathbf{B}$ to $\mathbf{A}$; specifically, product rule 6 states that

$$\nabla\cdot(\mathbf{A}\times\mathbf{B}) = \mathbf{B}\cdot(\nabla\times\mathbf{A}) - \mathbf{A}\cdot(\nabla\times\mathbf{B})$$

so

$$\mathbf{A}\cdot(\nabla\times\mathbf{B}) = B^2 - \nabla\cdot(\mathbf{A}\times\mathbf{B})$$

Consequently,

$$W = \frac{1}{2\mu_0}\left[\int B^2\,d\tau - \int \nabla\cdot(\mathbf{A}\times\mathbf{B})\,d\tau\right]$$
$$= \frac{1}{2\mu_0}\left[\int_{\text{volume}} B^2\,d\tau - \oint_{\text{surface}} (\mathbf{A}\times\mathbf{B})\cdot d\mathbf{a}\right] \tag{7.34}$$

Now, the integration (in 7.32) is to be taken over the *entire volume occupied by the current.* But any region *larger* than this will do just as well, for **J** is zero out there anyway. In the form (7.34), the larger the region we pick, the greater is the contribution from the B^2 integral and therefore the smaller is that of the surface integral (this makes sense: as the surface gets farther from the current, both **A** and **B** decrease). In particular, if we agree to integrate over *all* space, then the surface integral goes to zero, and we are left with

$$\boxed{W = \frac{1}{2\mu_0}\int_{\text{all space}} B^2\,d\tau} \tag{7.35}$$

In view of this result, we say the energy is "stored in the magnetic field," in the amount $(1/2\mu_0)(B^2)$ per unit volume. This is a nice way to think of it, though someone looking at equation (7.32) might prefer to say that the energy is stored in the *current distribution,* in the amount $\frac{1}{2}(\mathbf{A}\cdot\mathbf{J})$ per unit volume. The distinction is one of bookkeeping; the important quantity is the total energy W, and we shall not worry about where (if anywhere) the energy is "located".

You might find it strange that it takes work to set up a magnetic field—after all, magnetic fields *themselves* do no work. The point is that producing a magnetic field where before there was none requires a *changing* field, and a changing **B**-field, according to Faraday, induces an *electric* field. The latter, of course, *can* do work. In the beginning there is no **E**, and at the end there is no **E**; but in between, while **B** is building up, there *is* an **E**, and it is against *this* that the work is done. (You see why I could not calculate the energy stored in a magnetostatic field back in Chapter 5.) In the light of this, it is extraordinary how similar the magnetic energy formulas are to their electrostatic counterparts:

$$W_{\text{elec}} = \frac{1}{2}\int (V\rho)\,d\tau = \frac{\epsilon_0}{2}\int E^2\,d\tau \qquad \text{(2.37 and 2.39)}$$

$$W_{\text{mag}} = \frac{1}{2}\int (\mathbf{A}\cdot\mathbf{J})\,d\tau = \frac{1}{2\mu_0}\int B^2\,d\tau \qquad \text{(7.32 and 7.35)}$$

Example 13

A long coaxial cable carries current I (the current flows down the surface of the inner cylinder, radius a, and back along the outer cylinder, radius b) as shown in Fig. 7.38. Find the energy stored in a section of length l.

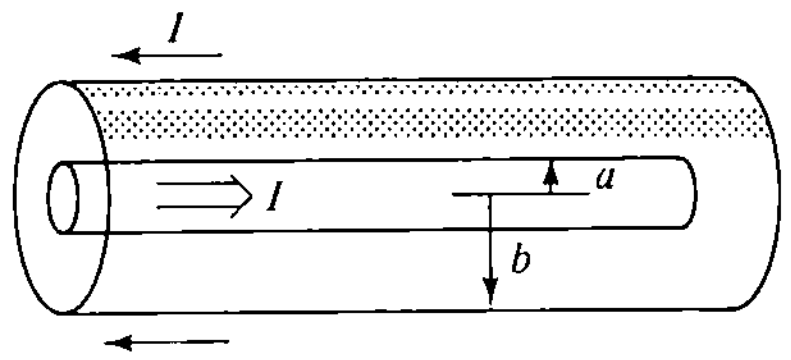

Figure 7.38

Solution: According to Ampère's law, the field between the cylinders is

$$\mathbf{B} = \frac{\mu_0 I}{2\pi r}\,\hat{\phi}$$

Elsewhere, the field is zero. Thus, the energy per unit volume is

$$\frac{1}{2\mu_0}\left(\frac{\mu_0 I}{2\pi r}\right)^2 = \frac{\mu_0 I^2}{8\pi^2 r^2}$$

The energy in a cylindrical shell of length l, radius r, and thickness dr, then, is

$$\left(\frac{\mu_0 I^2}{8\pi^2 r^2}\right) l\, 2\pi r\, dr = \frac{\mu_0 I^2 l}{4\pi}\frac{dr}{r}$$

Integrating from a to b, we have:

$$W = \frac{\mu_0 I^2 l}{4\pi}\ln\frac{b}{a}$$

By the way, this suggests a very simple way to calculate the self-inductance of the cable. According to (7.30), the energy can also be written as $\frac{1}{2}LI^2$. Comparing the two expressions,[8]

$$L = \frac{\mu_0 l}{2\pi}\ln\frac{b}{a}$$

(The self-inductance *per unit length* is $(\mu_0/2\pi)\ln(b/a)$.) This method of calculating the self-inductance is especially useful when the current is not confined to a single path but spreads over some surface or volume. In such cases different parts of the current may circle different amounts of flux, and it can be very difficult to get L from a formula like (7.25).

Problem 7.25 Find the energy stored in a section of length l of a long solenoid (radius R, current I, N turns per unit length),

(a) Using (7.30) (you found L in Problem 7.20).

(b) Using (7.31) (we worked out $\mathbf{A}$ in Example 12, Chapter 5).

[8]Notice the similarity to equation (7.27)—in a sense, the rectangular toroid is a short coaxial cable turned on its side.

(c) Using (7.35).
(d) Using (7.34): (take as your volume the cylindrical tube from radius $a < R$ out to radius $b > R$).

Problem 7.26 Calculate the energy stored in the toroidal coil of Example 11, by applying equation (7.35). Use the answer to check equation (7.27).

Problem 7.27 A long cable carries current in one direction uniformly distributed over its (circular) cross section. The current returns along the surface (there is a very thin insulating sheath separating the currents). Find the self-inductance per unit length.

Problem 7.28 Suppose the circuit in Fig. 7.39 has been connected for a long time when suddenly, at time $t = 0$, switch S is thrown, bypassing the battery.
(a) What is the current at any subsequent time t?
(b) What is the total energy delivered to the resistor?
(c) Show that this is equal to the energy originally stored in the inductor.

Problem 7.29 A certain transmission line is constructed from two thin metal "ribbons" of width w, a very small distance $s \ll w$ apart. The current travels down one strip and back along the other. In each case it spreads out uniformly over the surface of the ribbon.
(a) Find the capacitance per unit length, $\mathcal{C}$.
(b) Find the inductance per unit length, $\mathcal{L}$.
(c) What is the product $\mathcal{L}\mathcal{C}$, numerically? [$\mathcal{L}$ and $\mathcal{C}$ will, naturally, vary from one kind of transmission line to another, but their product is a universal constant—check, for example, the cable in Example 13—provided the space between the conductors is a vacuum. In the theory of transmission lines, this product is related to the speed with which an electromagnetic disturbance propagates down the line: $v = 1/\sqrt{\mathcal{L}\mathcal{C}}$.]
(d) If the strips are insulated from one another by a nonconducting material of permittivity ϵ and permeability μ, what then is the product $\mathcal{L}\mathcal{C}$? What is the propagation speed? (See Problem 7.24 and Example 6 of Chapter 4.)

Problem 7.30 Two tiny wire loops, of areas $\mathbf{a}_1$ and $\mathbf{a}_2$, are situated a distance r apart (Fig. 7.40).
(a) Find their mutual inductance. (*Hint*: Treat them as magnetic dipoles, and use equation (5.85)). Is your formula consistent with equation (7.23)?
(b) Suppose a current I_1 is flowing in loop 1, and we propose to turn on a current I_2 in loop 2. How much work must be done, against the mutually induced emf, to keep the current I_1 flowing in loop 1? (In the light of this result, comment on equation (6.31)).

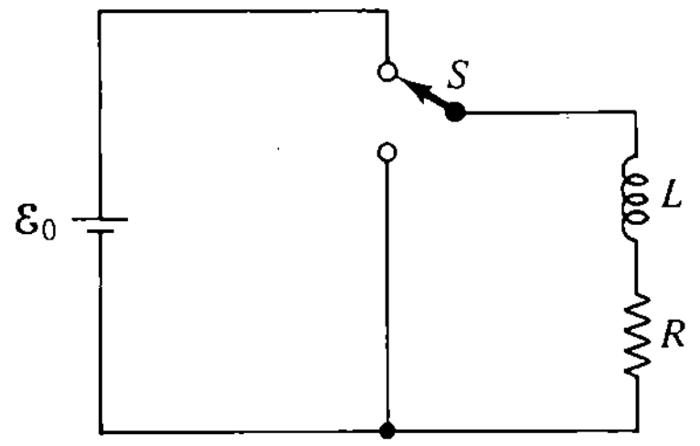

Figure 7.39

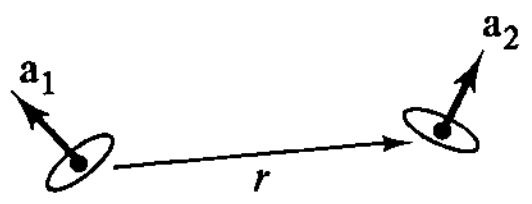

Figure 7.40

7.3 MAXWELL'S EQUATIONS

7.3.1 Electrodynamics Before Maxwell

So far, we have encountered the following laws, specifying the divergence and curl of electric and magnetic fields:

$$\begin{aligned}
&\text{(i)} \quad \nabla\cdot\mathbf{E} = \frac{1}{\epsilon_0}\rho && \text{(Gauss's law)}\\
&\text{(ii)} \quad \nabla\cdot\mathbf{B} = 0 && \text{(no name)}\\
&\text{(iii)} \quad \nabla\times\mathbf{E} = -\frac{\partial\mathbf{B}}{\partial t} && \text{(Faraday's law)}\\
&\text{(iv)} \quad \nabla\times\mathbf{B} = \mu_0\mathbf{J} && \text{(Ampère's law)}
\end{aligned} \tag{7.36}$$

These equations represent the state of electromagnetic theory over a century ago, when Maxwell began his work. They were not written in so compact a form in those days, but their physical content was familiar. Now, it happens there is a fatal inconsistency in these formulas. It has to do with the old rule that divergence of curl is always zero. If you apply the divergence to number (iii), everything works out:

$$\nabla\cdot(\nabla\times\mathbf{E}) = \nabla\cdot\left(-\frac{\partial\mathbf{B}}{\partial t}\right) = -\frac{\partial}{\partial t}(\nabla\cdot\mathbf{B})$$

The left side is zero because divergence of curl is zero; the right side is zero by virtue of equation (ii). But when you do the same thing to number (iv), you get into trouble:

$$\nabla\cdot(\nabla\times\mathbf{B}) = \mu_0(\nabla\cdot\mathbf{J}) \tag{7.37}$$

the left side must be zero, but the right side, in general, is *not*. For *steady* currents, the divergence of **J** is zero, but evidently when we go beyond magnetostatics Ampère's law cannot be right.

There's another way to see that Ampère's law is bound to fail for nonsteady currents. Suppose we're in the process of charging up a capacitor (Fig. 7.41). In integral form, Ampère's law reads

$$\oint \mathbf{B}\cdot d\mathbf{l} = \mu_0 I_{enc}$$

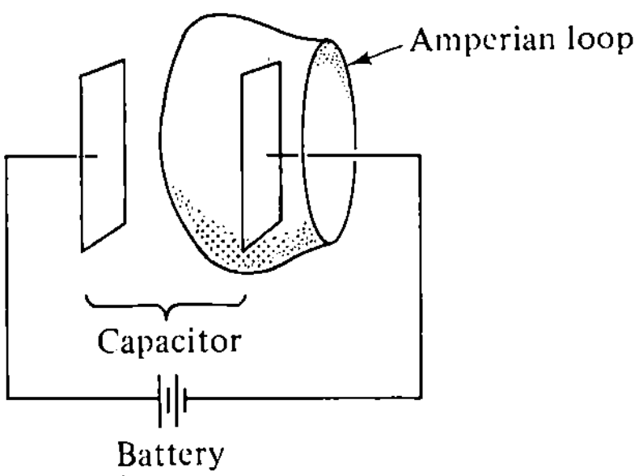

Figure 7.41

I want to apply it to the Amperian loop shown in the diagram. How do I determine I_{enc}? Well, it's the total current passing through the loop or, more precisely, the current passing through a surface which has the loop for its boundary. In this case, the *simplest* surface lies in the plane of the loop—the wire passes through this surface, so $I_{\text{enc}} = I$. Fine—but what if I draw instead the balloon-shaped surface in Fig. 7.41? *No* current passes through *this* surface, and I conclude that $I_{\text{enc}} = 0$! We never had this problem in magnetostatics because the conflict arises only when charge is piling up somewhere (in this case, on the capacitor plate). But *for nonsteady currents* (such as this one) *"the current enclosed by a loop" is an ill-defined notion,* since it depends entirely on what surface you use. (If this seems pedantic to you—"obviously one should use the plane surface"—remember that the Amperian loop could be some contorted shape that doesn't even lie in a plane.)

Of course, we had no right to *expect* Ampère's law to hold outside of magnetostatics; after all, we derived it from the Biot-Savart law. However, in Maxwell's time there was no *experimental* reason to doubt that Ampère's law was of wider validity. The flaw was a purely theoretical one, and Maxwell fixed it by purely theoretical arguments.

7.3.2 How Maxwell Fixed Up Ampère's Law

The problem is in the right side of equation (7.37), which *should be* zero but *isn't*. Applying the continuity equation (5.25) and Gauss's law, the offending term can be rewritten:

$$\nabla\cdot\mathbf{J} = -\frac{\partial\rho}{\partial t} = -\frac{\partial}{\partial t}(\epsilon_0\nabla\cdot\mathbf{E}) = -\nabla\cdot\left(\epsilon_0\frac{\partial\mathbf{E}}{\partial t}\right)$$

It might occur to you that if we were to add the quantity $\epsilon_0(\partial\mathbf{E}/\partial t)$ to $\mathbf{J}$, in Ampère's law, it would be just right to kill off the extra divergence:

$$\boxed{\nabla\times\mathbf{B} = \mu_0\mathbf{J} + \mu_0\epsilon_0\frac{\partial\mathbf{E}}{\partial t}} \tag{7.38}$$

(Maxwell himself had other reasons for wanting to add this quantity to Ampère's law. To him the rescue of the continuity equation was a happy dividend rather than a primary motive. But today we recognize this argument as a far more compelling one than Maxwell's, which was based on a now-discredited model of the ether.)[9]

Such a modification changes nothing insofar as magneto*statics* is concerned: When $\mathbf{E}$ is constant, we still have $\nabla\times\mathbf{B} = \mu_0\mathbf{J}$, as Ampère said. In fact, Maxwell's term is hard to detect in ordinary electromagnetic experiments, where it must compete for recognition with $\mathbf{J}$; that's why Faraday and the others never discovered it in the laboratory. However, it plays a crucial role in the propagation of electromagnetic waves, as we'll see in the next chapter.

[9]For the history of this subject, see A. M. Bork, *Am. J. Phys.* **31**, 854 (1963).

Apart from curing the defect in Ampère's law, Maxwell's term has a certain aesthetic appeal: Just as a changing *magnetic* field induces an *electric* field (Faraday's law), so a changing *electric* field induces a *magnetic* field. Of course, theoretical convenience and aesthetic consistency are only *suggestive*—there might, after all, be other ways to doctor up Ampère's law. The real confirmation of Maxwell's theory came in 1888 with Hertz's experiments on electromagnetic radiation.

Maxwell called his extra term the **displacement current:**

$$\mathbf{J}_d = \epsilon_0 \frac{\partial \mathbf{E}}{\partial t} \tag{7.39}$$

It's a misleading name, since $\epsilon_0(\partial \mathbf{E}/\partial t)$ has nothing to do with current, except that it adds to $\mathbf{J}$ in Ampère's law. Let's see now how the displacement current resolves the paradox of the charging capacitor (Fig. 7.41). If the capacitor plates are very close together (I didn't *draw* them that way, but the calculation is simpler if you assume this) then the field between them is

$$\mathbf{E} = \frac{1}{\epsilon_0}\sigma = \frac{1}{\epsilon_0}\frac{Q}{A}$$

where Q is the charge on the plate and A is its area. Thus, between the plates

$$\frac{\partial E}{\partial t} = \frac{1}{\epsilon_0 A}\frac{dQ}{dt} = \frac{1}{\epsilon_0 A} I$$

Now, equation (7.38) reads, in integral form,

$$\oint \mathbf{B}\cdot d\mathbf{l} = \mu_0 I_{\text{enc}} + \mu_0\epsilon_0 \int \left(\frac{\partial \mathbf{E}}{\partial t}\right)\cdot d\mathbf{a} \tag{7.40}$$

If we chose the *flat* surface, then $E = 0$ and $I_{\text{enc}} = I$. If, on the other hand, we use the balloon-shaped surface, then $I_{\text{enc}} = 0$, but $\int (\partial \mathbf{E}/\partial t)\cdot d\mathbf{a} = I/\epsilon_0$. So we get the same answer for either surface, though in the first case it comes from the genuine current and in the second from the displacement current.

Problem 7.31 A capacitor made from parallel circular plates, of radius a and separation s, is inserted into a long straight wire carrying current I (Fig. 7.42). As the capacitor charges up, find the induced magnetic field midway between the plates, at a distance r ($r < a$) from the center.

Problem 7.32 Refer to Problem 7.13, to which the correct answer was

$$\mathbf{E} = \frac{\mu_0 I_0 \omega}{2\pi} \sin \omega t \ln\left(\frac{R}{r}\right)\hat{\mathbf{z}}$$

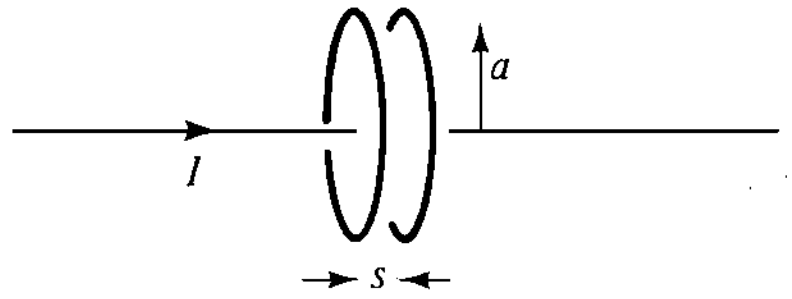

Figure 7.42

(a) Find the displacement current density, $\mathbf{J}_d$.
(b) Integrate it to get the total displacement current,

$$I_d = \int \mathbf{J}_d \cdot d\mathbf{a}$$

(c) Compare I_d and I (what's the ratio?). If the outer cylinder were, say, 2 mm in diameter, how high would the frequency have to be, for I_d to be 1% of I?
(This problem is designed to show you why Faraday never discovered displacement currents, and why it is ordinarily safe to ignore them unless the frequency is extremely high.)

7.3.3 Maxwell's Equations and Magnetic Charge

In the last section, we put the finishing touches on Maxwell's equations. In their final form, they read:

(i)	$\nabla \cdot \mathbf{E} = \frac{1}{\epsilon_0}\rho$	(Gauss's law)
(ii)	$\nabla \cdot \mathbf{B} = 0$	(no name)
(iii)	$\nabla \times \mathbf{E} = -\frac{\partial \mathbf{B}}{\partial t}$	(Faraday's law)
(iv)	$\nabla \times \mathbf{B} = \mu_0 \mathbf{J} + \mu_0\epsilon_0 \frac{\partial \mathbf{E}}{\partial t}$	(Ampère's law with Maxwell's correction)

(7.41)

Together with the Lorentz force law,[10]

$$\mathbf{F} = q(\mathbf{E} + \mathbf{v} \times \mathbf{B}) \tag{7.42}$$

they summarize the entire theoretical content of classical electrodynamics, save for some special properties of matter which we discussed in Chapters 4 and 6. Even the continuity equation,

$$\nabla \cdot \mathbf{J} = -\frac{\partial \rho}{\partial t} \tag{7.43}$$

which is the mathematical expression of conservation of charge, can be derived from Maxwell's equations by applying the divergence to number (iv).

[10]There's a kind of reciprocity here: Maxwell's equations tell you how charges produce fields, and the Lorentz force law tells you how fields affect charges. By the way, like any differential equations, Maxwell's must be supplemented by suitable *boundary conditions*. Because these are typically "obvious" from the context (e.g. **E** and **B** go to zero at large distances from a localized charge distribution), it is easy to forget that they play an essential role.

There is a certain pleasing symmetry about Maxwell's equations; it is particularly striking in free space, where ρ and $\mathbf{J}$ are zero:

$$\nabla \cdot \mathbf{E} = 0, \qquad \nabla \cdot \mathbf{B} = 0$$

$$\nabla \times \mathbf{E} = -\frac{\partial \mathbf{B}}{\partial t}, \qquad \nabla \times \mathbf{B} = \mu_0 \epsilon_0 \frac{\partial \mathbf{E}}{\partial t}$$

If you replace $\mathbf{E}$ by $\mathbf{B}$ and $\mathbf{B}$ by $-\mu_0\epsilon_0\mathbf{E}$, the first pair of equations turns into the second and vice versa. This symmetry[11] between $\mathbf{E}$ and $\mathbf{B}$ is spoiled, though, by the charge term in Gauss's law and the current term in Ampère's law. You can't help wondering why the corresponding terms are "missing" from $\nabla \cdot \mathbf{B} = 0$ and $\nabla \times \mathbf{E} = -\partial \mathbf{B}/\partial t$. What if we had

$$\nabla \cdot \mathbf{B} = \mu_0 \eta, \quad \text{corresponding to} \quad \nabla \cdot \mathbf{E} = \frac{1}{\epsilon_0}\rho$$

and

$$\nabla \times \mathbf{E} = -\mu_0 \mathbf{K} - \frac{\partial \mathbf{B}}{\partial t}, \quad \text{matching} \quad \nabla \times \mathbf{B} = \mu_0 \mathbf{J} + \mu_0 \epsilon_0 \frac{\partial \mathbf{E}}{\partial t}$$

Then η would represent the density of magnetic "charge," just as ρ is the density of electric charge, and $\mathbf{K}$ would be the current of magnetic charge, just as $\mathbf{J}$ is the current of electric charge. The magnetic charge would be conserved,

$$\nabla \cdot \mathbf{K} = -\frac{\partial \eta}{\partial t}$$

just as electric charge is,

$$\nabla \cdot \mathbf{J} = -\frac{\partial \rho}{\partial t}$$

The former follows by application of the divergence to the ($\nabla \times \mathbf{E}$) equation, just as the latter follows by application of the divergence to the ($\nabla \times \mathbf{B}$) equation.

In a sense, Maxwell's equations *beg* for magnetic charge to exist—it would fit in so nicely. And yet, in spite of diligent search, no one has ever found any. (See references in Section 5.3.4.) As far as is known, η is zero everywhere and so is $\mathbf{K}$. So $\mathbf{B}$ is *not* on an equal footing with $\mathbf{E}$: there exist stationary sources for $\mathbf{E}$ (electric charges) but none for $\mathbf{B}$. (This is reflected in the fact that magnetic multipole expansions have no monopole term, and magnetic dipoles consist of current loops, not separated north and south "poles.") Apparently God just didn't *make* any magnetic charge. (In the quantum theory of electrodynamics, by the way, it's a more than merely aesthetic shame that magnetic charge does not seem to exist: Dirac showed that the *existence* of *magnetic* charge would explain why *electric* charge is *quantized*. See Problem 7.52.)

[11]Don't be distracted by the pesky constants μ_0 and ϵ_0; these are present only because the SI system measures E and B in different units, and would not occur, for instance, in the Gaussian system.

Problem 7.33 Assuming that "Coulomb's law" for magnetic charges (s_1 and s_2) reads

$$\mathbf{F} = \frac{\mu_0}{4\pi}\frac{s_1 s_2}{\mathscr{r}^2}\hat{\mathscr{r}}$$

work out the "Lorentz force law" for a monopole s moving with velocity $\mathbf{v}$ through electric and magnetic fields $\mathbf{E}$ and $\mathbf{B}$.

7.3.4 Maxwell's Equations Inside Matter

Maxwell's equations in the form (7.41) are complete and correct as they stand. However, when you are working with materials which are subject to electric and magnetic polarization there is a more pertinent way to *write* them. The reason is that inside polarized matter there will be accumulations of "bound" charge and current over which we exert no direct control. I would like to reformulate Maxwell's equations in such a way as to make explicit reference only to those sources we control directly: the "free" charges and currents.

We have already learned, from the static case, that an electric polarization $\mathbf{P}$ results in an accumulation of bound charge (equation (4.12))

$$\rho_b = -\nabla\cdot\mathbf{P} \tag{7.44}$$

Likewise, a magnetic polarization (or "magnetization") $\mathbf{M}$ results in a bound current (eq. 6.13)

$$\mathbf{J}_b = \nabla\times\mathbf{M} \tag{7.45}$$

There's just one new feature to consider in the *non*static case: any *change* in the electric polarization involves a flow of (bound) charge (call it $\mathbf{J}_p$) which must be included in the total current. For suppose we examine a tiny chunk of polarized material, as shown in Fig. 7.43. The polarization introduces a charge density $\sigma_b = P$ at one end and $-\sigma_b$ at the other (equation (4.11)). If P now *increases* a bit, the charge on each end increases accordingly, giving a net current

$$dI = \frac{\partial\sigma_b}{\partial t}da_\perp = \frac{\partial P}{\partial t}da_\perp$$

The current density, therefore, is

$$\mathbf{J}_p = \frac{\partial\mathbf{P}}{\partial t} \tag{7.46}$$

This "polarization current" has nothing whatever to do with the *bound* current $\mathbf{J}_b$. The latter is associated with *magnetization* of the material and involves the spin and orbital motion of electrons. $\mathbf{J}_p$, by contrast, is the result of linear motion of

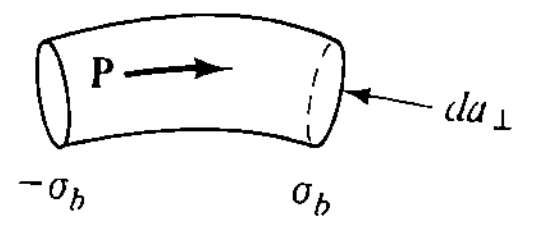

Figure 7.43

charge when the electric polarization changes. If $\mathbf{P}$ points to the right and is increasing, then each plus charge moves a bit to the right and each minus charge to the left; the cumulative effect is the polarization current $\mathbf{J}_p$. In this connection, we might just check that (7.46) is consistent with the continuity equation:

$$\nabla \cdot \mathbf{J}_p = \nabla \cdot \frac{\partial \mathbf{P}}{\partial t} = \frac{\partial}{\partial t}(\nabla \cdot \mathbf{P}) = -\frac{\partial \rho_b}{\partial t}$$

Yes: The continuity equation *is* satisfied; in fact, $\mathbf{J}_p$ is essential to account for the conservation of bound charge. (Incidentally, a changing *magnetization* does *not* lead to any analogous accumulation of charge or current. The bound current $\mathbf{J}_b = \nabla \times \mathbf{M}$ varies in response to changes in $\mathbf{M}$, to be sure, but that's about it.)

In view of all this, the total charge density can be separated into two parts:

$$\rho = \rho_f + \rho_b = \rho_f - \nabla \cdot \mathbf{P} \tag{7.47}$$

and the current density into *three* parts:

$$\mathbf{J} = \mathbf{J}_f + \mathbf{J}_b + \mathbf{J}_p = \mathbf{J}_f + \nabla \times \mathbf{M} + \frac{\partial \mathbf{P}}{\partial t} \tag{7.48}$$

Gauss's law can now be written as

$$\nabla \cdot \mathbf{E} = \frac{1}{\epsilon_0}(\rho_f - \nabla \cdot \mathbf{P})$$

or

$$\nabla \cdot \mathbf{D} = \rho_f \tag{7.49}$$

where $\mathbf{D}$, as in the static case, is given by

$$\mathbf{D} = \epsilon_0 \mathbf{E} + \mathbf{P} \tag{7.50}$$

Meanwhile, Ampère's law (with Maxwell's term) becomes

$$\nabla \times \mathbf{B} = \mu_0\left(\mathbf{J}_f + \nabla \times \mathbf{M} + \frac{\partial \mathbf{P}}{\partial t}\right) + \mu_0 \epsilon_0 \frac{\partial \mathbf{E}}{\partial t}$$

or

$$\nabla \times \mathbf{H} = \mathbf{J}_f + \frac{\partial \mathbf{D}}{\partial t} \tag{7.51}$$

where, as before,

$$\mathbf{H} = \frac{1}{\mu_0}\mathbf{B} - \mathbf{M} \tag{7.52}$$

Faraday's law and $\nabla \cdot \mathbf{B} = 0$ are not affected by our separation of charge and current into free and bound parts, since they do not involve ρ or $\mathbf{J}$.

In terms of *free* charges and currents, then, Maxwell's equations read

$$\boxed{\begin{array}{ll} \text{(i)} & \nabla \cdot \mathbf{D} = \rho_f \\ \text{(ii)} & \nabla \cdot \mathbf{B} = 0 \\ \text{(iii)} & \nabla \times \mathbf{E} = -\dfrac{\partial \mathbf{B}}{\partial t} \\ \text{(iv)} & \nabla \times \mathbf{H} = \mathbf{J}_f + \dfrac{\partial \mathbf{D}}{\partial t} \end{array}} \tag{7.53}$$

Many people regard these as the "true" Maxwell's equations, but please understand that they are not in the least more "general" than (7.41); they simply reflect a convenient division of charge and current into free and nonfree parts. And they have the disadvantage of hybrid notation, involving as they do both **E** and **D**, both **B** and **H**. They must be supplemented, therefore, by the defining relations (7.50) and (7.52). And even so, further equations are necessary to specify **P** and **M**. These depend on the nature of the material; for linear media they take the form

$$\left.\begin{array}{l} \mathbf{P} = \epsilon_0 \chi_e \mathbf{E}, \\ \mathbf{M} = \chi_m \mathbf{H}, \end{array}\right\} \quad \text{so that} \quad \left\{\begin{array}{l} \mathbf{D} = \epsilon \mathbf{E}, \\ \mathbf{H} = \dfrac{1}{\mu}\mathbf{B}. \end{array}\right. \tag{7.54}$$

Incidentally, you'll remember that **D** is called the electric "displacement"; this is why the second term in the Ampère/Maxwell equation (iv) is called the "displacement current":

$$\mathbf{J}_d = \frac{\partial \mathbf{D}}{\partial t} \tag{7.55}$$

Problem 7.34 Sea water at frequency $\nu = 4 \times 10^8$ Hz has permittivity $\epsilon = 81\,\epsilon_0$, permeability $\mu = \mu_0$, and resistivity $\rho = 0.23\ \Omega \cdot \text{m}$. What is the ratio of conduction current to displacement current? (*Hint:* Consider a parallel-plate capacitor immersed in sea water and driven by a voltage $V_0 \cos(2\pi\nu t)$.)

7.3.5 Boundary Conditions

In general, the fields **E**, **B**, **D**, and **H** will be discontinuous at a boundary between two different media or at a surface which carries charge density σ or current density **K**. The precise nature of these discontinuities can be deduced from Maxwell's equations (7.53), in their integral form

$$\left.\begin{array}{ll}\text{(i)} & \oint_S \mathbf{D}\cdot d\mathbf{a} = Q_{f_{\text{enc}}} \\ \text{(ii)} & \oint_S \mathbf{B}\cdot d\mathbf{a} = 0\end{array}\right\} \text{ over any closed surface } S.$$

$$\left.\begin{array}{ll}\text{(iii)} & \oint_L \mathbf{E}\cdot d\mathbf{l} = -\frac{d}{dt}\int_S \mathbf{B}\cdot d\mathbf{a} \\ \text{(iv)} & \oint_L \mathbf{H}\cdot d\mathbf{l} = I_{f_{\text{enc}}} + \frac{d}{dt}\int_S \mathbf{D}\cdot d\mathbf{a}\end{array}\right\} \text{ for any surface } S \text{ bounded by the closed loop } L.$$

Applying (i) to a tiny, wafer-thin, Gaussian pillbox extending just a hair into the material on either side of the boundary, we obtain (Fig. 7.44):[12]

$$\mathbf{D}_1\cdot\mathbf{a} - \mathbf{D}_2\cdot\mathbf{a} = \sigma_f a$$

(The edge of the wafer contributes nothing in the limit as the thickness goes to zero. Nor does any *volume* density of free charge contribute, in this limit.) Thus, the component of **D** which is perpendicular to the interface is discontinuous in the amount

$$\boxed{D_{1_\perp} - D_{2_\perp} = \sigma_f} \tag{7.57}$$

Identical reasoning, applied to equation (ii), yields

$$\boxed{B_{1_\perp} - B_{2_\perp} = 0} \tag{7.58}$$

Turning to (iii), a very thin Amperian loop straddling the surface (Fig. 7.45) gives

$$\mathbf{E}_1\cdot\mathbf{l} - \mathbf{E}_2\cdot\mathbf{l} = -\frac{d}{dt}\int_S \mathbf{B}\cdot d\mathbf{a}$$

[12]The positive direction for **a** and $D_\perp$ is *from* 2 *toward* 1.

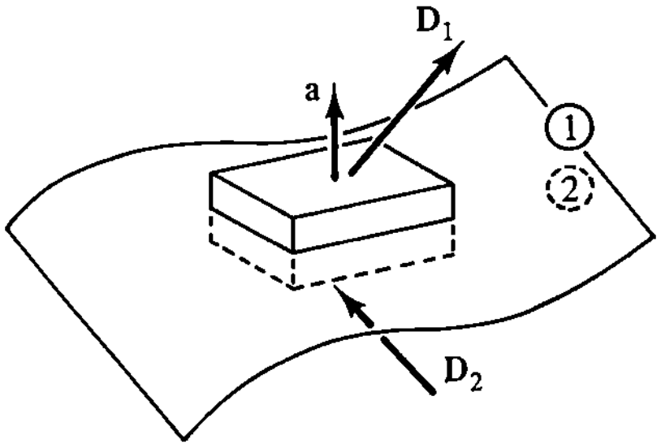

Figure 7.44

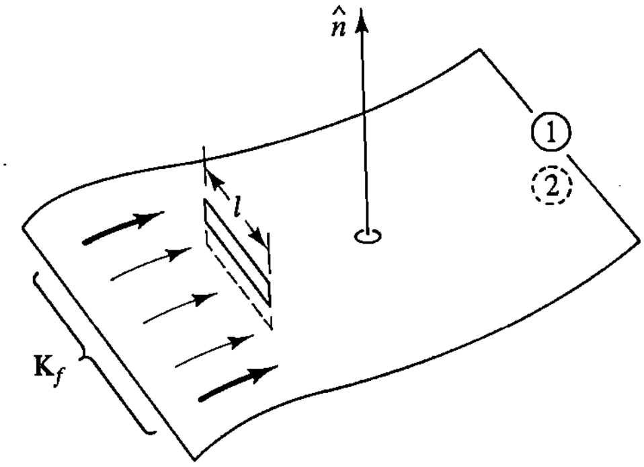

Figure 7.45

But in the limit as the width of the loop goes to zero, the flux vanishes. (I have already dropped the contribution of the two ends to $\oint \mathbf{E} \cdot d\mathbf{l}$, on the same grounds.) Therefore,

$$\boxed{\mathbf{E}_{1_\parallel} - \mathbf{E}_{2_\parallel} = 0} \tag{7.59}$$

That is, the components of **E** *parallel* to the interface are continuous across the boundary. By the same token, (iv) implies

$$\mathbf{H}_1 \cdot \mathbf{l} - \mathbf{H}_2 \cdot \mathbf{l} = I_{f_{\text{enc}}}$$

where $I_{f_{\text{enc}}}$ is the free current passing through the Amperian loop. No *volume* current density will contribute (in the limit of infinitesimal width) but a *surface* current can. In fact, if $\hat{n}$ is a unit vector perpendicular to the interface (pointing from 2 toward 1), so that $(\hat{n} \times \hat{l})$ is normal to the Amperian loop, then

$$I_{f_{\text{enc}}} = \mathbf{K}_f \cdot (\hat{n} \times \hat{l})l = (\mathbf{K}_f \times \hat{n}) \cdot \mathbf{l}$$

and hence

$$\boxed{\mathbf{H}_{1_\parallel} - \mathbf{H}_{2_\parallel} = \mathbf{K}_f \times \hat{n}} \tag{7.60}$$

So the *parallel* components of **H** are discontinuous by an amount proportional to the free surface current density.

Equations (7.57)-(7.60) constitute the general boundary conditions for electrodynamics. In the case of *linear* media, they can be expressed in terms of E and B alone:

$$\left.\begin{array}{llll} \text{(i)} & \epsilon_1 E_{1_\perp} - \epsilon_2 E_{2_\perp} = \sigma_f & \text{(iii)} & \mathbf{E}_{1_\parallel} - \mathbf{E}_{2_\parallel} = 0 \\ \text{(ii)} & B_{1_\perp} - B_{2_\perp} = 0 & \text{(iv)} & \dfrac{1}{\mu_1}\mathbf{B}_{1_\parallel} - \dfrac{1}{\mu_2}\mathbf{B}_{2_\parallel} = \mathbf{K}_f \times \hat{n} \end{array}\right\} \tag{7.61}$$

If there is no free charge or free current at the interface, then

$$\left.\begin{array}{llll}\text{(i)} & \epsilon_1 E_{1_\perp} - \epsilon_2 E_{2_\perp} = 0 & \text{(iii)} & \mathbf{E}_{1_\parallel} - \mathbf{E}_{2_\parallel} = 0 \\ \text{(ii)} & B_{1_\perp} - B_{2_\perp} = 0 & \text{(iv)} & \frac{1}{\mu_1}\mathbf{B}_{1_\parallel} - \frac{1}{\mu_2}\mathbf{B}_{2_\parallel} = 0\end{array}\right\} \tag{7.62}$$

As we shall see in the next chapter, these equations are the basis for the theory of reflection and refraction.

7.4 POTENTIAL FORMULATIONS OF ELECTRODYNAMICS

7.4.1 Scalar and Vector Potentials

In electrostatics, $\nabla \times \mathbf{E} = 0$ allowed us to write $\mathbf{E}$ as the gradient of a scalar potential: $\mathbf{E} = -\nabla V$. In electro*dynamics* this is no longer possible, because the curl of $\mathbf{E}$ is nonzero. But $\mathbf{B}$ remains divergenceless, so we can still write

$$\boxed{\mathbf{B} = \nabla \times \mathbf{A}} \tag{7.63}$$

as in magnetostatics. Putting this into Faraday's law yields

$$\nabla \times \mathbf{E} = -\frac{\partial}{\partial t}(\nabla \times \mathbf{A})$$

or

$$\nabla \times \left(\mathbf{E} + \frac{\partial \mathbf{A}}{\partial t}\right) = 0$$

Here is a quantity, unlike $\mathbf{E}$ alone, whose curl *does* vanish; it can therefore be written as the gradient of a scalar potential:

$$\mathbf{E} + \frac{\partial \mathbf{A}}{\partial t} = -\nabla V$$

In terms of V and $\mathbf{A}$, then,

$$\boxed{\mathbf{E} = -\nabla V - \frac{\partial \mathbf{A}}{\partial t}} \tag{7.64}$$

This reduces to the old form, of course, in the static case $(\partial \mathbf{A}/\partial t) = 0$.

The potential formulation (equations (7.63) and (7.64)) automatically fulfills the two homogeneous Maxwell equations, (ii) and (iii) (equation (7.41)). How about Gauss's law (i) and Ampère's law, with Maxwell's term, (iv)? Putting (7.64) into (i), we find that

$$\nabla^2 V + \frac{\partial}{\partial t}(\nabla \cdot \mathbf{A}) = -\frac{1}{\epsilon_0}\rho \tag{7.65}$$

This replaces Poisson's equation (to which it reduces in the static case). Putting (7.63) and (7.64) into (iv) yields

$$\nabla \times (\nabla \times \mathbf{A}) = \mu_0 \mathbf{J} - \mu_0\epsilon_0 \nabla\left(\frac{\partial V}{\partial t}\right) - \mu_0\epsilon_0 \frac{\partial^2 \mathbf{A}}{\partial t^2}$$

or, using the vector identity $\nabla \times (\nabla \times \mathbf{A}) = \nabla(\nabla \cdot \mathbf{A}) - \nabla^2 \mathbf{A}$ and rearranging terms a bit:

$$\left(\nabla^2 \mathbf{A} - \mu_0\epsilon_0 \frac{\partial^2 \mathbf{A}}{\partial t^2}\right) - \nabla\left(\nabla \cdot \mathbf{A} + \mu_0\epsilon_0 \frac{\partial V}{\partial t}\right) = -\mu_0 \mathbf{J} \tag{7.66}$$

Equations (7.65) and (7.66) carry all of the information in Maxwell's equations.

Example 14

Find the charge and current distributions that would give rise to the potentials $V = 0$,

$$\mathbf{A} = \begin{cases} \dfrac{\mu_0 \alpha}{4c}(ct - |x|)^2 \hat{k}, & \text{for } |x| < ct \\ 0, & \text{for } |x| > ct \end{cases}$$

(here α is some constant, and c is shorthand for $1/\sqrt{\epsilon_0\mu_0}$).

Solution: First we'll determine the electric and magnetic fields, using (7.63) and (7.64):

$$\mathbf{E} = -\frac{\partial \mathbf{A}}{\partial t} = -\frac{\mu_0 \alpha}{2}(ct - |x|)\hat{k}$$

$$\mathbf{B} = \nabla \times \mathbf{A} = -\frac{\mu_0 \alpha}{4c}\frac{\partial}{\partial x}(ct - |x|)^2 \hat{j} = \pm\frac{\mu_0 \alpha}{2c}(ct - |x|)\hat{j}$$

(plus for $x > 0$, minus for $x < 0$). These are for $|x| < ct$; when $|x| > ct$, $\mathbf{E} = \mathbf{B} = 0$ (Fig. 7.46).

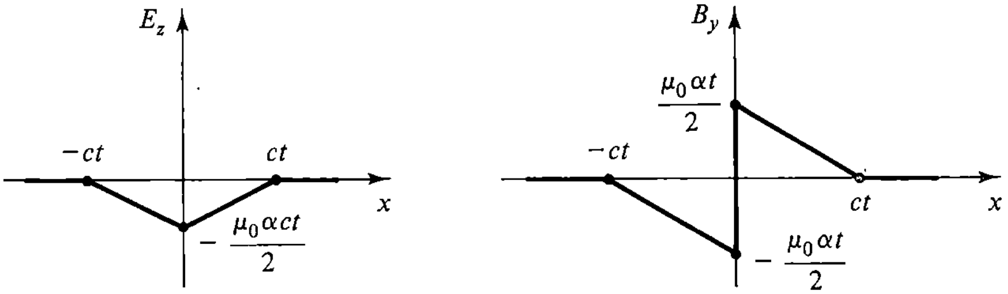

Figure 7.46

Calculating every derivative in sight, I find

$$\nabla \cdot \mathbf{E} = 0; \qquad \nabla \cdot \mathbf{B} = 0; \qquad \nabla \times \mathbf{E} = \mp\frac{\mu_0 \alpha}{2}\hat{j}; \qquad \nabla \times \mathbf{B} = -\frac{\mu_0 \alpha}{2c}\hat{k}$$

$$\frac{\partial \mathbf{E}}{\partial t} = -\frac{\mu_0 \alpha c}{2}\hat{k}; \qquad \frac{\partial \mathbf{B}}{\partial t} = \pm\frac{\mu_0 \alpha}{2}\hat{j}$$

As you can easily check, Maxwell's equations (7.41) are satisfied (the Ampère/Maxwell law holds in virtue of the condition $c^2 = 1/\mu_0\epsilon_0$), with ρ and $\mathbf{J}$ both *zero*. Notice, however, that $\mathbf{B}$ has a discontinuity at $x = 0$, and this signals the presence of a surface current $\mathbf{K}$ in the yz plane. Assuming there are no magnetic materials around, boundary condition (iv) (equation 7.61) gives

$$\alpha t\hat{j} = \mathbf{K} \times \hat{i}$$

and hence

$$\mathbf{K} = \alpha t\hat{k}$$

Evidently we have here a uniform surface current flowing in the z direction over the plane $x = 0$, which starts up at $t = 0$, and increases in proportion to t.

7.4.2 Gauge Transformations

Equations (7.65) and (7.66) are ugly, and you might be inclined at this stage to despair of the potential formulation altogether. However, we *have* succeeded in reducing six problems—finding $\mathbf{E}$ and $\mathbf{B}$ (three components each)—down to four: V (one component) and $\mathbf{A}$ (three more). Moreover, equations (7.63) and (7.64) do not uniquely define the potentials; we are free to impose extra conditions on V and $\mathbf{A}$, as long as nothing happens to $\mathbf{E}$ and $\mathbf{B}$. Let's work out precisely what this freedom entails. Suppose we have two sets of potentials, $(V, \mathbf{A})$ and $(V', \mathbf{A}')$, which correspond to the *same* electric and magnetic fields. By how much can they differ? Write

$$\mathbf{A}' = \mathbf{A} + \boldsymbol{\alpha} \quad \text{and} \quad V' = V + \beta$$

Since the two $\mathbf{A}$'s give the same $\mathbf{B}$, their curls must be equal (7.63), and hence

$$\nabla \times \boldsymbol{\alpha} = 0$$

We can therefore write $\boldsymbol{\alpha}$ as the gradient of some scalar:

$$\boldsymbol{\alpha} = \nabla\lambda$$

The two potentials also give the same $\mathbf{E}$, so (7.64) requires that

$$\nabla\beta + \frac{\partial\boldsymbol{\alpha}}{\partial t} = 0$$

or

$$\nabla\left(\beta + \frac{\partial\lambda}{\partial t}\right) = 0$$

The term in parentheses is therefore independent of position (it could, however, depend on time); call it $k(t)$:

$$\beta = -\frac{\partial\lambda}{\partial t} + k(t)$$

Actually, we may as well absorb $k(t)$ into λ, defining a new λ by adding $\int_0^t k(t')\,dt'$ to the old one. This will not affect the gradient of λ; it just adds $k(t)$ to $\partial\lambda/\partial t$. With this done,

$$\left.\begin{aligned} \mathbf{A}' &= \mathbf{A} + \nabla\lambda \\ V' &= V - \frac{\partial\lambda}{\partial t} \end{aligned}\right\} \tag{7.67}$$

Conclusion. For any old scalar function λ, we can with impunity add $\nabla\lambda$ to $\mathbf{A}$, provided we simultaneously subtract $\partial\lambda/\partial t$ from V. None of this will affect the physical quantities $\mathbf{E}$ and $\mathbf{B}$. Such changes in V and $\mathbf{A}$ are called **gauge transformations.** They can be exploited to adjust the divergence of $\mathbf{A}$, with a view to simplifying the "ugly" equations (7.65) and (7.66). In magnetostatics, it was best to choose $\nabla \cdot \mathbf{A} = 0$ (equation (5.55)); in electrodynamics the situation is not so clear-cut, and the most convenient gauge depends to some extent on the problem at hand. There are many famous gauges in the literature; I'll just show you the two most popular ones.

Problem 7.35 Find the charge and current distributions corresponding to

$$V(\mathbf{r}, t) = 0; \qquad \mathbf{A}(\mathbf{r}, t) = -\frac{1}{4\pi\epsilon_0}\frac{qt}{r^2}\hat{r}$$

Problem 7.36 Suppose $V = 0$ and $\mathbf{A} = A_0 \sin(\kappa x - \omega t)\hat{j}$, where A_0, ω, and κ are constants. Find $\mathbf{E}$ and $\mathbf{B}$, and check that they satisfy Maxwell's equations in vacuum. What condition must you impose on ω and κ?

Problem 7.37 Use the gauge function $\lambda = -(1/4\pi\epsilon_0)(qt/r)$ to transform the potentials in Problem 7.35, and comment on the result.

7.4.3 Coulomb Gauge and Lorentz Gauge

The Coulomb gauge. As in magnetostatics, we pick

$$\nabla \cdot \mathbf{A} = 0 \tag{7.68}$$

With this, equation (7.65) becomes

$$\nabla^2 V = -\frac{1}{\epsilon_0}\rho \tag{7.69}$$

This is just Poisson's equation. We know how to solve it: Setting $V = 0$ at infinity,

$$V = \frac{1}{4\pi\epsilon_0}\int \frac{\rho}{\mathscr{r}}\,d\tau \tag{7.70}$$

as in electrostatics. Don't be fooled, though—*unlike* electrostatics, V by itself doesn't tell you $\mathbf{E}$; you have to know $\mathbf{A}$ as well (see equation 7.64).

There is a peculiar thing about equation (7.70): it says that the potential everywhere is determined by the distribution of charge *right now.* If I move an electron in my laboratory, the potential V on the moon instantaneously records this change.

That sounds particularly odd in the light of special relativity, which allows no message to travel faster than the speed of light. The point is that V *by itself* is not a physically measurably quantity—all the man in the moon can measure is $\mathbf{E}$, and that involves $\mathbf{A}$ as well. Somehow it is built into the vector potential, in the Coulomb gauge, that whereas V instantaneously reflects all changes in ρ, the combination $-\nabla V - (\partial\mathbf{A}/\partial t)$ does *not*. $\mathbf{E}$ will change only after sufficient time has elapsed for the "news" to arrive.[13]

The *advantage* of the Coulomb gauge is that the *scalar* potential is particularly simple to calculate; the *disadvantage* (apart from the noncausal appearance of V mentioned above) is that $\mathbf{A}$ is particularly *difficult* to calculate. The differential equation for $\mathbf{A}$ (7.66) in the Coulomb gauge reads

$$\left(\nabla^2\mathbf{A} - \mu_0\epsilon_0\frac{\partial^2\mathbf{A}}{\partial t^2}\right) = -\mu_0\mathbf{J} + \mu_0\epsilon_0\nabla\left(\frac{\partial V}{\partial t}\right) \tag{7.71}$$

The Lorentz gauge. In the Lorentz gauge we pick

$$\nabla\cdot\mathbf{A} = -\mu_0\epsilon_0\frac{\partial V}{\partial t} \tag{7.72}$$

This is designed to eliminate the middle term in (7.66) entirely:

$$\nabla^2\mathbf{A} - \mu_0\epsilon_0\frac{\partial^2\mathbf{A}}{\partial t^2} = -\mu_0\mathbf{J} \tag{7.73}$$

Meanwhile, the equation for V, (7.65), becomes

$$\nabla^2 V - \mu_0\epsilon_0\frac{\partial^2 V}{\partial t^2} = -\frac{1}{\epsilon_0}\rho \tag{7.74}$$

The special virtue of the Lorentz gauge is that it treats V and $\mathbf{A}$ on an equal footing: The same differential operator,

$$\nabla^2 - \mu_0\epsilon_0\frac{\partial^2}{\partial t^2} \equiv \Box^2 \tag{7.75}$$

(called the **d'Alembertian**) occurs in both equations:

$$\left.\begin{aligned}\Box^2\mathbf{A} &= -\mu_0\mathbf{J},\\ \Box^2 V &= -\frac{1}{\epsilon_0}\rho.\end{aligned}\right\} \tag{7.76}$$

This democratic treatment of V and $\mathbf{A}$ is particularly nice in the context of special relativity, where the d'Alembertian plays somewhat the same role as the Laplacian in classical physics, and (7.76) may be regarded as a four-dimensional version of Poisson's equation. In this book we'll use the Lorentz gauge exclusively; the whole of electrodynamics reduces, then, to the problem of solving 7.76 for specified $\mathbf{J}$ and ρ—we'll take up this question in Chapter 9.

[13]See O. L. Brill and B. Goodman, *Am. J. Phys.* **35**, 832 (1967).

Problem 7.38 Which of the potentials in Example 14, Problem 7.35, and Problem 7.36 are in the Coulomb gauge? Which are in the Lorentz gauge? (Notice that these gauges are not mutually exclusive.)

Problem 7.39 In Chapter 5, I showed that it is always possible to pick a vector potential whose divergence is zero (Coulomb gauge). Show that it is always possible to choose $\nabla \cdot \mathbf{A} = -\mu_0\epsilon_0(\partial V/\partial t)$, as required for the Lorentz gauge, assuming you know how to solve equations of the form (7.76). Is it always possible to pick $V = 0$? How about $\mathbf{A} = 0$?

7.4.4 Lorentz Force Law in Potential Form

It is sometimes useful to express the Lorentz force law in terms of potentials:

$$\mathbf{F} = \frac{d\mathbf{p}}{dt} = q(\mathbf{E} + \mathbf{v} \times \mathbf{B}) = q\left[-\nabla V - \frac{\partial \mathbf{A}}{\partial t} + \mathbf{v} \times (\nabla \times \mathbf{A})\right]$$

Now, product rule 4 says

$$\nabla(\mathbf{v}\cdot\mathbf{A}) = \mathbf{v} \times (\nabla \times \mathbf{A}) + (\mathbf{v}\cdot\nabla)\mathbf{A}$$

for $\mathbf{v}$, the velocity of the particle, is a function of *time* but not *position*. Thus,

$$\frac{d\mathbf{p}}{dt} = -q\left[\frac{\partial \mathbf{A}}{\partial t} + (\mathbf{v}\cdot\nabla)\mathbf{A} + \nabla(V - \mathbf{v}\cdot\mathbf{A})\right]$$

The combination

$$\left[\frac{\partial \mathbf{A}}{\partial t} + (\mathbf{v}\cdot\nabla)\mathbf{A}\right]$$

is called the **convective derivative** of $\mathbf{A}$ and written $d\mathbf{A}/dt$ (*total* derivative). It represents the time rate of change of $\mathbf{A}$ *at the location of the moving particle.* For suppose that at time t the particle is at point $\mathbf{r}$ (Fig. 7.47), where the potential is $\mathbf{A}(\mathbf{r}, t)$, and a moment dt later it is at $(\mathbf{r} + \mathbf{v}\,dt)$, where the potential is $\mathbf{A}(\mathbf{r} + \mathbf{v}\,dt, t + dt)$. The *change* in $\mathbf{A}$, then, is

$$\begin{aligned} d\mathbf{A} &= \mathbf{A}(\mathbf{r} + \mathbf{v}\,dt, t + dt) - \mathbf{A}(\mathbf{r}, t) \\ &= \left(\frac{\partial \mathbf{A}}{\partial x}\right)(v_x\,dt) + \left(\frac{\partial \mathbf{A}}{\partial y}\right)(v_y\,dt) + \left(\frac{\partial \mathbf{A}}{\partial z}\right)(v_z\,dt) + \left(\frac{\partial \mathbf{A}}{\partial t}\right)dt \end{aligned}$$

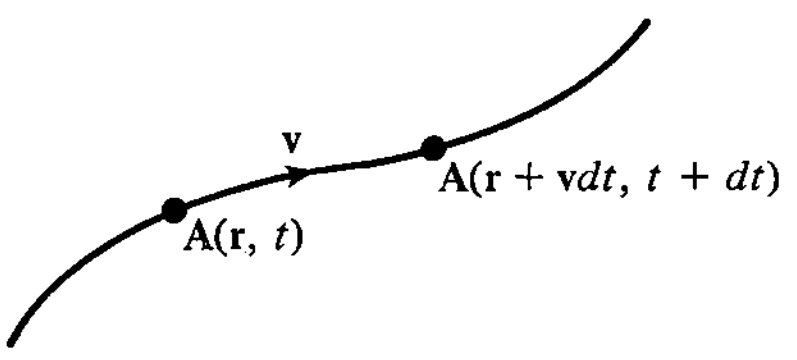

Figure 7.47

so that

$$\frac{d\mathbf{A}}{dt} = \frac{\partial \mathbf{A}}{\partial t} + (\mathbf{v}\cdot\nabla)\mathbf{A} \tag{7.77}$$

As the charge moves, the potential it "feels" changes for two reasons: first, because the potential itself varies with *time* and, second, because the particle is now in a new location where **A** is different because of its variation in *space*. Hence, the two terms in equation (7.77).

With the aid of the convective derivative, the Lorentz force law reads:

$$\frac{d}{dt}(\mathbf{p} + q\mathbf{A}) = -\nabla[q(V - \mathbf{v}\cdot\mathbf{A})] \tag{7.78}$$

This is reminiscent of the formula for the motion of a particle whose potential energy U is a specified function of position:

$$\frac{d\mathbf{p}}{dt} = -\nabla U$$

Playing the role of **p** is the **canonical momentum,**

$$\mathbf{p}_{\text{can}} = \mathbf{p} + q\mathbf{A} \tag{7.79}$$

while the part of U is taken by the velocity-dependent quantity

$$U = q(V - \mathbf{v}\cdot\mathbf{A}) \tag{7.80}$$

Problem 7.40 The vector potential for a uniform magnetostatic field is $\mathbf{A} = -\frac{1}{2}(\mathbf{r}\times\mathbf{B})$ (see Problem 5.27). Show that $d\mathbf{A}/dt = -\frac{1}{2}(\mathbf{v}\times\mathbf{B})$, in this case, and confirm that (7.78) yields the correct equation of motion.

7.5 ENERGY AND MOMENTUM IN ELECTRODYNAMICS

7.5.1 Newton's Third Law in Electrodynamics

Imagine a point charge q traveling in along the x axis at a constant speed v. Because it is moving, its electric field is *not* given by Coulomb's law; nevertheless, **E** still points radially outward from the instantaneous position of the charge (Fig. 7.48(a)), as we'll see in Chapter 9. Since, moreover, a moving point charge does not constitute a steady current, its magnetic field is *not* given by the Biot-Savart law. Nevertheless, it's a fact that **B** still circles around the axis in a manner suggested by the right-hand rule (Fig. 7.48(b)); again, the proof will come in Chapter 9.

Now suppose this charge encounters an identical one, proceeding in at the same speed along the y axis. Of course, the electromagnetic force between them would tend to drive them off the axes, but let's assume they're mounted on tracks, or something, so they're forced to maintain the same direction and the same speed (Fig. 7.49). The electrical force between them is repulsive, but how about the magnetic force? Well the magnetic field of q_1 points into the page (at the position of q_2), so the

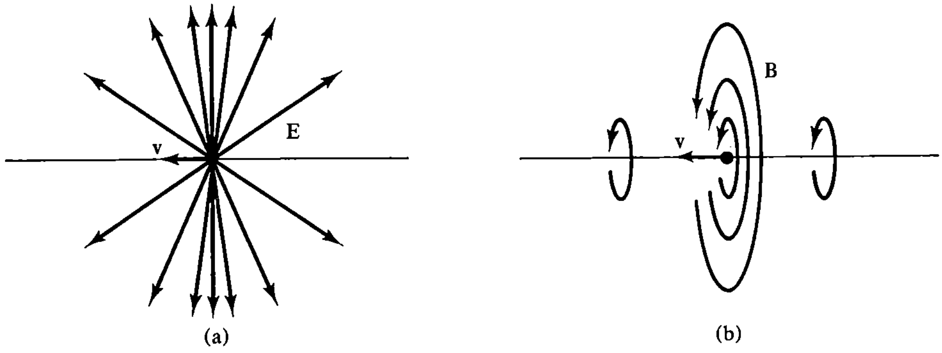

Figure 7.48

magnetic force on q_2 is toward the *right,* whereas the magnetic field of q_2 is *out* of the page (at the position of q_1), and the magnetic force on q_1 is *upward.* Now, here's the point: *The electromagnetic force of q_1 on q_2 is equal but not opposite to the force of q_2 on q_1, in violation of Newton's third law.* In electro*statics* and magneto*statics* the third law holds, but in electrodynamics it does *not.*

Well, that's an interesting curiosity, but then how often does one actually use the third law, in practice? Answer: All the time! For the proof of conservation of momentum rests on the cancellation of internal forces, which follows from the third law. When you tamper with the third law, then, you are placing conservation of momentum in jeopardy, and there is no principle in physics more sacred than *that.*

Momentum conservation is rescued in electrodynamics by the realization that *the fields themselves carry momentum.* This is not so surprising when you consider that we have already attributed *energy* to the fields. In the case of the two point charges in Fig. 7.49, whatever momentum is lost to the particles is gained by the fields. Only when the field momentum is added to the mechanical momentum of the charges is momentum conservation restored. My aim in Section 7.5 is to prove the laws of conservation of momentum and energy, and in the course of this to derive formulas for the energy, momentum and angular momentum stored in the fields.

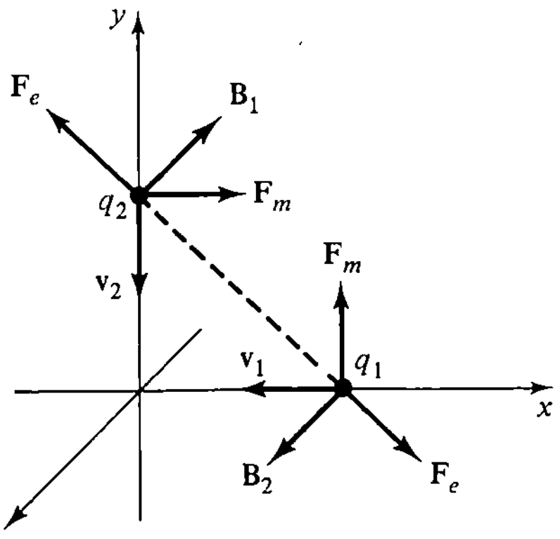

Figure 7.49

7.5.2 Poynting's Theorem

In Chapter 2, we found that the work necessary to assemble a static charge distribution (against the Coulomb repulsion of like charges) is (equation (2.39))

$$W_E = \frac{\epsilon_0}{2}\int E^2\,d\tau$$

where **E** is the resulting electric field. Likewise, the work required to get currents going (against the back emf) is (equation (7.35))

$$W_B = \frac{1}{2\mu_0}\int B^2\,d\tau$$

where **B** is the resulting magnetic field. Evidently, the total energy stored in electromagnetic fields is

$$W_{EB} = \frac{1}{2}\int\left(\epsilon_0 E^2 + \frac{1}{\mu_0}B^2\right)d\tau \tag{7.81}$$

I want to derive (7.81) more generally, now, in the context of the energy conservation law for electrodynamics.

Suppose we have some charge and current configuration which, at time t, produces fields **E** and **B**. In the next instant, dt, the charges move around a bit. *Question*: How much work, dW, is done by the electromagnetic forces on these charges in the interval dt? According to the Lorentz force law (5.2), the work done on an element of charge dq is

$$\mathbf{F}\cdot d\mathbf{l} = dq(\mathbf{E} + \mathbf{v}\times\mathbf{B})\cdot\mathbf{v}\,dt = \mathbf{E}\cdot\mathbf{v}\,dq\,dt$$

Now, $dq = \rho\,d\tau$ and $\rho\mathbf{v} = \mathbf{J}$, so the total work done on all the charges in some volume V is given by

$$\frac{dW}{dt} = \int_V (\mathbf{E}\cdot\mathbf{J})\,d\tau \tag{7.82}$$

Thus $\mathbf{E}\cdot\mathbf{J}$ is the work done per unit time, per unit volume—which is to say, the *power* delivered per unit volume. We can express this quantity in terms of the fields alone, using Ampère's law (with Maxwell's extra term) to eliminate **J**:

$$\mathbf{E}\cdot\mathbf{J} = \frac{1}{\mu_0}\mathbf{E}\cdot(\nabla\times\mathbf{B}) - \epsilon_0\mathbf{E}\cdot\frac{\partial\mathbf{E}}{\partial t}$$

From product rule 6,

$$\nabla\cdot(\mathbf{E}\times\mathbf{B}) = \mathbf{B}\cdot(\nabla\times\mathbf{E}) - \mathbf{E}\cdot(\nabla\times\mathbf{B})$$

Invoking Faraday's law ($\nabla\times\mathbf{E} = -\partial\mathbf{B}/\partial t$) it follows that

$$\mathbf{E}\cdot(\nabla\times\mathbf{B}) = -\mathbf{B}\cdot\frac{\partial\mathbf{B}}{\partial t} - \nabla\cdot(\mathbf{E}\times\mathbf{B})$$

Meanwhile,

$$\mathbf{B}\cdot\frac{\partial\mathbf{B}}{\partial t}=\frac{1}{2}\frac{\partial}{\partial t}(B^2),\qquad \mathbf{E}\cdot\frac{\partial\mathbf{E}}{\partial t}=\frac{1}{2}\frac{\partial}{\partial t}(E^2)$$

so that

$$\mathbf{E}\cdot\mathbf{J}=-\frac{1}{2}\frac{\partial}{\partial t}\left(\epsilon_0E^2+\frac{1}{\mu_0}B^2\right)-\frac{1}{\mu_0}\nabla\cdot(\mathbf{E}\times\mathbf{B}) \tag{7.83}$$

Putting (7.83) into (7.82) and applying the divergence theorem to the second term, we have

$$\frac{dW}{dt}=-\frac{d}{dt}\int_V\frac{1}{2}\left(\epsilon_0E^2+\frac{1}{\mu_0}B^2\right)d\tau-\frac{1}{\mu_0}\oint_S(\mathbf{E}\times\mathbf{B})\cdot d\mathbf{a}, \tag{7.84}$$

where S is the surface bounding V. This is **Poynting's theorem;** it is the work-energy theorem of electrodynamics. The first integral on the right is the total energy stored in the fields, W_{EB} (equation (7.81)). The second term evidently represents the rate at which energy is carried out of V, across its boundary surface, by the electromagnetic fields. Poynting's theorem says, then, that *the work done on the charges by the electromagnetic force is equal to the decrease in energy stored in the field, less the energy which flowed out through the surface.*

The *energy per unit time, per unit area,* transported by the fields is called the **Poynting vector:**

$$\boxed{\mathbf{S}=\frac{1}{\mu_0}(\mathbf{E}\times\mathbf{B})} \tag{7.85}$$

specifically, $\mathbf{S}\cdot d\mathbf{a}$ is the energy per unit time crossing the infinitesimal surface $d\mathbf{a}$—the energy *flux*, if you like (so $\mathbf{S}$ is the *energy flux density*). We will see many applications of the Poynting vector in the next two chapters, but for the moment I am mainly interested in using it to express Poynting's theorem more compactly:

$$\frac{dW}{dt}=-\frac{dW_{EB}}{dt}-\oint_S\mathbf{S}\cdot d\mathbf{a} \tag{7.86}$$

Of course, the work W done on the charges will increase their mechanical energy (kinetic, potential, or whatever). If we let U_M denote the mechanical energy density, so that

$$\frac{dW}{dt}=\frac{d}{dt}\int_V U_M\,d\tau \tag{7.87}$$

and use U_{EB} for the energy density of the fields,

$$\boxed{U_{EB} = \frac{1}{2}\left(\epsilon_0 E^2 + \frac{1}{\mu_0} B^2\right)} \tag{7.88}$$

then

$$\frac{d}{dt}\int_V (U_M + U_{EB})\, d\tau = -\oint_S \mathbf{S}\cdot d\mathbf{a} = -\int_V (\nabla\cdot\mathbf{S})\, d\tau$$

and hence

$$\boxed{\nabla\cdot\mathbf{S} = -\frac{\partial}{\partial t}(U_M + U_{EB})} \tag{7.89}$$

This is the differential version of Poynting's theorem. You might compare it with the continuity equation expressing conservation of charge:

$$\nabla\cdot\mathbf{J} = -\frac{\partial}{\partial t}\rho$$

The charge density is replaced by the total energy density (mechanical plus electromagnetic), whereas the current density is replaced by the Poynting vector. **S** describes the flow of *energy*, then, in the same way that **J** describes the flow of *charge*.[14]

Example 15

When current flows down a wire, electromagnetic work is done, which shows up as Joule heating of the wire (equation (7.6)). Though there are *easier* ways to do it, the energy per unit time delivered to the wire can be calculated using the Poynting vector. Assuming it's uniform, the electric field parallel to the wire is

$$E = \frac{V}{L}$$

where V is the potential difference between the ends and L is the length of the wire (Fig. 7.50). The magnetic field is "circumferential"; at the surface (radius R) it has the value

$$B = \frac{\mu_0 I}{2\pi R}$$

[14]In the presence of linear dielectric, diamagnetic, or paramagnetic materials, one is typically interested only in the work done on *free* charges and currents (see Section 4.4.3). In that case the appropriate energy density is $\frac{1}{2}(\mathbf{E}\cdot\mathbf{D} + \mathbf{B}\cdot\mathbf{H})$, and the Poynting vector becomes $(\mathbf{E}\times\mathbf{H})$. See J. D. Jackson, *Classical Electrodynamics*, 2nd ed. (New York: John Wiley, 1975, Section 6.9).

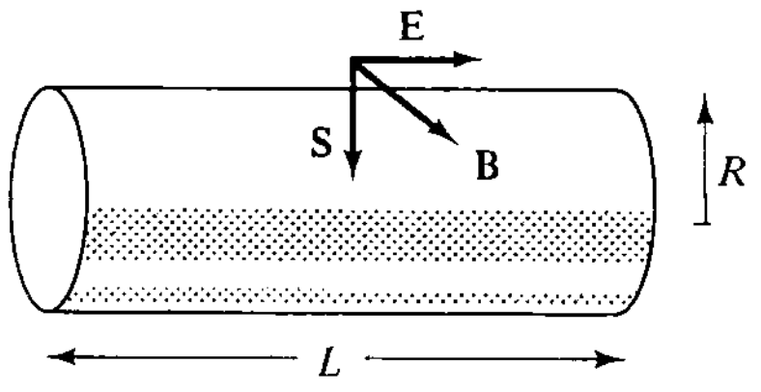

Figure 7.50

Accordingly, the Poynting vector is

$$S \doteq \frac{1}{\mu_0} \frac{V}{L} \frac{\mu_0 I}{2\pi R} = \frac{VI}{2\pi RL}$$

pointing radially inward. The energy per unit time passing in through the surface of the wire is therefore

$$\int \mathbf{S} \cdot d\mathbf{a} = S(2\pi RL) = VI$$

which is exactly what we concluded, on rather more direct grounds, in Section 7.1.

Problem 7.41 Calculate the power (energy per unit time) transported down the cables of Example 13 and Problem 7.29, assuming the two conductors are held at a potential difference V, and carry current I (down one and back up the other).

Problem 7.42 For the configuration in Example 14, consider a rectangular box of length l, width w, and height h, situated a distance d above the yz plane (Fig. 7.51).

(a) Find the energy in the box at time $t_1 = d/c$, and at $t_2 = (d + h)/c$.

(b) Find the Poynting vector, and determine the energy per unit time flowing into the box during the interval $t_1 < t < t_2$.

(c) Integrate the result in (b) from t_1 to t_2 and confirm that the increase in energy (a) equals the net influx.

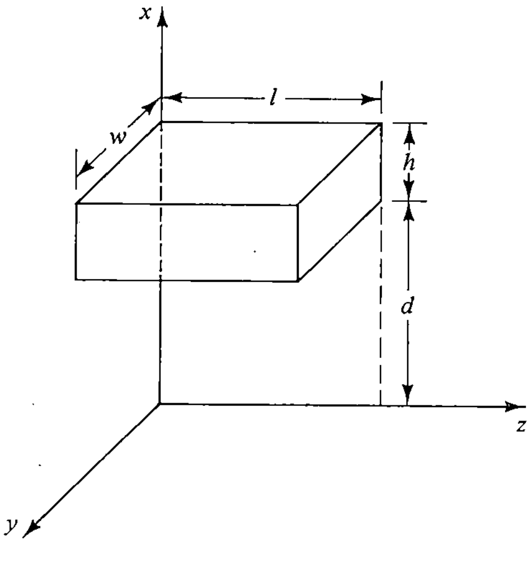

Figure 7.51

7.5.3 Maxwell's Stress Tensor

Let's calculate now the total electromagnetic force on the charges in volume V:

$$\mathbf{F} = \int_V (\mathbf{E} + \mathbf{v} \times \mathbf{B})\rho \, d\tau = \int_V (\rho\mathbf{E} + \mathbf{J} \times \mathbf{B}) \, d\tau \tag{7.90}$$

The *force per unit volume* is evidently

$$\mathfrak{F} = \rho\mathbf{E} + \mathbf{J} \times \mathbf{B} \tag{7.91}$$

As before, we propose to write this in terms of the fields alone, eliminating ρ and $\mathbf{J}$ by using Maxwell's equations (i) and (iv) (equation (7.41)):

$$\mathfrak{F} = \epsilon_0(\nabla \cdot \mathbf{E})\mathbf{E} + \left(\frac{1}{\mu_0}\nabla \times \mathbf{B} - \epsilon_0\frac{\partial \mathbf{E}}{\partial t}\right) \times \mathbf{B}$$

Now

$$\frac{\partial}{\partial t}(\mathbf{E} \times \mathbf{B}) = \frac{\partial \mathbf{E}}{\partial t} \times \mathbf{B} + \mathbf{E} \times \frac{\partial \mathbf{B}}{\partial t}$$

whereas Faraday's law gives

$$\frac{\partial \mathbf{B}}{\partial t} = -\nabla \times \mathbf{E}$$

so

$$\frac{\partial \mathbf{E}}{\partial t} \times \mathbf{B} = \frac{\partial}{\partial t}(\mathbf{E} \times \mathbf{B}) + \mathbf{E} \times (\nabla \times \mathbf{E})$$

Thus,

$$\mathfrak{F} = \epsilon_0[(\nabla \cdot \mathbf{E})\mathbf{E} - \mathbf{E} \times (\nabla \times \mathbf{E})] - \frac{1}{\mu_0}[\mathbf{B} \times (\nabla \times \mathbf{B})] - \epsilon_0\frac{\partial}{\partial t}(\mathbf{E} \times \mathbf{B})$$

Just to make things look more symmetrical, let's throw in a term $(\nabla \cdot \mathbf{B})\mathbf{B}$ within the second square bracket—since $\nabla \cdot \mathbf{B} = 0$, this costs us nothing. Meanwhile, product rule 4 says

$$\nabla(E^2) = 2(\mathbf{E} \cdot \nabla)\mathbf{E} + 2\mathbf{E} \times (\nabla \times \mathbf{E})$$

so

$$\mathbf{E} \times (\nabla \times \mathbf{E}) = \tfrac{1}{2}\nabla(E^2) - (\mathbf{E} \cdot \nabla)\mathbf{E}$$

and the same for $\mathbf{B}$. Therefore,

$$\mathfrak{F} = \epsilon_0[(\nabla \cdot \mathbf{E})\mathbf{E} + (\mathbf{E} \cdot \nabla)\mathbf{E}] + \frac{1}{\mu_0}[(\nabla \cdot \mathbf{B})\mathbf{B} + (\mathbf{B} \cdot \nabla)\mathbf{B}]$$
$$- \frac{1}{2}\nabla\left(\epsilon_0 E^2 + \frac{1}{\mu_0}B^2\right) - \epsilon_0\frac{\partial}{\partial t}(\mathbf{E} \times \mathbf{B}) \tag{7.91}$$

Ugly! But it can be simplified by introducing the **Maxwell stress tensor,**

$$T_{ij} \equiv \epsilon_0\left(E_iE_j - \frac{1}{2}\,\delta_{ij}E^2\right) + \frac{1}{\mu_0}\left(B_iB_j - \frac{1}{2}\,\delta_{ij}B^2\right) \tag{7.92}$$

The indices i and j refer to the coordinates x, y, and z, so the stress tensor has a total of nine components (T_{xx}, T_{yy}, T_{xz}, T_{yx}, and so on). The **Kronecker delta,** δ_{ij}, is defined to be 1 if the indices are the same ($\delta_{xx} = \delta_{yy} = \delta_{zz} = 1$) and zero otherwise ($\delta_{xy} = \delta_{xz} = \delta_{yz} = 0$). Thus,

$$T_{xx} = \frac{1}{2}\,\epsilon_0(E_x^2 - E_y^2 - E_z^2) + \frac{1}{2\mu_0}\,(B_x^2 - B_y^2 - B_z^2)$$

$$T_{xy} = \epsilon_0(E_xE_y) + \frac{1}{\mu_0}\,(B_xB_y)$$

and so on. Because it carries *two* indices, where a vector has only one, T_{ij} is sometimes written with a double arrow: $\overleftrightarrow{\mathbf{T}}$. One can form the dot product of $\mathbf{T}$ with a vector $\mathbf{a}$:

$$(\mathbf{a}\cdot\overleftrightarrow{\mathbf{T}})_j = \sum_{i=x,y,z} a_iT_{ij} \tag{7.93}$$

The resulting object is itself a vector, having one remaining index. In particular, the divergence of $\overleftrightarrow{\mathbf{T}}$ has for its jth component

$$(\nabla\cdot\overleftrightarrow{\mathbf{T}})_j = \epsilon_0\left[(\nabla\cdot\mathbf{E})E_j + (\mathbf{E}\cdot\nabla)E_j - \frac{1}{2}\,\nabla_jE^2\right]$$

$$+ \frac{1}{\mu_0}\left[(\nabla\cdot\mathbf{B})B_j + (\mathbf{B}\cdot\nabla)B_j - \frac{1}{2}\,\nabla_jB^2\right]$$

Thus the force per unit volume (7.91) can be written

$$\mathfrak{F} = \nabla\cdot\overleftrightarrow{\mathbf{T}} - \epsilon_0\mu_0\frac{\partial\mathbf{S}}{\partial t} \tag{7.94}$$

where $\mathbf{S}$ is the Poynting vector (7.85).

The *total* force on V (equation (7.90)) is therefore

$$\mathfrak{F} = \oint_S \overleftrightarrow{\mathbf{T}}\cdot d\mathbf{a} - \epsilon_0\mu_0\frac{d}{dt}\int_V \mathbf{S}\,d\tau \tag{7.95}$$

(I used the divergence theorem to convert the first term to a surface integral). In the *static* case (or, more generally, whenever $\int\mathbf{S}\,d\tau$ is independent of time), the second term drops out, and the electromagnetic force on the charge configuration can be expressed entirely in terms of the stress tensor at the boundary. Physically, $\overleftrightarrow{\mathbf{T}}$ is the force per unit area (or **stress**) acting on the surface. More precisely, T_{ij} is the force (per unit area) in the ith direction acting on an element of surface oriented in the jth direction—"diagonal" elements (T_{xx}, T_{yy}, T_{zz}) represent *pressures,* and "off-diagonal" elements (T_{xy}, T_{xz}, etc.) are *shears.*

Example 16

Determine the net force on the "northern" hemisphere of a uniformly charged solid sphere of radius R and charge Q (the same as Problem 2.44, only this time we'll use the Maxwell stress tensor and equation (7.95)).

Solution: The boundary surface consists of two parts—a hemispherical "bowl" at radius R, and a circular disc at $\theta = \pi/2$ (Fig. 7.52). For the bowl,

$$d\mathbf{a} = R^2 \sin\theta \, d\theta \, d\phi \, \hat{r}$$

and

$$\mathbf{E} = \frac{1}{4\pi\epsilon_0} \frac{Q}{R^2} \hat{r}$$

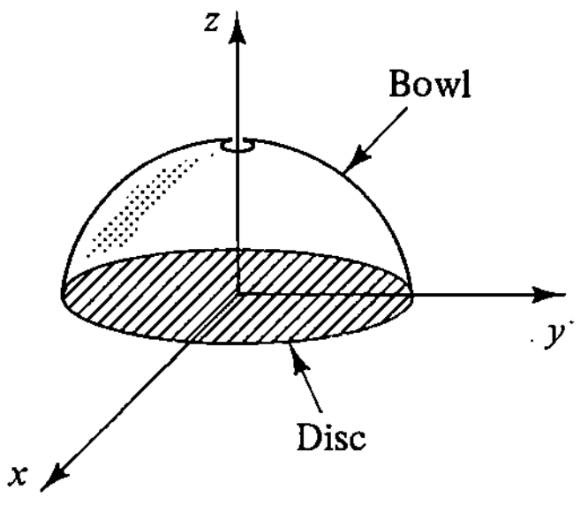

Figure 7.52

In Cartesian components,

$$\hat{r} = \sin\theta \cos\phi \, \hat{\imath} + \sin\theta \sin\phi \, \hat{\jmath} + \cos\theta \, \hat{k}$$

so

$$T_{zx} = \epsilon_0 E_z E_x = \epsilon_0 \left(\frac{Q}{4\pi\epsilon_0 R^2}\right)^2 \sin\theta \cos\theta \cos\phi$$

$$T_{zy} = \epsilon_0 E_z E_y = \epsilon_0 \left(\frac{Q}{4\pi\epsilon_0 R^2}\right)^2 \sin\theta \cos\theta \sin\phi \tag{7.96}$$

$$T_{zz} = \frac{\epsilon_0}{2}(E_z^2 - E_x^2 - E_y^2) = \frac{\epsilon_0}{2}\left(\frac{Q}{4\pi\epsilon_0 R^2}\right)^2 (\cos^2\theta - \sin^2\theta)$$

The net force is obviously in the z-direction, so it suffices to calculate

$$(\overleftrightarrow{\mathbf{T}} \cdot d\mathbf{a})_z = T_{zx} da_x + T_{zy} da_y + T_{zz} da_z = \frac{\epsilon_0}{2}\left(\frac{Q}{4\pi\epsilon_0 R}\right)^2 \sin\theta \cos\theta \, d\theta \, d\phi$$

The force on the "bowl" is therefore

$$F_{\text{bowl}} = \frac{\epsilon_0}{2}\left(\frac{Q}{4\pi\epsilon_0 R}\right)^2 2\pi \int_0^{\pi/2} \sin\theta\cos\theta\, d\theta = \frac{1}{4\pi\epsilon_0}\left(\frac{Q^2}{8R^2}\right)$$

Meanwhile, for the equatorial disc,

$$d\mathbf{a} = -r\, dr\, d\phi\, \hat{k} \tag{7.98}$$

and (since we are now *inside* the sphere)

$$\mathbf{E} = \frac{1}{4\pi\epsilon_0}\frac{Q}{R^3}\mathbf{r} = \frac{1}{4\pi\epsilon_0}\frac{Q}{R^3} r(\cos\phi\,\hat{\imath} + \sin\phi\,\hat{\jmath})$$

Thus

$$T_{zz} = \frac{\epsilon_0}{2}(E_z^2 - E_x^2 - E_y^2) = -\frac{\epsilon_0}{2}\left(\frac{Q}{4\pi\epsilon_0 R^3}\right)^2 r^2$$

and hence

$$(\overleftrightarrow{\mathbf{T}}\cdot d\mathbf{a})_z = \frac{\epsilon_0}{2}\left(\frac{Q}{4\pi\epsilon_0 R^3}\right)^2 r^3\, dr\, d\phi$$

The force on the disc is therefore

$$F_{\text{disc}} = \frac{\epsilon_0}{2}\left(\frac{Q}{4\pi\epsilon_0 R^3}\right)^2 2\pi \int_0^R r^3\, dr = \frac{1}{4\pi\epsilon_0}\left(\frac{Q^2}{16R^2}\right) \tag{7.99}$$

Combining (7.97) and (7.99), I conclude that the net force on the northern hemisphere is

$$F = \frac{1}{4\pi\epsilon_0}\left(\frac{3Q^2}{16R^2}\right) \tag{7.100}$$

Incidentally, in applying (7.95) *any* volume that encloses all of the charge in question (and no *other* charge) will do the job. For example, in the present case we could use the whole region $z > 0$. In the case the boundary surface consists of the entire xy plane (plus a hemisphere at $r = \infty$—but $E = 0$ out there anyway, so it contributes nothing). In place of the "bowl," we now have the outer portion of the plane ($r > R$). Here

$$T_{zz} = -\frac{\epsilon_0}{2}\left(\frac{Q}{4\pi\epsilon_0}\right)^2 \frac{1}{r^4}$$

(equation (7.96) with $\theta = \pi/2$ and $R \to r$), and $d\mathbf{a}$ is given by (7.98), so

$$(\overleftrightarrow{\mathbf{T}}\cdot d\mathbf{a})_z = \frac{\epsilon_0}{2}\left(\frac{Q}{4\pi\epsilon_0}\right)^2 \frac{1}{r^3}\, dr\, d\phi$$

and

$$F_{\text{plane}(r>R)} = \frac{\epsilon_0}{2}\left(\frac{Q}{4\pi\epsilon_0}\right)^2 2\pi \int_R^\infty \frac{1}{r^3}\, dr = \frac{1}{4\pi\epsilon_0}\left(\frac{Q^2}{8R^2}\right)$$

the same as for the bowl (7.97).

Problem 7.43 Calculate the force of magnetic attraction between the northern and southern hemispheres of a uniformly charged spinning spherical shell, with radius R, angular velocity ω, and surface charge density σ. (This is the same as Problem 5.52, but this time use the Maxwell stress tensor and equation (7.95).)

Problem 7.44

(a) Consider two equal point charges q, separated by a distance r. Construct the plane equidistant from the two charges. By integrating Maxwell's stress tensor over this plane, determine the force of one charge on the other.

(b) Do the same for charges that are opposite in sign.

7.5.4 Conservation of Momentum

According to Newton's second law, the force on an object is equal to the rate of charge of its momentum:

$$\mathbf{F} = \frac{d\mathbf{p}}{dt}$$

In electrodynamics, then, we have (equation 7.95)

$$\frac{d\mathbf{p}}{dt} = -\epsilon_0\mu_0 \frac{d}{dt}\int_V \mathbf{S}\, d\tau + \oint_S \overleftrightarrow{\mathbf{T}}\cdot d\mathbf{a} \tag{7.101}$$

This equation is similar in structure to Poynting's theorem (7.86), and it admits an analogous interpretation: the first integral represents *momentum stored in the electromagnetic fields themselves:*

$$\mathbf{p}_{EB} = \mu_0\epsilon_0 \int_V \mathbf{S}\, d\tau \tag{7.102}$$

while the second integral is the momentum per unit time flowing in through the surface. Thus equation (7.101) is the general statement of *conservation of momentum* in electrodynamics: Any increase in the *total* momentum (mechanical plus electromagnetic) is accounted for in the form of momentum carried in by the fields. (If V is *all* of space, then no momentum flows in or out, and $\mathbf{p} + \mathbf{p}_{EB}$ is constant.)

As in the case of conservation of charge and conservation of energy, conservation of momentum can be given a differential formulation. Let $\wp$ be the density of *mechanical* momentum, and $\wp_{EB}$ the density of momentum in the fields:

$$\boxed{\wp_{EB} = \mu_0\epsilon_0\, \mathbf{S}} \tag{7.103}$$

Then (7.101) is equivalent to

$$\boxed{\nabla \cdot (-\overleftrightarrow{\mathbf{T}}) = -\frac{\partial}{\partial t}(\wp + \wp_{EB}).} \tag{7.104}$$

Evidently $(-\overleftrightarrow{\mathbf{T}})$ is the **momentum flux density**, playing the role of **J** (charge flux density) in the continuity equation or **S** (energy flux density) in Poynting's theorem. Specifically, $-T_{ij}$ is the momentum if the i direction crossing a surface oriented in the j direction, per unit area per unit time. Notice that Poynting's vector has appeared in two quite different guises: **S** itself is the energy per unit area, per unit time, transported by the electromagnetic fields, while $\mu_0\epsilon_0\mathbf{S}$ is the momentum per unit volume stored in those fields. Similarly, $\overleftrightarrow{\mathbf{T}}$ plays a dual role: $\overleftrightarrow{\mathbf{T}}$ itself is the electromagnetic stress acting on a surface, and $-\overleftrightarrow{\mathbf{T}}$ describes the flow of momentum carried by the fields.

By this time, the electromagnetic fields (which started out as mediators of forces between charges) have taken on a life of their own. They carry *energy* (7.88)

$$U_{EB} = \frac{1}{2}\left(\epsilon_0 E^2 + \frac{1}{\mu_0}B^2\right),$$

and *momentum* (7.103)

$$\wp_{EB} = \mu_0\epsilon_0\mathbf{S} = \epsilon_0(\mathbf{E} \times \mathbf{B}),$$

and, for that matter, *angular* momentum:

$$\mathcal{L}_{EB} = \mathbf{r} \times \wp_{EB} = \epsilon_0\mathbf{r} \times (\mathbf{E} \times \mathbf{B}) \tag{7.105}$$

Even perfectly static fields carry momentum, as long as $\mathbf{E} \times \mathbf{B}$ is nonzero, and it is only when these field contributions are included that the classical conservation laws hold.

Example 17

Imagine a very long solenoid with radius R, N turns per unit length, and current I. Coaxial with the solenoid are two long cylindrical shells of length l—one, *inside* the solenoid at radius a, carries a charge $+Q$, uniformly distributed over its surface; the other, *outside* the solenoid at radius b, carries charge $-Q$ (see Fig. 7.53; l is supposed to be much greater than b). When the current in the solenoid is gradually reduced, the cylinders begin to rotate, as we found in Example 7. *Question:* Where does the angular momentum come from?[15]

[15]This is a variant on the "Feynman disk paradox" (R. P. Feynman, R. B. Leighton, and M. Sands, *The Feynman Lectures,* (Reading, Mass.: Addison-Wesley, 1964, vol 2, pp. 17-5) suggested by F. L. Boos, Jr. (*Am. J. Phys.* **52,** 756 (1984)). A similar model was proposed earlier by R. H. Romer (*Am. J. Phys.* **34,** 772 (1966)). For further references, see T.-C. E. Ma, *Am. J. Phys.* **54,** 949 (1986).

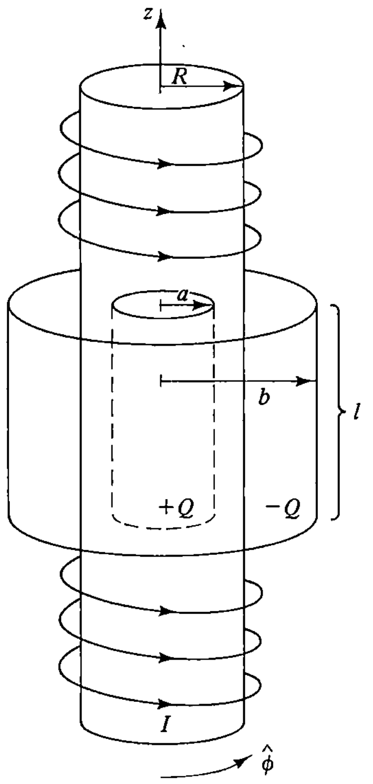

Figure 7.53

Solution: It was initially stored in the fields. Before the current was switched off, there was an electric field

$$\mathbf{E} = \frac{Q}{2\pi\epsilon_0 l}\frac{1}{r}\hat{\mathbf{r}}$$

in the region between the cylinders, and a magnetic field

$$\mathbf{B} = \mu_0 N I \hat{\mathbf{z}}$$

inside the solenoid. The momentum density (7.103) was therefore

$$\boldsymbol{\wp}_{EB} = \epsilon_0(\mathbf{E} \times \mathbf{B}) = -\frac{\mu_0 N I Q}{2\pi l r}\hat{\boldsymbol{\phi}}$$

in the region $a < r < R$. The *angular* momentum density was

$$\boldsymbol{\ell}_{EB} = \mathbf{r} \times \boldsymbol{\wp}_{EB} = -\frac{\mu_0 N I Q}{2\pi l}\hat{\mathbf{z}}$$

—a *constant,* as it turns out; to get the *total* anglar momentum in the fields, we simply multiply by the volume, $\pi(R^2 - a^2)l$:

$$\mathbf{L}_{EB} = -\frac{1}{2}\mu_0 N I Q(R^2 - a^2)\hat{\mathbf{z}} \tag{7.106}$$

When the current is turned off, the changing magnetic field induces a circumferential electric field, given by Faraday's law:

$$\mathbf{E} = \begin{cases} -\frac{1}{2}\mu_0 N\dot{I}\frac{R^2}{r}\hat{\phi}, & (r > R) \\ -\frac{1}{2}\mu_0 N\dot{I}r\hat{\phi}, & (r < R) \end{cases}$$

Thus the torque on the outer cylinder is

$$\mathbf{N}_b = \mathbf{r} \times (-Q\mathbf{E}) = \tfrac{1}{2}\mu_0 NQR^2\dot{I}\hat{z}$$

and it picks up an angular momentum

$$\mathbf{L}_b = \frac{1}{2}\mu_0 NQR^2\hat{z}\int_I^0 \frac{dI}{dt}\,dt = -\frac{1}{2}\mu_0 NIQR^2\hat{z}$$

Similarly, the torque on the inner cylinder is

$$\mathbf{N}_a = -\tfrac{1}{2}\mu_0 NQa^2\dot{I}\hat{z}$$

and its angular momentum increase is

$$\mathbf{L}_a = \tfrac{1}{2}\mu_0 NIQa^2\hat{z}$$

So it all works out: $\mathbf{L}_{EB} = \mathbf{L}_a + \mathbf{L}_b$. The angular momentum *lost* by the fields is precisely equal to the angular momentum *gained* by the cylinders, and the *total* angular momentum (fields plus matter) is conserved.

Problem 7.45 Consider an infinite parallel-plate capacitor, with the lower plate (at $z = -(d/2)$) carrying charge density $-\sigma$, and the upper plate at ($z = +(d/2)$) carrying charge density $+\sigma$.

(a) Determine all nine elements of the stress tensor, in the region between the plates. Display your answer as a 3×3 matrix:

$$\begin{pmatrix} T_{xx} & T_{xy} & T_{xz} \\ T_{yx} & T_{yy} & T_{yz} \\ T_{zx} & T_{zy} & T_{zz} \end{pmatrix}$$

(b) Use equation 7.95 to determine the force per unit area on the top plate. Compare equation 2.44.

(c) What is the momentum per unit area per unit time crossing the yz plane (or any other plane parallel to that one, between the plates)?

(d) At the plates this momentum is absorbed, and the plates recoil (unless there is some nonelectrical force holding them in position). Find the recoil force per unit area on the top plate, and compare your answer in (b). (*Note:* this is not an *additional* force, but rather an alternative way of calculating the *same* force—in (b) we got it from the Lorentz force law, and in (d) we did it by conservation of momentum.)

Problem 7.46 In Example 17, suppose that instead of turning off the *magnetic* field (by reducing I) we turn off the *electric* field, by connecting a weakly[16] conducting radial spoke between the cylinders. (We'll have to cut a slot in the solenoid, so the cylinders can still rotate freely.) From the magnetic force on the current $I'(t)$ in the spoke, determine the total angular momentum delivered to the cylinders, as they discharge (they are now rigidly connected, so they rotate together). Compare the initial angular momentum stored in the fields (7.106). (Notice that the *mechanism* by which angular momentum is transferred from the fields to the cylinders is entirely different in the two cases: In Example 17 it was Faraday's law, but here it is the Lorentz force law.)

Further Problems on Chapter 7

Problem 7.47 A perfectly conducting spherical shell of radius R rotates about the z axis with angular velocity ω, in a uniform magnetic field $\mathbf{B} = B_0\hat{k}$. Calculate the emf developed between the "north pole" and the equator. (*Answer*: $\frac{1}{2} B_0 \omega R^2$.)

Problem 7.48 Refer to Problem 7.12 (and use the result of Problem 5.58, if it helps):

(a) Does the square ring fall faster in the orientation shown (Fig. 7.19), or when rotated 45° about an axis coming out of the page? Find the ratio of the two terminal velocities. If you dropped the loop, which orientation would it assume in falling?
[*Answer*: $(\sqrt{2} - 2y/s)^2$, where s is the length of a side, and y is the height of the center above the edge of the magnetic field, in the rotated configuration.]

(b) How long does it take a *circular* ring to cross the bottom of the magnetic field, at its (changing) terminal velocity?

Problem 7.49 The current in a long solenoid is increasing linearly with time, so that the flux is proportional to t: $\Phi = \alpha t$. Two voltmeters are connected to diametrically opposite points (A and B), together with resistors (R_1 and R_2), as shown in Fig. 7.54. What is the reading on each voltmeter? Assume that these are *ideal* voltmeters, that draw negligible current (they have huge internal resistance), and that a voltmeter registers $\int_a^b \mathbf{E} \cdot d\mathbf{l}$ between the terminals and through the meter.
(*Answer*: $V_1 = \alpha R_1/(R_1 + R_2)$; $V_2 = -\alpha R_2/(R_1 + R_2)$. Notice that $V_1 \neq V_2$, even though they are connected to the same points. See R. H. Romer, *Am. J. Phys.* **50**, 1089 (1982).)

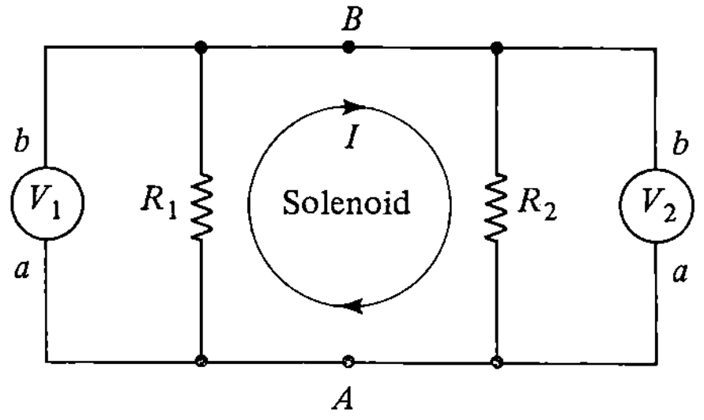

Figure 7.54

Problem 7.50

(a) Use the Neumann formula (7.22) to calculate the mutual inductance of the configuration in Fig. 7.34, assuming a is very small ($a \ll b$, $a \ll z$). Compare your answer to Problem 7.18.

[16]In Example 17 we turned the current off slowly, to keep things quasistatic; here we reduce the electric field slowly to keep the displacement current negligible.

(b) For the general case (*not* assuming a is small) show that

$$M = \frac{\mu_0 \pi}{2} \sqrt{ab\beta}\, \beta\left(1 + 3\beta + \frac{75}{8}\beta^2 + \cdots\right)$$

where

$$\beta = \frac{ab}{z^2 + (a + b)^2}$$

Problem 7.51 Suppose $\mathbf{J}(\mathbf{r})$ is constant in time but $\rho(\mathbf{r}, t)$ is *not*—conditions that might prevail, for instance, during the charging of a capacitor.

(a) Show that the charge density at any particular point is a linear function of time:

$$\rho(\mathbf{r}, t) = \rho(\mathbf{r}, 0) + \dot{\rho}(\mathbf{r}, 0)t$$

where $\dot{\rho}(\mathbf{r}, 0)$ is the time derivative of ρ at $t = 0$.

This is *not* an electrostatic or magnetostatic configuration; nevertheless—rather surprisingly—both Coulomb's law (in the form 2.6) and the Biot-Savart law (5.35) hold, as you can confirm by showing that they satisfy Maxwell's equations. In particular:

(b) Show that

$$\mathbf{B} = \frac{\mu_0}{4\pi} \int \frac{\mathbf{J} \times \hat{\boldsymbol{\imath}}}{\imath^2}\, d\tau$$

obeys Ampère's law *with Maxwell's displacement current term.*

(*Comment*: Some authors would regard this as a magnetostatic configuration—after all, $\mathbf{B}$ is independent of t. For them, the Biot-Savart law is a general rule of magnetostatics, but $\nabla \cdot \mathbf{J} = 0$ and $\nabla \times \mathbf{B} = \mu_0 \mathbf{J}$ apply only under the *additional* assumption that ρ is constant. In such a formulation Maxwell's displacement term would be *derived* from the Biot-Savart law, by the method you used in (b). The trouble is, if this current were *truly* steady, the charge would pile up without limit. To avoid such infinite accumulations of charge, we must turn the current on and off, in which case $\mathbf{J}$ is no longer constant. Of course, if we switch the current on and then wait a *reasonable period t,* we may hope that the change in $\mathbf{J}$ has been forgotten, at least as far as nearby points are concerned. Indeed, as we shall see in Chapter 9, the "news" of the change travels outward at the speed of light; it is within a sphere of radius ct, then, that magnetostatics in the broader sense may be said to hold. But in my view the situation has by now become so complicated that the small benefit to be derived from extending the domain of magnetostatics is not worth the price.)

! **Problem 7.52** Suppose you had an electric charge e and a magnetic charge g separated by a distance d. The field of the electric charge is

$$\mathbf{E} = \frac{1}{4\pi\epsilon_0} \frac{e}{r^2},$$

of course, and the field of the magnetic charge is

$$\mathbf{B} = \frac{\mu_0}{4\pi} \frac{g}{r^2}.$$

Find the total angular momentum stored in the resulting electromagnetic fields. (This system is known as **Thomson's monopole.** See I. Adawi, *Am. J. Phys.*, **44,** 762 (1976) and *Phys. Rev.* **D31,** 3301 (1985) for discussion and references.) [*Answer*: $(\mu_0/4\pi)\, eg$.]

(*Comment*: The answer is independent of the separation distance (!), and points in the direction from e toward g. In quantum mechanics angular momentum comes in half-integer multiples of $\hbar$, so this problem suggests that if magnetic monopoles exist, electric and magnetic charge must be quantized, according to the relation $\mu_0 eg/4\pi = n\hbar/2$, for $n = 1, 2, 3, \ldots$, an idea first proposed by Dirac in 1931.)

! **Problem 7.53**[17] Imagine an iron sphere of radius R which carries a charge Q and a uniform magnetization $\mathbf{M} = M\hat{\mathbf{k}}$. The sphere is initially at rest.

(a) Compute the angular momentum stored in the electromagnetic fields.

(b) Suppose the sphere is gradually (and uniformly) demagnetized (perhaps by heating it up through the Curie point). Use Faraday's law to determine the induced electric field, find the torque this field exerts on the sphere, and calculate the total angular momentum imparted to the sphere in the course of the demagnetization.

(c) Suppose instead of *demagnetizing* the sphere we *discharge* it, by connecting a grounding wire to the north pole. Assume the current flows over the surface in such a way that the charge density remains uniform. Use the Lorentz force law to determine the torque on the sphere, and calculate the total angular momentum imparted to the sphere in the course of the discharge. (The magnetic field is discontinuous at the surface . . . does this matter?)

(*Answer*: $\frac{2}{9}\mu_0 MQR^2$.)

Problem 7.54 The *magnetic* field outside a long straight wire carrying a steady current I is (of course)

$$\mathbf{B} = \frac{\mu_0}{2\pi}\frac{I}{r}\hat{\boldsymbol{\phi}}$$

The *electric field inside* the wire is uniform:

$$\mathbf{E} = \frac{I}{\pi a^2 \sigma}\hat{\mathbf{z}}$$

where σ is the conductivity and a is the radius (see Examples 1 and 3). *Question*: what is the electric field *outside* the wire? (This is a famous problem, first analyzed by Sommerfeld, and known in its most recent incarnation as "Merzbacher's puzzle."[18]) The answer depends on how you complete the circuit. Suppose the current returns along a perfectly conducting grounded coaxial cylinder of radius b (Fig. 7.55). In the region $a < r < b$, the potential $V(r, z)$ satisfies Laplace's equation, with the boundary conditions

$$\text{(i)}\quad V(a, z) = -\frac{Iz}{\pi a^2 \sigma}; \qquad \text{(ii)}\quad V(b, z) = 0$$

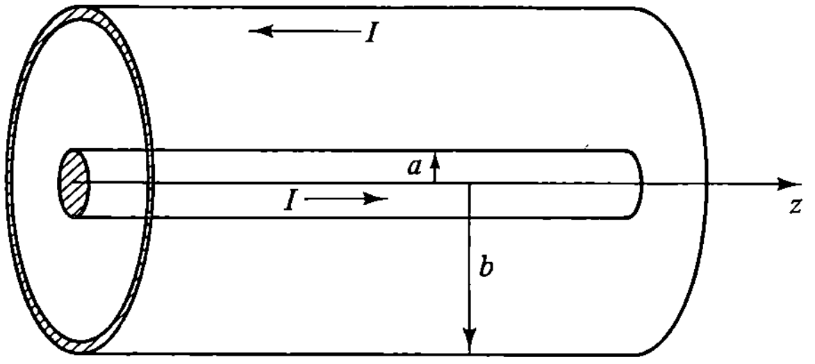

Figure 7.55

[17]This variation on the Feynman disk paradox was proposed by N. L. Sharma (*Am. J. Phys.* **56**, 420 (1988)); similar models were analysed by E. M. Pugh and G. E. Pugh, *Am. J. Phys.* **35**, 153 (1967) and by R. H. Romer, *Am. J. Phys.* **35**, 445 (1967).

[18]A. Sommerfeld, *Electrodynamics* (New York: Academic Press, 1952) p. 125; E. Merzbacher, *Am. J. Phys.* **48**, 104 (1980); further references in M. A. Heald, *Am. J. Phys.* **52**, 522 (1984).

Unfortunately, this does not suffice to determine the answer—we still need to specify boundary conditions at the two ends. In the literature it is customary to sweep this ambiguity under the rug by simply *asserting* (in so many words) that $V(r, z)$ is proportional to z:

$$V(r, z) = zf(r).$$

On this assumption,

(a) Determine $V(r, z)$.

(b) Find $\mathbf{E}(r, z)$.

(c) Calculate the surface charge density $\sigma(z)$ on the wire (not to be confused with the *conductivity*, σ).

(d) Find the Poynting vector, and describe the flow of energy.

[*Answer*: $V = (-Iz/\pi a^2\sigma)(\ln(r/b)/\ln(a/b))$. This is a *peculiar* result, since E_r and $\sigma(z)$ are *not* independent of z—as they should be for a truly *infinite* wire.]

Problem 7.55 Prove **Alfven's theorem**: In a perfectly conducting fluid (say, a gas of free electrons), the magnetic flux through any closed loop moving with the fluid is constant in time. (The magnetic field lines are, as it were, "frozen" into the fluid.)

(a) Use Ohm's law, in the form (7.2), together with Faraday's law, to prove that if $\sigma = \infty$ and $\mathbf{J}$ is finite, then

$$\frac{\partial \mathbf{B}}{\partial t} = \nabla \times (\mathbf{v} \times \mathbf{B}).$$

(b) Let S be a surface bounded by the loop (C) at time t, and S' a surface bounded by the loop in its new position (C') at time $t + dt$ (see Fig. 7.56). The change in flux is

$$d\Phi = \int_{S'} \mathbf{B}(t + dt) \cdot d\mathbf{a} - \int_{S} \mathbf{B}(t) \cdot d\mathbf{a}$$

Show that

$$\int_{S'} \mathbf{B}(t + dt) \cdot d\mathbf{a} + \int_{R} \mathbf{B}(t + dt) \cdot d\mathbf{a} = \int_{S} \mathbf{B}(t + dt) \cdot d\mathbf{a}$$

(where R is the "ribbon" joining C and C'), and hence that

$$d\Phi = dt \int_{S} \frac{\partial \mathbf{B}}{\partial t} \cdot d\mathbf{a} - \int_{R} \mathbf{B}(t + dt) \cdot d\mathbf{a}$$

(for infinitesimal dt). Use the method of Section 7.1.3 to rewrite the second integral as

$$dt \oint_{C} (\mathbf{B} \times \mathbf{v}) \cdot d\mathbf{l}$$

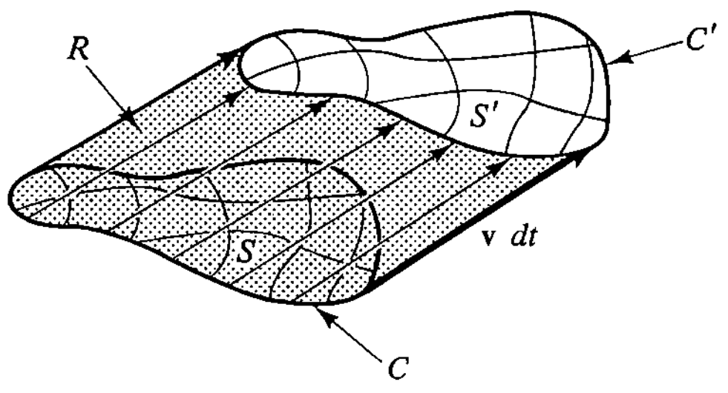

Figure 7.56

and invoke Stokes' theorem to conclude that

$$\frac{d\Phi}{dt} = \int_{S} \left(\frac{\partial \mathbf{B}}{\partial t} - \nabla \times (\mathbf{v} \times \mathbf{B}) \right) \cdot d\mathbf{a}$$

Together with the result in (a), this proves the theorem.

Problem 7.56 In the discussion of motional emf (Section 7.1.3) I assumed that the wire loop (Fig. 7.12) has a resistance R; the current generated is then $I = vBh/R$. But what if the wire is made of perfectly conducting material, so that R is essentially *zero*? In that case the current is limited only by the back emf associated with the self-inductance of the loop (which would ordinarily be negligible in comparison with IR). Show that in this regime the loop executes simple harmonic motion, and find its frequency.[19]

[*Answer*: $\omega = Bh/\sqrt{mL}$.]

Problem 7.57 Suppose

$$\mathbf{E}(\mathbf{r}, t) = \frac{q}{r^2}\theta(vt - |\mathbf{r}|)\hat{\mathbf{r}}; \qquad \mathbf{B}(\mathbf{r}, t) = 0$$

(the theta function is defined in Problem 1.44(b)). Show that these fields satisfy all of Maxwell's equations, and determine ρ and $\mathbf{J}$. Describe the physical situation that gives rise to these fields.

Problem 7.58 A transformer (Problem 7.22) takes an input AC voltage of amplitude V_1, and delivers an output voltage of amplitude V_2, which is determined by the turns ratio $(V_2/V_1 = N_2/N_1)$. If $N_2 > N_1$ the output voltage is greater than the input voltage. Why doesn't this violate conservation of energy? *Answer*: Power is the product of voltage and current; evidently if the voltage goes *up*, the current must come *down*. The purpose of this problem is to see exactly how this works out, in a simplified model.

(a) In an ideal transformer the same flux passes through all turns of the primary and of the secondary. Show that in this case $M^2 = L_1L_2$, where M is the mutual inductance of the coils, and L_1, L_2 are their individual self-inductances.

(b) Suppose the primary is driven with an AC voltage $V_{in} = V_1\cos(\omega t)$, and the secondary is connected to a resistor, R. Show that the two currents satisfy the relations

$$L_1(dI_1/dt) - M(dI_2/dt) = V_1\cos(\omega t); \qquad L_2(dI_2/dt) - M(dI_1/dt) + I_2R = 0$$

(c) Using the result in (a), solve these equations for $I_1(t)$ and $I_2(t)$. (Assume I_1 has no DC component.)

(d) Show that the output voltage ($V_{out} = I_2R$) divided by the input voltage (V_{in}) is equal to the turns ratio: $V_{out}/V_{in} = N_2/N_1$.

(e) Calculate the input power ($P_{in} = V_{in}I_1$) and the output power ($P_{out} = V_{out}I_2$), and show that their averages over a full cycle are equal.

Problem 7.59 In a **perfect conductor**, the conductivity is infinite, so $\mathbf{E} = 0$ (equation 7.3), and any net charge resides on the surface (just as it does for an *im*perfect conductor, in electro*statics*).

(a) Show that the magnetic field is constant (i.e., $\partial\mathbf{B}/\partial t = 0$), inside a perfect conductor.

A **superconductor** is a perfect conductor with the *additional* property that this constant $\mathbf{B}$ is always *zero*. (This "flux exclusion" is known as the **Meissner effect.**)

[19]For a collection of related problems, see W. M. Saslow, *Am. J. Phys.* **55**, 986 (1987).

(b) Show that the current in a superconductor is confined to the surface.
Superconductivity is lost above a certain critical temperature (T_c), which varies from one material to another.
(c) Suppose you had a sphere (radius R) above its critical temperature, and you held it in a uniform magnetic field $B_0\hat{z}$ while cooling it below T_c. Find the induced surface current density $\mathbf{K}$, as a function of the polar angle θ.
(d) Suppose you made a loop of perfectly conducting wire, and a single magnetic monopole g (Problem 7.52) passed through it. What is the resulting current in the loop?

Problem 7.60 Picture the electron as a uniformly charged spherical shell, with charge e and radius R, spinning at angular velocity ω.
(a) Calculate the total energy contained in the electromagnetic fields.
(b) Calculate the total angular momentum contained in the fields.
(c) According to the Einstein formula ($E = mc^2$), the energy in the fields should contribute to the mass of the electron. Lorentz and others speculated that the *entire* mass of the electron might be accounted for in this way: $W_{EB} = m_e c^2$. Suppose moreover that the electron's spin angular momentum is entirely attributable to the electromagnetic fields: $L_{EB} = \hbar/2$. On these two assumptions, determine the radius and angular velocity of the electron. What is their product, ωR? Does this classical model make sense? [See J. Higbie, *Am. J. Phys.* **56**, 378 (1988).]

Problem 7.61 A very long solenoid of radius a, with N turns per unit length, carries a current I_s. Coaxial with the solenoid, at radius $b \gg a$, is a circular ring of wire, with resistance R. When the current in the solenoid is (gradually) decreased, a current I_r is induced in the ring.
(a) Calculate I_r, in terms of dI_s/dt.
(b) The power ($I_r^2 R$) delivered to the ring must have come from the solenoid. Confirm this by calculating the Poynting vector just outside the solenoid (the *electric* field is due to the changing flux in the solenoid; the *magnetic* field is due to the current in the ring). Integrate over the entire surface of the solenoid, and check that you recover the correct total power. [For extensive discussion, see M. A. Heald, *Am. J. Phys.* **56**, 540 (1988).]

Problem 7.62 The magnetic field of an infinite straight wire carrying a steady current I can be obtained from the *displacement* current term in the Ampere/Maxwell law, as follows: Picture the current as consisting of a uniform line charge λ moving along the x axis at speed v (so that $I = \lambda v$), with a tiny gap of length ϵ, which reaches the origin at time $t = 0$. In the next instant (up to $t = \epsilon/v$) there is no *real* current passing through a circular Amperian loop in the yz plane, but there *is* a *displacement* current, due to the "missing" charge in the gap.
(a) Use Coulomb's law to calculate the x component of the electric field, for points in the yz plane a distance r from the origin, due to a segment of wire with uniform charge density $-\lambda$ extending from $x = \epsilon - vt$ to $x = vt$.
(b) Determine the flux of this electric field through a circle of radius R in the yz plane.
(c) Find the displacement current through this circle. Show that I_d is equal to I, in the limit as the gap width (ϵ) goes to zero.
[For a slightly different approach to the same problem, see W. K. Terry, *Am. J. Phys.* **50**, 742 (1982).]

8

ELECTROMAGNETIC WAVES

8.1 THE WAVE EQUATION

8.1.1 Introduction

In regions of space where there is no charge or current, Maxwell's equations read

$$\left.\begin{array}{llll}\text{(i)} & \nabla\cdot\mathbf{E}=0 & \text{(iii)} & \nabla\times\mathbf{E}=-\dfrac{\partial\mathbf{B}}{\partial t}\\[2ex] \text{(ii)} & \nabla\cdot\mathbf{B}=0 & \text{(iv)} & \nabla\times\mathbf{B}=\mu_0\epsilon_0\dfrac{\partial\mathbf{E}}{\partial t}\end{array}\right\} \tag{8.1}$$

They constitute a set of coupled, first-order, partial differential equations for **E** and **B**. They can be *de*coupled by applying the curl to (iii) and (iv):

$$\nabla\times(\nabla\times\mathbf{E})=\nabla(\nabla\cdot\mathbf{E})-\nabla^2\mathbf{E}=\nabla\times\left(-\frac{\partial\mathbf{B}}{\partial t}\right)$$

$$=-\frac{\partial}{\partial t}(\nabla\times\mathbf{B})=-\mu_0\epsilon_0\frac{\partial^2\mathbf{E}}{\partial t^2}$$

$$\nabla\times(\nabla\times\mathbf{B})=\nabla(\nabla\cdot\mathbf{B})-\nabla^2\mathbf{B}=\nabla\times\left(\mu_0\epsilon_0\frac{\partial\mathbf{E}}{\partial t}\right)$$

$$=\mu_0\epsilon_0\frac{\partial}{\partial t}(\nabla\times\mathbf{E})=-\mu_0\epsilon_0\frac{\partial^2\mathbf{B}}{\partial t^2}$$

Or, since $\nabla \cdot \mathbf{E} = 0$ and $\nabla \cdot \mathbf{B} = 0$,

$$\boxed{\nabla^2 \mathbf{E} = \mu_0 \epsilon_0 \frac{\partial^2 \mathbf{E}}{\partial t^2}, \qquad \nabla^2 \mathbf{B} = \mu_0 \epsilon_0 \frac{\partial^2 \mathbf{B}}{\partial t^2}} \tag{8.2}$$

We now have *separate* equations for **E** and **B**, but they are of *second* order; that's the price you pay for decoupling them.

In vacuum, then, the components of **E** and **B** satisfy the equation

$$\nabla^2 f = \frac{1}{v^2} \frac{\partial^2 f}{\partial t^2} \tag{8.3}$$

This is called the (classical) **wave equation**; as we shall see in the next section, it describes waves traveling with a velocity v. According to Maxwell's equations, then, empty space supports the propagation of electromagnetic waves at a speed

$$v = \frac{1}{\sqrt{\epsilon_0 \mu_0}} = 3.00 \times 10^8 \text{ m/s} \tag{8.4}$$

which is precisely the speed of light, c! The implication is stunning: Perhaps light *is* an electromagnetic wave.[1] Unfortunately, this conclusion does not surprise anyone today, but imagine what a triumph it was in Maxwell's time. Remember how ϵ_0 and μ_0 came into the theory in the first place: They were constants in Coulomb's law and the Biot-Savart law, respectively. You measure them in experiments involving charged pith balls, batteries, and wires—experiments having nothing whatever to do with light. And yet, according to Maxwell's theory these two constants are related in a beautifully simple manner to the speed of light. Notice also the crucial role played by Maxwell's term in Ampère's law ($\mu_0 \epsilon_0 \, \partial \mathbf{E}/\partial t$); without it, the wave equation would not emerge.

Inside matter, but in regions where there is no *free* charge or *free* current, Maxwell's equations become

$$\left.\begin{array}{llll} \text{(i)} & \nabla \cdot \mathbf{D} = 0 & \text{(iii)} & \nabla \times \mathbf{E} = -\dfrac{\partial \mathbf{B}}{\partial t} \\ \text{(ii)} & \nabla \cdot \mathbf{B} = 0 & \text{(iv)} & \nabla \times \mathbf{H} = \dfrac{\partial \mathbf{D}}{\partial t} \end{array}\right\} \tag{8.5}$$

If the medium is *linear,*

$$\mathbf{D} = \epsilon \mathbf{E}, \qquad \mathbf{H} = \frac{1}{\mu} \mathbf{B} \tag{8.6}$$

and *homogeneous,* so that ϵ and μ do not vary from point to point, then equations (8.5) reduce to

[1] As Maxwell himself put it, "We can scarcely avoid the inference that light consists in the transverse undulations of the same medium which is the cause of electric and magnetic phenomena." See Ivan Tolstoy, *James Clerk Maxwell, A Biography* (Chicago: University of Chicago Press, 1983).

$$\left.\begin{array}{ll}\text{(i)}\quad \nabla\cdot\mathbf{E}=0 & \text{(iii)}\quad \nabla\times\mathbf{E}=-\dfrac{\partial\mathbf{B}}{\partial t}\\[2ex] \text{(ii)}\quad \nabla\cdot\mathbf{B}=0 & \text{(iv)}\quad \nabla\times\mathbf{B}=\mu\epsilon\dfrac{\partial\mathbf{E}}{\partial t}\end{array}\right\} \tag{8.7}$$

which differ from (8.1) only in the replacement of $\mu_0\epsilon_0$ by $\mu\epsilon$. Evidently, electromagnetic waves propagate through a linear homogeneous medium at the speed

$$v=\frac{1}{\sqrt{\epsilon\mu}} \tag{8.8}$$

For transparent media, μ is typically very close to 1 and ϵ is *greater* than 1; evidently, then, light should travel *slower* through matter—a fact that is, of course, well known from optics:

$$v=\frac{c}{n} \tag{8.9}$$

where n is the index of refraction. It follows that n is related to the electric and magnetic properties of the material by the equation

$$n=\sqrt{\frac{\epsilon\mu}{\epsilon_0\mu_0}} \tag{8.10}$$

Now, the index of refraction typically varies somewhat according to the wavelength (color) of the incident light—that's what accounts for the *dispersion* of light by a prism or by drops of water in the formation of a rainbow. This, too, finds a natural electromagnetic explanation, for we shall see that the electric polarizability of a molecule, and hence also the permittivity ϵ, depends on how rapidly the electric field oscillates.

My purpose in this chapter is to develop the theory of electromagnetic waves (including reflection, refraction, polarization, absorption, and dispersion) as an application of classical electrodynamics—to bridge the gap between Maxwell's equations and the laws of optics. But before we get into this, I must tell you more about the wave equation itself. I shall do this in the simplest possible context: transverse waves on a stretched string.

8.1.2 The Wave Equation in One Dimension

Suppose you have a very long string held taut by a tension T. In equilibrium it coincides with the x axis, but if you shake it a wave will propagate down the line. The net transverse force on the segment of string between x and $x+\Delta x$ (Fig. 8.1) is

$$F=T\sin\theta'-T\sin\theta,$$

where θ' is the angle the string makes with the x-direction at point $x+\Delta x$, and θ is the corresponding angle at point x. Provided that the distortion of the string is not too great, these angles are small (the figure is exaggerated, obviously), and we can re-

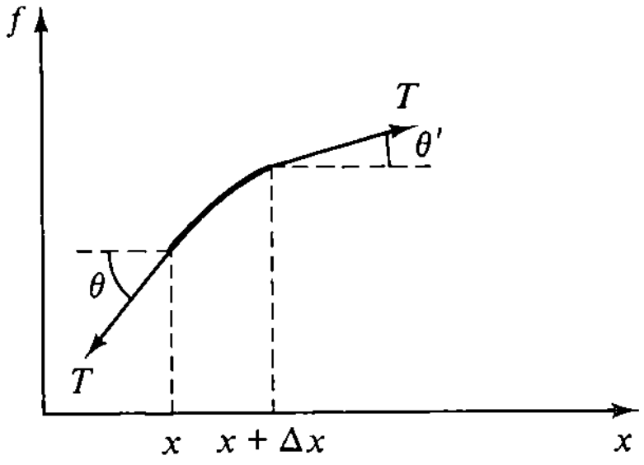

Figure 8.1

place the sine by the tangent. If $f(x, t)$ is the transverse displacement of the string,

$$F \cong T(\tan\theta' - \tan\theta) = T\left(\left.\frac{\partial f}{\partial x}\right|_{x+\Delta x} - \left.\frac{\partial f}{\partial x}\right|_{x}\right) \cong T\frac{\partial^2 f}{\partial x^2}\Delta x$$

If the mass per unit length is μ, Newton's second law gives

$$F = \mu(\Delta x)\frac{\partial^2 f}{\partial t^2}$$

and hence

$$\frac{\partial^2 f}{\partial x^2} = \frac{\mu}{T}\frac{\partial^2 f}{\partial t^2}$$

So small disturbances on the string obey the wave equation in one dimension:

$$\frac{\partial^2 f}{\partial x^2} = \frac{1}{v^2}\frac{\partial^2 f}{\partial t^2} \tag{8.11}$$

with

$$v = \sqrt{\frac{T}{\mu}} \tag{8.12}$$

Equation (8.11) admits as a solution any function of the form

$$f(x, t) = g(x - vt) \tag{8.13}$$

that is, any function which depends on the variables x and t in the special combination $z \equiv x - vt$. For

$$\frac{\partial f}{\partial x} = \frac{dg}{dz}\frac{\partial z}{\partial x} = \frac{dg}{dz}, \qquad \frac{\partial f}{\partial t} = \frac{dg}{dz}\frac{\partial z}{\partial t} = -v\frac{dg}{dz}$$

and

$$\frac{\partial^2 f}{\partial x^2} = \frac{\partial}{\partial x}\left(\frac{dg}{dz}\right) = \frac{d^2 g}{dz^2}\frac{\partial z}{\partial x} = \frac{d^2 g}{dz^2}$$

$$\frac{\partial^2 f}{\partial t^2} = -v\frac{\partial}{\partial t}\left(\frac{dg}{dz}\right) = -v\frac{d^2 g}{dz^2}\frac{\partial z}{\partial t} = v^2\frac{d^2 g}{dz^2}$$

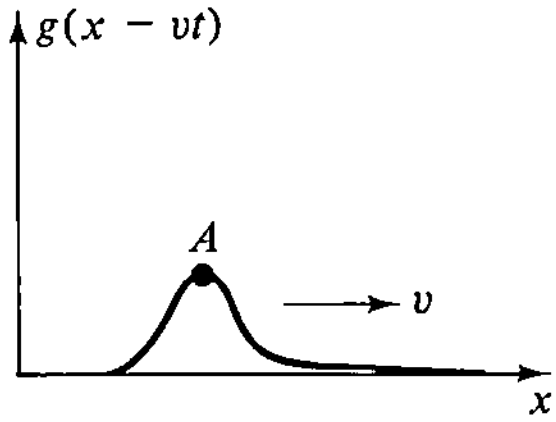

Figure 8.2

so that

$$\frac{d^2g}{dz^2} = \frac{\partial^2 f}{\partial x^2} = \frac{1}{v^2}\frac{\partial^2 f}{\partial t^2}$$

Note that $g(z)$ can be *any* (differentiable) *function whatever.* Physically, equation (8.13) represents a wave of constant shape $g(z)$ traveling in the positive x-direction at speed v. A fixed point on the wave (say, the peak A in Fig. 8.2) corresponds to a fixed value of the argument z:

$$z = x - vt = \text{constant}$$

from which it follows that such a point moves with speed $dx/dt = v$.

Of course, the string can also support waves traveling to the *left,*

$$f(x, t) = h(x + vt) \tag{8.14}$$

as you can readily check for yourself. Now the wave equation is **linear**, in the sense that the *sum* of any two solutions is itself a solution:

$$\frac{\partial^2}{\partial x^2}(f_1 + f_2) = \frac{\partial^2 f_1}{\partial x^2} + \frac{\partial^2 f_2}{\partial x^2} = \frac{1}{v^2}\frac{\partial^2 f_1}{\partial t^2} + \frac{1}{v^2}\frac{\partial^2 f_2}{\partial t^2} = \frac{1}{v^2}\frac{\partial^2}{\partial t^2}(f_1 + f_2) \tag{8.15}$$

Therefore, the superposition of a wave going to the left and a wave going to the right,

$$f(x, t) = g(x - vt) + h(x + vt) \tag{8.16}$$

satisfies (8.11) for arbitrary functions g and h.

As a matter of fact, (8.16) represents the *most general* solution to the wave equation in one dimension. For suppose you give the string an initial shape $f(x, 0)$, and an initial transverse velocity $\dot{f}(x, 0)$. Then

$$\left.\begin{aligned} f(x, 0) &= g(x) + h(x) \\ \dot{f}(x, 0) &= -vg'(x) + vh'(x) \end{aligned}\right\} \tag{8.17}$$

Integrating the latter with respect to x, we have[2]

$$\frac{1}{v}\int_0^x \dot{f}(\bar{x}, 0)\, d\bar{x} = -g(x) + h(x) \tag{8.18}$$

[2]Since only the *sum* of g and h appears in (8.16), you are at liberty to add any constant to g, provided you simultaneously subtract the same from h. I have used this flexibility to eliminate the arbitrary constant of integration in (8.18) by requiring $g(0) = h(0)$.

Combining (8.18) and (8.17), we find that the initial conditions are satisfied if we choose

$$\left.\begin{aligned} g(x) &= \frac{1}{2}\left[f(x, 0) - \frac{1}{v}\int_0^x \dot{f}(\bar{x}, 0)\, d\bar{x}\right] \\ h(x) &= \frac{1}{2}\left[f(x, 0) + \frac{1}{v}\int_0^x \dot{f}(\bar{x}, 0)\, d\bar{x}\right] \end{aligned}\right\} \tag{8.19}$$

Example 1

Roughly speaking, a *plucked* string (as on a guitar or a harpsichord) has some initial displacement but no initial velocity; a *hammered* string (as on a piano or a dulcimer) has an initial velocity but no initial displacement. Suppose a long string is plucked at its center into a triangular shape, as shown in Fig. 8.3, and released. What is the shape of the string at any subsequent time?

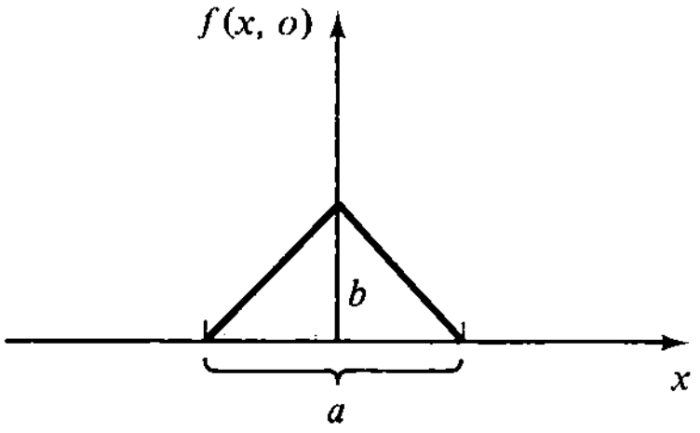

Figure 8.3

Solution: Here $\dot{f}(x, 0) = 0$, so $g(x)$ and $h(x)$ are each $\frac{1}{2}f(x, 0)$—that is, triangles of height $b/2$. As time passes, g moves off to the right and h to the left, at speed v, so the string has the shape shown in Fig. 8.4.

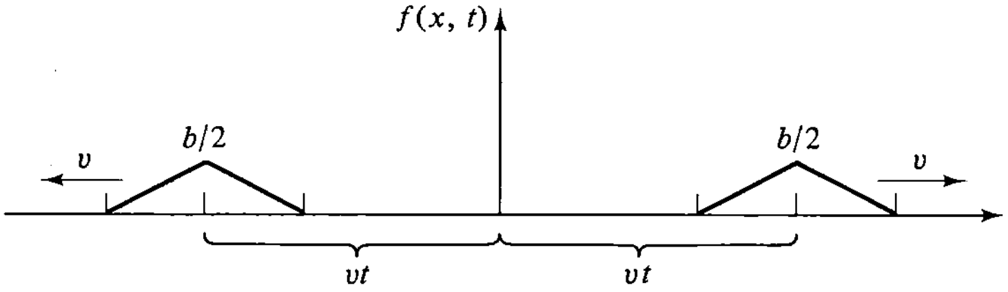

Figure 8.4

Problem 8.1 Suppose the string is *hammered* at time $t = 0$, in such a way that $f(x, 0) = 0$ and

$$\dot{f}(x, 0) = \begin{cases} -w, & -a < x < 0 \\ +w, & 0 < x < +a \\ 0, & \text{otherwise} \end{cases} \qquad \text{where } w \text{ is some constant}$$

(a) Determine $g(x)$ and $h(x)$.
(b) Find the shape of the string at $t = a/2v$, and sketch it.
(c) Do the same for $t = 3a/2v$.
(d) Describe the subsequent shape of the string.

Problem 8.2

(a) Combine (8.19) and (8.16) to obtain **d'Alembert's solution** to the wave equation:

$$f(x, t) = \frac{1}{2}\left[f(x + vt, 0) + f(x - vt, 0) + \frac{1}{v}\int_{x-vt}^{x+vt} \dot{f}(\bar{x}, 0)\, d\bar{x}\right]$$

(b) Suppose a transverse velocity

$$\dot{f}(x, 0) = Axe^{-x^2}$$

is imparted to the string, which is initially straight ($f(x, 0) = 0$). Find the subsequent shape of the string, and sketch it.

8.1.3 Sinusoidal Waves

(i) Terminology. Of all possible wave forms, the sinusoidal one

$$f(x, t) = A\cos[\kappa(x - vt) + \delta] \tag{8.20}$$

is (for good reason) the most familiar. Fig. 8.5 displays this function at time $t = 0$. A is called the **amplitude** of the wave; it is a positive number, representing the maximum displacement from equilibrium. δ is called the **phase constant**. Notice that at $x = vt - \delta/\kappa$, the argument of the cosine is zero; let's call this the "central maximum." If $\delta = 0$, the central maximum passes the origin at time $t = 0$; more generally, δ/κ *is the distance by which the central maximum* (and therefore the entire wave) *is "delayed."* For this reason δ is also called the "phase delay." κ is called the **wave number**; it is related to the **wavelength** λ by the equation

$$\lambda = \frac{2\pi}{\kappa} \tag{8.21}$$

for when x advances by $2\pi/\kappa$ the cosine goes through one complete cycle.

As time passes, the entire wave train travels to the right at speed v. At any fixed point x, the string vibrates up and down, undergoing one full cycle in a time,

$$T = \frac{2\pi}{\kappa v} \tag{8.22}$$

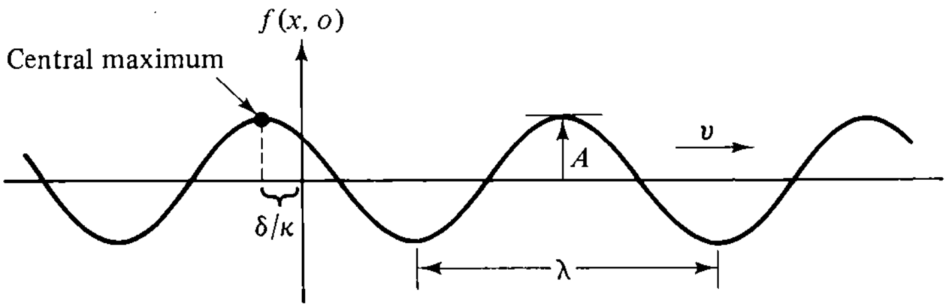

Figure 8.5

T is called the **period** of the oscillation (as t advances by this amount, the cosine executes one cycle). The **frequency** ν (number of oscillations per unit time) is therefore

$$\nu = \frac{1}{T} = \frac{\kappa v}{2\pi} = \frac{v}{\lambda} \tag{8.23}$$

For our purposes, a more convenient unit is the **angular frequency** ω, so-called because in the analogous case of uniform circular motion it represents the number of radians swept out per unit time:

$$\omega = 2\pi\nu = \kappa v \tag{8.24}$$

Ordinarily, I prefer to write the sinusoidal wave (8.20) in terms of ω rather than ν:

$$f(x, t) = A\cos(\kappa x - \omega t + \delta) \tag{8.25}$$

A sinusoidal oscillation of wave number κ and (angular) frequency ω traveling to the *left* would be written

$$f(x, t) = A\cos(\kappa x + \omega t - \delta) \tag{8.26}$$

The sign of the phase constant is chosen for consistency with our previous convention that δ/κ shall represent the distance by which the wave is "delayed"—since the wave is now moving to the *left,* a delay means a shift to the *right.* At $t = 0$, the wave looks like Fig. 8.6. Because the cosine is an *even* function, we could as well write (8.26) thus:

$$f(x, t) = A\cos(-\kappa x - \omega t + \delta) \tag{8.27}$$

Comparison with (8.25) shows that, in effect, *we simply switch the sign of* κ to produce a wave with the same amplitude, phase, frequency, and wavelength traveling in the opposite direction.

(ii) Complex notation. In view of Euler's famous formula,

$$e^{i\theta} = \cos\theta + i\sin\theta \tag{8.28}$$

the sinusoidal wave (8.25) can be written

$$f(x, t) = \mathrm{Re}(Ae^{i(\kappa x - \omega t + \delta)}) \tag{8.29}$$

where Re(w) denotes the real part of w. Suppose we introduce the *complex* wave function

$$\tilde{f}(x, t) \equiv \tilde{A}e^{i(\kappa x - \omega t)} \tag{8.30}$$

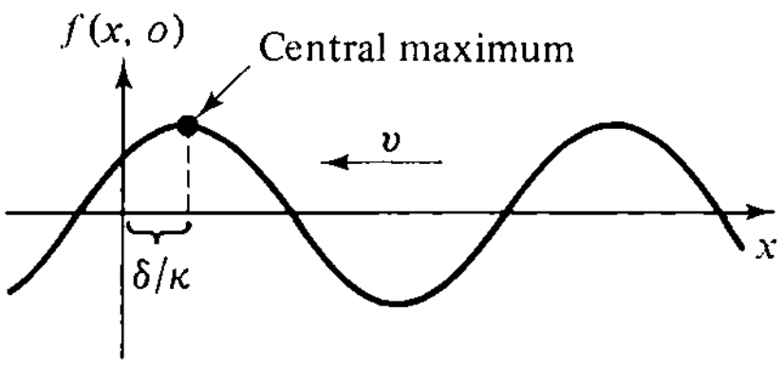

Figure 8.6

with the *complex amplitude* $\tilde{A} \equiv Ae^{i\delta}$ absorbing the phase constant. The *actual* (real) wave is the real part of $\tilde{f}$:

$$f(x, t) = \text{Re}\,\tilde{f}(x, t) \tag{8.31}$$

If you know $\tilde{f}$, it is a simple matter to find f; the *virtue* of the complex notation is that exponentials are much easier to manipulate than sines and cosines.

Example 2

Suppose you want to add two sinusoidal waves:

$$f_3 = f_1 + f_2 = \text{Re}(\tilde{f}_1) + \text{Re}(\tilde{f}_2) = \text{Re}(\tilde{f}_1 + \tilde{f}_2) = \text{Re}(\tilde{f}_3)$$

with $\tilde{f}_3 = \tilde{f}_1 + \tilde{f}_2$. Evidently you simply add the corresponding *complex* waves and then take the real part. To see how much easier this is than adding the waves directly, let's combine two waves of identical frequency and wavelength but different amplitude and phase:

$$\begin{aligned} f_3(x, t) &= A_1 \cos(\kappa x - \omega t + \delta_1) + A_2 \cos(\kappa x - \omega t + \delta_2) \\ &= A_1[\cos(\kappa x - \omega t)\cos\delta_1 - \sin(\kappa x - \omega t)\sin\delta_1] \\ &\quad + A_2[\cos(\kappa x - \omega t)\cos\delta_2 - \sin(\kappa x - \omega t)\sin\delta_2] \\ &= [A_1 \cos\delta_1 + A_2 \cos\delta_2]\cos(\kappa x - \omega t) \\ &\quad - [A_1 \sin\delta_1 + A_2 \sin\delta_2]\sin(\kappa x - \omega t) \end{aligned}$$

Now this is an ugly expression, and I'd forgive you if you did not happen to notice that it can be simplified by defining

$$\begin{aligned} A_3 &= \sqrt{A_1^2 + A_2^2 + 2A_1A_2\cos(\delta_1 - \delta_2)} \\ \delta_3 &= \tan^{-1}\left(\frac{A_1 \sin\delta_1 + A_2 \sin\delta_2}{A_1 \cos\delta_1 + A_2 \cos\delta_2}\right) \end{aligned} \tag{8.32}$$

Since

$$\begin{aligned} (A_1 \sin\delta_1 + A_2 \sin\delta_2)^2 &+ (A_1 \cos\delta_1 + A_2 \cos\delta_2)^2 \\ &= (A_1^2 \sin^2\delta_1 + 2A_1A_2 \sin\delta_1 \sin\delta_2 + A_2^2 \sin^2\delta_2) \\ &\quad + (A_1^2 \cos^2\delta_1 + 2A_1A_2 \cos\delta_1 \cos\delta_2 + A_2^2 \cos^2\delta_2) \\ &= A_1^2 + A_2^2 + 2A_1A_2\cos(\delta_1 - \delta_2) = A_3^2, \end{aligned}$$

it follows (Fig. 8.7) that

$$A_3 \cos\delta_3 = A_1 \cos\delta_1 + A_2 \cos\delta_2$$

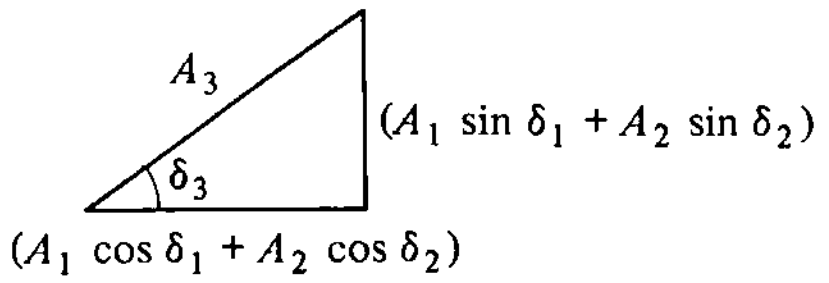

Figure 8.7

and

$$A_3 \sin \delta_3 = A_1 \sin \delta_1 + A_2 \sin \delta_2$$

and therefore

$$\begin{aligned} f_3(x, t) &= (A_3 \cos \delta_3)\cos(\kappa x - \omega t) - (A_3 \sin \delta_3)\sin(\kappa x - \omega t) \\ &= A_3 \cos(\kappa x - \omega t + \delta_3) \end{aligned}$$

The combined wave form, then, is equivalent to a *single* sinusoidal wave whose amplitude and phase constant are given by (8.32).

If you got lost in all that algebra, *good*! Because my whole point is to show you that this is a rotten way to treat the problem. Suppose we did it in the complex notation:

$$\begin{aligned} \tilde{f}_3 = \tilde{f}_1 + \tilde{f}_2 &= \tilde{A}_1 e^{i(\kappa x - \omega t)} + \tilde{A}_2 e^{i(\kappa x - \omega t)} \\ &= (\tilde{A}_1 + \tilde{A}_2) e^{i(\kappa x - \omega t)} = \tilde{A}_3 e^{i(\kappa x - \omega t)} \end{aligned}$$

So

$$f_3(x, t) = A_3 \cos(\kappa x - \omega t + \delta_3)$$

For waves of the same frequency and wavelength, then, you simply *add the complex amplitudes*:

$$\tilde{A}_3 = \tilde{A}_1 + \tilde{A}_2 \quad \text{or} \quad A_3 e^{i\delta_3} = A_1 e^{i\delta_1} + A_2 e^{i\delta_2} \tag{8.33}$$

From here it is an easy exercise to obtain (8.32).

(iii) Linear combinations of sinusoidal waves. Although the sinusoidal function (8.25) as it stands represents only one very special wave form, the fact is that *any* wave can be expressed as a linear combination of sinusoidal ones. That is, if we add together sinusoidal wave functions of different wavelengths (remembering that the sum of two or more solutions is itself a solution) it is possible to select the various amplitudes and phases so as to fit *any* solution to the wave equation. Since the wavelength (or, more conveniently, the wave number $\kappa = 2\pi/\lambda$) can take on any value, the most general linear combination has the form of an *integral*:

$$\tilde{f}(x, t) = \int_{-\infty}^{\infty} \tilde{A}(\kappa) e^{i(\kappa x - \omega t)}\, d\kappa \tag{8.34}$$

I have allowed κ to run over negative values in order to include waves going both directions.[3]

The actual formula for $\tilde{A}(\kappa)$, in terms of the initial conditions $f(x, 0)$ and $\dot{f}(x, 0)$, can be obtained from the theory of Fourier transforms, (see Problem 8.31) but the details are not relevant to my purpose here. The *point* is that any wave can be written as a linear combination of sinusoidal waves, and therefore if you know how sinusoidal

[3]This does not mean that λ and ω are negative—wavelength and frequency are *intrinsically* positive qualities. If we allow negative wave numbers, then (8.21) and (8.24) should really be written

$$\lambda = 2\pi/|\kappa| \text{ and } \omega = |\kappa| v$$

waves behave, you know in principle how *any* wave behaves. So, from now on, we shall limit our attention to sinusoidal waves.

Problem 8.3 Derive equation (8.32) from (8.33).

Problem 8.4 Obtain equation (8.34) from equation (8.11) by the method of separation of variables.

8.1.4 Polarization

The waves that travel down a string when you shake it are called **transverse**, because the displacement is perpendicular (transverse) to the direction of propagation. If the string is reasonably elastic, it is also possible to stimulate compression waves by giving the string little tugs. Compression waves are hard to see on a string, but if you try it with a slinky they're quite noticeable (Fig. 8.8). These waves are called **longitudinal**, because the displacement from equilibrium is along the direction of propagation. Sound waves, which are nothing but compression waves in air, are longitudinal; electromagnetic waves, as we shall see, are transverse.

Now there are, of course, *two* dimensions perpendicular to any given line of propagation. Accordingly, transverse waves occur in two independent states of **polarization**: you can shake the string up-and-down ("vertical" polarization—Fig. 8.9(a)),

$$\tilde{\mathbf{f}}_v(x, t) = \tilde{A}e^{i(\kappa x-\omega t)}\hat{j} \tag{8.35}$$

or left-and-right ("horizontal" polarization—Fig. 8.9(b)),

$$\tilde{\mathbf{f}}_h(x, t) = \tilde{A}e^{i(\kappa x-\omega t)}\hat{k} \tag{8.36}$$

or along any other direction in the yz plane (Fig. 8.9(c)):

$$\tilde{\mathbf{f}}(x, t) = \tilde{A}e^{i(\kappa x-\omega t)}\hat{n} \tag{8.37}$$

The **polarization vector** $\hat{n}$ defines the plane of vibration.[4] Because the waves are transverse, $\hat{n}$ is perpendicular to the direction of propagation:

$$\hat{n}\cdot\hat{i} = 0 \tag{8.38}$$

In terms of the **polarization angle** θ,

$$\hat{n} = \cos\theta\hat{j} + \sin\theta\hat{k} \tag{8.39}$$

Thus, the wave pictured in Fig. 8.9(c) can be considered a superposition of two waves—one horizontally polarized, the other vertically:

$$\tilde{\mathbf{f}}(x, t) = (\tilde{A}\cos\theta)e^{i(\kappa x-\omega t)}\hat{j} + (\tilde{A}\sin\theta)e^{i(\kappa x-\omega t)}\hat{k} \tag{8.40}$$

[4]Notice that you can always switch the *sign* of $\hat{n}$, provided you simultaneously advance the phase constant by 180°, since both operations change the sign of the wave.

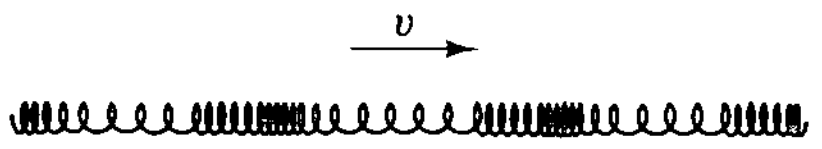

Figure 8.8

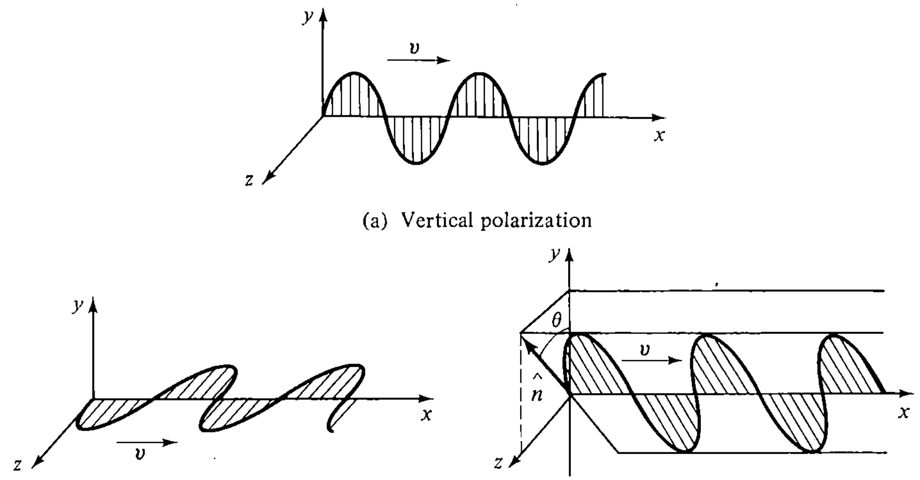

(a) Vertical polarization

(b) Horizontal polarization

(c) Polarization vector

Figure 8.9

Problem 8.5 Equation (8.37) describes the most general **linearly** polarized wave on a string. Linear (or "plane") polarization (so-called because the displacement is parallel to a fixed vector $\hat{n}$) results from the combination of horizontally and vertically polarized waves of the *same phase* (equation (8.40)). If the two components are of equal amplitude, but *out of phase* by 90° (say, $\delta_v = 0$, $\delta_h = 90°$), the result is a *circularly* polarized wave:

(a) At a fixed point x, show that the string moves in a circle about the x axis. Does it go *clockwise* or *counterclockwise*, as you look down the axis toward the origin? How would you construct a wave circling the *other* way? (In optics, the clockwise case is called **right circular polarization**, and the counterclockwise, **left circular polarization.**)

(b) Sketch the string at time $t = 0$.

(c) How would you shake the string in order to produce a circularly polarized wave?

8.1.5 Boundary Conditions: Reflection and Transmission

So far I have assumed the string is infinitely long—or, at any rate, long enough so that we don't have to worry about what happens to a wave when it reaches the end. As a matter of fact, what happens depends a lot on how the string is *attached* at the end—that is, on the specific boundary conditions to which the wave is subject. Suppose, for instance, that the string is simply tied onto the end of a *second* string. The tension T is the same for both strings, but the mass per unit length μ presumably is not, and hence the wave velocities v_1 and v_2 are different (remember, $v = \sqrt{T/\mu}$). Let's say, for convenience, that the knot occurs at $x = 0$. The **incident** wave[5]

$$\tilde{f}_I = \tilde{A}_I e^{i(\kappa_1 x - \omega t)}, \qquad (x < 0) \tag{8.41}$$

[5]Polarization is irrelevant, for the moment, so I shall treat f as a scalar.

coming in from the left, gives rise to a **reflected** wave

$$\tilde{f}_R = \tilde{A}_R e^{i(-\kappa_1 x - \omega t)}, \qquad (x < 0) \tag{8.42}$$

traveling *back* along string 1 (hence the minus sign in front of κ_1), in addition to a **transmitted** wave

$$\tilde{f}_T = \tilde{A}_T e^{i(\kappa_2 x - \omega t)}, \qquad (x > 0) \tag{8.43}$$

which continues on to the right in string 2.

Please note: The incident wave $f_I(x, t)$ is a sinusoidal oscillation that extends (in principle) all the way back to $x = -\infty$, and has been doing so for all of time. The same goes for f_R and f_T (except that the latter, of course, extends to $x = +\infty$). *All parts of the system are oscillating at the same frequency* ω (a frequency determined by the person at $x = -\infty$, who is shaking the string in the first place). Since the wave velocities are different in the two strings, the wavelengths and wave numbers are also different:

$$\frac{\lambda_1}{\lambda_2} = \frac{\kappa_2}{\kappa_1} = \frac{v_1}{v_2} \tag{8.44}$$

Of course, this situation is pretty artificial—what's more, with incident and reflected waves of infinite extent traveling on the same piece of string, it's hard for a spectator to tell them apart. You might therefore prefer to consider an incident wave of *finite* extent—say, the pulse shown in Fig. 8.10(a). You can work out the details for yourself (see Problem 8.6). The *trouble* with this approach is that no *finite* pulse is truly sinusoidal. The waves of Fig. 8.10 may *look* like sine functions to you, but they're *not*: they're little snippets of sines, joined onto an entirely *different* function (namely, zero). Like any other waves, they can be built up as *linear combinations* of true sinusoidal functions (8.34), but only by putting together a whole range of frequencies and wavelengths. If you want a *single* incident frequency (as we shall in the electromagnetic case), you must let your waves extend to infinity. In practice, if you use a very *long* pulse with many oscillations, it will be *close* to the ideal of a single frequency.

For a sinusoidal incident wave, then, the net disturbance of the string is given by:

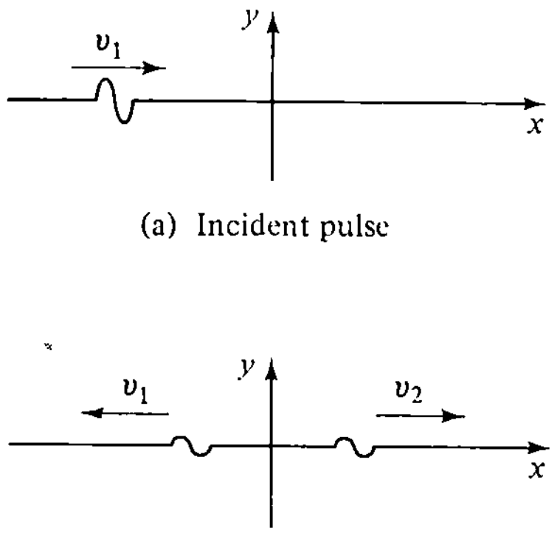

(b) Reflected and transmitted pulses **Figure 8.10**

$$\tilde{f}(x, t) = \begin{cases} \tilde{A}_I e^{i(\kappa_1 x - \omega t)} + \tilde{A}_R e^{i(-\kappa_1 x - \omega t)}, & \text{for } x < 0 \\ \tilde{A}_T e^{i(\kappa_2 x - \omega t)}, & \text{for } x > 0 \end{cases} \tag{8.45}$$

At the join ($x = 0$), the displacement just slightly to the left ($x = 0^-$) must equal the displacement slightly to the right ($x = 0^+$), else there would be a break between the two strings. Mathematically, $f(x, t)$ *is a continuous function at* $x = 0$:

$$f(0^-, t) = f(0^+, t) \tag{8.46}$$

If the knot itself is of negligible mass, then the *derivative* of f must *also* be continuous:

$$\left.\frac{\partial f}{\partial x}\right|_{0-} = \left.\frac{\partial f}{\partial x}\right|_{0+} \tag{8.47}$$

Otherwise there would be a net force on the knot and, consequently, an infinite acceleration (Fig. 8.11). These boundary conditions, which hold for all times t, apply directly to the *real* wave function $f(x, t)$. But since the imaginary part of $\tilde{f}$ differs from the real part only in the replacement of cosine by sine (8.28), it follows that the complex wave function $\tilde{f}(x, t)$ obeys the same rules:

$$\tilde{f}(0^-, t) = \tilde{f}(0^+, t), \qquad \left.\frac{\partial \tilde{f}}{\partial x}\right|_{0-} = \left.\frac{\partial \tilde{f}}{\partial x}\right|_{0+} \tag{8.48}$$

When applied to equation (8.45), these boundary conditions determine the outgoing amplitudes ($\tilde{A}_R$ and $\tilde{A}_T$) in terms of the incoming one ($\tilde{A}_I$):

$$\tilde{A}_I + \tilde{A}_R = \tilde{A}_T, \qquad \kappa_1(\tilde{A}_I - \tilde{A}_R) = \kappa_2 \tilde{A}_T$$

from which it follows that

$$\tilde{A}_R = \left(\frac{\kappa_1 - \kappa_2}{\kappa_1 + \kappa_2}\right)\tilde{A}_I, \qquad \tilde{A}_T = \left(\frac{2\kappa_1}{\kappa_1 + \kappa_2}\right)\tilde{A}_I \tag{8.49}$$

Or, in terms of the velocities (8.44):

$$\tilde{A}_R = \left(\frac{v_2 - v_1}{v_2 + v_1}\right)\tilde{A}_I, \qquad \tilde{A}_T = \left(\frac{2v_2}{v_2 + v_1}\right)\tilde{A}_I \tag{8.50}$$

The *real* amplitudes and phases, then, are related by

$$A_R e^{i\delta_R} = \left(\frac{v_2 - v_1}{v_2 + v_1}\right)A_I e^{i\delta_I}, \qquad A_T e^{i\delta_T} = \left(\frac{2v_2}{v_2 + v_1}\right)A_I e^{i\delta_I} \tag{8.51}$$

If the second string is *lighter* than the first, $\mu_2 < \mu_1$, so that $v_2 > v_1$, then all three

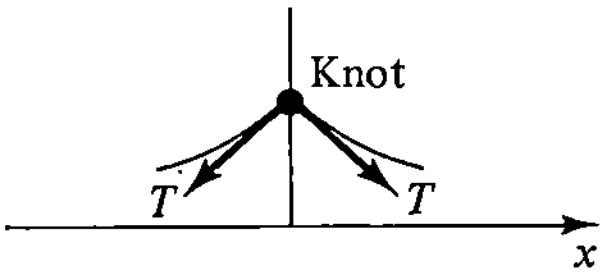

(a) Discontinuous slope; force on knot

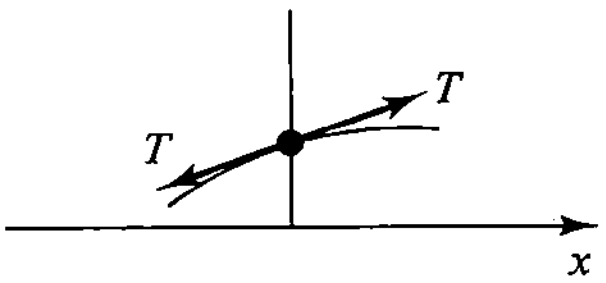

(b) Continuous slope; no force on knot

Figure 8.11

waves have the same phase angle ($\delta_R = \delta_T = \delta_I$), and the outgoing amplitudes are

$$A_R = \left(\frac{v_2 - v_1}{v_2 + v_1}\right)A_I, \qquad A_T = \left(\frac{2v_2}{v_2 + v_1}\right)A_I \tag{8.52}$$

If the second string is *heavier* than the first, so that $v_2 < v_1$, then the reflected wave is out of phase by 180° ($\delta_R + \pi = \delta_T = \delta_I$). In other words, since

$$\cos(-\kappa_1 x - \omega t + \delta_I - \pi) = -\cos(-\kappa_1 x - \omega t + \delta_I)$$

the reflected wave is "upside down." The amplitudes in this case are

$$A_R = \left(\frac{v_1 - v_2}{v_2 + v_1}\right)A_I \quad \text{and} \quad A_T = \left(\frac{2v_2}{v_2 + v_1}\right) \tag{8.53}$$

In particular, if the second string is *infinitely* massive—or, what amounts to the same thing, if the first string is simply *nailed down* at the end—then

$$A_R = A_I \quad \text{and} \quad A_T = 0$$

Problem 8.6 Suppose you send an incident wave of *arbitrary* shape, $g_I(x - v_1 t)$, down string number 1. It gives rise to a reflected wave, $h_R(x + v_1 t)$, and a transmitted wave, $g_T(x - v_2 t)$. By imposing the boundary conditions (8.46) and (8.47), find h_R and g_T.

Problem 8.7

(a) Formulate an appropriate boundary condition, to replace (8.47), for the case of two strings joined by a knot of mass M.

(b) Find the amplitude and phase of the reflected and transmitted waves for the case where the knot has a mass M, and the second string is massless.

! **Problem 8.8** Suppose string 2 is embedded in a viscous medium (such as molasses) which imposes a frictional force, on any section of length Δx, that is proportional to its (transverse) speed:

$$F_{\text{friction}} = -\gamma \frac{\partial f}{\partial t} \Delta x$$

(a) Derive the modified wave equation describing the motion of the string.

(b) Solve this equation, assuming the string oscillates at the incident frequency ω. That is, look for solutions of the form $\tilde{f}(x, t) = e^{-i\omega t}\tilde{F}(x)$.

(c) Show that the waves are **attenuated** (that is, their amplitude decreases with increasing x). Find the characteristic penetration distance, at which the amplitude is $1/e$ of its original value, in terms of γ, T, μ, and ω.

(d) If a wave of amplitude A_I, phase $\delta_I = 0$, and frequency ω is incident from the left (string 1), find the reflected wave's amplitude and phase.

8.2 ELECTROMAGNETIC WAVES IN NONCONDUCTING MEDIA

8.2.1 Monochromatic Plane Waves in Vacuum

Section 8.1 was largely a digression—I began with the theory of waves on a string because (a) they are easier to "picture" than electromagnetic waves, and (b) I want it to be clear which features are common to *all* wave motion and which are peculiar to electrodynamics. At any rate, we return now to the case of **E** and **B** in empty space, which, you will recall from Section 8.1.1, satisfy the three-dimensional wave equation

$$\nabla^2\mathbf{E} = \frac{1}{c^2}\frac{\partial^2\mathbf{E}}{\partial t^2}, \qquad \nabla^2\mathbf{B} = \frac{1}{c^2}\frac{\partial^2\mathbf{B}}{\partial t^2} \tag{8.54}$$

where $c = 1/\sqrt{\epsilon_0\mu_0}$ is the speed of light in vacuum. For reasons discussed in Section 8.1.3, we may confine our attention to sinusoidal waves of frequency ω. Since frequency corresponds perceptually to *color,* such waves are called **monochromatic** (Table 8.1). Suppose for the moment that the waves are traveling in the x-direction and have no y- or z-dependence; these are called **plane waves,**[6] because the fields are

TABLE 8.1 THE ELECTROMAGNETIC SPECTRUM

Frequency (Hz)	Name of radiation	Wavelength (m)
10^{22}—	Gamma rays	—10^{-13}
10^{21}—		—10^{-12}
10^{20}—		—10^{-11}
10^{19}—	X rays	—10^{-10}
10^{18}—		—10^{-9}
10^{17}—	Ultraviolet	—10^{-8}
10^{16}—		—10^{-7}
10^{15}—	Visible	—10^{-6}
10^{14}—	Infrared	—10^{-5}
10^{13}—		—10^{-4}
10^{12}—		—10^{-3}
10^{11}—	Microwave	—10^{-2}
10^{10}—		—10^{-1}
10^{9} —	TV, FM	—1
10^{8} —		—10
10^{7} —	Standard broadcast	—10^{2}
10^{6} —		—10^{3}
10^{5} —	Radiofrequency	—10^{4}
10^{4} —		—10^{5}
10^{3} —		

Color	Wavelength (m)	Frequency (Hz)
Near ultraviolet	3.0×10^{-7}	10.0×10^{14}
Shortest visible blue	4.0×10^{-7}	7.5×10^{14}
Blue	4.6×10^{-7}	6.5×10^{14}
Green	5.4×10^{-7}	5.6×10^{14}
Yellow	5.9×10^{-7}	5.1×10^{14}
Orange	6.1×10^{-7}	4.9×10^{14}
Longest visible red	7.6×10^{-7}	3.9×10^{14}
Near infrared	10.0×10^{-7}	3.0×10^{14}

[6]For a discussion of *spherical* waves, at this level, see J. R. Reitz, F. J. Milford, and R. W. Christy, *Foundations of Electromagnetic Theory,* 3d ed. Section 17-5, (Reading, MA: Addison-Wesley, 1979). Or work Problem 8.32. Of course, over small enough regions *any* wave is essentially plane, as long as the wavelength is much less than the radius of curvature of the wave front.

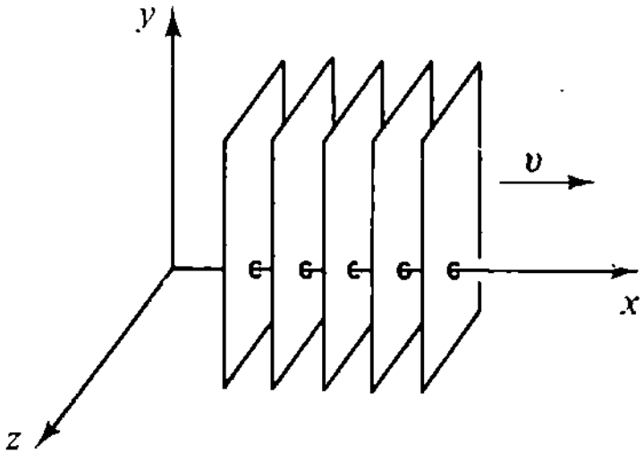

Figure 8.12

uniform over every plane perpendicular to the direction of propagation (Fig. 8.12):

$$\tilde{\mathbf{E}}(x, t) = \tilde{\mathbf{E}}_0 e^{i(\kappa x - \omega t)}, \qquad \tilde{\mathbf{B}}(x, t) = \tilde{\mathbf{B}}_0 e^{i(\kappa x - \omega t)} \tag{8.55}$$

$\tilde{\mathbf{E}}_0$ and $\tilde{\mathbf{B}}_0$ are the (complex) amplitudes of the electric and magnetic fields—the *physical* fields are the real parts of $\tilde{\mathbf{E}}$ and $\tilde{\mathbf{B}}$.

Now, the wave equations (8.54) were derived from Maxwell's equations. However, whereas every solution to Maxwell's equations (in empty space) must obey the wave equation, the converse is *not* true; Maxwell's equations impose special constraints on $\tilde{\mathbf{E}}_0$ and $\tilde{\mathbf{B}}_0$. In particular, since $\nabla \cdot \mathbf{E} = 0$ and $\nabla \cdot \mathbf{B} = 0$, it follows[7] that

$$(\tilde{E}_0)_x = (\tilde{B}_0)_x = 0 \tag{8.56}$$

That is, ***electromagnetic waves are transverse***: the electric and magnetic fields are perpendicular to the direction of propagation. Moreover, Faraday's law, $\nabla \times \mathbf{E} = -\partial \mathbf{B}/\partial t$, implies a relation between the electric and magnetic amplitudes, to wit:

$$-\kappa(\tilde{E}_0)_z = \omega(\tilde{B}_0)_y, \qquad \kappa(\tilde{E}_0)_y = \omega(\tilde{B}_0)_z \tag{8.57}$$

or, more compactly:

$$\tilde{\mathbf{B}}_0 = \frac{\kappa}{\omega}(\hat{\imath} \times \tilde{\mathbf{E}}_0) \tag{8.58}$$

Evidently, **E** and **B** are ***in phase*** and ***mutually perpendicular***; their (real) amplitudes are related by

$$B_0 = \frac{\kappa}{\omega} E_0 = \frac{1}{c} E_0 \tag{8.59}$$

The fourth of Maxwell's equations, $\nabla \times \mathbf{B} = \mu_0 \epsilon_0 (\partial \mathbf{E}/\partial t)$, does not yield an independent condition; it simply reproduces (8.57).

Example 3

If **E** points in the y-direction, then according to (8.58) **B** points in the z-direction (Fig. 8.13):

$$\tilde{\mathbf{E}}(x, t) = \tilde{\mathbf{E}}_0 e^{i(\kappa x - \omega t)} \hat{\jmath}, \qquad \tilde{\mathbf{B}}(x, t) = \frac{1}{c} \tilde{\mathbf{E}}_0 e^{i(\kappa x - \omega t)} \hat{k}$$

[7] Because the real part of $\tilde{\mathbf{E}}$ differs from the imaginary part only in the replacement of sine by cosine, if the former obeys Maxwell's equations, so does the latter, and hence $\tilde{\mathbf{E}}$ as well.

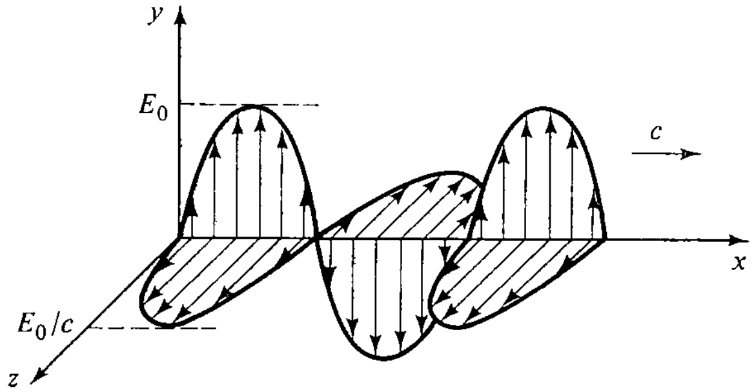

Figure 8.13

or

$$\mathbf{E}(x, t) = E_0 \cos(\kappa x - \omega t + \delta)\hat{j}, \qquad \mathbf{B}(x, t) = \frac{1}{c} E_0 \cos(\kappa x - \omega t + \delta)\hat{k} \tag{8.60}$$

The wave as a whole is said to be polarized along the y-direction: By convention, we use the direction of $\mathbf{E}$ to specify the polarization of an electromagnetic wave.

There is nothing special about the x-direction, of course—we can easily generalize to monochromatic plane waves traveling in an arbitrary direction. The notation is facilitated by introduction of the **propagation** (or **wave**) **vector** $\boldsymbol{\kappa}$, pointing in the direction of propagation, whose magnitude is the wave number κ. The scalar product $(\boldsymbol{\kappa} \cdot \mathbf{r})$ is then the appropriate generalization of κx (Fig. 8.14), and so

$$\tilde{\mathbf{E}}(\mathbf{r}, t) = \tilde{E}_0 e^{i(\boldsymbol{\kappa} \cdot \mathbf{r} - \omega t)} \hat{n} \tag{8.61}$$

$$\tilde{\mathbf{B}}(\mathbf{r}, t) = \frac{1}{c} \tilde{E}_0 e^{i(\boldsymbol{\kappa} \cdot \mathbf{r} - \omega t)} (\hat{\kappa} \times \hat{n}) = \frac{1}{c} \hat{\kappa} \times \tilde{\mathbf{E}} \tag{8.62}$$

where $\hat{n}$ is the polarization vector. Because $\mathbf{E}$ is transverse,

$$\hat{n} \cdot \hat{\kappa} = 0 \tag{8.63}$$

The transversality of $\mathbf{B}$ follows automatically from (8.62). The actual (real) electric and magnetic fields in a monochromatic plane wave with propagation vector $\boldsymbol{\kappa}$ and polarization $\hat{n}$ are therefore

$$\mathbf{E}(\mathbf{r}, t) = E_0 \cos(\boldsymbol{\kappa} \cdot \mathbf{r} - \omega t + \delta)\hat{n} \tag{8.64}$$

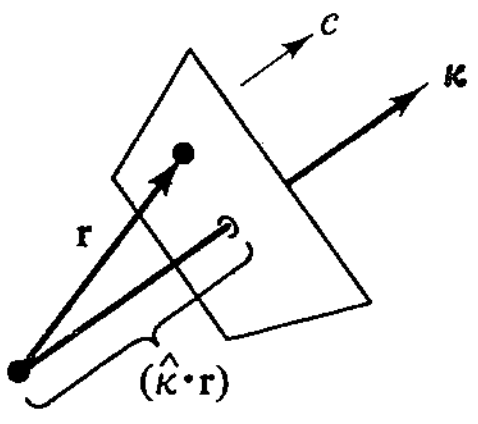

Figure 8.14

$$\mathbf{B}(\mathbf{r}, t) = \frac{1}{c} E_0 \cos(\boldsymbol{\kappa} \cdot \mathbf{r} - \omega t + \delta)(\hat{\kappa} \times \hat{n}) \tag{8.65}$$

Problem 8.9 Write down the (real) electric and magnetic fields for a monochromatic plane wave of amplitude E_0, frequency ω, and phase angle zero that is
(a) traveling in the negative y-direction and polarized in the x-direction;
(b) traveling in the direction from the origin to the point (1, 1, 1), with polarization parallel to the xy plane.
In each case, sketch the wave, and give the explicit Cartesian components of $\boldsymbol{\kappa}$ and $\hat{n}$.

8.2.2 Energy and Momentum of Electromagnetic Waves

According to equation (7.88), the energy per unit volume stored in an electromagnetic field is

$$U = \frac{1}{2}\left(\epsilon_0 E^2 + \frac{1}{\mu_0} B^2\right) \tag{8.66}$$

In the case of a monochromatic plane wave, (8.60),

$$B^2 = \frac{1}{c^2} E^2 = \mu_0 \epsilon_0 E^2 \tag{8.67}$$

so the *electric and magnetic contributions are equal*:

$$U = \epsilon_0 E^2 = \epsilon_0 E_0^2 \cos^2(\kappa x - \omega t + \delta) \tag{8.68}$$

As the wave propagates, it carries this energy along with it. The energy flux density (energy per unit area, per unit time) transported by the fields is given by the Poynting vector (7.85):

$$\mathbf{S} = \frac{1}{\mu_0}(\mathbf{E} \times \mathbf{B}) \tag{8.69}$$

For monochromatic plane waves,

$$\mathbf{S} = c\epsilon_0 E_0^2 \cos^2(\kappa x - \omega t + \delta)\hat{\imath} = cU\hat{\imath} \tag{8.70}$$

Observe that $\mathbf{S}$ is the energy density (U) times the velocity of the waves ($c\hat{\imath}$)—as it *should* be. For in a time Δt, a length $c\,\Delta t$ of wave passes through area A (Fig. 8.15), carrying with it an energy $UAc\,\Delta t$. The energy per unit time, per unit area, transported by the wave is therefore Uc.

Electromagnetic fields not only carry *energy,* they also carry *momentum.* In fact, we found in equation (7.103) that the momentum density stored in the fields is

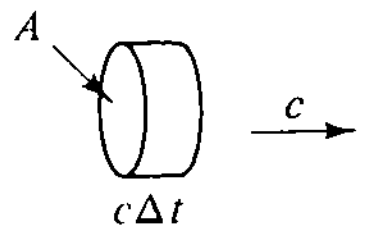

Figure 8.15

given by

$$\mathbf{p} = \frac{1}{c^2}\mathbf{S} \tag{8.71}$$

For monochromatic plane waves, then,

$$\mathbf{p} = \frac{1}{c}\epsilon_0 E_0^2 \cos^2(\kappa x - \omega t + \delta)\hat{i} = \frac{1}{c}U\hat{i} \tag{8.72}$$

Notice the connection between the energy and the momentum carried by the wave:

$$U = \mathrm{p}c \tag{8.73}$$

In the case of *light*, the wavelength is so short (5×10^{-7} m) and the period so brief (10^{-15} s) that any macroscopic measurement will encompass many cycles. Typically, therefore, we're not interested in the fluctuating cosine-squared term in the energy and momentum densities; all we want is its *average* value. Now, the average of $\cos^2$ over a complete cycle[8] is $\frac{1}{2}$, so

$$\langle U \rangle = \frac{1}{2}\epsilon_0 E_0^2 \tag{8.74}$$

$$\langle \mathbf{S} \rangle = \frac{1}{2}c\epsilon_0 E_0^2\hat{i} = c\langle U \rangle\hat{i} \tag{8.75}$$

$$\langle \mathbf{p} \rangle = \frac{1}{2}\frac{1}{c}\epsilon_0 E_0^2\hat{i} = \frac{1}{c}\langle U \rangle\hat{i} \tag{8.76}$$

I use brackets, $\langle\ \rangle$, to denote the average over a complete cycle (or *many* cycles, if you prefer[9]). The average power per unit area transported by an electromagnetic wave is called the **intensity**:

$$I = \langle S \rangle \tag{8.77}$$

Problem 8.10 The intensity of sunlight hitting the earth is about 1300 W/m^2. Find the amplitude of the electric and magnetic fields. How far from a stationary electron would you have to be to get a comparable electric field? (Actually, the sun's radiation is neither monochromatic nor linearly polarized, so don't take these numbers too seriously.) If the sunlight falls on a perfect absorber (which soaks up all the incident momentum), what pressure does it exert? How about a perfect reflector? What fraction of atmospheric pressure does this amount to?

Problem 8.11 In the complex notation there is a cute trick to handle the time averaging of U

[8] $\sin^2\theta + \cos^2\theta = 1$, but over a complete cycle the average of $\sin^2\theta$ is equal to the average of $\cos^2\theta$, so $\langle \sin^2 \rangle = \langle \cos^2 \rangle = \frac{1}{2}$. More formally,

$$\frac{1}{T}\int_0^T \cos^2\left(\kappa x - \frac{2\pi}{T}t + \delta\right)dt = \frac{1}{2}$$

[9] If you use *many* cycles, it need not even be a whole number of them, since the extra bit makes a negligible contribution to the average.

and **S**: Take the complex conjugate of the second field and multiply by $\frac{1}{2}$. Thus (from 8.66 and 8.69),

$$\langle U\rangle = \frac{1}{4}\left(\epsilon_0 \tilde{\mathbf{E}}\cdot\tilde{\mathbf{E}}^* + \frac{1}{\mu_0}\tilde{\mathbf{B}}\cdot\tilde{\mathbf{B}}^*\right)$$

$$\langle \mathbf{S}\rangle = \frac{1}{2\mu_0}(\tilde{\mathbf{E}}\times\tilde{\mathbf{B}}^*)$$

Check that these expressions reproduce 8.74 and 8.75. Show that this device works even if **E** and **B** are out of phase, provided you take the real part of the result.

Problem 8.12 Find all elements of the Maxwell stress tensor for a monochromatic plane wave traveling in the x-direction and linearly polarized in the y-direction (equation (8.60)). Does your answer make sense? (Remember that **T** represents the momentum flux density.) How is the momentum flux density related to the energy density?

8.2.3 Propagation Through Linear Media

So far, I have assumed the electromagnetic waves are traveling through *empty space*, but it is easy to generalize these results to *any linear medium*:[10] You simply change ϵ_0 *to* ϵ *and* μ_0 *to* μ. Thus the speed of propagation becomes

$$v = \frac{1}{\sqrt{\epsilon\mu}} = \frac{c}{n} \tag{8.78}$$

(as we found in Section 8.1.1), where

$$n = \sqrt{\frac{\epsilon\mu}{\epsilon_0\mu_0}} \tag{8.79}$$

is the "index of refraction" of the material. The energy density is

$$U = \frac{1}{2}\left(\epsilon E^2 + \frac{1}{\mu}B^2\right) \tag{8.80}$$

and the Poynting vector is

$$\mathbf{S} = \frac{1}{\mu}(\mathbf{E}\times\mathbf{B}) \tag{8.81}$$

(see Section 7.5.2). In particular, the *intensity* of the wave is

$$I = \frac{1}{2}\epsilon v E_0^2 \tag{8.82}$$

The interesting question is this: What happens when a wave passes from one such medium into another—air to water, say, or glass to plastic? As in the case of

[10]By "linear" I mean to imply not only that **D** and **H** are proportional to **E** and **B**, respectively, but that the proportionality constants ϵ and μ are independent of position and direction; in other words, the medium is *homogenous, isotropic,* and *linear.* I also assume there are no *free* charges or currents in the material.

waves on a string, we can expect to get a reflected wave and a transmitted wave. The details depend on the exact nature of the electrodynamic boundary conditions, which we derived in Chapter 7 (equation (7.62)):

$$\left.\begin{array}{llll} \text{(i)} & \epsilon_1 E_{1_\perp} = \epsilon_2 E_{2_\perp} & \text{(iii)} & \mathbf{E}_{1_\parallel} = \mathbf{E}_{2_\parallel} \\ \text{(ii)} & B_{1_\perp} = B_{2_\perp} & \text{(iv)} & \dfrac{1}{\mu_1}\mathbf{B}_{1_\parallel} = \dfrac{1}{\mu_2}\mathbf{B}_{2_\parallel} \end{array}\right\} \tag{8.83}$$

These equations relate the electric and magnetic fields just to the left and just to the right of the interface between two linear media. In the following sections we use them to deduce the laws governing reflection and refraction of electromagnetic waves.

8.2.4 Reflection and Transmission at Normal Incidence

Suppose the yz plane forms the boundary between two linear media. A plane wave of frequency ω, traveling in the x direction and polarized in the y direction, approaches the interface from the left (Fig. 8.16):

$$\left.\begin{aligned} \tilde{\mathbf{E}}_I(x,t) &= \tilde{E}_{0_I} e^{i(\kappa_1 x - \omega t)}\hat{j} \\ \tilde{\mathbf{B}}_I(x,t) &= \frac{1}{v_1}\tilde{E}_{0_I} e^{i(\kappa_1 x - \omega t)}\hat{k} \end{aligned}\right\} \tag{8.84}$$

It gives rise to a reflected wave

$$\left.\begin{aligned} \tilde{\mathbf{E}}_R(x,t) &= \tilde{E}_{0_R} e^{i(-\kappa_1 x - \omega t)}\hat{j} \\ \tilde{\mathbf{B}}_R(x,t) &= -\frac{1}{v_1}\tilde{E}_{0_R} e^{i(-\kappa_1 x - \omega t)}\hat{k} \end{aligned}\right\} \tag{8.85}$$

which travels back to the left in medium (1), and a transmitted wave

$$\left.\begin{aligned} \tilde{\mathbf{E}}_T(x,t) &= \tilde{E}_{0_T} e^{i(\kappa_1 x - \omega t)}\hat{j} \\ \tilde{\mathbf{B}}_T(x,t) &= \frac{1}{v_2}\tilde{E}_{0_T} e^{i(\kappa_2 x - \omega t)}\hat{k} \end{aligned}\right\} \tag{8.86}$$

which continues on to the right in medium (2). Note the minus sign in $\tilde{\mathbf{B}}_R$, as re-

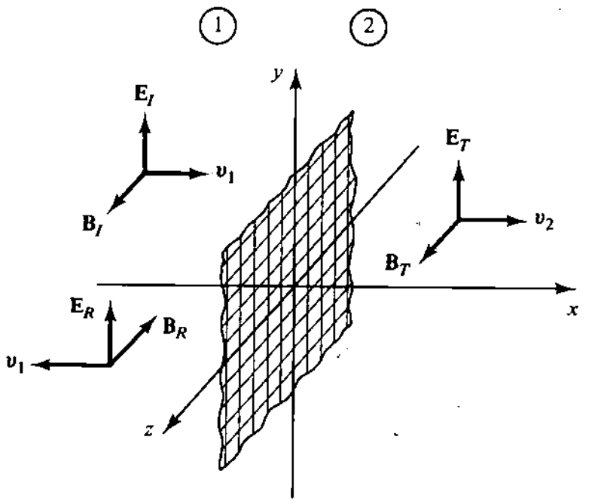

Figure 8.16

quired by (8.62)—or, if you prefer, by the fact that the Poynting vector must aim in the direction of propagation.

At $x = 0$, the combined fields to the left, $\tilde{\mathbf{E}}_I + \tilde{\mathbf{E}}_R$ and $\tilde{\mathbf{B}}_I + \tilde{\mathbf{B}}_R$, must join the fields to the right, $\tilde{\mathbf{E}}_T$ and $\tilde{\mathbf{B}}_T$, in accordance with the boundary conditions (8.83). In this case there are no components perpendicular to the surface so (i) and (ii) are trivial. However, (iii) requires that

$$\tilde{E}_{0_I} + \tilde{E}_{0_R} = \tilde{E}_{0_T} \tag{8.87}$$

whereas (iv) says

$$\frac{1}{\mu_1}\left(\frac{1}{v_1}\tilde{E}_{0_I} - \frac{1}{v_1}\tilde{E}_{0_R}\right) = \frac{1}{\mu_2}\left(\frac{1}{v_2}\tilde{E}_{0_T}\right)$$

or

$$\tilde{E}_{0_I} - \tilde{E}_{0_R} = \beta\tilde{E}_{0_T} \tag{8.88}$$

where

$$\beta = \frac{\mu_1 v_1}{\mu_2 v_2} \tag{8.89}$$

Equations (8.87) and (8.88) are easily solved for the outgoing amplitudes, in terms of the incident amplitude:

$$\tilde{E}_{0_R} = \left(\frac{1-\beta}{1+\beta}\right)\tilde{E}_{0_I}, \qquad \tilde{E}_{0_T} = \left(\frac{2}{1+\beta}\right)\tilde{E}_{0_I} \tag{8.90}$$

These results are strikingly similar to those describing waves on a string (8.49). Indeed, if the permittivities μ are close to their values in vacuum (as, in fact, they *are* for most media), then $\beta = v_1/v_2$, and we have

$$\tilde{E}_{0_R} = \left(\frac{v_2 - v_1}{v_2 + v_1}\right)\tilde{E}_{0_I}, \qquad \tilde{E}_{0_T} = \left(\frac{2v_2}{v_2 + v_1}\right)\tilde{E}_{0_I} \tag{8.91}$$

which are *identical* to (8.50). As before, the reflected wave is *in phase* (right side up) if $v_2 > v_1$ and *out of phase* (upside down) if $v_2 < v_1$; the real amplitudes are related by

$$E_{0_R} = \left|\frac{v_2 - v_1}{v_2 + v_1}\right| E_{0_I}, \qquad E_{0_T} = \left(\frac{2v_2}{v_2 + v_1}\right) E_{0_I} \tag{8.92}$$

or, in terms of the index of refraction,

$$E_{0_R} = \left|\frac{n_1 - n_2}{n_1 + n_2}\right| E_{0_I}, \qquad E_{0_T} = \left(\frac{2n_1}{n_1 + n_2}\right) E_{0_I} \tag{8.93}$$

What fraction of the incident energy is reflected and what fraction transmitted? According to (8.82), the intensity (average power per unit area) is given by

$$I = \tfrac{1}{2}\epsilon v E_0^2$$

If (again) $\mu_1 = \mu_2 = \mu_0$, then the ratio of the reflected intensity to the incident intensity is

$$R = \frac{I_R}{I_I} = \left(\frac{E_{0_R}}{E_{0_I}}\right)^2 = \left(\frac{n_1 - n_2}{n_1 + n_2}\right)^2 \tag{8.94}$$

whereas the ratio of the transmitted intensity to the incident intensity is

$$T = \frac{I_T}{I_I} = \frac{\epsilon_2 v_2}{\epsilon_1 v_1}\left(\frac{E_{0_T}}{E_{0_I}}\right)^2 = \frac{n_2}{n_1}\left(\frac{2n_1}{n_1 + n_2}\right)^2 \tag{8.95}$$

R is called the **reflection coefficient** and T the **transmission coefficient** at the surface; they measure the fraction of the incident energy that is reflected and transmitted, respectively. Notice that

$$R + T = 1 \tag{8.96}$$

as conservation of energy, of course, requires. For instance, when light passes from air ($n_1 = 1$) into glass ($n_2 = 1.5$), $R = 0.04$ and $T = 0.96$. Not surprisingly, most of the light is transmitted.

Problem 8.13 Calculate the *exact* reflection and transmission coefficients *without* assuming $\mu_1 = \mu_2 = \mu_0$. Confirm that $R + T = 1$.

Problem 8.14 In writing equations (8.85) and (8.86), I implicitly assumed that the reflected and transmitted waves have the same polarization as the incident wave—along the y-direction. Prove that this *must* be so. (Let the polarization vectors of the transmitted and reflected waves be

$$\hat{n}_T = \cos\theta_T \hat{j} + \sin\theta_T \hat{k}, \qquad \hat{n}_R = \cos\theta_R \hat{j} + \sin\theta_R \hat{k}$$

and prove from the boundary conditions that $\theta_T = \theta_R = 0$.)

8.2.5 Reflection and Transmission at Oblique Incidence

In the last section I treated reflection and transmission at *normal* incidence, that is, when the incoming wave hits the interface head-on. We now turn to the more general case of *oblique* incidence, in which the incoming wave meets the boundary at a glancing angle θ_I (Fig. 8.17). Of course, normal incidence is really just a special case of

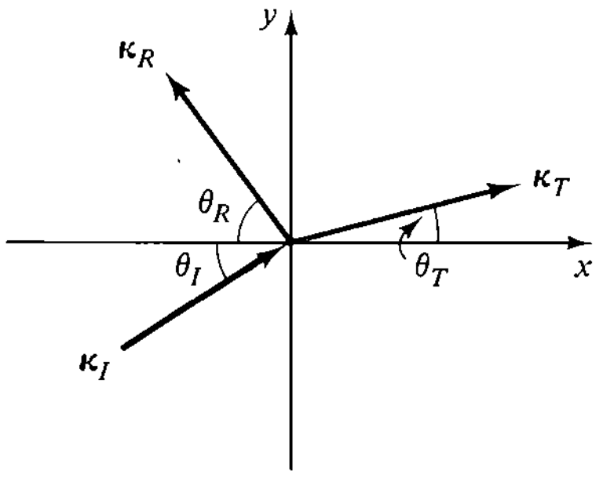

Figure 8.17

oblique incidence, with $\theta_I = 0$, but I wanted to treat it separately, as a kind of warm-up because the algebra is now going to get a little heavy.

Suppose, then, that a monochromatic plane wave

$$\tilde{\mathbf{E}}_I(\mathbf{r}, t) = \tilde{\mathbf{E}}_{0_I} e^{i(\boldsymbol{\kappa}_I \cdot \mathbf{r} - \omega t)}, \qquad \tilde{\mathbf{B}}_I(\mathbf{r}, t) = \frac{1}{v_1}(\hat{\boldsymbol{\kappa}}_I \times \tilde{\mathbf{E}}_I) \tag{8.97}$$

approaches from the left, giving rise to a reflected wave,

$$\tilde{\mathbf{E}}_R(\mathbf{r}, t) = \tilde{\mathbf{E}}_{0_R} e^{i(\boldsymbol{\kappa}_R \cdot \mathbf{r} - \omega t)}, \qquad \tilde{\mathbf{B}}_R(\mathbf{r}, t) = \frac{1}{v_1}(\hat{\boldsymbol{\kappa}}_R \times \tilde{\mathbf{E}}_R) \tag{8.98}$$

and a transmitted wave,

$$\tilde{\mathbf{E}}_T(\mathbf{r}, t) = \tilde{\mathbf{E}}_{0_T} e^{i(\boldsymbol{\kappa}_T \cdot \mathbf{r} - \omega t)}, \qquad \tilde{\mathbf{B}}_T(\mathbf{r}, t) = \frac{1}{v_2}(\hat{\boldsymbol{\kappa}}_T \times \tilde{\mathbf{E}}_T) \tag{8.99}$$

All three waves have the same *frequency* ω—that is determined once and for all at the source (the flashlight, or whatever, that is producing the incident beam). The wave numbers are related by (8.24):

$$\kappa_I v_1 = \kappa_R v_1 = \kappa_T v_2 = \omega, \qquad \text{or} \quad \kappa_I = \kappa_R = \frac{v_2}{v_1}\kappa_T = \frac{n_1}{n_2}\kappa_T \tag{8.100}$$

The combined fields in medium (1), $\tilde{\mathbf{E}}_I + \tilde{\mathbf{E}}_R$, and $\tilde{\mathbf{B}}_I + \tilde{\mathbf{B}}_R$, must now be joined to the fields in medium (2), in accordance with the boundary conditions (8.83). All these boundary conditions have the general form

$$(\quad)e^{i(\boldsymbol{\kappa}_I \cdot \mathbf{r} - \omega t)} + (\quad)e^{i(\boldsymbol{\kappa}_R \cdot \mathbf{r} - \omega t)} = (\quad)e^{i(\boldsymbol{\kappa}_T \cdot \mathbf{r} - \omega t)} \qquad \text{at} \quad x = 0 \tag{8.101}$$

I'll fill in the parentheses in a moment; for now, the important thing to notice is that the y-, z-, and t-dependence is confined to the exponents. *Because the boundary conditions must hold at all points on the plane, and for all times, these exponential factors must be equal.* Otherwise, a slight change in y, say, would destroy the equality (Problem 8.15). Of course, the time factors are *already* equal (in fact, you could regard this as an independent confirmation that the transmitted and reflected frequencies must match the incident one). As for the spatial terms, evidently

$$\boldsymbol{\kappa}_I \cdot \mathbf{r} = \boldsymbol{\kappa}_R \cdot \mathbf{r} = \boldsymbol{\kappa}_T \cdot \mathbf{r}, \qquad \text{when} \quad x = 0 \tag{8.102}$$

or, more explicitly,

$$(\kappa_I)_y y + (\kappa_I)_z z = (\kappa_R)_y y + (\kappa_R)_z z = (\kappa_T)_y y + (\kappa_T)_z z \tag{8.103}$$

for all y and all z.

Now, (8.103) holds if and only if the components are separately equal, for if $y = 0$, we get

$$(\kappa_I)_z = (\kappa_R)_z = (\kappa_T)_z \tag{8.104}$$

while $z = 0$ gives

$$(\kappa_I)_y = (\kappa_R)_y = (\kappa_T)_y \tag{8.105}$$

We may as well orient our axes so that κ_I lies in the xy plane ($(\kappa_I)_z = 0$); according to (8.104), so too will κ_R and κ_T. Conclusion:

(A) *The incident, reflected, and transmitted wave vectors form a plane* (called the **plane of incidence**), *which also includes the normal to the surface* (here, the x axis).

Meanwhile, (8.105) implies that

$$\kappa_I \sin\theta_I = \kappa_R \sin\theta_R = \kappa_T \sin\theta_T \tag{8.106}$$

where θ_I is the **angle of incidence**, θ_R is the **angle of reflection**, and θ_T is the angle of transmission, more commonly known as the **angle of refraction** (Fig. 8.17). In view of (8.100), then, we have

(B)
$$\theta_I = \theta_R \tag{8.107}$$

(*the angle of incidence is equal to the angle of reflection*), and

(C)
$$\frac{\sin\theta_T}{\sin\theta_I} = \frac{n_1}{n_2} \tag{8.108}$$

which is **Snell's law.**[11]

(A), (B), and (C) are fundamental laws of geometrical optics. It is astonishing how little actual *electrodynamics* goes into them: We have yet to invoke any *specific* boundary conditions—only the general format (8.101) was used. Any *other* waves (water waves, for instance, or sound waves) can be expected to obey the same "optical" laws when they pass from one medium into another.

Now that we have taken care of the exponential factors in (8.101)—they cancel, given (8.102)—the boundary conditions (8.83) become:

$$\left.\begin{aligned}
&\text{(i)} && \epsilon_1[\tilde{\mathbf{E}}_{0_I} + \tilde{\mathbf{E}}_{0_R}]_x = \epsilon_2(\tilde{\mathbf{E}}_{0_T})_x\\
&\text{(ii)} && [\tilde{\mathbf{B}}_{0_I} + \tilde{\mathbf{B}}_{0_R}]_x = (\tilde{\mathbf{B}}_{0_T})_x\\
&\text{(iii)} && [\tilde{\mathbf{E}}_{0_I} + \tilde{\mathbf{E}}_{0_R}]_{y,z} = (\tilde{\mathbf{E}}_{0_T})_{y,z}\\
&\text{(iv)} && \frac{1}{\mu_1}[\tilde{\mathbf{B}}_{0_I} + \tilde{\mathbf{B}}_{0_R}]_{y,z} = \frac{1}{\mu_2}(\tilde{\mathbf{B}}_{0_T})_{y,z}
\end{aligned}\right\} \tag{8.109}$$

where $\tilde{\mathbf{B}}_0 = (\hat{\kappa} \times \tilde{\mathbf{E}}_0)/v$. (The last two represent *pairs* of equations, one for the y-component and one for the z-component.)

Let's suppose that the polarization of the incident wave is *parallel* to the plane of incidence (the xy plane in Fig. 8.18). (I shall leave for you the case of polarization *perpendicular* to the plane of incidence—see Problem 8.16.) Then (i) reads

$$\epsilon_1[-\tilde{E}_{0_I}\sin\theta_I + \tilde{E}_{0_R}\sin\theta_R] = \epsilon_2(-\tilde{E}_{0_T}\sin\theta_T) \tag{8.110}$$

[11] If $n_2 < n_1$ and θ_I is greater than the **critical angle** $\theta_c = \sin^{-1}(n_2/n_1)$, Snell's law predicts an impossible angle of refraction: $\sin\theta_T > 1$. In this case we obtain "total internal reflection," with *no* transmitted wave. For details, see P. Lorrain and D. R. Corson, *Electromagnetic Fields and Waves*, 2d ed. (San Francisco: W. A. Freeman, 1970), sec. 12.4.1.

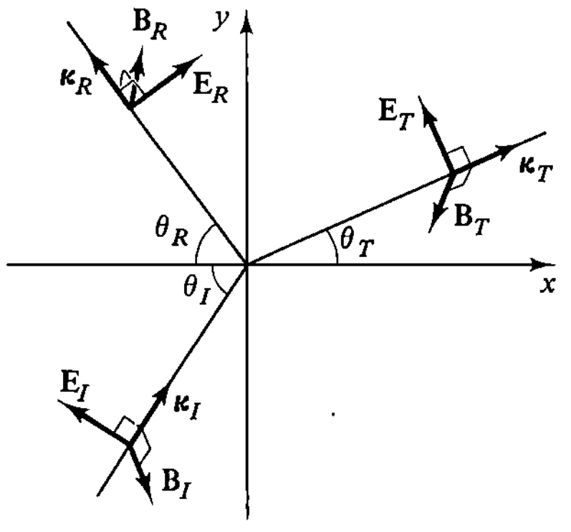

Figure 8.18

(ii) gives nothing (0 = 0); (iii) becomes

$$(\tilde{E}_{0_I} \cos\theta_I + \tilde{E}_{0_R} \cos\theta_R) = \tilde{E}_{0_T} \cos\theta_T \tag{8.111}$$

and (iv) says

$$\frac{1}{\mu_1 v_1}[\tilde{E}_{0_I} - \tilde{E}_{0_R}] = \frac{1}{\mu_2 v_2}\tilde{E}_{0_T} \tag{8.112}$$

Given the laws of reflection and refraction (equations (8.107) and (8.108)), (8.110) and (8.112) both reduce to

$$\tilde{E}_{0_I} - \tilde{E}_{0_R} = \beta\tilde{E}_{0_T} \tag{8.113}$$

where

$$\beta = \frac{\mu_1 v_1}{\mu_2 v_2} = \frac{\mu_1 n_2}{\mu_2 n_1} \tag{8.114}$$

as before, and (8.111) becomes

$$\tilde{E}_{0_I} + \tilde{E}_{0_R} = \alpha\tilde{E}_{0_T} \tag{8.115}$$

where

$$\alpha = \frac{\cos\theta_T}{\cos\theta_I} \tag{8.116}$$

Solving (8.113) and (8.115) for the reflected and transmitted amplitudes, we obtain

$$\tilde{E}_{0_R} = \left(\frac{\alpha - \beta}{\alpha + \beta}\right)\tilde{E}_{0_I}, \qquad \tilde{E}_{0_T} = \left(\frac{2}{\alpha + \beta}\right)\tilde{E}_{0_I} \tag{8.117}$$

These are known as **Fresnel's equations**, for the case of polarization in the plane of incidence. (There are two other Fresnel equations, giving the reflected and transmitted amplitudes when the polarization is *perpendicular* to the plane of incidence—see Problem 8.16.) Notice that the transmitted wave is always *in phase* with the incident one; the reflected wave is either in phase ("right side up"), if $\alpha > \beta$, or 180° out of phase ("upside down"), if $\alpha < \beta$.

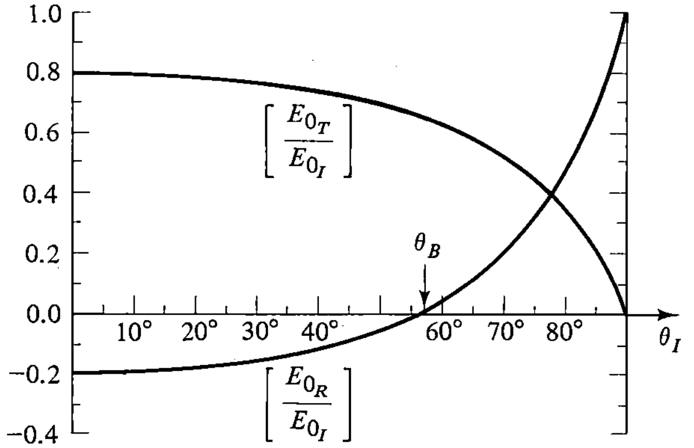

Figure 8.19

The amplitudes of the transmitted and reflected waves depend on the angle of incidence, because α is a function of θ_I:

$$\alpha = \frac{\sqrt{1-\sin^2\theta_T}}{\cos\theta_I} = \frac{\sqrt{1-\left(\frac{n_1}{n_2}\sin\theta_I\right)^2}}{\cos\theta_I}. \tag{8.118}$$

In the case of normal incidence ($\theta_I = 0$), $\alpha = 1$, and we recover (8.90). At grazing incidence ($\theta_I = 90°$), α diverges, and the wave is totally reflected (a fact that is painfully familiar to anyone who has driven at night on a wet road). Interestingly, there is an intermediate angle, θ_B (called **Brewster's angle**) at which the reflected wave is completely extinguished.[12] According to (8.117), this occurs when $\alpha = \beta$, or, using (8.118):

$$\sin^2\theta_B = \frac{1-\beta^2}{(n_1/n_2)^2 - \beta^2} \tag{8.119}$$

For the typical case $\mu_1 \approx \mu_2$, so that $\beta = n_2/n_1$, $\sin^2\theta_B = \beta^2/(1+\beta^2)$, and hence

$$\tan\theta_B = \left(\frac{n_2}{n_1}\right) \tag{8.120}$$

Figure 8.19 shows a plot of the transmitted and reflected amplitudes (8.117) as functions of θ_I, for light incident on glass ($n_2 = 1.5$) from air ($n_1 = 1$). (On the graph, a *negative* number indicates that the wave is 180° out of phase with the incident beam—the amplitude itself is the absolute value.)

The power per unit area striking the interface is given by $\mathbf{S}\cdot\hat{n}$, where $\hat{n}$ is the normal unit vector, in this case $\hat{x}$. Thus the incident intensity is

$$I_I = \tfrac{1}{2}\epsilon_1 v_1 E_{0_I}^2 \cos\theta_I \tag{8.121}$$

[12]Because waves polarized *perpendicular* to the plane of incidence exhibit no corresponding quenching of the reflected component, an arbitrary beam incident at Brewster's angle yields a reflected beam that is *totally* polarized parallel to the interface. That's why Polaroid glasses, with the transmission axis vertical, help to reduce glare off a horizontal surface.

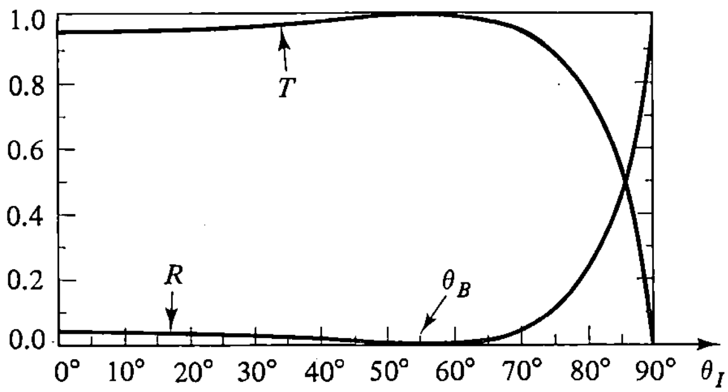

Figure 8.20

while the reflected and transmitted intensities are

$$I_R = \frac{1}{2}\epsilon_1 v_1 E_{0_R}^2 \cos\theta_R, \qquad I_T = \frac{1}{2}\epsilon_2 v_2 E_{0_T}^2 \cos\theta_T \tag{8.122}$$

(The cosines are present because we are speaking now of the average power per unit area of *interface,* and the interface is at an angle to the wave front.) The reflection and transmission coefficients for waves polarized parallel to the plane of incidence are

$$R \equiv \frac{I_R}{I_I} = \left(\frac{E_{0_R}}{E_{0_I}}\right)^2 = \left(\frac{\alpha - \beta}{\alpha + \beta}\right)^2 \tag{8.123}$$

$$T \equiv \frac{I_T}{I_I} = \frac{\epsilon_2 v_2}{\epsilon_1 v_1}\left(\frac{E_{0_T}}{E_{0_I}}\right)^2 \frac{\cos\theta_T}{\cos\theta_I} = \alpha\beta\left(\frac{2}{\alpha + \beta}\right)^2 \tag{8.124}$$

They are plotted as functions of the angle of incidence in Fig. 8.20 (for the air/glass interface). R is the fraction of the incident energy that is reflected—naturally, it goes to zero at Brewster's angle; T is the fraction transmitted—it goes to 1 at θ_B. Note that $R + T = 1$, as required by conservation of energy: the energy per unit time *reaching* a particular patch of area on the surface is equal to the energy per unit time *leaving* the patch.

Problem 8.15 Suppose $Ae^{iay} + Be^{iby} = Ce^{icy}$, for some nonzero constants A, B, C, a, b, c and for all y. Prove that $a = b = c$.

! **Problem 8.16** Analyze the case of polarization *perpendicular* to the plane of incidence. Impose the boundary conditions (8.109), and obtain the Fresnel equations for $\tilde{E}_{0_R}$ and $\tilde{E}_{0_T}$. Sketch (E_{0_R}/E_{0_I}) and (E_{0_T}/E_{0_I}) as functions of θ_I, for the case $\beta = n_2/n_1 = 1.5$. (Note that for this β the reflected wave is *always* 180° out of phase.) Show that there is no Brewster's angle for this case—E_{0_R} is *never* zero (unless, of course, $n_1 = n_2$, in which case the two media are optically indistinguishable). Confirm that your Fresnel equations reduce to the proper forms at normal incidence. Compute the reflection and transmission coefficients, and check that they add up to 1.

Problem 8.17 The index of refraction of diamond is 2.42. Construct the graph analogous to Fig. 8.19 for the air/diamond interface. (Assume $\mu_1 = \mu_2 = \mu_0$.) In particular, calculate

(a) the amplitudes at normal incidence,
(b) Brewster's angle,
(c) the "crossover" angle, at which the reflected and transmitted amplitudes are equal.

8.3 ELECTROMAGNETIC WAVES IN CONDUCTORS

8.3.1 The Modified Wave Equation

When I derived the wave equation for electromagnetic fields in Section 8.1.1, I stipulated that the free charge density ρ_f and the free current density $\mathbf{J}_f$ be zero, and everything we have done since then is predicated on that assumption. Such a restriction is perfectly reasonable when you're talking about wave propagation through a vacuum or through dielectric materials such as glass or (pure) water. But in the case of conductors (sea water, say, or metals) we cannot control the flow of charge, and in general $\mathbf{J}_f$ is certainly *not* zero. In fact, according to Ohm's law, the (free) current density in a conductor is proportional to the electric field:

$$\mathbf{J}_f = \sigma \mathbf{E} \tag{8.125}$$

With this, Maxwell's equations for linear media assume the form

$$\left.\begin{array}{llll} \text{(i)} & \nabla \cdot \mathbf{E} = \dfrac{1}{\epsilon} \rho_f & \text{(iii)} & \nabla \times \mathbf{E} = -\dfrac{\partial \mathbf{B}}{\partial t} \\ \text{(ii)} & \nabla \cdot \mathbf{B} = 0 & \text{(iv)} & \nabla \times \mathbf{B} = \mu \sigma \mathbf{E} + \mu \epsilon \dfrac{\partial \mathbf{E}}{\partial t} \end{array}\right\} \tag{8.126}$$

Now the continuity equation for the free current,

$$\nabla \cdot \mathbf{J}_f = -\frac{\partial \rho_f}{\partial t} \tag{8.127}$$

together with Ohm's law and Gauss's law (i), gives

$$\frac{\partial \rho_f}{\partial t} = -\sigma(\nabla \cdot \mathbf{E}) = -\frac{\sigma}{\epsilon} \rho_f$$

from which it follows that

$$\rho_f(t) = e^{-(\sigma/\epsilon)t} \rho_f(0) \tag{8.128}$$

Thus any initial free charge density $\rho_f(0)$ dissipates in a characteristic time $\tau = (\epsilon/\sigma)$. This reflects the familiar fact that if you put some free charge on a conductor, it will flow out to the edges.[13] At present we're not interested in this kind of transient behavior—we'll wait for any accumulated free charge to disappear; from then on $\rho_f = 0$, and we have

[13]The time constant τ affords a measure of how "good" a conductor is: For a "perfect" conductor $\sigma = \infty$ and $\tau = 0$; for a "good" conductor, τ is much less than the other relevant times in the problem (in oscillatory systems, that means $\tau \ll 1/\omega$); for a "poor" conductor, τ is *greater* than the characteristic times in the problem ($\tau \gg 1/\omega$).

$$\left.\begin{array}{llll}\text{(i)} & \nabla\cdot\mathbf{E}=0 & \text{(iii)} & \nabla\times\mathbf{E}=-\dfrac{\partial\mathbf{B}}{\partial t}\\ \text{(ii)} & \nabla\cdot\mathbf{B}=0 & \text{(iv)} & \nabla\times\mathbf{B}=\mu\epsilon\dfrac{\partial\mathbf{E}}{\partial t}+\mu\sigma\mathbf{E}\end{array}\right\} \tag{8.129}$$

These differ from the corresponding equations for *non*conducting media (8.7) only in the addition of the last term in (iv).

Applying the curl to (iii) and (iv), as before, we obtain modified wave equations for **E** and **B**:

$$\nabla^2\mathbf{E}=\mu\epsilon\frac{\partial^2\mathbf{E}}{\partial t^2}+\mu\sigma\frac{\partial\mathbf{E}}{\partial t},\qquad \nabla^2\mathbf{B}=\mu\epsilon\frac{\partial^2\mathbf{B}}{\partial t^2}+\mu\sigma\frac{\partial\mathbf{B}}{\partial t} \tag{8.130}$$

These equations again admit plane-wave solutions,

$$\tilde{\mathbf{E}}(x,t)=\tilde{\mathbf{E}}_0 e^{i(\kappa x-\omega t)},\qquad \tilde{\mathbf{B}}(x,t)=\tilde{\mathbf{B}}_0 e^{i(\kappa x-\omega t)} \tag{8.131}$$

only this time the wave number κ is complex:

$$\kappa^2=\mu\epsilon\omega^2+i\mu\sigma\omega \tag{8.132}$$

as you can easily check by putting (8.131) into (8.130). Taking the square root,[14] we find

$$\kappa=\kappa_+ + i\kappa_- \tag{8.133}$$

with

$$\kappa_\pm=\omega\sqrt{\frac{\epsilon\mu}{2}}\left[\sqrt{1+\left(\frac{\sigma}{\epsilon\omega}\right)^2}\pm 1\right]^{1/2} \tag{8.134}$$

The imaginary part of κ results in an **attenuation** of the wave (decreasing amplitude with increasing x):

$$\tilde{\mathbf{E}}(x,t)=\tilde{\mathbf{E}}_0 e^{-\kappa_- x}e^{i(\kappa_+ x-\omega t)},\qquad \tilde{\mathbf{B}}(x,t)=\tilde{\mathbf{B}}_0 e^{-\kappa_- x}e^{i(\kappa_+ x-\omega t)} \tag{8.135}$$

The distance it takes to reduce the amplitude by a factor of $1/e$ (about a third) is called the **skin depth**:

$$d=\left(\frac{1}{\kappa_-}\right) \tag{8.136}$$

It is a measure of the depth to which an electromagnetic wave penetrates into the conductor. Meanwhile, the real part of κ determines the wavelength, the propagation speed, and the index of refraction in the usual way:

$$\lambda=\frac{2\pi}{\kappa_+},\qquad v=\frac{\omega}{\kappa_+},\qquad n=\frac{c\kappa_+}{\omega} \tag{8.137}$$

[14] Equation (8.132) has, of course, *two* square roots. We want the one that reduces to the previous answer, $\kappa=\omega\sqrt{\epsilon\mu}$ when $\sigma\to 0$.

For a "poor" conductor,

$$\sigma \ll \omega\epsilon \tag{8.138}$$

In this case

$$\kappa_+ \cong \omega\sqrt{\epsilon\mu}, \qquad \kappa_- \cong \frac{\sigma}{2}\sqrt{\frac{\mu}{\epsilon}} \tag{8.139}$$

and the skin depth is independent of frequency. For a "good" conductor,

$$\sigma \gg \omega\epsilon \tag{8.140}$$

κ_+ and κ_- are nearly equal:

$$\kappa_+ \cong \kappa_- \cong \sqrt{\frac{\omega\sigma\mu}{2}} \tag{8.141}$$

and the skin depth decreases with increasing frequency. (In this regime the notion of "wavelength" loses its geometrical significance, since the wave only penetrates about a sixth of a wavelength:

$$d \cong \frac{\lambda}{2\pi} \tag{8.142}$$

it's attenuated away before it gets a chance to complete a single cycle.) Remember that the terms "good" and "poor" conductor depend on the frequency you have in mind—the same substance can be a good conductor at low frequencies and a poor one at high frequencies.[15]

Example 4

For sea water, $\mu = \mu_0 = 4\pi \times 10^{-7}\ \mathrm{N/A^2}$, $\epsilon \cong 70\epsilon_0 \cong 6 \times 10^{-10}\ \mathrm{C^2/N \cdot m^2}$, and $\sigma \cong 5\ (\Omega \cdot \mathrm{m})^{-1}$. Consequently,

$$\frac{\sigma}{\epsilon} \cong 10^{10}/\mathrm{s}$$

Sea water is a poor conductor, then, for frequencies substantially above 10^{10} Hz (including the visible region), and here equation (8.136) gives

$$d \cong \frac{2}{\sigma}\sqrt{\frac{\epsilon}{\mu}} \cong 1\ \mathrm{cm}$$

which is patently incorrect (see footnote 15).

In the radio frequency range sea water is a good conductor. The skin depth is quite short; to reach a depth of 10 m, say, for communication with submarines, you would need a frequency in the neighborhood of

$$\nu = \frac{\omega}{2\pi} = \frac{1}{\pi\sigma\omega d^2} \cong 500\ \mathrm{Hz}$$

[15] Moreover, σ itself is generally a function of frequency, as we shall see in Section 8.4.

or a wavelength (in air) of about 600 km. This calls for gigantic ground antennas and effectively rules out any possibility of a two-way conversation with a submerged submarine.

Example 5

For a typical metal, σ is on the order of 10^7 $(\Omega \cdot \text{m})^{-1}$, μ is 10^{-6} N/A^2, and ϵ is around 10^{-11} $\text{C}^2/\text{N}\cdot\text{m}^2$. Therefore, $(\sigma/\epsilon) \cong 10^{18}/\text{s}$, and the material is a good conductor well into the ultraviolet region. The skin depth at optical frequencies ($\omega \cong 10^{15}/\text{s}$) is roughly 10^{-8} m, which helps to explain why metals are opaque. (This argument is naïve, however, because it ignores the frequency dependence of σ. See Example 8.)

Problem 8.18

(a) Suppose you imbedded some free charge in a piece of glass. About how long would it take for the charge to flow to the surface?

(b) Silver is an excellent conductor, but it's expensive. Suppose you were designing a microwave experiment to operate at a frequency of 10^{10} Hz. How thick would you make the silver coatings?

(c) Find the wavelength and propagation speed in copper for radio waves at 1 MHz. Compare the corresponding values in air (or vacuum).

Problem 8.19 It may have occurred to you, in studying Example 4, that some *higher* frequency might give you a greater skin depth. Show, on the contrary, that if ϵ, σ, and μ are all constant, then d is a monotonically decreasing function of ω; the *only* way to increase the skin depth is to use very long wavelengths. Graph d as a function of ω. (However: see footnote 15.)

8.3.2 Monochromatic Plane Waves in Conducting Media

The attenuated plane waves (8.135) satisfy the modified wave equation (8.130) for *any* $\tilde{\mathbf{E}}_0$ and $\tilde{\mathbf{B}}_0$. But Maxwell's equations (8.129) impose further constraints, which serve to determine the relative amplitudes, phases, and polarizations of **E** and **B**. As before, (i) and (ii) rule out any x-components: the fields are *transverse*. We may as well orient our axes so that **E** is polarized along the y-direction:

$$\tilde{\mathbf{E}}(x, t) = \tilde{E}_0 e^{-\kappa_- x} e^{i(\kappa_+ x - \omega t)} \hat{\jmath} \tag{8.143}$$

Then (iii) gives

$$\tilde{\mathbf{B}}(x, t) = \left(\frac{\kappa}{\omega}\right) \tilde{E}_0 e^{-\kappa_- x} e^{i(\kappa_+ x - \omega t)} \hat{k} \tag{8.144}$$

(Equation (iv) says the same thing.) Once again, the electric and magnetic fields are mutually perpendicular.

Like any complex number, κ can be expressed in terms of its modulus and phase:

$$\kappa = \kappa_+ + i\kappa_- = |\kappa|e^{i\phi} \tag{8.145}$$

with

$$|\kappa| = \sqrt{\kappa_+^2 + \kappa_-^2} = \omega\sqrt{\epsilon\mu\sqrt{1 + \left(\frac{\sigma}{\epsilon\omega}\right)^2}} \tag{8.146}$$

and

$$\phi = \tan^{-1}\left(\frac{\kappa_-}{\kappa_+}\right) \tag{8.147}$$

According to (8.143) and (8.144), the complex amplitudes $\tilde{E}_0 = E_0 e^{i\delta_E}$ and $\tilde{B}_0 = B_0 e^{i\delta_B}$ are related by

$$B_0 e^{i\delta_B} = \frac{|\kappa|e^{i\phi}}{\omega} E_0 e^{i\delta_E} \tag{8.148}$$

Evidently, the electric and magnetic fields are no longer in phase; in fact,

$$\delta_B - \delta_E = \phi \tag{8.149}$$

Since δ represents a *delay* (see Fig. 8.5) this means that the magnetic field in a conductor *lags behind* the electric field. Meanwhile, the (real) amplitudes of **E** and **B** satisfy the equation

$$B_0 = \frac{|\kappa|}{\omega} E_0 = \sqrt{\epsilon\mu\sqrt{1 + \left(\frac{\sigma}{\epsilon\omega}\right)^2}}\, E_0 \tag{8.150}$$

The (real) electric and magnetic fields are, finally,

$$\left.\begin{aligned} \mathbf{E}(x, t) &= E_0 e^{-\kappa_- x} \cos(\kappa_+ x - \omega t + \delta_E)\hat{j} \\ \mathbf{B}(x, t) &= \frac{|\kappa|}{\omega} E_0 e^{-\kappa_- x} \cos(\kappa_+ x - \omega t + \delta_E + \phi)\hat{k} \end{aligned}\right\} \tag{8.151}$$

These fields are shown in Fig. 8.21.

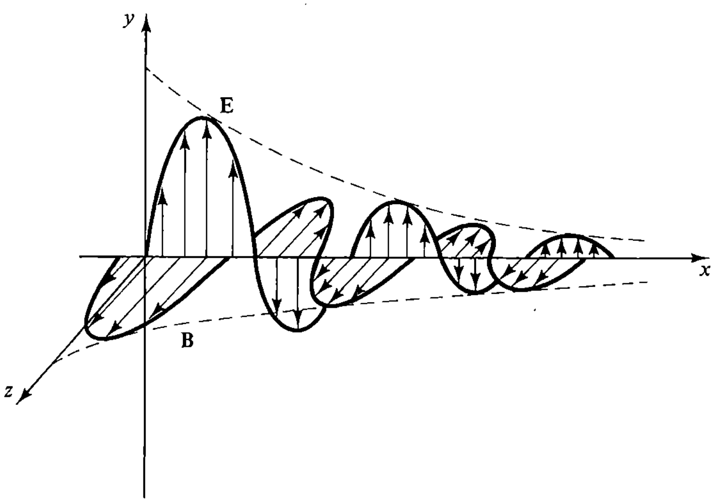

Figure 8.21

In a conductor, the energy is *not* equally shared between the electric and magnetic fields:

$$U = \frac{1}{2}\left(\epsilon E^2 + \frac{1}{\mu}B^2\right)$$
$$= \frac{1}{2}E_0^2 e^{-2\kappa_- x}\left\{\epsilon\cos^2(\kappa_+ x - \omega t + \delta_E) + \frac{|\kappa^2|}{\mu\omega^2}\cos^2(\kappa_+ x - \omega t + \delta_E + \phi)\right\} \tag{8.152}$$

Averaging over a full cycle, we get

$$\langle U\rangle = \frac{1}{4}\epsilon E_0^2 e^{-2\kappa_- x}\left[1 + \sqrt{1 + \left(\frac{\sigma}{\epsilon\omega}\right)^2}\right] \tag{8.153}$$

Notice that the magnetic contribution (the second term) always dominates; in fact, for a "good" conductor the energy is almost exclusively magnetic:

$$\langle U\rangle \cong \frac{1}{4}\frac{\sigma}{\omega}E_0^2 e^{-2\kappa_- x} \tag{8.154}$$

Meanwhile, the energy flux is given by the Poynting vector:

$$\mathbf{S} = \frac{1}{\mu}(\mathbf{E}\times\mathbf{B})$$
$$= \frac{|\kappa|}{\mu\omega}E_0^2 e^{-2\kappa_- x}[\cos(\kappa_+ x - \omega t + \delta_E)\cos(\kappa_+ x - \omega t + \delta_E + \phi)]\hat{\imath} \tag{8.155}$$

Because they differ in phase, the average of the product of the cosines is no longer $\frac{1}{2}$; rather,

$$\frac{1}{2\pi}\int_0^{2\pi}\cos\theta\cos(\theta + \phi)\,d\theta = \frac{1}{2}\cos\phi \tag{8.156}$$

Since $|\kappa|\cos\phi = \kappa_+$, therefore,

$$\langle\mathbf{S}\rangle = \frac{1}{2}\frac{\kappa_+}{\mu\omega}E_0^2 e^{-2\kappa_- x}\hat{\imath} \tag{8.157}$$

Example 6

As the wave progresses, the energy density decreases, because of the attenuation factor $e^{-2\kappa_- x}$ in (8.154). Where does this "lost" energy go? *Answer*: It heats up the conductor. In its most general form (7.82), the Joule heating law states that the power delivered to a volume V is

$$P = \int_V (\mathbf{E}\cdot\mathbf{J})\,d\tau$$

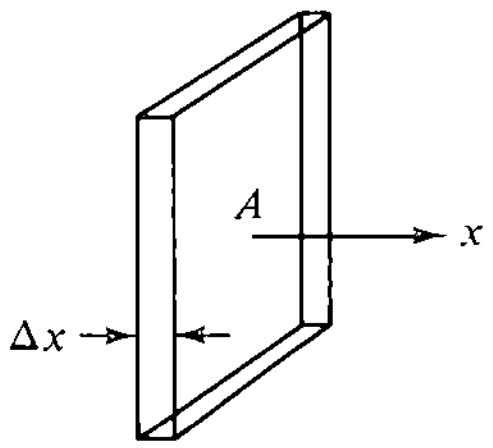

Figure 8.22

Applied to a slab of conductor, with infinitesimal thickness Δx and area A (Fig. 8.22), we obtain

$$P = \sigma E^2 A \, \Delta x$$

In the case of plane waves (8.152), the average rate of heating of the slab is

$$\langle P \rangle = \tfrac{1}{2}\sigma E_0^2 e^{-2\kappa_- x} A \, \Delta x$$

Is this gain in heat energy by the slab consistent with the loss of energy by the fields, as determined by the Poynting vector? According to (8.158) the average power flowing *into* the slab from the left is

$$\frac{1}{2}\frac{\kappa_+}{\mu\omega}E_0^2 e^{-2\kappa_- x} A$$

whereas the power flowing *out*, at the right, is given by the same formula, evaluated at $x + \Delta x$. The power dissipated in the slab is the *difference*:

$$\langle P \rangle = \frac{1}{2}\frac{\kappa_+}{\mu\omega}E_0^2 A\left(-\frac{d}{dx}e^{-2\kappa_- x}\right)\Delta x = \frac{\kappa_+\kappa_-}{\mu\omega}E_0^2 e^{-2\kappa_- x} A \, \Delta x$$

Since $\kappa_+\kappa_- = \frac{1}{2}\sigma\mu\omega$, the two expressions agree: the heat generated in the slab is equal to the energy lost by the fields. Whichever way you calculate it, the average power dissipated per unit volume is given by

$$\tfrac{1}{2}\sigma E_0^2 e^{-2\kappa_- x} \tag{8.158}$$

Problem 8.20 Show that in a "good" conductor the magnetic field lags the electric field by 45°, and determine the ratio of their amplitudes.

Problem 8.21 Consider a monochromatic electromagnetic wave incident on a thick slab of conducting material.

(a) What percent of the total energy delivered to the slab is deposited within the skin depth?

(b) Confirm that the total power dissipated in the slab is equal to power crossing the surface (8.157 with $x = 0$).

8.3.3 Reflection and Transmission at a Conducting Surface

The boundary conditions (8.83) that we used to analyze reflection and refraction at an interface between dielectrics do not hold in the presence of free charges and currents. Instead, we have the more general relations (7.61):

$$\left.\begin{array}{llll} \text{(i)} & \epsilon_1 E_{1_\perp} - \epsilon_2 E_{2_\perp} = \sigma_f & \text{(iii)} & \mathbf{E}_{1_\parallel} = \mathbf{E}_{2_\parallel} \\ \text{(ii)} & B_{1_\perp} = B_{2_\perp} & \text{(iv)} & \dfrac{1}{\mu_1}\mathbf{B}_{1_\parallel} - \dfrac{1}{\mu_2}\mathbf{B}_{2_\parallel} = \mathbf{K}_f \times \hat{n} \end{array}\right\} \tag{8.159}$$

where σ_f is the free surface charge and $\mathbf{K}_f$ the free surface current at the boundary. The unit vector $\hat{n}$ is perpendicular to the surface, pointing from medium (2) into medium (1). For ohmic conductors ($\mathbf{J}_f = \sigma\mathbf{E}$) there can be no free surface current, since this would require an infinite field at the boundary: $\mathbf{K}_f = 0$.

Suppose now that the yz plane forms the boundary between a nonconducting linear medium (1) and a conductor (2). A monochromatic plane wave, traveling in the x-direction and polarized in the y-direction, approaches from the left, as in Fig. 8.16:

$$\tilde{\mathbf{E}}_I(x,t) = \tilde{E}_{0_I} e^{i(\kappa_1 x - \omega t)}\hat{j}, \qquad \tilde{\mathbf{B}}_I(x,t) = \frac{1}{v_1}\tilde{E}_{0_I} e^{i(\kappa_1 x - \omega t)}\hat{k} \tag{8.160}$$

This incident wave gives rise to a reflected wave,

$$\tilde{\mathbf{E}}_R(x,t) = \tilde{E}_{0_R} e^{i(-\kappa_1 x - \omega t)}\hat{j}, \qquad \tilde{\mathbf{B}}_R(x,t) = -\frac{1}{v_1}\tilde{E}_{0_R} e^{i(-\kappa_1 x - \omega t)}\hat{k} \tag{8.161}$$

propagating back to the left in medium (1), and a transmitted wave

$$\tilde{\mathbf{E}}_T(x,t) = \tilde{E}_{0_T} e^{i(\kappa_2 x - \omega t)}\hat{j}, \qquad \tilde{\mathbf{B}}_T(x,t) = \frac{\kappa_2}{\omega}\tilde{E}_{0_T} e^{i(\kappa_2 x - \omega t)}\hat{k} \tag{8.162}$$

which is attenuated (κ_2 has an imaginary part, in accordance with (8.133)) as it penetrates into the conductor.

At $x = 0$, the combined wave in medium (1) must join the wave in medium (2), pursuant to the boundary conditions (8.159). Since $E_\perp = 0$ on both sides, boundary condition (i) yields $\sigma_f = 0$. Since $B_\perp = 0$, (ii) is automatically satisfied. Meanwhile, (iii) gives

$$\tilde{E}_{0_I} + \tilde{E}_{0_R} = \tilde{E}_{0_T} \tag{8.163}$$

and (iv) requires that

$$\frac{1}{\mu_1 v_1}(\tilde{E}_{0_I} - \tilde{E}_{0_R}) - \frac{1}{\mu_2}\frac{\kappa_2}{\omega}\tilde{E}_{0_T} = 0 \tag{8.164}$$

or

$$\tilde{E}_{0_I} - \tilde{E}_{0_R} = \beta\tilde{E}_{0_T} \tag{8.165}$$

where, in this case,

$$\beta = \left(\frac{\mu_1 v_1 \kappa_2}{\mu_2 \omega}\right) \tag{8.166}$$

It follows from (8.163) and (8.165) that

$$\tilde{E}_{0_R} = \left(\frac{1-\beta}{1+\beta}\right)\tilde{E}_{0_I}, \qquad \tilde{E}_{0_T} = \left(\frac{2}{1+\beta}\right)\tilde{E}_{0_I} \tag{8.167}$$

These results are formally identical to those which apply at the boundary between *non*conductors (8.90), but the resemblance is deceptive since β, through κ_2, is now a complex number.

For a *perfect* conductor ($\sigma = \infty$), β is infinite, and

$$\tilde{E}_{0_R} = -\tilde{E}_{0_I}, \qquad \tilde{E}_{0_T} = 0 \tag{8.168}$$

In this case the wave is totally reflected, with a 180° phase shift. (That's why excellent conductors, such as silver, make good mirrors. In practice, you paint a thin coating of silver onto the back of a piece of glass—the glass has nothing to do with the *reflection,* as such; it's just there to support the silver and to keep it from tarnishing. Since the skin depth at optical frequencies is on the order of 100 Å (Example 5), you don't need a very thick layer.) For a *good* conductor ($\sigma \gg \omega\epsilon_2$, but not infinite) $|\beta|$ is very large, and we may expand in powers of $1/\beta$:

$$\left(\frac{1-\beta}{1+\beta}\right) = -\frac{(1-1/\beta)}{(1+1/\beta)} \cong -\left(1-\frac{1}{\beta}\right)^2 \cong \frac{2}{\beta} - 1$$

In this approximation the reflection coefficient becomes

$$R = \frac{|\tilde{E}_{0_R}|^2}{|\tilde{E}_{0_I}|^2} \cong \left|\left(1-\frac{2}{\beta}\right)\right|^2 \cong 1 - 2\frac{(\beta+\beta^*)}{|\beta|^2}$$

or (referring to (8.166), (8.133), and (8.139)):

$$R \cong 1 - \sqrt{8\frac{\mu_2}{\mu_1}\left(\frac{\omega\epsilon_1}{\sigma}\right)} \tag{8.169}$$

Example 7

For air-to-silver ($\mu_1 = \mu_2 = \mu_0$, $\epsilon_1 = \epsilon_0$, $\sigma = 6 \times 10^7\ (\Omega\cdot\text{m})^{-1}$), at optical frequencies ($\omega = 4 \times 10^{15}$/s), the reflection coefficient is 0.93. In other words, 93% of the energy is reflected and 7% absorbed by the silver. At lower frequencies, the reflection is even more complete; at 1 MHz, for example, $R = 0.999994$. (However: See Example 8.)

Problem 8.22 Use the Poynting vector (8.157) to calculate the transmission coefficient at the surface of a conductor (express your answer in terms of β). Confirm that $T + R = 1$ (do *not* assume the conductor is a "good" one).

Problem 8.23 Calculate the average radiation pressure on a "good" conductor for an incident wave in air of amplitude E_0 and frequency ω. (*Hint*: First find the incident wave's momentum and that of the reflected wave. How much momentum is delivered to the conductor in a time Δt?) Account for this pressure (qualitatively) in terms of the forces on free electrons in the conductor.

8.4 DISPERSION

8.4.1 The Frequency Dependence of ϵ, μ, and σ

In the preceding sections, we have seen that the propagation of electromagnetic waves through matter is governed by three properties of the material, which we took to be constants: the permittivity ϵ, the permeability μ, and the conductivity σ. Actually, each of these parameters depends to some extent on the frequency of the waves you are considering. Indeed, if ϵ and μ were *truly* constant, then the speed of waves in a nonconductor,

$$v = \frac{1}{\sqrt{\epsilon\mu}}$$

and the index of refraction,

$$n = \frac{c}{v}$$

would likewise be constant. And yet, it is well known from optics that n is a function of ω (Fig. 8.23 shows the graph for a typical glass). Thus a prism bends blue light more sharply than red and spreads white light out into a rainbow of colors. This phenomenon is called **dispersion**. By extension, whenever the speed of a wave varies with its frequency, the supporting medium is called **dispersive**.[16] (Unfortunately, the

[16]Conductors, for example, are dispersive: See equations (8.134) and (8.137).

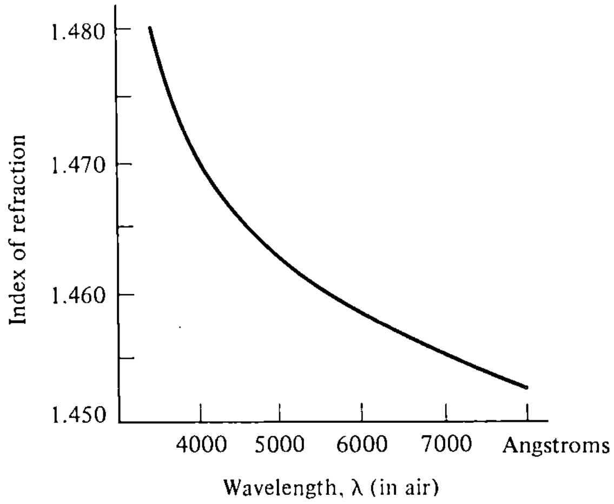

Figure 8.23

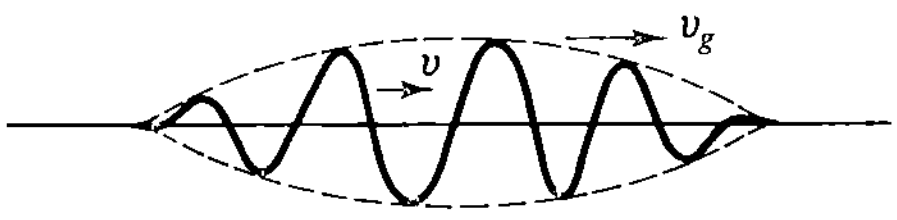

Figure 8.24

simple model with which I began this chapter—transverse waves on a stretched string—will not serve to illustrate dispersion, for the propagation speed $\sqrt{T/\mu}$ is independent of frequency. Dispersion *does* occur in water waves, however, and in many other contexts—see Problem 8.24.)

Because waves of different frequency travel at different speeds in a dispersive medium, a wave form that incorporates a range of frequencies will change shape as it progresses. A sharply peaked wave typically flattens out as it goes along, and whereas each sinusoidal component travels at the ordinary *wave* (or *phase*) *velocity*,

$$v = \frac{\omega}{\kappa} \tag{8.170}$$

the packet as a whole (the "envelope") travels at the so-called **group velocity**[17]

$$v_g = \frac{d\omega}{d\kappa} \tag{8.171}$$

You can demonstrate this by dropping a rock into the nearest pond and watching the waves that form: while the disturbance as a whole spreads out in a circle, moving at speed v_g, the ripples that go to make it up will be seen to travel *twice* as fast ($v = 2v_g$ for these waves). They appear at the back end of the packet, growing as they move forward to the center, then shrinking again and fading away at the front (Fig. 8.24). We shall not concern ourselves with these matters—we'll stick to monochromatic waves, for which the problem does not arise. But please note that the *energy* carried by a wave packet in a dispersive medium ordinarily travels at the *group* velocity, not the phase velocity. Don't be too alarmed, therefore, if in some circumstances v comes out greater than c.[18]

When you stop to think about it, it is not unreasonable that ϵ, μ, and σ should be functions of ω. After all, they describe the response of the material to electric and magnetic fields, and this response can depend on the driving frequency, much as the response of a swing depends on the frequency with which you push it. In practice, μ is very close to μ_0, for most materials, and its variation with ω is insignificant. This leaves ϵ and σ. My purpose in the following sections is to develop a model for the behavior of electrons inside matter, from which we can deduce the functions $\epsilon(\omega)$ and $\sigma(\omega)$. Like all classical models of atomic-scale phenomena, it is at best an *approximation* to the truth; nevertheless, it describes the observed behavior of matter in a qualitatively satisfactory manner.

[17] See A. P. French, *Vibrations and Waves*, (New York: W.W. Norton & Co., 1971), p. 230, or F. S. Crawford, Jr., *Waves*, (New York: McGraw-Hill, 1968) sec. 6.2.

[18] Even the group velocity can exceed c in some circumstances—see P. C. Peters, *Am. J. Phys.* **56**, 129 (1988).

Problem 8.24

(a) Shallow water is nondispersive; the waves travel at a speed which is proportional to the square root of the depth. In deep water, however, the waves can't "feel" all the way down to the bottom—they behave as though the depth were proportional to λ. (Actually, the distinction between "shallow" and "deep" itself depends on the wavelength: if the depth is less than λ the water is "shallow"; if it is substantially greater than λ the water is "deep.") Show that the wave velocity of deep water waves is *twice* the group velocity.

(b) In quantum mechanics, a free particle of mass m traveling in the x-direction is described by the wave function

$$\psi(x, t) = e^{i(px-Et)/\hbar},$$

where p is the momentum, and $E = p^2/2m$ is the kinetic energy. Calculate the group velocity and the wave velocity. Which one corresponds to the classical speed of the particle? Note that the wave velocity is *half* the group velocity.

8.4.2 Dispersion in Nonconductors

The electrons in a nonconductor are attached to specific molecules by binding forces whose detailed structure may be extremely complicated (especially in a large molecule) and would, in any event, require quantum mechanics for a proper treatment. But we shall picture the electron as attached to the end of a spring, of force constant k (Fig. 8.25):

$$F_{\text{binding}} = -ky = -m\omega_0^2 y \tag{8.172}$$

where y is displacement from equilibrium, m is the electron's mass, and ω_0 is the natural oscillation frequency, $\sqrt{k/m}$. If this strikes you as an implausible model for the binding force, refer to Example 1 of Chapter 4, where an admittedly crude atomic model led to a force of precisely this form. As a matter of fact, practically *any* binding force can be approximated this way for sufficiently small displacements from equilibrium, as you can see by expanding the potential energy in a Taylor series about the equilibrium point:

$$U(y) = U(0) + \left(\left.\frac{dU}{dy}\right|_0\right)y + \frac{1}{2}\left(\left.\frac{d^2U}{dy^2}\right|_0\right)y^2 + \cdots$$

The first term is a constant, with no dynamical significance (you can always adjust the zero of potential energy so that $U(0) = 0$). The second term automatically vanishes, since $dU/dy = -F$, and by the nature of an equilibrium the force at that point

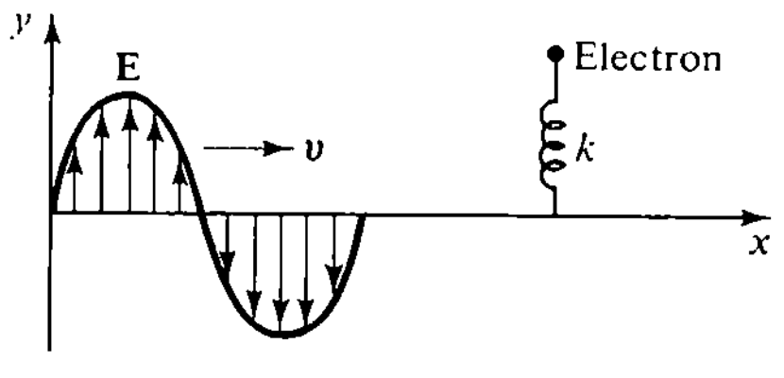

Figure 8.25

is zero. The third term is precisely the potential energy of a spring, with force constant $k = d^2U/dy^2|_0$ (the second derivative is necessarily *positive* for a point of *stable* equilibrium). As long as the displacements are small, so that the higher terms in the series can be neglected, the binding energy is that of a simple harmonic oscillator.[19]

When the electron oscillates, there will presumably be some damping force, which I shall assume is proportional to the velocity:

$$F_{\text{damping}} = -m\gamma \frac{dy}{dt} \tag{8.173}$$

Once again, I have chosen the simplest possible form that such a force can take: The damping must be opposite in direction to the velocity, and making it *proportional* to the velocity is the easiest way to accomplish this. The *cause* of the damping does not concern us here—among other things, an oscillating charge *radiates,* and the radiation carries away energy at the expense of the oscillation energy. (We will calculate this "radiation damping" in Chapter 9.)

In the presence of an electromagnetic wave of frequency ω, the electron is subject to a driving force

$$F_{\text{driving}} = qE = qE_0 \cos \omega t \tag{8.174}$$

where q is the charge of the electron, and E_0 is the amplitude of the wave at the point x where the electron is situated. (Since we're only interested in one fixed point, I have reset the clock so that maximum E occurs at $t = 0$.) Putting all this into Newton's second law gives

$$m \frac{d^2y}{dt^2} = F_{\text{tot}} = F_{\text{binding}} + F_{\text{damping}} + F_{\text{driving}}$$

or

$$m \frac{d^2y}{dt^2} + m\gamma \frac{dy}{dt} + m\omega_0^2 y = qE_0 \cos \omega t \tag{8.175}$$

Our model, then, describes the electron as a damped harmonic oscillator, driven at frequency ω. (We shall assume that the much more massive nuclei remain at rest.)

Equation (8.175) is easier to handle if we regard it as the real part of a *complex* equation:

$$\frac{d^2\tilde{y}}{dt^2} + \gamma \frac{d\tilde{y}}{dt} + \omega_0^2 \tilde{y} = \frac{q}{m} E_0 e^{-i\omega t} \tag{8.176}$$

In the steady state, the system oscillates at the driving frequency:

$$\tilde{y}(t) = \tilde{y}_0 e^{-i\omega t} \tag{8.177}$$

[19]In principle, the second derivative might *also* be zero. For instance, $U(y) = y^4$ has a point of stable equilibrium at the origin, but it is *not* approximately simple harmonic, even for small y. However, such potential energy functions are extremely rare in practice. Geometrically, what my argument amounts to is that almost any function can be fit near a minimum by a suitable parabola.

Inserting this into (8.176), I find that

$$\tilde{y}_0 = \frac{(q/m)}{(\omega_0^2 - \omega^2) - i\gamma\omega} E_0 \tag{8.178}$$

The dipole moment associated with the motion of this electron is therefore the real part of

$$\tilde{p}(t) = q\tilde{y}(t) = \frac{(q^2/m)}{(\omega_0^2 - \omega^2) - i\gamma\omega} E_0 e^{-i\omega t} \tag{8.179}$$

The imaginary term in the denominator indicates that p is *out of phase* with E—lagging behind by an angle $\tan^{-1}[\gamma\omega/(\omega_0^2 - \omega^2)]$ that is very small when $\omega \ll \omega_0$ and rises to π when $\omega \gg \omega_0$.

In general, differently situated electrons within a given molecule exhibit different natural frequencies and damping coefficients. Let's say there are f_j electrons with frequency ω_j and damping γ_j in each molecule. If there are N molecules per unit volume, the polarization $\mathbf{P}$ is given by[20] the real part of

$$\tilde{\mathbf{P}} = \frac{Nq^2}{m}\left(\sum_j \frac{f_j}{(\omega_j^2 - \omega^2) - i\gamma_j\omega}\right)\tilde{\mathbf{E}} \tag{8.180}$$

Now, we defined the electric susceptibility as the proportionality constant between $\mathbf{P}$ and $\mathbf{E}$ (more precisely, $\mathbf{P} = \epsilon_0\chi_e\mathbf{E}$). In the present case $\mathbf{P}$ is *not* proportional to $\mathbf{E}$ (this is not, strictly speaking, a linear medium) because of the difference in phase. However, the *complex* polarization $\tilde{\mathbf{P}}$ *is* proportional to the *complex* field $\tilde{\mathbf{E}}$, and this suggests that we introduce a *complex* susceptibility:

$$\tilde{\mathbf{P}} = \epsilon_0\chi_e\tilde{\mathbf{E}} \tag{8.181}$$

All of the manipulations we went through before carry over, on the understanding that the physical polarization is the real part of $\tilde{\mathbf{P}}$, just as the physical field is the real part of $\tilde{\mathbf{E}}$. In particular, the proportionality between $\tilde{\mathbf{D}}$ and $\tilde{\mathbf{E}}$ is the *complex* permittivity $\epsilon = \epsilon_0(1 + \chi_e)$, which in this model takes the form

$$\epsilon = \epsilon_0\left[1 + \frac{Nq^2}{m\epsilon_0}\sum_j \frac{f_j}{(\omega_j^2 - \omega^2) - i\gamma_j\omega}\right] \tag{8.182}$$

Ordinarily, the imaginary term is insignificant; however, when ω is very close to one of the resonant frequencies ω_j it plays an important role.

In a dispersive medium the wave equation for a given frequency reads

$$\nabla^2\tilde{\mathbf{E}} = \epsilon\mu_0\frac{\partial^2\tilde{\mathbf{E}}}{\partial t^2} \tag{8.183}$$

with ϵ now a complex function of ω; it admits plane-wave solutions, as before,

$$\tilde{\mathbf{E}}(x, t) = \tilde{\mathbf{E}}_0 e^{i(\kappa x - \omega t)} \tag{8.184}$$

[20]This applies directly to the case of a dilute gas; for denser materials the theory is modified slightly, in accordance with the Clausius-Mossotti equation (see Section 4.4.4). Incidentally, don't confuse the "polarization" of a medium, $\mathbf{P}$, with the "polarization" of a *wave*—same *word*, but the two uses have no connection.

However, the wave number

$$\kappa = \sqrt{\epsilon\mu_0}\,\omega \tag{8.185}$$

is complex, because ϵ is. Writing κ in terms of its real and imaginary parts,

$$\kappa = \kappa_+ + i\kappa_- \tag{8.186}$$

equation (8.184) becomes

$$\tilde{\mathbf{E}}(x, t) = \tilde{\mathbf{E}}_0 e^{-\kappa_- x} e^{i(\kappa_+ x - \omega t)} \tag{8.187}$$

Evidently, κ_- measures the *attenuation* of the wave. Because the *intensity* is proportional to E^2 (and hence to $e^{-2\kappa_- x}$), the quantity

$$\alpha = 2\kappa_- \tag{8.188}$$

is called the **absorption coefficient**. Meanwhile, the wave velocity is ω/κ_+, and therefore the index of refraction is

$$n = \frac{c}{\omega}\kappa_+ \tag{8.189}$$

I have deliberately used notation reminiscent of Section 8.3.1. However, in the present case κ_+ and κ_- have nothing to do with conductivity; rather they are determined by the parameters of our damped harmonic oscillator. For gases, the second term in (8.182) is typically small, and we can approximate the square root by using $\sqrt{1+x} \cong 1 + \frac{1}{2}x$. Then (8.185) gives

$$\kappa = \frac{\omega}{c}\sqrt{\frac{\epsilon}{\epsilon_0}} \cong \frac{\omega}{c}\left[1 + \frac{Nq^2}{2m\epsilon_0}\sum_j \frac{f_j}{(\omega_j^2 - \omega^2) - i\gamma_j\omega}\right] \tag{8.190}$$

so that

$$n = \frac{c}{\omega}\kappa_+ \cong 1 + \frac{Nq^2}{2m\epsilon_0}\sum_j \frac{f_j(\omega_j^2 - \omega^2)}{(\omega_j^2 - \omega^2)^2 + \gamma_j^2\omega^2} \tag{8.191}$$

and

$$\alpha = 2\kappa_- \cong \frac{Nq^2\omega^2}{m\epsilon_0 c}\sum_j \frac{f_j\gamma_j}{(\omega_j^2 - \omega^2)^2 + \gamma_j^2\omega^2} \tag{8.192}$$

In Fig. 8.26 I have plotted the index of refraction and the absorption coefficient near one of the resonances. *Most* of the time the index of refraction *rises* gradually with increasing frequency, consistent with our experience from optics (Fig. 8.23). However, in the immediate neighborhood of a resonance, the index of refraction *drops* sharply. Because this behavior is atypical, it is called **anomalous dispersion**. Notice that the region of anomalous dispersion ($\omega_1 < \omega < \omega_2$, in the figure) coincides with the region of maximum absorption; in fact, the material may be practically opaque in this frequency range. The reason is that we are now driving the electrons at their "favorite" frequency; the amplitude of their oscillation is relatively large, and a correspondingly large amount of energy is dissipated by the damping mechanism.

In Fig. 8.26, n runs below 1 above the resonance, indicating that the wave speed exceeds c. As I mentioned in Section 8.4.1, this is of no great consequence, since

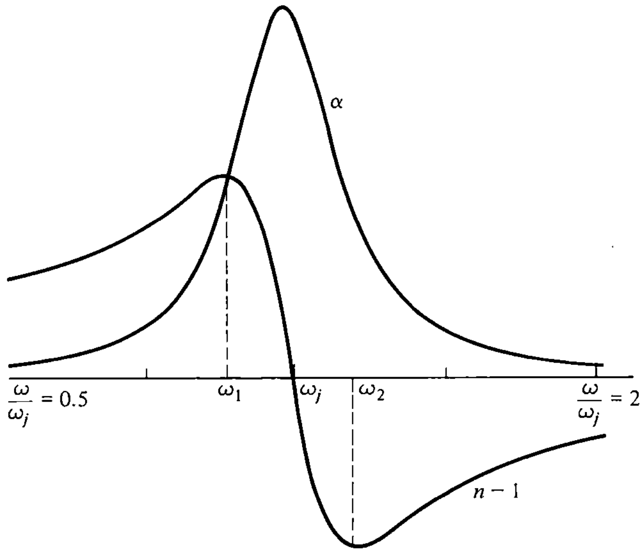

Figure 8.26

energy does not travel at the wave speed (see Problem 8.27). Moreover, the graph does not include the contributions of other terms in the sum, which add a relatively constant "background" that, in some cases, keeps $n > 1$ on both sides of the resonance.

If you agree to stay away from the resonances, the damping can be ignored, and the formula for the index of refraction simplifies somewhat:

$$n = 1 + \frac{Nq^2}{2m\epsilon_0} \sum_j \frac{f_j}{(\omega_j^2 - \omega^2)} \tag{8.193}$$

For most substances the natural frequencies ω_j are scattered all over the spectrum in a rather chaotic fashion. For transparent materials, the nearest significant resonances generally lie in the ultraviolet, so that $\omega < \omega_j$. In this circumstance,

$$\frac{1}{(\omega_j^2 - \omega^2)} = \frac{1}{\omega_j^2}\left(1 - \frac{\omega^2}{\omega_j^2}\right)^{-1} \cong \frac{1}{\omega_j^2}\left(1 + \frac{\omega^2}{\omega_j^2}\right)$$

and (8.193) assumes the form

$$n = 1 + \left(\frac{Nq^2}{2m\epsilon_0} \sum_j \frac{f_j}{\omega_j^2}\right) + \omega^2\left(\frac{Nq^2}{2m\epsilon_0} \sum_j \frac{f_j}{\omega_j^4}\right) \tag{8.194}$$

Or, in terms of the wavelength in vacuum ($\lambda = 2\pi c/\omega$):

$$n = 1 + A\left(1 + \frac{B}{\lambda^2}\right) \tag{8.195}$$

This is known as **Cauchy's equation**; the constant A is called the **coefficient of refraction** and B is called the **coefficient of dispersion**. Cauchy's equation applies reasonably well to most gases, in the optical region.

What I have described in this section is certainly not the complete story of dispersion in nonconducting media. Nevertheless, it does indicate how the damped harmonic motion of electrons can account for the observed frequency dependence of the index of refraction, and it explains why n is ordinarily an increasing function of ω. In the following section I'll develop a similar model, in the same spirit, for the frequency dependence of σ in conductors.

Problem 8.25 If you took the model in Example 1, Chapter 4, at face value, what natural frequency would you get? (Put in the actual numbers—where, in the electromagnetic spectrum, does this lie, assuming the radius of the atom is 0.5 Å?) Find the coefficients of refraction and dispersion and compare them with those for hydrogen at 0°C and atmospheric pressure: $A = 1.36 \times 10^{-4}$, $B = 7.7 \times 10^{-15}\ \mathrm{m}^2$.

Problem 8.26 Find the width of the anomalous dispersion region for the case of a single resonance at frequency ω_0. Assume $\gamma \ll \omega_0$. Show that the index of refraction assumes its maximum and minimum values at points where the absorption coefficient is at half-maximum.

Problem 8.27 Assuming negligible damping ($\gamma_j = 0$), calculate the group velocity ($v_g = d\omega/d\kappa_+$) of the waves described by (8.187) and (8.190). Show that $v_g < c$, even when $v > c$.

8.4.3 Free Electrons in Conductors and Plasmas

The free electrons in a conductor are not bound to any particular atom or molecule; nevertheless, essentially the same model can be applied to them, if we set the binding force equal to zero and understand that the damping factor γ, attributable now to the cumulative effect of many collisions, is relatively large. The equation of motion (8.176) becomes

$$\frac{d^2\tilde{y}}{dt^2} + \gamma\frac{d\tilde{y}}{dt} = \frac{q}{m}E_0 e^{-i\omega t} \tag{8.196}$$

with the solution

$$\tilde{y}(t) = \tilde{y}_0 e^{-i\omega t} \tag{8.197}$$

where

$$\tilde{y}_0 = -\frac{(q/m)}{\omega^2 + i\gamma\omega}E_0 \tag{8.198}$$

This time we are not interested in the *polarization* associated with the electrons but rather the *current* they generate:

$$J = (Nfq)\frac{dy}{dt} \tag{8.199}$$

in which f represents the number of *free* electrons per molecule, and N, again, is the number of molecules per unit volume. Evidently, J is the real part of

$$\tilde{\mathbf{J}} = \left(\frac{Nfq^2/m}{\gamma - i\omega}\right)\tilde{\mathbf{E}} \tag{8.200}$$

The imaginary term in the denominator of (8.200) means that J is out of phase with E—the current is *not* proportional to the field, and this material does not, strictly speaking, obey Ohm's law. However, the fact that the *complex* current is proportional to the *complex* field invites us to introduce a *complex* conductivity:

$$\tilde{\mathbf{J}} = \sigma \tilde{\mathbf{E}} \tag{8.201}$$

In the present model

$$\sigma = \frac{(Nfq^2/m)}{\gamma - i\omega} \tag{8.202}$$

At low frequencies the imaginary term is negligible and σ is independent of ω, but for sufficiently large frequencies the phase difference between J and E cannot be ignored. In effect, the current cannot "keep up" when the driving force oscillates too rapidly.

Example 8

Let's calculate the frequency at which the imaginary term in σ might be expected to become significant, for a typical metal.[21] According to Table 7.1, the conductivity of copper, say, at zero frequency, is about $6 \times 10^7/\Omega \cdot \text{m}$; its density is about $9 \times 10^3\ \text{kg/m}^3$, and let's say it has one free electron per atom. Then

$$N = \frac{(\text{atoms})}{(\text{volume})} = \frac{(\text{mass/volume})}{(\text{mass/mole})}\frac{(\text{atoms})}{(\text{mole})}$$

$$= \frac{(\text{density})}{(\text{atomic weight})}(\text{Avogadro's number})$$

$$= \left(\frac{9 \times 10^3\ \text{kg/m}^3}{60 \times 10^{-3}\ \text{kg}}\right)(6 \times 10^{23}) = 9 \times 10^{28}/\text{m}^3$$

It follows that

$$\gamma = \frac{Nfq^2}{\sigma m} = \frac{(9 \times 10^{28}/\text{m}^3)(1.6 \times 10^{-19}\ \text{C})^2}{(6 \times 10^7/\Omega \cdot \text{m})(9 \times 10^{-31}\ \text{kg})} = 4 \times 10^{13}/\text{s}$$

Evidently, the imaginary term becomes significant in the infrared region. Much above this, the phase difference between J and E goes to 90°, and the conductivity drops off in inverse proportion to the frequency. (That's why the use of dc conductivities in Examples 5 and 7 was naïve.)

Naturally, all this affects the behavior of electromagnetic waves in a conductor at high frequencies. But the modifications are relatively easy to make: Simply insert

[21] The application of this model to metals is to be taken with a grain of salt, since it ignores the forces between the electrons themselves, to say nothing of quantum effects. The quantitative predictions of Section 8.4 are valid for dilute substances only—gases and plasmas, in particular.

the complex conductivity into the general formula for the wave number (8.132),

$$\kappa^2 = \mu\epsilon\omega^2 + i\sigma\mu\omega \tag{8.203}$$

One special case is of particular interest: In a dilute plasma (ionized gas) the damping is negligible, and σ is purely imaginary,

$$\sigma = i\left(\frac{Nfq^2}{m\omega}\right) \tag{8.204}$$

Since, moreover, $\mu \cong \mu_0$ and $\epsilon \cong \epsilon_0$ in a plasma, (8.203) becomes

$$\kappa^2 = \frac{1}{c^2}(\omega^2 - \omega_p^2) \tag{8.205}$$

where

$$\omega_p = q\sqrt{\frac{Nf}{m\epsilon_0}} \tag{8.206}$$

is the **plasma frequency**. For frequencies in excess of ω_p, the wave number is real, and waves propagate without attenuation at a wave speed

$$v = \frac{\omega}{\kappa} = c[1 - (\omega_p/\omega)^2]^{-1/2} \tag{8.207}$$

(which, again, is greater than c). The index of refraction of the plasma is

$$n = \sqrt{1 - \left(\frac{\omega_p}{\omega}\right)^2} \tag{8.208}$$

For frequencies *below* ω_p, on the other hand, κ is purely imaginary, and the waves are attenuated away,

$$\tilde{\mathbf{E}}(x, t) = \tilde{\mathbf{E}}_0 e^{-(1/c)\sqrt{\omega_p^2-\omega^2}x} e^{-i\omega t} \tag{8.209}$$

with a low frequency skin depth of (c/ω_p). Accordingly, the plasma is opaque to waves of frequency less than ω_p and transparent to those above ω_p.

Example 9 The ionosphere

In the altitude range between about 50 and 1000 km, the atmosphere is partially ionized by ultraviolet rays from the sun. The density of electrons[22] in this region (called the **ionosphere**) is typically around $10^{11}/\text{m}^3$. With this value for Nf, the plasma frequency is about 3 MHz, just above the AM radio band. Lower frequency radio waves cannot penetrate the ionosphere; they are reflected back to earth. That's why at night, when, in the absence of ionizing radiation from the sun, the ionosphere recedes to a higher altitude, you can often pick up very distant stations.

---

[22]Actually, the density of *molecules* in the ionosphere ranges from 10^{15} to 10^{22} per cubic meter, but only a small fraction of them are ionized.

Problem 8.28 Calculate the group velocity of waves in a plasma for $\omega > \omega_p$, and check that it is less than the speed of light.

Problem 8.29 By combining (7.5)—the formula for conductivity at low frequencies—with (8.202), show that the damping γ is related to the mean free time (average time between collisions). Find the specific relation, and estimate the mean free time for copper on the basis of Example 8. Assuming thermal velocities ($\frac{1}{2}mv^2 = \frac{3}{2}kT$) estimate the mean free path of an electron in copper at room temperature.

Problem 8.30 Find approximate formulas for the skin depth and index of refraction of a conductor at frequencies substantially greater than γ.

8.5 GUIDED WAVES

8.5.1 Wave Guides

So far, we have dealt with plane waves of infinite extent; now we consider electromagnetic waves confined to the interior of a hollow cylindrical pipe, or **wave guide** (Fig. 8.27). We'll assume the wave guide is a perfect conductor, so that $\mathbf{E} = 0$ and $\mathbf{B} = 0$ inside the material itself, and hence the boundary conditions at the inner wall are[23]

$$\left.\begin{array}{ll}\text{(i)} & \mathbf{E}_{\parallel} = 0 \\ \text{(ii)} & \mathbf{B}_{\perp} = 0\end{array}\right\} \tag{8.210}$$

We are interested in monochromatic waves that propagate down the tube, so $\mathbf{E}$ and $\mathbf{B}$ have the general form

$$\left.\begin{array}{ll}\text{(i)} & \tilde{\mathbf{E}}(x, y, z, t) = \tilde{\mathbf{E}}_0(y, z)e^{i(\kappa x - \omega t)} \\ \text{(ii)} & \tilde{\mathbf{B}}(x, y, z, t) = \tilde{\mathbf{B}}_0(y, z)e^{i(\kappa x - \omega t)}\end{array}\right\} \tag{8.211}$$

The electric and magnetic fields must, of course, satisfy Maxwell's equations, in the interior of the wave guide:

[23] See eq. (8.159). In a perfect conductor $\mathbf{E} = 0$, and hence (by Faraday's law) $\partial\mathbf{B}/\partial t = 0$. Assuming the magnetic field *started out* zero, then, it will *remain* so.

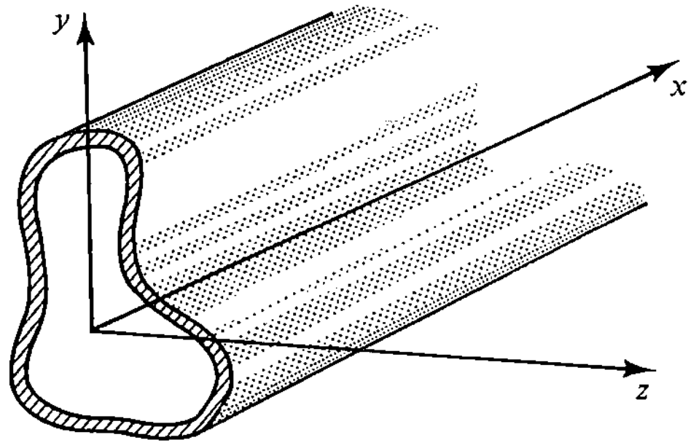

Figure 8.27

$$\left.\begin{array}{llll}\text{(i)} & \nabla\cdot\mathbf{E}=0 & \text{(iii)} & \nabla\times\mathbf{E}=-\dfrac{\partial\mathbf{B}}{\partial t}\\[2ex] \text{(ii)} & \nabla\cdot\mathbf{B}=0 & \text{(iv)} & \nabla\times\mathbf{B}=\dfrac{1}{c^2}\dfrac{\partial\mathbf{E}}{\partial t}\end{array}\right\} \tag{8.212}$$

The problem, then, is to find functions $\tilde{\mathbf{E}}_0$ and $\tilde{\mathbf{B}}_0$ such that the fields (8.211) obey the differential equations (8.212), subject to the boundary conditions (8.210).

As we shall soon see, *confined* waves are *not* (in general) transverse; in order to fit the boundary conditions we shall have to include longitudinal components (E_x and B_x):[24]

$$\tilde{\mathbf{E}}_0 = E_x\hat{\imath} + E_y\hat{\jmath} + E_z\hat{k} \qquad \tilde{\mathbf{B}}_0 = B_x\hat{\imath} + B_y\hat{\jmath} + B_z\hat{k} \tag{8.213}$$

where each of the components is a function of y and z. Putting this into Maxwell's equations (iii) and (iv), we obtain (Problem 8.31(a))

$$\left.\begin{array}{llll}\text{(i)} & \dfrac{\partial E_z}{\partial y}-\dfrac{\partial E_y}{\partial z}=i\omega B_x & \text{(iv)} & \dfrac{\partial B_z}{\partial y}-\dfrac{\partial B_y}{\partial z}=-\dfrac{i\omega}{c^2}E_x\\[2ex] \text{(ii)} & \dfrac{\partial E_x}{\partial z}-i\kappa E_z=i\omega B_y & \text{(v)} & \dfrac{\partial B_x}{\partial z}-i\kappa B_z=-\dfrac{i\omega}{c^2}E_y\\[2ex] \text{(iii)} & i\kappa E_y-\dfrac{\partial E_x}{\partial y}=i\omega B_z & \text{(vi)} & i\kappa B_y-\dfrac{\partial B_x}{\partial y}=-\dfrac{i\omega}{c^2}E_z\end{array}\right\} \tag{8.214}$$

Equations (ii), (iii), (v), and (vi) can be solved for E_y, E_z, B_y, and B_z:

$$\left.\begin{array}{lll}\text{(i)} & E_y=\dfrac{i}{(\omega/c)^2-\kappa^2}\left(\kappa\dfrac{\partial E_x}{\partial y}+\omega\dfrac{\partial B_x}{\partial z}\right)\\[2ex] \text{(ii)} & E_z=\dfrac{i}{(\omega/c)^2-\kappa^2}\left(\kappa\dfrac{\partial E_x}{\partial z}-\omega\dfrac{\partial B_x}{\partial y}\right)\\[2ex] \text{(iii)} & B_y=\dfrac{i}{(\omega/c)^2-\kappa^2}\left(\kappa\dfrac{\partial B_x}{\partial y}-\dfrac{\omega}{c^2}\dfrac{\partial E_x}{\partial z}\right)\\[2ex] \text{(iv)} & B_z=\dfrac{i}{(\omega/c)^2-\kappa^2}\left(\kappa\dfrac{\partial B_x}{\partial z}+\dfrac{\omega}{c^2}\dfrac{\partial E_x}{\partial y}\right)\end{array}\right\} \tag{8.215}$$

It suffices, then, to determine the longitudinal components E_x and B_x—if we knew those, we could quickly calculate all the others, using (8.215). Meanwhile, inserting (8.215) into the remaining Maxwell equations (Problem 8.31(b)) yields uncoupled equations for E_x and B_x themselves:

[24] To avoid cumbersome notation I leave the subscript and tilde off the individual components.

$$\left.\begin{aligned}&\text{(i)}\quad \left[\frac{\partial^2}{\partial y^2}+\frac{\partial^2}{\partial z^2}+(\omega/c)^2-\kappa^2\right]E_x=0\\&\text{(ii)}\quad \left[\frac{\partial^2}{\partial y^2}+\frac{\partial^2}{\partial z^2}+(\omega/c)^2-\kappa^2\right]B_x=0\end{aligned}\right\}\tag{8.216}$$

If $E_x = 0$ we call these **TE** ("transverse electric") **waves**; if $B_x = 0$ they are called **TM** ("transverse magnetic") **waves**; if both $E_x = 0$ and $B_x = 0$, we call them **TEM waves.**[25] I shall now prove that TEM waves cannot occur in a hollow wave guide. If $E_x = 0$, Gauss's law (8.212(i)) says that

$$\frac{\partial E_y}{\partial y}+\frac{\partial E_z}{\partial z}=0$$

and if $B_x = 0$, Faraday's law (8.214(i)) requires

$$\frac{\partial E_z}{\partial y}-\frac{\partial E_y}{\partial z}=0$$

Indeed, the vector $\tilde{\mathbf{E}}_0$ in eq. (8.213) has zero divergence and zero curl. It can therefore be written as the gradient of a scalar potential which satisfies Laplace's equation. But the boundary condition on **E** (8.210(i)) implies that the surface is an equipotential, and since Laplace's equation admits no local maxima or minima (Section 3.1.4), this means that the potential is a constant throughout, and hence the electric field is *zero*—no wave at all. Notice that this argument applies only to a *hollow* pipe—if you run a separate conductor down the middle, the potential at *its* surface need not be the same as on the outer wall, and hence a nontrivial potential is possible. We'll see an example of this in Section 8.5.3.

Problem 8.31

(a) Derive equations (8.214), and from these obtain (8.215).

(b) Put (8.215) into (i) and (ii) of (8.212) to obtain (8.216). Check that you get the same results using (i) and (iv) of (8.214).

8.5.2 TE Waves in a Rectangular Wave Guide

Suppose we have a wave guide of rectangular shape (Fig. 8.28), with height a and width b, and we are interested in the propagation of TE waves. The problem is to solve equation (8.216(ii)), subject to the boundary condition (8.210(ii)). We'll do it by separation of variables (Section 3.3.1). Let

$$B_x(y,z)=Y(y)Z(z)$$

so that

$$Z\frac{d^2Y}{dy^2}+Y\frac{d^2Z}{dz^2}+[(\omega/c)^2-\kappa^2]YZ=0$$

[25] In the case of TEM waves (including the unconfined plane waves of Section 8.2), $\kappa=\omega/c$, equations (8.215) are indeterminate, and we have to go back to (8.214).

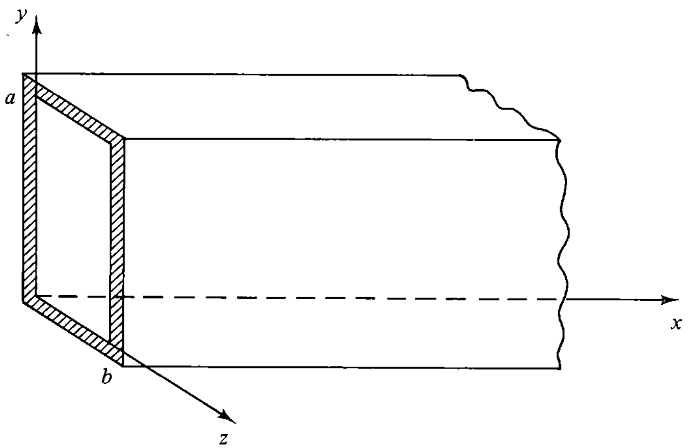

Figure 8.28

Divide by YZ and note that the y- and z-dependent terms must each be constant:

$$\text{(i)} \quad \frac{1}{Y}\frac{d^2Y}{dy^2} = -k_y^2 \qquad \text{(ii)} \quad \frac{1}{Z}\frac{d^2Z}{dz^2} = -k_z^2 \tag{8.217}$$

with

$$-k_y^2 - k_z^2 + (\omega/c)^2 - \kappa^2 = 0 \tag{8.218}$$

The general solution to (8.217(i)) is

$$Y = A\sin(k_y y) + B\cos(k_y y)$$

But the boundary conditions require that B_y—and hence also (see 8.215(ii)) dY/dy—vanishes at $y = 0$ and $y = a$. So $A = 0$, and

$$k_y = m\pi/a \qquad (m = 0, 1, 2, \ldots) \tag{8.219}$$

The same goes for Z, with

$$k_z = n\pi/b \qquad (n = 0, 1, 2, \ldots) \tag{8.220}$$

and we conclude that

$$B_x = B_0\cos(m\pi y/a)\cos(n\pi z/b) \tag{8.221}$$

This solution is called the TE_{mn} mode. (The first index is conventionally associated with the *larger* dimension, so we assume $a \geq b$. By the way, at least *one* of the indices must be nonzero—see Problem 8.32.) The wave number (κ) is obtained by putting (8.219) and (8.220) into (8.218):

$$\kappa = \sqrt{(\omega/c)^2 - \pi^2[(m/a)^2 + (n/b)^2]} \tag{8.222}$$

If

$$\omega < c\pi\sqrt{(m/a)^2 + (n/b)^2} \equiv \omega_{mn} \tag{8.223}$$

the wave number runs imaginary, and instead of a traveling wave we have exponentially attenuated fields (see equation (8.211)). For this reason ω_{mn} is called the "cutoff frequency" for the mode in question. The *lowest* cutoff frequency for a given wave guide occurs for the mode T_{10}:

$$\omega_{10} = c\pi/a \tag{8.224}$$

Frequencies less than this will not propagate down the wave guide at all.

The wave number (8.222) can be written more simply in terms of the cutoff frequency:

$$\kappa = \frac{1}{c}\sqrt{\omega^2 - \omega_{mn}^2} \tag{8.225}$$

The wave velocity (8.170) is

$$v = \frac{\omega}{k} = \frac{c}{\sqrt{1 - (\omega/\omega_{mn})^2}} \tag{8.226}$$

—which is greater than c. However (see Problem 8.34), the energy carried by the wave travels at the *group velocity* (8.171):

$$v_g = \frac{1}{(dk/d\omega)} = c\sqrt{1 - (\omega/\omega_{mn})^2} \tag{8.227}$$

There's another way to think about the propagation of an electromagnetic wave in a rectangular pipe, which serves to illuminate many of these results. Consider an ordinary *plane* wave, traveling at an angle θ to the x axis, and reflecting perfectly off each conducting surface (Fig. 8.29). In the y and z directions the (multiply reflected) waves interfere to form standing wave patterns, of wavelength $\lambda_y = 2a/m$ and $\lambda_z = 2b/n$ (hence wave number $k_y = 2\pi/\lambda_y = \pi m/a$ and $k_z = \pi n/b$), respectively. Meanwhile, in the x direction there remains a traveling wave, of wave number $k_x = \kappa$. The propagation vector for the "original" plane wave is therefore

$$\mathbf{k} = \kappa\hat{i} + \frac{\pi m}{a}\hat{j} + \frac{\pi n}{b}\hat{k}$$

and the frequency is

$$\omega = c|\mathbf{k}| = c\sqrt{\kappa^2 + \pi^2[(m/a)^2 + (n/b)^2]} = \sqrt{(c\kappa)^2 + (\omega_{mn})^2}$$

Only certain angles will lead to one of the allowed standing wave patterns:

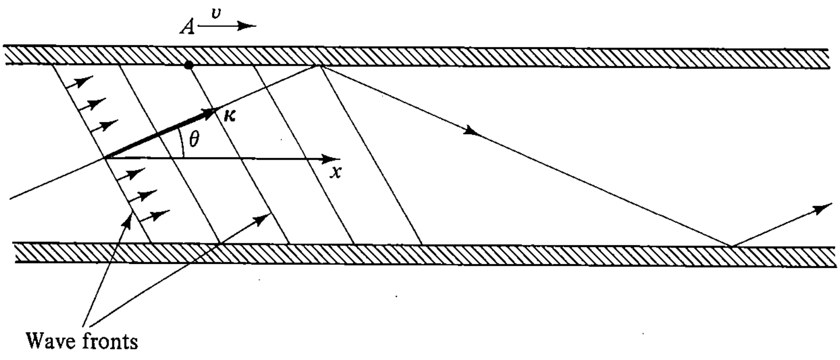

Figure 8.29

$$\cos\theta = \frac{\kappa}{|\mathbf{k}|} = \sqrt{1 - (\omega_{mn}/\omega)^2}$$

The plane wave travels at speed c, but because it is going at an angle θ to the x axis, its net velocity down the wave guide is

$$v_g = c\cos\theta = c\sqrt{1 - (\omega_{mn}/\omega)^2}$$

The *wave* velocity, on the other hand, is the speed of the wave fronts (say, A in Fig. 8.29) with respect to the pipe. Like the intersection of a line of breakers with the beach, they can move much faster than the waves themselves—in fact

$$v = \frac{c}{\cos\theta} = \frac{c}{\sqrt{1 - (\omega_{mn}/\omega)^2}}$$

Problem 8.32 Show that the mode TE_{00} cannot occur in a rectangular wave guide. (*Note*: In this case $\omega/c = \kappa$ (8.222), so equations (8.215) are indeterminate, and you must go back to (8.214). Show that these equations, together with (8.212(ii)), imply that $B_0 = 0$, and hence that this would be a TEM mode.)

Problem 8.33 Consider a rectangular wave guide with dimensions 2.28 cm $\times$ 1.01 cm. What TE modes will propagate in this wave guide, if the driving frequency is 1.70×10^{10} Hz? Suppose you wanted to excite only *one* TE mode; what range of frequencies could you use? What are the corresponding wavelengths (in open space)?

Problem 8.34 Confirm that the energy in a rectangular wave guide travels at the group velocity (8.227), as follows: Find the time averaged Poynting vector $\langle \mathbf{S} \rangle$ and energy density $\langle \mathbf{U} \rangle$ (use Problem 8.11 if you wish). Integrate over the cross section of the wave guide to get the energy per unit time and per unit length carried by the wave, and take their ratio.

Problem 8.35 Work out the theory of TM modes for a rectangular wave guide. In particular, find the longitudinal electric field, the cutoff frequencies, and the wave and group velocities. Find the ratio of the lowest TM cutoff frequency to the lowest TE cutoff frequency, for a given wave guide. (*Caution*: What is the lowest TM mode?).

8.5.3 The Coaxial Transmission Line

In Section 8.5.1 I showed that a *hollow* wave guide cannot support TEM waves. But a coaxial transmission line, consisting of a long straight wire surrounded by a cylindrical conducting sheath of radius a (Fig. 8.30) *does* admit modes with $E_x = 0$ and $B_x = 0$. In this case Maxwell's equations (in the form (8.214)) yield

$$\kappa = \omega/c \tag{8.228}$$

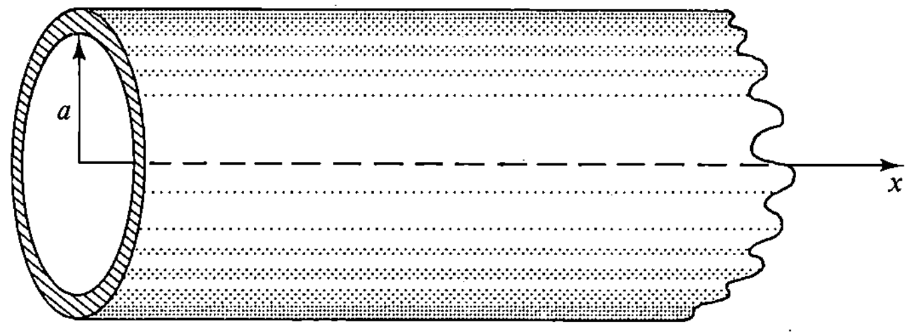

Figure 8.30

(so the waves travel at speed c, and are nondispersive),

$$cB_z = E_y \qquad \text{and} \qquad cB_y = -E_z \tag{8.229}$$

(so **E** and **B** are mutually perpendicular), and

$$\left.\begin{aligned} \frac{\partial E_y}{\partial y} + \frac{\partial E_z}{\partial z} = 0 \qquad \frac{\partial E_z}{\partial y} - \frac{\partial E_y}{\partial z} = 0 \\ \frac{\partial B_y}{\partial y} + \frac{\partial B_z}{\partial z} = 0 \qquad \frac{\partial B_z}{\partial y} - \frac{\partial B_y}{\partial z} = 0 \end{aligned}\right\} \tag{8.230}$$

These are precisely the equations of *electrostatics* and *magnetostatics*, for empty space, in two dimensions—the solution with cylindrical symmetry can be borrowed directly from the case of an infinite line charge and an infinite straight current, respectively:

$$\mathbf{E}_0 = E_0 \frac{1}{r}\hat{r} \qquad \mathbf{B}_0 = \frac{E_0}{c}\frac{1}{r}\hat{\phi} \tag{8.231}$$

Putting these into (8.211), and taking the real part, we have

$$\left.\begin{aligned} \mathbf{E} &= E_0 \frac{\cos(\kappa x - \omega t)}{r}\hat{r} \\ \mathbf{B} &= \frac{E_0}{c}\frac{\cos(\kappa x - \omega t)}{r}\hat{\phi} \end{aligned}\right\} \tag{8.232}$$

Problem 8.36

(a) Show directly that (8.232) satisfy Maxwell's equations (8.212) and the boundary conditions (8.210).

(b) Find the charge density, $\lambda(x, t)$, and the current, $I(x, t)$, on the inner conductor. Also find the net line charge density and the net current on the outer conductor.

Further Problems on Chapter 8

! Problem 8.37 The "inversion theorem" for Fourier transforms states that

$$f(x) = \int_{-\infty}^{\infty} F(\kappa)\, e^{i\kappa x}\, d\kappa \quad \Leftrightarrow \quad F(\kappa) = \frac{1}{2\pi}\int_{-\infty}^{\infty} f(x) e^{-i\kappa x}\, dx$$

Use this to determine $A(\kappa)$, in equation (8.34), in terms of $f(x, 0)$ and $\dot{f}(x, 0)$.

$$\left(\textit{Answer}: \frac{1}{2\pi}\int_{-\infty}^{\infty}\left[f(x, 0) + \frac{i}{\omega}\dot{f}(x, 0)\right] e^{-i\kappa x}\, dx\right)$$

! Problem 8.38 Suppose

$$\mathbf{E} = E_0\left(\sin\frac{\theta}{r}\right)\left[\cos(\kappa r - \omega t) - \left(\frac{1}{\kappa r}\right)\sin(\kappa r - \omega t)\right]\hat{\phi}, \qquad \text{with} \quad \frac{\omega}{\kappa} = c.$$

(This is, incidentally, the simplest possible *spherical* wave. For notational simplicity, set $(\kappa r - \omega t) = u$ in your calculations.)

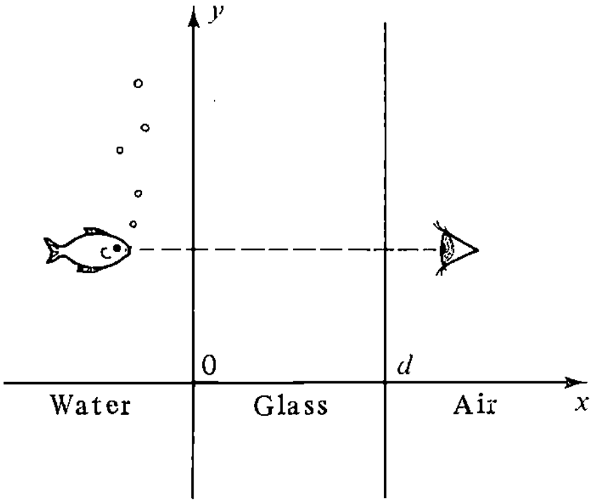

Figure 8.31

(a) Show that **E** obeys all four of Maxwell's equations, in vacuum, and find the associated magnetic field.
(b) Calculate the Poynting vector. Average **S** over a full cycle to get the intensity vector **I**. (Does it point in the expected direction? Does it fall off with r as anticipated?)
(c) Integrate **I** over a sphere of radius R ($\int \mathbf{I} \cdot d\mathbf{a}$) to get the total (average) power radiated.

$$\left(\textit{Answer}: \frac{4}{3}\pi\frac{E_0^2}{\mu_0 c}\right)$$

! **Problem 8.39** Find the transmission coefficient for waves passing through a pane of glass, of thickness d, at normal incidence. (*Note*: To the *left,* there is an incident wave and a reflected wave; to the *right* there is a transmitted wave; *inside the glass* there is a wave going to the right and a wave going to the left. Express each of these waves in terms of its complex amplitude, and relate the amplitudes by imposing suitable boundary conditions at the two edges. Neglect dispersion, and assume $\mu = \mu_0$.)

$$\left(\textit{Answer}: T^{-1} = 1 + \left[\frac{(n^2-1)}{2n}\sin\kappa d\right]^2, \text{ where } \kappa \text{ is the wave number in glass}\right)$$

Problem 8.40 A microwave antenna radiating at 10 GHz is to be protected from the environment by a plastic shield of dielectric constant 2.5. What is the minimum thickness of this shielding that will allow perfect transmission (assuming normal incidence)? (*Hint*: Use the result of Problem 8.33.)

! **Problem 8.41** The light emerging from an aquarium (Fig. 8.31) passes from water ($n = \frac{4}{3}$) through a pane of glass ($n = \frac{3}{2}$) into air ($n = 1$). Assuming it's a monochromatic plane wave and strikes the glass at normal incidence, calculate the transmission coefficient—that is, the ratio of the intensity of the beam that reaches your eye to the intensity of the beam that left the fish. Assume $\mu = \mu_0$ in all three media. You can see the fish pretty well. How well can it see you?

! **Problem 8.42** Consider the **resonant cavity** produced by closing off the two ends of the rectangular wave guide, at $x = 0$ and at $x = d$, making a perfectly conducting empty box. Show that the resonant frequencies for both TE and TM modes are given by

$$\omega_{lmn} = c\pi\sqrt{(l/d)^2 + (m/a)^2 + (n/b)^2}$$

for integers l, m, and n.

9

ELECTROMAGNETIC RADIATION

9.1 DIPOLE RADIATION

9.1.1 Retarded Potentials

In Chapter 8 we discussed the propagation of plane electromagnetic waves through various media—the vacuum, linear dielectrics, conductors, and dilute plasmas. But I did not address the question of how the waves got started in the first place. Like all electromagnetic fields, their source is some arrangement of electric charge. But a charge at rest does not generate electromagnetic waves; nor does a charge moving at constant velocity. It takes an *accelerating* charge, as we shall see. My purpose in this chapter, then, is to show you how accelerating electric charges produce electromagnetic waves—that is, how they *radiate.*

In order to do this, we must calculate the electric and magnetic fields of moving charges. As usual, it is easiest to begin with the potentials and then get the fields from (7.63) and (7.64):

$$\mathbf{E} = -\nabla V - \frac{\partial \mathbf{A}}{\partial t} \tag{9.1}$$

$$\mathbf{B} = \nabla \times \mathbf{A} \tag{9.2}$$

In the Lorentz gauge, defined by the condition (7.72)

$$\nabla \cdot \mathbf{A} = -\frac{1}{c^2}\frac{\partial V}{\partial t} \tag{9.3}$$

the potentials are determined by (7.73) and (7.74):

$$\nabla^2 V - \frac{1}{c^2}\frac{\partial^2 V}{\partial t^2} = -\frac{1}{\epsilon_0}\rho \tag{9.4}$$

$$\nabla^2 \mathbf{A} - \frac{1}{c^2}\frac{\partial^2 \mathbf{A}}{\partial t^2} = -\mu_0 \mathbf{J} \tag{9.5}$$

Apart from the "source" terms on the right, these look just like the wave equation (8.3). They are called **inhomogeneous wave equations**; the source-free versions are called **homogeneous wave equations**.

In the static case, (9.4) and (9.5) reduce to Poisson's equation,

$$\nabla^2 V = -\frac{1}{\epsilon_0}\rho, \qquad \nabla^2 \mathbf{A} = -\mu_0 \mathbf{J}$$

with the familiar solutions

$$V(\mathbf{r}) = \frac{1}{4\pi\epsilon_0}\int \frac{\rho(\mathbf{r}')}{\mathscr{r}}\,d\tau, \qquad \mathbf{A}(\mathbf{r}) = \frac{\mu_0}{4\pi}\int \frac{\mathbf{J}(\mathbf{r}')}{\mathscr{r}}\,d\tau \tag{9.6}$$

Here $\mathscr{r} = |\mathbf{r} - \mathbf{r}'|$ is the distance from the source to the point P at which the potentials are evaluated (Fig. 9.1). Now, electromagnetic "news" travels at the speed of light. In the *nonstatic* case, therefore, it's not the condition of the source *right now* that matters, but rather its condition at some earlier time t_r (called the **retarded time**) when the "message" left. Since this message must travel a distance $\mathscr{r}$ to reach P, the delay is $\mathscr{r}/c$:

$$t_r = t - \frac{\mathscr{r}}{c} \tag{9.7}$$

The appropriate generalization of (9.6) for nonstatic sources is therefore

$$\boxed{V(\mathbf{r}, t) = \frac{1}{4\pi\epsilon_0}\int \frac{\rho(\mathbf{r}', t_r)}{\mathscr{r}}\,d\tau, \qquad \mathbf{A}(\mathbf{r}, t) = \frac{\mu_0}{4\pi}\int \frac{\mathbf{J}(\mathbf{r}', t_r)}{\mathscr{r}}\,d\tau} \tag{9.8}$$

Here $\rho(\mathbf{r}', t_r)$ is the charge density that prevailed at point $\mathbf{r}'$ at the retarded time t_r. Because the integrands are evaluated at the retarded time, these are called **retarded potentials**. (I speak of "the" retarded time, but of course the most distant parts of the

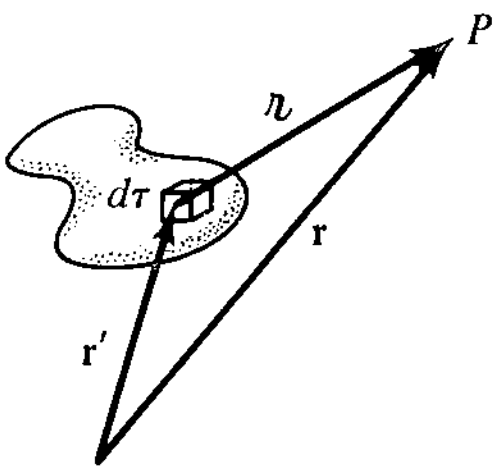

Figure 9.1

charge distribution have earlier retarded times than nearby ones. It's just like the night sky: the light we see now left each star at the retarded time corresponding to that star's distance from the earth.) Note that the retarded potentials reduce properly to (9.6) in the static case, for which ρ and $\mathbf{J}$ are independent of time.

Well, that all sounds very *reasonable*—and delightfully simple. But are we sure it's *right*? I didn't actually *derive* these formulas for V and $\mathbf{A}$; all I did was invoke a heuristic argument ("electromagnetic news travels at the speed of light") to make them *plausible*. To *prove* them, we must show that they satisfy (9.4) and (9.5) and meet the Lorentz condition (9.3). In case you think I'm being fussy, let me warn you that if you applied the same argument to the *fields*[1] you'd get entirely the *wrong* answer (see Problem 9.33):

$$\mathbf{E}(\mathbf{r}, t) \neq \frac{1}{4\pi\epsilon_0}\int \frac{\rho(\mathbf{r}', t_r)}{\mathscr{r}^2}\hat{\mathscr{r}}\, d\tau, \qquad \mathbf{B}(\mathbf{r}, t) \neq \frac{\mu_0}{4\pi}\int \frac{\mathbf{J}(\mathbf{r}', t_r)\times\hat{\mathscr{r}}}{\mathscr{r}^2}\, d\tau$$

as you would suppose, if the same "logic" worked for Coulomb's law and the Biot-Savart law. Let's stop and check, then, that the retarded scalar potential satisfies (9.4);[2] the same argument would serve to establish that the retarded vector potential obeys (9.5). I shall leave it for you (Problem 9.1) to check that the retarded potentials meet the Lorentz condition.

In calculating the Laplacian of $V(\mathbf{r}, t)$, the crucial point to notice is that the integrand depends on $\mathbf{r}$ in *two* places: *explicitly*, in the denominator ($\mathscr{r} = |\mathbf{r} - \mathbf{r}'|$), and *implicitly*, through $t_r = t - \mathscr{r}/c$, in the numerator. Thus

$$\nabla V = \frac{1}{4\pi\epsilon_0}\int\left[(\nabla\rho)\frac{1}{\mathscr{r}} + \rho\nabla\left(\frac{1}{\mathscr{r}}\right)\right]d\tau \tag{9.9}$$

and

$$\nabla\rho = \dot{\rho}(\nabla t_r) = -\frac{1}{c}\dot{\rho}(\nabla\mathscr{r}) \tag{9.10}$$

(the dot denotes differentiation with respect to t_r). Now $\nabla\mathscr{r} = \hat{\mathscr{r}}$ and $\nabla(1/\mathscr{r}) = -\hat{\mathscr{r}}/\mathscr{r}^2$ (Problem 1.13), so

$$\nabla V = \frac{1}{4\pi\epsilon_0}\int\left[-\frac{1}{c}\dot{\rho}\frac{\hat{\mathscr{r}}}{\mathscr{r}} - \rho\frac{\hat{\mathscr{r}}}{\mathscr{r}^2}\right]d\tau \tag{9.11}$$

Taking the divergence, we obtain

$$\nabla^2 V = \frac{1}{4\pi\epsilon_0}\int\left\{-\frac{1}{c}\left[\dot{\rho}\left(\nabla\cdot\frac{\hat{\mathscr{r}}}{\mathscr{r}}\right) + \frac{\hat{\mathscr{r}}}{\mathscr{r}}\cdot(\nabla\dot{\rho})\right] - \left[\rho\left(\nabla\cdot\frac{\hat{\mathscr{r}}}{\mathscr{r}^2}\right) + \frac{\hat{\mathscr{r}}}{\mathscr{r}^2}\cdot(\nabla\rho)\right]\right\}d\tau$$

[1]The same goes for *potentials*, in any other gauge. For example, the scalar potential in the *Coulomb* gauge *instantaneously* responds to changes in the source (see equation (7.70) and the paragraph following).

[2]I'll give you the straightforward but cumbersome proof; for a clever indirect argument see J. B. Marion and M. A. Heald, *Classical Electromagnetic Radiation*, 2d ed. (New York: Academic Press, 1980), p. 200.

But

$$\nabla \dot{\rho} = -\frac{1}{c}\ddot{\rho}(\nabla \mathscr{r})$$

as in (9.10), and

$$\nabla \cdot \left(\frac{\hat{\mathscr{r}}}{\mathscr{r}}\right) = \frac{1}{\mathscr{r}^2}$$

(Problem 1.62), whereas

$$\nabla \cdot \left(\frac{\hat{\mathscr{r}}}{\mathscr{r}^2}\right) = 4\pi\delta^3(\mathscr{r})$$

(equation (1.79)). Thus

$$\nabla^2 V = \frac{1}{4\pi\epsilon_0}\int \left[\frac{1}{c^2}\frac{\ddot{\rho}}{\mathscr{r}} - 4\pi\rho\delta^3(\mathscr{r})\right] d\tau = \frac{1}{c^2}\frac{\partial^2}{\partial t^2}V - \frac{1}{\epsilon_0}\rho$$

confirming that the retarded potential (9.8) satisfies the inhomogeneous wave equation (9.4).

Incidentally, this proof applies equally well to the **advanced potentials**,

$$V_a(\mathbf{r}, t) = \frac{1}{4\pi\epsilon_0}\int \frac{\rho(\mathbf{r}', t_a)}{\mathscr{r}}\, d\tau, \qquad \mathbf{A}_a(\mathbf{r}, t) = \frac{\mu_0}{4\pi}\int \frac{\mathbf{J}(\mathbf{r}', t_a)}{\mathscr{r}}\, d\tau \tag{9.12}$$

in which the charge and current densities are evaluated at the "advanced" time

$$t_a = t + \frac{\mathscr{r}}{c} \tag{9.13}$$

A few signs are changed, but the final result is unaffected. Although the advanced potentials are entirely consistent with Maxwell's equations, they violate the most sacred tenet in all of physics: the principle of **causality**. They suggest that the potentials *now* depend on what the charge and current distribution *will* be at some time in the future—the effect, in other words, precedes the cause. Although the advanced potentials are of some theoretical interest, they are without direct physical significance.

Example 1

An infinite straight wire carries the current

$$I(t) = \begin{cases} 0, & \text{for } t \le 0 \\ I_0, & \text{for } t > 0 \end{cases}$$

(that is, a constant current I_0 is turned on abruptly at $t = 0$).[3] Find the resulting electric and magnetic fields.

[3] See footnote 7, p. 290.

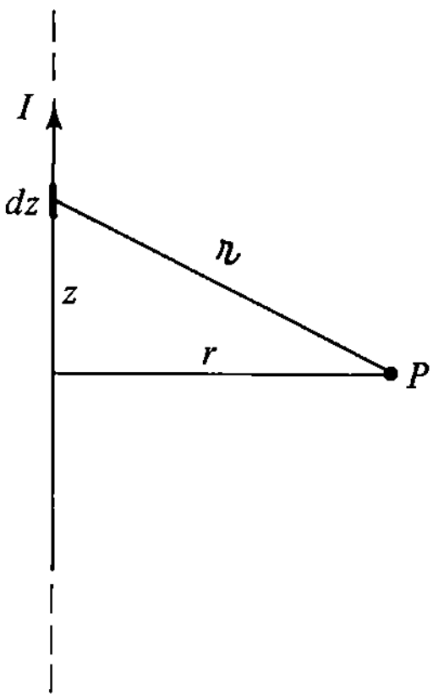

Figure 9.2

Solution: The wire is presumably electrically neutral, so the scalar potential is zero. Let the wire lie along the z axis (Fig. 9.2); the retarded vector potential at point P is

$$\mathbf{A}(r, t) = \frac{\mu_0}{4\pi} \hat{k} \int_{-\infty}^{\infty} \frac{I(t_r)}{\mathscr{r}} dz$$

For $t < r/c$, the "news" has not yet reached P, and the potential is zero. For $t > r/c$ only points out to

$$z = \pm\sqrt{(ct)^2 - r^2} \tag{9.14}$$

contribute (beyond this, t_r is negative, so $I(t_r) = 0$); thus

$$\mathbf{A}(r, t) = \left(\frac{\mu_0 I_0}{4\pi} \hat{k}\right) 2 \int_0^{\sqrt{(ct)^2 - r^2}} \frac{dz}{\sqrt{r^2 + z^2}} = \frac{\mu_0 I_0}{2\pi} \hat{k} \ln(\sqrt{r^2 + z^2} + z)\Big|_0^{\sqrt{(ct)^2 - r^2}}$$

$$= \frac{\mu_0 I_0}{2\pi} \ln\left(\frac{ct + \sqrt{(ct)^2 - r^2}}{r}\right) \hat{k}$$

The electric field is

$$\mathbf{E}(r, t) = -\frac{\partial \mathbf{A}}{\partial t} = -\frac{\mu_0 I_0 c}{2\pi\sqrt{(ct)^2 - r^2}} \hat{k}$$

and the magnetic field (evaluating the curl in cylindrical coordinates) is

$$\mathbf{B}(r, t) = \nabla \times \mathbf{A} = -\frac{\partial A_z}{\partial r} \hat{\phi} = \frac{\mu_0 I_0}{2\pi r} \frac{ct}{\sqrt{(ct)^2 - r^2}} \hat{\phi}$$

Notice that as $t \to \infty$ we recover the static case: $\mathbf{E} = 0$, $\mathbf{B} = (\mu_0 I_0 / 2\pi r)\hat{\phi}$.

Problem 9.1 Confirm that the retarded potentials satisfy the Lorentz gauge condition. [*Hint*: First show that

$$\nabla \cdot \left(\frac{\mathbf{J}}{\mathscr{r}}\right) = \frac{1}{\mathscr{r}} (\nabla \cdot \mathbf{J}) + \frac{1}{\mathscr{r}} (\nabla' \cdot \mathbf{J}) - \nabla' \cdot \left(\frac{\mathbf{J}}{\mathscr{r}}\right)$$

where ∇ denotes derivative with respect to $\mathbf{r}$, and ∇' denotes derivatives with respect to $\mathbf{r}'$. Next, noting that $\mathbf{J}(\mathbf{r}', t - \mathscr{r}/c)$ depends on $\mathbf{r}'$ both explicitly and through $\mathscr{r}$, whereas it depends on $\mathbf{r}$ only through $\mathscr{r}$, confirm that

$$\nabla \cdot \mathbf{J} = -\frac{1}{c}\frac{\partial \mathbf{J}}{\partial t_r}\cdot(\nabla \mathscr{r}), \qquad \nabla' \cdot \mathbf{J} = -\frac{\partial \rho}{\partial t} - \frac{1}{c}\frac{\partial \mathbf{J}}{\partial t_r}\cdot(\nabla' \mathscr{r})$$

Use this to calculate the divergence of $\mathbf{A}$ (equation 9.8) . . .]

Problem 9.2

(a) Suppose the wire in Example 1 carries a linearly increasing current

$$I(t) = \alpha t$$

for $t > 0$. Find the electric and magnetic fields generated.

(b) Do the same for the case of a sudden burst of current:

$$I(t) = q_0 \delta(t)$$

Problem 9.3 A piece of wire bent into a loop, as shown in Fig. 9.3, carries a current that increases linearly with time:

$$I(t) = \alpha t$$

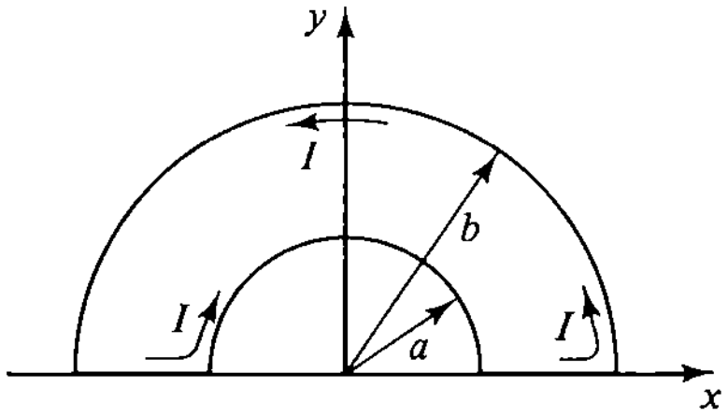

Figure 9.3

Calculate the retarded vector potential $\mathbf{A}$ at the center. Find the electric field at the center. Why does this (neutral) wire produce an *electric* field? (Why can't you determine the *magnetic* field from this expression for $\mathbf{A}$?)

The rest of this chapter amounts to a series of applications of the retarded potentials (9.8). Specifically, we shall use them to determine the electric and magnetic fields of an oscillating electric dipole, an oscillating magnetic dipole, and an accelerating point charge. Through these examples we will develop the theory of electromagnetic radiation.

9.1.2 Electric Dipole Radiation

Imagine two tiny metal spheres separated by a distance s and connected by a fine wire (Fig. 9.4). Assume the system as a whole is electrically neutral, so if at time t the charge on the upper sphere is $q(t)$, then the charge on the lower sphere is $-q(t)$. Suppose further that we somehow contrive to drive the charge back and forth through the wire, from one end to the other, at a frequency ω:

$$q(t) = q_0 \cos \omega t \tag{9.15}$$

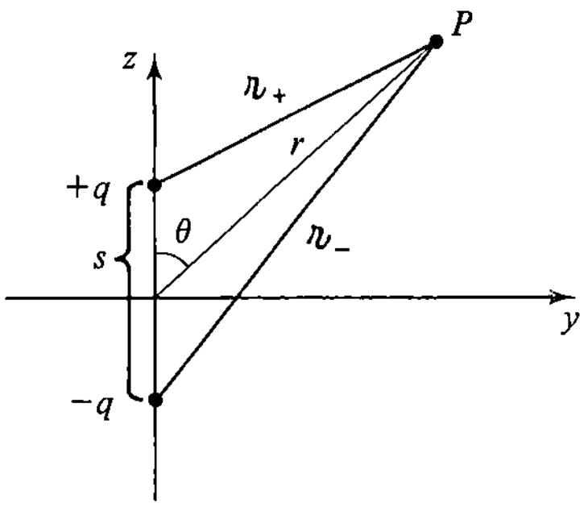

Figure 9.4

What I have described is a simple model for an oscillating electric dipole:[4]

$$\mathbf{p}(t) = q(t)\mathbf{s} = p_0 \cos \omega t \hat{k} \tag{9.16}$$

where

$$p_0 = q_0 s$$

is the maximum value of the dipole moment.

The retarded potential at P is given by

$$V(\mathbf{r}, t) = \frac{1}{4\pi\epsilon_0}\left[\frac{q_0 \cos \omega(t - \imath_+/c)}{\imath_+} - \frac{q_0 \cos \omega(t - \imath_-/c)}{\imath_-}\right] \tag{9.17}$$

where, by the law of cosines,

$$\imath_\pm = \sqrt{r^2 \mp rs\cos\theta + (s/2)^2} \tag{9.18}$$

Now, to make this *physical* dipole into a "perfect" dipole, we want the separation distance to be extremely small. In particular,

$$s \ll r \qquad (\textit{Approximation 1}) \tag{9.19}$$

Of course, if s is *zero* we get no potential at all; what we require is an *expansion* carried to *first order in s*. Thus,

$$\imath_\pm = r\left(1 \mp \frac{1}{2}\frac{s}{r}\cos\theta\right) \tag{9.20}$$

so that

$$\frac{1}{\imath_\pm} = \frac{1}{r}\left(1 \pm \frac{1}{2}\frac{s}{r}\cos\theta\right) \tag{9.21}$$

and

$$\begin{aligned}\cos \omega(t - \imath_\pm/c) &= \cos\left[\omega(t - r/c) \pm \frac{\omega s}{2c}\cos\theta\right] \\ &= \cos \omega(t - r/c)\cos\left(\frac{\omega s}{2c}\cos\theta\right) \mp \sin \omega(t - r/c)\sin\left(\frac{\omega s}{2c}\cos\theta\right)\end{aligned}$$

[4]It might strike you that a more natural model would consist of equal and opposite charges mounted on a spring, say, so that q is constant while s oscillates, instead of the other way around. Such a model would lead to the same result, but there is a subtle problem in calculating the retarded potentials of a moving point charge, which I would prefer to save for Section 9.2.

In the perfect dipole limit we have, further,

$$s \ll \frac{c}{\omega} \qquad \textit{(Approximation 2)} \tag{9.22}$$

(Since waves of frequency ω have a wavelength $\lambda = 2\pi c/\omega$, Approximation 2 amounts to the requirement $s \ll \lambda$.) Under these conditions

$$\cos\omega(t - \imath_{\pm}/c) = \cos\omega(t - r/c) \mp \frac{\omega s}{2c}\cos\theta \sin\omega(t - r/c) \tag{9.23}$$

Putting (9.21) and (9.23) into (9.17), then, we obtain

$$V(r, \theta, t) = \frac{p_0 \cos\theta}{4\pi\epsilon_0 r}\left[-\frac{\omega}{c}\sin\omega(t - r/c) + \frac{1}{r}\cos\omega(t - r/c)\right] \tag{9.24}$$

In the static limit ($\omega \rightarrow 0$) the second term properly reproduces the old formula for the potential of a stationary dipole (3.94):

$$V = \frac{p_0 \cos\theta}{4\pi\epsilon_0 r^2}$$

This is not, however, the term that concerns us now; we are interested in the fields which survive at *large distances from the source,* in the so-called **radiation zone:**[5]

$$r \gg \frac{c}{\omega} \qquad \textit{(Approximation 3)} \tag{9.25}$$

(Or, in terms of wavelength, $r \gg \lambda$.) In this region the potential reduces to

$$\boxed{V(r, \theta, t) = -\frac{p_0\omega}{4\pi\epsilon_0 c}\left(\frac{\cos\theta}{r}\right)\sin\omega(t - r/c)} \tag{9.26}$$

Meanwhile, the *vector* potential is determined by the current flowing in the wire:

$$\mathbf{I}(t) = \frac{dq}{dt}\hat{k} = -q_0\omega \sin\omega t\, \hat{k} \tag{9.27}$$

Referring to Fig. 9.5,

$$\mathbf{A}(\mathbf{r}, t) = \frac{\mu_0}{4\pi}\int_{-s/2}^{s/2} \frac{-q_0\omega \sin\omega(t - \imath/c)\hat{k}}{\imath}\, dz \tag{9.28}$$

Because the integration itself introduces a factor of s, we can, to first order, replace the integrand by its value at the center:

$$\boxed{\mathbf{A}(r, \theta, t) = -\frac{\mu_0 p_0 \omega}{4\pi r}\sin\omega(t - r/c)\hat{k}} \tag{9.29}$$

[5]Note that Approximations 2 and 3 together subsume Approximation 1.

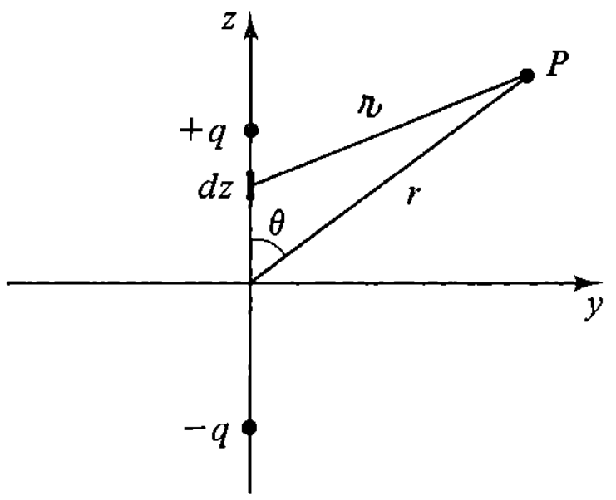

Figure 9.5

In spherical coordinates, of course,

$$\hat{k} = \cos\theta\,\hat{r} - \sin\theta\,\hat{\theta} \tag{9.30}$$

Notice that whereas I implicitly used Approximations 1 and 2, in keeping only the first order in s, (9.29) is *not* subject to Approximation 3.

From the potentials, it is a straightforward matter to compute the fields:

$$\begin{aligned}
\nabla V &= \frac{\partial V}{\partial r}\hat{r} + \frac{1}{r}\frac{\partial V}{\partial\theta}\hat{\theta} \\
&= -\frac{p_0\omega}{4\pi\epsilon_0 c}\left\{\cos\theta\left[-\frac{1}{r^2}\sin\omega(t - r/c) - \frac{\omega}{rc}\cos\omega(t - r/c)\right]\hat{r}\right. \\
&\quad \left. - \frac{\sin\theta}{r^2}\sin\omega(t - r/c)\hat{\theta}\right\} \\
&= \frac{p_0\omega^2}{4\pi\epsilon_0 c^2}\left(\frac{\cos\theta}{r}\right)\cos\omega(t - r/c)\hat{r}
\end{aligned}$$

(I dropped the first and last terms, in accordance with Approximation 3.) Likewise,

$$\frac{\partial \mathbf{A}}{\partial t} = -\frac{\mu_0 p_0\omega^2}{4\pi r}\cos\omega(t - r/c)(\cos\theta\,\hat{r} - \sin\theta\,\hat{\theta})$$

and therefore

$$\boxed{\mathbf{E} = -\nabla V - \frac{\partial \mathbf{A}}{\partial t} = -\frac{\mu_0 p_0\omega^2}{4\pi}\left(\frac{\sin\theta}{r}\right)\cos\omega(t - r/c)\hat{\theta}} \tag{9.31}$$

Meanwhile,

$$\begin{aligned}
\nabla \times \mathbf{A} &= \frac{1}{r}\left[\frac{\partial}{\partial r}(rA_\theta) - \frac{\partial A_r}{\partial\theta}\right]\hat{\phi} \\
&= -\frac{\mu_0 p_0\omega}{4\pi r}\left[\frac{\omega}{c}\sin\theta\cos\omega(t - r/c) + \frac{\sin\theta}{r}\sin\omega(t - r/c)\right]\hat{\phi}
\end{aligned}$$

The second term is again eliminated by Approximation 3, so that

$$\boxed{\mathbf{B} = \nabla \times \mathbf{A} = -\frac{\mu_0 p_0 \omega^2}{4\pi c}\left(\frac{\sin\theta}{r}\right)\cos\omega(t - r/c)\hat{\boldsymbol{\phi}}} \tag{9.32}$$

Equations (9.31) and (9.32) represent monochromatic waves of frequency ω traveling in the radial direction at the speed of light. **E** and **B** are in phase, mutually perpendicular, and transverse; the ratio of their amplitudes is given by $E_0/B_0 = c$. All of which is precisely what we expect for electromagnetic waves in free space. (These are actually *spherical* waves, not plane waves, and their amplitude decreases like $1/r$ as they progress. But for large r, they are approximately plane over small regions—just as the surface of the earth is reasonably flat, locally.)

The energy radiated by an oscillating electric dipole is determined by the Poynting vector:

$$\mathbf{S} = \frac{1}{\mu_0}(\mathbf{E} \times \mathbf{B}) = \frac{\mu_0}{c}\left[\frac{p_0\omega^2}{4\pi}\left(\frac{\sin\theta}{r}\right)\cos\omega(t - r/c)\right]^2 \hat{\mathbf{r}} \tag{9.33}$$

The intensity is obtained by averaging (in time) over a complete cycle:

$$\langle \mathbf{S} \rangle = \left(\frac{\mu_0 p_0^2 \omega^4}{32\pi^2 c}\right)\frac{\sin^2\theta}{r^2}\hat{\mathbf{r}} \tag{9.34}$$

Notice that there is no radiation along the axis of the dipole (here $\sin\theta = 0$); the profile of intensity looks like a donut, with its maximum in the equatorial plane (Fig. 9.6). The total power radiated is found by integrating **S** over a sphere of radius r:

$$\begin{aligned}\langle P \rangle &= \int \langle \mathbf{S} \rangle \cdot d\mathbf{a} = \left(\frac{\mu_0 p_0^2 \omega^4}{32\pi^2 c}\right)\int \frac{\sin^2\theta}{r^2} r^2 \sin\theta \, d\theta \, d\phi \\ &= \left(\frac{\mu_0 p_0^2}{12\pi c}\right)\omega^4 = \frac{1}{4\pi\epsilon_0}\frac{p_0^2\omega^4}{3c^3}\end{aligned} \tag{9.35}$$

This quantity is independent of the radius of the sphere, as one would expect from conservation of energy.

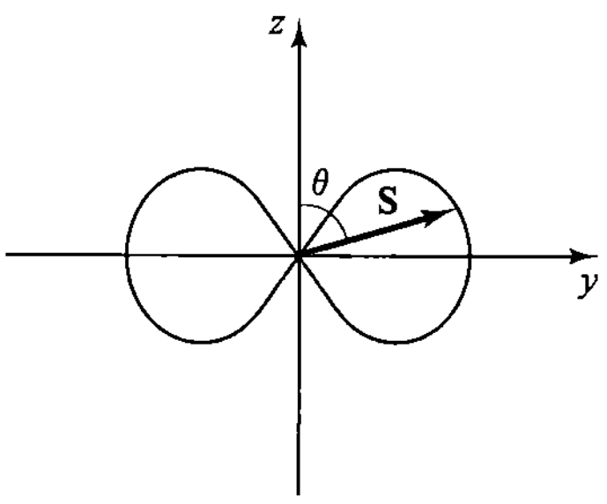

Intensity profile for dipole radiation **Figure 9.6**

Example 2

The sharp frequency dependence of the power formula is what accounts for the blueness of the sky. Sunlight passing through the atmosphere stimulates atoms to oscillate as tiny dipoles. The incident solar radiation covers a broad range of frequencies (white light), but the energy absorbed and reradiated by the atmospheric dipoles is stronger at the higher frequencies because of the ω^4 in (9.35). It is more intense in the blue, then, than in the red. It is this reradiated light that you see when you look up in the sky—unless, of course, you're looking directly at the sun.

Because electromagnetic waves are transverse, the dipoles oscillate in a direction perpendicular to the sun's rays. In the celestial arc perpendicular to these rays, where the blueness is most pronounced, dipoles oscillating along the line of sight send no radiation to the observer (because of the $\sin^2\theta$ in equation (9.34)); light received at this angle is therefore polarized perpendicular to the sun's rays (Fig. 9.7).

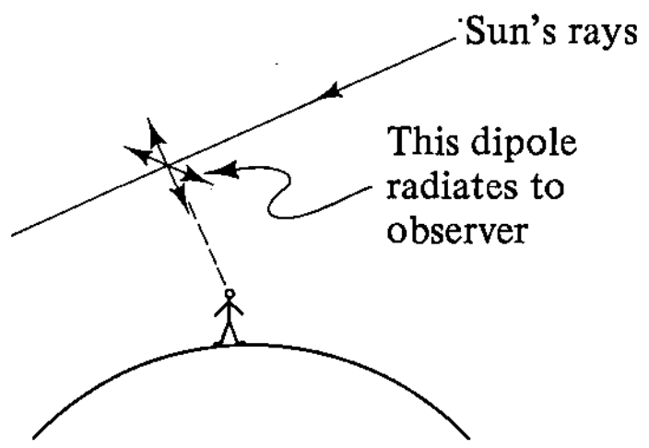

Figure 9.7

The redness of sunset is the other side of the same coin: sunlight coming in at a tangent to the earth's surface must pass through a much longer stretch of atmosphere than sunlight coming from overhead (Fig. 9.8). Accordingly, a much higher fraction of the blue has been removed by scattering from atmospheric dipoles.

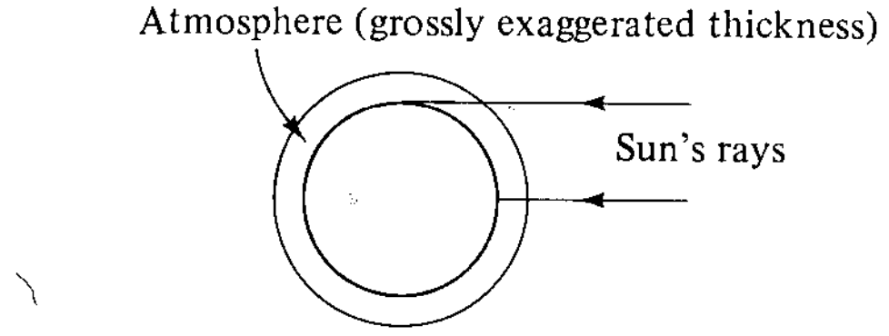

Figure 9.8

Problem 9.4 Check that the retarded potentials of an oscillating dipole, (9.24) and (9.29), satisfy the Lorentz gauge condition. Do not assume Approximation 3.

Problem 9.5 Equation (9.26) can be expressed in "coordinate-free" form by writing $p_0 \cos\theta = \mathbf{p}_0 \cdot \hat{r}$. Do so, and do the same for (9.29), (9.31), (9.32), and (9.34).

Problem 9.6 Find the **radiation resistance** of the wire joining the two ends of the dipole. (This is the resistance that would give the same average power loss—to heat—as the oscillating dipole in *fact* puts out in the form of radiation.) Show that $R = 790\,(s/\lambda)^2\,\Omega$,

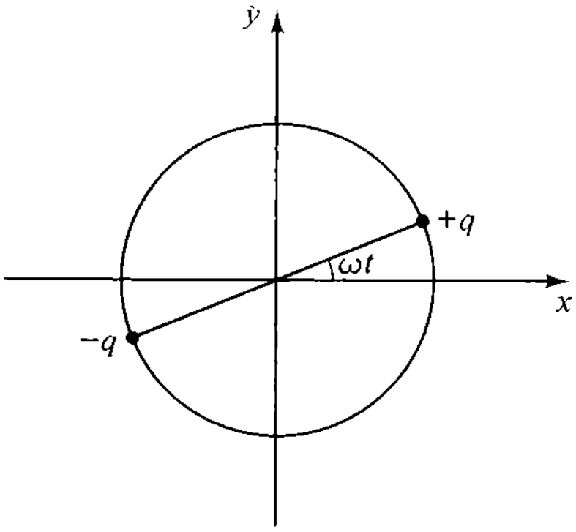

Figure 9.9

where λ is the wavelength of the radiation. For the wires in an ordinary radio (say, s = 5 cm), should you worry about the radiative contribution to the total resistance?

! Problem 9.7 A *rotating* electric dipole can be thought of as the superposition of two oscillating dipoles, one along the x axis, and the other along the y axis (Fig. 9.9), with the latter out of phase by 90°:

$$\mathbf{p} = p_0(\cos \omega t\, \hat{i} + \sin \omega t\, \hat{j})$$

Using the principle of superposition and equations (9.31) and (9.32) (perhaps in the form of Problem 9.5), find the fields of a rotating dipole. Also find the Poynting vector and the intensity of the radiation. Sketch the intensity profile as a function of the polar angle θ, and calculate the total power radiated. Does the answer seem reasonable? (Note that power, being *quadratic* in the fields, does *not* satisfy the superposition principle. In this instance, however, it *seems* to. Any explanation?)

9.1.3 Magnetic Dipole Radiation

Suppose now that we have a wire loop of radius a (Fig. 9.10), around which we drive a sinusoidally varying current, at frequency ω:

$$I(t) = I_0 \cos \omega t \tag{9.36}$$

This is a model for an oscillating *magnetic* dipole,

$$\mathbf{m}(t) = \pi a^2 I(t)\hat{k} = m_0 \cos \omega t\, \hat{k} \tag{9.37}$$

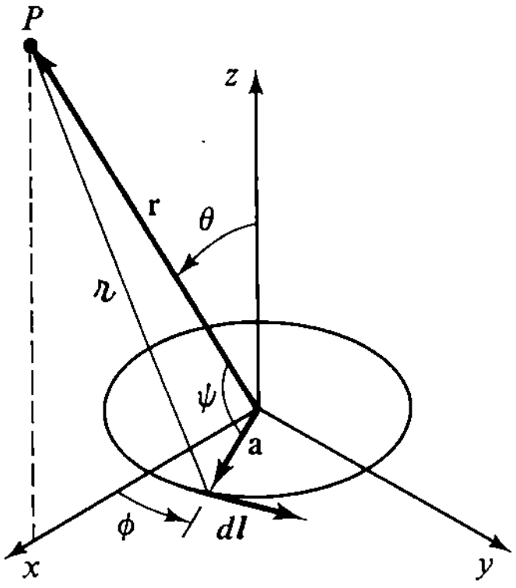

Figure 9.10

where

$$m_0 = \pi a^2 I_0 \tag{9.38}$$

is the maximum value of the magnetic dipole moment.

The loop is uncharged, so the scalar potential is zero. The retarded vector potential is

$$\mathbf{A}(\mathbf{r}, t) = \frac{\mu_0}{4\pi} \int \frac{I_0 \cos \omega(t - \mathscr{r}/c)}{\mathscr{r}} \, d\mathbf{l} \tag{9.39}$$

For a point P directly above the x axis (Fig. 9.10), $\mathbf{A}$ must aim in the y-direction; x-components from symmetrically placed points on either side of the x axis will cancel. Thus,

$$\mathbf{A}(\mathbf{r}, t) = \frac{\mu_0}{4\pi} I_0 a \hat{\jmath} \int_0^{2\pi} \frac{\cos \omega(t - \mathscr{r}/c)}{\mathscr{r}} \cos \phi \, d\phi \tag{9.40}$$

(the $\cos \phi$ serves to pick out the y-component of $d\mathbf{l}$). By the law of cosines,

$$\mathscr{r} = \sqrt{r^2 + a^2 - 2ra \cos \psi}$$

where ψ is the angle between the vectors $\mathbf{r}$ and $\mathbf{a}$:

$$\mathbf{r} = r \sin \theta \, \hat{\imath} + r \cos \theta \, \hat{k}, \qquad \mathbf{a} = a \cos \phi \, \hat{\imath} + a \sin \phi \, \hat{\jmath}$$

Thus, $ra \cos \psi = \mathbf{r} \cdot \mathbf{a} = ra \sin \theta \cos \phi$, and therefore

$$\mathscr{r} = \sqrt{r^2 + a^2 - 2ra \sin \theta \cos \phi} \tag{9.41}$$

For a "perfect" dipole, we require that the loop be extremely small. In particular,

$$a \ll r \qquad (\textit{Approximation 1}) \tag{9.42}$$

To first order in a, then,

$$\mathscr{r} = r\left(1 - \frac{a}{r} \sin \theta \cos \phi\right)$$

so that

$$\frac{1}{\mathscr{r}} = \frac{1}{r}\left(1 + \frac{a}{r} \sin \theta \cos \phi\right) \tag{9.43}$$

and

$$\cos \omega(t - \mathscr{r}/c) = \cos\left[\omega(t - r/c) + \frac{\omega a}{c} \sin \theta \cos \phi\right]$$

$$= \cos \omega(t - r/c) \cos\left(\frac{\omega a}{c} \sin \theta \cos \phi\right) - \sin \omega(t - r/c) \sin\left(\frac{\omega a}{c} \sin \theta \cos \phi\right)$$

As before, we also assume the radius is much less than the wavelength generated:

$$a \ll \frac{c}{\omega} \qquad (\textit{Approximation 2}) \tag{9.44}$$

In this case,

$$\cos\omega\left(t-\frac{\imath}{c}\right)=\cos\omega\left(t-\frac{r}{c}\right)-\frac{\omega a}{c}\sin\theta\cos\phi\sin\omega\left(t-\frac{r}{c}\right) \quad (9.45)$$

Inserting (9.43) and (9.45) into the equation for **A** (9.40), and dropping the second-order term:

$$\mathbf{A}(\mathbf{r},t)=\frac{\mu_0 I_0 a}{4\pi r}\hat{j}\int_0^{2\pi}\left\{\cos\omega(t-r/c)+a\sin\theta\cos\phi\left[\frac{1}{r}\cos\omega(t-r/c)-\frac{\omega}{c}\sin\omega(t-r/c)\right]\right\}\cos\phi\,d\phi$$

The first term integrates to zero:

$$\int_0^{2\pi}\cos\phi\,d\phi=0$$

The second term involves the integral of cosine squared:

$$\int_0^{2\pi}\cos^2\phi\,d\phi=\pi$$

Putting this in, and noting that in general **A** points in the $\hat{\phi}$-direction, we find

$$\mathbf{A}(r,\theta,t)=\frac{\mu_0 m_0}{4\pi}\left(\frac{\sin\theta}{r}\right)\left[\frac{1}{r}\cos\omega\left(t-\frac{r}{c}\right)-\frac{\omega}{c}\sin\omega\left(t-\frac{r}{c}\right)\right]\hat{\phi} \quad (9.46)$$

In the static limit ($\omega=0$) we recover the familiar formula for the potential of a magnetic dipole (5.83)

$$\mathbf{A}(r,\theta)=\frac{\mu_0}{4\pi}\frac{m_0\sin\theta}{r^2}\hat{\phi}$$

In the radiation zone,

$$r\gg\frac{c}{\omega} \qquad \textit{(Approximation 3)} \quad (9.47)$$

the first term in **A** is negligible, and so

$$\boxed{\mathbf{A}(r,\theta,t)=-\frac{\mu_0 m_0\omega}{4\pi c}\left(\frac{\sin\theta}{r}\right)\sin\omega(t-r/c)\hat{\phi}} \quad (9.48)$$

From this potential, we obtain the fields

$$\boxed{\mathbf{E}=-\frac{\partial\mathbf{A}}{\partial t}=\frac{\mu_0 m_0\omega^2}{4\pi c}\left(\frac{\sin\theta}{r}\right)\cos\omega(t-r/c)\hat{\phi}} \quad (9.49)$$

and

$$\boxed{\mathbf{B} = \nabla \times \mathbf{A} = -\frac{\mu_0 m_0 \omega^2}{4\pi c^2}\left(\frac{\sin\theta}{r}\right)\cos\omega(t - r/c)\hat{\theta}} \tag{9.50}$$

(I used Approximation 3 in the latter.) These fields are in phase, mutually perpendicular, and transverse to the direction of propagation $\hat{r}$. The ratio of their amplitudes ($E_0/B_0 = c$) is likewise correct for electromagnetic waves. They are, in fact, remarkably similar in structure to the fields of an oscillating *electric* dipole, equations (9.31) and (9.32), only this time it is $\mathbf{B}$ that points in the $\hat{\theta}$ direction and $\mathbf{E}$ in the $\hat{\phi}$, whereas for electric dipoles it's the other way around.

The energy flux for magnetic dipole radiation is

$$\mathbf{S} = \frac{1}{\mu_0}(\mathbf{E} \times \mathbf{B}) = \frac{\mu_0}{c}\left[\frac{m_0\omega^2}{4\pi c}\left(\frac{\sin\theta}{r}\right)\cos\omega(t - r/c)\right]^2 \hat{r} \tag{9.51}$$

and the intensity is

$$\langle \mathbf{S} \rangle = \frac{\mu_0 m_0^2 \omega^4}{32\pi^2 c^3}\left(\frac{\sin^2\theta}{r^2}\right)\hat{r} \tag{9.52}$$

The total radiated power is therefore

$$\langle P \rangle = \left(\frac{\mu_0 m_0^2}{12\pi c^3}\right)\omega^4 = \frac{1}{4\pi\epsilon_0}\frac{m_0^2\omega^4}{3c^5} \tag{9.53}$$

Once again, the intensity profile has the shape of a donut, and the power radiated goes as the fourth power of the frequency. There is, however, one important difference between electric and magnetic dipole radiation: For configurations with comparable dimensions, the power radiated electrically is enormously greater. Comparing (9.35) and (9.53),

$$\frac{P_{\text{magnetic}}}{P_{\text{electric}}} = \left(\frac{m_0}{p_0 c}\right)^2 \tag{9.54}$$

where $m_0 = \pi a^2 I_0$, and $p_0 = q_0 s$. The amplitude of the current in the electrical case (9.27) was $I_0 = q_0\omega$. Setting $s = \pi a$, just for the sake of comparison, I get

$$\frac{P_{\text{magnetic}}}{P_{\text{electric}}} = \left(\frac{a\omega}{c}\right)^2 \tag{9.55}$$

Now, $(a\omega/c)$ is precisely the quantity we assumed was very small in Approximation 2, and here it appears *squared.* Ordinarily, then, one should expect electric dipole radiation to dominate. Only when the system is carefully arranged to exclude any electric contribution (as in the case just treated) will the magnetic dipole radiation reveal itself.

Problem 9.8 Calculate the electric and magnetic fields of an oscillating magnetic dipole without using Approximation 3. (Do they look familiar? Compare Problem 8.32.) Find the Poynting vector, and show that the intensity of the radiation is the same as we got using Approximation 3. (Why *should* it be?)

Problem 9.9 Find the radiation resistance (see Problem 9.6) of the oscillating magnetic dipole in Fig. 9.10. Express your answer in terms of λ and a, and compare the radiation resistance of the *electric* dipole. (*Answer*: $3 \times 10^5 (a/\lambda)^4\ \Omega$)

9.1.4 Radiation from an Arbitrary Distribution of Charges and Currents

In the previous sections we studied the radiation produced by two specific systems: oscillating electric dipoles and oscillating magnetic dipoles. Now I want to apply the same procedures to a configuration of charge and current that is perfectly arbitrary, except that it is localized within some small region near the origin (Fig. 9.11).

The scalar potential is

$$V(\mathbf{r}, t) = \frac{1}{4\pi\epsilon_0}\int \frac{\rho(\mathbf{r}', t - \boldsymbol{\imath}/c)}{\boldsymbol{\imath}}\, d\tau \tag{9.56}$$

where

$$\boldsymbol{\imath} = \sqrt{r^2 + r'^2 - 2\mathbf{r}\cdot\mathbf{r}'} \tag{9.57}$$

As before, we shall assume that the point of observation P is far away, in comparison to the size of the source distribution:

$$r' \ll r \qquad (\textit{Approximation 1}) \tag{9.58}$$

(Actually, r' is a variable of integration; Approximation 1 holds provided the *maximum* value of r', as it ranges over the source, is much less than r.) On this assumption,

$$\boldsymbol{\imath} = r\left(1 - \frac{\mathbf{r}\cdot\mathbf{r}'}{r^2}\right) \tag{9.59}$$

so that

$$\frac{1}{\boldsymbol{\imath}} = \frac{1}{r}\left(1 + \frac{\mathbf{r}\cdot\mathbf{r}'}{r^2}\right) \tag{9.60}$$

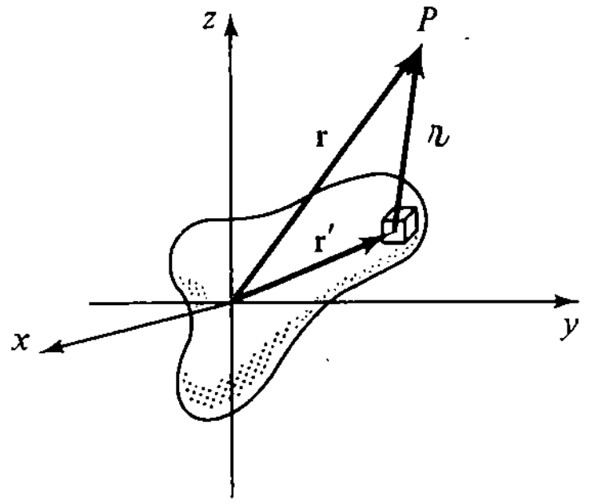

Figure 9.11

and

$$\rho\left(\mathbf{r}', t - \frac{\imath}{c}\right) = \rho\left(\mathbf{r}', t - \frac{r}{c} + \frac{\hat{r}\cdot\mathbf{r}'}{c}\right)$$

Expanding ρ as a Taylor series in the variable t, about the retarded time at the origin,

$$t_0 = t - \frac{r}{c} \tag{9.61}$$

we have

$$\rho\left(\mathbf{r}', t - \frac{\imath}{c}\right) = \rho(\mathbf{r}', t_0) + \dot{\rho}(\mathbf{r}', t_0)\left(\frac{\hat{r}\cdot\mathbf{r}'}{c}\right) + \cdots \tag{9.62}$$

where the dot signifies differentiation with respect to time. The next terms in the series would be

$$\frac{1}{2}\ddot{\rho}\left(\frac{\hat{r}\cdot\mathbf{r}'}{c}\right)^2, \quad \frac{1}{3!}\dddot{\rho}\left(\frac{\hat{r}\cdot\mathbf{r}'}{c}\right)^3, \ldots$$

We can afford to drop them, provided

$$r' \ll \frac{c}{|\dot{\rho}/\ddot{\rho}|}, \frac{c}{|\dot{\rho}/\dddot{\rho}|^{1/2}}, \frac{c}{|\dot{\rho}/\ddddot{\rho}|^{1/3}}, \ldots \quad \text{(Approximation 2)} \tag{9.63}$$

For an oscillating system each of these ratios is (c/ω); in the general case it's more difficult to interpret this approximation, but as a *procedural* matter Approximations 1 and 2 amount to *keeping only first-order terms in* r'.

Using (9.60) and (9.62) in the formula for V (9.56), and again discarding the second-order term:

$$V(\mathbf{r}, t) = \frac{1}{4\pi\epsilon_0 r}\left[\int \rho(\mathbf{r}', t_0)\, d\tau + \frac{\hat{r}}{r}\cdot\int \mathbf{r}'\rho(\mathbf{r}', t_0)\, d\tau + \frac{\hat{r}}{c}\cdot\frac{d}{dt}\int \mathbf{r}'\rho(\mathbf{r}', t_0)\, d\tau\right]$$

The first integral is simply the total charge at time t_0. Because charge is conserved, however, Q is actually *independent* of time. The other two integrals represent the electric dipole moment at time t_0. Thus,

$$V(\mathbf{r}, t) = \frac{1}{4\pi\epsilon_0}\left[\frac{Q}{r} + \frac{\hat{r}\cdot\mathbf{p}(t_0)}{r^2} + \frac{\hat{r}\cdot\dot{\mathbf{p}}(t_0)}{rc}\right] \tag{9.64}$$

In the static case, the first two terms in (9.64) are the monopole and dipole contributions to the multipole expansion for V; the third term, of course, would not be present.

Meanwhile, the vector potential is

$$\mathbf{A}(\mathbf{r}, t) = \frac{\mu_0}{4\pi}\int \frac{\mathbf{J}(\mathbf{r}', t - \imath/c)}{\imath}\, d\tau \tag{9.65}$$

As you'll see in a moment, to first order in r' it suffices to replace $\imath$ by r in the integrand:

$$\mathbf{A}(\mathbf{r}, t) = \frac{\mu_0}{4\pi r}\int \mathbf{J}(\mathbf{r}', t_0)\, d\tau \tag{9.66}$$

According to Problem 5.7, the integral of **J** is the time derivative of the dipole moment, so

$$\mathbf{A}(\mathbf{r}, t) = \frac{\mu_0}{4\pi}\frac{\dot{\mathbf{p}}(t_0)}{r} \tag{9.67}$$

Now you see why it was unnecessary to carry the approximation of $\imath$ beyond the zeroth order ($\imath = r$): **p** is *already* first order in r', and any refinements would be corrections of *second* order.

Next we must calculate the fields. Once again, we are interested in the radiation zone (that is, in the fields which survive at large distances from the source), so we keep only those terms that go like $1/r$:

$$\text{discard } 1/r^2 \text{ terms in } \mathbf{E} \text{ and } \mathbf{B} \qquad (\textit{Approximation } 3) \tag{9.68}$$

For instance, the Coulomb field,

$$\mathbf{E} = \frac{1}{4\pi\epsilon_0}\frac{Q}{r^2}\hat{r}$$

coming from the first term in (9.64), does not contribute to the electromagnetic radiation. In fact, the radiation comes entirely from those terms in which we differentiate the argument t_0:

$$\nabla V = \frac{1}{4\pi\epsilon_0}\left[\frac{\hat{r}}{rc}\cdot\ddot{\mathbf{p}}(t_0)\right]\nabla t_0 = -\frac{1}{4\pi\epsilon_0 c^2}\frac{(\hat{r}\cdot\ddot{\mathbf{p}}(t_0))}{r}\hat{r} \tag{9.69}$$

$$\frac{\partial \mathbf{A}}{\partial t} = \frac{\mu_0}{4\pi}\frac{\ddot{\mathbf{p}}(t_0)}{r}, \tag{9.70}$$

$$\nabla\times\mathbf{A} = \frac{\mu_0}{4\pi r}\left\{-\frac{\partial}{\partial r}[\dot{p}_\phi(t_0)]\hat{\theta} + \frac{\partial}{\partial r}[\dot{p}_\theta(t_0)]\hat{\phi}\right\} = -\frac{\mu_0}{4\pi c r}(-\ddot{p}_\phi\hat{\theta} + \ddot{p}_\theta\hat{\phi}) \tag{9.71}$$

So

$$\boxed{\mathbf{E}(\mathbf{r}, t) = \frac{\mu_0}{4\pi r}[(\hat{r}\cdot\ddot{\mathbf{p}}(t_0))\hat{r} - \ddot{\mathbf{p}}(t_0)]} \tag{9.72}$$

and

$$\boxed{\mathbf{B}(\mathbf{r}, t) = -\frac{\mu_0}{4\pi r c}[\hat{r}\times\ddot{\mathbf{p}}(t_0)]} \tag{9.73}$$

In particular, if we choose spherical polar coordinates, with the z axis in the direction of $\ddot{\mathbf{p}}(t_0)$, then

$$\left.\begin{aligned} \mathbf{E}(r, \theta, t) &= \frac{\mu_0}{4\pi}\ddot{p}(t_0)\left(\frac{\sin\theta}{r}\right)\hat{\theta} \\ \mathbf{B}(r, \theta, t) &= \frac{\mu_0}{4\pi c}\ddot{p}(t_0)\left(\frac{\sin\theta}{r}\right)\hat{\phi} \end{aligned}\right\} \tag{9.74}$$

The Poynting vector takes the form

$$\mathbf{S} = \frac{1}{\mu_0}(\mathbf{E} \times \mathbf{B}) = \frac{\mu_0}{16\pi^2 c}(\ddot{p}(t_0))^2\frac{\sin^2\theta}{r^2}\hat{r} \tag{9.75}$$

and the total power radiated through a sphere of radius r is

$$P = \int \mathbf{S}\cdot d\mathbf{a} = \frac{\mu_0}{6\pi c}(\ddot{p}(t_0))^2 = \frac{1}{4\pi\epsilon_0}\frac{2\ddot{p}^2}{3c^3} \tag{9.76}$$

Notice how the r's cancel out. If (contrary to 9.68) we kept the terms in **E** and **B** that go like $1/r^2$, they would yield contributions to P proportional to inverse powers of r—which vanish as $r \to \infty$. *By "radiation" we mean fields that transport energy out to infinity*; that is why only fields that go like $1/r$ constitute radiation. Notice also that **E** and **B** are mutually perpendicular, transverse to the direction of propagation $\hat{r}$, and in the ratio $E/B = c$, as always for radiation fields.

Example 3

(a) In the case of an oscillating electric dipole,

$$p(t) = p_0 \cos\omega t, \qquad \ddot{p}(t) = -\omega^2 p_0 \cos\omega t$$

and all these formulas reduce to the results of Section 9.1.2.

(b) For a single point charge q, the dipole moment is

$$\mathbf{p}(t) = q\mathbf{s}(t)$$

where **s** is its position with respect to the origin. Accordingly,

$$\ddot{\mathbf{p}}(t) = q\mathbf{a}(t)$$

where **a** is the acceleration of the charge. In this case the power radiated (9.76) is

$$\boxed{P = \frac{1}{4\pi\epsilon_0}\frac{2}{3}\frac{q^2a^2}{c^3}} \tag{9.77}$$

This is known as the **Larmor formula**; we'll derive it again, by rather different means, in Section 9.2. Notice the power radiated by a point charge is proportional to the *square* of its *acceleration*.

What I have done in this section amounts to a multipole expansion of the retarded potentials, carried to the lowest order in r' which is capable of producing electromagnetic radiation (fields that go like $1/r$). This turns out to be the electric dipole term. Because charge is conserved, an electric *monopole* does not radiate—if charge were *not* conserved, the first term in (9.64) would read

$$V_{\text{mono}} = \frac{1}{4\pi\epsilon_0}\frac{Q(t_0)}{r}$$

and we *would* get a monopole field proportional to $1/r$:

$$\mathbf{E}_{\text{mono}} = \frac{1}{4\pi\epsilon_0 c}\frac{\dot{Q}(t_0)}{r}\hat{r}$$

You might think that a charged sphere whose radius oscillates in and out would radiate, but it *doesn't*—the field outside, according to Gauss's law, is exactly $Q\hat{r}/(4\pi\epsilon_0 r^2)$, regardless of the fluctuations in size. (In the acoustical analog, by the way, monopoles *do* radiate: witness the croak of a bullfrog.)

If the electric dipole moment should happen to vanish (or, at any rate, if its second time derivative is zero) then there is no electric dipole radiation, and one must look to the next term: the one of *second* order in r'. As it happens, this term can be separated into two parts, one of which is related to the *magnetic* dipole moment of the source, the other to its electric *quadrupole* moment. The former is a generalization of the magnetic dipole radiation we considered in Section 9.1.3. If the magnetic dipole and electric quadrupole contributions vanish, the $(r')^3$ term must be considered. This yields magnetic quadrupole and electric octopole radiation . . . and so it goes.

Problem 9.10 Apply (9.75) and (9.76) to the rotating dipole of Problem 9.7. Explain any apparent discrepancies with your previous answer.

Problem 9.11 An insulating circular ring (radius R) lies in the xy plane, centered at the origin. It carries a linear charge density $\lambda = \lambda_0 \sin\phi$, where λ_0 is constant and ϕ is the usual azimuthal angle. The ring is now set spinning at constant angular velocity ω about the z axis. Calculate the power radiated.

Problem 9.12 An electron is released from rest and falls under the influence of gravity. In the first centimeter, what fraction of the potential energy lost is radiated away?

! **Problem 9.13** As a model for electric quadrupole radiation, consider two oppositely oriented oscillating electric dipoles, separated by a distance d, as shown in Fig. 9.12. (Use the results of Section 9.1.2 for the potentials of each dipole, but note that they are *not* located at the origin.) Keeping only terms of first order in d,

(a) Find the scalar and vector potentials.

(b) Find the electric and magnetic fields.

(c) Find the Poynting vector and the power radiated. Sketch the intensity profile as a function of θ.

! **Problem 9.14** A current $I(t)$ flows around the circular ring in Fig. 9.10. Derive the general

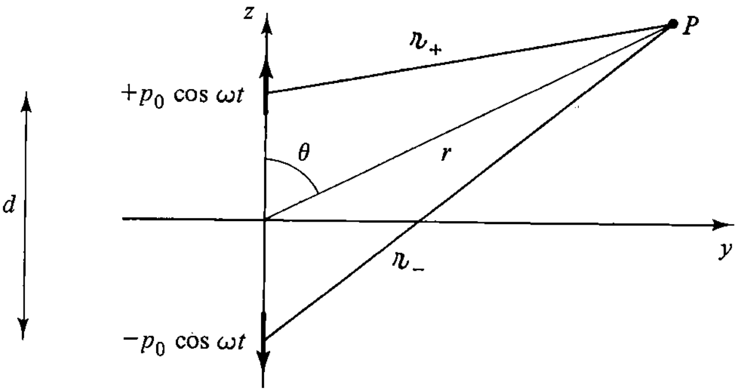

Figure 9.12

formula for the power radiated (analogous to 9.76), expressing your answer in terms of the magnetic dipole moment (m) of the loop.

$\left(\textit{Answer}: P = \dfrac{1}{4\pi\epsilon_0}\dfrac{2\ddot{m}^2}{3c^5}\right)$

9.2 RADIATION FROM A POINT CHARGE

9.2.1 Liénard-Wiechert Potentials

Our next project is to calculate the fields of a point charge q that is moving on a specified trajectory

$$\mathbf{w}(t) = \text{position of } q \text{ at time } t \tag{9.78}$$

The retarded potentials $V(\mathbf{r}, t)$ and $\mathbf{A}(\mathbf{r}, t)$ do not, of course, depend on the status of the charge at time t, but rather at the earlier time t_r when the "message" left the particle—a message that, traveling at the speed of light, reaches P at time t. This "retarded time" is determined implicitly by the equation

$$|\mathbf{r} - \mathbf{w}(t_r)| = c(t - t_r) \tag{9.79}$$

for the left side is the distance the "news" must travel, and $(t - t_r)$ is the time it takes to make the trip. I shall call $\mathbf{w}(t_r)$ the **retarded position** of the charge, and let $\boldsymbol{\mathscr{r}}$ be the vector from the retarded position to P (see Fig. 9.13):

$$\boldsymbol{\mathscr{r}} = \mathbf{r} - \mathbf{w}(t_r) \tag{9.80}$$

It is important to note that at most *one* point on the trajectory is "in communication" with P at any particular time t. For suppose there were *two* such points, with retarded times t_1 and t_2:

$$\mathscr{r}_1 = c(t - t_1) \quad \text{and} \quad \mathscr{r}_2 = c(t - t_2)$$

Then $\mathscr{r}_1 - \mathscr{r}_2 = c(t_2 - t_1)$, so the average velocity of the particle in the direction of P would have to be c, and that's not to mention whatever velocity the charge might have in the perpendicular direction. Since no charged particle can travel at the speed of

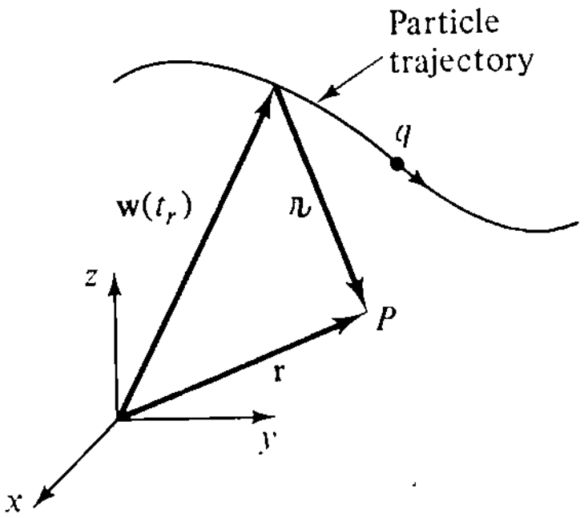

Figure 9.13

light, it follows that only *one retarded point contributes to the potentials at P, at any given moment.*[6]

Now, a naive reading of the formula

$$V(\mathbf{r}, t) = \frac{1}{4\pi\epsilon_0} \int \frac{\rho(\mathbf{r}', t_r)}{\mathscr{r}} \, d\tau \tag{9.81}$$

might suggest to you that the retarded potential of a point charge is simply

$$\frac{1}{4\pi\epsilon_0} \frac{q}{\mathscr{r}}$$

(the same as in the static case, only with the understanding that $\mathscr{r}$ is the distance to the *retarded* position of the charge). But this is wrong, for a very subtle reason: It is true that for a point source the denominator $\mathscr{r}$ comes outside the integral (9.81), but what remains,

$$\int \rho(\mathbf{r}', t_r) \, d\tau \tag{9.82}$$

is *not* equal to the charge of the particle. To get the total charge of a configuration, you must integrate ρ over the entire distribution at *one instant of time,* but here the retardation, $t_r = t - \mathscr{r}/c$, obliges us to evaluate ρ at *different times* for different parts of the configuration. If the source is moving, this will give a distorted picture of the total charge. You might think that this problem would disappear for *point* charges, but it doesn't. In Maxwell's electrodynamics, formulated as it is in terms of charge and current *densities,* a point charge must be regarded as the limit of an extended charge, as the size goes to zero. And as we'll see in the next paragraph, for an extended particle, however small, the retardation in equation (9.82) throws in a factor $(1 - \hat{\mathscr{r}} \cdot \mathbf{v}/c)^{-1}$, where $\mathbf{v}$ is the velocity of the charge at the retarded time:

$$\int \rho(\mathbf{r}', t_r) \, d\tau = q\left(1 - \frac{\hat{\mathscr{r}} \cdot \mathbf{v}}{c}\right)^{-1} \tag{9.83}$$

[6] For the same reason an observer at *P sees* the particle in only one place at a time. By contrast, it is possible in principle to *hear* an object in two places at once. Consider a bear who growls at you and then runs toward you at the speed of sound and growls again: You'd hear both growls at the same time, coming from different locations, but there's only one bear.

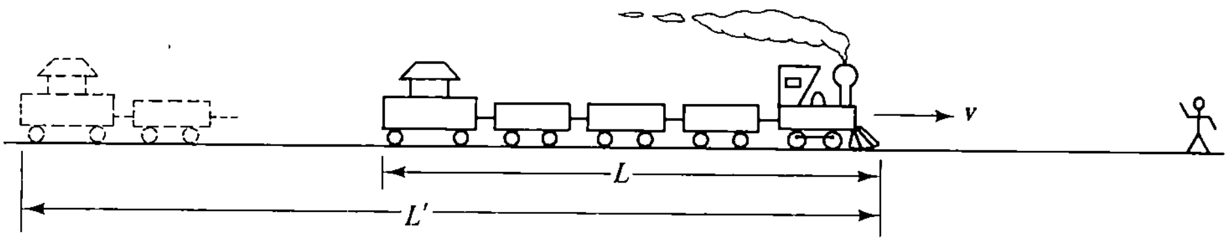

Figure 9.14

You will not have noticed it, for obvious reasons, but the fact is that a train coming toward you looks a little longer than it really is, because the light you receive from the caboose left earlier than the light you receive simultaneously from the engine, and at that earlier time the train was farther away (Fig. 9.14). In the interval it takes light from the caboose to travel the extra distance L', the train moves a distance $L' - L$:

$$\frac{L'}{c} = \frac{L' - L}{v} \quad \text{or} \quad L' = \frac{L}{(1 - v/c)}$$

So approaching trains appear *longer*, by a factor $(1 - v/c)^{-1}$. By contrast, a train going *away* from you looks *shorter*,[7] by a factor $(1 + v/c)^{-1}$. In general, if the train's velocity makes an angle θ with your line of sight,[8] the extra distance light from the caboose must cover is $L'\cos\theta$ (Fig. 9.15). In the time $L'\cos\theta/c$, then, the train moves a distance $(L' - L)$:

$$\frac{L'\cos\theta}{c} = \frac{L' - L}{v}, \quad \text{or} \quad L' = \frac{L}{1 - \dfrac{v}{c}\cos\theta}$$

Notice that this effect does *not* distort dimensions perpendicular to the motion (the height and width of the train): Never mind that the light from the far side of the train is delayed in reaching you (relative to light from the near side)—since there's no *motion* in that direction, they'll still look the same distance apart. The apparent *volume* τ' of the train, then, is related to the *actual* volume τ by

$$\tau' = \frac{\tau}{\left(1 - \dfrac{\hat{\boldsymbol{\imath}}\cdot\mathbf{v}}{c}\right)} \tag{9.84}$$

where $\hat{\boldsymbol{\imath}}$ is a unit vector from the train to the observer.

In case the connection between moving trains and retarded potentials escapes you, the point is this: When you do an integral of the type (9.82), in which the integrand is evaluated at the retarded time, the effective volume of the charge is modified by the factor (9.84), for exactly the same reason as was the apparent volume of the train. At each point you are putting in the charge density, *not* as it is now but as it *was*

[7]Please note that this has nothing whatever to do with special relativity or Lorentz contraction—it's a purely *geometrical* matter.

[8]I assume the train is far enough away or (more to the point) *short* enough so that rays from the caboose and engine can be considered parallel.

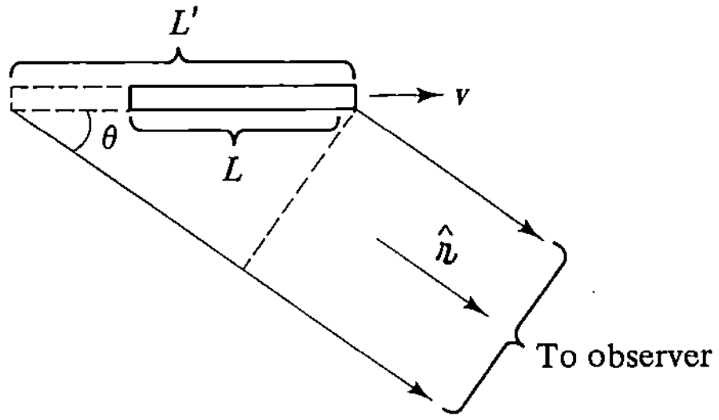

Figure 9.15

at the retarded time appropriate to that point. Because the "correction" factor makes no reference to the *size* of the particle, it is every bit as significant for a point charge as for an extended charge:

$$\int \rho(\mathbf{r}', t_r)\, d\tau = \frac{q}{\left(1 - \dfrac{\hat{\mathbf{\mathscr{r}}}\cdot\mathbf{v}}{c}\right)} \tag{9.85}$$

It follows, then, that

$$\boxed{V(\mathbf{r}, t) = \frac{1}{4\pi\epsilon_0}\frac{q}{\mathscr{r}\left(1 - \dfrac{\hat{\mathbf{\mathscr{r}}}\cdot\mathbf{v}}{c}\right)}} \tag{9.86}$$

where $\mathbf{v}$ is the velocity of the charge at the retarded time, and $\mathbf{\mathscr{r}}$ is the vector from the retarded position to the point of observation P. Meanwhile, since the current density of a rigid object is $\rho\mathbf{v}$, we also have

$$\mathbf{A}(\mathbf{r}, t) = \frac{\mu_0}{4\pi}\int \frac{\rho(\mathbf{r}', t_r)\mathbf{v}(t_r)}{\mathscr{r}}d\tau = \frac{\mu_0}{4\pi}\frac{\mathbf{v}}{\mathscr{r}}\int \rho(\mathbf{r}', t_r)\, d\tau$$

or

$$\boxed{\mathbf{A}(\mathbf{r}, t) = \frac{\mu_0}{4\pi}\frac{q\mathbf{v}}{\mathscr{r}\left(1 - \dfrac{\hat{\mathbf{\mathscr{r}}}\cdot\mathbf{v}}{c}\right)} = \frac{\mathbf{v}}{c^2}V(\mathbf{r}, t)} \tag{9.87}$$

Equations (9.86) and (9.87) are the famous **Liénard-Wiechert potentials** for a moving point charge.[9]

[9]There are many ways to obtain the Liénard-Wiechert potentials. I have tried to emphasize the *geometrical* origin of the factor $(1 - \hat{\mathbf{\mathscr{r}}}\cdot\mathbf{v}/c)^{-1}$; for illuminating commentary see W. K. H. Panofsky and M. Phillips, *Classical Electricity and Magnetism,* 2d ed., (Reading, MA: Addison-Wesley (1962)), pp. 342–3. A more rigorous derivation is provided by J. R. Reitz, F. J. Milford, and R. W. Christy, *Foundations of Electromagnetic Theory,* 3d ed., (Reading, MA: Addison-Wesley, 1979), sec. 21.1.

Example 4

Find the potential of a point charge moving with constant velocity.

Solution: For convenience, say the particle passes through the origin at time $t = 0$, so that

$$\mathbf{w}(t) = \mathbf{v}t$$

We first compute the retarded time, using (9.79):

$$|\mathbf{r} - \mathbf{v}t_r| = c(t - t_r)$$

or, squaring:

$$r^2 - 2\mathbf{r}\cdot\mathbf{v}t_r + v^2t_r^2 = c^2(t^2 - 2tt_r + t_r^2)$$

Solving for t_r, by the quadratic formula, I find that

$$t_r = \frac{(c^2t - \mathbf{r}\cdot\mathbf{v}) \pm \sqrt{(c^2t - \mathbf{r}\cdot\mathbf{v})^2 + (c^2 - v^2)(r^2 - c^2t^2)}}{(c^2 - v^2)} \tag{9.88}$$

To fix the sign, consider the case $v = 0$:

$$t_r = t \pm \frac{r}{c}$$

In this case the charge is at rest at the origin, and the retarded time should be $(t - r/c)$; evidently we want the *minus* sign. Now, from (9.79) and (9.80),

$$\mathscr{r} = c(t - t_r), \qquad \text{and} \quad \hat{\mathscr{r}} = \frac{\mathbf{r} - \mathbf{v}t_r}{c(t - t_r)}$$

so

$$\mathscr{r}\left(1 - \frac{\hat{\mathscr{r}}\cdot\mathbf{v}}{c}\right) = c(t - t_r)\left[1 - \frac{\mathbf{v}}{c}\cdot\frac{(\mathbf{r} - \mathbf{v}t_r)}{c(t - t_r)}\right] = c(t - t_r) - \frac{\mathbf{v}\cdot\mathbf{r}}{c} + \frac{v^2}{c}t_r$$

$$= \frac{1}{c}\left[(c^2t - \mathbf{r}\cdot\mathbf{v}) - t_r(c^2 - v^2)\right]$$

$$= \frac{1}{c}\sqrt{(c^2t - \mathbf{r}\cdot\mathbf{v})^2 + (c^2 - v^2)(r^2 - c^2t^2)}$$

(I used (9.88), with the minus sign, in the last step.) Therefore,

$$V(\mathbf{r}, t) = \frac{1}{4\pi\epsilon_0}\frac{qc}{\sqrt{(c^2t - \mathbf{r}\cdot\mathbf{v})^2 + (c^2 - v^2)(r^2 - c^2t^2)}} \tag{9.89}$$

Problem 9.15 A particle of charge q moves in a circle of radius R at constant angular velocity ω. (Assume that the circle lies in the xy plane, centered at the origin, and at time $t = 0$ the charge is at $(R, 0)$, on the positive x axis.) Find the Liénard-Wiechert potentials for points on the z axis.

Problem 9.16 Show that the potential of a point charge moving with constant velocity (9.89) can be written equivalently as

$$V(\mathbf{r}, t) = \frac{1}{4\pi\epsilon_0} \frac{q}{R\sqrt{1 - \frac{v^2}{c^2}\sin^2\theta}}$$

where $\mathbf{R} = (\mathbf{r} - \mathbf{v}t)$ is the vector from the *present* (!) position of the particle to point P, and θ is the angle between $\mathbf{R}$ and $\mathbf{v}$ (Fig. 9.16). Evidently, for nonrelativistic velocities ($v^2 \ll c^2$),

$$V(\mathbf{r}, t) = \frac{1}{4\pi\epsilon_0} \frac{q}{R}$$

Problem 9.17 In comments following equation (9.80), I showed that *at most one* point on the particle trajectory communicates with P at any given time. In some cases there may be *no* such point (an observer at P would not yet see the particle). As an example, consider "hyperbolic" motion along the x axis:

$$\mathbf{w}(t) = \sqrt{b^2 + (ct)^2}\,\hat{i} \qquad (-\infty < t < \infty)$$

(This is the trajectory of a particle subject to a constant force $F = mc^2/b$, according to special relativity.) Sketch the graph of w versus t. At four or five representative points on the curve, draw the trajectory of a light signal emitted by the particle at that point—both in the plus x-direction and in the minus x-direction. What region on your graph corresponds to points and times (x, t) from which the particle cannot be seen? At what time does someone at point x first see the particle? (Prior to this the potential at x is evidently zero.) Is it possible for a particle ever to *disappear* from view?

! **Problem 9.18** Find the Liénard-Wiechert potentials for a charge in hyperbolic motion (see Problem 9.17). Assume the point P is on the x axis and to the right of the charge.

9.2.2 The Fields of a Point Charge in Motion

We are now in a position to calculate the electric and magnetic fields of a point charge in arbitrary motion, using the Liénard-Wiechert potentials, which I write for convenience in the form

$$V(\mathbf{r}, t) = \frac{1}{4\pi\epsilon_0} \frac{qc}{(\imath c - \boldsymbol{\imath} \cdot \mathbf{v})}, \qquad \mathbf{A}(\mathbf{r}, t) = \frac{\mathbf{v}}{c^2} V(\mathbf{r}, t) \tag{9.90}$$

and the equations

$$\mathbf{E} = -\nabla V - \frac{\partial \mathbf{A}}{\partial t}, \qquad \mathbf{B} = \nabla \times \mathbf{A}$$

The differentiation is not as easy as it looks, however, because

$$\boldsymbol{\imath} = \mathbf{r} - \mathbf{w}(t_r) \quad \text{and} \quad \mathbf{v} = \dot{\mathbf{w}}(t_r) \tag{9.80}$$

are both evaluated at the retarded time, and t_r—defined implicitly by the equation

$$|\mathbf{r} - \mathbf{w}(t_r)| = c(t - t_r) \tag{9.79}$$

—is *itself* a function of $\mathbf{r}$ and t.[10] So hang on: The next two pages are rough going . . . but the answer is worth the effort. Let's begin with the gradient of V:

$$\nabla V = \frac{qc}{4\pi\epsilon_0} \frac{-1}{(\mathscr{r}c - \boldsymbol{\mathscr{r}}\cdot\mathbf{v})^2} \nabla(\mathscr{r}c - \boldsymbol{\mathscr{r}}\cdot\mathbf{v}) \tag{9.91}$$

Since $\mathscr{r} = c(t - t_r)$,

$$\nabla \mathscr{r} = -c\nabla t_r \tag{9.92}$$

As for the second term, product rule 4 gives

$$\nabla(\boldsymbol{\mathscr{r}}\cdot\mathbf{v}) = (\boldsymbol{\mathscr{r}}\cdot\nabla)\mathbf{v} + (\mathbf{v}\cdot\nabla)\boldsymbol{\mathscr{r}} + \boldsymbol{\mathscr{r}} \times (\nabla \times \mathbf{v}) + \mathbf{v} \times (\nabla \times \boldsymbol{\mathscr{r}}) \tag{9.93}$$

Evaluating these terms one at a time:

$$\begin{aligned}(\boldsymbol{\mathscr{r}}\cdot\nabla)\mathbf{v} &= \left(\mathscr{r}_x \frac{\partial}{\partial x} + \mathscr{r}_y \frac{\partial}{\partial y} + \mathscr{r}_z \frac{\partial}{\partial z}\right)\mathbf{v}(t_r) \\ &= \mathscr{r}_x \frac{d\mathbf{v}}{dt_r}\frac{\partial t_r}{\partial x} + \mathscr{r}_y \frac{d\mathbf{v}}{dt_r}\frac{\partial t_r}{\partial y} + \mathscr{r}_z \frac{d\mathbf{v}}{dt_r}\frac{\partial t_r}{\partial z} \\ &= \mathbf{a}(\boldsymbol{\mathscr{r}}\cdot\nabla t_r)\end{aligned} \tag{9.94}$$

where $\mathbf{a} = \dot{\mathbf{v}}$ is the *acceleration* of the particle at the retarded time. Now

$$(\mathbf{v}\cdot\nabla)\boldsymbol{\mathscr{r}} = (\mathbf{v}\cdot\nabla)\mathbf{r} - (\mathbf{v}\cdot\nabla)\mathbf{w}, \tag{9.95}$$

and

$$\begin{aligned}(\mathbf{v}\cdot\nabla)\mathbf{r} &= \left(v_x \frac{\partial}{\partial x} + v_y \frac{\partial}{\partial y} + v_z \frac{\partial}{\partial z}\right)(x\hat{\imath} + y\hat{\jmath} + z\hat{k}) \\ &= v_x\hat{\imath} + v_y\hat{\jmath} + v_z\hat{k} = \mathbf{v}\end{aligned} \tag{9.96}$$

while

$$(\mathbf{v}\cdot\nabla)\mathbf{w} = \mathbf{v}(\mathbf{v}\cdot\nabla t_r)$$

(same reasoning as (9.94)). Moving on to the third term in (9.93),

$$\begin{aligned}\nabla \times \mathbf{v} &= \hat{\imath}\left(\frac{\partial v_z}{\partial y} - \frac{\partial v_y}{\partial z}\right) + \hat{\jmath}\left(\frac{\partial v_x}{\partial z} - \frac{\partial v_z}{\partial x}\right) + \hat{k}\left(\frac{\partial v_y}{\partial x} - \frac{\partial v_x}{\partial y}\right) \\ &= \hat{\imath}\left(\frac{dv_z}{dt_r}\frac{\partial t_r}{\partial y} - \frac{dv_y}{dt_r}\frac{\partial t_r}{\partial z}\right) + \hat{\jmath}\left(\frac{dv_x}{dt_r}\frac{\partial t_r}{\partial z} - \frac{dv_z}{dt_r}\frac{\partial t_r}{\partial x}\right) + \hat{k}\left(\frac{dv_y}{dt_r}\frac{\partial t_r}{\partial x} - \frac{dv_x}{dt_r}\frac{\partial t_r}{\partial y}\right) \\ &= -\mathbf{a} \times \nabla t_r\end{aligned} \tag{9.97}$$

Finally,

$$\nabla \times \boldsymbol{\mathscr{r}} = \nabla \times \mathbf{r} - \nabla \times \mathbf{w} \tag{9.98}$$

[10]The following calculation is done here by the most direct, "brute force" method. For a more efficient approach—which, however, relies on a certain amount of trickery—see J. D. Jackson, *Classical Electrodynamics*, 2d ed. (New York: John Wiley, 1975), sec. 14.

but $\nabla \times \mathbf{r} = 0$, while, by the same argument as (9.97),

$$\nabla \times \mathbf{w} = -(\mathbf{v} \times \nabla t_r) \tag{9.99}$$

Putting all this back into (9.93), and using the "BAC-CAB" rule to reduce the triple cross products,

$$\begin{aligned}\nabla(\boldsymbol{\mathscr{r}}\cdot\mathbf{v}) &= \mathbf{a}(\boldsymbol{\mathscr{r}}\cdot\nabla t_r) + \mathbf{v} - \mathbf{v}(\mathbf{v}\cdot\nabla t_r) - \boldsymbol{\mathscr{r}} \times (\mathbf{a} \times \nabla t_r) + \mathbf{v} \times (\mathbf{v} \times \nabla t_r)\\ &= \mathbf{v} + ((\boldsymbol{\mathscr{r}}\cdot\mathbf{a}) - v^2)\nabla t_r \end{aligned} \tag{9.100}$$

Collecting (9.92) and (9.100) together, we have

$$\nabla V = \frac{qc}{4\pi\epsilon_0}\frac{1}{(\mathscr{r}c - \boldsymbol{\mathscr{r}}\cdot\mathbf{v})^2}[\mathbf{v} + (c^2 - v^2 + (\boldsymbol{\mathscr{r}}\cdot\mathbf{a}))\nabla t_r] \tag{9.101}$$

To complete the calculation, we need to know ∇t_r. This can be found by taking the gradient of the defining equation (9.79)—as we have already done in (9.92)—and expanding out $\nabla \mathscr{r}$:

$$\begin{aligned}-c\nabla t_r = \nabla\mathscr{r} = \nabla\sqrt{\boldsymbol{\mathscr{r}}\cdot\boldsymbol{\mathscr{r}}} &= \frac{1}{2\sqrt{\boldsymbol{\mathscr{r}}\cdot\boldsymbol{\mathscr{r}}}}\nabla(\boldsymbol{\mathscr{r}}\cdot\boldsymbol{\mathscr{r}})\\ &= \frac{1}{\mathscr{r}}[(\boldsymbol{\mathscr{r}}\cdot\nabla)\boldsymbol{\mathscr{r}} + \boldsymbol{\mathscr{r}} \times (\nabla \times \boldsymbol{\mathscr{r}})]\end{aligned} \tag{9.102}$$

But

$$(\boldsymbol{\mathscr{r}}\cdot\nabla)\boldsymbol{\mathscr{r}} = \boldsymbol{\mathscr{r}} - \mathbf{v}(\boldsymbol{\mathscr{r}}\cdot\nabla t_r)$$

(same idea as (9.95)), while (from (9.98) and (9.99))

$$\nabla \times \boldsymbol{\mathscr{r}} = (\mathbf{v} \times \nabla t_r)$$

Thus

$$-c\nabla t_r = \frac{1}{\mathscr{r}}[\boldsymbol{\mathscr{r}} - \mathbf{v}(\boldsymbol{\mathscr{r}}\cdot\nabla t_r) + \boldsymbol{\mathscr{r}} \times (\mathbf{v} \times \nabla t_r)] = \frac{1}{\mathscr{r}}(\boldsymbol{\mathscr{r}} - (\boldsymbol{\mathscr{r}}\cdot\mathbf{v})\nabla t_r)$$

and hence

$$\nabla t_r = \frac{-\boldsymbol{\mathscr{r}}}{(\mathscr{r}c - \boldsymbol{\mathscr{r}}\cdot\mathbf{v})} \tag{9.103}$$

Incorporating this result into (9.101), I conclude that

$$\nabla V = \frac{qc}{4\pi\epsilon_0}\frac{1}{(\mathscr{r}c - \boldsymbol{\mathscr{r}}\cdot\mathbf{v})^3}\{(\mathscr{r}c - \boldsymbol{\mathscr{r}}\cdot\mathbf{v})\mathbf{v} - [c^2 - v^2 + (\boldsymbol{\mathscr{r}}\cdot\mathbf{a})]\boldsymbol{\mathscr{r}}\} \tag{9.104}$$

A similar calculation, which I shall leave for you (Problem 9.19) yields

$$\begin{aligned}\frac{\partial\mathbf{A}}{\partial t} = \frac{qc}{4\pi\epsilon_0}\frac{1}{(\mathscr{r}c - \boldsymbol{\mathscr{r}}\cdot\mathbf{v})^3}\bigg\{&(\mathscr{r}c - \boldsymbol{\mathscr{r}}\cdot\mathbf{v})\left(-\mathbf{v} + \frac{\mathscr{r}}{c}\mathbf{a}\right)\\ &+ \frac{\mathscr{r}}{c}[c^2 - v^2 + (\boldsymbol{\mathscr{r}}\cdot\mathbf{a})]\mathbf{v}\bigg\}\end{aligned} \tag{9.105}$$

When you put these together to form **E**, you will find that the result can be simplified by introducing the vector

$$\mathbf{u} \equiv c\hat{\boldsymbol{\mathscr{r}}} - \mathbf{v} \tag{9.106}$$

With this,

$$\boxed{\mathbf{E}(\mathbf{r}, t) = \frac{q}{4\pi\epsilon_0} \frac{\mathscr{r}}{(\boldsymbol{\mathscr{r}} \cdot \mathbf{u})^3} [\mathbf{u}(c^2 - v^2) + \boldsymbol{\mathscr{r}} \times (\mathbf{u} \times \mathbf{a})]} \tag{9.107}$$

Meanwhile,

$$\nabla \times \mathbf{A} = \frac{1}{c^2} \nabla \times (\mathbf{v}V) = \frac{1}{c^2} [V(\nabla \times \mathbf{v}) - \mathbf{v} \times (\nabla V)]$$

We have already calculated $(\nabla \times \mathbf{v})$ in (9.97) and ∇V in (9.104). Putting it together, I find that

$$\nabla \times \mathbf{A} = -\frac{1}{c} \frac{q}{4\pi\epsilon_0} \frac{1}{(\mathbf{u} \cdot \boldsymbol{\mathscr{r}})^3} \boldsymbol{\mathscr{r}} \times [\mathbf{v}(c^2 - v^2) + \mathbf{v}(\boldsymbol{\mathscr{r}} \cdot \mathbf{a}) + \mathbf{a}(\boldsymbol{\mathscr{r}} \cdot \mathbf{u})]$$

The quantity in brackets is remarkably similar to the one in (9.107)—the only difference is that we have **v**'s instead of **u**'s in the first two terms. In fact, since it's all crossed into $\boldsymbol{\mathscr{r}}$ *anyway,* we can with impunity *change* these **v**'s into −**u**'s; the extra term proportional to $\hat{\boldsymbol{\mathscr{r}}}$ disappears in the cross product. It follows that

$$\boxed{\mathbf{B} = \frac{1}{c} \hat{\boldsymbol{\mathscr{r}}} \times \mathbf{E}} \tag{9.108}$$

Evidently, *the magnetic field of a point charge is always perpendicular to the electric field and to the vector from the retarded point.*

The first term in **E** (the one involving $\mathbf{u}(c^2 - v^2)$) falls off as the inverse *square* of the distance from the particle. If the velocity and acceleration are both zero, this term alone survives and reduces to the old electrostatic result

$$\mathbf{E} = \frac{1}{4\pi\epsilon_0} \frac{q}{\mathscr{r}^2} \hat{\boldsymbol{\mathscr{r}}}$$

For this reason, the first term in **E** is sometimes referred to as the **generalized Coulomb field.** (Because it does not depend on the acceleration, it is also known as the **velocity field.**) The second term (the one involving $\boldsymbol{\mathscr{r}} \times (\mathbf{u} \times \mathbf{a})$) falls off as the inverse *first* power of $\mathscr{r}$ and is therefore dominant at large distances. It is this term that is responsible for electromagnetic radiation; accordingly, it is called the **radiation field**—or, since it is proportional to a, the **acceleration field.** The same terminology applies to the magnetic field. Notice that the electric and magnetic radiation fields are mutually perpendicular, transverse to the direction of propagation $\hat{\boldsymbol{\mathscr{r}}}$, and stand in the ratio $E_{\text{rad}}/B_{\text{rad}} = c$, all of which is as it should be for electromagnetic radiation.

Back in Chapter 2, I commented that if we could only write down the formula for the force one charge exerts on another, we would be done with electrodynamics, in principle. That, together with the superposition principle, would tell us the force exerted on a test charge Q by any configuration whatever. Well . . . here we are: (9.107) and (9.108) give us the fields, and the Lorentz Force Law determines the resulting force:

$$\mathbf{F} = \frac{qQ}{4\pi\epsilon_0}\frac{\mathscr{r}}{(\boldsymbol{\mathscr{r}}\cdot\mathbf{u})^3}\Big\{[\mathbf{u}(c^2 - v^2) + \boldsymbol{\mathscr{r}} \times (\mathbf{u}\times\mathbf{a})]$$

$$+ \frac{\mathbf{V}}{c}\times\left[\hat{\boldsymbol{\mathscr{r}}}\times[\mathbf{u}(c^2 - v^2) + \boldsymbol{\mathscr{r}}\times(\mathbf{u}\times\mathbf{a})]\right]\Big\}$$

where $\mathbf{V}$ is the velocity of Q, and $\boldsymbol{\mathscr{r}}$, $\mathbf{u}$, $\mathbf{v}$, and $\mathbf{a}$ are all evaluated at the retarded time. The entire theory of classical electrodynamics is contained in that equation . . . but you see why I preferred to start out with Coulomb's law.

Example 5

Calculate the electric and magnetic fields of a point charge moving with constant velocity.

Solution: Putting $\mathbf{a} = 0$ in (9.107),

$$\mathbf{E} = \frac{q}{4\pi\epsilon_0}\frac{(c^2 - v^2)}{(\boldsymbol{\mathscr{r}}\cdot\mathbf{u})^3}\mathscr{r}\mathbf{u}$$

In this case, using $\mathbf{w} = \mathbf{v}t$,

$$\mathscr{r}\mathbf{u} = c\boldsymbol{\mathscr{r}} - \mathscr{r}\mathbf{v} = c(\mathbf{r} - \mathbf{v}t_r) - c(t - t_r)\mathbf{v} = c(\mathbf{r} - \mathbf{v}t)$$

In Example 4, we found that

$$(\mathscr{r}c - \boldsymbol{\mathscr{r}}\cdot\mathbf{v}) = (\boldsymbol{\mathscr{r}}\cdot\mathbf{u}) = \sqrt{(c^2t - \mathbf{r}\cdot\mathbf{v})^2 + (c^2 - v^2)(r^2 - c^2t^2)}$$

In Problem 9.16, you showed that this radical could be written as

$$Rc\sqrt{1 - \frac{v^2}{c^2}\sin^2\theta}$$

where

$$\mathbf{R} = \mathbf{r} - \mathbf{v}t$$

and θ is the angle between $\mathbf{R}$ and $\mathbf{v}$ (Fig. 9.16). Thus

$$\boxed{\mathbf{E}(\mathbf{r}, t) = \frac{q}{4\pi\epsilon_0}\frac{\hat{\mathbf{R}}}{R^2}\frac{(1 - v^2/c^2)}{\left(1 - \frac{v^2}{c^2}\sin^2\theta\right)^{3/2}}} \tag{9.109}$$

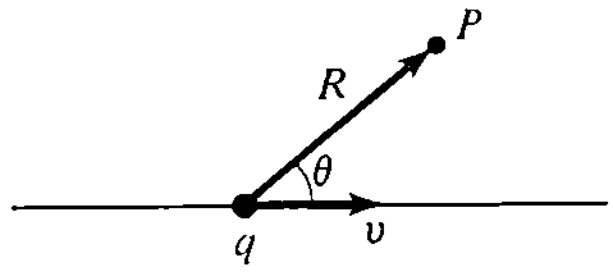

Figure 9.16

Notice that **E** points along the vector from the *present* position of the particle—an *extraordinary* coincidence, since the "message" came from the *retarded* position. The field of a fast-moving charge is flattened out like a pancake in the direction perpendicular to the motion (Fig. 9.17). In the forward and backward directions **E** is *reduced* by a factor $(1 - v^2/c^2)$ relative to the field of a charge at rest; in the perpendicular direction it is *enhanced* by a factor $1/\sqrt{1 - v^2/c^2}$.

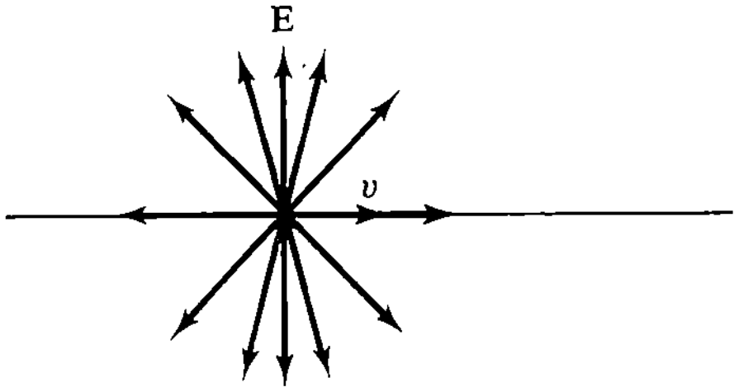

Figure 9.17

As for **B**, we have

$$\hat{\mathscr{r}} = \frac{\mathbf{r} - \mathbf{v}t_r}{\mathscr{r}} = \frac{(\mathbf{r} - \mathbf{v}t) + \mathbf{v}(t - t_r)}{\mathscr{r}} = \frac{\mathbf{R}}{\mathscr{r}} + \frac{\mathbf{v}}{c}$$

and therefore

$$\boxed{\mathbf{B} = \frac{1}{c}(\hat{\mathscr{r}} \times \mathbf{E}) = \frac{1}{c^2}(\mathbf{v} \times \mathbf{E})} \tag{9.110}$$

Lines of **B** circle around the charge, as shown in Fig. 9.18.

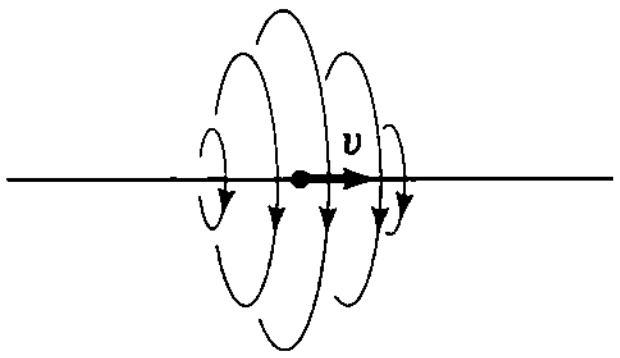

Figure 9.18

Problem 9.19 Derive equation (9.105). First show that

$$\frac{\partial t_r}{\partial t} = \frac{\mathscr{r}c}{(\boldsymbol{\mathscr{r}} \cdot \mathbf{u})}$$

Problem 9.20 Suppose a point charge q is constrained to move along the x axis. Show that the fields at points on the axis to the *right* of the charge are given by

$$\mathbf{E} = \frac{q}{4\pi\epsilon_0} \frac{1}{\imath^2} \left(\frac{c+v}{c-v}\right)\hat{i}, \qquad \mathbf{B} = 0$$

What are the fields on the axis, to the *left* of the charge?

Problem 9.21 Show that (9.107) can be expressed more compactly as

$$\mathbf{E} = \frac{q}{4\pi\epsilon_0} \frac{1}{(\imath \cdot \mathbf{u})} \frac{\partial}{\partial t_r} \left(\frac{\imath \mathbf{u}}{\imath \cdot \mathbf{u}}\right)$$

(The derivative $\partial/\partial t_r$ is with respect to the *explicit* dependence on t_r, treating $\mathbf{r}$ and t as constants.)

Problem 9.22

(a) Use equation (9.109) to calculate the electric field a distance d from an infinite straight wire carrying a uniform line charge λ, moving at a constant speed v down the wire.

(b) Use (9.110) to find the *magnetic* field of this wire.

Problem 9.23 For the configuration in Problem 9.15, find the electric and magnetic fields at the center. From your formula for $\mathbf{B}$, determine the magnetic field at the center of a circular loop carrying a steady current I, and compare your answer with the result of Example 6 in Chapter 5.

9.2.3 Power Radiated by a Point Charge

The energy flux associated with the fields of a point charge is given by

$$\mathbf{S} = \frac{1}{\mu_0}(\mathbf{E} \times \mathbf{B}) = \frac{1}{\mu_0 c}[\mathbf{E} \times (\hat{\imath} \times \mathbf{E})] = \frac{1}{\mu_0 c}[\hat{\imath}E^2 - \mathbf{E}(\hat{\imath} \cdot \mathbf{E})] \tag{9.111}$$

However, not all of this energy flux represents true *radiation*; some of it is just field energy carried along by the particle as it moves. The *radiated* energy is the stuff that, in effect, *detaches* itself from the charge and propagates off to infinity. To calculate the total power radiated by the particle at time t_r, we draw a huge sphere of radius $\imath$ (Fig. 9.19), centered at the position of the particle (at that instant), wait the appropriate interval

$$t - t_r = \frac{\imath}{c}$$

for the radiation to reach the sphere, and at that moment integrate the Poynting vector over the surface of the sphere. (I have used the notation t_r because, of course, this *is* the retarded time for all points on the sphere at time t.) Now, the area of the sphere is proportional to $\imath^2$, so any term in $\mathbf{S}$ which goes like $1/\imath^2$ will yield a finite

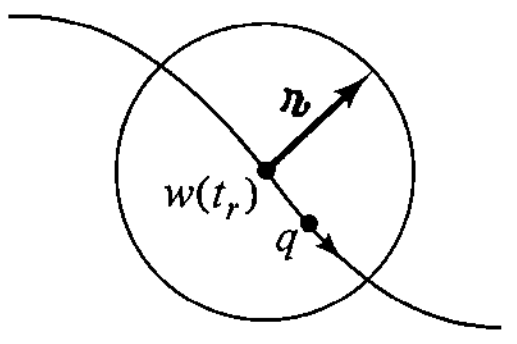

Figure 9.19

answer, but terms like $1/\mathscr{r}^3$ or $1/\mathscr{r}^4$ will contribute nothing in the limit $\mathscr{r} \to \infty$. For this reason only the *acceleration* fields represent true radiation:

$$\mathbf{E}_{\text{rad}} = \frac{q}{4\pi\epsilon_0} \frac{\mathscr{r}}{(\boldsymbol{\mathscr{r}} \cdot \mathbf{u})^3} (\boldsymbol{\mathscr{r}} \times (\mathbf{u} \times \mathbf{a})) \tag{9.112}$$

(The velocity fields carry energy, and as the charge moves this energy is dragged along—but it's not *radiation.* In particular, a charge must *accelerate* in order to radiate.) Now the radiation fields are perpendicular to $\hat{\boldsymbol{\mathscr{r}}}$, so the second term in (9.111) is zero:

$$\mathbf{S}_{\text{rad}} = \frac{1}{\mu_0 c} E_{\text{rad}}^2 \hat{\boldsymbol{\mathscr{r}}} \tag{9.113}$$

If the charge is instantaneously at *rest* (at time t_r), then $\mathbf{u} = c\hat{\boldsymbol{\mathscr{r}}}$ (9.106), and

$$\begin{aligned} \mathbf{E}_{\text{rad}} &= \frac{q}{4\pi\epsilon_0 c^2} \frac{1}{\mathscr{r}} [\hat{\boldsymbol{\mathscr{r}}} \times (\hat{\boldsymbol{\mathscr{r}}} \times \mathbf{a})] \\ &= \frac{q}{4\pi\epsilon_0 c^2} \frac{1}{\mathscr{r}} [\hat{\boldsymbol{\mathscr{r}}}(\hat{\boldsymbol{\mathscr{r}}} \cdot \mathbf{a}) - \mathbf{a}] \end{aligned} \tag{9.114}$$

In this case

$$\begin{aligned} \mathbf{S}_{\text{rad}} &= \frac{1}{\mu_0 c} \left(\frac{q}{4\pi\epsilon_0 c^2}\right)^2 \frac{1}{\mathscr{r}^2} (a^2 - (\hat{\boldsymbol{\mathscr{r}}} \cdot \mathbf{a})^2) \hat{\boldsymbol{\mathscr{r}}} \\ &= \frac{1}{4\pi\epsilon_0} \left(\frac{q^2}{4\pi c^3}\right) \frac{a^2 \sin^2\theta}{\mathscr{r}^2} \hat{\boldsymbol{\mathscr{r}}} \end{aligned} \tag{9.115}$$

where θ is the angle between $\hat{\boldsymbol{\mathscr{r}}}$ and $\mathbf{a}$. No power is radiated in the forward or backward direction—rather, it is emitted in a donut about the direction of instantaneous acceleration (Fig. 9.20).

The total power radiated is evidently

$$P = \int \mathbf{S}_{\text{rad}} \cdot d\mathbf{a} = \frac{1}{\epsilon_0 c} \left(\frac{q}{4\pi c}\right)^2 a^2 \int \frac{\sin^2\theta}{\mathscr{r}^2} \mathscr{r}^2 \sin\theta \, d\theta \, d\phi$$

or

$$\boxed{P = \frac{1}{4\pi\epsilon_0} \frac{2}{3} \frac{q^2 a^2}{c^3}} \tag{9.116}$$

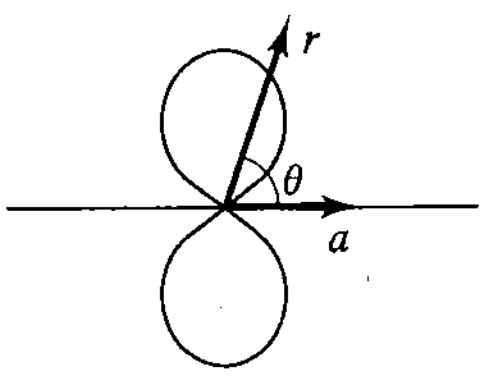

Figure 9.20

This, again, is the Larmor formula, which we obtained earlier by different means (equation (9.77)).

Although I *derived* them on the assumption that $v = 0$, equations (9.114) and (9.115) hold to good approximation provided $v \ll c$. An exact treatment of the case $v \neq 0$ is more difficult,[11] both for the obvious reason that $\mathbf{E}_{\text{rad}}$ is more complicated, and also for the more subtle reason that $\mathbf{S}_{\text{rad}}$, the rate at which energy passes through the sphere, is *not* the same as the rate at which energy left the particle. Suppose someone is firing a stream of bullets out the window of a moving car (Fig. 9.21). The rate N_t at which the bullets strike a stationary target is not the same as the rate N_g at which they left the gun, because of the motion of the car.[12] In fact, you can easily check that $N_g = (1 - v/c)N_t$, if the car is moving toward the target, and

$$N_g = \left(1 - \frac{\hat{\boldsymbol{\mathscr{r}}} \cdot \mathbf{v}}{c}\right) N_t$$

for arbitrary directions (here $\mathbf{v}$ is the velocity of the car, c is that of the bullets—relative to the ground—and $\hat{\boldsymbol{\mathscr{r}}}$ a unit vector from car to target). In our case, if dW/dt is the rate at which energy passes through the sphere at radius $\mathscr{r}$, then the rate at which energy left the charge was

$$\frac{dW}{dt_r} = \frac{dW}{dt} \bigg/ \left(\frac{\partial t_r}{\partial t}\right) = \frac{(\boldsymbol{\mathscr{r}} \cdot \mathbf{u})}{\mathscr{r} c} \frac{dW}{dt} \tag{9.117}$$

(I used the result of Problem 9.19 to express $\partial t_r / \partial t$.) Note that

$$\left(\frac{\boldsymbol{\mathscr{r}} \cdot \mathbf{u}}{\mathscr{r} c}\right) = \left(1 - \frac{\hat{\boldsymbol{\mathscr{r}}} \cdot \mathbf{v}}{c}\right)$$

which is the same factor as relates N_g to N_t, and for the same reason. The power radiated by the particle into a patch of area $\mathscr{r}^2 \sin\theta \, d\theta \, d\phi = \mathscr{r}^2 \, d\Omega$ on the sphere is therefore given by

$$\frac{dP}{d\Omega} = \left(\frac{\boldsymbol{\mathscr{r}} \cdot \mathbf{u}}{\mathscr{r} c}\right) \frac{1}{\mu_0 c} E_{\text{rad}}^2 \mathscr{r}^2 = \frac{1}{4\pi\epsilon_0} \left(\frac{q^2}{4\pi}\right) \frac{[\hat{\boldsymbol{\mathscr{r}}} \times (\mathbf{u} \times \mathbf{a})]^2}{(\hat{\boldsymbol{\mathscr{r}}} \cdot \mathbf{u})^5} \tag{9.118}$$

($d\Omega = \sin\theta \, d\theta \, d\phi$ is the solid angle into which this power is radiated.) Integrating this over θ and ϕ to get the total power radiated is no picnic, though the *answer* is reasonably simple:

$$P = \frac{1}{4\pi\epsilon_0} \frac{2}{3} \frac{q^2}{c^3} \gamma^6 \left[a^2 - \left(\frac{\mathbf{v}}{c} \times \mathbf{a}\right)^2\right] \tag{9.119}$$

[11] In the context of special relativity, the condition $v = 0$ simply represents an astute choice of reference system, with no essential loss of generality. If you can decide how P transforms, you can *deduce* the general (Liénard) result from the $v = 0$ (Larmor) formula. (See Problem 10.56.)

[12] The same idea underlies the Doppler effect.

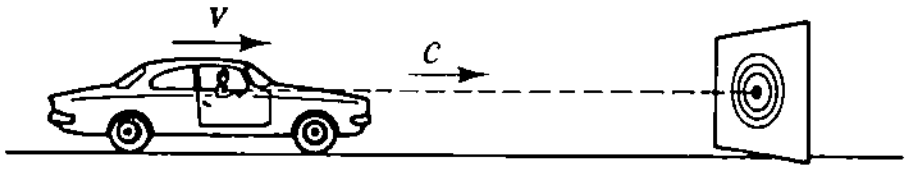

Figure 9.21

where $\gamma = 1/\sqrt{1 - v^2/c^2}$. This is Liénard's generalization of the Larmor formula (to which it reduces when $v = 0$). The factor γ^6 means that the radiated power increases enormously as the particle velocity approaches the speed of light.

Example 6

Suppose $\mathbf{v}$ and $\mathbf{a}$ are instantaneously collinear (at time t_r), as, for example, in straight-line motion. Find the angular distribution of the radiation (9.118) and the total power emitted.

Solution: In this case $(\mathbf{u} \times \mathbf{a}) = (c\hat{\boldsymbol{\mathscr{r}}} \times \mathbf{a})$, so

$$\frac{dP}{d\Omega} = \frac{1}{4\pi\epsilon_0}\left(\frac{q^2}{4\pi}\right)c^2\frac{[\hat{\boldsymbol{\mathscr{r}}} \times (\hat{\boldsymbol{\mathscr{r}}} \times \mathbf{a})]^2}{(c - \hat{\boldsymbol{\mathscr{r}}}\cdot\mathbf{v})^5}$$

Now

$$\hat{\boldsymbol{\mathscr{r}}} \times (\hat{\boldsymbol{\mathscr{r}}} \times \mathbf{a}) = \hat{\boldsymbol{\mathscr{r}}}(\hat{\boldsymbol{\mathscr{r}}}\cdot\mathbf{a}) - \mathbf{a}, \quad \text{so } (\hat{\boldsymbol{\mathscr{r}}} \times (\hat{\boldsymbol{\mathscr{r}}} \times \mathbf{a}))^2 = a^2 - (\hat{\boldsymbol{\mathscr{r}}}\cdot\mathbf{a})^2$$

In particular, if we let the z axis point along $\mathbf{v}$, then

$$\frac{dP}{d\Omega} = \frac{1}{4\pi\epsilon_0}\left(\frac{q^2}{4\pi c^3}\right)\frac{a^2\sin^2\theta}{(1 - \beta\cos\theta)^5} \tag{9.120}$$

where $\beta = v/c$. This is consistent, of course, with (9.115), in the case $v = 0$. However, for very large v ($\beta \approx 1$) the donut of radiation (Fig. 9.20) is stretched out and pushed forward by the factor $(1 - \beta\cos\theta)^{-5}$, as shown in Fig. 9.22. Although there is still no radiation in *precisely* the forward direction, most of it is concentrated within an increasingly narrow cone *about* the forward direction (see Problem 9.26).

The total power emitted is found by integration of (9.120) over all angles:

$$P = \int \frac{dP}{d\Omega}\,d\Omega$$
$$= \frac{1}{4\pi\epsilon_0}\frac{q^2a^2}{4\pi c^3}\int \frac{\sin^2\theta}{(1 - \beta\cos\theta)^5}\sin\theta\,d\theta\,d\phi$$

The ϕ integral gives 2π; the θ integral is simplified by the substitution $x = \cos\theta$. With this,

$$P = \frac{1}{4\pi\epsilon_0}\frac{q^2a^2}{2c^3}\int_{-1}^{+1}\frac{(1 - x^2)}{(1 - \beta x)^5}\,dx$$

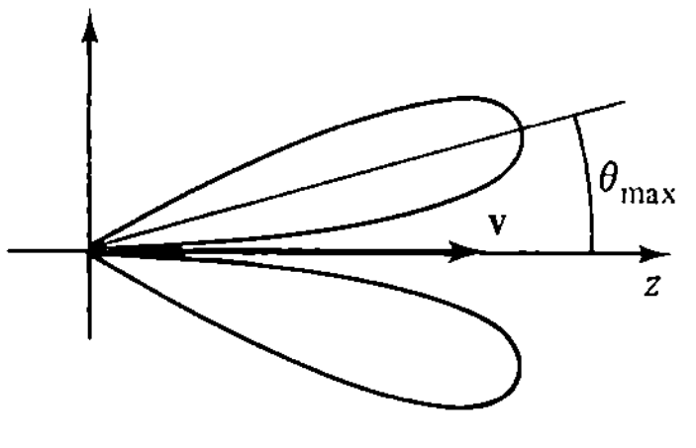

Figure 9.22

Integration by parts yields $\frac{4}{3}(1-\beta^2)^{-3}$ for this integral, and I conclude that

$$P = \frac{1}{4\pi\epsilon_0}\frac{2}{3}\left(\frac{q^2a^2}{c^3}\right)\gamma^6 \tag{9.121}$$

(since $\gamma = 1/\sqrt{1-\beta^2}$). This result is consistent with the Liénard formula (9.119) for the case of collinear **v** and **a**. Notice that the distribution of radiation is the same whether the particle is *accelerating* or *decelerating*; it only depends on the *square* of a. When a high-speed electron hits a metal target it rapidly decelerates, giving off what is called **bremsstrahlung**, or "braking radiation." What I have described in this example is essentially the classical theory of bremsstrahlung.

Problem 9.24

(a) Suppose an electron decelerated at a constant rate a from some initial velocity v_0 down to zero. What fraction of its initial kinetic energy is lost to radiation? (The rest is absorbed by whatever mechanism keeps the acceleration constant.) Assume $v_0 \ll c$ so that the Larmor formula can be used.

(b) To get a sense of the numbers involved, suppose the initial velocity is thermal (around 10^5 m/s) and the distance the electron goes is 30 Å. What can you conclude about radiation losses for the electrons in an ordinary conductor?

Problem 9.25 In Bohr's theory of hydrogen, the electron in its ground state was supposed to travel in a circular orbit of radius 0.5 Å, with an angular momentum $\hbar$ (Planck's constant). Show that $v \ll c$ for such an electron, so that you can safely use the Larmor formula. Find the power radiated, according to classical electrodynamics. At this rate, about how long would it take the electron to spiral in and hit the nucleus (say the radius of the nucleus is 2×10^{-16} m)?

Problem 9.26 Find the angle θ_{max} at which the maximum radiation is emitted, in Example 6 (see Fig. 9.22.) Show that for ultrarelativistic speeds ($v \approx c$), $\theta_{max} \approx \sqrt{(1-\beta)/2}$. What is the intensity of the radiation in this maximal direction (in the ultrarelativistic case), in proportion to the same quantity for a particle instantaneously at rest? Give your answer in terms of γ.

! **Problem 9.27** In Example 6 we assumed the velocity and acceleration were (instantaneously, at least) *collinear*. Carry out the same analysis for the case where they are *perpendicular*. Choose your axes so that **v** lies along the z axis and **a** along the x axis (Fig. 9.23). Then

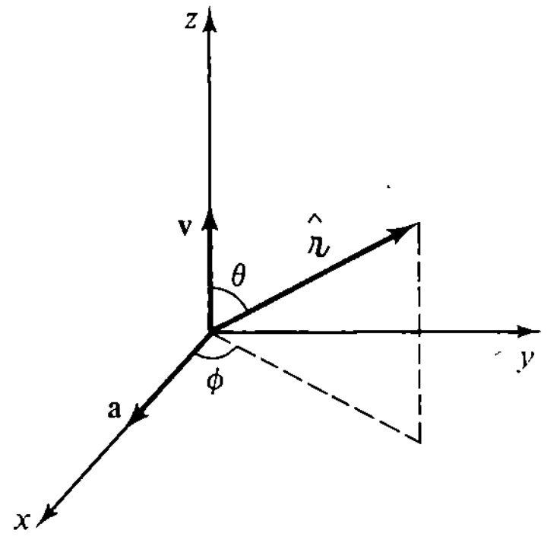

Figure 9.23

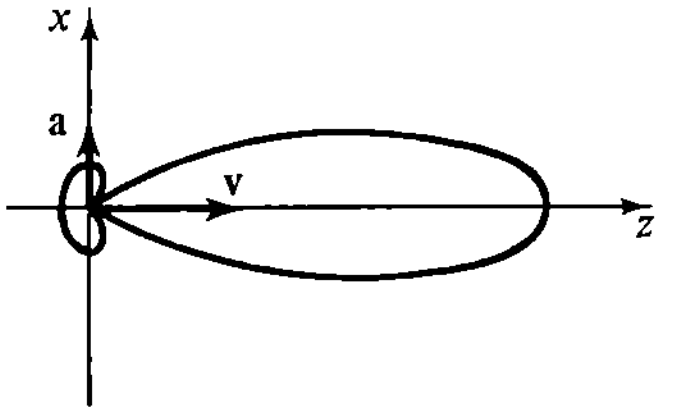

Figure 9.24

$$\mathbf{v} = v\hat{k}$$

$$\mathbf{a} = a\hat{\imath}$$

$$\hat{\boldsymbol{\mathscr{r}}} = \sin\theta\cos\phi\,\hat{\imath} + \sin\theta\sin\phi\,\hat{\jmath} + \cos\theta\,\hat{k}$$

Check that P is consistent with the Liénard formula.

$$\left(\textit{Answers}: \frac{dP}{d\Omega} = \frac{1}{4\pi\epsilon_0}\frac{q^2a^2}{4\pi c^3}\frac{[(1-\beta\cos\theta)^2 - (1-\beta^2)\sin^2\theta\cos^2\phi]}{(1-\beta\cos\theta)^5},\right.$$

$$\left. P = \frac{1}{4\pi\epsilon_0}\frac{2}{3}\frac{q^2a^2}{c^3}\gamma^4\right)$$

[*Comment*: For relativistic velocities ($\beta \approx 1$) the radiation is again sharply peaked in the forward direction (Fig. 9.24). The most important application of these formulas is to *circular* motion—in this case the radiation is called **synchrotron radiation**. For a relativistic electron the radiation sweeps around like a locomotive's headlight as the electron circles.]

9.3 RADIATION REACTION

9.3.1 The Abraham-Lorentz Formula

According to the laws of classical electrodynamics, an accelerating charge radiates. This radiation carries off energy, which must come at the expense of the particle's kinetic energy. Under the influence of a given force, therefore, a charged particle accelerates *less* than a neutral one of the same mass. The radiation evidently exerts a force ($\mathbf{F}_{\text{rad}}$) back on the charge—a *recoil* force, rather like that of a bullet on a gun. In this section we'll derive this **radiation reaction** force from conservation of energy. Then in the next section I'll show you the actual mechanism responsible, and derive it again in the context of a simple model.

For a nonrelativistic particle ($v \ll c$) the total power radiated is given by the Larmor formula (9.116):

$$P = \frac{1}{4\pi\epsilon_0}\frac{2}{3}\frac{q^2a^2}{c^3} \tag{9.122}$$

Conservation of energy suggests that this is also the rate at which the *particle loses* energy, under the influence of the radiation reaction force $\mathbf{F}_{\text{rad}}$:

$$\mathbf{F}_{\text{rad}}\cdot\mathbf{v} = -\frac{1}{4\pi\epsilon_0}\frac{2}{3}\frac{q^2a^2}{c^3} \tag{9.123}$$

I say "suggests" advisedly because this equation is actually *incorrect.* For we calculated the radiated power by integrating the Poynting vector over a sphere of "infinite" radius; in this calculation the *velocity* fields played no part, since they fall off too rapidly as a function of r to make any contribution. But the velocity fields *do* carry energy—they just don't transport it out to infinity. And as the particle accelerates and decelerates energy is pumped back and forth between the charge and the velocity fields, at the same time as energy is irretrievably radiated away by the acceleration fields. Equation (9.123) accounts only for the latter, but if we want to know the recoil force exerted by the fields on the charge, we need to consider the *total* power lost at any instant, not just the portion that eventually escapes in the form of radiation. (The term "radiation reaction" is a misnomer. We should really call it the *field reaction.* In fact, we'll soon see that $\mathbf{F}_{\text{rad}}$ is determined by the *time derivative* of the acceleration and can be nonzero even when the acceleration itself is instantaneously zero, so that the particle is not radiating.)

The energy lost by the particle in any interval of time, then, must equal the energy carried away by the radiation *plus* whatever extra energy has been pumped into the velocity fields.[13] Now, if we agree to consider intervals at the end of which the system has returned to its initial state, then the energy in the velocity fields is the same at either end and the only *net* loss is in the form of radiation. Thus, equation (9.123), while incorrect *instantaneously,* is valid on the *average*:

$$\int_{t_1}^{t_2} \mathbf{F}_{\text{rad}} \cdot \mathbf{v}\, dt = -\frac{1}{4\pi\epsilon_0}\frac{2}{3}\frac{q^2}{c^3}\int_{t_1}^{t_2} a^2\, dt \tag{9.124}$$

with the *proviso* that the *state of the system is identical at t_1 and t_2*. In the case of periodic motion, for instance, we must integrate over an integral number of full cycles.[14] Now, the right side of (9.124) can be integrated by parts:

$$\int_{t_1}^{t_2} a^2\, dt = \int_{t_1}^{t_2}\left(\frac{d\mathbf{v}}{dt}\right)\cdot\left(\frac{d\mathbf{v}}{dt}\right) dt = \left(\mathbf{v}\cdot\frac{d\mathbf{v}}{dt}\right)\Bigg|_{t_1}^{t_2} - \int_{t_1}^{t_2}\frac{d^2\mathbf{v}}{dt^2}\cdot\mathbf{v}\, dt$$

The boundary term drops out, since the velocities and accelerations are identical at t_1 and t_2, so (9.124) can be written equivalently as

$$\int_{t_1}^{t_2}\left[\mathbf{F}_{\text{rad}} - \frac{1}{4\pi\epsilon_0}\frac{2}{3}\frac{q^2}{c^3}\dot{\mathbf{a}}\right]\cdot\mathbf{v}\, dt = 0 \tag{9.125}$$

[13] Actually, while the total field is the sum of velocity and acceleration fields, $\mathbf{E} = \mathbf{E}_v + \mathbf{E}_a$, the *energy* is proportional to $E^2 = E_v^2 + 2\mathbf{E}_v\cdot\mathbf{E}_a + E_a^2$ and contains *three* terms: energy stored in the velocity fields alone (E_v^2), energy radiated away (E_a^2), and a *cross* term $\mathbf{E}_v\cdot\mathbf{E}_a$. For the sake of simplicity, I'm referring to the *combination* ($E_v^2 + 2\mathbf{E}_v\cdot\mathbf{E}_a$) as "energy stored in the velocity fields." These terms go like $1/\mathscr{r}^4$ and $1/\mathscr{r}^3$, respectively, so neither one contributes to the radiation.

[14] For *non*periodic motion the condition that the energy in the velocity fields be the same at t_1 and t_2 is more difficult to achieve. It is not enough that the instantaneous velocities and accelerations be equal, since the fields farther out depend on v and a at *earlier* times. In *principle,* then, v and *a and all higher derivatives* must be identical at t_1 and t_2. *In practice,* since the velocity fields fall off rapidly with $\mathscr{r}$, it is sufficient that v and a be the same over a brief interval prior to t_1 and t_2.

This equation will certainly be satisfied if

$$\boxed{\mathbf{F}_{\text{rad}} = \frac{1}{4\pi\epsilon_0}\frac{2}{3}\frac{q^2}{c^3}\dot{\mathbf{a}}} \tag{9.126}$$

Equation (9.126) is called the **Abraham-Lorentz formula** for the radiation reaction force.

Of course, (9.125) doesn't *prove* (9.126). It tells you nothing whatever about the component of $\mathbf{F}_{\text{rad}}$ *perpendicular* to $\mathbf{v}$; and it only tells you the *time average* of the parallel component—the average, moreover, over very special time intervals. As we'll see in the next section, there are other reasons for believing in the Abraham-Lorentz formula, but for now the best that can be said is that it represents the *simplest* form the radiation reaction force could take, consistent with conservation of energy.

The Abraham-Lorentz formula has disturbing implications, which are not entirely understood 80 years after the law was first introduced. For suppose a particle is subject to *no external* forces; then Newton's second law says

$$F_{\text{rad}} = \frac{1}{4\pi\epsilon_0}\frac{2}{3}\frac{q^2}{c^3}\dot{a} = ma$$

from which it follows that

$$a = a_0 e^{t/\tau} \tag{9.127}$$

where

$$\tau = \frac{1}{4\pi\epsilon_0}\frac{2}{3}\frac{q^2}{mc^3} \tag{9.128}$$

(In the case of the electron, $\tau = 6 \times 10^{-24}$ s.) The acceleration spontaneously *increases* exponentially with time! This absurd conclusion can be avoided if $a_0 = 0$, of course, but it turns out that the systematic elimination of such "runaway" solutions has an equally unpleasant consequence: if you *do* apply an external force, the particle starts to respond *before the force acts*!! (See Problem 9.30.) This acausal "preacceleration" jumps the gun by only a short time τ; nevertheless, it is philosophically repugnant that the theory should allow it at all.[15]

Example 7

Calculate the **radiation damping** of a charged particle attached to a spring of natural frequency ω_0, driven at frequency ω.

Solution: The equation of motion is

$$\begin{aligned} m\ddot{x} &= F_{\text{tot}} = F_{\text{spring}} + F_{\text{rad}} + F_{\text{driving}} \\ &= -m\omega_0^2 x + m\tau\dddot{x} + F_{\text{driving}} \end{aligned}$$

[15] See J. D. Jackson, *Classical Electrodynamics*, 2d ed. (New York: John Wiley, 1975), Chap. 17, for details. These difficulties persist in the relativistic version of the Abraham-Lorentz formula, which can be derived by starting with Liénard's formula instead of Larmor's.

With the system oscillating at frequency ω,

$$x(t) = x_0 \cos(\omega t + \delta)$$

so that

$$\dddot{x} = -\omega^2 \dot{x}$$

Therefore

$$m\ddot{x} + m\gamma\dot{x} + m\omega_0^2 x = F_{\text{driving}} \tag{9.129}$$

with the damping factor γ given by

$$\gamma = \omega^2 \tau \tag{9.130}$$

(When I wrote $F_{\text{damping}} = -\gamma m v$, back in Chapter 8 (equation (8.175)), I assumed for simplicity that the damping was proportional to the velocity. We now know that *radiation* damping, at least, is proportional to $\ddot{v}$. But it hardly matters: for sinusoidal oscillations *any* even number of derivatives of v would do, since they're all proportional to v.)

Problem 9.28

(a) A particle of charge q moves in a circle of radius R at constant speed v. To sustain the motion, you must, of course, provide a centripetal force mv^2/R; what *additional* force ($\mathbf{F}_e$) must you exert, in order to counteract the radiation reaction? (It's easiest to express the answer in terms of the instantaneous velocity $\mathbf{v}$.) What power (P_e) does this extra force deliver? Compare P_e with the power radiated (Larmor formula).

(b) Repeat part (a) for a particle in simple harmonic motion with amplitude A and angular frequency ω ($\mathbf{w}(t) = A\cos(\omega t)\,\hat{k}$). Explain the discrepancy.

(c) Consider the case of a particle in free fall (constant acceleration g). What is the radiation reaction force? What is the power radiated? Comment on these results.

Problem 9.29

(a) Assuming that γ is entirely attributable to radiation damping (equation (9.130)), show that for optical dispersion the damping is "small" ($\gamma \ll \omega_0$). (For optical dispersion the relevant resonances lie in or near the optical frequency range.)

(b) Using your results from Problem 8.26, estimate the width of the anomalous dispersion region, for the model in Problem 8.25.

! **Problem 9.30** With the inclusion of the radiation reaction (9.126), Newton's second law for a charged particle becomes

$$a = \tau\dot{a} + \frac{F}{m}$$

where F is the external force acting on the particle.

(a) In contrast to the case of an *uncharged* particle ($a = F/m$), acceleration (like position and velocity) must now be a *continuous* function of time, even if the force changes abruptly. (Physically, the radiation reaction damps out any rapid change in a). *Prove* that a is continuous at time t_0, by integrating the equation of motion above from $(t_0 - \epsilon)$ to $(t_0 + \epsilon)$ and taking the limit $\epsilon \to 0$.

(b) A particle is subjected to a constant force F, beginning at time $t = 0$ and lasting until

time T. Find the most general solution $a(t)$ to the equation of motion in each of the three periods: (I) $t < 0$; (II) $0 < t < T$; (iii) $t > T$.
(c) Impose the continuity condition (a) at $t = 0$ and $t = T$. Show that you can *either* eliminate the runaway in region (III) *or* avoid preacceleration in region (I), *but not both.*
(d) If you choose to eliminate the runaway, what is the velocity as a function of time (it must, of course, be continuous at $t = 0$ and $t = T$). Assume the particle was originally at rest: $v(-\infty) = 0$.
(e) Plot $a(t)$ and $v(t)$, both for an *uncharged* particle and for a (nonrunaway) charged particle, subject to this force.

9.3.2 The Physical Origin of the Radiation Reaction

In the last section I derived the Abraham-Lorentz formula for the radiation reaction, using conservation of energy. I made no attempt to identify the actual *mechanism* responsible for this force, except to point out that it must be a recoil effect of the particle's own fields acting back on the charge. Unfortunately, the fields of a *point* charge *blow up* right at the particle, so it's hard to see how one can calculate the force they exert.[16] Let's avoid this problem by considering an *extended* charge distribution, for which the field is finite everywhere; at the end, we'll take the limit as the size of the charge goes to zero. Now, it turns out that for an *extended* charge, the force of one part (A) on another part (B) is *not* equal and opposite to the force of B on A (Fig. 9.25). If the distribution is divided up into infinitesimal chunks, and the imbalances are added up for all such pairs, the result is a *net force of the charge* on itself. It is this *self-force*, resulting from the breakdown of Newton's third law within the structure of the particle, which accounts for the radiation reaction.

Lorentz originally calculated the electromagnetic self-force using a *spherical* charge distribution, which seems reasonable but makes the mathematics rather cumbersome.[17] Because I am only trying to elucidate the *mechanism* involved, I shall use a less "realistic" model: a "dumbbell" in which the total charge q is divided into two halves separated by a fixed distance d (Fig. 9.26). This is the simplest possible arrangement of the charge which permits the essential mechanism (imbalance of internal electromagnetic forces) to function. Never mind that it's an unlikely model for an elementary particle: In the point limit ($d \to 0$) *any* model must yield the Abraham-Lorentz formula, to the extent that conservation of energy alone dictates that answer.

Let's assume the dumbbell moves in the x direction, and is (instantaneously) at rest at the retarded time. The electric field at (1) due to (2) is

[16]It *can* be done by a suitable *averaging* of the field, but it's not easy. (See T. H. Boyer, *Am. J. Phys.* **40**, 1843 (1972), and references cited therein.)

[17]See J. D. Jackson, *Classical Electrodynamics,* 2d ed. (New York: John Wiley, 1975), sec. 17-3, for details.

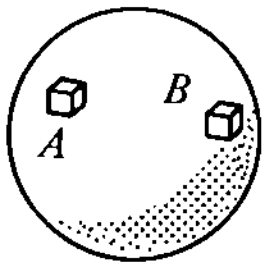

Figure 9.25

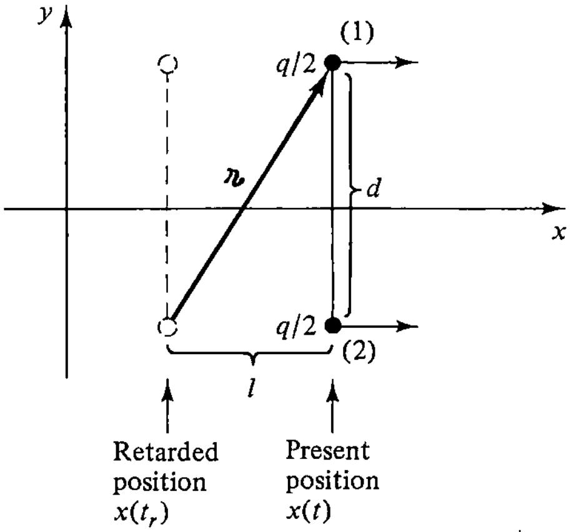

Figure 9.26

$$\mathbf{E}_{(1)} = \frac{(q/2)}{4\pi\epsilon_0} \frac{\mathscr{r}}{(\boldsymbol{\mathscr{r}}\cdot\mathbf{u})^3}[\mathbf{u}(c^2 - \boldsymbol{\mathscr{r}}\cdot\mathbf{a}) - \mathbf{a}(\boldsymbol{\mathscr{r}}\cdot\mathbf{u})] \tag{9.131}$$

(equation (9.107)), where

$$\mathbf{u} = c\hat{\mathscr{r}} \quad \text{and} \quad \boldsymbol{\mathscr{r}} = l\hat{i} = d\hat{j} \tag{9.132}$$

so that

$$\boldsymbol{\mathscr{r}}\cdot\mathbf{u} = c\mathscr{r}, \qquad \boldsymbol{\mathscr{r}}\cdot\mathbf{a} = la, \qquad \text{and} \quad \mathscr{r} = \sqrt{l^2 + d^2} \tag{9.133}$$

Actually, we're only interested in the x-component of $\mathbf{E}_{(1)}$, since the y-components will cancel when we add the forces on the two ends (for the same reason, we don't need to worry about the magnetic forces). Now

$$u_x = cl/\mathscr{r} \tag{9.134}$$

and hence

$$E_{(1)_x} = \frac{(q/2)}{4\pi\epsilon_0} \frac{(l - ad^2/c^2)}{(l^2 + d^2)^{3/2}} \tag{9.135}$$

By symmetry, $E_{(2)_x} = E_{(1)_x}$; evidently the net force on the dumbbell is

$$\mathbf{F}_{\text{self}} = \frac{q}{2}(\mathbf{E}_{(1)} + \mathbf{E}_{(2)}) = \frac{2(q/2)^2}{4\pi\epsilon_0} \frac{(l - ad^2/c^2)}{(l^2 + d^2)^{3/2}}\hat{i} \tag{9.136}$$

So far everything is *exact*. The idea now is to *expand in powers of d*; when the size of the "particle" goes to zero, all *positive* powers will disappear. Using Taylor's theorem,

$$x(t) = x(t_r) + \dot{x}(t_r)(t - t_r) + \frac{1}{2}\ddot{x}(t_r)(t - t_r)^2 + \frac{1}{3!}\dddot{x}(t_r)(t - t_r)^3 + \cdots \tag{9.137}$$

we have,

$$l = x(t) - x(t_r) = \frac{1}{2}aT^2 + \frac{1}{6}\dot{a}T^2 + \cdots \tag{9.138}$$

where $T = t - t_r$, for short. T is determined by the retarded time condition

$$l^2 + d^2 = (cT)^2 \tag{9.139}$$

so that

$$T = \frac{d}{c} + (\quad)d^3 + \cdots \tag{9.140}$$

and hence

$$l = \frac{a}{2c^2}d^2 + \frac{\dot{a}}{6c^3}d^3 + (\quad)d^4 + \cdots \tag{9.141}$$

Putting this into equation (9.136), I conclude that

$$\mathbf{F}_{\text{self}} = \frac{q^2}{4\pi\epsilon_0}\left(-\frac{a}{4c^2d} + \frac{\dot{a}}{12c^3} + (\quad)d + \cdots\right)\hat{\imath} \tag{9.142}$$

Here a and $\dot{a}$ are evaluated at the *retarded* time (t_r), but it's easy to rewrite the result in terms of the *present* time (t):

$$a(t_r) = a(t) + \dot{a}(t)(t_r - t) + \cdots = a(t) - \dot{a}(t)T + \cdots = a(t) - \dot{a}(t)\frac{d}{c} + \cdots \tag{9.143}$$

so that

$$\mathbf{F}_{\text{self}} = \frac{q^2}{4\pi\epsilon_0}\left[-\frac{a(t)}{4c^2d} + \frac{\dot{a}(t)}{3c^3} + (\quad)d + \cdots\right]\hat{\imath} \tag{9.144}$$

The first term on the right is proportional to the acceleration of the charge; if we pull it over to the other side of Newton's second law, it simply adds to the dumbbell's mass. In effect, the total inertia of the charged dumbbell is

$$m = 2m_0 + \frac{1}{4\pi\epsilon_0}\frac{q^2}{4dc^2} \tag{9.145}$$

where m_0 is the mass of either end alone. In the context of special relativity it is not surprising that the electrical repulsion of the charges should enhance the mass of the dumbbell. For the potential energy of this configuration (in the static case) is

$$\frac{1}{4\pi\epsilon_0}\frac{(q/2)^2}{d} \tag{9.146}$$

and according to the Einstein formula $E = mc^2$, this energy should be reflected in the mass of the object.[18]

[18]The fact that the actual *numbers* work out perfectly is a lucky feature of this particular configuration. If you do the same calculation for the dumbbell in *longitudinal* motion, the mass correction is only *half* of what it "should" be (there's a 2, instead of a 4, in equation 9.145), and for a sphere it's off by a factor of $\frac{3}{4}$. This notorious paradox has been the subject of much debate over the years. (See D. J. Griffiths and R. E. Owen, *Am. J. Phys.* 51, 1120 (1983), and references therein).

The second term in (9.144) is the radiation reaction proper:

$$F_{\text{rad}}^{\text{int}} = \frac{1}{4\pi\epsilon_0}\frac{q^2}{3c^3}\dot{a} \tag{9.147}$$

It alone (apart from the mass correction[19]) survives in the "point dumbbell" limit $d \to 0$. Unfortunately, it differs from the Abraham-Lorentz formula by a factor of 2. But then, this is only the self-force associated with the *interaction* between (1) and (2)—hence, the superscript "int." There remains the force of *each end on itself.* When the latter is included (see Problem 9.31) the result is

$$F_{\text{rad}} = \frac{1}{4\pi\epsilon_0}\frac{2q^2}{3c^3}\dot{a} \tag{9.148}$$

reproducing the Abraham-Lorentz formula exactly. *Conclusion: The radiation reaction is due to the force of the charge on itself*—or, more elaborately, the net force exerted by the fields generated by different parts of the charge distribution acting on one another.

Problem 9.31 Deduce 9.148 from 9.147, as follows:

(a) Use the Abraham-Lorentz formula to determine the radiation reaction on each end of the dumbbell; add this to the interaction term (9.147) to obtain (9.148).

Method (a) has the defect that it *uses* the Abraham-Lorentz formula—the very thing we were trying to *derive*; we can avoid this by method (b).

(b) Smear out the charge along a strip of length L oriented perpendicular to the motion (the charge density, then, is $\lambda = q/L$); find the cumulative interaction force for all pairs of segments, using 9.147 (with the correspondence $q/2 \to \lambda dy_1$, at one end and $q/2 \to \lambda dy_2$ at the other). Make sure you don't count the same pair twice.

Further Problems on Chapter 9

Problem 9.32 Suppose you take a plastic ring of radius R and glue charge on it, so that the line charge density is $\lambda_0|\sin(\theta/2)|$. Then you spin the loop about its axis at an angular velocity ω. Find the (exact) scalar and vector potentials at the center of the ring.

$$\left(\textit{Answer}: \mathbf{A} = -\frac{\mu_0\lambda_0\omega R}{3\pi}\left\{\sin\left[\omega\left(t - \frac{R}{c}\right)\right]\hat{\imath} + \cos\left[\omega\left(t - \frac{R}{c}\right)\right]\hat{\jmath}\right\}.\right)$$

Problem 9.33

(a) By applying the curl to Faraday's law and to Ampère's law (with Maxwell's term), show that $\mathbf{E}$ and $\mathbf{B}$ satisfy the inhomogeneous wave equation, with appropriate sources.

(b) We know how to solve the inhomogeneous wave equation ((9.8) does it for (9.4) and (9.5)). With this in mind, write the general formulas for $\mathbf{E}$ and $\mathbf{B}$ in terms of ρ and $\mathbf{J}$.

(c) Show that these equations reduce to Coulomb's law and the Biot-Savart law in the static case.

$$\left(\textit{Answer}: \mathbf{E} = \frac{1}{4\pi\epsilon_0}\int\frac{[-\epsilon_0\mu_0\dot{\mathbf{J}} - \nabla\rho](\mathbf{r}', t_r)}{\mathscr{r}}\,d\tau;\ \mathbf{B} = \frac{\mu_0}{4\pi}\int\frac{[(\nabla\times\mathbf{J})](\mathbf{r}', t_r)}{\mathscr{r}}\,d\tau\right)$$

[19]Of course, the limit $d \to 0$ has an embarrassing effect on the mass term. In a sense, it doesn't matter, since only the *total* mass m is observable; maybe m_0 somehow has a compensating infinity, so that m comes out finite. Or perhaps this means that we should not take the idea of a point charge too literally in classical electrodynamics.

Problem 9.34 Figure 2.35 summarizes the laws of *electrostatics* in a "triangle diagram" relating the *source* (ρ), the *field* (**E**), and the *potential* (V). Figure 5.46 does the same for *magnetostatics,* where the source is **J**, the field is **B**, and the potential is **A**. Construct the analogous diagram for *electrodynamics,* with sources ρ and **J** (constrained by the continuity equation), fields **E** and **B**, and potentials V and **A** (constrained by the Lorentz gauge condition). Include the results of Problem 9.33, but skip the formulas for V and **A** in terms of **E** and **B**.

Problem 9.35 A particle of mass m and charge q is attached to a spring with force constant k, hanging from the ceiling (Fig. 9.27). Its equilibrium position is a distance h above the floor. It is pulled down a distance d below equilibrium and released at time $t = 0$.

(a) Under the usual assumptions ($d \ll \lambda \ll h$), calculate the intensity of the radiation hitting the floor, as a function of the distance R from the point directly below q. (*Note*: Intensity is the average power per unit area of *floor.*) At what R is the radiation most intense? (Neglect the radiative damping of the oscillator.)

(b) As a check on your formula, assume the floor is of infinite extent, and calculate the average energy per unit time striking the entire floor. Is it what you'd expect?

(c) Because it is losing energy in the form of radiation, the amplitude of the oscillation will gradually decrease. After what time τ has the amplitude been reduced to d/e? (Assume the fraction of the total energy lost in one cycle is very small.)

$$\left(\textit{Answer}: I(R) = \left(\frac{\mu_0 q^2 d^2 \omega^4}{32\pi^2 c}\right)\frac{R^2 h}{(R^2 + h^2)^{5/2}}\right)$$

Problem 9.36 As you know, the magnetic north pole of the earth does not coincide with the geographic north pole—in fact, it's off by about 11°. Relative to the fixed axis of rotation, therefore, the magnetic dipole moment vector of the earth is changing with time, and the earth must be giving off magnetic dipole radiation.

(a) Find the formula for the total power radiated, in terms of the following parameters:

Ψ (the angle between the geographic and magnetic north poles)
M (the magnitude of the earth's magnetic dipole moment)
ω (the angular velocity of rotation of the earth)

(b) Using the fact that the earth's magnetic field is about $\frac{1}{2}$ gauss at the equator, estimate the magnetic dipole moment M of the earth.

(c) Find the power radiated *in watts.*

(d) Pulsars are thought to be rotating neutron stars, with a typical radius of 10 km, a rotational period of 10^{-3} s, and a surface magnetic field of 10^8 T. What sort of radiated power would you expect from such a star? (See J. P. Ostriker and J. E. Gunn, *Astrophys. J.*, **157**, 1395 (1969).) (*Answers*: (c) 4×10^{-5} W; (d) 2×10^{36} W)

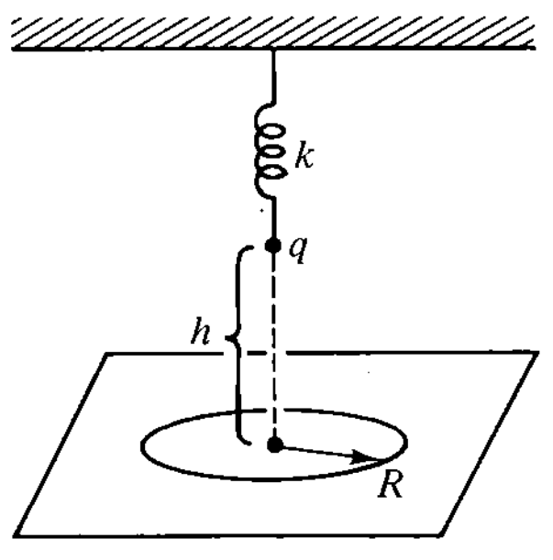

Figure 9.27

Problem 9.37 One particle, of charge q_1, is held at rest at the origin. Another particle, of charge q_2, moves in along the x axis in "hyperbolic" motion:

$$x(t) = \sqrt{b^2 + (ct)^2}$$

(It reaches the point of closest approach, b, at time $t = 0$ and then returns out to infinity.)
(a) What is the force F_2 on q_2 (due to q_1) at time t?
(b) What total impulse ($I_2 = \int_{-\infty}^{\infty} F_2\, dt$) is delivered to q_2 by q_1?
(c) What is the force F_1 on q_1 (due to q_2) at time t?
(d) What total impulse ($I_1 = \int_{-\infty}^{\infty} F_1\, dt$) is delivered to q_1 by q_2? (It might help to look over Problem 9.17 before doing this integral.)

$$\left(\textit{Answer}: I_2 = -I_1 = \frac{q_1 q_2}{4\epsilon_0 bc}\right)$$

Problem 9.38 Suppose the (electrically neutral) yz plane carries a time-dependent but uniform surface current $K(t)\hat{k}$.
(a) Find the electric and magnetic fields at a height x above the plane if
(i) a constant current is turned on at $t = 0$:

$$K(t) = \begin{Bmatrix} 0, & t \leq 0 \\ K_0, & t > 0 \end{Bmatrix}$$

(ii) a linearly increasing current is turned on at $t = 0$:

$$K(t) = \begin{Bmatrix} 0, & t \leq 0 \\ \alpha t, & t > 0 \end{Bmatrix}$$

(b) Show that the retarded vector potential can be written in the form

$$\mathbf{A}(x, t) = \frac{\mu_0 c}{2} \hat{k} \int_0^\infty K\left(t - \frac{x}{c} - u\right) du$$

and from this determine **E** and **B**. Show that the total power radiated per unit area of surface is

$$\frac{\mu_0 c}{2} [K(t)]^2$$

(For details, and related problems, see T. A. Abbott and D. J. Griffiths, *Am. J. Phys.* **53**, 1203 (1985).)

! Problem 9.39
(a) Find the radiation reaction force on a particle moving with *arbitrary* velocity in a straight line, by reconstructing the argument in Section 9.3.2 *without* assuming $v(t_r) = 0$.

$$\left(\textit{Answer}: F_{\text{rad}} = \frac{1}{4\pi\epsilon_0} \frac{2q^2}{3c^3} \gamma^4 \left(\dot{a} + \frac{3\gamma^2 a^2 v}{c^2}\right)\right)$$

(b) Show that this result is consistent (in the sense of equation (9.124)) with the power radiated by such a particle (9.121).

! Problem 9.40
(a) Does a particle in hyperbolic motion radiate? (Use the exact formula (9.121) to calculate the power radiated.)

(b) Does a particle in hyperbolic motion experience a radiation reaction? (Use the exact formula (Problem 9.39) to determine the reaction force.)
[*Comment*: These famous questions carry important implications for the Principle of Equivalence. See T. Fulton and F. Rohrlich, *Annals of Physics* **9**, 499 (1960), Chapter 8 of R. Peierls, *Surprises in Theoretical Physics* (Princeton: Princeton University Press, 1979) and the article by P. Pearle in *Electromagnetism: Paths to Research*, ed. D. Teplitz, (New York: Plenum Press, 1982).]

10

ELECTRODYNAMICS AND RELATIVITY

10.1 THE SPECIAL THEORY OF RELATIVITY

10.1.1 Einstein's Postulates

Aristotle taught that a moving object, left to itself, will eventually come to rest. I don't suppose there is any law of physics that sounds more reasonable, more consistent with everyday experience, than that. If I put a brick on the table and give it a shove, it certainly soon comes to a stop. But progress is sometimes made by questioning the obvious. Galileo observed that the more you polish the surface of the brick, the farther it goes. Moreover, if the table *itself* is moving, the brick comes to rest with respect to the table, not the floor. Galileo concluded that the brick stops because of *friction* between the table and the brick—if you could eliminate *all* friction by somehow polishing the surfaces to perfect smoothness, then the brick would keep going forever. Newton, of course, incorporated this idea into his first law of motion. Aristotle was wrong: Left *completely* to themselves, objects do *not* come to rest—they keep moving. In practice, there is always some friction at work; however, this does not bring the object to an "absolute" state of rest, but merely to rest with respect to whatever it's rubbing against.

In fact, Galileo's discovery raises a deeper question: Is there *any such thing* as an "absolute" state of rest, or must we always speak of rest *with respect to some specified object?* Do the laws of physics, as Aristotle held, presuppose the existence of a unique stationary reference system, or could we just as well work in a moving system? Imagine, for example, that we load a billiard table onto a freight car (we'll assume the track is smooth and straight, and the speed of the train is constant). Could we, by studying the progress of the game, determine that the train is in motion? Would the players be obliged to "correct" their shots to account for the speed of the train, or would the game proceed as normal?

Suppose, to begin with, that a ball is rolling forward at speed u' relative to the train and the train's speed is v relative to the ground. The speed of the ball relative to the *ground* is then (according to classical physics)

$$u = u' + v \tag{10.1}$$

Ignoring friction, Newton's first law says that

$$u \text{ is constant}$$

As long as the train's speed, v, is a constant, it follows that

$$u' \text{ is constant}$$

In other words, Newton's first law applies just as well in the train system as it does in the ground system. Of course, the actual *numbers* involved may be entirely different (the ball could be moving 20 mi/h relative to the train and 110 mi/h relative to the ground) but the *physical law* describing the motion (constant velocity) is the same in either system. By watching a freely rolling ball, therefore, you could *not* tell whether the train was moving. Notice, by contrast, that you could *easily* tell if the train *accelerated:* The ball would appear to slow down or, if the train rounded a corner, to follow a curved trajectory—and you yourself would lurch backward or to one side. A *non*accelerated reference frame, in which Newton's first law holds, is called an **inertial** system.

If *freely* moving objects provide no means for establishing an "absolute" rest system, what about objects that *interact?* Suppose, for instance, one billiard ball, A, on the train collides head-on with another, B (Fig. 10.1). We'll assume the collision is elastic, so that the total kinetic energy is the same before and after, but we won't assume the masses are identical—in fact, we'll even allow that in the course of the collision some mass rubs off A and sticks onto B, so that the outgoing objects, C and D, do not necessarily have the same masses as A and B, though the *total* is, of course, unchanged:

$$m_A + m_B = m_C + m_D \qquad \textit{(conservation of mass)} \tag{10.2}$$

The collision is governed by the laws of conservation of momentum,

$$m_A u_A + m_B u_B = m_C u_C + m_D u_D \qquad \textit{(conservation of momentum)} \tag{10.3}$$

and kinetic energy,

$$\tfrac{1}{2}m_A u_A^2 + \tfrac{1}{2}m_B u_B^2 = \tfrac{1}{2}m_C u_C^2 + \tfrac{1}{2}m_D u_D^2 \qquad \textit{(conservation of energy)} \tag{10.4}$$

which together determine the outgoing velocities in terms of the incoming ones and the various masses. I have written the conservation laws from the point of view of an

Figure 10.1

observer on the *ground* (that is, I used velocities relative to ground), because I know it is safe to apply the laws of mechanics in this system. The question is, could I just as well have used the train as my reference system? The answer is *yes,* for if we use (10.1) to rewrite (10.3),

$$m_A(u_A' + v) + m_B(u_B' + v) = m_C(u_C' + v) + m_D(u_D' + v)$$

and exploit conservation of mass (10.2) to cancel the terms in u, we are left with

$$m_A u_A' + m_B u_B' = m_C u_C' + m_D u_D'$$

which is precisely conservation of momentum relative to the train. (I'll leave it for you to check conservation of energy.) If momentum and energy are conserved relative to ground, they must *also* be conserved relative to the train—again, the actual *numbers* may be quite different in the two systems, but the *physical laws* are identical.

In fact, *all* the laws of classical mechanics hold as well in uniformly moving systems as "stationary" ones. I put "stationary" in quotation marks because, of course, a corollary of this result is that there is no way, in the context of classical mechanics, to decide whether a given inertial system *is* stationary; there *is* no absolute state of rest. To be quite precise about it,

The laws of classical mechanics apply in all inertial reference systems.

This is called **Galileo's principle of relativity.** It was well known long before Einstein—in fact, it played a crucial role in vindicating the Copernican theory of planetary motion. (The opponents of Copernicus had argued that the earth must be stationary, else a falling object would not descend in a straight line but would appear to curve backward, lagging behind as the earth moved out from under it. Galileo is supposed to have demolished this objection by dropping a rock from the mast of a moving ship: It landed at the base of the mast, just as it would for a ship at rest. The fact that an object falls straight down tells you nothing at all about the speed of the system in which you perform the experiment; only on an *accelerating* ship would the lag be observed.)

Insofar as classical mechanics is concerned, then, it doesn't matter whether we define positions (and velocities) with respect to a coordinate system fixed on the ground or one attached to the train; the same physical laws describe the process in either system. But what about electrodynamics: Do we have the same freedom there? At first glance the answer would seem to be *no.* After all, a charge in motion produces a magnetic field, whereas a charge at rest does not. A charge carried along by the train would generate a magnetic field, according to an observer on the ground, but a man on the train, if he were to apply the laws of electrodynamics in his system, would predict no magnetic field. In fact, many of the equations of electrodynamics, such as the Lorentz force law and the Biot-Savart law, make explicit reference to "the" velocity of the charge. It certainly appears, therefore, that classical electromagnetic theory presupposes the existence of a unique stationary reference frame, with respect to which all velocities are to be measured.

And yet there is an extraordinary coincidence which has to make you wonder. Suppose we mount a wire loop in our freight car and have the train pass between the poles of a giant magnet (Fig. 10.2). As the loop rides through the magnetic field, a

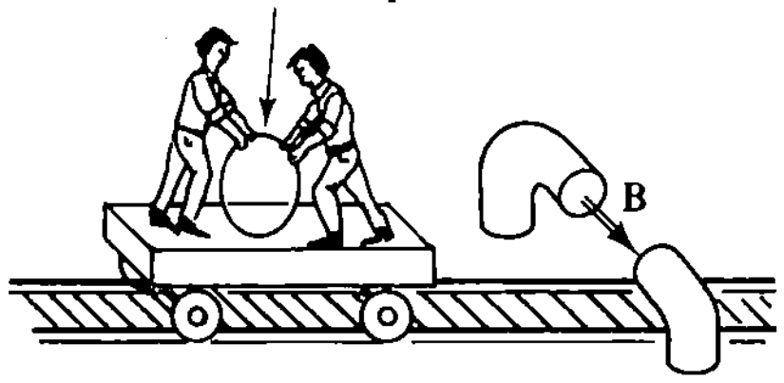

Figure 10.2

motional emf is established; according to the flux rule (7.13),

$$\mathcal{E} = -\frac{d\Phi}{dt}$$

This emf, remember, is due to the magnetic force on charges in the wire loop, which are moving along with the train. On the other hand, if a man on the train naively applied the laws of electrodynamics in *his* system, what would he get? He would expect no *magnetic* force because to him the loop is at rest. But as the magnet flies by, the magnetic field in his freight car will change, and a changing magnetic field should induce an electric field, by Faraday's law. The resulting *electric* force would generate an emf in the loop given by (7.14):

$$\mathcal{E} = -\frac{d\Phi}{dt}$$

Because Faraday's law and the flux rule predict the same emf, the man on the train will get the right answer, *even though his physical interpretation of the process is entirely wrong.*

Or *is* it? Einstein could not believe this was a mere coincidence; he took it, rather, as a clue that electromagnetic phenomena, like mechanical ones, obey the principle of relativity. In his view the analysis by the observer on the train is just as valid as that of the observer on the ground. If their *interpretations* differ (one calling the process electric, the other magnetic) so be it; their actual *predictions* are in agreement. Here's what he wrote on the first page of his 1905 paper introducing the Special Theory:

On the Electrodynamics of Moving Bodies by A. Einstein

It is known that Maxwell's electrodynamics—as usually understood at the present time—when applied to moving bodies, leads to asymmetries which do not appear to be inherent in the phenomena. Take, for example, the reciprocal electrodynamic action of a magnet and a conductor. The observable phenomenon here depends only on the relative motion of the conductor and the magnet, whereas the customary view draws a sharp distinction between the two cases in which either the one or the other of these bodies is in motion. For if the magnet is in motion and the conductor at rest, there arises in the neighborhood of the magnet an electric field . . . producing a current at the places where parts of the conductor are situated. But if the magnet is stationary and the conductor in motion, no electric field arises in the neighborhood of the magnet. In the conductor, however, we find an electromotive force . . . which gives rise—assuming equality of relative motion in the two cases discussed—to electric currents of the same path and intensity as those produced by the electric forces in the former case.

> Examples of this sort, together with the unsuccessful attempts to discover any motion of the earth relative to the "light medium," suggest that the phenomena of electrodynamics as well as of mechanics possess no properties corresponding to the idea of absolute rest.

But I'm getting ahead of the story. To Einstein's predecessors the equality of the two emf's was just a lucky accident—they had no doubt that one observer was right and the other was wrong. It's not that they believed in an absolute rest system, though, nor that they rejected the principle of relativity as such. Rather, orthodox opinion at the time regarded electric and magnetic fields as strains in an invisible jellylike medium called **ether,** which permeated all of space; and in this view the speed of a charge is to be measured with *respect to the ether—only then are the laws of electrodynamics valid.* The train observer is wrong because he is moving with respect to the ether. But wait: How do we know the *ground* observer isn't moving relative to the ether, too? Suddenly it becomes a matter of crucial importance to *find* the ether frame, experimentally, else *all* our calculations will be invalid.

When you stop to think about it, it's most unlikely that the ground should be at rest with respect to the ether. After all, the earth rotates on its axis once a day and revolves around the sun once a year; the solar system circulates around the galaxy, and for all I know the galaxy itself may be moving at a high speed through the cosmos. All told, we should be traveling at well over 50 km/s with respect to the ether. Like a motorcycle rider on the open road, we face an "ether wind" of high velocity—unless by some miraculous coincidence we just happen to find ourselves in a tailwind of precisely the right strength, or the earth has some sort of "windshield" and drags its local supply of ether along with it.

The problem, then, is to detect our motion through the ether—to measure the speed and direction of the "ether wind." How shall we do it? At first glance you might suppose that practically *any* electromagnetic experiment would suffice: If Maxwell's equations are valid only with respect to the ether frame, any discrepancy between the experimental result and the theoretical prediction (using the earth system) should be attributable to the ether wind. Unfortunately, as nineteenth century physicists soon realized, the anticipated error in a typical experiment is of order v^2/c^2 or smaller,[1] where v is our velocity through the ether. So unless the ether wind is blowing at close to the speed of light (and there's no reason to expect *that*) it's going to take a delicate experiment to detect it at all.

Now, among the results of classical electrodynamics is the prediction that electromagnetic waves travel through the vacuum at a speed

$$\frac{1}{\sqrt{\epsilon_0\mu_0}} = 3.00 \times 10^8 \text{ m/s}$$

relative (presumably) *to the ether.* In principle, then, one should be able to detect the ether wind by simply measuring the speed of light in various directions. Like a motor-

[1] In the example discussed earlier, involving induced emf, the error is actually zero. In other cases, where nonzero errors appeared, we now know that the fault lay with the laws and procedures of classical mechanics, not with the principle of relativity.

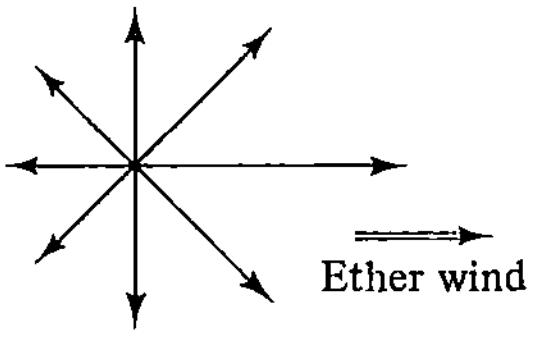

Figure 10.3

boat on a river, the net speed "downstream" should be a maximum, for here the light is swept along by the ether; in the opposite direction, where it is bucking the current, the speed should be a minimum (Fig. 10.3). While the *idea* of this experiment could not be simpler, its *execution* is another matter because light travels so inconveniently fast. If it weren't for that "technical problem" you could do it all with a flashlight and a stopwatch. As it was, an elaborate and lovely experiment was devised by Michelson and Morley, using an optical interferometer of fantastic precision. I shall not go into the details here, because I do not want to distract your attention from the two essential points: (1) all Michelson and Morley were trying to do was compare the speed of light in different directions, and (2) what they in fact *discovered* was that this speed is *exactly the same in all directions.*

Nowadays, when students are taught in high school to snicker at the naïveté of the ether model, it takes some imagination to realize how utterly perplexing this result must have been at the time. All other waves (water waves, sound waves, waves on a string) travel at a prescribed speed *relative to the propagating medium* (the stuff that does the waving), and if this medium is in motion with respect to the observer, the net speed is always greater "downstream" than "upstream." Over the next 20 years a series of improbable schemes were concocted in an effort to explain why this did *not* occur with light. Michelson and Morley themselves interpreted their experiment as confirmation of the "ether drag" hypothesis, which held that the earth sweeps the ether along with it. But this was found to be inconsistent with other observations, notably the aberration of starlight.[2] Various so-called "emission" theories were proposed, according to which the speed of electromagnetic waves is governed by the motion of the *source*—as it would be in a corpuscular theory (conceiving of light as a stream of particles). Such theories called for implausible modifications in Maxwell's equations, but in any event they were discredited by experiments using extraterrestrial light sources. Meanwhile, Fitzgerald and Lorentz suggested that the ether wind physically compresses all matter (including the Michelson-Morley apparatus itself) in just the right way to compensate for, and thereby conceal, the variation in speed with direction. As it turns out, there is a grain of truth in this, although their idea of the *reason* for the contraction was quite wrong.

At any rate, it was not until Einstein that anyone took the Michelson-Morley result at face value and suggested that the speed of light is a universal constant, the same in all directions, regardless of the motion of the observer or the source. There *is* no ether wind because there is no ether. *Any* inertial system is a suitable reference frame for the application of Maxwell's equations, and the velocity of a charge is to be measured *not* with respect to a (nonexistent) absolute rest frame nor with respect to a

[2]A discussion of the Michelson-Morley experiment and related matters is to be found in R. Resnick's *Introduction to Special Relativity,* Chap. 1 (New York: John Wiley, 1968).

(nonexistent) ether, but simply with respect to the particular reference system you happen to have chosen.

Inspired, then, both by internal theoretical hints (the fact that the laws of electrodynamics are such as to give the right answer even when applied in the "wrong" system) and by external empirical evidence (the Michelson-Morley experiment[3]), Einstein proposed his two famous postulates:

1. **The principle of relativity.** The laws of physics apply in all inertial reference systems.
2. **The universal speed of light.** The speed of light in vacuum is the same for all inertial observers, regardless of the motion of the source, the observer, or any assumed medium of propagation.

The special theory of relativity derives from these two postulates. The first elevates Galileo's observation about classical mechanics to the status of a general law, applying to *all* of physics. It states that there is no absolute rest system. The second might be considered Einstein's interpretation of the Michelson-Morley experiment. It asserts that there is no ether. (Some authors consider Einstein's second postulate redundant—no more than a special case of the first. They maintain that the very existence of ether would violate the principle of relativity, in the sense that it would define a unique stationary reference frame. I think this is nonsense. The existence of air as a medium for sound propagation does not invalidate the theory of relativity. Ether is no more an absolute rest system than the water in a goldfish bowl—which is a *special* system, if you happen to be the goldfish, but scarcely "absolute.")[4]

Unlike the principle of relativity, which had roots going back several centuries, the universal speed of light was radically new and, on the face of it, in drastic conflict with the dictates of common sense. For if I walk 5 mi/h down the corridor of a train going 60 mi/h, my net speed relative to the ground is "obviously" 65 mi/h—the speed of A (me) with respect to C (ground) is equal to the speed of A relative to B (train) plus the speed of B relative to C:

$$v_{AC} = v_{AB} + v_{BC} \tag{10.5}$$

And yet, if A is a *light* signal (whether it comes from a flashlight on the train or a lamp on the ground or a star in the sky) Einstein would have us believe that its speed is c relative to the train *and* c relative to the ground:

$$v_{AC} = v_{AB} = c \tag{10.6}$$

Evidently, equation (10.5), which we now call **Galileo's velocity addition rule** (no one before Einstein would have bothered to give it a name at all) is incompatible with the

[3]Actually, Einstein appears to have been only dimly aware of the Michelson-Morley experiment at the time. For him, the theoretical argument alone was decisive.

[4]I put it this way in an effort to dispel some misunderstanding as to what constitutes an absolute rest frame. In 1977, it became possible for the first time to measure the speed of the earth through the 3 K background radiation left over from the "big bang." Does this mean we have found an absolute rest system, in violation of relativity? Of course not.

second postulate. In special relativity, as we shall see, it is replaced by **Einstein's velocity addition rule:**

$$\boxed{v_{AC} = \frac{v_{AB} + v_{BC}}{1 + \dfrac{v_{AB}v_{BC}}{c^2}}} \tag{10.7}$$

For "ordinary" speeds ($v_{AB} \ll c$, $v_{BC} \ll c$), the denominator is so close to 1 that the discrepancy between Galileo's formula and Einstein's is negligible. On the other hand, (10.7) has the desired property that if $v_{AB} = c$, then automatically $v_{AC} = c$, consistent with (10.6):

$$v_{AC} = \frac{c + v_{BC}}{1 + \dfrac{cv_{BC}}{c^2}} = c$$

But how can Galileo's rule, which relies on nothing but the most primitive geometrical inference, possibly be wrong? And if it *is* wrong, what does this do to all of classical physics? (As a matter of fact, I used this very rule in equation (10.1), to support the principle of relativity itself!) The answer is that special relativity compels us to alter our notions of space and time themselves, and therefore also of such derived quantities as velocity, momentum, and energy. Although it developed historically out of Einstein's contemplation of electrodynamics, the special theory is not limited to any particular class of phenomena—rather, it is a description of the space-time "arena" in which *all* physical phenomena take place. And in spite of the reference to the speed of light in the second postulate, relativity has nothing to do with light: c is evidently a fundamental velocity, and it happens that light (also neutrinos and, presumably, gravitons) travel at that speed, but it is possible to conceive of a universe in which there are no electric charges, and hence no electromagnetic fields or waves, and yet relativity would still prevail. Because relativity defines the structure of space and time, it claims authority not merely over all presently known phenomena, but over those yet to be discovered. It is, as Kant would say, a "prolegomenon to any future physics."

Problem 10.1

(a) What's the percent error introduced when you use Galileo's rule, instead of Einstein's, with $v_{AB} = 5$ mi/h and $v_{BC} = 60$ mi/h?

(b) Suppose you could run at half the speed of light down the corridor of a train going three-quarters the speed of light. What would your speed be relative to ground?

(c) Prove, using equation (10.7), that if $v_{AB} < c$ and $v_{BC} < c$ then $v_{AC} < c$. Interpret this result.

Problem 10.2 As the outlaws escape in their getaway car, which goes $\frac{3}{4}c$, the police officer fires a bullet from the pursuit car, which only goes $\frac{1}{2}c$. (Fig. 10.4). The muzzle velocity (speed relative to gun) of the bullet is $\frac{1}{3}c$. Does the bullet reach its target (a) according to Galileo, (b) according to Einstein?

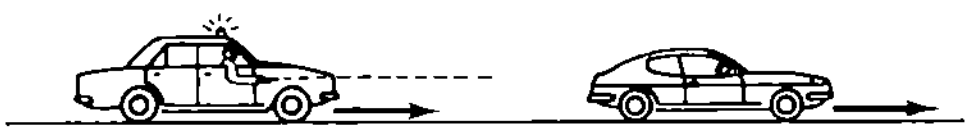

Figure 10.4

10.1.2 The Geometry of Relativity

In this section I present a series of *gedanken* (thought) experiments which serve to introduce the three most striking geometrical consequences of Einstein's postulates: time dilation, Lorentz contraction, and the relativity of simultaneity. In Section 10.1.3 the same results are derived more systematically by using the Lorentz transformations.

(i) The relativity of simultaneity. Imagine a freight car, traveling at constant speed along a smooth, straight track (Fig. 10.5). In the very center of the car there hangs a light bulb. When someone switches it on, the light spreads out in all directions at the speed c. Because the lamp is equidistant from the two ends, an observer on the train will find that the light reaches the front end at the same instant as it reaches the back end: the two events in question—(a) light reaches the front end and (b) light reaches the back end—occur *simultaneously*. However, to an observer on the *ground* these same two events are *not* simultaneous. For as the light travels out from the bulb, the train itself moves forward, so the beam going to the back end has a shorter distance to travel than the one going forward (Fig. 10.6). According to this observer, therefore, event (b) happens *before* event (a). (An observer passing by on an express train, meanwhile, would report that (a) preceded (b).) *Conclusion*:

Two events which are simultaneous in one inertial system are not, in general, simultaneous in another.

This is a purely relativistic matter, undreamt of in classical physics. It derives from Einstein's second postulate: I assumed the speed of light is the same in both directions for each observer, an assumption anyone before Einstein would have considered preposterous. Naturally, the train has to be going awfully fast before the discrepancy becomes detectable—that's why you don't notice it all the time.

Of course, it was *always* possible for a naive witness to be *misled* about simultaneity: you hear the thunder *after* you see the lightning, and a child might infer that the source of the light was not simultaneous with the source of the sound. But this is a trivial error, having nothing to do with moving observers or relativity—*obviously*, you must correct for the time the signal (sound, light, carrier pigeon, or whatever) takes to reach you. When I speak of an "observer," I mean someone having the sense to make this correction, and an "observation" is what an observer records *after* doing

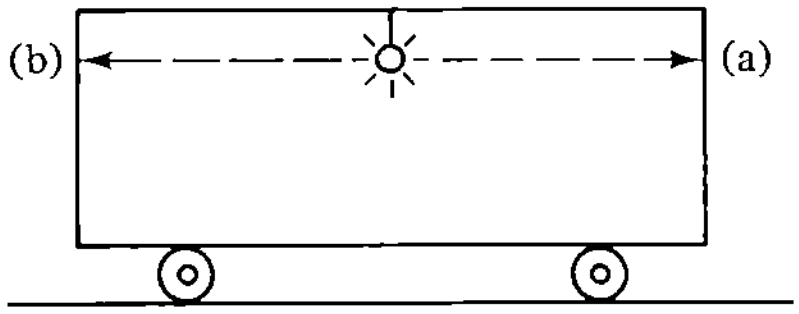

Figure 10.5

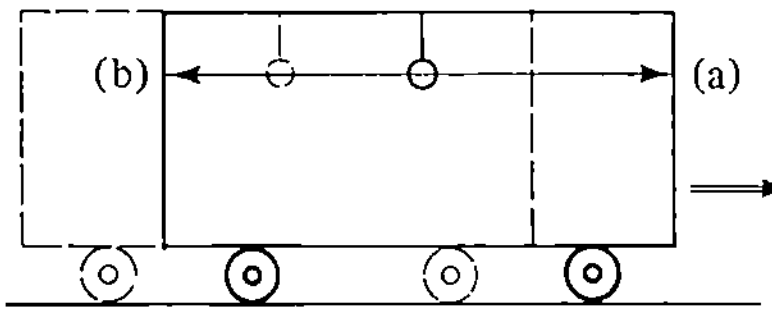

Figure 10.6

so. What you *see*, therefore, is not the same as what you *observe*. An observation cannot be made with a camera—it's an artificial reconstruction made after the fact, when all the data are in. In fact, a wise observer will avoid the problem by stationing assistants at strategic locations, each equipped with a watch synchronized to a master clock, so that time measurements can be made right at the scene. I belabor this point in order to emphasize that the relativity of simultaneity is a genuine discrepancy between measurements made by competent observers in relative motion, not a simple mistake arising from a failure to account for the travel time of light signals.

Problem 10.3 Synchronized clocks are stationed at regular intervals, 1 million km apart, along a straight line. When the clock next to you reads 12 noon,

(a) What time do you *see* on the 90th clock down the line?

(b) What time do you *observe* on that clock?

Problem 10.4 Every 2 years, more or less, *The New York Times* publishes an article in which some astronomer reports that he has found an object traveling faster than light. Many of these claims result from a failure to distinguish what is *seen* from what is *observed*—that is, from a failure to account for light travel time. Here's an example.

A star is traveling with speed v at an angle θ to the line of sight (Fig. 10.7). What is its apparent speed across the sky? (Suppose the light signal from b reaches earth a time Δt after the signal from a, and the star has meanwhile advanced a distance Δs across the celestial sphere; by "apparent speed" I mean $\Delta s/\Delta t$.) What angle θ gives the maximum apparent speed? Show that the apparent speed can be much greater than c, even if v itself is less than c.

(ii) Time dilation. Now let's consider a light ray that leaves the bulb and strikes the floor of the car directly below. *Question*: How long does it take the light to make this trip? From the point of view of an observer on the train, the answer is easy: If the height of the car is h, the time is

$$\Delta t' = \frac{h}{c} \tag{10.8}$$

(I'll use a prime to denote measurements made on the train.) On the other hand, as observed from the ground this same ray must travel farther because the train itself is moving. From Fig. 10.8, this distance is $\sqrt{h^2 + (v\,\Delta t)^2}$, and so

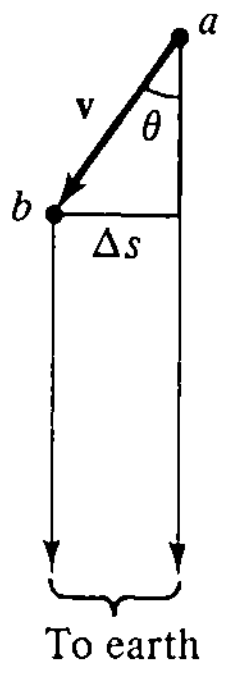

Figure 10.7

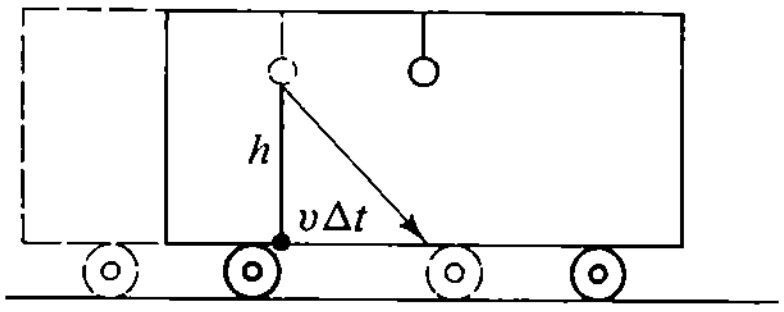

Figure 10.8

$$\Delta t = \frac{\sqrt{h^2 + (v\Delta t)^2}}{c}$$

Solving for Δt, we have

$$\Delta t = \frac{h}{c}\frac{1}{\sqrt{1 - v^2/c^2}}$$

so that

$$\boxed{\Delta t' = \sqrt{1 - v^2/c^2}\,\Delta t} \tag{10.9}$$

Evidently, the time elapsed between the *same two events*—(a) light leaves bulb, and (b) light strikes center of floor—is different for the two observers. In fact, the interval recorded on the train clock, $\Delta t'$, is *shorter* by the factor

$$\boxed{\gamma = \frac{1}{\sqrt{1 - v^2/c^2}}} \tag{10.10}$$

Conclusion:

Moving clocks run slow.

This effect is called **time dilation.** It doesn't have anything to do with the mechanics of clocks; it's a statement about the nature of time, which applies to *all* properly functioning timepieces.

Of all Einstein's predictions, none has received more spectacular and persuasive confirmation than time dilation. Most elementary particles are unstable: They disintegrate after a characteristic lifetime[5] that varies from one species to the next. The lifetime of a neutron is 15 min, of a muon, 2×10^{-6} s, of a neutral pion, 9×10^{-17} s. But these are the lifetimes of particles at *rest.* When particles are moving at speeds close to c, they in fact last much longer, for their internal clocks (whatever it is that tells them when their time is up) are running slow, in accordance with Einstein's time dilation formula.

Example 1

A muon is traveling through the laboratory at three-fifths the speed of light. How long does it last?

Solution: In this case,

$$\gamma = \frac{1}{\sqrt{1 - (\frac{3}{5})^2}} = \frac{5}{4}$$

[5]Actually, an individual particle may last longer or shorter than this. Particle disintegration is a random process, and I should really speak of the *average* lifetime for the species. But to avoid irrelevant complication I shall pretend that every particle disintegrates after precisely the average lifetime.

so it lives longer (than at rest) by a factor of $\frac{5}{4}$:

$$\tfrac{5}{4} \times (2 \times 10^{-6})\,\text{s} = 2.5 \times 10^{-6}\,\text{s}$$

It may strike you that time dilation is plainly inconsistent with the principle of relativity. For if the ground observer says the train clock runs slow, the train observer can with equal justice claim that the *ground* clock runs slow—after all, from the train's point of view it is the *ground* that is in motion. Who is right? *Answer*: They're *both* right! On closer inspection the "contradiction," which seems so stark, evaporates. Let me explain: In order to check the rate of the train clock, the ground observer uses *two* of his own clocks (Fig. 10.9): one to compare times at the beginning of the interval, when the train clock passes point A, the other to compare times at the end of the interval, when the train clock passes point B. Of course, he must be careful to synchronize his clocks before the experiment. What he finds is that while the train clock ticked off, say, 3 minutes, the interval between his two clock readings was 5 minutes. He concludes that the *train* clock runs slow.

Meanwhile, the observer on the train is checking the rate of the ground clock by the same procedure: He uses two carefully synchronized train clocks, and compares times with a single ground clock as it passes by each of them in turn (Fig. 10.10). He finds that while the ground clock ticks off 3 minutes, the interval between his train clocks is 5 minutes, and concludes that the *ground* clock runs slow. Is there a contradiction? *No,* for the two observers have measured *different things.* The ground observer compared *one* train clock with *two* ground clocks; the train observer compared one *ground* clock with two *train* clocks. Each followed a sensible and correct procedure, comparing a single moving clock with two stationary ones. "So what," you say, "the stationary clocks were synchronized in each instance, so it cannot matter that they used two different ones." But there's the rub. *Clocks that are properly synchronized in one system will not be synchronized when observed from another system.* They *can't* be, for to say that two clocks are synchronized is to say that they read 12 noon *simultaneously,* and we have already learned that what's simultaneous to one observer is *not* simultaneous to another. So whereas each observer conducted a perfectly sound measurement, from his own point of view, the *other* observer, watching him in action, considers that he made the most elementary blunder: He used two unsynchronized clocks. That's how, in spite of the fact that "*his*" clocks "*actually*" run slow, he manages to conclude that "*mine*" are running slow.

Because moving clocks are not synchronized, it is essential when checking time dilation to focus attention on a *single* moving clock. *All* moving clocks run slow by

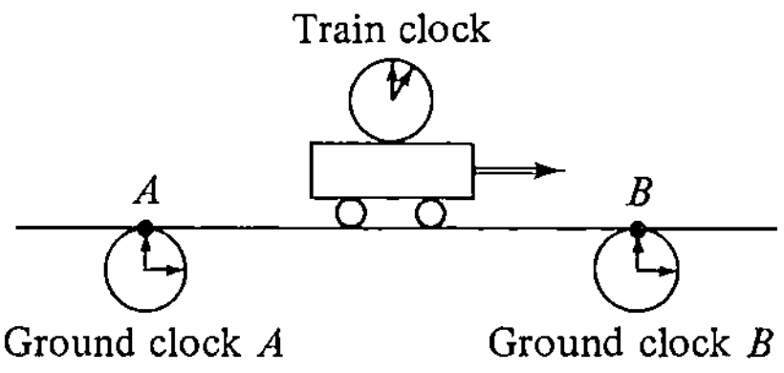

Figure 10.9

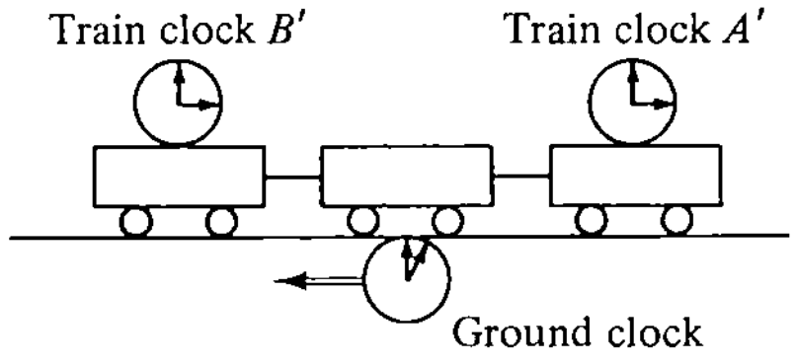

Figure 10.10

the same factor, but you can't start the timing on one clock and then switch to another because they weren't in step to begin with. But you can use as many *stationary* clocks (stationary with respect to you, the observer) as you please, for they *are* properly synchronized (a moving observer would dispute this, but that's *his* problem).

Example 2 The twin paradox

On her 21st birthday, an astronaut takes off in a rocket ship at a speed of $\frac{12}{13}c$. After 5 years have elapsed on her watch, she turns around and heads back at the same speed to rejoin her twin brother, who stayed at home. *Question*: How old is each twin at their reunion?

Solution: The traveling twin has aged 10 years (5 years out, 5 years back); she arrives home just in time to celebrate her 31st birthday. However, as viewed from earth, the moving clock has been running slow by a factor

$$\gamma = \frac{1}{\sqrt{1-(\frac{12}{13})^2}} = \frac{13}{5}$$

The time elapsed on earthbound clocks is $\frac{13}{5} \times 10 = 26$, and her brother will therefore be celebrating his 47th birthday—he is now 16 years older than his twin sister! But don't be deceived: This is no fountain of youth for the traveling twin, for though she may die later than her brother, she will not have lived any *more*—she's just done it *slower*. During the flight, all her biological processes—metabolism, pulse, thought, and speech—are subject to the same time dilation that affects her watch.

The so-called **twin paradox** arises when you try to tell this story from the point of view of the *traveling* twin. She sees the *earth* fly off at $\frac{12}{13}c$, turn around after 5 years, and return. From her point of view, it would seem, *she's* at rest, whereas her *brother* is in motion, and hence it is *he* who should be younger at the reunion. An enormous amount has been written about the twin paradox, but the truth is there's really no paradox here at all: This second analysis is simply *wrong*. The two twins are *not* equivalent. The traveling twin experiences *acceleration* when she turns around to head home, but her brother does *not*. To put it in fancier language, the traveling twin is not in an inertial system—more precisely, she's in *one* inertial system on the way out and a completely different one on the way back. I'll show you in Section 10.1.3 how to analyze this problem *correctly* from her point of view, but as far as the resolution of the "paradox" is concerned, it is enough to note that the *traveling twin cannot claim to be a stationary observer* because you can't undergo acceleration and remain stationary.

Problem 10.5 In a laboratory experiment a muon is observed to travel 800 m before disintegrating. A graduate student looks up the lifetime of a muon (2×10^{-6} s) and concludes that its speed was

$$v = \frac{800 \text{ m}}{2 \times 10^{-6} \text{ s}} = 4 \times 10^8 \text{ m/s}$$

Faster than light! Identify the student's error, and find the *actual* speed of this muon.

Problem 10.6 A rocket ship leaves earth at a speed of $\frac{3}{5}c$. When a clock on the rocket says 1 hour has elapsed, the rocket sends a light signal back to earth.

(a) According to *earth* clocks, when was the signal sent?

(b) According to *earth* clocks, how long after the rocket left did the signal arrive back on earth?

(c) According to the *rocket* observer, how long after the rocket left did the signal arrive back on earth?

(iii) Lorentz contraction. For the third gedanken experiment you must imagine that we have set up a lamp at one end of the boxcar and a mirror at the other, so that a light signal can be sent down and back (Fig. 10.11). *Question*: How long does the signal take to complete the round trip? To an observer on the train, the answer is

$$\Delta t' = 2\frac{\Delta x'}{c} \tag{10.11}$$

where $\Delta x'$ is the length of the car, and primes, as before, denote measurements made on the train. To an observer on the ground the process is more complicated because of the motion of the train. If Δt_1 is the time for the light signal to reach the front end and Δt_2 is the return time, then (see Fig. 10.12):

$$\Delta t_1 = \frac{\Delta x + v\Delta t_1}{c}, \qquad \Delta t_2 = \frac{\Delta x - v\Delta t_2}{c}$$

or, solving for Δt_1 and Δt_2:

$$\Delta t_1 = \frac{\Delta x}{c - v}, \qquad \Delta t_2 = \frac{\Delta x}{c + v}$$

So the round-trip time is

$$\Delta t = \Delta t_1 + \Delta t_2 = 2\frac{\Delta x}{c}\frac{1}{(1 - v^2/c^2)} \tag{10.12}$$

Meanwhile, these same intervals are related by the time dilation formula, (10.9):

$$\Delta t' = \sqrt{(1 - v^2/c^2)}\,\Delta t$$

Applying this to (10.11) and (10.12), I conclude that

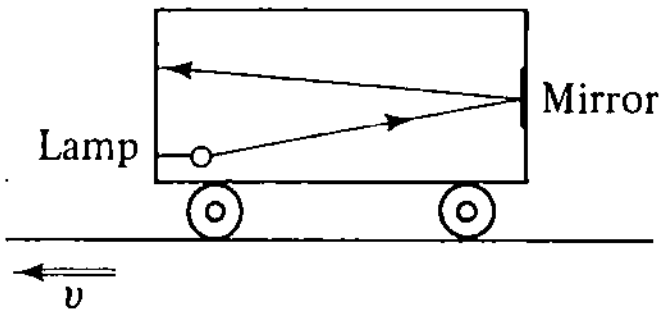

Figure 10.11

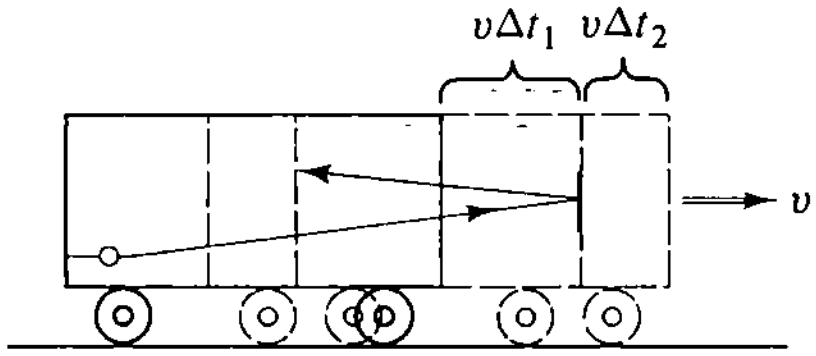

Figure 10.12

$$\boxed{\Delta x' = \frac{1}{\sqrt{1 - v^2/c^2}}\,\Delta x} \tag{10.13}$$

The length of the boxcar is not the same when measured by an observer on the ground, as it is when measured by an observer on the train—from the ground point of view it is somewhat *shorter*. *Conclusion*:

Moving objects are shortened.

We call this **Lorentz contraction.** Notice that the same factor,

$$\gamma = \frac{1}{\sqrt{1 - v^2/c^2}}$$

appears in both the time dilation formula and the Lorentz contraction formula. This makes it all very easy to remember: Moving clocks run slow, moving sticks are shortened, and the factor is always γ.

Of course, the person on the train doesn't think his car is shortened—his meter sticks are shortened by the same factor, so all his measurements come out the same as when the train was standing in the station. In fact, from *his* point of view, it is objects on the *ground* that are shortened. This raises again a paradoxical problem: If A says B's sticks are short, and B says A's sticks are short, who is right? *Answer*: They *both* are! But to reconcile the rival claims we must study carefully the actual process by which length is measured. Suppose you want to find the length of a board. If it's at rest (with respect to you) you simply lay your ruler down next to the board and record the reading at each end. The difference is the length of the board (Fig. 10.13). (If you're very clever, you will line up the left end of the ruler against the left end of the board—then you only have to read *one* number.) But what if the board is moving? Same story, only this time, of course, you must take care to read the two ends *at the same instant of time.* If you don't, the board will move in the course of the measurement, and obviously you'll get the wrong answer. But therein lies the problem: Because of the relativity of simultaneity the two observers disagree as to what constitutes "the same instant of time." When the person on the ground measures the length of the boxcar, he reads the positions of the two ends at the same instant *in his system.* But the person on the train, watching him do it, complains that he read the front end first, then waited a moment before reading the back end. *Naturally,* he came out short, in spite of the fact that (to the train observer) he was using undersized meter sticks, which would otherwise have yielded a number too *large.* Each observer measures lengths correctly (from the point of view of his own inertial system) and each finds the other's sticks to be shortened. Yet there is no inconsistency, for they are measuring in different ways, and each considers the other's method improper.

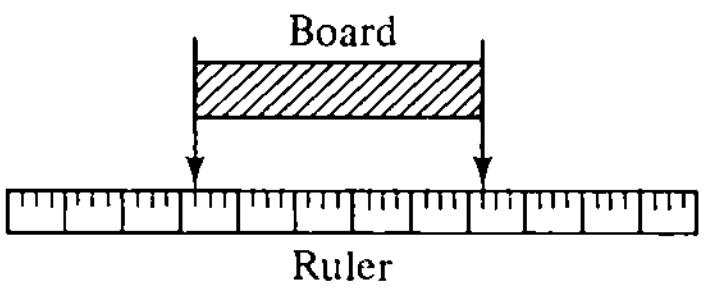

Figure 10.13

Example 3 The barn and ladder paradox

Unlike time dilation, there is no direct experimental confirmation of Lorentz contraction, simply because it's too difficult to get an object of measurable size going anywhere near the speed of light. The following parable illustrates how bizarre the world would be if the speed of light were more accessible.

There was once a farmer who had a ladder too long to store in his barn (Fig. 10.14(a)). He chanced one day to read some relativity, and a solution to his problem suggested itself. He instructed his son to run with the ladder as fast as he could—the moving ladder having Lorentz-contracted to a size the barn could easily accommodate, the son was to rush through the door, whereupon the farmer would slam it behind him, capturing the ladder inside (Fig. 10.14(b)). The son, however, was a lazy fellow, and moreover he had read somewhat farther in the relativity book. He argued that from *his* point of view the *barn,* not the ladder, would contract, and the fit would be even worse than it was with the two at rest (Fig. 10.14(c)).

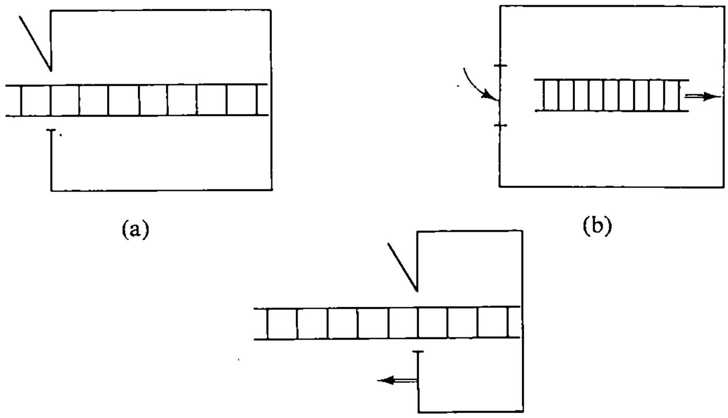

(a) (b) (c)

Figure 10.14

Question: Who's right? Will the ladder fit inside the barn, or won't it?

Answer: They're *both* right! When you say "the ladder is in the barn", you mean that all parts of it are inside *at one instant of time,* but in view of the relativity of simultaneity, *that's* a condition that depends on the observer. Think of it this way: There are really *two* relevant events here:

A. Back end of ladder makes it in the door;
B. Front end of ladder hits far wall of barn.

The farmer says (A) occurs before (B), so there *is* a time when the ladder is entirely within the barn; the son says (B) precedes (A), so there is *not. Contradiction?* Nope—just a difference in perspective.

"But *come* now," you say, "when it's all over and the dust clears, either the ladder is inside the barn, or it isn't. There can be no dispute about *that.*" Quite so, but now you're introducing a new element into the story: What happens *as the ladder is brought to a stop?* Suppose the farmer grabs the last rung of the ladder firmly with one hand, while he slams the door with the other. Assuming it remains intact, the ladder must now stretch out to its rest length. Evidently, the front end keeps going, even after the rear end has been stopped! Expanding like a telescope, the front end of the ladder smashes into the far side of the barn. (The whole notion of a rigid object loses its meaning in relativity, for when it changes its speed, different parts do not in general accelerate simultaneously—in this way the material stretches or shrinks to reach the length appropriate to its new velocity.) But to return to the question at hand: When the ladder finally comes to a stop, is it inside the barn or not? The answer is indeterminate. When the front end of the ladder hits the far side of the barn, something has to give, and the farmer is left either with a broken ladder inside the barn or with the ladder intact poking through a hole in the wall. In any event, he is unlikely to be pleased with the outcome.

One final matter concerning Lorentz contraction. A moving object is shortened *only along the direction of its motion:*

Dimensions perpendicular to the velocity are not contracted.

Indeed, in deriving the time dilation formula I took it for granted that the *height* of the train is the same for both observers. I'll now justify this, using a lovely gedanken experiment presented by Taylor and Wheeler.[6] Imagine that we build a wall beside the railroad tracks, and 1 m above the rails (*as measured on the ground*) we paint a horizontal blue line. When the train goes by, a passenger leans out the window holding a wet paintbrush 1 m above the rails *as measured on the train,* leaving a horizontal *red* line on the wall. *Question*: Does the passenger's red line lie above or below our blue one? If the rule were that perpendicular dimensions contract, then the person on the ground would predict that the *red* line is lower, while the one on the train would say it's the *blue* one (to the latter, of course, the *ground* is moving). The principle of relativity says that both observers are equally justified, but they cannot both be right. No subtleties of simultaneity or synchronization can rationalize this contradiction; either the blue line is higher or the red one is—*unless they exactly coincide,* which is the inescapable conclusion. There *cannot* be a law of contraction (or, for that matter, *expansion*) of perpendicular dimensions, for it would lead to irreconcilably inconsistent predictions.

Problem 10.7 A Lincoln Continental is twice as long as a VW Beetle, when they are at rest. As the Continental overtakes the VW, going through a speed trap, a (stationary) policeman observes that they both have the same length. The VW is going at half the speed of light. How fast is the Lincoln going? (Leave your answer as a multiple of c.)

Problem 10.8 A sailboat is manufactured so that the mast leans at an angle θ with respect to

[6] E. F. Taylor and J. A. Wheeler, *Spacetime Physics* (San Francisco: W. H. Freeman, 1966). A somewhat different version of the same argument is given in J. H. Smith, *Introduction to Special Relativity* (New York: Benjamin, 1967).

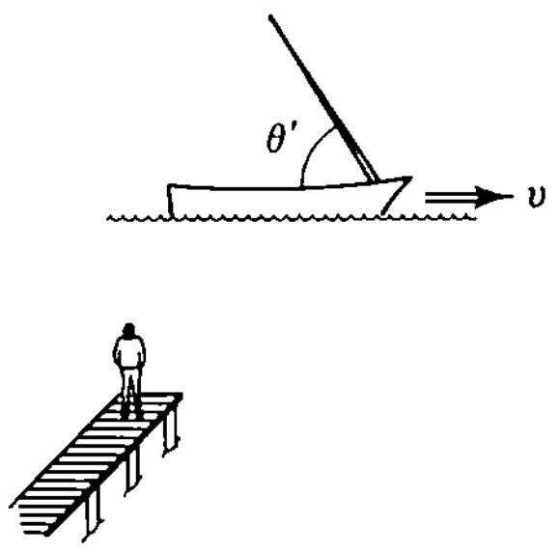

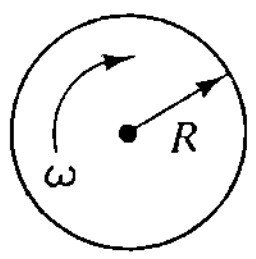

Figure 10.15 Figure 10.16

the deck. An observer standing on a dock sees the boat go by at speed v (Fig. 10.15). What angle does this *observer* say the mast makes?

Problem 10.9 A record turntable of radius R rotates at angular velocity ω (Fig. 10.16). The circumference is presumably Lorentz-contracted, but the radius (being perpendicular to the velocity) is *not*. What's the ratio of the circumference to the diameter, in terms of ω and R? According to the rules of ordinary geometry, that has to be π. What's going on here? (This is known as **Ehrenfest's paradox**; for discussion and references see H. Arzelies, *Relativistic Kinematics*, Chap. IX (Elmsford, N.Y.: Pergamon Press, Inc., 1966).)

10.1.3 The Lorentz Transformations

In the previous section I introduced Lorentz contraction, time dilation, and the relativity of simultaneity and synchronization by means of a series of parables. Notwithstanding their casual appearance, these gedanken experiments were skillfully designed by Einstein and others to make specific points while avoiding treacherous pitfalls. Each story involves the observation of some process from two different inertial systems—one which we chose to regard as "stationary" (the ground), the other "in motion" (the train). What I plan to do now is develop a systematic and rigorous machinery for comparing observations made in different inertial frames.

Every physical process consists of one or more **events.** An "event," in the sense that we shall use the word, is something that takes place at a specific location (x, y, z), at a precise time (t). The explosion of a firecracker, for example, is an event; a trip to the moon is not. Suppose, now, that we know the coordinates (x, y, z, t) of a particular event E in *one* inertial system S, and we would like to calculate the coordinates (x', y', z', t') of that *same event* in some other inertial system S', which is moving with respect to S. What we need is a "dictionary" for translating from the language of S to the language of S'.

We may as well choose our axes as shown in Fig. 10.17, so that S' slides along the x axis at speed v. If we "start the clock" ($t = 0$) at the moment the origins coincide, then at time t, O' will be a distance vt from O, and hence

$$x = d + vt \tag{10.14}$$

where d is the distance from O' to A' at time t. (A' is that point on the x' axis which is even with E when the event occurs.) Before Einstein, anyone would have said

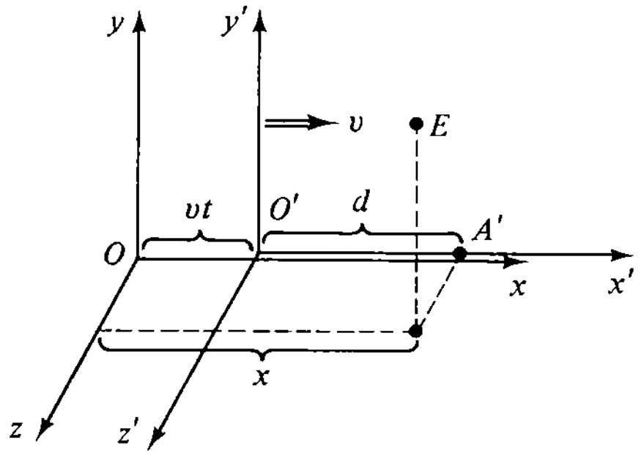

Figure 10.17

immediately that

$$d = x' \tag{10.15}$$

and thus constructed the "dictionary"

$$\left.\begin{aligned}&\text{(i)}\ \ x' = x - vt\\&\text{(ii)}\ \ y' = y\\&\text{(iii)}\ \ z' = z\\&\text{(iv)}\ \ t' = t\end{aligned}\right\} \tag{10.16}$$

These are now called the **Galilean transformations**, though they scarcely deserve so fine a title—the last one, in particular, went without saying, since everyone assumed the flow of time was the same for all observers. In the context of special relativity, however, we must expect (iv) to be replaced by a rule that incorporates time dilation, the relativity of simultaneity, and the nonsynchronization of moving clocks. Likewise, there will be a modification in (i) to account for Lorentz contraction. As for (ii) and (iii), they, at least, remain unchanged, for we have already seen that there can be no contraction of lengths perpendicular to the motion.

But where does the classical derivation of (i) break down? *Answer*: In equation (10.15). For d is the distance from O' to A' *as measured in* S, whereas x' is the distance from O' to A' *as measured in* S'. Because O' and A' are at rest in S', x' is the "moving stick," which appears contracted to S:

$$d = \frac{1}{\gamma}x' \tag{10.17}$$

When this is inserted in (10.14), we obtain the relativistic version of (i):

$$x' = \gamma(x - vt) \tag{10.18}$$

Of course, we could have run the same argument from the point of view of S'. The diagram (Fig. 10.18) looks similar, but in this case it depicts the scene *at time* t', whereas Fig. 10.17 showed the scene *at time* t. (t and t' represent the same physical instant *at* E, but not at other places, because of the relativity of simultaneity.) If we assume that S' also starts his clock when the origins coincide, then at time t', O will

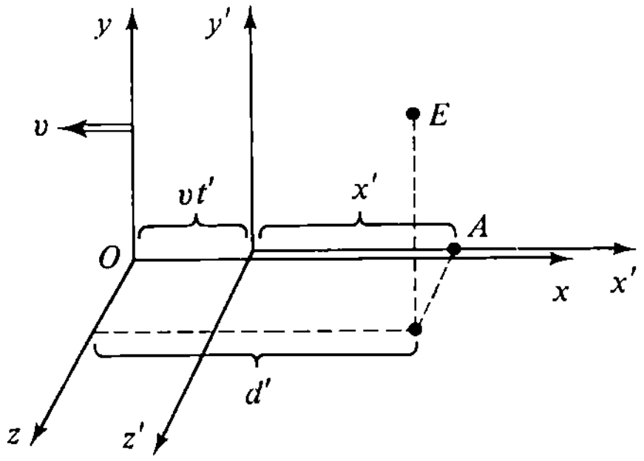

Figure 10.18

be a distance vt' from O', and so

$$x' = d' - vt' \tag{10.19}$$

where d' is the distance from O to A at time t', and A is that point on the x axis which is even with E when the event occurs. The classical physicist would have said that $x = d'$, and (using (iv)) recovered (i). But, as before, relativity demands that we observe a subtle distinction: x is the distance from O to A *in S*, whereas d' is the distance from O to A *in S'*. Because O and A are at rest in S, x is the "moving stick," and

$$d' = \frac{1}{\gamma}x \tag{10.20}$$

It follows that

$$x = \gamma(x' + vt') \tag{10.21}$$

This last equation comes as no surprise, for the symmetry of the situation dictates that the formula for x, in terms of x' and t', should be identical to the formula for x' in terms of x and t (equation (10.18)), except for a switch in the sign of v. (If S' is going to the *right* at speed v, with respect to S, then S is going to the *left* at speed v, with respect to S'.) Nevertheless, this is a useful result, for if we substitute x' from equation (10.18) and solve for t', we complete the relativistic "dictionary":

$$\boxed{\begin{aligned} &\text{(i)}\ \ x' = \gamma(x - vt) \\ &\text{(ii)}\ \ y' = y \\ &\text{(iii)}\ z' = z \\ &\text{(iv)}\ t' = \gamma\left(t - \frac{v}{c^2}x\right) \end{aligned}} \tag{10.22}$$

These are the famous **Lorentz transformations,** with which Einstein replaced the Galilean ones. They contain all of the kinematic information in the special theory, as the following examples illustrate. The reverse dictionary, which carries you from S' back to S, can be obtained algebraically by solving (i) and (iv) for x and t, or, more simply, by switching the sign of v:

$$\left.\begin{array}{ll} \text{(i}') & x = \gamma(x' + vt') \\ \text{(ii}') & y = y' \\ \text{(iii}') & z = z' \\ \text{(iv}') & t = \gamma\left(t' + \dfrac{v}{c^2}x'\right) \end{array}\right\} \quad (10.23)$$

Example 4 Simultaneity, synchronization, and time dilation

Suppose event A occurs at $x = 0$, $t = 0$, and event B occurs at $x = a$, $t = 0$. The two events are simultaneous in S: they both take place at the same instant, $t = 0$. But they are *not* simultaneous in S', for according to (10.22) the coordinates of A and B in S' are $x' = 0$, $t' = 0$ and $x' = \gamma a$, $t' = -\gamma(v/c^2)a$, respectively. According to the S' clocks, B occurred *before* A. This is nothing *new*, of course—just the relativity of simultaneity. But I wanted you to see how it follows from the Lorentz transformations.

Now, suppose that at time $t = 0$ observer S decides to examine *all* the clocks in S'. He finds that they read *different* times, depending on their location, to wit, from (iv):

$$t' = -\gamma \frac{v}{c^2} x$$

Those to the left of the origin (negative x) are *ahead*, and those to the right are *behind*, by an amount that increases in proportion to their distance (Fig. 10.19). Only the master clock at the origin reads $t' = 0$. Thus, the nonsynchronization of moving clocks, too, follows directly from the Lorentz transformations. Of course, from the S' viewpoint it is the S clocks which are out of synchronization, as you can check by putting $t' = 0$ into equation (iv').

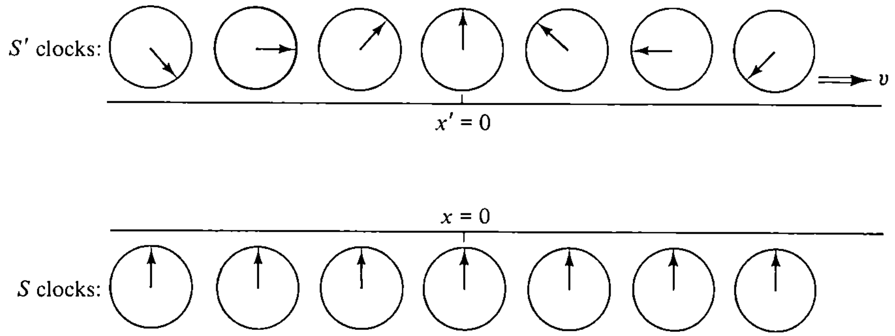

Figure 10.19

Finally, suppose S focuses his attention on a single clock in the S' frame (say, the one at $x' = b$) and watches it over some interval Δt. How much time passes on the moving clock? Because x' is fixed, (iv′) gives $\Delta t = \gamma \Delta t'$, or

$$\Delta t' = \frac{1}{\gamma} \Delta t$$

That's the old time dilation formula, derived now from the Lorentz transformations. Please note that it's x' we hold fixed, here, because we're watching *one moving clock*. If you hold x fixed, then you're watching a whole series of different S' clocks as they pass by, and that won't tell you whether any one of them is running slow.

Example 5 Lorentz contraction

Imagine a stick moving to the right at speed v. Its rest length (that is, its length as measured in S') is $\Delta x' = x_r' - x_l'$, where the subscripts denote the right and left ends of the stick. If an observer in S were to measure the stick, he would subtract the positions of the two ends *at one instant of time t*: $\Delta x = x_r - x_l$. According to (i), then,

$$\Delta x = \frac{1}{\gamma} \Delta x'$$

This is the old Lorentz contraction formula, derived now from the Lorentz transformations. Please note that it's t we hold fixed, here, because we're talking about a measurement made by S, and he marks off the two ends at the same instant of his time. (S' doesn't have to be so fussy, since the stick is at rest in his frame.)

Example 6 Einstein's velocity addition rule

Suppose a particle moves a distance dx (in S) in a time dt. Its velocity u is then

$$u = \frac{dx}{dt}$$

In S', meanwhile, it has moved a distance

$$dx' = \gamma(dx - v\,dt)$$

(as we see from (i) in equation (10.22)), in a time given by (iv):

$$dt' = \gamma\left(dt - \frac{v}{c^2}\,dx\right)$$

The velocity in S' is therefore

$$u' = \frac{dx'}{dt'} = \frac{\gamma(dx - v\,dt)}{\gamma\left(dt - \dfrac{v}{c^2}\,dx\right)} = \frac{\left(\dfrac{dx}{dt} - v\right)}{1 - \dfrac{v}{c^2}\dfrac{dx}{dt}} = \frac{u - v}{1 - \dfrac{uv}{c^2}} \tag{10.24}$$

If the derivation seems slippery to you, think of it this way: There are *two* events involved in describing the motion,

1. (x_i, t_i): particle leaves initial point
2. (x_f, t_f): particle arrives at final point

The velocity is

$$u = \frac{x_f - x_i}{t_f - t_i} = \frac{dx}{dt}$$

In S' the same two events are labeled by (x_i', t_i') and (x_f', t_f'), and the velocity is

$$u' = \frac{x_f' - x_i'}{t_f' - t_i'} = \frac{dx'}{dt'}$$

The primed coordinates of each event are now related to the unprimed coordinates by Lorentz transformations and we obtain (10.24).)

Equation (10.24) is the **Einstein velocity addition rule,** derived now from the Lorentz transformations. To recover the more transparent notation of equation (10.7), let A be the particle, B be S, and C be S'; then $u = v_{AB}$, $u' = v_{AC}$, and $v = v_{CB} = -v_{BC}$, so (10.24) becomes

$$v_{AC} = \frac{v_{AB} + v_{BC}}{1 + \dfrac{v_{AB}v_{BC}}{c^2}} \tag{10.7}$$

as before.

Problem 10.10 Solve equations (10.22) for x, y, z, t in terms of x', y', z', t', and check that you recover (10.23).

Problem 10.11 Sophie Zabar, clairvoyante, cried out in pain at precisely the instant her twin brother, 500 km away, hit his thumb with a hammer. A skeptical scientist observed both events (brother's accident, Sophie's cry) from an airplane traveling at $\frac{12}{13}c$ to the right (see Fig. 10.20). Which event occurred first, according to the scientist? How *much* earlier was it, in seconds?

Problem 10.12

(a) In Example 6 we found how velocities *in the x-direction* transform when you go from

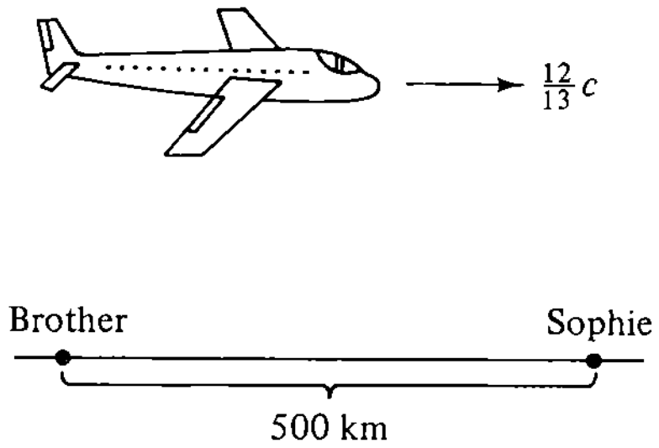

Figure 10.20

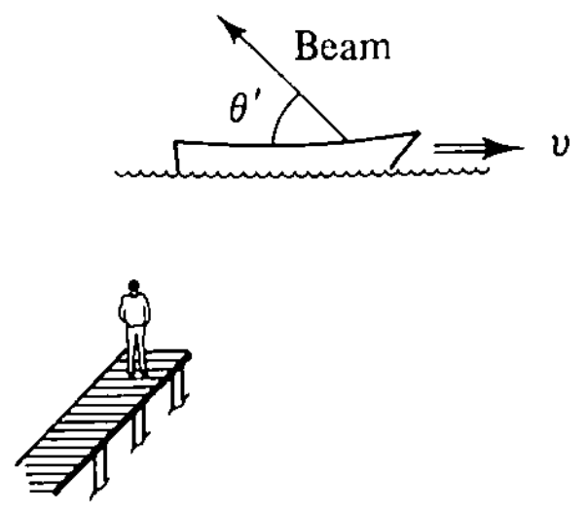

Figure 10.21

S to S'. Derive the analogous formulas for velocities in the y- and z-directions (that is, directions perpendicular to the relative motion of S and S').

(b) A spotlight is mounted on a boat so that its beam makes an angle θ with the deck (Fig. 10.21). If this boat is then set in motion at speed v, what angle θ' does an observer on the *dock* say the beam makes with the deck? Compare Problem 10.8, and explain the difference.

Problem 10.13 You probably did Problem 10.2 from the point of view of an observer on the *ground.* Now do it from the point of view of the police car, the outlaws, and the bullet. That is, fill in the gaps in the following table:

Relative to ↓ / Speed of →	GROUND	POLICE	OUTLAWS	BULLET	DO THEY ESCAPE?
GROUND	0	$\frac{1}{2}c$	$\frac{3}{4}c$		
POLICE				$\frac{1}{3}c$	
OUTLAWS					
BULLET					

Problem 10.14 (Twin paradox reconsidered). On their 21st birthday, one twin gets on a moving sidewalk, which carries her out to star X at a speed $\frac{4}{5}c$; the other twin stays home. When the traveling twin gets to star X, she immediately jumps onto the returning moving sidewalk and comes back to earth, again at speed $\frac{4}{5}c$. She arrives on her 39th birthday (as determined by *her* watch).

(a) How old is her twin brother (who stayed at home)?

(b) How far away is star X? (Give your answer in light years.)

Call the outbound sidewalk system S' and the inbound one S'' (the earth system is S). All three systems set their master clocks, and choose their origins, so that $x = x' = x'' = 0$, $t = t' = t'' = 0$ at the moment of departure.

(c) What are the coordinates (x, t) of the jump (from outbound to inbound sidewalk) in S?

(d) What are the coordinates (x', t') of the jump in S'?

(e) What are the coordinates (x'', t'') of the jump in S''?

(f) If the traveling twin wants her watch to agree with the clocks in S'', what must she do to her watch immediately after the jump? If she *does* this, what will her watch read when she gets home? (This won't change her *age,* of course—she's still 39—it'll just make her watch agree with the standard synchronization in S''.)

(g) If the traveling twin is asked the question, "How old is your brother *right now,* and which of you is younger?", what is the correct reply (i) just *before* she makes the jump, (ii) just *after* she makes the jump? (Nothing dramatic happens to her brother during the split second between (i) and (ii), of course—what *does* change radically is his sister's notion of what "right now" means.)

10.1.4 The Structure of Spacetime

(i) Four-vectors. The Lorentz transformations take on a simpler appearance when expressed in terms of the new quantities

$$x^0 = ct, \qquad \beta = \frac{v}{c} \tag{10.25}$$

Using x^0 (instead of t) and β (instead of v) amounts to changing the unit of time from the *second* to the *meter*—1 meter of x^0 corresponds to the time it takes light to travel 1 meter (in vacuum). If, at the same time, we number the x, y, z coordinates, so that

$$x^1 = x, \qquad x^2 = y, \qquad x^3 = z \tag{10.26}$$

then the Lorentz transformations read

$$\left.\begin{aligned} (x^0)' &= \gamma(x^0 - \beta x^1) \\ (x^1)' &= \gamma(x^1 - \beta x^0) \\ (x^2)' &= x^2 \\ (x^3)' &= x^3 \end{aligned}\right\} \tag{10.27}$$

Or, in matrix form:

$$\begin{pmatrix} x^{0\prime} \\ x^{1\prime} \\ x^{2\prime} \\ x^{3\prime} \end{pmatrix} = \begin{pmatrix} \gamma & -\gamma\beta & 0 & 0 \\ -\gamma\beta & \gamma & 0 & 0 \\ 0 & 0 & 1 & 0 \\ 0 & 0 & 0 & 1 \end{pmatrix} \begin{pmatrix} x^0 \\ x^1 \\ x^2 \\ x^3 \end{pmatrix} \tag{10.28}$$

Letting the Greek indices run from 0 to 3, this can be distilled into a single equation:

$$(x^\mu)' = \sum_{\nu=0}^{3} (\Lambda^\mu_\nu) x^\nu \tag{10.29}$$

where Λ is the **Lorentz transformation matrix** in equation (10.28) (the superscript μ labels the row, the subscript ν labels the column). One virtue of writing things in this abstract manner is that we can handle in the same format a more general transformation, in which the relative motion is *not* along a common x-x' axis; the matrix Λ would be more complicated, but the structure of equation (10.29) remains unchanged.

If this reminds you of the *rotations* we studied in Chapter 1, it's no accident.

There we were concerned with the change in components when you switch to a *rotated* coordinate system; here we are interested in the change of components when you go to a *moving* system. The former is governed by a rotation matrix R, which mixes together the three spatial terms (x^1, x^2, x^3); the latter is governed by a Lorentz transformation matrix Λ, which reshuffles all four components (x^0, x^1, x^2, x^3). Indeed, we may now contemplate a combined transformation, incorporating both a rotation and a "pure" Lorentz transformation into a single 4×4 matrix Λ. In Chapter 1 we defined a (3-) vector as any set of three components which transform under rotations the same way (x^1, x^2, x^3) do; by extension, we now define a **4-vector** as any set of *four* components which transform in the same manner as (x^0, x^1, x^2, x^3) under generalized Lorentz transformations:

$$(a^\mu)' = \sum_{\nu=0}^{3} \Lambda^\mu_\nu a^\nu \tag{10.30}$$

For the particular case of a transformation along the x axis:

$$\left.\begin{aligned} (a^0)' &= \gamma(a^0 - \beta a^1) \\ (a^1)' &= \gamma(a^1 - \beta a^0) \\ (a^2)' &= a^2 \\ (a^3)' &= a^3 \end{aligned}\right\} \tag{10.31}$$

There is a 4-vector analog to the scalar product $(\mathbf{A} \cdot \mathbf{B}) = A_x B_x + A_y B_y + A_z B_z$, but it's not just the sum of the products of like components. Rather, the zeroth components have a minus sign:

$$-a^0 b^0 + a^1 b^1 + a^2 b^2 + a^3 b^3$$

You should check for yourself (Problem 10.15) that the resulting expression has the same value in the primed system:

$$-a^0 b^0 + a^1 b^1 + a^2 b^2 + a^3 b^3 = -a^{0\prime} b^{0\prime} + a^{1\prime} b^{1\prime} + a^{2\prime} b^{2\prime} + a^{3\prime} b^{3\prime} \tag{10.32}$$

Just as the ordinary dot product is *invariant* (unchanged) under rotations, this combination is invariant under Lorentz transformations. To keep track of the minus sign it is convenient to introduce the **covariant** vector a_μ, which differs from the **contravariant** a^μ only in the sign of the zeroth component:

$$a_\mu = (a_0, a_1, a_2, a_3) = (-a^0, a^1, a^2, a^3) \tag{10.33}$$

(You must be scrupulously careful about the placement of indices in this business: *Upper* indices designate *contravariant* vectors; *lower* indices are for *covariant* vectors. Raising or lowering the temporal index costs a minus sign ($a_0 = -a^0$); raising or lowering a spatial index changes nothing ($a_1 = a^1$, $a_2 = a^2$, $a_3 = a^3$).) The scalar product can now be written with the summation symbol,

$$\sum_{\mu=0}^{3} a_\mu b^\mu \tag{10.34}$$

or, more compactly still,

$$a_\mu b^\mu \tag{10.35}$$

Summation is implied whenever a Greek index is repeated in a product—once as a covariant index and once as contravariant. This is called the **Einstein summation convention,** after its inventor, who regarded it as one of his most important contributions. Of course, we could as well take care of the minus sign by switching to covariant b:

$$a_\mu b^\mu = a^\mu b_\mu = -a^0b^0 + a^1b^1 + a^2b^2 + a^3b^3 \tag{10.36}$$

• **Problem 10.15** Check equation 10.32, using 10.31. (This only proves the invariance of the scalar product for transformations along the x-direction. But (10.32) is also invariant under *rotations,* since the first term is not affected at all, and the last three constitute the dot product $\mathbf{a}\cdot\mathbf{b}$. By a suitable rotation, the x-direction can be aimed any way you please, so (10.32) is actually invariant under *any* Lorentz transformation.)

Problem 10.16

(a) Write out the matrix that describes a *Galilean* transformation (10.16).

(b) Write out the matrix describing a Lorentz transformation along the y axis.

(c) Find the matrix describing a Lorentz transformation with velocity v along the x axis followed by a Lorentz transformation with velocity $\bar{v}$ along the y axis. Does it matter in what order the transformations are carried out?

Problem 10.17 The parallel between rotations and Lorentz transformation is even more striking if we define

$$\frac{v}{c} = \tanh\theta \tag{10.37}$$

(a) Write the Lorentz transformation matrix Λ (equation (10.28)) in terms of θ, and compare the rotation matrix (equation (1.19)). The quantity θ is called **rapidity;** in some respects it is a more natural way to describe the motion than the velocity v. For one thing, it ranges from $-\infty$ to $+\infty$, instead of $-c$ to $+c$. More significantly, rapidities add, whereas velocities do not:

(b) Express the Einstein velocity addition law in terms of rapidity.

(See E. F. Taylor and J. A. Wheeler, *Spacetime Physics* (San Francisco: W. H. Freeman, 1966) for more on rapidity.)

(ii) The invariant interval. Suppose event A occurs at $(x_A^0, x_A^1, x_A^2, x_A^3)$, and event B at $(x_B^0, x_B^1, x_B^2, x_B^3)$. The difference,

$$\Delta x^\mu = x_A^\mu - x_B^\mu \tag{10.38}$$

is a 4-vector—the **displacement 4-vector.** The invariant product of Δx^μ with itself is a quantity of special importance; we call it the **interval** between the two events:

$$\begin{aligned} I &= (\Delta x)_\mu (\Delta x)^\mu \\ &= -(\Delta x^0)^2 + (\Delta x^1)^2 + (\Delta x^2)^2 + (\Delta x^3)^2 = -c^2t^2 + d^2 \end{aligned} \tag{10.39}$$

where t is the time interval between the two events and d their spatial separation.

When you transform to a moving system, the *time* between A and B is altered ($t' \neq t$), and so is the *spatial separation* ($d' \neq d$), but the interval I remains the same.

Notice that, depending on the two events in question, the interval can be positive, negative, or zero:

1. If $I < 0$ we call the interval **timelike,** for this is the sign we get when the two occur at the *same place* ($d = 0$), and are separated only temporally.
2. If $I > 0$ we call the interval **spacelike,** for this is the sign we get when the two occur at the *same time* ($t = 0$) and are separated only spatially.
3. If $I = 0$ we call the interval **lightlike,** for this is the relation that holds when the two events are connected by a signal traveling at the speed of light.

In fact, if the interval between two events is timelike, there exists an inertial system (accessible by Lorentz transformation) in which they occur at the same point. For if I hop on a train going from (A) to (B) at the speed $v = d/t$, leaving event A when it occurs, I shall be just in time to pass B when *it* occurs. In the train system, A and B take place at the same point. You cannot do this for a *spacelike* interval, of course, because v would have to be greater than c, and no observer can exceed the speed of light (if he did, γ would be imaginary and the Lorentz transformations would be nonsense). On the other hand, if the interval is spacelike then there exists a system in which the two events occur at the same time (see Problem 10.19).

Problem 10.18

(a) Event A happens at point ($x = 5, y = 3, z = 0$) and at time t given by $ct = 15$; event B occurs at (10, 8, 0) and $ct = 5$, both in system S.

(i) What is the invariant interval between A and B?

(ii) Is there an inertial system in which they occur *simultaneously?* If so, find its velocity (magnitude and direction) relative to S.

(iii) Is there an inertial system in which they occur at the same point? If so, find its velocity relative to S.

(b) Repeat part a for $A = (2, 0, 0)$, $ct = 1$; and $B = (5, 0, 0)$, $ct = 3$.

Problem 10.19 The coordinates of event A are (x_A, t_A), and the coordinates of event B are (x_B, t_B). Assuming the interval between them is spacelike, find the velocity of the system in which they occur at the same time.

(iii) Space-time diagrams. If you wish to represent the motion of a particle graphically, the normal practice is to plot position versus time (that is, x runs vertically and t horizontally). On such a graph, the velocity can be read off as the slope of the curve. For some reason the convention is reversed in relativity: Everyone plots position horizontally and time (or, better, $x^0 = ct$) vertically. Velocity is then given by the *reciprocal* of the slope. A particle at rest is represented by a vertical line; a photon, traveling at the speed of light, is described by a 45° line; and a rocket going at some intermediate speed follows a line of slope $c/v = 1/\beta$. The trajectory of a particle on such a space-time diagram is called its **world line** (Fig. 10.22).

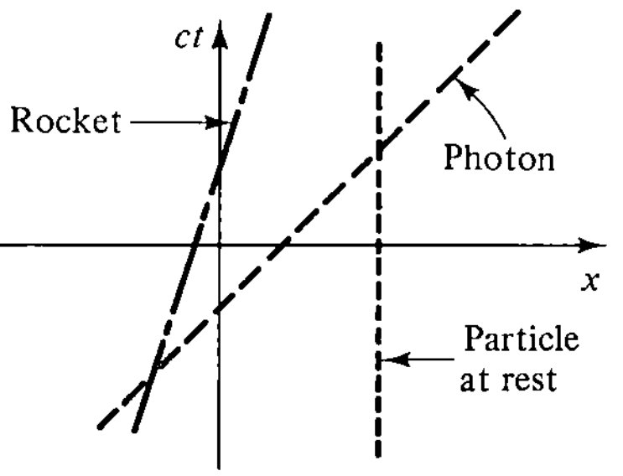

Figure 10.22

Suppose, now, that you set out from the origin, $x = 0$, at time $t = 0$. Because no material object can travel faster than light, your world line can never have a slope less than 1. Accordingly, your motion is restricted to the wedge-shaped region bounded by the two 45° lines. We call this your "future," in the sense that it is the locus of all points accessible to you when you first start out. Of course, as time goes on, and you move along your chosen world line, the options open to you progressively narrow: Your "future" at any moment is the forward "wedge" constructed at whatever point you find yourself (Fig. 10.23). Meanwhile, the *backward* wedge represents your "past," in the sense that it is the locus of all points from which you might have come. As for the rest (the region outside the forward and backward wedges) this is the generalized "present." You can't *get* there, and you didn't *come* from there. In fact, there's no way you can influence any event in the present (the message would have to travel faster than light)—it's a vast expanse of spacetime which is absolutely inaccessible to you.

I've been ignoring the y- and z-directions. If we include a y axis coming out of the page, the "wedges" become cones—and, with an undrawable z axis, hypercones. Because their boundaries are the trajectories of light rays, we call them the **forward light cone** and the **backward light cone.** Your future, in other words, lies within your forward light cone, your past within your backward light cone.

Notice that the slope of the line connecting two events on a space-time diagram tells you at a glance whether the invariant interval between them is timelike (slope

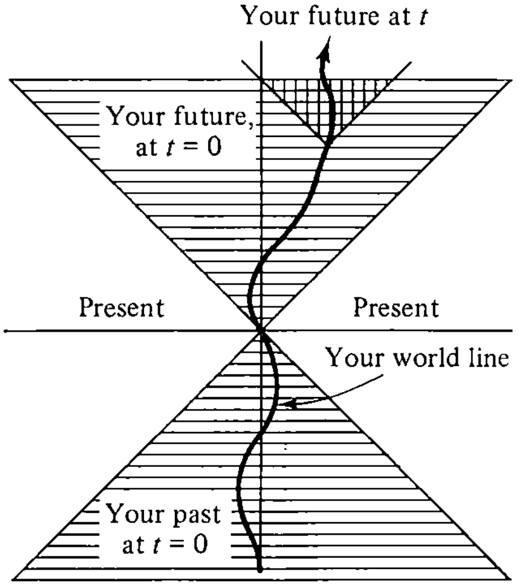

Figure 10.23

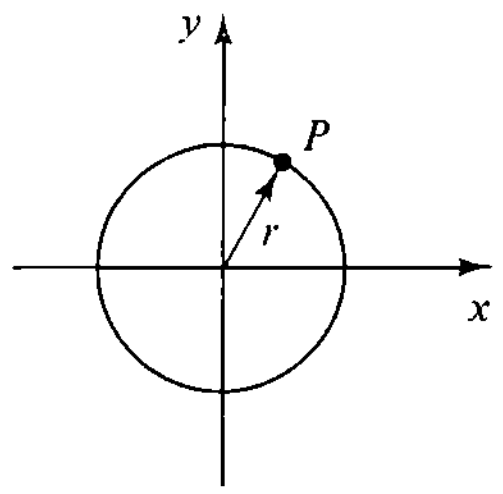

Figure 10.24

greater than 1), spacelike (slope less than 1), or lightlike (slope 1). For example, all points in the past and future are timelike with respect to your present location, whereas points in the present are spacelike, and points on the light cone are lightlike.

Hermann Minkowski, who was the first to recognize the full geometrical significance of special relativity, began a classic paper with the words, "Henceforth space by itself, and time by itself, are doomed to fade away into mere shadows, and only a kind of union of the two will preserve an independent reality." It is a lovely thought, but you must be careful not to read too much into it. For it is not at all the case that time is "just another coordinate, on the same footing with x, y, and z" (except that for obscure reasons we measure it on clocks instead of rulers). *No:* Time is *utterly different* from the others, and the mark of its distinction is precisely the minus sign in the invariant interval. That minus sign imparts to space-time a hyperbolic geometry that is much richer than the circular geometry of 3-space.

Under rotations about the z axis, a point P in the xy plane describes a *circle:* the locus of all points a fixed distance $r = \sqrt{x^2 + y^2}$ from the origin (Fig. 10.24). Under Lorentz transformations, however, it is the interval $I = (x^2 - c^2t^2)$ which is preserved, and the locus of all points with a given value of I is a *hyperbola*—or, if we include the y axis, a *hyperboloid of revolution.* When the interval is *timelike,* it's a "hyperboloid of two sheets" (Fig. 10.25(a)); when the interval is *spacelike,* it's a "hyperboloid of one sheet" (Fig. 10.25(b)). When you perform a Lorentz transformation (that is, when you go into a moving inertial system), the coordinates (x, t) of a given event will change to (x', t'), but these new coordinates *will lie on the same hyperbola* as (x, t). By appropriate combinations of Lorentz transformation and rotation, a

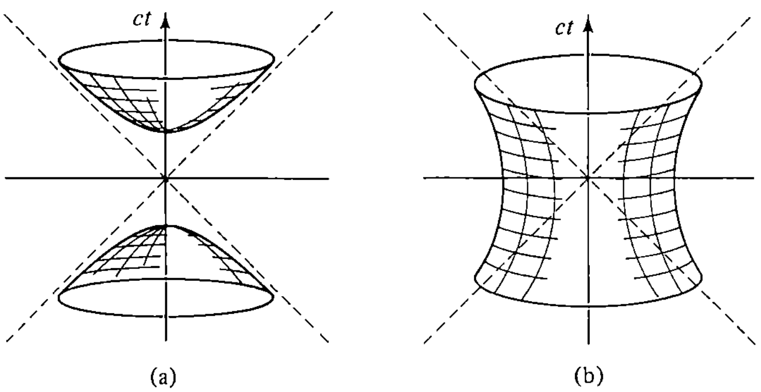

Figure 10.25

spot can be moved around at will over the surface of a given hyperboloid, but no amount of transformation will carry it, say, from the upper sheet of the timelike hyperboloid to the lower sheet or to a spacelike hyperboloid.

When we were discussing simultaneity I pointed out that the time ordering of two events can, at least in certain cases, be reversed, simply by going into a moving system. But we now see that this is not *always* possible: ***If the invariant interval between two events is timelike, then their ordering is absolute; if the interval is spacelike, their ordering depends on the inertial system from which they are observed.*** In terms of the space-time diagram, an event on the upper sheet of a timelike hyperboloid *definitely* occurred *after* (0, 0), and one on the lower sheet certainly occurred *before;* but an event on a spacelike hyperboloid occurred at positive t, or negative t, depending on your reference frame. This is not an idle curiosity, for it rescues the notion of causality, on which all physics is based. If it were *always* possible to reverse the order of two events, then we could never say "A caused B," since a rival observer would retort that *B preceded A*. This embarrassment is avoided, provided the two events are timelike-separated. And causally related events *are* timelike-separated, otherwise no influence could travel from one to the other. Conclusion: ***The invariant interval between causally related events is always timelike, and their temporal ordering is the same for all inertial observers.***

Problem 10.20

(a) Draw a space-time diagram representing a game of catch (or a conversation) between two people at rest, 10 ft apart. How is it possible for them to communicate, given that their separation is spacelike?

(b) There's an old limerick that runs as follows:

There once was a girl named Ms. Bright,
Who could travel much faster than light.
She departed one day,
The Einsteinian way,
And returned on the previous night.

What do you think? Even if she *could* travel faster than light, could she return before she set out? Draw a space-time diagram representing this trip.

Problem 10.21 Inertial system S' moves in the x-direction at speed $\frac{3}{5}c$ relative to system S. (The x' axis slides along the x axis, and the origins coincide at $t = t' = 0$, as usual.)

(a) On graph paper set up a Cartesian coordinate system with axes ct and x. Carefully draw in the lines representing $x' = -3, -2, -1, 0, 1, 2$, and 3. Also draw in the lines corresponding to $ct' = -3, -2, -1, 0, 1, 2$, and 3. Label your lines clearly.

(b) In S', a free particle is observed to travel from the point $x' = -2$ at time $ct' = -2$ to the point $x' = 2$ at $ct' = +3$. Indicate this displacement on your graph. From the slope of this line, determine the particle's speed in S.

(c) Use the velocity addition rule to determine the velocity in S algebraically, and check that your answer is consistent with the graphical solution in (b).

10.2 RELATIVISTIC MECHANICS

10.2.1 Proper Time and Proper Velocity

As you progress along your world line, your watch runs slow, relative to a fixed inertial system. The dilation factor is large or small in accordance with your speed[7] u at the moment in question—specifically, while the "stationary" clocks tick off an interval dt, your watch only ticks off $d\tau$:

$$d\tau = \sqrt{1 - u^2/c^2}\, dt \tag{10.40}$$

τ, the time your *own* watch registers, is called the **proper time.** (The word suggests a mistranslation of the French *propre,* meaning "own.") In some cases τ may be a more relevant or useful quantity than t. For one thing, the proper time is obviously invariant, whereas "ordinary" time depends on the particular reference frame you have in mind.

Now, imagine you're on a flight to Los Angeles, and the pilot announces that the plane's speed is $\frac{4}{5}c$. What precisely does he mean by "speed"? Well, of course, he means the distance traveled per unit time: As a vector,

$$\mathbf{u} = \frac{d\mathbf{l}}{dt} \tag{10.41}$$

and, since he is presumably talking about the speed relative to ground, *both* l *and* t *are to be measured by the ground observer.* That's the important number to know, if you're concerned about being on time for an appointment in Los Angeles, but if you're wondering whether you'll be hungry on arrival, you might be more interested in the distance covered per unit *proper* time:

$$\boldsymbol{\eta} = \frac{d\mathbf{l}}{d\tau} \tag{10.42}$$

This hybrid quantity—distance measured on the ground over time measured in the airplane—is called **proper velocity;** for contrast, I'll call **u** the **ordinary velocity.** The two are related: from (10.40), (10.41), and (10.42),

$$\boldsymbol{\eta} = \frac{1}{\sqrt{1 - u^2/c^2}}\,\mathbf{u} \tag{10.43}$$

For speeds much less than c, of course, the difference between ordinary and proper velocity is negligible.

From a theoretical standpoint, proper velocity has an enormous advantage over ordinary velocity: It transforms simply, when you go from one inertial system to another. In fact, $\boldsymbol{\eta}$ is the spatial part of a 4-vector,

$$\eta^\mu = \frac{dx^u}{d\tau} \tag{10.44}$$

[7]I'll use u for the velocity of a particular object and save v for the relative velocity of two inertial systems.

whose zeroth component is

$$\eta^0 = \frac{dx^0}{d\tau} = c\frac{dt}{d\tau} = \frac{c}{\sqrt{1 - u^2/c^2}} \tag{10.45}$$

For the numerator dx^μ is a displacement 4-vector, while the denominator $d\tau$ is invariant. Thus, for instance, when you go from system S to system S', moving at speed v along the common x-x' axis,

$$\left.\begin{aligned} \eta^{0\prime} &= \gamma(\eta^0 - \beta\eta^1) \\ \eta^{1\prime} &= \gamma(\eta^1 - \beta\eta^0) \\ \eta^{2\prime} &= \eta^2 \\ \eta^{3\prime} &= \eta^3 \end{aligned}\right\} \tag{10.46}$$

and, more generally,

$$\eta^{\mu\prime} = \Lambda^\mu_\nu \eta^\nu \tag{10.47}$$

η^μ is called the **proper velocity 4-vector** or, more simply, the **4-velocity.**

By contrast, the transformation rule for *ordinary* velocities is extremely cumbersome, as we found in Example 6 and Problem 10.12:

$$\left.\begin{aligned} u_x' &= \frac{dx'}{dt'} = \frac{u_x - v}{1 - vu_x/c^2} \\ u_y' &= \frac{dy'}{dt'} = \frac{1}{\gamma}\left(\frac{u_y}{1 - vu_x/c^2}\right) \\ u_z' &= \frac{dz'}{dt'} = \frac{1}{\gamma}\left(\frac{u_z}{1 - vu_x/c^2}\right) \end{aligned}\right\} \tag{10.48}$$

The *reason* for the added complexity in this case is plain: We're obliged to transform both the numerator *dl and the denominator dt*, whereas for *proper* velocity the denominator $d\tau$ is invariant, so the ratio inherits the transformation rule of the numerator alone.

Problem 10.22

(a) Equation (10.43) defines proper velocity in terms of ordinary velocity. Invert that equation to get a formula for **u** in terms of $\boldsymbol{\eta}$.

(b) What is the relation between proper velocity and *rapidity?* (Rapidity was defined in Problem 10.17.) Assume the velocity is along the x-direction, and find η as a function of θ.

Problem 10.23 A car is traveling along the 45° line as shown in Fig. 10.26, at (ordinary) speed $(2/\sqrt{5})c$.

(a) Find the components u_x and u_y of the (ordinary) velocity.

(b) Find the components η_x and η_y of the proper velocity.

(c) Find the zeroth component of the 4-velocity, η^0.

A system S' is moving in the x-direction with (ordinary) speed $\sqrt{2/5}\,c$. By using the appropriate transformation laws:

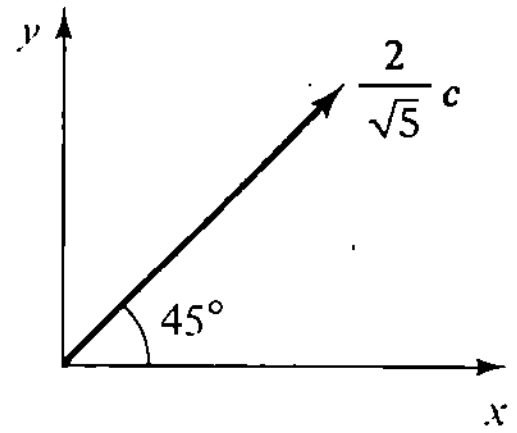

Figure 10.26

(d) Find the (ordinary) velocity components u_x' and u_y' in S'.
(e) Find the proper velocity components η_x' and η_y' in S'.
(f) As a consistency check, verify that

$$\eta' = \frac{\mathbf{u}'}{\sqrt{1 - u'^2/c^2}}$$

Problem 10.24 Find the invariant product of the 4-velocity with itself, $\eta^\mu \eta_\mu$.

Problem 10.25 For a particle in "hyperbolic" motion,

$$x(t) = \sqrt{b^2 + (ct)^2}, \qquad y = z = 0$$

(a) Find the proper time τ as a function of t, assuming the clocks are set so that $\tau = 0$ when $t = 0$. (*Note*: you must *integrate* equation (10.40).)
(b) Find x and v (ordinary velocity) as functions of τ.
(c) Find η^μ (proper velocity) as a function of t.

10.2.2 Relativistic Energy and Momentum

In classical mechanics momentum is defined as mass times velocity. I would like to extend this notion to the relativistic domain, but immediately a question arises: Should I use *ordinary* velocity or *proper* velocity? In classical physics τ and t are identical, so there is no *a priori* reason to favor one over the other. However, in the context of relativity it is essential that we use proper velocity, for the law of conservation of momentum would be inconsistent with the principle of relativity if we were to define momentum as $m\mathbf{u}$. Let me explain.

Remember how, at the beginning of the chapter, I discussed the classical kinematics of a collision between two objects (Fig. 10.27). Momentum was conserved in this process:

$$m_A u_A + m_B u_B = m_C u_C + m_D u_D \tag{10.49}$$

We then examined the same collision from the point of view of a moving observer, using Galileo's velocity addition rule,

$$u' = u - v \tag{10.50}$$

and concluded that momentum is conserved in this system as well,

$$m_A u_A' + m_B u_B' = m_C u_C' + m_D u_D' \tag{10.51}$$

Figure 10.27

provided that mass itself is conserved:

$$m_A + m_B = m_C + m_D \tag{10.52}$$

In fact, I used this example to lend credibility to the principle of relativity.

Unfortunately, when Galileo's rule is replaced by Einstein's (10.24):

$$u' = \frac{u - v}{1 - uv/c^2} \tag{10.53}$$

the argument falls apart. Solving for u gives,[8]

$$u = \frac{u' + v}{1 + u'v/c^2} \tag{10.54}$$

Equation (10.49) is a mess when written in terms of the primed velocities:

$$m_A\left(\frac{u_A' + v}{1 + u_A'v/c^2}\right) + m_B\left(\frac{u_B' + v}{1 + u_B'v/c^2}\right) = m_C\left(\frac{u_C' + v}{1 + u_C'v/c^2}\right) + m_D\left(\frac{u_D' + v}{1 + u_D'v/c^2}\right) \tag{10.55}$$

and conservation of momentum does *not* hold in the moving system. *Conclusion*: If you define momentum as $m\mathbf{u}$, then conservation of momentum is an impossible physical law—if true in one inertial frame, it is certainly false in another.

Now let's see how the argument goes when we use *proper* velocity. Assume the momentum so defined is conserved in the original system:

$$m_A\eta_A + m_B\eta_B = m_C\eta_C + m_D\eta_D \tag{10.56}$$

According to equation (10.46),

$$\eta' = \gamma\eta - \gamma\beta\eta^0$$

or

$$\eta = \frac{1}{\gamma}\eta' + \beta\eta^0$$

(since we're only dealing with the x-direction, I won't bother with the superscript 1 on η). So (10.56) becomes

$$m_A\left(\frac{1}{\gamma}\eta_A' + \beta\eta_A^0\right) + m_B\left(\frac{1}{\gamma}\eta_B' + \beta\eta_B^0\right)$$

$$= m_C\left(\frac{1}{\gamma}\eta_C' + \beta\eta_C^0\right) + m_D\left(\frac{1}{\gamma}\eta_D' + \beta\eta_D^0\right) \tag{10.57}$$

It follows that conservation of momentum holds in the new system,

$$m_A\eta_A' + m_B\eta_B' = m_C\eta_C' + m_D\eta_D'$$

[8]There's a handy mnemonic for Einstein's rule, which enables you to write down (10.54) directly without resort to (10.53): First write the formula for the velocity in question as *Galileo* would have; then put in the denominator, which is always 1 ± the product of the velocities in the numerator (+ if they have the same sign, − if they have opposite signs) divided by c^2.

provided that

$$m_A\eta_A^0 + m_B\eta_B^0 = m_C\eta_C^0 + m_D\eta_D^0 \tag{10.58}$$

But what does this last equation mean? It arose in exactly the same way as conservation of *mass*, in the classical argument, and yet it's *not* quite conservation of mass because of the factor $\eta^0 = c/\sqrt{1 - u^2/c^2}$. Evidently, if $m\eta$ is conserved, then mass is *not;* rather, the quantity

$$\frac{m}{\sqrt{1 - u^2/c^2}} \tag{10.59}$$

(which *reduces* to m in the nonrelativistic limit), is conserved. Einstein called this the **relativistic mass** (m itself was then called the **rest mass**) but modern usage discards this term in favor of **relativistic energy**, defined by

$$E = \frac{mc^2}{\sqrt{1 - u^2/c^2}} \tag{10.60}$$

(which differs from "relativistic mass" only by the constant factor c^2). Of course, you can *call* it anything you like—the *physics* resides in the fact that it is *conserved.* But there *is* a reason for the terminology: In the classical domain $u \ll c$ the denominator can be expanded in powers of u^2/c^2, giving

$$E = mc^2 + \frac{1}{2}mu^2 + \frac{3}{8}m\frac{u^4}{c^2} + \cdots \tag{10.61}$$

The first term is a constant, but the second is precisely the classical kinetic energy. In relativity we call E the **(total) energy**, mc^2 the **rest energy** (that's the energy of a particle at rest), and the remainder ($E - mc^2$) the **kinetic energy** (that's the energy attributable to *motion*). *Conclusion*: If you define momentum as $m\eta$, then conservation of momentum is consistent with the principle of relativity, provided the relativistic energy (10.60) is *also* conserved. Accordingly,[9] we define the relativistic momentum and energy as

$$\boxed{\begin{aligned} \mathbf{p} &= m\boldsymbol{\eta} = \frac{m\mathbf{u}}{\sqrt{1 - u^2/c^2}} \\ E &= mc\eta^0 = \frac{mc^2}{\sqrt{1 - u^2/c^2}} \end{aligned}} \tag{10.62}$$

Of course, the fact that a law is consistent with relativity does not prove it's *true*—that is an *experimental* matter, to be decided in the laboratory. Suffice it to say that few laws of physics have been confirmed more conclusively than **conservation of relativistic momentum and energy:**

[9]There are other ways to motivate the definitions of relativistic energy and momentum, but I find this the most compelling one, even though it is rather subtle and abstract.

In every closed[10] system, the total relativistic energy and momentum are conserved.

"Relativistic mass" (if you care to use that term) is *also* conserved—but this is equivalent to conservation of energy. *Rest* mass is *not* conserved—a fact that has been painfully familiar to everyone since 1945, though the so-called conversion of mass into energy is really a conversion of *rest* energy into *kinetic* energy. Incidentally, please observe the distinction between an **invariant** quantity (same value in all inertial systems) and a **conserved** quantity (same value before as after same process). Mass is invariant, but not conserved; energy is conserved, but not invariant; electric charge (as we shall see) is both conserved *and* invariant; velocity is neither conserved *nor* invariant.

Since η is the spatial part of a 4-vector, η^μ, the momentum too is part of a 4-vector (m being invariant):

$$p^\mu = m\eta^\mu \tag{10.63}$$

Because its zeroth component,

$$p^0 = m\eta^0 = \frac{E}{c} \tag{10.64}$$

is proportional to the energy, p^μ is known as the **energy-momentum 4-vector.** It behaves like any other 4-vector under Lorentz transformations:

$$\left.\begin{aligned} p^{0\prime} &= \gamma(p^0 - \beta p^1), \\ p^{1\prime} &= \gamma(p^1 - \beta p^0) \\ p^{2\prime} &= p^2 \\ p^{3\prime} &= p^3 \end{aligned}\right\} \tag{10.65}$$

or, more generally,

$$p^{\mu} = \Lambda^{\mu}_{\nu} p^{\nu}$$

The invariant quantity constructed from this 4-vector is

$$p^\mu p_\mu = -(p^0)^2 + (\mathbf{p}\cdot\mathbf{p}) = -m^2c^2$$

as you can easily check from equation (10.62). In terms of energy, then,

$$\boxed{E^2 - p^2c^2 = m^2c^4} \tag{10.66}$$

This result can be very useful, for it enables you to calculate E (if you know p), or p (knowing E), without ever having to determine the velocity.

Problem 10.26 Show that conservation of (relativistic) energy is consistent with the principle of relativity. That is: if it is conserved in one inertial system, it's conserved in any other.

[10]If there are *external* forces at work, then (just as in the classical case) the energy and momentum of the system itself will *not*, in general, be conserved.

Problem 10.27 If a particle's kinetic energy is n times its rest energy, what is its speed?

Problem 10.28 Suppose you have a collection of particles, all moving in the x-direction, with energies $(E_1, E_2, E_3, \ldots)$ and momenta $(p_1, p_2, p_3, \ldots)$. Find the velocity of the **center of momentum** frame, in which the total momentum is zero.

10.2.3 Relativistic Kinematics

I'm going to pause now to show you some applications of the conservation laws to particle decays and collisions.

Example 7

Two lumps of clay, each of (rest) mass m, collide head-on at $\frac{3}{5}c$ (Fig. 10.28). They stick together. *Question*: what is the mass (M) of the composite lump?

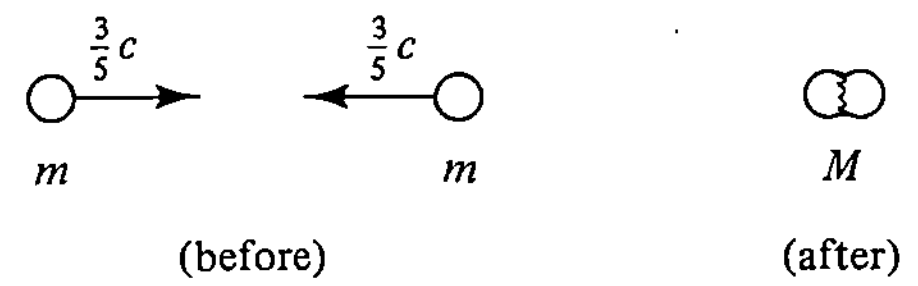

Figure 10.28

Solution: In this case conservation of momentum is trivial: zero before, zero after. The energy of each lump prior to the collision is

$$\frac{mc^2}{\sqrt{1-(\frac{3}{5})^2}} = \frac{5}{4}mc^2$$

and the energy of the composite lump after the collision is Mc^2 (since it's at rest). So conservation of energy says

$$\frac{5}{4}mc^2 + \frac{5}{4}mc^2 = Mc^2$$

and hence

$$M = \frac{5}{2}m$$

Notice that this is *greater* than the sum of the initial masses! Mass was *not* conserved in this collision; kinetic energy was converted into rest energy, so the mass increased.

In the *classical* analysis of such a collision, we say that kinetic energy was converted into *thermal* energy—the composite lump is *hotter* than the two constituents. This is, of course, true in the relativistic picture too. But what *is* thermal energy? It's the sum total of the random kinetic and potential energies of all the atoms and molecules in the substance. What relativity says is that these microscopic energies are reflected in the *mass* of the object: a hot potato is *heavier* than a cold potato, and a compressed spring is *heavier* than a relaxed spring. Not by *much*, it's true—internal energy (U) contributes an amount U/c^2 to the mass, and c^2 is a very large number, by everyday standards. You could never get two lumps of clay going anywhere *near* fast enough to detect the nonconservation of mass in their collision—and if you *could,* the whole thing would vaporize. But in the realm of elementary particles, the effect can be very striking. For example, when the neutral pi meson (mass 2.4×10^{-28} kg) decays spontaneously into two photons (mass zero), the rest energy is converted *entirely* into kinetic energy—the mass disappears completely.

You may have been puzzled by my claim that the photon has mass zero. In classical mechanics there's no such thing as a massless particle—its kinetic energy ($\frac{1}{2}mu^2$) and its momentum ($m\mathbf{u}$) would be zero, you couldn't apply a force to it ($F = ma$), and hence (by Newton's third law) *it* couldn't exert a force on anything else—it's a cipher, as far as physics is concerned. You might at first assume that the same is true in relativity; after all, $\mathbf{p}$ and E are still proportional to m. However, a closer inspection of (10.62) reveals a loophole worthy of a congressman: If $u = c$, then the zero in the numerator is balanced by a zero in the denominator, leaving $\mathbf{p}$ and E indeterminate (zero over zero). It is conceivable, therefore, that a massless particle could carry energy and momentum, *provided it always traveled at the speed of light.* Although (10.62) would no longer serve to specify E and p, equation (10.66) suggests that the two should be related by[11]

$$E = pc \tag{10.67}$$

Personally, I would regard this reasoning as laughable, or at best a curiosity, were it not for the fact that at least two kinds of massless particles are known to exist in nature: photons and neutrinos.[12] They *do* travel at the speed of light, and they obey equation (10.67). They force us to take the "loophole" seriously. (By the way, you might ask what distinguishes a photon with a *lot* of energy from one with very little—after all, they have the same mass (zero) and the same speed (c). Relativity offers no answer to this question; strangely enough, quantum mechanics *does:* According to the Planck formula, $E = h\nu$, where h is Planck's constant and ν is the *frequency.* A *blue* photon is more energetic than a *red* one.)

Example 8

A pion at rest decays into a muon and a neutrino (Fig. 10.29). Find the momentum of the outgoing muon, in terms of the two masses, m_π and m_μ ($m_\nu = 0$).

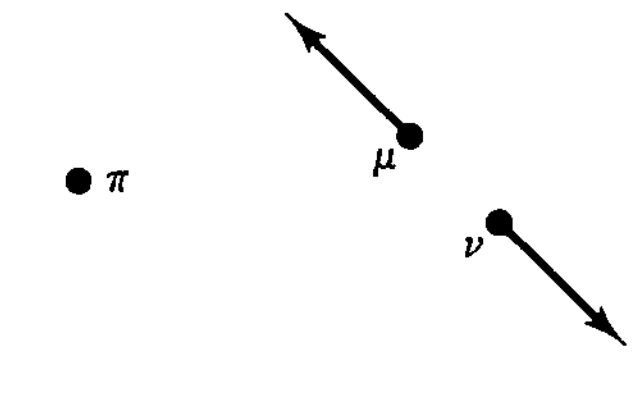

Figure 10.29

[11]Compare equation (8.73), where the same relation was found to hold for electromagnetic waves—of which the photon is the quantum.

[12]Neutrinos may actually carry a (very small) mass—this is a subject of experimental investigation at the present time. But for the moment we assume they are massless, as are the more hypothetical gluons and gravitons.

Solution:

$$E_{\text{before}} = m_\pi c^2, \qquad \mathbf{p}_{\text{before}} = 0$$
$$E_{\text{after}} = E_\mu + E_\nu, \qquad \mathbf{p}_{\text{after}} = \mathbf{p}_\mu + \mathbf{p}_\nu$$

Conservation of momentum requires that $\mathbf{p}_\nu = -\mathbf{p}_\mu$. Conservation of energy says that

$$E_\mu + E_\nu = m_\pi c^2$$

Now, $E_\nu = p_\nu c$, by equation (10.67), whereas $E_\mu = \sqrt{m_\mu^2 c^4 + p_\mu^2 c^2}$, by equation (10.66), so

$$\sqrt{m_\mu^2 c^4 + p_\mu^2 c^2} + p_\mu c = m_\pi c^2$$

from which it follows that

$$p_\mu = |\mathbf{p}_\mu| = \frac{c(m_\pi^2 - m_\mu^2)}{2m_\pi}$$

In a classical collision process, momentum and mass are always conserved, whereas kinetic energy, in general, is not. A "sticky" collision generates heat at the expense of kinetic energy; an "explosive" collision generates kinetic energy at the expense of chemical energy (or some other kind). If the kinetic energy *is* conserved, as in the ideal collision of two billiard balls, we call the process *elastic*. In the relativistic case, momentum and total energy are always conserved but mass and kinetic energy, in general, are not. Once again, we call the process **elastic** if kinetic energy is conserved. In such a case the *rest* energy (being the total minus the kinetic) is *also* conserved, and therefore so too is the mass. Examples 7 and 8 are *inelastic* processes; the following case is an *elastic* process.

Example 9 Compton scattering

A photon of energy E_0 "bounces" off an electron, initially at rest. Find the energy E of the outgoing photon, as a function of the "scattering angle" θ (see Fig. 10.30).

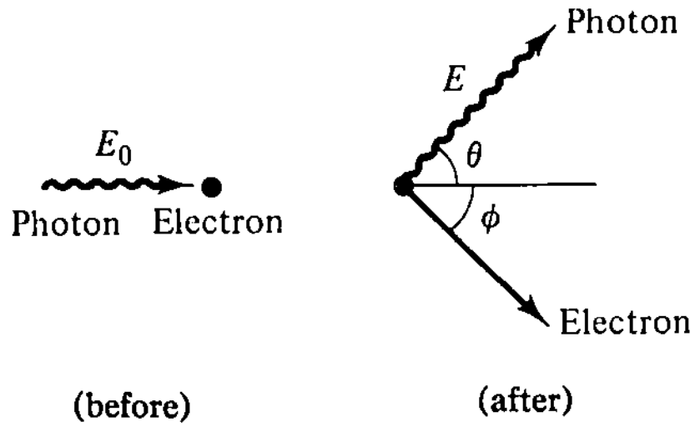

Figure 10.30

Solution: Conservation of momentum in the "vertical" direction gives

$$p_e \sin\phi = p_p \sin\theta, \quad \text{or, since } p_p = \frac{E}{c}, \quad \sin\phi = \frac{E}{p_e c}\sin\theta$$

Conservation of momentum in the "horizontal" direction gives

$$\frac{E_0}{c} = p_p\cos\theta + p_e\cos\phi = \frac{E}{c}\cos\theta + p_e\sqrt{1-\left(\frac{E}{p_e c}\sin\theta\right)^2}$$

or

$$p_e^2c^2 = (E_0 - E\cos\theta)^2 + E^2\sin^2\theta = E_0^2 - 2E_0E\cos\theta + E^2$$

Finally, conservation of energy says that

$$\begin{aligned} E_0 + mc^2 &= E + E_e = E + \sqrt{m^2c^4 + p_e^2c^2} \\ &= E + \sqrt{m^2c^4 + E_0^2 - 2E_0E\cos\theta + E^2} \end{aligned}$$

Solving for E, I find that

$$E = \frac{1}{\dfrac{(1-\cos\theta)}{mc^2} + \dfrac{1}{E_0}} \tag{10.68}$$

The answer looks nicer when expressed in terms of photon *wavelengths:*

$$E = h\nu = \frac{hc}{\lambda}$$

so

$$\lambda = \lambda_0 + \left(\frac{h}{mc}\right)(1-\cos\theta) \tag{10.69}$$

The quantity (h/mc) is called the **Compton wavelength** of the electron.

Problem 10.29 Find the energy and velocity of the muon in Example 8.

Problem 10.30 A particle of mass m whose total energy is twice its rest energy collides with an identical particle at rest. If they stick together, what is the mass of the resulting composite particle? What is its velocity?

Problem 10.31 A neutral pion of (rest) mass m and (relativistic) momentum $p = \frac{3}{4}mc$ decays into two photons (both massless). One of the photons is emitted in the same direction as the original pion, and the other in the opposite direction. Find the (relativistic) energy of each photon.

Problem 10.32 In the past, most experiments in particle physics were of the following kind: One particle (usually a proton or an electron) was accelerated to a high energy E, and collided with a target particle at rest:

Such experiments are limited by our ability to accelerate the incoming particle. Far higher relative energies are obtainable if we accelerate *both* particles to energy E, and have them collide:

We call these "colliding beam" experiments. *Classically,* the energy E' of one particle, relative to the other, is just 4E (why?)—not much of a gain (only a factor of 4). But *relativistically* the gain can be enormous. Assuming the two particles have the same mass, M, show that

$$E' = \frac{2E^2}{Mc^2} - Mc^2 \tag{10.70}$$

Suppose we use protons ($Mc^2 = 1$ GeV) and $E = 30$ GeV. What E' does this give? What multiple of E does this amount to? (1 GeV $= 10^9$ electron volts.)

Problem 10.33 In a "pair annihilation" experiment, an electron (mass m) with momentum p_0 hits a positron (same mass, but opposite charge) at rest. They annihilate, producing two photons. (Why couldn't they produce just *one* photon?) If one of the photons emerges at 60° to the incident electron direction, what is its energy?

10.2.4 Relativistic Dynamics

Newton's *first* law is built into the principle of relativity. His second law, in the form

$$\boxed{\mathbf{F} = \frac{d\mathbf{p}}{dt}} \tag{10.71}$$

retains its validity in relativistic mechanics, ***provided we use the relativistic momentum for* p**. Work, as always, is the line integral of the force:

$$W = \int \mathbf{F}\cdot d\mathbf{l} \tag{10.72}$$

With F and W defined this way, the work-energy theorem—the work done on a particle equals the increase in its energy—holds relativistically:

$$W = \int \frac{d\mathbf{p}}{dt}\cdot d\mathbf{l} = \int \frac{d\mathbf{p}}{dt}\cdot\frac{d\mathbf{l}}{dt}\,dt = \int \frac{d\mathbf{p}}{dt}\cdot\mathbf{u}\,dt$$

while

$$\frac{d\mathbf{p}}{dt}\cdot\mathbf{u} = \frac{d}{dt}\left(\frac{m\mathbf{u}}{\sqrt{1-u^2/c^2}}\right)\cdot\mathbf{u}$$

$$= \frac{m\mathbf{u}}{(1-u^2/c^2)^{3/2}}\cdot\frac{d\mathbf{u}}{dt} = \frac{d}{dt}\left(\frac{mc^2}{\sqrt{1-u^2/c^2}}\right) = \frac{dE}{dt} \tag{10.73}$$

so that

$$W = \int \frac{dE}{dt}\,dt = E_{\text{final}} - E_{\text{initial}} \tag{10.74}$$

Example 10 Motion under the influence of a constant force

Suppose a particle of mass m is subjected to a constant force F, and it starts from rest at the origin at time $t = 0$. Find the position x, as a function of time.

Solution:

$$\frac{dp}{dt} = F \quad \text{yields } p = Ft + \text{constant}$$

but since $p = 0$ at $t = 0$, the constant must be zero, and hence

$$p = \frac{mu}{\sqrt{1 - u^2/c^2}} = Ft$$

Solving for u, we obtain

$$u = \frac{(F/m)t}{\sqrt{1 + (Ft/mc)^2}} \tag{10.75}$$

The numerator, of course, is the classical answer—it's approximately right, if $(F/m)t \ll c$. On the other hand, the relativistic denominator ensures that u never exceeds c; in fact, as $t \to \infty$, $u \to c$.

To complete the problem we must integrate again:

$$\begin{aligned} x &= \frac{F}{m}\int_0^t \frac{t'}{\sqrt{1 + (Ft'/mc)^2}}\,dt' \\ &= \frac{mc^2}{F}\sqrt{1 + (Ft'/mc)^2}\,\Bigg|_0^t = \frac{mc^2}{F}\left[\sqrt{1 + \left(\frac{Ft}{mc}\right)^2} - 1\right] \end{aligned} \tag{10.76}$$

In place of the classical parabola, $x = (F/2m)t^2$, we have a *hyperbola* (see Fig. 10.31); for this reason, motion under a constant force is often called **hyperbolic motion.** It occurs, for example, when a charged particle is placed in a uniform electric field.

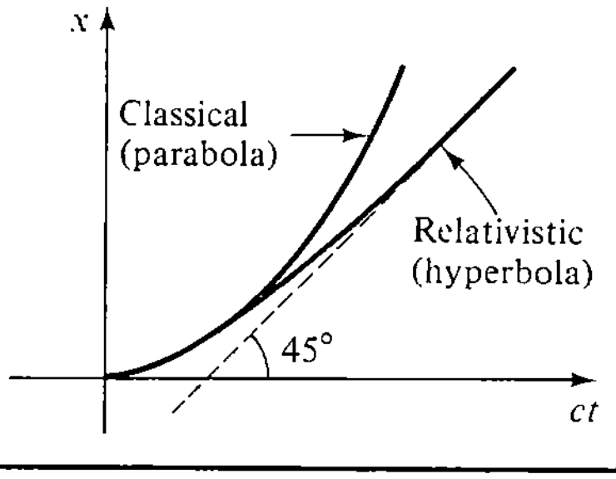

Figure 10.31

Example 11 Cyclotron motion

The typical trajectory of a charged particle in a uniform *magnetic* field is "cyclotron" motion (Fig. 10.32). The magnetic force pointing toward the center,

$$F = quB$$

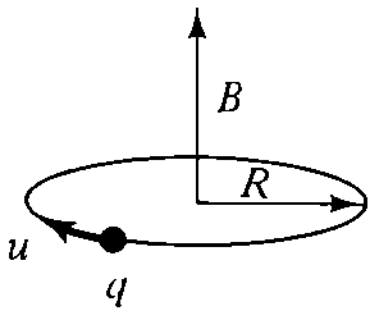

Figure 10.32

provides the centripetal acceleration necessary to sustain circular motion. Beware, however—in special relativity the centripetal force is *not* mu^2/R, as in classical mechanics. Rather, as you can see from Fig. 10.33, $dp = p\, d\theta$, so that

$$F = \frac{dp}{dt} = p\frac{d\theta}{dt} = p\frac{u}{R}$$

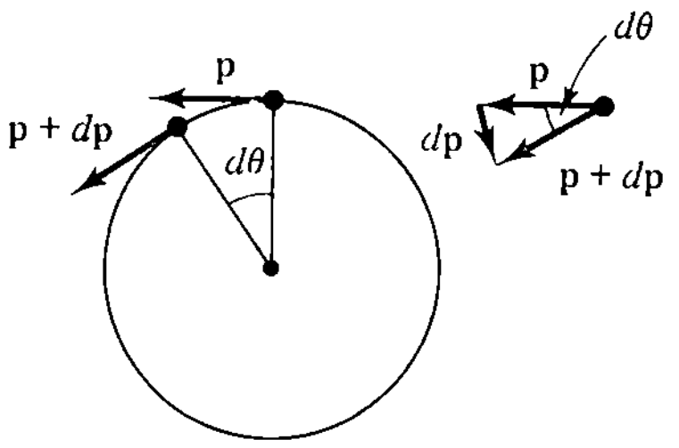

Figure 10.33

(Classically, $p = mu$, so $F = mu^2/R$.) Thus,

$$quB = p\frac{u}{R}$$

or

$$p = qBR \tag{10.77}$$

In this form the relativistic cyclotron formula is identical to the nonrelativistic one, (5.3)—the only difference is that p is now the relativistic momentum.

In contrast to the first two, Newton's *third* law does *not*, in general, extend to the relativistic domain. Indeed, if the two objects in question are separated in space, the third law is incompatible with the relativity of simultaneity. For suppose the force of A on B at some instant t is $\mathbf{F}(t)$, and the force of B on A at the same instant is $-\mathbf{F}(t)$; then the third law applies, *in this reference frame*. But a moving observer will report that these equal and opposite forces occurred *at different times;* in his system, therefore, the third law is *violated.* Only in the case of contact forces, where the two are applied at the *same physical point,* can the third law be retained.

Because $\mathbf{F}$ is the derivative of momentum with respect to *ordinary* time, it shares the ugly behavior of (ordinary) velocity, when you go from one inertial system to another: both the numerator *and the denominator* must be transformed. Thus,[13]

$$F'_y = \frac{dp'_y}{dt'} = \frac{dp_y}{\gamma\, dt - \dfrac{\gamma\beta}{c}\, dx} = \frac{(dp_y/dt)}{\gamma\left(1 - \dfrac{\beta}{c}\dfrac{dx}{dt}\right)} = \frac{F_y}{\gamma\left(1 - \dfrac{\beta u_x}{c}\right)} \tag{10.78}$$

and similarly for the z-component:

$$F'_z = \frac{F_z}{\gamma\left(1 - \dfrac{\beta u_x}{c}\right)}$$

The x-component is even worse:

$$F'_x = \frac{dp'_x}{dt'} = \frac{\gamma\, dp_x - \gamma\beta\, dp^0}{\gamma\, dt - \dfrac{\gamma\beta}{c}\, dx} = \frac{\dfrac{dp_x}{dt} - \beta\dfrac{dp^0}{dt}}{1 - \dfrac{\beta}{c}\dfrac{dx}{dt}} = \frac{F_x - \dfrac{\beta}{c}\left(\dfrac{dE}{dt}\right)}{\left(1 - \dfrac{\beta u_x}{c}\right)}$$

We calculated dE/dt in equation (10.73); putting that in,

$$F'_x = \frac{F_x - \dfrac{\beta}{c}(\mathbf{u}\cdot\mathbf{F})}{\left(1 - \dfrac{\beta u_x}{c}\right)} \tag{10.79}$$

Only in one special case are these equations reasonably tractable: *If the particle is (instantaneously) at rest in S*, so that $\mathbf{u} = 0$, then

$$\mathbf{F}'_\perp = \frac{1}{\gamma}\mathbf{F}_\perp, \qquad F'_\parallel = F_\parallel \tag{10.80}$$

That is, the component of $\mathbf{F}$ *parallel* to the motion of S' is unchanged, whereas components perpendicular are divided by γ.

It has perhaps occurred to you that we could avoid the bad transformation behavior of $\mathbf{F}$ by introducing a "proper" force, analogous to the proper velocity, which would be the derivative of momentum with respect to *proper* time:

$$K^\mu = \frac{dp^\mu}{d\tau} \tag{10.81}$$

This quantity is known as the **Minkowski force**; it is plainly a 4-vector, since p^μ is a 4-vector and proper time is invariant. The spatial components of K^μ are related to the

[13]Remember: γ and β pertain to the motion of S' with respect to S—they are *constants;* $\mathbf{u}$ is the velocity of the *particle* with respect to S.

"ordinary" force by

$$\mathbf{K} = \left(\frac{dt}{d\tau}\right)\frac{d\mathbf{p}}{dt} = \frac{1}{\sqrt{1 - u^2/c^2}}\mathbf{F} \tag{10.82}$$

while the zeroth component,

$$K^0 = \frac{dp^0}{d\tau} = \frac{1}{c}\frac{dE}{d\tau} \tag{10.83}$$

is, apart from the $1/c$, the (proper) rate at which the energy of the particle increases—in other words, the (proper) *power* delivered to the particle. Relativistic dynamics can be formulated in terms of the ordinary force *or* the Minkowski force. The latter version is generally much *neater,* but since in the long run we are interested in the particle's trajectory as a function of *ordinary* time, the former is often more useful. When we wish to generalize some classical force law, such as Lorentz's, to the relativistic domain, the question arises: Does the classical formula correspond to the *ordinary* force or to the Minkowski force? In other words, should we write

$$\mathbf{F} = q(\mathbf{E} + \mathbf{u} \times \mathbf{B})$$

or should it rather be

$$\mathbf{K} = q(\mathbf{E} + \mathbf{u} \times \mathbf{B})$$

Since proper time and ordinary time are identical in classical physics, there is no way at this stage to decide the issue. The Lorentz force, as it turns out, is an *ordinary* force—you'll see later on why this is so, and how to construct the electromagnetic Minkowski force.

Problem 10.34 Newton's second law in classical mechanics can be written in the more familiar form $\mathbf{F} = m\mathbf{a}$ (provided the mass is constant). The relativistic equation, $\mathbf{F} = d\mathbf{p}/dt$, *cannot* be so simply expressed. Show, rather, that

$$\mathbf{F} = \frac{m}{\sqrt{1 - u^2/c^2}}\left[\mathbf{a} + \frac{\mathbf{u}(\mathbf{u} \cdot \mathbf{a})}{(c^2 - u^2)}\right] \tag{10.84}$$

where $\mathbf{a} = d\mathbf{u}/dt$.

Problem 10.35 Show that it is possible to outrun a light ray, if you're given a sufficient head start, and your feet exert a constant force.

Problem 10.36 Define the "proper acceleration" in the obvious way:

$$\alpha^\mu = \frac{d\eta^\mu}{d\tau} = \frac{d^2x^\mu}{d\tau^2} \tag{10.85}$$

(a) Find α^0 and $\boldsymbol{\alpha}$ in terms of $\mathbf{u}$ and $\mathbf{a}$ (the "ordinary" acceleration).
(b) Express $\alpha_\mu \alpha^\mu$ in terms of $\mathbf{u}$ and $\mathbf{a}$.
(c) Show that $\eta^\mu \alpha_\mu = 0$.
(d) Write the Minkowski version of Newton's second law, equation (10.81) in terms of α^μ. Evaluate the invariant product $K^\mu \eta_\mu$.

Problem 10.37 Show that

$$K_\mu K^\mu = \left[\frac{1 - (u^2/c^2)\cos^2\theta}{1 - u^2/c^2}\right] F^2$$

where θ is the angle between $\mathbf{u}$ and $\mathbf{F}$.

Problem 10.38 Show that the acceleration of a particle of mass m and charge q, under the influence of electromagnetic fields $\mathbf{E}$ and $\mathbf{B}$, is given by

$$\mathbf{a} = \frac{q}{m}\sqrt{1 - u^2/c^2}\left[\mathbf{E} + \mathbf{u} \times \mathbf{B} - \frac{1}{c^2}\mathbf{u}(\mathbf{u} \cdot \mathbf{E})\right]$$

where $\mathbf{u}$ is the particle's velocity. (*Hint*: Use equation (10.84).)

10.3 RELATIVISTIC ELECTRODYNAMICS

10.3.1 Magnetism as a Relativistic Phenomenon

Unlike Newtonian mechanics, classical electrodynamics is *already* consistent with special relativity. Maxwell's equations and the Lorentz force law can be applied legitimately in any inertial system. Of course, what one observer interprets as an electrical process another may regard as magnetic, but the actual particle motions they predict will be identical. To the extent that this did *not* work out for Lorentz and others, who studied the question in the late nineteenth century, the fault lay with the nonrelativistic mechanics they used, not with the electrodynamics. Having corrected Newtonian mechanics, we are now in a position to develop a complete and consistent formulation of relativistic electrodynamics. But I emphasize that we will not be changing the rules of electrodynamics in the slightest—rather, we will be *expressing* these rules in a notation which exposes and illuminates their relativistic character. As we go along, I shall pause now and then to rederive, using the Lorentz transformations, results obtained earlier by more laborious means. But the main purpose of this section is to provide you with a deeper understanding of the structure of electrodynamics—laws that had seemed arbitrary and unrelated before take on a kind of coherence and inevitability when approached from the point of view of relativity.

To begin with I'd like to show you why there *had* to be such a thing as magnetism, given electrostatics and relativity, and how, in particular, you can calculate the magnetic force between a current-carrying wire and a moving charge without ever invoking the laws of magnetism.[14] Suppose you had a string of positive charges moving along to the right at speed v. I'll assume the charges are close enough together so that we may regard them as a continuous line charge λ. Superimposed on this positive string is a negative one, $-\lambda$, proceeding to the left at the same speed v. We have, then, a net current to the right, of magnitude

$$I = 2\lambda v \tag{10.86}$$

Meanwhile, a distance r away there is a point charge q traveling to the right at speed

[14]This and several other arguments in this section are adapted from E. M. Purcell's *Electricity and Magnetism*, 2d ed. (New York: McGraw-Hill, 1985).

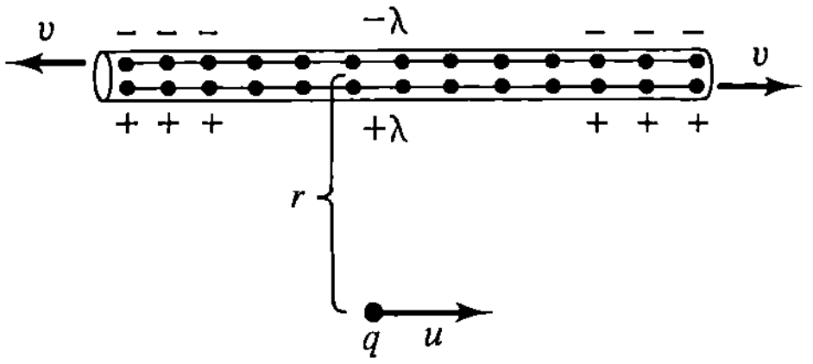

Figure 10.34

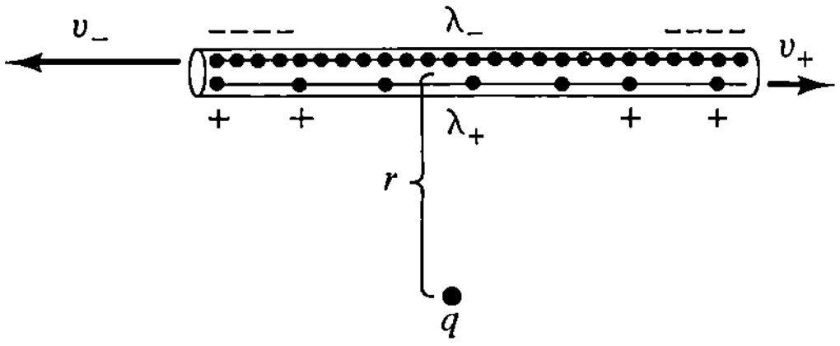

Figure 10.35

u (see Fig. 10.34). Because the two line charges cancel, there is *no electrical force on* q in this system S.

However, let's look at the same situation from the point of view of system S', in which q is at rest (S' moves to the right with speed u, relative to S—see Fig. 10.35). By the Einstein velocity addition rule, the velocities of the positive and negative lines are now

$$v_{\pm} = \frac{v \mp u}{1 \mp vu/c^2} \tag{10.87}$$

Because v_- is greater than v_+, the Lorentz contraction of the spacing between negative charges is more severe than that between positive charges; *in this frame,* therefore, ***the wire carries a net negative charge!*** In fact,

$$\lambda_{\pm} = \pm\gamma_{\pm}\lambda_0 \tag{10.88}$$

where

$$\gamma_{\pm} = \frac{1}{\sqrt{1 - v_{\pm}^2/c^2}} \tag{10.89}$$

and λ_0 is the charge density of the positive line in its own rest system. That's not the same as λ, of course—in S they're already moving at speed v, so

$$\lambda = \gamma\lambda_0 \tag{10.90}$$

where

$$\gamma = \frac{1}{\sqrt{1 - v^2/c^2}} \tag{10.91}$$

It takes a little algebra to put $\gamma_{\pm}$ into simple form:

$$\gamma_{\pm} = \frac{1}{\sqrt{1 - \frac{1}{c^2}(v \mp u)^2\left(1 \mp \frac{vu}{c^2}\right)^{-2}}} = \frac{c^2 \mp uv}{\sqrt{(c^2 \mp uv)^2 - c^2(v \mp u)^2}}$$

$$= \frac{c^2 \mp uv}{\sqrt{(c^2 - v^2)(c^2 - u^2)}} = \gamma\frac{(1 \mp uv/c^2)}{\sqrt{1 - u^2/c^2}} \tag{10.92}$$

Evidently, then, the net line charge in S' is

$$\lambda_{\text{tot}} = \lambda_+ + \lambda_- = \lambda_0(\gamma_+ - \gamma_-) = \frac{-2\lambda uv}{c^2\sqrt{1 - u^2/c^2}} \tag{10.93}$$

Conclusion: As a result of unequal Lorentz contraction of the positive and negative lines, a current-carrying wire which is electrically neutral in one system will be charged in another.

Now, a line charge λ_{tot} sets up an *electric* field

$$E = \frac{\lambda_{tot}}{2\pi\epsilon_0 r}$$

so *there is an electrical force on q in S'*, to wit:

$$F' = qE = -\frac{\lambda v}{\pi\epsilon_0 c^2 r}\frac{qu}{\sqrt{1-u^2/c^2}} \tag{10.94}$$

But if there's a force on q in S', there must be one S—we can *calculate* it by using the transformation rules for forces. Since q is at rest in S', and F' is perpendicular to u, the force in S is given by equation (10.80):

$$F = \sqrt{1-u^2/c^2}\,F' = -\frac{\lambda v}{\pi\epsilon_0 c^2}\frac{qu}{r} \tag{10.95}$$

The charge is attracted toward the wire by a force that is purely electrical in S' (where the wire is charged, and q is at rest), but distinctly *non*electrical in S (where the wire is neutral). Taken together, then, electrostatics and relativity imply the existence of *another* force. This "other force" is, of course, *magnetic.* In fact, we can put equation (10.95) into more familiar form by using $c^2 = (\epsilon_0\mu_0)^{-1}$ and expressing λv in terms of the current (equation (10.86)):

$$F = -\left(\frac{\mu_0 I}{2\pi r}\right)qu \tag{10.96}$$

The term in parentheses is the magnetic field of a long, straight wire, and the force is precisely what we would have obtained by using the Lorentz force law in system S.

10.3.2 How the Fields Transform

We have learned, in various special cases, that one person's electric field is another's magnetic field. What we need now are the *general* transformation rules for electromagnetic fields, so that we can answer the question: Given the fields in S, what are the fields in S'? Your first guess might be that **E** is the spatial part of one 4-vector and **B** the spatial part of another. If so, your intuition is wrong—it's more complicated than that. Let me begin by making explicit an assumption that was already used implicitly in Section 10.3.1, to wit, that *charge itself is invariant.* Like mass, but unlike energy, the charge of a particle is a fixed number, independent of how fast it happens to be going. We shall assume also that the transformation rules are the same no matter how the fields were produced—electric fields generated by moving currents transform the same way as those set up by stationary charges. Were this not the case we'd have to abandon the field formulation altogether, for it is the essence of a field theory that the fields at a given point tell you *all there is to know,* electromagnetically, about that point; you do *not* have to append extra information regarding their source.

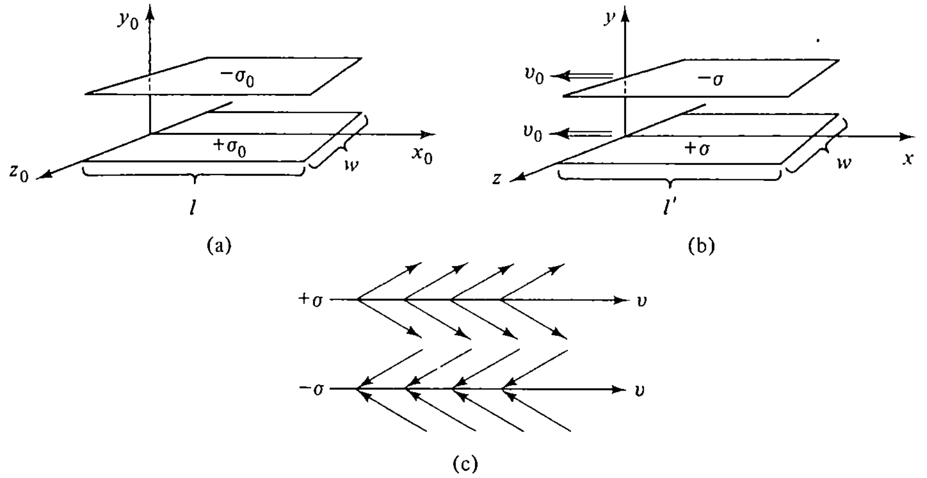

Figure 10.36

With this in mind, let's consider the *simplest possible* electric field: the uniform field in the region between the plates of a large parallel-plate capacitor (Fig. 10.36(a)). Say the capacitor is at rest in S_0 and carries surface charges $\pm\sigma_0$. Then

$$\mathbf{E}_0 = \frac{\sigma_0}{\epsilon_0}\hat{j} \tag{10.97}$$

Now, what if we examine this same capacitor from system S, moving to the right at speed v_0, as in Fig. 10.36(b)? In this system the plates are moving to the left, but the field still takes the form

$$\mathbf{E} = \frac{\sigma}{\epsilon_0}\hat{j} \tag{10.98}$$

the only difference is the surface charge σ. (Wait a minute. *Is* that the only difference? The formula $E = \sigma/\epsilon_0$ for a parallel plate capacitor came from Gauss's law, and Gauss's law relies on symmetry. How can we be sure the field is still perpendicular to the plates? What if the field of a moving plane *tilts*, say, in the direction of motion, as in Fig. 10.36(c)? Well, *even if it did* (it *doesn't*), the field between the plates, being the superposition of the $+\sigma$ field and the $-\sigma$ field, would nevertheless run perpendicular to the plates. For the $-\sigma$ field would aim as shown in Fig. 10.36(c) (changing the sign of the charge merely reverses the direction of the arrows), and the vector sum kills off the parallel components.)

Now, the total charge on each plate is invariant, and the *width* (w) is unchanged, but the *length* (l) is Lorentz-contracted by a factor

$$\frac{1}{\gamma_0} = \sqrt{1 - v_0^2/c^2} \tag{10.99}$$

so the charge per unit area is *increased* by a factor γ_0:

$$\sigma = \gamma_0\sigma_0 \tag{10.100}$$

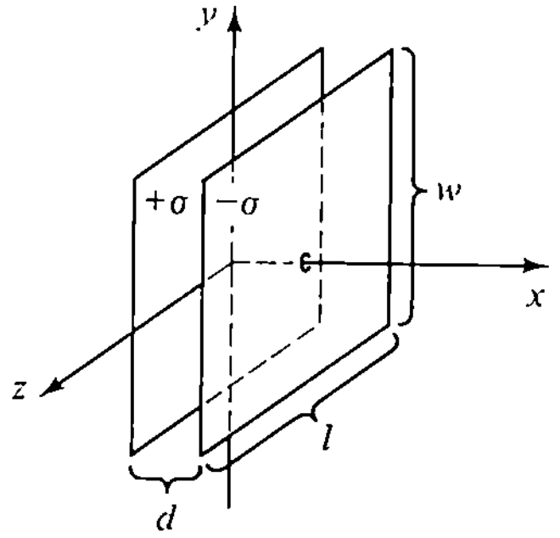

Figure 10.37

Accordingly,

$$E_{\perp} = \gamma_0 E_{0_{\perp}} \tag{10.101}$$

I have put in the subscript $\perp$ to make it clear that this rule pertains to components of **E** which are *perpendicular* to the direction of motion of S. To extract the rule for *parallel* components, consider the capacitor lined up with the yz plane (Fig. 10.37). This time it is the plate separation (d) that is Lorentz-contracted, whereas l and w are the same in both frames. Since the field does not depend on d, it follows that

$$E_{\parallel} = E_{0_{\parallel}} \tag{10.102}$$

Example 12 Electric field of a point charge in uniform motion

A point charge q is at rest at the origin in system S_0. *Question*: What is the electric field of this same charge in system S, moving to the right at speed v_0 relative to S_0?

Solution: In S_0 the field is

$$\mathbf{E}_0 = \frac{1}{4\pi\epsilon_0}\frac{q}{r_0^2}\hat{r}_0$$

or

$$E_{x_0} = \frac{1}{4\pi\epsilon_0}\frac{qx_0}{(x_0^2 + y_0^2 + z_0^2)^{3/2}}$$

$$E_{y_0} = \frac{1}{4\pi\epsilon_0}\frac{qy_0}{(x_0^2 + y_0^2 + z_0^2)^{3/2}}$$

$$E_{z_0} = \frac{1}{4\pi\epsilon_0}\frac{qz_0}{(x_0^2 + y_0^2 + z_0^2)^{3/2}}$$

From the transformation rules (10.101) and (10.102), we have

$$E_x = E_{x_0} = \frac{1}{4\pi\epsilon_0}\frac{qx_0}{(x_0^2 + y_0^2 + z_0^2)^{3/2}}$$

$$E_y = \gamma_0 E_{y0} = \frac{1}{4\pi\epsilon_0} \frac{q\gamma_0 y_0}{(x_0^2 + y_0^2 + z_0^2)^{3/2}}$$

$$E_z = \gamma_0 E_{z0} = \frac{1}{4\pi\epsilon_0} \frac{q\gamma_0 z_0}{(x_0^2 + y_0^2 + z_0^2)^{3/2}}$$

These are still expressed in terms of the S_0 coordinates (x_0, y_0, z_0) of the point P; I'd prefer to write them in terms of the S coordinates of the same point. From the Lorentz transformations (or, actually, the inverse transformations),

$$x_0 = \gamma_0(x + v_0 t) = \gamma_0 \mathscr{r}_x$$

$$y_0 = y = \mathscr{r}_y$$

$$z_0 = z = \mathscr{r}_z$$

where $\mathscr{r}$ is the vector from q to P (see Fig. 10.38). Thus, (dropping the subscript 0):

$$\begin{aligned} \mathbf{E} &= \frac{1}{4\pi\epsilon_0} \frac{\gamma q \mathscr{r}}{(\gamma^2 \mathscr{r}^2 \cos^2\theta + \mathscr{r}^2 \sin^2\theta)^{3/2}} \\ &= \frac{1}{4\pi\epsilon_0} \frac{q(1 - v^2/c^2)}{[1 - (v^2/c^2)\sin^2\theta]^{3/2}} \frac{\hat{\mathscr{r}}}{\mathscr{r}^2} \end{aligned} \quad (10.103)$$

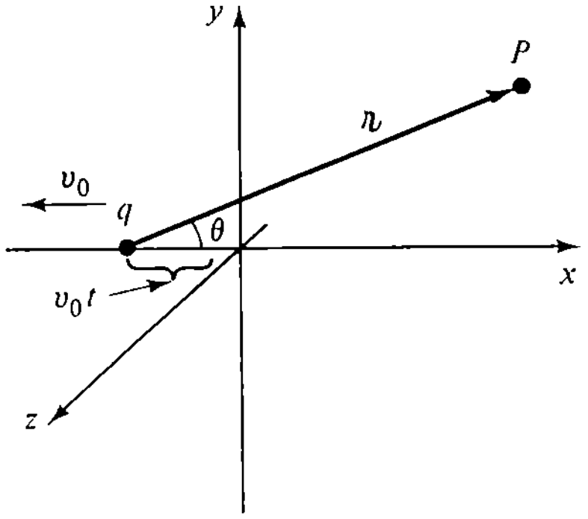

Figure 10.38

This, then, is the field of a charge in uniform motion, which we obtained in Chapter 9 from the retarded potentials. The present derivation is far more efficient and sheds some light on the remarkable fact that the field points away from the instantaneous (as opposed to the retarded) position of the charge: E_x gets a factor of γ_0 from the Lorentz transformation of the *coordinates*; E_y and E_z pick up theirs from the transformation of the *field*. It's the balancing of these two γ_0's that leaves $\mathbf{E}$ parallel to $\mathscr{r}$.

Equations (10.101) and (10.102) are not the *general* transformation laws we seek, for while they tell us how the components of $\mathbf{E}$ transform, they say nothing about the components of $\mathbf{B}$. The reason is simple: We began with a system S_0 in which the charges were at rest and where, consequently, there *was* no magnetic field.

To derive the *general* rules we must start out in a system with both electric and magnetic fields. For this purpose S itself will do the job. In addition to the electric field

$$E_y = \frac{\sigma}{\epsilon} \tag{10.104}$$

there is a *magnetic* field due to the surface currents (Fig. 10.36(b))

$$\mathbf{K}_\pm = \mp\sigma v_0 \hat{i} \tag{10.105}$$

By the right-hand rule, this field points in the negative z-direction; its magnitude is given by Ampère's law:

$$B_z = -\mu_0 \sigma v_0 \tag{10.106}$$

In a *third* system, S', traveling to the right with speed v relative to S (Fig. 10.39), the fields would be

$$E_y' = \frac{\sigma'}{\epsilon_0}, \qquad B_z' = -\mu_0 \sigma' v' \tag{10.107}$$

where v' is the velocity of S' relative to S_0:

$$v' = \frac{v + v_0}{1 + v v_0/c^2}, \qquad \gamma' = \frac{1}{\sqrt{1 - v'^2/c^2}} \tag{10.108}$$

and

$$\sigma' = \gamma' \sigma_0 \tag{10.109}$$

It remains only to express $\mathbf{E}'$ and $\mathbf{B}'$ (equation (10.107)), in terms of $\mathbf{E}$ and $\mathbf{B}$ (equations (10.104) and (10.106)). In view of (10.100), and (10.109), we have

$$E_y' = \left(\frac{\gamma'}{\gamma_0}\right)\frac{\sigma}{\epsilon_0}, \qquad B_z' = -\left(\frac{\gamma'}{\gamma_0}\right)\mu_0 \sigma v' \tag{10.110}$$

With a little algebra, you will find that

$$\frac{\gamma'}{\gamma_0} = \frac{\sqrt{1 - v_0^2/c^2}}{\sqrt{1 - v'^2/c^2}} = \frac{(1 + v v_0/c^2)}{\sqrt{1 - v^2/c^2}} = \gamma\left(1 + \frac{v v_0}{c^2}\right) \tag{10.111}$$

where

$$\gamma = \frac{1}{\sqrt{1 - v^2/c^2}} \tag{10.112}$$

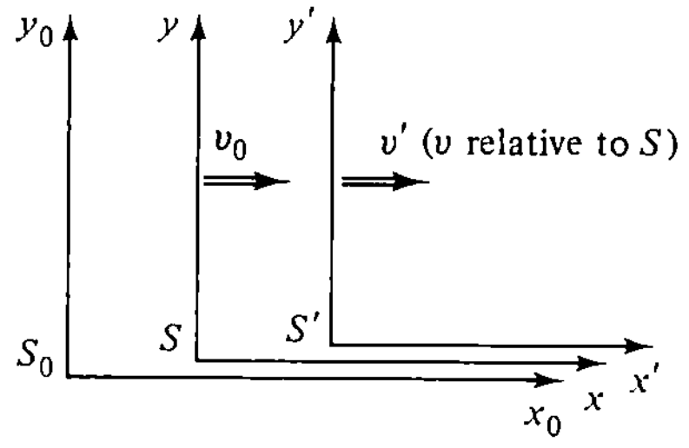

Figure 10.39

as always. Thus,

$$E'_y = \gamma\left(1 + \frac{vv_0}{c^2}\right)\frac{\sigma}{\epsilon_0} = \gamma\left(E_y - \frac{v}{c^2\epsilon_0\mu_0}B_z\right)$$

whereas

$$B'_z = -\gamma\left(1 + \frac{vv_0}{c^2}\right)\mu_0\sigma\left(\frac{v + v_0}{1 + vv_0/c^2}\right) = \gamma(B_z - \mu_0\epsilon_0 v E_y)$$

Or, since $\mu_0\epsilon_0 = 1/c^2$,

$$\left.\begin{aligned} E'_y &= \gamma(E_y - vB_z) \\ B'_z &= \gamma\left(B_z - \frac{v}{c^2}E_y\right) \end{aligned}\right\} \tag{10.113}$$

This tells us how E_y and B_z transform—to do E_z and B_y we simply align the same capacitor parallel to the xy plane instead of the xz plane. The fields in S are then (Fig. 10.40)

$$E_z = \frac{\sigma}{\epsilon_0}, \qquad B_y = \mu_0\sigma v_0$$

(Check the right-hand rule to get the sign of B_y.) The rest of the argument is identical—everywhere we had E_y before, read E_z, and everywhere we had B_z, read $-B_y$:

$$\left.\begin{aligned} E'_z &= \gamma(E_z + vB_y) \\ B'_y &= \gamma\left(B_y + \frac{v}{c^2}E_z\right) \end{aligned}\right\} \tag{10.114}$$

As for the x-components, we have already seen (by orienting the capacitor parallel to the yz plane) that

$$E'_x = E_x \tag{10.115}$$

Since in this configuration there is no accompanying magnetic field, we cannot deduce the transformation rule for B_x. But another device will permit us to work this out: Imagine a long *solenoid* aligned parallel to the x axis (Fig. 10.41) and at rest in

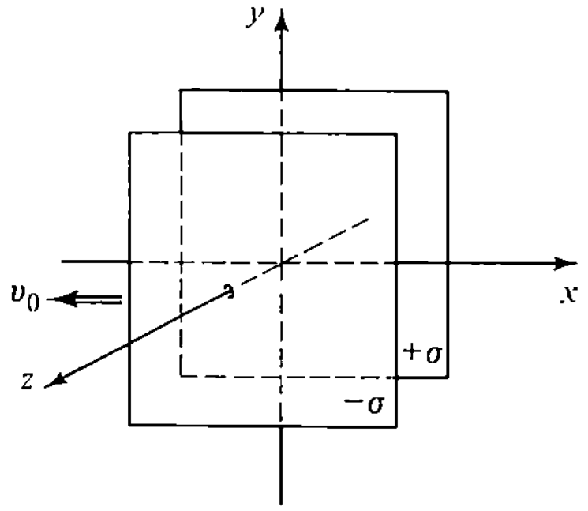

Figure 10.40

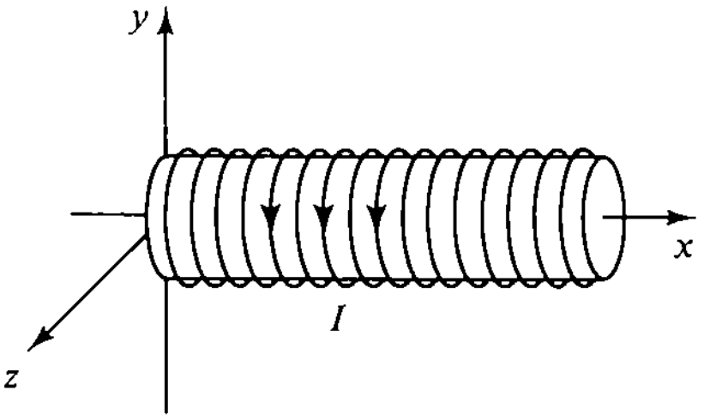

Figure 10.41

S. The magnetic field within the coil is

$$B_x = \mu_0 n I \tag{10.116}$$

where n is the number of turns per unit length, and I is the current. In system S', the length contracts, so *n increases:*

$$n' = \gamma n \tag{10.117}$$

On the other hand, time *dilates:* The S clock, which rides along with the solenoid, runs slow, so the current (charge *per unit time*) in S' is given by

$$I' = \frac{1}{\gamma} I \tag{10.118}$$

The two factors of γ exactly cancel, and we conclude that

$$B'_x = B_x$$

Like **E**, the component of **B** *parallel* to the motion is unchanged.

Let's now collect the complete set of transformation rules:

$$\boxed{\begin{array}{lll} E'_x = E_x, & E'_y = \gamma(E_y - vB_z), & E'_z = \gamma(E_z + vB_y) \\ B'_x = B_x, & B'_y = \gamma\left(B_y + \frac{v}{c^2}E_z\right), & B'_z = \gamma\left(B_z - \frac{v}{c^2}E_y\right) \end{array}} \tag{10.119}$$

Two special cases warrant particular attention:

1. If $\mathbf{B} = 0$ in S, then

$$\mathbf{B}' = \gamma \frac{v}{c^2}(E_z\hat{j} - E_y\hat{k}) = \frac{v}{c^2}(E'_z\hat{j} - E'_y\hat{k})$$

or, since $\mathbf{v} = v\hat{i}$,

$$\boxed{\mathbf{B}' = -\frac{1}{c^2}(\mathbf{v} \times \mathbf{E}')} \tag{10.120}$$

2. If $\mathbf{E} = 0$ in S, then

$$\mathbf{E}' = -\gamma v(B_z\hat{j} - B_y\hat{k}) = -v(B'_z\hat{j} - B'_y\hat{k})$$

or

$$\boxed{\mathbf{E}' = (\mathbf{v} \times \mathbf{B}')} \tag{10.121}$$

In other words, if either **E** or **B** is zero (at a particular point) in *one* system, then in any other system the fields (at that point) are very simply related by (10.120) or (10.121).

Example 13

Find the *magnetic* field of a point charge in uniform motion.

Solution: In the particle's *rest* frame (S_0) the magnetic field is zero (everywhere), so in a system S moving to the right at speed v,

$$\mathbf{B} = -\frac{1}{c^2}(\mathbf{v} \times \mathbf{E})$$

We calculated the *electric* field in Example 12; see equation (10.103). The magnetic field, then, is

$$\mathbf{B} = \frac{\mu_0}{4\pi} \frac{qv(1 - v^2/c^2)\sin\theta}{[1 - (v^2/c^2)\sin^2\theta]^{3/2}} \frac{\hat{\boldsymbol{\phi}}}{\mathscr{r}^2} \tag{10.122}$$

where $\hat{\boldsymbol{\phi}}$ aims counterclockwise as you face the oncoming charge. Incidentally, in the nonrelativistic limit ($v^2 \ll c^2$), equation (10.122) reduces to

$$\mathbf{B} = \frac{\mu_0}{4\pi} q \frac{\mathbf{v} \times \hat{\mathscr{r}}}{\mathscr{r}^2}$$

which is exactly what you would get by naïve application of the Biot-Savart law to a point charge (equation (5.36)).

Problem 10.39 A parallel-plate capacitor, at rest in S_0 and tilted at a 45° angle to the x_0 axis, carries charge densities $\pm\sigma_0$ on the two plates (Fig. 10.42). System S is moving to the right at speed v relative to S_0.

(a) Find $\mathbf{E}_0$, the field in S_0.

(b) Find $\mathbf{E}$, the field in S.

(c) What angle do the plates make with the x axis?

(d) Is the field perpendicular to the plates in S?

Problem 10.40

(a) Check that Gauss's law, $\int \mathbf{E} \cdot d\mathbf{a} = (1/\epsilon_0)Q_{\text{enc}}$, is obeyed by the field of a point charge in uniform motion, by integrating the field over a sphere of radius R centered on the charge.

(b) Find the Poynting vector for a point charge in uniform motion. (Say the charge is going in the z-direction at speed v, and calculate $\mathbf{S}$ at the instant q passes the origin.)

Problem 10.41 Charge q_A is at rest at the origin in system S; charge q_B flies by at speed v on a trajectory parallel to the x axis but at $y = d$.

(a) What is the electromagnetic force on B as it crosses the y axis?

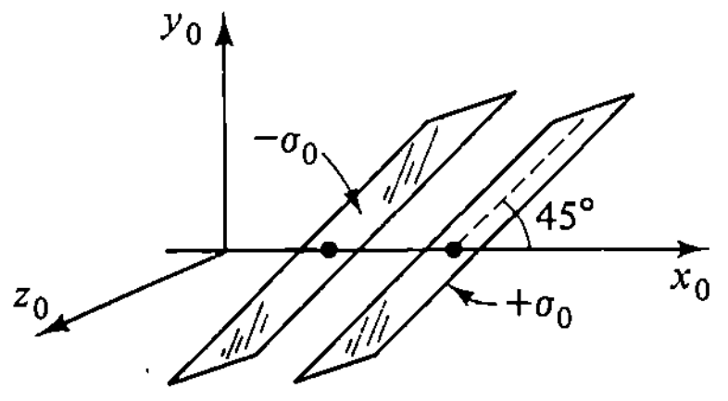

Figure 10.42

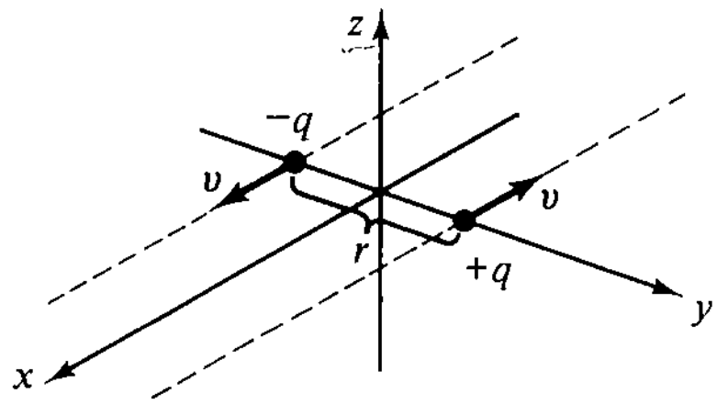

Figure 10.43

Now study the same problem from system S', which moves to the right with speed v:
(b) What is the force on B when A passes the y' axis? [Do it two ways: (i) by using your answer to (a) and transforming the force; (ii) by computing the fields in S' and using the Lorentz force law.]

Problem 10.42 Two charges, $\pm q$, are on parallel trajectories a distance r apart, moving with equal speeds v in opposite directions. We're interested in the force on $+q$ due to $-q$ at the instant they cross (Fig. 10.43). Fill in the following table, doing all the consistency checks you can think of as you go along.

	System A (Fig. 10.43)	System B ($+q$ at rest)	System C ($-q$ at rest)
E at $+q$ due to $-q$:			
B at $+q$ due to $-q$:			
F on $+q$ due to $-q$:			

Problem 10.43 Two charges $\pm q$ approach the origin at constant velocity from opposite directions along the x axis. They collide and stick together, forming a neutral particle at rest. Sketch the electric field before and shortly after the collision (remember that electromagnetic "news" travels at the speed of light). How would you interpret the field after the collision, physically?[15]

Problem 10.44
(a) Show that $(\mathbf{E} \cdot \mathbf{B})$ is relativistically invariant.
(b) Show that $(E^2 - c^2B^2)$ is relativistically invariant.
(c) Suppose the *magnetic* field at some point is *zero* in one system. Is it possible to find another system in which the *electric* field at that point is zero?

Problem 10.45 A charge q is released from rest at the origin, in the presence of a uniform electric field $\mathbf{E} = E_0\hat{k}$ and a uniform magnetic field $\mathbf{B} = B_0\hat{i}$. Find the trajectory of the particle by transforming to a system in which $\mathbf{E} = 0$, finding the path in that system and then transforming back to the original system. Assume $E_0 < cB_0$. Compare your result with Example 2 of Chapter 5.

10.3.3 The Field Tensor

As equation (10.119) shows, **E** and **B** certainly do *not* transform like the spatial parts of two 4-vectors. In fact, *all six* components of **E** and **B** are stirred together when you

[15]See E. M. Purcell, *Electricity and Magnetism*, 2d ed. (New York: McGraw-Hill, 1985), sec. 5.7, and R. Y. Tsien, *Am. J. Phys.* **40**, 46 (1972).

go from one inertial system to another. What sort of an object is this, which has six components and transforms according to equation (10.119)? *Answer*: It's an antisymmetric, second-rank tensor. Remember that a 4-vector transforms by the rule

$$a^{\mu\prime} = \Lambda^{\mu}_{\nu} a^{\nu} \tag{10.123}$$

(summation over ν implied), where Λ is the Lorentz transformation matrix. If S' is moving in the x-direction at speed v, Λ has the form

$$\Lambda = \begin{pmatrix} \gamma & -\gamma\beta & 0 & 0 \\ -\gamma\beta & \gamma & 0 & 0 \\ 0 & 0 & 1 & 0 \\ 0 & 0 & 0 & 1 \end{pmatrix} \tag{10.124}$$

Λ^{μ}_{ν} is the entry in row μ, column ν. A (second-rank) tensor is an object with *two* indices, which transforms with *two* factors of Λ (one for each index):

$$t^{\mu\nu\prime} = \Lambda^{\mu}_{\lambda}\Lambda^{\nu}_{\sigma} t^{\lambda\sigma} \tag{10.125}$$

A tensor (in 4 dimensions) has $4 \times 4 = 16$ components, which we can display in a 4×4 array:

$$t^{\mu\nu} = \begin{pmatrix} t^{00} & t^{01} & t^{02} & t^{03} \\ t^{10} & t^{11} & t^{12} & t^{13} \\ t^{20} & t^{21} & t^{22} & t^{23} \\ t^{30} & t^{31} & t^{32} & t^{33} \end{pmatrix}$$

However, the 16 elements need not all be different. For instance, a *symmetric* tensor has the property

$$t^{\mu\nu} = t^{\nu\mu} \qquad (\textit{symmetric tensor}) \tag{10.126}$$

In this case there are 10 distinct components; 6 of the 16 are repeats ($t^{01} = t^{10}$, $t^{02} = t^{20}$, $t^{03} = t^{30}$, $t^{12} = t^{21}$, $t^{13} = t^{31}$, $t^{23} = t^{32}$). Similarly, an *anti*symmetric tensor obeys

$$t^{\mu\nu} = -t^{\nu\mu} \qquad (\textit{antisymmetric tensor}) \tag{10.127}$$

Such an object has just 6 distinct elements—of the original 16, six are repeats (the same ones as before, only this time with a minus sign) and four are zero (t^{00}, t^{11}, t^{22}, and t^{33}). Thus, the general antisymmetric tensor has the form

$$t^{\mu\nu} = \begin{pmatrix} 0 & t^{01} & t^{02} & t^{03} \\ -t^{01} & 0 & t^{12} & t^{13} \\ -t^{02} & -t^{12} & 0 & t^{23} \\ -t^{03} & -t^{13} & -t^{23} & 0 \end{pmatrix}$$

FUNDAMENTAL EQUATIONS OF ELECTRODYNAMICS

MAXWELL'S EQUATIONS

In vacuum:

$$\begin{cases} \nabla \cdot \mathbf{E} = \dfrac{1}{\epsilon_0}\rho \\ \nabla \times \mathbf{E} = -\dfrac{\partial \mathbf{B}}{\partial t} \\ \nabla \cdot \mathbf{B} = 0 \\ \nabla \times \mathbf{B} = \mu_0 \mathbf{J} + \mu_0 \epsilon_0 \dfrac{\partial \mathbf{E}}{\partial t} \end{cases}$$

In matter:

$$\begin{cases} \nabla \cdot \mathbf{D} = \rho_f \\ \nabla \times \mathbf{E} = -\dfrac{\partial \mathbf{B}}{\partial t} \\ \nabla \cdot \mathbf{B} = 0 \\ \nabla \times \mathbf{H} = \mathbf{J}_f + \dfrac{\partial \mathbf{D}}{\partial t} \end{cases}$$

AUXILLARY FIELDS

Definitions:

$$\begin{cases} \mathbf{D} = \epsilon_0 \mathbf{E} + \mathbf{P} \\ \mathbf{H} = \dfrac{1}{\mu_0}\mathbf{B} - \mathbf{M} \end{cases}$$

In linear media:

$$\begin{cases} \mathbf{P} = \epsilon_0 \chi_e \mathbf{E}, & \mathbf{D} = \epsilon \mathbf{E} \\ \mathbf{M} = \chi_m \mathbf{H}, & \mathbf{H} = \dfrac{1}{\mu}\mathbf{B} \end{cases}$$

POTENTIALS: $\mathbf{E} = -\nabla V - \dfrac{\partial \mathbf{A}}{\partial t}, \quad \mathbf{B} = \nabla \times \mathbf{A}$

LORENTZ FORCE LAW: $\mathbf{F} = q(\mathbf{E} + \mathbf{v} \times \mathbf{B})$

ENERGY, MOMENTUM, AND POWER

Energy: $$W = \frac{1}{2}\int \left(\epsilon_0 E^2 + \frac{1}{\mu_0} B^2\right) d\tau$$

Momentum: $$\mathbf{P} = \epsilon_0 \int (\mathbf{E} \times \mathbf{B})\, d\tau$$

Poynting vector: $$\mathbf{S} = \frac{1}{\mu_0}(\mathbf{E} \times \mathbf{B})$$

Larmor formula: $$P = \frac{1}{4\pi\epsilon_0}\frac{2}{3}\frac{q^2 a^2}{c^3}$$